Lecture Notes in Computer Science 516

Edited by G. Goos and J. Hartmanis

S. Kaplan M. Okada (Eds.)

Conditional and Typed Rewriting Systems

2nd International CTRS Workshop
Montreal, Canada, June 11-14, 1990
Proceedings

Springer-Verlag
Berlin Heidelberg New York
London Paris Tokyo
Hong Kong Barcelona
Budapest

Series Editors

Gerhard Goos
GMD Forschungsstelle
Universität Karlsruhe
Vincenz-Priessnitz-Straße 1
W-7500 Karlsruhe, FRG

Juris Hartmanis
Department of Computer Science
Cornell University
Upson Hall
Ithaca, NY 14853, USA

Volume Editors

Stéphane Kaplan
Laboratoire de Recherche en Informatique, Université Paris-11
Centre d'Orsay, Bâtiment 490, F-91405 Orsay Cedex, France

Mitsuhiro Okada
Department of Computer Science, Concordia University
1455 de Maisonneuve Quest, Montreal, Quebec H3G 1M8, Canada

CR Subject Classification (1991): F.4.1-2, D.3.1, I.2.3

ISBN 3-540-54317-1 Springer-Verlag Berlin Heidelberg New York
ISBN 0-387-54317-1 Springer-Verlag New York Berlin Heidelberg

Printed in Germany

Typesetting: Camera ready by author
Printing and binding: Druckhaus Beltz, Hemsbach/Bergstr.
2145/3140-543210 - Printed on acid-free paper

In Memoriam

Dr. Stéphane Kaplan

As this volume was going to press, we heard the very tragic news of the untimely passing of Stéphane Kaplan, co-editor of this volume. The death of this brilliant young scientist is a great loss to the theoretical computer science community.

Beyond other distinguished work in such domains as classical term rewriting, abstract datatypes, software development, and database theory, Stéphane played a most important role in the domain of this volume, conditional and typed rewriting systems. He introduced the notion of simplifying conditional rewriting systems, which has become a new research paradigm and already resulted in very rich research activities in the field. His pioneering work on negative conditions in conditional rewriting and on infinite terms have propelled further research by his colleagues.

His work exhibits the powerful potential of the use of extended framework of the traditional term rewriting theory, which gave birth to the spirit of the CTRS workshop series. Indeed it was he who conceived of the CTRS workshops and served as co-organizer of both the first and second meetings.

For me, his death represents not only the loss of an excellent research colleague but also of a close friend. His intelligent and friendly style of talking always pleased and delighted me. Many scenes of delightful conversation with Stéphane still remain clear in my mind.

July 1991

Mitsuhiro Okada
Co-editor and
Co-organizer of CTRS 90

CTRS '90 Organization

Program Committee:

- R. V. Book, *University of California-Santa Barbara*
- N. Dershowitz, *University of Illinois-Urbana Champaign*
- K. Futatsugi, *Electrotechnical Lab, Tsukuba*
- H. Ganzinger, *Universität Dortmund*
- J. P. Jouannaud, *Université de Paris-Sud*
- S. Kaplan, *Bar Ilan University / Université de Paris-Sud**
- J. W. Klop, *CWI, Amsterdam*
- J. Meseguer, *SRI International, Menlo Park*
- M. Okada, *Concordia University**
- D. Plaisted, *University of North Carolina-Chapel Hill*
- J. L. Rémy, *Université de Nancy*

* Co-editor

CRTS '90 Secretary:
Michael Assels, *Concordia University*

Co-sponsored by

- Natural Science and Engineering Research Council of Canada
- Centre de recherche informatique de Montréal
- Concordia University

Preface

In recent years, extensions of rewriting techniques that go beyond the traditional untyped algebraic rewriting framework have been investigated and developed. Among these extensions, conditional and typed systems are particularly important, as are higher-order systems, graph rewriting systems, etc. The international CTRS (Conditional and Typed Rewriting Systems) workshops are intended to offer a forum for researchers on such extensions of rewriting techniques.

The first CTRS workshop (focused on conditional rewriting systems) was held in July 1987 at Université de Paris XI in Orsay, France, and was very successful. (The proceedings of the first workshop are available as Vol. 308 of the Springer Lecture Notes in Computer Science.) To further research in these and related areas, the Second International Workshop was held June 11–14, 1990 at Concordia University in Montréal, Canada. Topics included the theory of conditional and typed rewriting and its application to programming languages, automated deduction, and other related extensions of rewriting techniques, such as graph rewriting, combinator based languages, and their application to parallel architectures, parallel computation models, and compilation techniques.

In particular, the second CTRS workshop contributed to discussion and evaluation of new directions of research. Through the presentation sessions and informal discussion sessions, several important directions for extensions of rewriting techniques were stressed, which are reflected in the organization of the chapters in this volume.

The program committee received 54 submissions in time; 26 papers were accepted as regular papers, and 11 others were also presented as short communication papers. 7 invited keynote lectures (by J. W. Klop and R. de Vrijer, H. Aida and J. Meseguer, L. Bachmair and H. Ganzinger, J. Hsiang, V. Breazu-Tannen and J. Gallier, A. R. Meyer, and J. L. Rémy) were also given. After the workshop, the program committee reevaluated the short paper presentations for inclusion in this volume. This volume contains 26 regular papers, 7 short papers, and 4 invited papers.

The workshop was co-sponsored by the Natural Science and Engineering Research Council of Canada, Centre de recherche informatique de Montréal and Concordia University. The third CTRS workshop is scheduled for 1992 in Nancy, France.

Concordia University
June 1991

Mitsuhiro Okada
Co-editor

Chapter 1

Theory of Conditional and Horn Clause Systems

Contents

Papers with '(*)' are "short papers". Papers without '(*)' are "regular papers".

Conditional Rewriting in Focus

Francois Bronsard
Uday S. Reddy
Department of Computer Science
The University of Illinois at Urbana-Champaign
Urbana, IL 61801.

Abstract

We discuss the conditional rewriting aspects of the Focus program transformation/synthesis system. The term-rewriting induction principle, presented earlier for unconditional rewrite systems, is generalized for conditional systems. We also present a novel deduction method for first-order clauses called *contextual deduction* which may be considered to be the first-order analogue of the rewriting operation. Like rewriting, contextual deduction uses pattern matching instead of unification and has the property of finite termination. We identify a sufficient condition for clausal theories, called *saturation*, under which contextual deduction is complete for refuting ground goal clauses.

1 Introduction

Focus is an interactive program transformation/synthesis system based on term rewriting techniques [Red88]. A formal description of the synthesis procedure was described in [Red89] as an adaptation of the Knuth-Bendix completion procedure. This presentation used a language of unconditional equations over algebraic terms. In this paper, we describe the extension of this procedure for conditional equations over terms in a functional logic language [Red87, BR90].

The two main component of the synthesis procedure are *rewriting* and *term rewriting induction.* Our goal here is to adapt these techniques to conditional equations.

Section 3 presents an inductive proof procedure, inspired by term rewriting induction, for conditional equations.

Rewriting for conditional equations has been defined before (see [Siv88] for a review). However multiple restrictions make this operation less powerful than one would desire. Consider rewriting a conditional equation of the form

$$p : t = u$$

While rewriting t and u, the condition p can be *assumed*, i.e., it can be used to satisfy the conditions of other rewrite rules which may be applicable. So, the rewriting problem in Focus is different from the word problem [KR87]. A *contextual rewriting* operation was defined by Zhang *et.al.* [ZK88] which addresses this situation. The conditions p are expressed as equations, or rewrite rules, but all their free variables are treated as constants. Now, the rewriting of t and u is performed in the *context* of the additional rules derived from p. However, as noted in [ZK88], conditional rewriting is not a complete deduction mechanism. For example, given the rewrite rules

$$q(y) = \text{False} : \quad p(y) \;\rightarrow\; \text{True}$$
$$q(z) = \text{True} : \quad p(z) \;\rightarrow\; \text{True}$$

it is not possible to prove the formula $p(x) = \text{True}$ using contextual rewriting.

A conditional rewrite rule $p_1 \wedge \cdots \wedge p_n : q \rightarrow \text{True}$ can be thought of as a clause $q \vee \neg p_1 \vee \cdots \vee \neg p_n$. The termination requirements of conditional rewriting are often satisfied by insisting that the conditions p_i be smaller than q by a reduction order $>$. In that case, conditional rewrite rules form a special case of clauses where (1) there is a single positive literal, and (2) the positive literal is maximal in the clause. Stated in this way, the limitations of conditional rewriting are apparent. It is not possible to have multiple maximal literals. Secondly, it is not possible to derive new information from the conditions in the context because negative literals are not permitted to be maximal. Atoms like $q(y) = \text{False}$ do not exactly correspond to negative literals because the equational representation does not capture the information that formulas can only be true or false.

To circumvent these limitations, we have defined a generalization of the contextual rewriting operation called *contextual deduction*. It is presented in Section 4. We believe this operation aptly generalizes the rewriting concept to the context of first-order clauses. In particular this operation enjoys the efficiency and the termination properties of rewriting. Section 5 discusses the completeness issue of contextual deduction.

The techniques discussed here will be illustrated with the problem of synthesizing a program to find the maximum of a list. But first the section 2 will give a short introduction to the notation used here.

2 Terms and Formulas

The language used by a user for interaction with Focus is a functional logic language [Red87, BR90], called FLL.

The *terms* in our language include the usual functional terms. Since we would like to use equational reasoning with formulas whenever possible, we also embed propositional formulas in terms. These are given by the constants True and False, and terms of the form $t_1 \; eq \; t_2$, $t_1 \wedge t_2$, $t_1 \vee t_2$ and $\neg t$. We assume a sorted language with at least the following type distinctions: sort *Data* for data values and sort *Bool* for boolean values. The function $\mathbf{if}_S : Bool \times S \times S \rightarrow S$ allow propositional formulas to be embedded in terms. $\mathbf{if}_S(p, t_1, t_2)$ is written as $\mathbf{if} \; p \; \mathbf{then} \; t_1 \; \mathbf{else} \; t_2$.

A *conditional equation* is a formula of the form $c : t = u$. A *conditional rewrite rule*, with respect to a reduction order $\succ$, is a rule of the form $c : t \rightarrow u$ where $t \succ c$ and $t \succ u$. A clause is a formula of the form $c_1 \vee \cdots \vee c_n$ where c_i are "literal" terms, i.e., terms without the boolean connectives "$\wedge$" and "$\vee$". We also write clauses in sequent notation $c_1 \wedge \ldots \wedge c_k \Rightarrow c_{k+1} \vee \ldots \vee c_n$. Conditional equations can also be formulated as clauses.

The functional logic language that Focus supports includes other kinds of terms. In contrast to first-order terms these terms can be partial or inconsistent. A term is *partial* if it has the possibility of denoting no value and *inconsistent* if it has the possibility of denoting more than one value. Inconsistent terms are illegitimate. The user should guarantee, by means of axioms, that terms are consistent. Partial terms, on the other hand, are legitimate. They can be combined with other partial terms to make total terms which are then treated like ordinary first-order terms.

The fragment of FLL we use in this paper includes a constant Undef, a choice operator denoted by ";" and a definite description construct denoted by "the ". Undef denotes the completely undefined value. The choice operator is defined as follows

$$x ;_s y \;=\; \begin{cases} x & \text{if } y = \text{Undef} \\ y & \text{if } x = \text{Undef} \\ x & \text{if } x = y \neq \text{Undef} \\ \text{inconsistent} & \text{otherwise} \end{cases}$$

The definite description construct of the form $\mathbf{the} \; t[\bar{x}] : p$ binds the variables $\bar{x}$ which may occur free in t and p. (So, it may be thought of as a quantifier). Its value is the *unique* value of t for the instantiation of variables $\bar{x}$ which satisfy the condition p. If there is no instantiation of $\bar{x}$, the term denotes Undef. If there are multiple instantiations which give distinct values to t, then the term is *inconsistent*. The following axioms are used to simplify expressions involving the

$$\begin{aligned}
(\text{the } t[] : p) &\rightarrow \text{if } p \text{ then } t \text{ else Undef} \\
(\text{the } t[x, \bar{y}] : x \; eq \; u \wedge p) &\rightarrow (\text{the } t^u_x[\bar{y}] : p^u_x) \quad \text{if } x \notin FV(u) \\
(\text{the } t[\bar{x}] : q \vee p) &\rightarrow (\text{the } t[\bar{x}] : q) \, ; (\text{the } t[\bar{x}] : p), \\
(\text{the } t[\bar{x}] : \text{False}) &\rightarrow \text{Undef.}
\end{aligned}$$

$$\frac{(\text{the } \; x[x] : p) = t}{p = (t \; eq \; x)}$$

Example 1 Consider the following axioms and properties needed for the specification of the function max which computes the maximum element of a list. The **Signature:** is $\{0 : Nat, S : Nat \rightarrow Nat, \geq : Nat \times Nat \rightarrow Bool, \text{Nil} : List, -.- : Nat \times List \rightarrow List, - \in - : Nat \times List \rightarrow Bool, \text{bounded} : Nat \times List \rightarrow Bool\}$. The ordering is a Recursive Path Ordering [Der82] based on the following precedence ordering:

$$\text{bounded} \succ \in \succ \geq \succ \; . \succ S \succ \text{Nil} \succ 0.$$

Axioms:

$$\begin{aligned}
x \geq 0 &\rightarrow \text{True} \\
S(x) \geq S(z) &\rightarrow x \geq z \\
0 \geq S(z) &\rightarrow \text{False} \\
x \in \text{Nil} &\rightarrow \text{False} \\
x \in a.l &\rightarrow x \; eq \; a \vee x \in l \\
\text{bounded}(x, \text{Nil}) &\rightarrow \text{True} \\
\text{bounded}(x, a.l) &\rightarrow x \geq a \wedge \text{bounded}(x, l)
\end{aligned}$$

Properties:

$$x \geq y \wedge y \geq z \Rightarrow x \geq z$$

Nil denotes the empty list and the predicate bounded(x, l) express that x is greater or equal than any element of l.

We express the transitivity of $\geq$ as a clause because there does not seem to be any satisfactory way to express it as an equation.

The specification of max is then:

$$l \neq \text{Nil} : \max(l) = \text{the } z[z] : z \in l \wedge \text{bounded}(z, l).$$

□

3 Induction

An induction method called term-rewriting induction is presented in [Red90] and it is shown that completion-based inductive proof procedures ([Mus80], [Fri86], and others) construct term-rewriting induction proofs. In [Red89] this method was used for program synthesis. In this section we extend it to the case of conditional equations. This procedure may be contrasted with the completion based inductive proof procedure for conditional equation given by Kaplan in [Kap87].

Definition 1 Let E be a set of conditional equations. An equation P is an inductive consequence of E if it holds in the initial FLL-model[1] of E.

[1]A FLL-model assigns the standard interpretations to boolean connectives and other FLL operators.

As usual, an equation P is an inductive consequence if all its ground instances are logical consequences. Unlike in algebraic theories, the converse does not hold because quantifiers are involved. A ground instance may involve quantifiers which only range over the elements of the initial model.

Since the set of ground terms is infinite, we use representative sets of terms called *cover sets* [Red90]. For the conditional case these are defined as follows.

Definition 2 Let E a set of conditional equations and $\succ$ a reduction ordering.

1. A *conditional cover set* for a sort S is a set of pairs $\{(t_i, c_i)\}_i$ such that for every ground term g of sort S there is an i and a substitution γ such that $g =_E t_i\gamma$, $g \succeq t_i\gamma$, and $c_i\gamma =_E$ True.

2. A conditional cover set for a set of variables, $\{x_1^{S_1}, \ldots, x_k^{S_k}\}$, is a set of pairs $\{(\sigma_i, c_i)\}_i$ such that for every ground instances of the variables, say $(g_1, \ldots, g_k)$, there is an i and a substitution γ such that $c_i\gamma =_E$ True, and for all j, $g_j =_E x_j\sigma_i\gamma$ and $g_j \succeq x_j\sigma_i\gamma$.

3. If e is a term, a set of pairs $\{(e\sigma_i, c_i)\}_i$ is called a conditional cover set of *cases* of e if $\{(\sigma_i, c_i)\}_i$ is a conditional cover set for the free variables in e.

4. A cover set of cases of e is said to be *reducible* if each $e\sigma_i$ is reducible in the context c_i.

Example 2 Consider the following axioms defining the membership relation for ordered lists, with type *OList*. The signature is the one given in the previous example plus the predicate, omember : $Nat \times OList \rightarrow Bool$.

Axioms:

$$
\begin{array}{rrcl}
 & \text{omember}(x, \text{Nil}) & \rightarrow & \text{False} \\
x \geq a: & \text{omember}(x, a.l) & \rightarrow & x \ eq \ a \vee \text{omember}(x, l) \\
a \geq x: & \text{omember}(x, a.l) & \rightarrow & x \ eq \ a
\end{array}
$$

A conditional cover set for the sort *OList* is: $\{(\text{Nil}, \text{True}), (a.l, \text{True})\}$. A conditional cover set for the variables $\{x, l\}$ is :

$$\{(\{x \rightarrow x, l \rightarrow \text{Nil}\}, \text{True}), (\{x \rightarrow x, l \rightarrow a.l'\}, x \geq a), (\{x \rightarrow x, l \rightarrow a.l'\}, a \geq x)\}.$$

With this conditional cover set we can make the following reducible conditional cover set of cases of the expression omember(x, l)

$$\{(\text{omember}(x, Nil), \text{True}), (\text{omember}(x, a.l'), x \geq a), (\text{omember}(x, a.l'), a \geq x)\}$$

□

Our inductive inference is based on *term-rewriting induction:*

Proposition 3 Let R be a rewrite system and $P \equiv p : e \rightarrow e'$ be an inductive proposition, all oriented by a reduction ordering $\succ$. P is an inductive consequence of R if

1. $\{(e\sigma_i, c_i)\}_i$ is a reducible conditional cover set of cases of e,

2. for each case, i, we have terms a and b such that $e\sigma_i \rightarrow_R a \rightarrow^*_{RUP} b \leftarrow^*_{RUP} e'\sigma_i$ in the context $p\sigma_i \wedge c_i$.

In [Red90] a procedure called *inductive completion* is presented which applies the term-rewriting induction principle repeatedly to construct hierarchical induction proofs. The heart of this procedure is an operation called *Expand*. We adapt the Expand operation below for conditional equations.

Let E be a set of equations to be proved, H a set of inductive hypotheses available for proving E via term-rewriting induction. The Expand operation is defined by the inference rule (used backwards):

$$\text{Expand} \quad \frac{(E \cup E', H \cup \{p : e \rightarrow e'\})}{(E \cup \{p : e = e'\}, H)}$$

under the conditions that $e \succ e', e \succ p$, $\{(e\sigma_i, c_i)\}_i$ is a reducible conditional cover set of cases of e, $e\sigma_i \succ c_i$ for each i, and $E' = \{p\sigma_i \wedge c_i : a = e'\sigma_i \mid p\sigma_i \wedge c_i : e\sigma_i \rightarrow_R a\}_i$.

The correctness of this procedure, can be shown by an argument using noetherian induction relative to the ordering $\succ$. The proof can be done in two steps, first, one shows that the set of ground proofs is the same throughout the procedure, and, second, that proof steps using the inductive hypothesis can be replaced by "simpler" proof steps. This second part is shown below.

Proposition 4 Let $\{(e\sigma_i, c_i)\}_i$ be a reducible conditional cover set of cases of e and E' be the set $\{p\sigma_i \wedge c_i : a = e'\sigma_i \mid p\sigma_i \wedge c_i : e\sigma_i \rightarrow_R a\}_i$. For every ground proof step $t \leftrightarrow_P u$, there is a proof $t \leftrightarrow^*_{R \cup E'} u$ such that every E' step in the latter is for terms smaller than $\{t, u\}$ by the multiset extension of the ordering $\succ$.

Proof:

1. For any ground instance $P\gamma$ there is an i and a γ' such that $c_i\gamma'$ is true and $e\gamma \rightarrow^*_R e\sigma\gamma'$, hence $p\gamma : e\gamma \rightarrow^*_R a\gamma'$. Whenever $p\gamma \rightarrow^*_R True$, we can replace $e\gamma \leftrightarrow_P e'\gamma$ by $e\gamma \rightarrow^*_R a\gamma' \leftrightarrow_{E'} e'\gamma$.

2. Moreover in the new proof the E'-steps are smaller than $e\gamma$ since $e \succ e'$ and $e\sigma_i \succ a$. □

The inference rule Expand places an important constraint on the choice of cover sets, viz. each $e\sigma_i$ must be reducible by R. In the equational framework this can be achieved by showing that the expanded term is inductively reducible[2] and using a complete set of narrowings for this term. Techniques for proving inductive reducibility may be found in [KNZ86, JK86, HH80]. The case of conditional equations is more complex because inductive reducibility is in general undecidable. Gaudel and Kaplan [GK86] present syntactic sufficient conditions to establish inductive reducibility. We use similar techniques and then use complete sets of narrowings to generate conditional cover sets.

The inductive completion procedure is used for program synthesis as follows: Given a specification $c : t = e$, we want to derive a program for t which would be inductively equivalent to the specification. The ordering must be defined such that $e \succ t$ (otherwise we could simply use e as the program for t) and $e \succ c$ (by definition of rewrite rules). We repeatedly apply the inductive completion procedure until we obtain a set E of equations which can be accepted as a program for t. In addition to the Expand rule above, we also use the following weaker version which involves case reasoning, but no induction.

$$\text{Cases} \quad \frac{(E \cup \{p\sigma_i \wedge c_i : e\sigma_i = e'\sigma_i\}_i, H)}{(E \cup \{p : e = e'\}, H)}$$

if $\{(e\sigma_i, c_i)\}_i$ is a conditional cover set of cases of e, and $e\sigma_i \succ c_i$ for each i.

Example 3 Let us derive a recursive program for max specified in Example 1. The specification is:

$$l \neq \text{Nil} : \text{max}(l) = \text{the } z[z] : z \in l \wedge \text{bounded}(z, l).$$

We expand this equation (by narrowing $z \in l$, we obtain reducible cover set of cases: $\{(\{l \rightarrow \text{Nil}\}, \text{True}), (\{l \rightarrow a.l'\}, \text{True})\}$). The inductive hypothesis is:

$$l \neq \text{Nil} : \ \text{max}(l) \ \leftarrow \ \text{the } z[z] : z \in l \wedge \text{bounded}(z, l)$$

[2] A term e is inductively reducible by R if every ground instance of e is reducible by R.

From the axioms given earlier, we deduce from this hypothesis the rule:

$$l \neq \text{Nil}: \ (z \in l \wedge \text{bounded}(z,l)) \ \rightarrow \ \max(l) = z)$$

In the derivation below, the successive lines are the results of simplifications:

$$\begin{array}{lll}
\text{case } l = \text{Nil} & & \\
\text{Nil} \neq \text{Nil}: & \dots & \\
\text{False}: & \dots & \\
\hline
\text{case } l = a.l' & & \\
a.l' \neq \text{Nil}: & \max(a.l') = & \text{the } z[z] : (z = a \vee z \in l') \wedge \text{bounded}(z, a.l') \\
\text{True}: & \max(a.l') = & \text{the } z[z] : (z = a \wedge \text{bounded}(z, a.l')) \vee \\
 & & (z \in l' \wedge \text{bounded}(z, a.l')) \\
 & \max(a.l') = & (\text{the } z[z] : z = a \wedge z \geq a \wedge \text{bounded}(z, l')); \\
 & & (\text{the } z[z] : z \in l' \wedge z \geq a \wedge \text{bounded}(z, l')) \\
 & \max(a.l') = & (\text{if bounded}(a, l') \text{ then } a); \\
 & & (\text{the } z[z] : z \in l' \wedge z \geq a \wedge \text{bounded}(z, l'))
\end{array}$$

Using the inference rule Cases with the conditional cover set $\{(\{\}, l' = \text{Nil}), (\{\}, l' \neq \text{Nil})\}$.

$$\begin{array}{lll}
\text{case } l' = \text{Nil} & & \\
l' = \text{Nil}: & \max(a.l') = & (\text{if True then } a); \\
 & & (\text{the } z[z] : \text{False} \wedge z \geq a \wedge \text{bounded}(z, l')) \\
 & \max(a.\text{Nil}) = & a \\
\hline
\text{case } l' \neq \text{Nil} & & \\
l' \neq \text{Nil}: & \max(a.l') = & (\text{if bounded}(a, l') \text{ then } a); \\
 & & (\text{the } z[z] : z = \max(l') \wedge z \geq a) \\
l' \neq \text{Nil}: & \max(a.l') = & (\text{if bounded}(a, l') \text{ then } a); \\
 & & (\text{if } \max(l') \geq a \text{ then } \max(l'))
\end{array}$$

This would give the program:

$$\begin{array}{rcl}
\max(a.\text{Nil}) & \rightarrow & a \\
l' \neq \text{Nil} : \max(a.l') & \rightarrow & (\text{if bounded}(a, l') \text{ then } a); \\
 & & (\text{if } \max(l') \geq a \text{ then } \max(l'))
\end{array}$$

Using the contextual deduction techniques presented in the next section, it is possible to prove the inductive theorem

$$l' \neq \text{Nil} : \text{bounded}(a, l') \rightarrow a \geq \max(l').$$

By a further use of Cases, the second rule can be transformed to

$$l' \neq \text{Nil} : \max(a.l') \rightarrow \text{if } \max(l') \geq a \text{ then } \max(l') \text{ else } a.$$

4 Contextual deduction

As remarked in the introduction *contexts* arise during the rewriting of conditional equations. In simplifying an equation of the form

$$p : t = u$$

the condition p needs to be used in simplifying t and u. Similarly, in simplifying

$$\text{if } p \text{ then } t \text{ else } u$$

the term t needs to be simplified in the context of p and u needs to be simplified in the context of $\neg p$. Simplifying the terms involves similar contexts. Simplification of propositional formulas involves contexts as well. For instance, in simplifying $p \wedge q$, p is the context for the simplification of q and vice versa.

Suppose we want to simplify a term t in the context p and we have a conditional rewrite rule $c : l \rightarrow r$ such that $l\sigma = t$. For the rule to be applicable, the formula $p \Rightarrow c\sigma$ must be proved. Note that all the variables in the formula are universally quantified and, so, they may be treated as constants. We call this the *first-order word problem*: given a set of theory clauses and a ground goal clause, decide if the goal clause follows from theory clauses.

Conditional rewriting is not adequate for solving first-order word problem as shown by the counter-example in the introduction. The solution we propose is to use a refutational proof procedure which is similar to rewriting in that it only uses instances of theory clauses obtained by pattern matching. There are however, two crucial differences from the rewriting operation. The first is caused by the fact that clauses can have multiple maximal literals, in contrast to rewrite rules which have a left hand side larger than the right hand side. Ordered rewriting, which generalizes rewriting to unoriented equations, still satisfies the crucial property that it does not need to introduce new variables. This is not the case with clauses. To use a clause of the form $p(x) \vee q(y)$, where both the literals are maximal, we may resolve a ground goal clause on the literal $p(x)$ and derive a nonground goal clause with the literal $q(y)$. Thus, the first-order analogue of rewriting must be able to derive nonground goal clauses. Secondly, such derived clauses may be further resolved with other ground goal clauses. This means that the first-order analogue of rewriting must be nonlinear. With these intuitions, we propose the following mechanism for rewrite-like reasoning with first-order clauses:

Definition 5 An ordering $\succ$ over terms is said to be *transparent*, if whenever $t \succ u$, all the free variables of u occur free in t.

Definition 6 Let $\mathcal{T}$ be a set of clauses ("theory" clauses) and $\succ$ a transparent reduction ordering. A *contextual deduction* step (using $\mathcal{T}$) is an inference step of the form

$$\frac{p \vee A}{A \vee B\sigma} \qquad \text{if } q \vee B \in \mathcal{T},\ q\sigma = \neg p \text{ and } q\sigma \text{ is maximal in } q\sigma \vee B\sigma$$

or of the form

$$\frac{p \vee A \qquad q \vee B}{A \vee B\sigma} \qquad \text{if } q\sigma = \neg p \text{ and } q\sigma \text{ is maximal in } q\sigma \vee B\sigma$$

A *contextual factoring* step is of the form

$$\frac{p \vee q \vee A}{p \vee A\sigma} \qquad \text{if } q\sigma = p \text{ and } q\sigma \text{ is maximal in } p \vee q\sigma \vee A\sigma$$

A *contextual proof* of a clause C from a set of goal clauses G using $\mathcal{T}$ ($G \vdash_{\mathcal{T}} C$) is a tree whose nodes are labeled by clauses such that the clause of a node is the consequent of its children by one of the inference rules given above, the root is labeled by C and the leaf nodes by clauses from G.

As an example of contextual refutation, consider proving $p(x)$ from the clauses

$$p(y) \vee q(y)$$
$$p(z) \vee \neg q(z)$$

Assume that $p > q$. The skolemized negation of $p(x)$ is $\neg p(A)$ where A is a new constant. The following contextual refutation shows the unsatisfiability of $\neg p(A)$:

$$\frac{\dfrac{\neg p(A)}{q(A)} \text{ by } p(y) \vee q(y) \qquad \dfrac{\neg p(A)}{\neg q(A)} \text{ by } p(z) \vee \neg q(z)}{\Box}$$

Example 4 Consider the theory made of the three clauses

$$x \in l \;\Rightarrow\; max(l) \geq x \qquad (1)$$
$$x \geq y \wedge y \geq z \;\Rightarrow\; x \geq z \qquad (2)$$
$$x \geq y \;\vee\; y \geq x \qquad (3)$$

We would like to prove

$$\neg(max(l) \geq a) \Rightarrow (x \in l \Rightarrow a \geq x).$$

The skolemized negation of the goal is

$$\neg(max(L) \geq A) \qquad (4)$$
$$X \in L \qquad (5)$$
$$\neg(A \geq X) \qquad (6)$$

The contextual refutation goes as follows:

$$A \geq max(L) \quad \text{from 4 using 3} \qquad (7)$$
$$max(L) \geq z \Rightarrow A \geq z \quad \text{from 7 using 2} \qquad (8)$$
$$max(L) \geq X \quad \text{from 5 using 1} \qquad (9)$$
$$A \geq X \quad \text{from 8 and 9} \qquad (10)$$
$$\Box \quad \text{from 10 and 6.} \qquad (11)$$

5 Completeness of contextual deduction

The contextual refutation procedure defined above is not in general complete. But neither is rewriting! The interesting question is: "Under what conditions is contextual deduction complete?"

Rewriting is complete if, under some ordering, the set of equations forms a confluent and noetherian set of rules. We show below that, under some additional restriction on the ordering, if the theory is saturated (i.e. contains all the ordered resolvents and factors) then contextual rewriting is complete (a stronger result for Horn Clauses is given in [Rus89]). However, the analogy with rewriting suggests that there may exist a smaller set, containing only the "irreducible" resolvents, which would be sufficient for completeness. Further investigation is needed for finding such a set.

First, recall the following result from [Bac89, HR86].

Definition 7 Assume $\succ$ is a complete reduction ordering.[3] *Ordered resolution* is an inference step of the form

$$\frac{p \vee A \qquad \neg q \vee B}{A\sigma \vee B\sigma} \quad \text{if } \sigma = mgu(p,q),\ p\sigma \text{ is maximal in } p\sigma \vee A\sigma, \text{ and } \neg q\sigma \text{ is maximal in } \neg q\sigma \vee B\sigma.$$

Ordered factoring is an inference step of the form

$$\frac{p \vee q \vee A}{p\sigma \vee A\sigma} \quad \text{if } \sigma = mgu(p,q), \text{ and } p\sigma \text{ is maximal in } p\sigma \vee q\sigma \vee A\sigma.$$

Proposition 8 If C is an unsatisfiable set of clauses, then there is an ordered refutation from C.

For defining saturated sets, we need the following stronger version of ordered resolution.

[3] A complete reduction ordering is one that can be extended to a total ordering on ground terms.

Definition 9 Strong (ordered) resolution is the inference step

$$\frac{p \vee A \qquad \neg q \vee B}{A\sigma \vee B\sigma} \qquad \text{if } \sigma = mgu(p,q),\ p \text{ is maximal in } p \vee A, \text{ and } \neg q \text{ is maximal in } \neg q \vee B.$$

Strong (ordered) factoring is the inference step

$$\frac{p \vee q \vee A}{p\sigma \vee A\sigma} \qquad \text{if } \sigma = mgu(p,q), \text{ and } p \text{ and } q \text{ are maximal in } p \vee q \vee A.$$

A set of clauses which is closed under strong resolution and factoring is said to be *saturated.* Suppose $\mathcal{T}$ is a saturated theory. Consider proving a boolean formula F (a formula using boolean connectives, but no quantifiers). We can obtain a set of clauses $\mathcal{G}$ by negating the formula, skolemizing it and taking its conjunctive normal form. Note that, since the variables in F are (implicitly) universally quantified, its skolemized negation is ground (variable-free). So, $\mathcal{G}$ is a set of ground clauses. We state the following result for refuting such clauses.

Proposition 10 Let $\mathcal{T}$ be a saturated theory with respect to a transparent and complete reduction ordering, and $\mathcal{G}$ a set of ground clauses. If $\mathcal{T} \cup \mathcal{G}$ is unsatisfiable, there exists an ordered refutation from $\mathcal{T} \cup \mathcal{G}$ such that, in every proof step, at least one of the two unified literals is ground.

Proof: Treat a refutation as a tree whose root is labeled by the empty clause and every internal node is labeled by the consequent of a proof step and its children labeled by the antecedents of the proof step. We use $\alpha, \beta, \ldots$ to refer to proof steps in the proof tree. Let "theory clause" refer to a member of $\mathcal{T}$, and "goal clause" refer to a member of $\mathcal{G}$ or the consequent of a proof step.

By proposition 8, we have an ordered refutation from $\mathcal{T} \cup \mathcal{G}$. First note that the refutation need not have any resolution steps between theory clauses or a factoring step on a theory clause (because the theory contains all its resolvents and factors). Any proof step in which both the the unified literals are nonground is called a *violation.* Consider an innermost violation α in the refutation of $\mathcal{T} \cup \mathcal{G}$. We show that there is another refutation where the subproof rooted at α is replaced by a proof without any violations. The result follows by induction on the address of α.

Case 1: α is a goal-theory resolution step (say between clauses G and T) and the resolved goal literal (say g) is not ground. Since all the input goal clauses are ground, the literal g must have been introduced in an inner goal-theory resolution step β (say between G' and T') where some generalization g' of g belongs to T'. Note that g' must be maximal in T'. (If it is not, then g is bounded above by one or more literals which are resolved away between the steps β and α. Since α is the innermost violation, all such literals must be ground. By definition of transparent reduction order, g must be ground as well). If g' is maximal in T', then the resolvent of the theory clauses T and T' (on the literal g') belongs to the theory. The inner resolution step β can then be modified to use the resolvent clause and the violating step at α eliminated.

Case 2: α is a goal-goal resolution step (say between G_1 and G_2) resolving nonground goal literals g_1 and g_2. As in case 1, there must be inner goal-theory resolution steps β_i for $i = 1,2$ (between some G'_i and T'_i) where some generalization g'_i of g_i belongs to T'_i. Further g'_i must be maximal in T'_i. So, the resolvent of T'_1 and T'_2 on the literals g'_1 and g'_2 belongs to the theory. Let the resolvent be T'. Transform the proof tree $\alpha(\ldots\beta_1, \ldots\beta_2)$ as follows. Replace the steps from β_2 to $\alpha.2$ by an equivalent proof using T' in place of T'_2. The final clause of this proof segment is $G_3 \equiv (G_2 \backslash g_2)\sigma \vee (T'_1 \backslash g'_1)\sigma$ for some substitution σ. Duplicate the proof steps from β_1 to $\alpha.1$ using G_3 in place of T'_1. The resulting proof does not have any violations because the violating step at α is eliminated and all the inner proof steps use clauses more specific than the corresponding ones in the original proof.

Case 3: α is a goal factoring step on two nonground literals g_1 and g_2. The proof transformation for this is similar to case 2 above. □

This result does not hold if saturation is defined using ordered inferences (rather than strong inferences). For example, consider the clauses

$$p(x) \vee q(x,y)$$
$$\neg q(x,0)$$

and the recursive path ordering generated by $p \succ q$. This set of clauses is saturated by ordered resolvents and factors. $q(x,0)$ is not maximal in $p(x) \vee q(x,0)$. But, there is no contextual refutation for the goal clause $\neg p(0)$. Therefore, the strong resolvents and factors seem necessary for contextual deduction to be complete.

Another question that needs to be addressed is whether there exist smaller sets than saturated sets for which contextual deduction is complete. To draw the analogy with equational reasoning, saturated sets correspond to rewrite systems which are closed under critical pairs. In contrast, complete sets of rewrite rules are smaller sets satisfying the property that all their critical pairs are joinable. So it is natural to ask whether a set of clauses whose ordered resolvents (and factors) are provable by contextual refutation is adequate for contextual deduction to be complete. However, there are subtle issues involved in this. Consider the clauses

$$p(x,0,x)$$
$$p(x,y,z) \quad \Rightarrow \quad p(y,x,z)$$

Assume a lexicographic path ordering. The two clauses have an ordered resolvent:

$$p(0,x,x).$$

To see if the resolvent is provable by contextual refutation, we consider its skolemized negation, viz., $\neg p(0,X,X)$. If we assume $0 \succ X$, it is easy enough to find a contextual refutation:

1. $\neg p(0,X,X)$
2. $\neg p(X,0,X)$ using the second clause.
3. $\square$ using the first clause.

However, if we assume $X \succ 0$ there is no contextual refutation. It would be necessary to consider all possible orderings over skolem constants before we can decide if a clause is needed in the "complete" set.

An alternative approach which seems promising is to define a notion of "reductive" proof. A contextual proof is ***reductive*** if it is obtainable by replacing all occurrences of "maximal" in definition 6 by "the largest literal." Reductive proofs have some of the properties of the rewriting operation. For instance, we have been able to verify the following generalized form of Newman's lemma:

Lemma 11 Let C be a set of clauses such that all its strong resolvents and factors are provable by reductive refutation using C. If $p \vee A$ and $\neg p \vee B$ are ground clauses provable by reductive refutation using C, then $A \vee B$ is provable by reductive refutation using C.

Let Sat($\mathcal{T}$) be the saturation of a set of clauses $\mathcal{T}$. Let Comp($\mathcal{T}$) be the smallest superset of $\mathcal{T}$ such that all its strong resolvents and factors are provable by reductive refutation. Then, it is easy to show, using the lemma, that every clause in Sat($\mathcal{T}$) is either in Comp($\mathcal{T}$) or is provable by reductive refutation using Comp($\mathcal{T}$). This corresponds to the Church-Rosser result in equational reasoning.

We conjecture that contextual deduction using Com($\mathcal{T}$) is complete for refuting ground goal clauses. We have not yet been able to verify this claim.

6 Conclusion

We have defined here an induction method for conditional equations over terms including propositional formulas. The method is based on term rewriting induction which allows inductive hypotheses to be used freely as rewrite rules and the smaller instance requirement is satisfied by the form of the proof.

We have also defined an efficient simplification / proof method for universally quantified boolean formulas in the context of theories of clauses. The method retains the advantages of the rewriting operation as it only uses pattern matching. We have also identified a sufficient condition, viz., saturation of the theory of clauses, for contextual deduction to be a complete refutational mechanism. It remains an open problem whether weaker conditions, analogous to complete sets of reductions, would be adequate for this purpose as well.

References

[Bac89] L. Bachmair. Proof normalization for resolution and paramodulation. In N. Dershowitz, editor, *Rewriting Techniques and Appl.*, pages 15–28, Springer-Verlag, Berlin, 1989. (Lecture Notes in Comp. Science, Vol 355).

[BR90] F. Bronsard and U. S. Reddy. An axiomatization of a functional logic language. In *Algebraic and Logic Programming*, pages 101–116, Springer-Verlag, Oct 1990. (LNCS Vol. 463).

[Der82] N. Dershowitz. Orderings for term-rewriting systems. *Theoretical Computer Science*, 17(3):279–301, 1982.

[Fri86] L. Fribourg. A strong restriction of the inductive completion procedure. In *Intern. Conf. Aut., Lang. and Program.*, pages 105–115, Rennes, France, July 1986. (Springer Lecture Notes in Computer Science, Vol. 226).

[GK86] M.-C. Gaudel and S. Kaplan. *How to build meaningful algebraic specification.* ESPRIT Meteor Project Technical report, Orsay-France, 1986.

[HH80] G. Huet and J.-M. Hullot. Proofs by induction in equational theories with constructors. In *Symp. on Foundations of Computer Science*, pages 96–107, IEEE, Lake Placid, NY, October 1980.

[HR86] Jieh Hsiang and Michaël Rusinowitch. A new method for establishing refutational completeness in theorem proving. In J. H. Siekmann, editor, *Proceedings of the Eighth International Conference on Automated Deduction*, pages 141–152, Oxford, England, July 1986. Vol. 230 of *Lecture Notes in Computer Science*, Springer, Berl in.

[JK86] J.-P. Jouannaud and E. Kounalis. Automatic proofs by induction in equational theories without constructors. In *Symp. on Logic in Computer Science*, pages 358–366, IEEE, Cambridge, MA., June 1986.

[Kap87] S. Kaplan. Simplifying conditional term rewriting systems: Unification, termination and confluence. *J. of Symbolic Computation*, 4:295–334, 1987.

[KNZ86] D. Kapur, P. Narendran, and H. Zhang. Proof by induction using test sets. In *Conf. on Automated Deduction*, Oxford, U.K., 1986.

[KR87] E. Kounalis and M. Rusinowitch. On word problems in Horn theories. In S. Kaplan and J.-P. Jouannaud, editors, *Conditional Term Rewriting Systems*, pages 144–160, Springer-Verlag, Berlin, 1987. (LNCS Vol 308).

[Mus80] D. R. Musser. On proving inductive properties of abstract data types. In *ACM Symp. on Princ. of Program. Languages*, pages 154–162, ACM, Las Vegas, 1980.

[Red87] U. S. Reddy. Functional logic languages, Part I. In *Graph Reduction*, pages 401–425, Springer-Verlag, 1987. (Lecture Notes in Computer Science, Vol 279).

[Red88] U. S. Reddy. Transformational derivation of programs using the Focus system. In *Symp. Practical Software Development Environments*, pages 163–172, ACM, December 1988.

[Red89] U. S. Reddy. Rewriting techniques for program synthesis. In N. Dershowitz, editor, *Rewriting Techniques and Appl.*, pages 388–403, Springer-Verlag, 1989. (LNCS Vol. 355).

[Red90] U. S. Reddy. Term rewriting induction. In M. Stickel, editor, *Conf. on Automated Deduction*, pages 162–177, Springer-Verlag, 1990. (Lecture Notes in Artificial Intelligence, Vol. 449).

[Rus89] Michaël Rusinowitch. *Démonstration Automatique: Techniques de réécriture.* InterEditions, Paris, France, 1989.

[Siv88] Sivakumar. *Conditional Rewriting.* PhD thesis, UIUC, 1988.

[ZK88] H. Zhang and D. Kapur. First-order theorem proving using conditional rewrite rules. In E. Lusk and R. Overbeek, editors, *Conf. on Automated Deduction*, pages 1–20, Springer-Verlag, Berlin, 1988.

A Maximal-Literal Unit Strategy for Horn Clauses*

Nachum Dershowitz
Department of Computer Science
University of Illinois at Urbana-Champaign
1304 West Springfield Avenue
Urbana, IL 61801, U.S.A.
nachum@cs.uiuc.edu

Abstract

A new positive-unit theorem-proving procedure for equational Horn clauses is presented. It uses a term ordering to restrict paramodulation to potentially maximal sides of equations. Completeness is shown using proof orderings.

1. Introduction

A *conditional equation* is a universally-quantified Horn clause in which the only predicate symbol is equality. We write such a clause in the form

$$e_1 \wedge \cdots \wedge e_n \Rightarrow s \simeq t$$

($n \geq 0$), meaning that the equality $s \simeq t$ holds whenever all the equations e_i, called *conditions*, hold. If $n = 0$, then the (positive unit) clause, $s \simeq t$, will be called an *unconditional equation*. Conditional equations are important for specifying abstract data types and expressing logic programs with equations. Our interest here is in procedures for proving validity of equations in all models of a given finite set E of conditional equations. Note that a conditional equation $e_1 \wedge \cdots \wedge e_n \Rightarrow s \simeq t$ is valid for E iff $s \simeq t$ is valid for $E \cup \{e_1, \ldots, e_n\}$. Hence, proving validity of conditional equations reduces to proving validity of unconditional ones.

The completeness of positive-unit resolution for Horn clauses is well-known. An advantage of positive-unit resolution is that the number of conditions never grows; it suffers from the disadvantage of being a bottom-up method. Ordinary Horn clauses

$$p_1 \wedge \cdots \wedge p_n \Rightarrow p_{n+1}$$

where the p_i are not equality literals, can be expressed as conditional equations, by turning each literal p_i into a Boolean equation $p_i \simeq T$, for the truth constant T. Ordered resolution, in which the literals of each clause are arranged in a linear order $>$, and only the largest literal may serve as a resolvent, is also complete for Horn clauses (see Boyer, 1971).

Positive-unit resolution can be expressed by means of the following inference rule:

$$\frac{E \cup \left\{\begin{array}{rcl} q \wedge s \simeq T & \Rightarrow & u \simeq T, \\ & & l \simeq T \end{array}\right\}}{E \cup \left\{\begin{array}{rcl} q \wedge s \simeq T & \Rightarrow & u \simeq T, \\ & & l \simeq T, \\ q\sigma & \Rightarrow & u\sigma \simeq T \end{array}\right\}}$$

*This research supported in part by the National Science Foundation under Grant CCR-9007195.

where σ is the most general unifier (*mgu*) of l and s. Here, the positive unit clause $l \simeq T$ is resolved with the negative literal $s \simeq T$ in the clause $q \land s \simeq T \Rightarrow u \simeq T$, and produces a new Horn clause $q\sigma \Rightarrow u\sigma \simeq T$. The new clause is a logical consequence of the two given clauses, since $s\sigma = l\tau\sigma$, where τ renames variables in l so that it shares none with s. Any unit clause that is a logical consequence of a set of Horn clauses E is an instance of a unit clause producible by repeated application of this rule of inference.

Horn clauses with both equality and non-equality literals can be expressed as conditional equations with equality literals only. The equality axioms, including functional reflexivity, are also Horn clauses. Positive-unit resolution, or any other complete variation of resolution, could be used to prove theorems in equational Horn theories, but the cost of treating equality axioms like any other clause is prohibitively high. For this reason, special inference mechanisms for equality, notably paramodulation (Robinson and Wos, 1969), have been devised. In the Horn case, a unit strategy can be combined with paramodulation (Henschen and Wos, 1974; Furbach, 1987).

In this paper, we describe a new complete theorem-proving method for equational Horn theories. It utilizes orderings of terms and atoms to restrict inferences, and is a generalization of *ordered completion* (Hsiang and Rusinowitch, 1987; Bachmair, *et al.*, 1989), an "unfailing" extension of the "completion procedure" in Knuth and Bendix (1970) for unconditional equational inference. Completion operates on asymmetrical equations, that is, on *rewrite rules*, and has as its goal the production of confluent (Church-Rosser) systems of rules that can be used to decide validity. For a survey of rewriting, see Dershowitz and Jouannaud (1990).

Brown (1975) and Lankford (1975) first suggested combining completion for oriented equations, with paramodulation for unorientable ones and resolution for non-equality atoms. Paul (1986) studied the application of completion to sets of Horn clauses with equality. Completion was extended to conditional equations by Kaplan (1987) and Ganzinger (1987). Unit strategies, such the one given here, do not seem to be appropriate for completion. Recently, several restrictions of paramodulation based on term orderings have been proposed for the full first-order case, including Zhang and Kapur (1988) and Rusinowitch (1989). Kounalis and Rusinowitch (1987) and Bachmair, *et al.* (1989) improved upon the earlier Horn-clause methods in various ways.

Our method severely restricts resolution with paramodulation by incorporating an ordering on (atoms and) terms. Inferences are limited in the following ways:

- The functional reflexive axioms are not needed and, at the same time, paramodulation into variables is avoided (as for some versions of paramodulation).
- For all (resolution and paramodulation) inferences, at least one of the equations must be unconditional (as in positive unit resolution and positive unit paramodulation).
- Unless an equation is unconditional only its conditional part is used for paramodulation (analogous to positive-unit resolution).
- Only maximal terms (with respect to a given ordering) are used (analogous to ordered resolution).

Unlike Kounalis and Rusinowitch (1987), we use only unit clauses when paramodulating into conditions; unlike Bachmair, *et al.* (1989), all our inference rules use only the maximal side of an equation. Thus, our method is the first to combine a unit strategy with one based on maximal terms. It also allows for (virtually unrestricted) simplification (demodulation) by unconditional equations. Since some of the rules we consider delete or simplify antecedent clauses, the above format for inference rules, with the equations that participated in the inference also appearing as part of the consequent, is advantageous.

Limiting inference partially controls growth; keeping clauses fully simplified stunts growth even further. Such restrictions are of paramount importance in any practical theorem prover, but their (refutational) completeness has been difficult to establish. For our completeness proof, we adapt the proof-ordering method of Bachmair, *et al.* (1986) to conditional proofs. Proof orderings allow us to limit narrowing to negative literals, something that appears impossible with the recent transfinite-tree proof method of Hsiang and Rusinowitch (1986). The crux of our method is the observation that any conditional equational proof not in "normal form" must either have an unconditional "peak", that is, two applications of unconditional equations such that the middle term is the largest of the three involved, or an unconditional "drop", that is, an application of an unconditional equation (or reflexivity of equals) to an instance of a condition. The proof procedure is designed to eliminate peaks and drops, thereby reducing the complexity assigned to the proof. The refutational completeness of this strategy, but not the more general proof normalization result, follows from concurrent work of Bachmair and Ganzinger (1990) on first-order proofs. A unit Horn-clause strategy with simplification is also proved complete in Bachmair (1991).

2. Orderings

Let $\mathcal{T}$ be a set of (first-order) terms, with variables taken from a set $\mathcal{X}$, and $\mathcal{G}$ be its subset of *ground* (variable-free) terms. If t is a term in $\mathcal{T}$, by $t|_\pi$ signifies the subterm of t rooted at position π; then by $t[s]_\pi$, for some term s, we denote the term t with its subterm $t|_\pi$ replaced by s.

Term orderings are of central importance in the proposed method. A total ordering $>$ on ground terms $\mathcal{G}$ is called a *complete simplification ordering* if it has (a) the "replacement property", $s > t$ implies that any term $u[s]_\pi$, with subterm s located at some position π, is greater under $>$ than the term $u[t]_\pi$ with that occurrence of s replaced by t, and (b) the "subterm property", $t \geq t|_\pi$ for all subterms $t|_\pi$ of t. Such a ground-term ordering must be a well-ordering (Dershowitz, 1982). A *completable simplification ordering* on all terms $\mathcal{T}$ is a well-founded *partial* ordering $\succ$ (c) that can be extended to a complete simplification ordering $>$ on ground terms, such that (d) $s \succ t$ implies that $s\sigma > t\sigma$ for all ground substitutions σ. Furthermore, we will assume (e) that the constant T is minimal in $\succ$.

With a total ordering of atoms and with no equations, *per se*, the method of the next section is just selected positive-unit resolution, in which the largest negative literal is chosen. The appropriate inference rule would be expressed as:

$$\frac{E \cup \left\{ \begin{array}{rcl} q \wedge s \simeq T & \Rightarrow & u \simeq T, \\ & & l \simeq T \end{array} \right\}}{E \cup \left\{ \begin{array}{rcl} q \wedge s \simeq T & \Rightarrow & u \simeq T, \\ & & l \simeq T, \\ q\sigma & \Rightarrow & u\sigma \simeq T \end{array} \right\}}$$

where $\sigma = mgu(s, l)$, and would only be applied when $s > q$, by which we mean that s is the largest negative literal in its clause. A total simplification ordering on non-ground literals is not actually possible (which is why the ordering of the parent clause is inherited in ordered resolution), but can be approximated by a partial ordering. If only a partial ordering $\succ$ is given, we resolve negative literals that are potentially maximal. That is, we apply the above rule if $s\sigma \not\prec q\sigma$, or, in other words, if the instance $s\sigma$ of s created by resolution is not necessarily smaller than the other instantiated negative literals.

Suppose E is a set of Horn clauses in conditional equation form. To handle equality literals we need to unify at subterms of conditions, not just at the literal level. Suppose $l \simeq r$ is an equation in E. Note that whenever we refer to equations in a set, we mean that it, or the symmetric equation (with l and r exchanged), or a variant with variables renamed uniformly, actually appears in the set. With that in mind, if l unifies with a non-variable subterm $s|_\pi$ of a maximal term s in a condition $s \simeq t$ of a conditional equation $q \wedge s \simeq t \Rightarrow u \simeq v$, then a new Horn clause is created by applying the most general unifying substitution σ to the conditional equation, and then replacing $l\sigma$ with $r\sigma$, as per the unit clause $l \simeq r$. More precisely, we infer the clause $q\sigma \wedge s\sigma[r\sigma]_\pi \simeq t\sigma \Rightarrow u\sigma \simeq v\sigma$, provided that $s\sigma$ is not smaller under $\succ$ than the other side of the condition $t\sigma$, or of either side of the other conditions $q\sigma$, or of the new term $s\sigma[r\sigma]_\pi$. Thus, the conditions ensure that $s\sigma$ is the (potentially) larger side of the condition that is being paramodulated into and that the replacement yields a (potentially) smaller condition.

Of course, the empty ordering is completable. But the strength of the method, both in minimizing possible inferences and maximizing potential simplifications, is brought to bear by employing more complete orderings. In practice, any efficiently computable ordering should be better than uncontrolled paramodulation. The polynomial and path orderings commonly used in rewrite-based theorem provers (see Dershowitz, 1987) are completable. In particular, the lexicographic variant of the recursive path ordering (Kamin and Lévy, 1980) has decidability properties (Comon, 1990) that make it ideal for this purpose. Choosing an ordering that takes the goal (theorem) into account can impart a top-down flavor to the otherwise bottom-up procedure.

3. Unit Strategy

We formulate our theorem-proving procedure as an inference system operating on a set of conditional equations, and parameterized by a completable ordering $\succ$.

The rules may be classified into three "expansion" rules and four "contraction" rules. The first expansion rule applies to unit clauses:

$$\textbf{Superpose}: \quad \frac{E \cup \left\{ \begin{array}{c} u \simeq v, \\ l \simeq r \end{array} \right\}}{E \cup \left\{ \begin{array}{c} u \simeq v, \\ l \simeq r, \\ u\sigma[r\sigma]_\pi \simeq v\sigma \end{array} \right\}} \quad \text{if} \left\{ \begin{array}{l} u|_\pi \notin \mathcal{X} \\ \sigma = mgu(u|_\pi, l) \\ u\sigma \not\prec v\sigma, u\sigma[r\sigma]_\pi \end{array} \right.$$

Superposition (i.e. oriented paramodulation of positive equational literals) is performed only at non-variable positions ($u|_\pi \notin \mathcal{X}$). Either side of an equation may be used for superposition, but only if, in the context of the paramodulation, it is potentially the largest term involved ($u\sigma \not\prec v\sigma, u\sigma[r\sigma]_\pi$). Note that the two equations, $u \simeq v$ and $l \simeq r$, may actually be the same (except for renaming).

The second rule applies a unit equation to a negative literal:

$$\textbf{Narrow}: \quad \frac{E \cup \left\{ \begin{array}{rcl} q \wedge s \simeq t & \Rightarrow & u \simeq v, \\ & & l \simeq r \end{array} \right\}}{E \cup \left\{ \begin{array}{rcl} q \wedge s \simeq t & \Rightarrow & u \simeq v, \\ & & l \simeq r, \\ q\sigma \wedge s\sigma[r\sigma]_\pi \simeq t\sigma & \Rightarrow & u\sigma \simeq v\sigma \end{array} \right\}} \quad \text{if} \left\{ \begin{array}{l} s|_\pi \notin \mathcal{X} \\ \sigma = mgu(s|_\pi, l) \\ s\sigma \not\prec q\sigma, t\sigma, s\sigma[r\sigma]_\pi \end{array} \right.$$

Whenever this or subsequent rules refer to a conditional equation like $q \wedge s \simeq t \Rightarrow u \simeq v$, the intent is that $s \simeq t$ is any one of the conditions and u is either side of the implied equation.

The last expansion rule in effect resolves a maximal negative literal with reflexivity of equals ($x \simeq x$):

$$\textbf{Reflect:}\quad \frac{E\cup\left\{\; q\wedge s\simeq t \;\Rightarrow\; u\simeq v \;\right\}}{E\cup\left\{\begin{array}{rcl} q\wedge s\simeq t & \Rightarrow & u\simeq v,\\ q\sigma & \Rightarrow & u\sigma\simeq v\sigma \end{array}\right\}} \quad \text{if} \left\{\begin{array}{l}\sigma = mgu(s,t)\\ s\sigma \not\succ q\sigma\end{array}\right.$$

The remaining, contraction rules all simplify the set of conditional equations. The first deletes trivial conditional equations:

$$\textbf{Delete:}\quad \frac{E\cup\left\{\; q \;\Rightarrow\; u\simeq u \;\right\}}{E}$$

Here and later, when a rule refers to a clause of the form $q \Rightarrow u \simeq v$, an unconditional equation $(u \simeq v)$ is also intended.

The next rule allows for deletion of conditions that are trivially true:

$$\textbf{Condense:}\quad \frac{E\cup\left\{\; q\wedge s\simeq s \;\Rightarrow\; u\simeq v \;\right\}}{E\cup\left\{\; q \Rightarrow u\simeq v \;\right\}}$$

The last two contraction rules use unit clauses to simplify other clauses. One rule simplifies conditions; the other applies to the equation part. In both cases, the original clause is *replaced* by a version that is equivalent but strictly smaller under $\succ$.

$$\textbf{Simplify:}\quad \frac{E\cup\left\{\begin{array}{rcl} q & \Rightarrow & u\simeq v,\\ & l\simeq r & \end{array}\right\}}{E\cup\left\{\begin{array}{rcl} q[r\sigma]_\pi & \Rightarrow & u\simeq v,\\ & l\simeq r & \end{array}\right\}} \quad \text{if} \left\{\begin{array}{l} q|_\pi = l\sigma\\ q\succ q[r\sigma]_\pi\end{array}\right.$$

$$\textbf{Compose:}\quad \frac{E\cup\left\{\begin{array}{rcl} q & \Rightarrow & u\simeq v,\\ & l\simeq r & \end{array}\right\}}{E\cup\left\{\begin{array}{rcl} q & \Rightarrow & u[r\sigma]_\pi\simeq v,\\ & l\simeq r & \end{array}\right\}} \quad \text{if} \left\{\begin{array}{l} u|_\pi = l\sigma\\ u\succ u[r\sigma]_\pi\\ q\neq T \vee v\succ u \vee u \blacktriangleright l\end{array}\right.$$

By $q \neq T$ we mean that the equation has at least one condition; by $u \blacktriangleright l$ we mean that u is strictly larger than l in the *encompassment* ordering in which a term is larger than its proper subterms and smaller than its proper instances.

Ordered completion, which deals just with unconditional equations, uses the rules, superpose, delete, and compose.

As a simple example of our unit strategy, consider the following three clauses:

$$0 < c(0) \simeq T$$
$$c(y) < c(z) \simeq y < z$$
$$x < y \simeq T \wedge y < z \simeq T \;\Rightarrow\; x < z \simeq T$$

Using a (left-to-right) lexicographic path ordering, they generate an infinite number of clauses to which contraction rules cannot be applied:

$$0 < c^i(0) \simeq T \qquad (i \geq 1) \tag{1}$$
$$c(y) < c(z) \simeq y < z \tag{2}$$
$$x < c^j(y) \simeq T \wedge y < z \simeq T \;\Rightarrow\; x < c^j(z) \simeq T \qquad (j \geq 0) \tag{3}$$
$$0 < z \simeq T \;\Rightarrow\; 0 < c^k(z) \simeq T \qquad (k \geq 1) \tag{4}$$
$$x < c^j(0) \simeq T \;\Rightarrow\; x < c^{i+j}(0) \simeq T \qquad (j \geq 0) \tag{5}$$
$$x < y \simeq T \wedge c^k(y) < z \simeq T \;\Rightarrow\; c^k(x) < z \simeq T \qquad (k \geq 1) \tag{6}$$

There are no possible reflections, since no condition has unifiable sides. No superposition inferences between clauses of type (1) and (2), or narrowing inferences between (2) and (4), are allowed, because $c(y)$ does not unify with 0.

Unit clauses of type (1) do not unify with the second condition of (6) for the same reason. Narrowing at the first condition of (6) is also not possible. That would produce an instance of (6) in which the chosen condition is not maximal, since, in

$$0 < c^i(0) \simeq T \wedge c^{k+i}(0) < z \simeq T \;\Rightarrow\; c^k(0) < z \simeq T$$

the term $c^{k+i}(0)$ is larger than 0.

Unifying $0 < c^i(0)$ from (1) with the condition of (4) and narrowing yields:

$$T \simeq T \;\Rightarrow\; 0 < c^{i+k}(0)$$

which condenses to a new equation of type (1):

$$0 < c^{i+k}(0)$$

Similarly, narrowing (5) with (1) generates an equation of type (1).

Unifying $0 < c^i(0)$ from (1) with the first condition $x < c^j(y)$ of (3) can only succeed if $i \geq j$, giving

$$0 < c^i(0) \simeq T \wedge c^{i-j}(0) < z \simeq T \;\Rightarrow\; 0 < c^i(z) \simeq T$$

as the relevant instance of (3). For $i > j$, the second condition is larger than the first and the inference is not performed. For $i = j$, we get the new clause

$$T \simeq T \wedge 0 < z \simeq T \;\Rightarrow\; 0 < c^i(z) \simeq T$$

which condenses to a clause of type (4).

Unifying $0 < c^i(0)$ with the second condition of (3) gives (after condensation)

$$x < c^i(0) \simeq T \;\Rightarrow\; x < c^{i+j}(0) \simeq T$$

which is of type (5).

Unifying (2) with the first condition of (6) gives

$$c(x) < c(y) \simeq T \wedge c^{k+1}(y) < z \simeq T \;\Rightarrow\; c^{k+1}(x) < z \simeq T$$

which simplifies to another type (6) clause:

$$x < y \simeq T \wedge c^{k+1}(y) < z \simeq T \;\Rightarrow\; c^{k+1}(x) < z \simeq T$$

Unifying (2) with the second condition of (6) gives (after simplification):

$$x < y \simeq T \wedge c^{k-1}(y) < z \simeq T \;\Rightarrow\; c^k(x) < c(z) \simeq T$$

which after composition turns out to be an already existing clause:

$$x < y \simeq T \wedge c^{k-1}(y) < z \simeq T \;\Rightarrow\; c^{k-1}(x) < z \simeq T$$

Similarly, narrowing (3) with (2) generates bigger clauses of type (3), while narrowing (5) with (2) gives smaller clauses of type (5).

Note that clauses like $c(0) < c(c(c(0))) \simeq T$ are not generated; all the same, the complete set of unit clauses, (1) and (2), will reduce any equation $c^j(0) < c^{i+j}(0) \not\simeq T$ to the contradiction $T \not\simeq T$.

4. Completeness

Let $>$ be any complete simplification ordering extending the given partial ordering $\succ$. We define a symmetric binary relation $\leftrightarrow$, for a particular set of conditional equations E, as the smallest relation satisfying $t[l\sigma]_\pi \leftrightarrow t[r\sigma]_\pi$ for all $u_1 \simeq v_1 \wedge \cdots \wedge u_n \simeq v_n \Rightarrow l \simeq r$ in E such that $u_i\sigma \leftrightarrow^* v_i\sigma$ for each i, where $\leftrightarrow^*$ is the reflexive-transitive closure of $\leftrightarrow$. This relation corresponds to "substitution of equals" according to the axioms in E. We also define a *rewrite* relation on ground terms $\mathcal{G}$ as the intersection of $>$ and $\leftrightarrow$. That is, $t[l\sigma]_\pi \rightarrow t[r\sigma]_\pi$ if $u_1 \simeq v_1 \wedge \cdots \wedge u_n \simeq v_n \Rightarrow l \simeq r$ is in E, $t[l\sigma]_\pi > t[r\sigma]_\pi$, and $u_i\sigma \leftrightarrow^* v_i\sigma$ for each i. We use $\leftarrow$ for the inverse of $\rightarrow$, and $\rightarrow^*$ and $\leftarrow^*$ for the reflexive-transitive closures of $\rightarrow$ and $\leftarrow$, respectively.

A *proof* of an equation $s \simeq t$ between *ground* terms (any variables in s and t may be treated as Skolem constants) is a "derivation"

$$\begin{array}{ccccccccc} s = t_1 & \overset{\pi_1}{\underset{e_1\sigma_1}{\longleftrightarrow}} & t_2 & \overset{\pi_2}{\underset{e_2\sigma_2}{\longleftrightarrow}} & \cdots & \overset{\pi_m}{\underset{e_m\sigma_m}{\longleftrightarrow}} & t_{m+1} = t \\ & | & & | & & | & \\ & P_1 & & P_2 & & P_m & \end{array}$$

of $m+1$ terms ($m \geq 0$), each step $t_k \leftrightarrow t_{k+1}$ of which is either *trivial* ($t_{k+1} = t_k$), or else is justified by a conditional equation e_k in E, a position π_k in t_k, a substitution σ_k for variables in the equation, and subproofs P_k for the conditions $p_k\sigma_k$ of the applied instance $e_k\sigma_k$. Steps employing an unconditional equation do not have subproofs as part of their justification. By the completeness of positive-unit resolution for Horn clauses, any equation $s \simeq t$ that is valid for a set E of conditional equations is amenable to such an equational proof.

We use the notation $E \vdash E'$ to denote one inference step, applying any of the seven rules to a set E of conditional equations to obtain a new set E'. The inference rules are evidently sound, in that the class of provable theorems is unchanged by an inference step.

By a *peak*, we mean a proof segment of the form $s \leftarrow u \rightarrow t$; by a *valley*, we mean a proof segment of the form $u \rightarrow^* w \leftarrow^* t$; by a *drop*, we mean a step $s \rightarrow t$ with valley subproofs; a *plateau* is a trivial subproof of form $s \leftrightarrow s$. The *depth* of a proof is the maximum number of nestings of subproofs; it is one more than the maximum depth of its subproofs. A *normal-form* proof is a valley proof of depth 0. That is the same as saying that a normal-form proof has no peaks, no drops, and no plateaus. Normal-form proofs may be thought of as "direct" proofs; in a refutational framework the existence of such a proof for $s \simeq t$ means that demodulation of s and t using positive unit equations suffices to derive a contradiction between the Skolemized negation $s' \not\simeq t'$ of the given theorem and $x \simeq x$.

The above inference rules are designed to allow any equational proof to be tranformed into normal form. A strategy based on these rules is complete if we can show that, with enough inferences, any theorem has a normal-form proof. We call an inference "fair" if all persistent superpositions, narrowings, and reflections have been considered:

Definition. An inference sequence $E_0 \vdash E_1 \vdash \cdots$ is *fair* if

$$\mathbf{exp}(E_\infty) \subseteq \bigcup_{i\geq 0} E_i,$$

where E_∞ is the set $\liminf_j E_j = \cup_{i\geq 0} \cap_{j\geq i} E_j$ of *persisting* conditional equations and $\mathbf{exp}(E_\infty)$ is the set of conditional equations that may be inferred from persisting equations by one application of an expansion rule (**superpose, narrow,** or **reflect**).

Our goal is to demonstrate that for any proof $s \leftrightarrow^* t$ of $s \simeq t$ in E_0, there eventually exists an unconditional valley proof $s \rightarrow^* w \leftarrow^* t$. Were it not for contraction rules, it would be relatively easy to show that **narrow** and **reflect** eventually provide an unconditional proof of $s \simeq t$, and that **superpose** eventually turns that into a valley.

Theorem (Normalization). *If an inference sequence $E_0 \vdash E_1 \vdash \cdots$ is fair, then for any proof of $s \simeq t$ in E_0, there is a normal-form proof of $s \simeq t$ in E_∞.*

This is shown by transfinite induction on proofs. Proofs are measured in the following way: For each step

$$t_k = w[u\sigma]_\pi \underset{\substack{e\sigma \\ | \\ P}}{\overset{\pi}{\longleftrightarrow}} w[v\sigma]_\pi = t_{k+1}$$

in a proof, where e is a conditional equation $q_1 \wedge \cdots \wedge q_n \Rightarrow u \simeq v$, we consider the quintuple

$$\langle n,\ q_1\sigma \wedge \cdots \wedge q_n\sigma,\ t_k,\ u,\ t_{k+1}\rangle,$$

where (with loss of generality) we are assuming $t_k \geq t_{k+1}$ in the complete ordering $>$. Quintuples are compared left-to-right lexicographically, with the first component (the number of conditions n) compared in the natural ordering of natural numbers, the second (the instantiated conditions $q_1\sigma \wedge \cdots \wedge q_n\sigma$), third ($t_k$), and fifth ($t_{k+1}$) components under the complete ordering $>$, and the fourth component (u) in the encompassment ordering $\blacktriangleright$. An unconditional step has 0 as its first component and T as its second. Finally, proofs are compared by comparing multisets consisting of quintuples for all steps in their top-level proofs *or* subproofs, using the extension to finite multisets of the above ordering on quintuples. We use $\gg$ to denote this proof ordering. It can be shown by standard arguments (Dershowitz and Manna, 1979) that $\gg$ is well-founded.

Intuitively, the first component is designed to decrease with each application of **reflect** or **condense**, the second with applications of **narrow** or **simplify**, the third with **superpose**, the fourth and fifth cater to **compose**. The multiset structure of the proof ordering ensures that decreasing the complexity of subproofs decreases the complexity of the whole proof (and also takes care of **delete**). We need to show that inferences never increase the complexity of proofs and, furthermore, that there are always inferences that can decrease the complexity of non-normal proofs. Then, by induction with respect to $\gg$, the eventual existence of a normal-form proof follows.

Lemma 1. *If $E \vdash E'$, then for any proof P in E of an equation $s \simeq t$, there exists a proof P' in E' of $s \simeq t$, such that $P \gg P'$ or $P = P'$.*

This is established by consideration of the effects of each contracting inference rule that deletes or replaces equations, since for expansion rules, $E \subseteq E'$, and we can take $P' = P$.

Consider any ground proof and look at a step $t_k = w[u\sigma]_\pi \leftrightarrow^{\pi}_{e\sigma} w[v\sigma]_\pi = t_{k+1}$.

Delete can prevent a proof from employing a clause $p \Rightarrow u \simeq u$. There is, however, an alternative proof that splices out the step $t_k \leftrightarrow t_{k+1} = t_k$, leaving just t_k, and omits all its subproofs. This strictly decreases the complexity of the whole proof, by eliminating one or more quintuples from the multiset measure.

Condense erases a condition $s \simeq s$ from a clause. If e is such a clause, there is a new proof using the condensed clause instead, which omits any (unnecessary) proofs of that condition. The quintuple associated with the step experiences a decrease in its first component.

Simplify changes the second component of the step's cost from $q[l\sigma]$ to the smaller $q[r\sigma]$. Though the cost of the subproofs is increased on account of an additional unconditional step $q[l\sigma] \to q[r\sigma]$ needed to establish $q[r\sigma]$ (given proofs of $q[l\sigma]$), that unconditional step (with 0 in its first component) is dominated by the simplified conditional one (with $m > 0$ in its first component).

Compose replaces a step $t_k = w[l\tau] \leftrightarrow_{p\sigma \Rightarrow u\sigma \simeq v\sigma} t_{k+1}$ with a two-step proof $t_k = w[l\tau] \to w[r\tau] \leftrightarrow t_{k+1}$. The cost of the replaced step is $\langle n, p\sigma, t_k, u, t_{k+1}\rangle$ if $t_k > t_{k+1}$; otherwise, it is $\langle n, p\sigma, t_{k+1}, v, t_k\rangle$. If $t_k > t_{k+1}$, then the cost of $w[r\tau] \leftrightarrow t_{k+1}$ is smaller in the third component; if $t_k \leq t_{k+1}$, it is smaller in the last. (The first two components are in any case unchanged.) The cost $\langle 0, T, t_k, l, w[r\tau]\rangle$ of the unconditional step $t_k \to w[r\tau]$ is smaller than that of the replaced step in the first, third, or fourth component, depending on which of the enabling conditions of the inference rule is satisfied: if the replaced step was conditional ($q \neq T$), then it is smaller in the first; if $q = T$, but $v > u$, then (by the replacement and substitution properties) $t_{k+1} > t_k$, and the reduction is in the third; if $q = T$ and $v \leq u$, but $u \blacktriangleright l$, then the first three components are the same, but the fourth is smaller in the encompassment ordering.

Lemma 2. *If P is a non-normal-form proof in E, then there exists a proof P' in $E \cup \mathbf{exp}(E)$ such that $P \gg P'$.*

The argument depends on a distinction between "non-critical" subproofs, for which there is a proof P' in E itself, and "critical" subproofs, for which equations in $\mathbf{exp}(E)$ are needed. A peak $t_{k-1} \leftarrow^{\pi}_{p\sigma \Rightarrow l\sigma \simeq r\sigma} t_k \rightarrow^{\rho}_{q\tau \Rightarrow u\tau \simeq v\tau} t_{k+1}$, where $t_k = w[l\sigma]_\pi[u\tau]_\rho$, is *critical* if the position π is at or below the position ρ in w at which $u \simeq v$ is applied, but not at or below a position corresponding to any variable in u, or (symmetrically) if ρ falls within the non-variable part of the occurrence of l in w. Similarly, a drop $t_k \rightarrow^{\pi}_{q\sigma \Rightarrow e\sigma} t_{k+1}$ is *critical* if the first or last step of one of the subproofs for $q\sigma$ takes place within the non-variable part of the condition q.

Since any proof must have at least one subproof of depth 0, any non-normal proof must have a plateau, an *unconditional* peak, or a drop of depth 1 with (unconditional) valley subproofs. Thus, we need not worry about peaks involving a conditional rule, nor drops in which the proof of some condition is not unconditional. All plateaus of depth 0 can be spliced out. Critical unconditional peaks, critical drops with non-empty unconditional valley subproofs, and drops with empty proofs of conditions can each be replaced by a smaller proof, using the conditional equation generated by a **superpose**, **narrow**, or **reflect** inference, respectively. Superposition replaces two steps with one that is smaller in the third component (the first two are unchanged); narrowing results in a step that is smaller in the second component (and removes a step from the subproof); reflection causes a decrease in the first component. Narrowing can be restricted to the maximal side of the maximal condition, since a drop with non-empty subproofs must have a step emanating from the larger side of its largest condition.

Non-critical unconditional peaks $t_{k-1} \leftarrow t_k \rightarrow t_{k+1}$ have alternative, smaller proofs $t_{k-1} \rightarrow^* t_k \leftarrow^* t_{k+1}$ in E by the version of the Critical Pair Lemma of (Knuth and Bendix, 1970) in (Lankford, 1975). Consider a non-critical drop $w[u\sigma] \leftrightarrow_{q\sigma \Rightarrow u\sigma \simeq v\sigma} w[v\sigma]$, with unconditional subproof $p\sigma \rightarrow p' \rightarrow^* p'' \leftarrow^* p'''$, where $p\sigma$ is no smaller than any other term in the subproof $q\sigma$. Suppose p has a variable x at position π and the first step applies within the variable part $p|_\pi$. That is, $p\sigma = p\sigma[x\sigma]_\pi \rightarrow p\sigma[r]_\pi = p'$. Let τ be the same substitution as σ except that $\tau : x \mapsto r$. There is a smaller proof (smaller, vis-a-vis $\gg$) in E: $w[u\sigma] \leftarrow^* w[u\tau] \leftrightarrow_{q\sigma \Rightarrow u\sigma \simeq v\sigma} w[v\tau] \rightarrow^* w[v\sigma]$. The new conditional step $w[u\tau] \leftrightarrow w[v\tau]$ is cheaper (in the second component) than the original, since $q\tau$ must be strictly smaller than $q\sigma$. The steps $w[u\sigma] \leftarrow^* w[u\tau]$ and $w[v\tau] \rightarrow^* w[v\sigma]$ are also cheaper than $w[u\sigma] \leftrightarrow w[v\sigma]$, since they are unconditional (hence smaller in the first component). Also any rewrites $x\sigma \rightarrow r$ that need to be added to turn a proof of $q\sigma$ into a proof of $q\tau$ are unconditional.

The Normalization Theorem follows. If $s \simeq t$ is provable in E_0, then (by Lemma 1) it has a proof P in the limit E_∞. If P is non-normal, then (by Lemma 2) it admits a smaller proof P' using (in addition to E_∞) a finite number of equations in $\mathbf{exp}(E_\infty)$. By fairness, each of those equations appeared at least once along the way. Subsequent inferences (by Lemma 1) can only decrease the complexity of the proof of such an equation once it appears in a set E_j (and has a one-step proof). Thus, each equation needed in P' has a proof of no greater complexity in E_∞ itself, and hence (by the multiset nature of the proof measure), there is a proof of $s \simeq t$ in E_∞ that is strictly smaller than P. Since the ordering on proofs is well-founded, by induction there must be a normal proof in E_∞.

5. Extensions

In the above method, the same ordering is used for simplification as for choosing the maximal literal. In fact, a different selection strategy can be used for choosing the literal to narrow, as in (Ganzinger, 1987; Sivakumar, 1989), but then the term ordering must be used to choose the larger side of the equality.

We used only unconditional equations for simplification and composition. Conditional equations can also be used—but only in those cases where the proof ordering shows a decrease anyway. An alternative is to design an inference system that distinguishes between different kinds of non-unit clauses. An instance $p\sigma \Rightarrow u\sigma \simeq v\sigma$ of a conditional equation is "decreasing" (in the terminology of Dershowitz and Okada, 1990) if $u\sigma \succ v\sigma, p\sigma$ in the completable ordering. This is the same condition as imposed on conditional rewrite rules by the completion-like procedures of Kaplan (1987) and Ganzinger (1987). In these methods, superposition is used when the left-hand side is larger than the conditions; narrowing, when a condition dominates the left-hand side. As theorem provers, however, they are refutationally *in*complete, since they make no provision for "unorientable" equations $s \simeq t$ such that $s \not\succ t$ and $t \not\succ s$. For a complete method, the inference rules given here must be modified to use the largest positive *or* negative clause in each expansion, and to treat decreasing equations like unit equations. In particular, superposition is needed between decreasing conditional rules. We must redefine a normal-form proof of $s \simeq t$ to be a valley proof in which each subproof is also in normal form and each term in a subproof is smaller than the larger of s and t; see (Dershowitz and Okada, 1988). (The normal forms of the previous section are a special case.) Any non-normal-form proof has a peak made from decreasing instances with normal-form subproofs, or else has a non-decreasing step with a drop.

References

[1] L. Bachmair, *Canonical Equational Proofs*. Boston: Birkhäuser, 1991. To appear.

[2] L. Bachmair, N. Dershowitz, and J. Hsiang, "Orderings for equational proofs," in *Proceedings of the IEEE Symposium on Logic in Computer Science*, (Cambridge, MA), pp. 346–357, June 1986.

[3] L. Bachmair, N. Dershowitz, and D. A. Plaisted, "Completion without failure," in *Resolution of Equations in Algebraic Structures 2: Rewriting Techniques* (H. Aït-Kaci and M. Nivat, eds.), ch. 1, pp. 1–30, New York: Academic Press, 1989.

[4] L. Bachmair and H. Ganzinger, "On restrictions of ordered paramodulation with simplification," in *Proceedings of the Second International Workshop on Conditional and Typed Rewriting Systems* (M. Okada, ed.), (Montreal, Canada), June 1990. *Lecture Notes in Computer Science*, Springer, Berlin.

[5] R. S. Boyer, *Locking: A restriction of resolution*. PhD thesis, University of Texas at Austin, Austin, TX, 1971.

[6] T. C. Brown, Jr., *A structured design-method for specialized proof procedures*. PhD thesis, California Institute of Technology, Pasadena, CA, 1975.

[7] H. Comon, "Solving inequations in term algebras (Preliminary version)," in *Proceedings of the Fifth Annual IEEE Symposium on Logic in Computer Science* (Philadelphia, PA), pp. 62—69, June 1990.

[8] N. Dershowitz, "Orderings for term-rewriting systems," *Theoretical Computer Science*, vol. 17, pp. 279–301, March 1982.

[9] N. Dershowitz, "Termination of rewriting," *J. of Symbolic Computation*, vol. 3, pp. 69–115, February/April 1987. Corrigendum: *4*, 3 (December 1987), 409–410.

[10] N. Dershowitz and J.-P. Jouannaud, "Rewrite systems," in *Handbook of Theoretical Computer Science B: Formal Methods and Semantics*, (J. van Leeuwen, ed.), ch. 6, Amsterdam: North-Holland, 1990.

[11] N. Dershowitz and Z. Manna, "Proving termination with multiset orderings," *Communications of the ACM*, vol. 22, pp. 465–476, August 1979.

[12] N. Dershowitz and M. Okada, "Proof-theoretic techniques and the theory of rewriting," in *Proceedings of the Third IEEE Symposium on Logic in Computer Science* (Edinburgh, Scotland), pp. 104–111, July 1988.

[13] N. Dershowitz and M. Okada, "A rationale for conditional equational programming," *Theoretical Computer Science*, vol. 75, pp. 111—138, 1990.

[14] U. Furbach, "Oldy but goody: Paramodulation revisited," in *Proceedings of the GI Workshop on Artificial Intelligence* (Morik, ed.), pp. 195–200, 1987. Vol. 152 of *Informatik Fachberichte.*

[15] H. Ganzinger, "A completion procedure for conditional equations," in *Proceedings of the First International Workshop on Conditional Term Rewriting Systems* (S. Kaplan and J.-P. Jouannaud, eds.), (Orsay, France), pp. 62–83, July 1987. Vol. 308 of *Lecture Notes in Computer Science*, Springer, Berlin (1988).

[16] L. Henschen and L. Wos, "Unit refutations and Horn sets," *J. of the Association for Computing Machinery*, vol. 21, pp. 590–605, 1974.

[17] J. Hsiang and M. Rusinowitch, "A new method for establishing refutational completeness in theorem proving," in *Proceedings of the Eighth International Conference on Automated Deduction* (J. H. Siekmann, ed.), (Oxford, England), pp. 141–152, July 1986. Vol. 230 of *Lecture Notes in Computer Science*, Springer, Berlin.

[18] J. Hsiang and M. Rusinowitch, "On word problems in equational theories," in *Proceedings of the Fourteenth EATCS International Conference on Automata, Languages and Programming* (T. Ottmann, ed.), (Karlsruhe, West Germany), pp. 54–71, July 1987. Vol. 267 of *Lecture Notes in Computer Science*, Springer, Berlin.

[19] S. Kamin and J.-J. Lévy, "Two generalizations of the recursive path ordering," Unpublished note, Department of Computer Science, University of Illinois, Urbana, IL, February 1980.

[20] S. Kaplan, "Simplifying conditional term rewriting systems: Unification, termination and confluence," *J. of Symbolic Computation*, vol. 4, pp. 295–334, December 1987.

[21] D. E. Knuth and P. B. Bendix, "Simple word problems in universal algebras," in *Computational Problems in Abstract Algebra* (J. Leech, ed.), pp. 263–297, Oxford, U. K.: Pergamon Press, 1970. Reprinted in *Automation of Reasoning 2*, Springer, Berlin, pp. 342–376 (1983).

[22] E. Kounalis and M. Rusinowitch, "On word problems in Horn theories," in *Proceedings of the First International Workshop on Conditional Term Rewriting Systems* (S. Kaplan and J.-P. Jouannaud, eds.), (Orsay, France), pp. 144–160, July 1987. Vol. 308 of *Lecture Notes in Computer Science*, Springer, Berlin (1988).

[23] D. S. Lankford, "Canonical inference," Memo ATP-32, Automatic Theorem Proving Project, University of Texas, Austin, TX, December 1975.

[24] E. Paul, "On solving the equality problem in theories defined by Horn clauses," *Theoretical Computer Science*, vol. 44, no. 2, pp. 127–153, 1986.

[25] G. Robinson and L. Wos, "Paramodulation and theorem-proving in first order theories with equality," in *Machine Intelligence 4* (B. Meltzer and D. Michie, eds.), pp. 135–150, Edinburgh, Scotland: Edinburgh University Press, 1969.

[26] M. Rusinowitch, *Démonstration Automatique: Techniques de réécriture.* Paris, France: InterEditions, 1989.

[27] G. Sivakumar, *Proofs and Computations in Conditional Equational Theories.* PhD thesis, Department of Computer Science, University of Illinois, Urbana, IL, 1989.

[28] H. Zhang and D. Kapur, "First-order theorem proving using conditional equations," in *Proceedings of the Ninth International Conference on Automated Deduction* (E. Lusk and R. Overbeek, eds.), (Argonne, Illinois), pp. 1–20, May 1988. Vol. 310 of *Lecture Notes in Computer Science*, Springer, Berlin.

EXTENDED TERM REWRITING SYSTEMS

Jan Willem Klop

CWI, Kruislaan 413, 1098 SJ Amsterdam, The Netherlands;
Dept. of Math. and Comp. Sci., Free University,
De Boelelaan 1081a, 1081 HV Amsterdam, The Netherlands;
jwk@cwi.nl

Roel de Vrijer

Dept. of Philosophy, University of Amsterdam,
Nieuwe Doelenstraat 15, 1012 CP Amsterdam, The Netherlands;
Dept. of Math. and Comp. Sci., Free University,
De Boelelaan 1081a, 1081 HV Amsterdam, The Netherlands;
rdv@cs.vu.nl

Abstract

In this paper we will consider some extensions of the usual term rewrite format, namely: term rewriting with conditions, infinitary term rewriting and term rewriting with bound variables. Rather than aiming at a complete survey, we discuss some aspects of these three extensions.

Contents

Note: Research of the first author is partially supported by ESPRIT BRA projects 3020: Integration and 3074: Semagraph.

INTRODUCTION

The aim of the present Workshop is to focus upon conditional term rewriting, typed term rewriting and other extended forms of term rewriting. Accordingly, in this paper we have set out to discuss three of these extensions, viz. infinite term rewriting, conditional rewriting and term rewriting with bound variables. Our discussion will be largely of an introductory nature. The subject of the second part, conditional term rewriting, is already well established and widely studied; we will develop some of the basic theory and then focus on a proof-theoretic application of Conditional Term Rewriting Systems that seems not to be generally known yet. The first part, which is rendered in a rather informal style following the corresponding talk at this Workshop, does present some recently obtained insights about infinite term rewriting. The extension discussed in section 3, sometimes called Combinatory Reduction Systems, gives a framework incorporating not only (ordinary) TRSs, but also rewrite systems with bound variables, as in the Lambda Calculus. We discuss some of the 'classical' theorems (Church-Rosser etc.) for the subclass of orthogonal CRSs. More introductory remarks and motivations can be found at the beginning of each of the three sections.

Our notation and terminology follow Klop [1987, 1991].

1. INFINITE TERM REWRITING

In this section we explain some recent work on infinite term rewriting as reported in Kennaway, Klop, Sleep & de Vries [1990ab, 1991] where the formal treatment including full proofs can be found. This work was stimulated by earlier studies of infinite rewriting by Dershowitz, Kaplan & Plaisted [1989] and Farmer & Watro [1989].

There is ample motivation for a theoretical study of infinite rewriting, in view of the facility that several lazy functional programming languages such as Miranda (Turner [1985]) and Haskell (Hudak [1988]) have enabling them to deal with (potentially) infinite terms, representing e.g. the list of all primes. Another motivation is the correspondence between infinite rewriting and rewriting of term graphs: a theory for infinite rewriting provides much of a foundation for a theory of term graph rewriting, since a cyclic term graph yields after unwinding an infinite term. Indeed, this correspondence has been the starting point for the work of Farmer & Watro [1989].

Our starting point is an ordinary TRS (Σ, R), where Σ is the signature and R is the set of rewrite rules. In fact, we will suppose throughout that our TRSs are orthogonal. Now it is obvious that the rules of the TRS (Σ, R) just as well apply to infinite terms as to the usual finite ones. First, let us explain the notion of infinite term that we have in mind. Let $Ter(\Sigma)$ be the set of finite Σ-terms. Then $Ter(\Sigma)$ can be equipped with a distance function d such that for $t, s \in Ter(\Sigma)$, we have $d(t, s) = 2^{-n}$ if the n-th level of the terms s, t (viewed as labelled trees) is the first level where a difference appears, in case s and t are not identical; furthermore, $d(t, t) = 0$. It is well-known that this construction yields $(Ter(\Sigma), d)$ as a metric space. Now infinite terms are obtained by taking the completion of this metric space, and they are represented by infinite trees. We will refer to the complete metric space arising in this way as $(Ter^{\infty}(\Sigma), d)$, where $Ter^{\infty}(\Sigma)$ is the set of finite and infinite terms over Σ.

A natural consequence of this construction is the emergence of the notion of Cauchy convergence as a possible basis for infinite reductions which have a limit: we say that $t_0 \rightarrow t_1 \rightarrow t_2 \rightarrow \ldots$ is an infinite reduction sequence with limit t, if t is the limit of the sequence $t_0, t_1, \ldots$ in the usual sense of Cauchy convergence. See Figure 1.1 for an example, based on a rewrite rule $F(x) \rightarrow P(x, F(S(x)))$ in the presence of a constant 0.

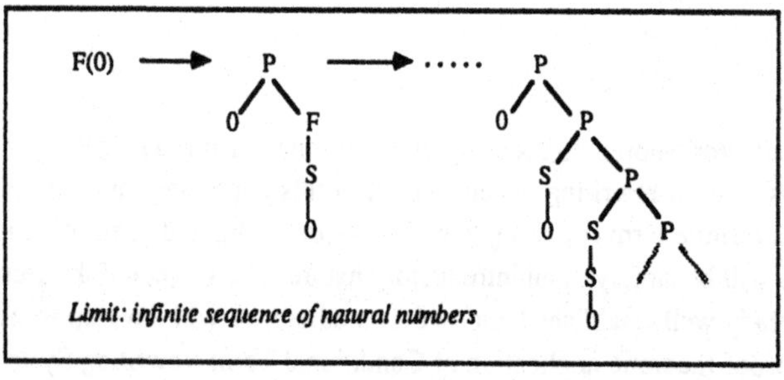

Figure 1.1

In fact, this notion of converging reduction sequence is the starting point for Dershowitz e.a. [1989]. In the sequel we will however adopt a stronger notion of converging reduction sequence which turns out to have better properties. First, let us argue that it makes sense to consider not only reduction sequences of length ω, but even reduction sequences of length α for arbitrary ordinals α. Given a notion of convergence, and limits, we may iterate reduction sequences beyond length ω and consider e.g. $t_0 \rightarrow t_1 \rightarrow t_2 \rightarrow \ldots \rightarrow t_n \rightarrow \ldots s_0 \rightarrow s_1 \rightarrow s_2 \rightarrow s_3 \rightarrow \ldots$ r where $\lim_{n\to\infty} t_n = s_0$ and $\lim_{n\to\infty} s_n = r$. See Figure 1.2 for such a reduction sequence of length $\omega + \omega$, which may arise by evaluating first the left part of the term at hand, and next the right part. Of course, in this example a 'fair' evaluation is possible in only ω many reduction steps, but we do not want to impose fairness requirements at the start of the theory development—even though we may (and will) consider it to be a desirable feature that reductions of length α could be 'compressed' to reductions of length not exceeding ω steps, yielding the same 'result'.

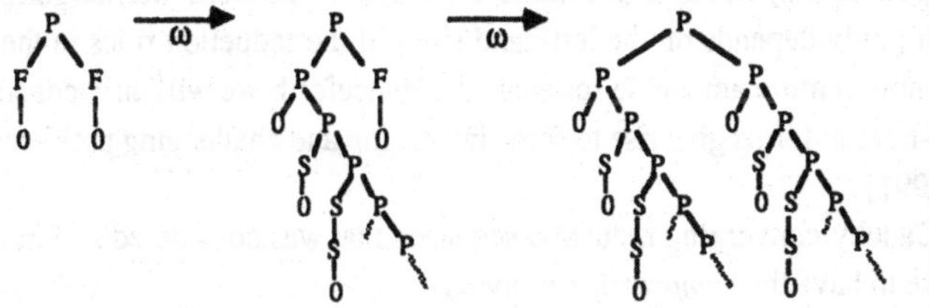

Transfinite reduction sequence of length $\omega + \omega$

Figure 1.2

We will give a formal definition now.

1.1. DEFINITION. Let (Σ, R) be a TRS. A *(Cauchy-) convergent* R-*reduction sequence of length* α (an ordinal) is a sequence $\langle t_\beta \mid \beta \leq \alpha \rangle$ of terms in $Ter^\infty(\Sigma)$, such that
(i) $t_\beta \rightarrow_R t_{\beta+1}$ for all $\beta < \alpha$,
(ii) $t_\lambda = \lim_{\beta<\lambda} t_\beta$ for every limit ordinal $\lambda \leq \alpha$.
Here (ii) means: $\forall n\ \exists \mu < \lambda\ \forall \nu\ (\mu \leq \nu \leq \lambda \Rightarrow d(t_\nu, t_\lambda) \leq 2^{-n})$.

NOTATION: If $\langle t_\beta \mid \beta \leq \alpha \rangle$ is a Cauchy-convergent reduction sequence we write $t_0 \rightarrow_\alpha^c t_\alpha$ ('c' for 'Cauchy').

The notion of normal form as a final result has to be considered next. We simply generalize the old finitary notion of normal form to the present infinitary setting thus: a (possibly infinite) term is a normal form *when it contains no redexes*. The only difference with the finitary case is that here a redex may be itself an infinite term. But note that a redex is still so by virtue of a finite prefix, called the redex pattern—this is so because our rewrite rules are orthogonal and hence contain no repeated variables. This choice of 'normal form' deviates from that in Dershowitz e.a. [1989]: there a (possibly infinite) term t is said to be an *ω-normal form* if either t contains no redexes, or the only possible reduction of t is to itself: $t \rightarrow t$, in one step.

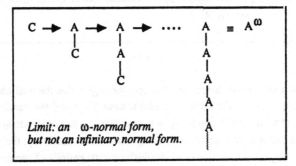

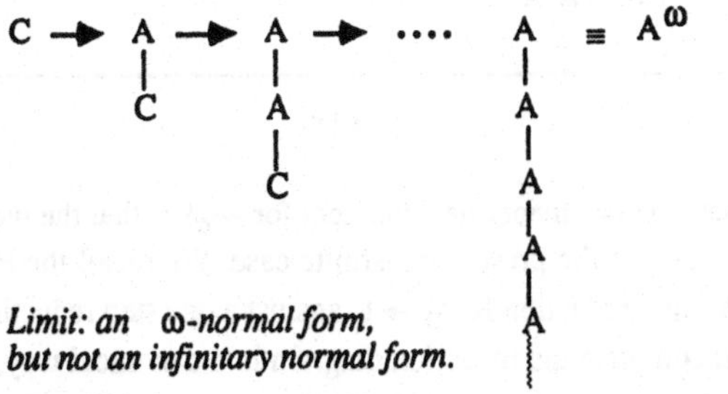

Limit: an ω*-normal form, but not an infinitary normal form.*

Figure 1.3

So, in Figure 1.3 we have, with as TRS $\{C \to A(C), A(x) \to x\}$, a (Cauchy-) converging reduction sequence with as limit the infinite term A(A(A(A..., abbreviated as A^ω; this limit is not a normal form in our sense but it is an ω-normal form, as A^ω only reduces to itself: $A^\omega \to A^\omega$. (Note that this step can be performed in infinitely many different ways, since every A in A^ω is the root of a redex.) Normal forms in our sense are shown in Figures 1.1, 1.2 as the rightmost terms (if no other reduction rules are present than the one mentioned above). Henceforth we will often drop the reference 'infinite' or 'infinitary'. Thus a term, or a normal form, may be finite or infinite. Note that the concept 'normal form', in contrast to that of 'ω-normal form', only depends on the left-hand sides of the reduction rules in the TRS (Σ, R), which makes the former notion more amenable for analysis. Henceforth we will only consider 'normal forms', but we note that ω-normal forms give rise to some interesting and challenging problems explicitly stated in Kennaway e.a. [1991].

The notion of Cauchy-converging reduction sequence that was considered so far, is not quite satisfactory. We would like to have the *compression property*:

$$t_0 \to_\alpha{}^c t_\alpha \;\Rightarrow\; t_0 \to_{\le\omega}{}^c t_\alpha.$$

That is, given a reduction $t_0 \to_\alpha{}^c t_\alpha$, of length α, the result t_α can already be found in at most ω many steps. ('At most', since it may happen that a transfinite reduction sequence can be compressed to finite length, but not to length ω.) Unfortunately, $\to_\alpha{}^c$ lacks this property:

1.2. COUNTEREXAMPLE. Consider the orthogonal TRS with rules $\{A(x) \to A(B(x)), B(x) \to E(x)\}$. Then $A(x) \to_\omega A(B^\omega) \to A(E(B^\omega))$, so $A(x) \to_{\omega+1} A(E(B^\omega))$. However, we do not have $A(x) \to_{\le\omega} A(E(B^\omega))$, as can easily be verified.

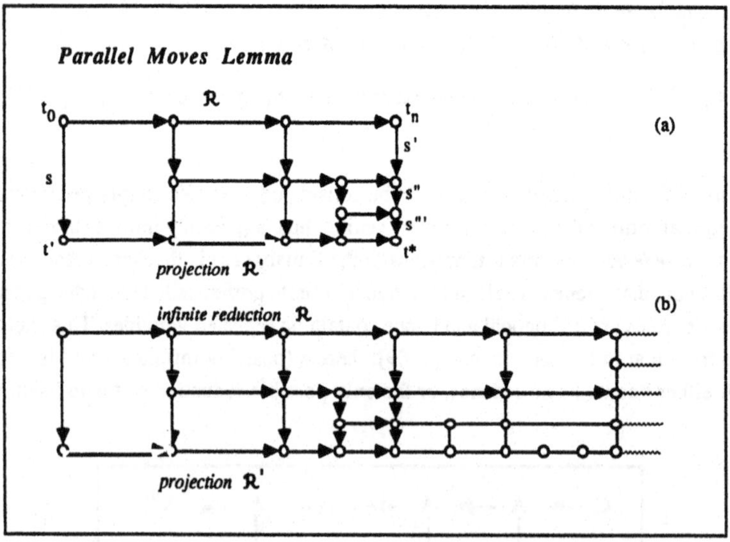

Figure 1.4

Another obstacle to a satisfactory theory development for $\to_\alpha{}^c$ is that the well-known Parallel Moves Lemma resists a generalization to the present transfinite case. We recall the Parallel Moves Lemma in Figure 1.4(a): setting out a finite reduction $\mathcal{R}$: $t_0 \twoheadrightarrow t_n$ against a one step reduction $t_0 \to_s t'$ (where s is the contracted redex), one can complete the reduction diagram in a canonical way, thereby obtaining as the right-hand side of the diagram a reduction $t_n \twoheadrightarrow t^*$ which consists entirely of contractions of all the *des-*

cendants of s *along* $\mathfrak{R}$. Furthermore, the reduction $\mathfrak{R}'$: t' ↠ t* arising as the lower side of this reduction diagram, is called the *projection of* $\mathfrak{R}$ *over the reduction step* $t_0 \to_s t'$. Notation: $\mathfrak{R}' = \mathfrak{R}/(t_0 \to_s t')$.

We would like to have a generalization of Parallel Moves Lemma where $\mathfrak{R}$ is allowed to be infinite, and converging to a limit. In this way we would have a good stepping stone towards establishing infinitary confluence properties. However, it is not clear at all how such a generalization can be established. The problem is shown in Figure 1.5. First note that we can without problem generalize the notion of 'projection' to infinite reductions, as in Figure 1.4(b): there $\mathfrak{R}'$ is the projection of the infinite $\mathfrak{R}$ over the displayed reduction step. This merely requires an iteration of the finitary Parallel Moves Lemma, no infinitary version is needed. Now consider the two rule TRS {A(x, y) → A(y, x), C → D}. Let $\mathfrak{R}$ be the infinite reduction A(C, C) → A(C, C) → A(C, C) → ... , in fact a reduction cycle of length 1. Note that $\mathfrak{R}$ is converging, with limit A(C, C). The projection $\mathfrak{R}'$ of $\mathfrak{R}$ over the step A(C, C) → A(D, C), however, is no longer converging. For, this is A(D, C) → A(C, D) → A(D, C) → ..., a 'two cycle'. So, the class of infinite converging reduction sequences is not closed under projection. This means that in order to get some decent properties of infinitary reduction in this sense, one has to impose further restrictions; Dershowitz e.a. [1989] chooses to impose these restrictions on the terms, thus ruling out e.g. terms as A(C, C) because they are not 'top-terminating'. (We will come back to this important notion later on.) Another road, the one taken here, is to strengthen the concept of converging reduction sequence—this option is also chosen in Farmer & Watro [1989].

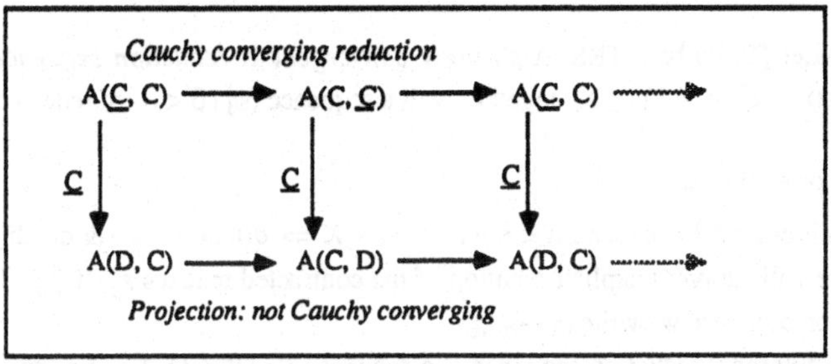

Figure 1.5

As the last example shows, there is a difficulty in that we loose the notion of descendants which is so clear and helpful in finite reductions. Indeed, after the infinite reduction A(C, C) → A(C, C) → A(C, C) → ... , with Cauchy limit A(C, C), what is the descendant of the original underlined redex C in the limit A(C, C)? There is no likely candidate.

We will now describe the stronger notion of converging reduction sequence that does preserve the notion of descendants in limits. If we have a converging reduction sequence $t_0 \to_{s_0} t_1 \to_{s_1} \ldots t$, where s_i is the redex contracted in the step $t_i \to t_{i+1}$ and t is the limit, we now moreover require that

$$\lim_{i \to \infty} \text{depth}(s_i) = \infty. \qquad (*)$$

Here depth(s_i), the depth of redex s_i, is the distance of the root of t_i to the root of the subterm s_i. If the converging reduction sequence satisfies this additional requirement (∗), it is called *strongly convergent*. The difference between the previous notion of (Cauchy-) converging reduction sequence and the present one, is suggested by Figure 1.6. The circles in that figure indicate the root nodes of the contracted redexes; the shaded part is that prefix part of the term that does not change anymore in the sequel of the reduction. The point of the additional requirement (∗) is that this growing non-changing prefix is required really to be non-changing, in the sense that no activity (redex contractions) in it may occur at all, even when this activity would by accident yield the same prefix.

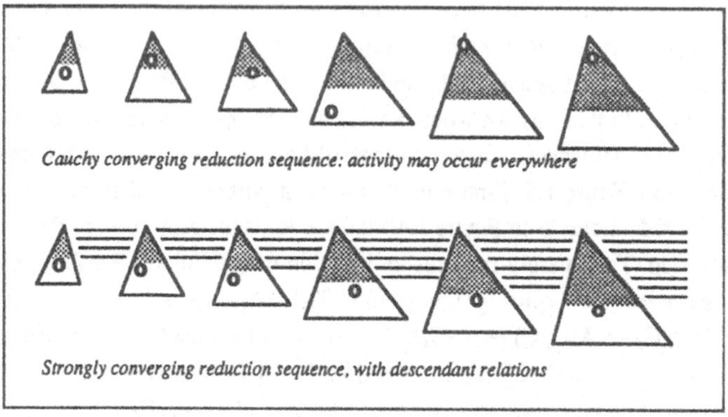

Figure 1.6

Note that there is now an obvious definition of descendants in the limit terms; the precise formulation is left to the reader.

In fact, we define strongly converging reductions of length α for every ordinal α, by imposing the additional condition (*) whenever a limit ordinal $\lambda \le \alpha$ is encountered. (It will turn out however that only countable ordinals may occur.) More formally:

1.3. DEFINITION. Let (Σ, R) be a TRS. A *strongly convergent* R-*reduction sequence of length* α is a sequence $\langle t_\beta \mid \beta \le \alpha \rangle$ of terms in $Ter^\infty(\Sigma)$, together with a sequence $\langle s_\beta \mid \beta < \alpha \rangle$ of redex occurrences s_β in t_β, such that

(i) $t_\beta \to_{s_\beta} t_{\beta+1}$ for all $\beta < \alpha$,

(ii) for every limit ordinal $\lambda \le \alpha$: $\forall n\ \exists \mu < \lambda\ \forall \nu\ (\mu \le \nu < \lambda \Rightarrow d(t_\nu, t_\lambda) \le 2^{-n}\ \&\ \text{depth}(s_\nu) \ge n)$.

Notation: Often we will suppress explicit mention of the contracted redexes s_β. If $\langle t_\beta \mid \beta \le \alpha \rangle$ is a strongly convergent reduction sequence we write $t_0 \to_\alpha t_\alpha$.

Furthermore, a *divergent* reduction sequence is a sequence $\langle t_\beta \mid \beta < \lambda \rangle$, λ some limit ordinal, such that every initial segment $\langle t_\beta \mid \beta \le \nu \rangle$ is strongly convergent, but there is no t_λ such that $\langle t_\beta \mid \beta \le \lambda \rangle$ is strongly convergent. (E.g. the infinite reduction $A(C, C) \to A(C, C) \to \ldots$ considered above, is divergent.)

Henceforth all our infinitary reductions will be strongly convergent. Now we can state the benefits of this notion; for the full proofs we refer to Kennaway e.a. [1990a].

1.4. COMPRESSION LEMMA. *In every orthogonal TRS:*

$$t \to_\alpha t' \;\Rightarrow\; t \to_{\le\omega} t'.$$

(Note that the counterexample 1.2 to compression for Cauchy converging reductions was not strongly converging.)

1.5. INFINITARY PARALLEL MOVES LEMMA. *In every orthogonal TRS:*

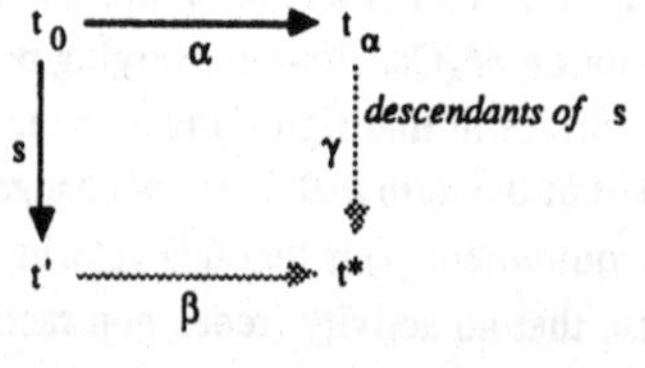

That is, whenever $t_0 \to_\alpha t_\alpha$ and $t_0 \to_s t'$, where s is the contracted redex (occurrence), there are infinitary reductions $t' \to_\beta t^*$ and $t_\alpha \to_\gamma t^*$. The latter reduction consists of contractions of all descendants of s along the reduction $t_0 \to_\alpha t_\alpha$.

Actually, by the Compression Lemma we can find $\beta, \gamma \leq \omega$.

As a side-remark, let us mention that in every TRS (even with uncountably many symbols and rules), all transfinite reductions, strongly convergent as well as divergent, have countable length. All countable ordinals can indeed occur as length of a strongly convergent reduction. (For ordinary Cauchy convergent reductions this is not so: the rewrite rule $C \to C$ yields arbitrarily long convergent reductions $C \to_\alpha^c C$. However, these are not strongly convergent.)

The infinitary Parallel Moves Lemma is "half of the infinitary confluence property". The question arises whether full infinitary confluence holds. That is, given $t_0 \to_\alpha t_1$, $t_0 \to_\beta t_2$, is there a t_3 such that $t_1 \to_\gamma t_3$, $t_2 \to_\delta t_3$ for some γ, δ? Using the Compression Lemma and the Parallel Moves Lemma all that remains to prove is: given $t_0 \to_\omega t_1$, $t_0 \to_\omega t_2$, is there a t_3 such that $t_1 \to_{\leq\omega} t_3$, $t_2 \to_{\leq\omega} t_3$? Surprisingly, the answer is negative: *full infinitary confluence for orthogonal rewriting does not hold.* The counterexample is in Figure 1.7, consisting of an orthogonal TRS with three rules, two of which are 'collapsing rules'. (A rule $t \to s$ is collapsing if s is a variable.) Indeed, in Figure 1.7(a) we have $C \to_\omega A^\omega$, $C \to_\omega B^\omega$ but A^ω, B^ω have no common reduct as they only reduce to themselves. Note that these reductions are indeed strongly convergent. (Figure 1.7(b) contains a rearrangement of these reductions that we need later on.)

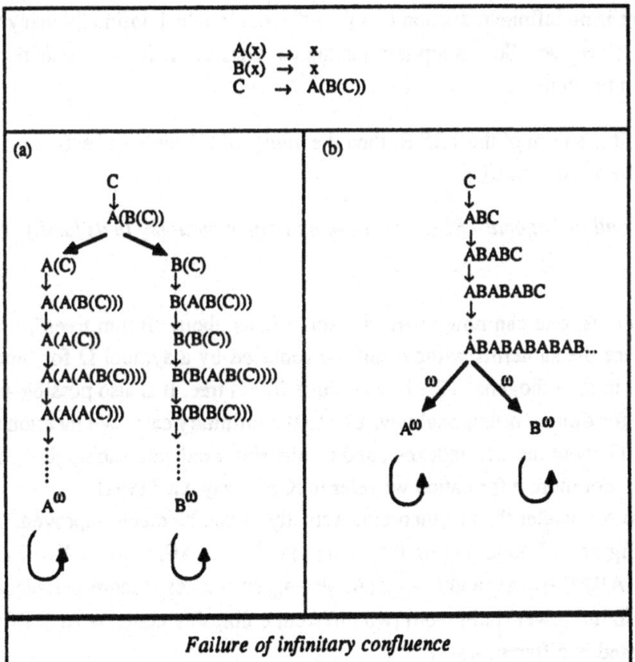

Failure of infinitary confluence

Figure 1.7

Yet, not all is lost: we do have unicity of (possibly infinite) normal forms.

1.6. THEOREM. *For all orthogonal TRSs: Let $t \to_\alpha t'$, $t \to_\beta t''$ where t', t'' are (possibly infinite) normal forms. Then $t' \equiv t''$.*

Here $\equiv$ denotes syntactical equality. Note that in the ABC counterexample in Figure 1.7 the terms A^ω and B^ω are not normal forms.

This Unique Normal Form property, by the way, also holds for Cauchy-converging reductions, that is, with $\rightarrow_\alpha$ replaced by $\rightarrow_\alpha{}^c$ and likewise for β. The reason is that we have:

$$t \rightarrow_\alpha{}^c t' \ \& \ t' \textit{ is a normal form} \quad \Rightarrow \quad t \rightarrow_{\leq\omega} t'.$$

(For $\alpha = \omega$ this is easy to prove; in fact a converging reduction of length ω to a normal form is already strongly convergent. For general α, the proof of the statement requires some work.)

We will now investigate the extent to which infinitary orthogonal rewriting lacks full confluence. It will turn out that non-confluence is only marginal, and that terms which display the bad behaviour are included in a very restricted class. The following definition is inspired by a classical notion in λ-calculus; see Barendregt [1981].

1.7. DEFINITION.
(i) The term t is in *head normal form* (hnf) if $t \equiv C[t_1, ..., t_n]$ where $C[\ , ..., \ ]$ is a non-empty context (prefix) such that no reduction of t can affect the prefix $C[\ , ..., \ ]$. More precisely, if $t \twoheadrightarrow s$ then $s \equiv C[s_1, ..., s_n]$ for some s_i $(i = 1, ..., n)$, and every redex of s is included in one of the s_i $(i = 1, ..., n)$.
(ii) t *has a hnf* if $t \twoheadrightarrow s$ and s is in hnf.

Actually, this definition is equivalent to one of Dershowitz e.a. [1989]; there a term t is called 'top-terminating' if there is no infinite reduction $t \rightarrow t' \rightarrow t'' \rightarrow ...$ in which infinitely many times a redex contraction *at the root* takes place. So: t is top-terminating $\Leftrightarrow$ t has a hnf. We need one more definition before formulating the next theorem.

1.8. DEFINITION. If t is a term of the TRS R, then the *family* of t is the set of subterms of reducts of t, i.e. $\{s \mid t \twoheadrightarrow_R C[s]$ for some context $C[\ ]\}$.

1.9. THEOREM. *For all orthogonal TRSs: Let* t *have no term without hnf in its family. Then* t *is infinitary confluent.*

Just as in λ-calculus, one can now formulate some facts about "Böhm trees", which are (possibly infinite) terms where the subterms without hnf are replaced by a symbol Ω for 'undefined'. As in λ-calculus, each term in an orthogonal TRS has a unique Böhm tree. It is also possible to generalize much of the usual theory for finitary orthogonal rewriting to the infinitary case. We mention the theory of Huet & Lévy [1979, 1991] about 'needed redexes', and results about reduction strategies (such as the parallel-outermost strategy). For more information we refer to Kennaway e.a. [1991].

Here we want to reconsider the last theorem. Actually, it can be much improved. Consider again the ABC example in Figure 1.7. Rearranging the reductions $C \rightarrow_\omega A^\omega$, $C \rightarrow_\omega B^\omega$ as in Figure 1.7(b) into reductions $C \rightarrow_\omega (AB)^\omega \rightarrow_\omega A^\omega$ and $C \rightarrow_\omega (AB)^\omega \rightarrow_\omega B^\omega$ makes it more perspicuous what is going on: $(AB)^\omega$ is an infinite 'tower' built from two different collapsing contexts A( ), B( ), and this infinite tower can be collapsed in different ways.

The ABC example (Figure 1.7) is not merely a pathological example; the same phenomenon (and therefore failure of infinitary confluence) occurs in Combinatory Logic, as in Figure 1.8, where an infinite tower built from the two different collapsing contexts K□K and K□S is able to collapse in two different ways. (Note that analogous to the situation in Figure 1.7, the middle term, built alternatingly from K□K and K□S, can be obtained after ω steps from a finite term which can easily be found by a fixed point construction.) Also for λ-calculus one can now easily construct a counterexample to infinitary confluence.

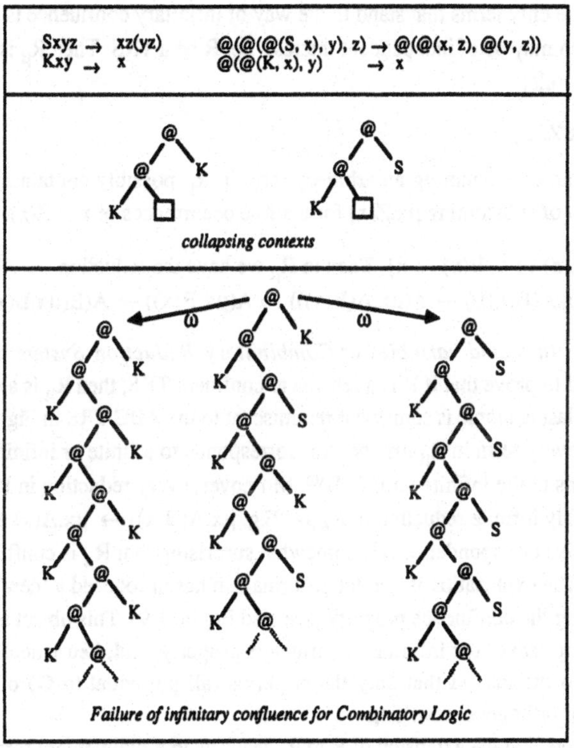

Figure 1.8

Remarkably, it turns out that the collapsing phenomenon is the *only* cause of failure of infinitary confluence. (The full proof is in Kennaway e.a. [1990b].) Thus we have:

1.10. THEOREM. (i) *Let the orthogonal TRS R have no collapsing rewrite rules* $t(x_1, ..., x_n) \to x_i$. *Then R is infinitary confluent.*
(ii) *If* R *is an orthogonal TRS with as only collapsing rule:* $I(x) \to x$, *then* R *is infinitary confluent.*

Call an infinite term $C_1[C_2[...C_n[...]...]]$, built from infinitely many non-empty collapsing contexts $C_i[\ ]$, a hereditarily collapsing (hc) term. (A context C[ ] is collapsing if C[ ] contains one hole □ and C[ ] $\twoheadrightarrow$ □.) Also a term reducing to a hc term is called a hc term. E.g. C from the ABC example in Figure 1.7 is a hc term. Clearly, hc terms do not have a hnf.

1.11. THEOREM. *Let* t *be a term in an orthogonal TRS, which has not a hc term in its family. Then* t *is infinitary confluent.*

This theorem can be sharpened somewhat, as follows. Let us introduce a new symbol **O** to denote hc terms, with the rewrite rule:

$t \to_O$ **O** *if* t *is a hc term.*

Of course this rule is not 'constructive', i.e. the reduction relation $\to_O$ may be undecidable (as it is in CL, Combinatory Logic). However, we now have that orthogonal reduction extended with $\to_O$ is infinitary confluent.

That hc terms are the only terms that stand in the way of infinitary confluence is also suggested by the following detour, which may be of independent interest. Let R be a TRS. Then R_μ is the TRS obtained by extending R with the μ-rule:

$$\mu x.Z(x) \to Z(\mu x.Z(x)).$$

Here Z(x) is a meta-variable, denoting an arbitrary term in R_μ possibly containing the variable x; and Z(μx.Z(x)) is the result of substituting μx.Z(x) for the free occurrences of x in Z(x).

EXAMPLE: Let R = {A(x) → x, B(x) → x}. Then in R_μ we have the reduction
$\mu x.A(B(x)) \to A(B(\mu x.A(B(x)))) \to A(\mu x.A(B(x))) \to A(\mu x.B(x)) \to A(B(\mu x.B(x))) \to \ldots$

In fact, R_μ is a *TRS with bound variables* or *Combinatory Reduction System* as we will present in Section 3. It is not hard to prove that if R is a left-linear confluent TRS, then R_μ is again confluent.

The TRS R_μ in the last example is somewhat reminiscent to the ABC TRS in Figure 1.7 (let us call this TRS: R_{ABC}). In fact, every term in R_μ *except* μx.x corresponds to a finite or infinite term in R_{ABC}. E.g. μx.A(B(x)) corresponds to the infinite term $(AB)^\omega$. Moreover, every reduction in R_μ not involving μx.x corresponds to a possibly infinite reduction in R_{ABC}. E.g. μx.A(B(x)) → μx.A(x) corresponds to $(AB)^\omega \to_\omega A^\omega$. In view of this correspondence it is somewhat surprising that R_μ is confluent, but R_{ABC} is not infinitarily confluent. The explanation is that the μ-formalism has introduced a 'new' object μx.x, without any meaning, but saving the confluence property (see also Figure 1.9). This object is of course the **O** that we introduced above, to 'save' the infinitary confluence property, and, heuristically, this consideration using R_μ has led us to the insight that only the hc terms (all put equal to **O**) obstruct the infinitary confluence property for orthogonal rewriting.

It is interesting to note that the extension of R to R_μ leads us to a rudimentary form of *graph rewriting*, as suggested in Figure 1.9(b), where graphs are given corresponding to the displayed μ-terms. It should also be noted however that the μ-formalism is not able to express all the 'sharing of subterms' that one wants in graph rewriting. To this end *systems of recursion equations* instead of μ-expressions are more expressible; e.g. μx.A(B(x)) 'reads' as a system of recursion equations: {ξ = A(η), η = B(ξ)} or equivalently {ξ = A(B(ξ))}, and μx.x as {ξ = ξ}. Here ξ, η are names of 'locations' (nodes). Now also a graph with sharing as in {ξ = F(η, η), η = G(ξ)} (here the subterm starting with G is shared) is in our scope, and this kind of sharing cannot be expressed by μ-expressions. A start of an exploration of properties of this kind of graph rewriting has been made in Farmer & Watro [1989].

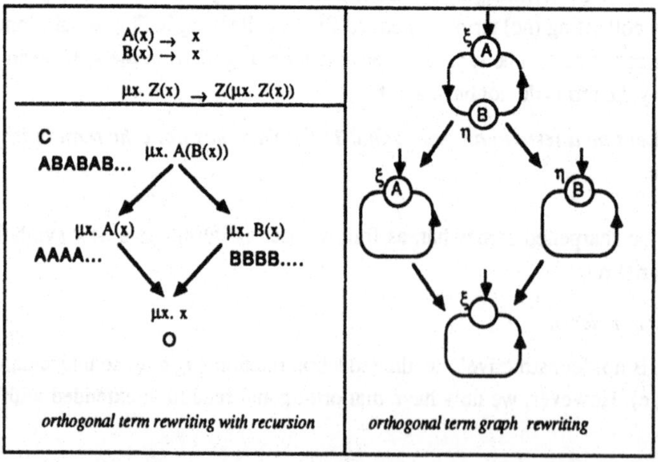

Figure 1.9

2. CONDITIONAL TERM REWRITING SYSTEMS

In this section we discuss another extension of the customary rewriting format, namely with conditions. As will be apparent from other papers in the present volume, a lot of the ongoing research in the field of term rewriting systems is at present devoted to *conditional* term rewriting systems (CTRSs). Conditional rewriting has important roots in Universal Algebra (Meinke & Tucker [1990]) and in the field of Algebraic Specifications (Ehrig & Mahr [1985]). Moreover, Conditional Term Rewriting Systems promise to be of value as a foundation for the integration of the disciplines of functional programming and logic programming (Dershowitz & Plaisted [1985, 1987], Dershowitz & Okada [1990]). Maybe less well-known, conditional rewriting has yet another origin. Out of the algebraic context, rewriting rules with conditions have been used as a proof-theoretic tool for establishing syntactic properties of unconditional rewriting systems and λ-calculus extensions in Klop [1980], de Vrijer [1987, 1989] and Klop & de Vrijer [1989].

In the short account in this section we will do mainly two things. First we sketch the general set up and discuss some of the fundamental definitions and results in the theory of CTRSs. Then we will describe in somewhat more detail one particular proof-theoretic application of CTRSs, namely as a tool for proving the property UN for non-confluent, non-leftlinear TRSs.

The general format we will use for conditional rewriting rules derives from the notation often used in logic programming (Apt [1990]). Then a definition of conditional rewriting can be given as follows.

2.1. DEFINITION. (i) A *conditional rewriting rule* has the following form (with m, n ≥ 0):

r: $t \to s \Leftarrow P_1(x_1, ..., x_m), ..., P_n(x_1, ..., x_m)$.

Here the rule r_u that remains when r is stripped of its conditions,

r_u: $t \to s$,

is supposed to be a usual (unconditional) rewriting rule, and $P_1(x_1, ..., x_m), ..., P_n(x_1, ..., x_m)$ are predicates on terms.

(ii) The instances of the conditional rule r are exactly those instances $t^\sigma \to s^\sigma$ of r_u that are obtained by a substitution σ such that $P_1(\sigma(x_1), ..., \sigma(x_m)) \wedge ... \wedge P_n(\sigma(x_1), ..., (x_m))$.

Of course, a CTRS will consist of a first order signature with a set of conditional rewriting rules, and all common TRS notions and notations immediately generalize. Observe that if n = 0 in Definition 2.1(i), the rewriting rule is unconditional; so the usual notion of TRS can be considered a special case of a CTRS.

2.2. EXAMPLES. (i) By way of a very simple example, observe that a non-leftlinear (unconditional) rule can always be seen as a special kind of conditional rewrite rule that is left-linear. E.g. the non-leftlinear rule r-e: $Dxx \to E$, test for syntactic identity in applicative notation, becomes in the format of conditional rewriting:

r-e: $Dxy \to e \Leftarrow x \equiv y$.

(ii) A natural rule for the transitivity of < could be the following (with T for *true*).

$$x < y \to T \Leftarrow x < z \twoheadrightarrow T, z < y \twoheadrightarrow T.$$

2.3. REMARK. In a rewrite rule $t \to s$ one requires that in s no new variables appear with respect to t. The same requirement is made for conditional rewrite rules $t \to s \Leftarrow C$. But, as observed in Dershowitz, Okada & Sivakumar [1988], for CTRSs it would make good sense to lift this requirement, as e.g. in the following perfectly natural conditional rewrite specification of the Fibonacci numbers. This more liberal format would introduce a considerable complication of the theory, however.

$$\mathrm{Fib}(0) \to \langle 0, 1\rangle,$$
$$\mathrm{Fib}(x + 1) \to \langle z, y + z\rangle \Leftarrow \mathrm{Fib}(x) \downarrow \langle y, z\rangle.$$

Special types of CTRS can be obtained by stipulating some restricted type or format of predicates to be used as conditions. But before we turn to special cases, already a quite general criterion for confluence of CTRSs can be given.

2.4. DEFINITION. (i) Let R be a CTRS. Then R_u, the *unconditional version* of R, is the TRS which arises from R by deleting all conditions (so each rule r is replaced by r_u as in Definition 2.1(i)).
(ii) The CTRS R is called *left-linear* if R_u is so; likewise for *ambiguous* and *orthogonal.*

2.5. DEFINITION. (i) Let R be a CTRS with rewrite relation $\to$, and let P be an n-ary predicate on the set of terms of R. Then P is *stable with respect to* $\to$ if for all terms t_i, t_i' such that $t_i \twoheadrightarrow t_i'$ ($i = 1, ..., n$):

$$P(t_1, ..., t_n) \Rightarrow P(t_1', ..., t_n').$$

(ii) Let R be a CTRS with rewrite relation $\to$. Then R is *stable* if all conditions appearing in some rule of R are stable with respect to $\to$.

2.6. THEOREM (O'Donnell [1977]). *Let* R *be an orthogonal CTRS which is stable. Then* R *is confluent.*

PROOF. A rather straightforward generalization of the confluence proof for orthogonal TRSs. □

Special types of CTRS: semi-equational, join and normal systems

Algebraically, conditional rewrite rules arise as implementations of equational specifications containing *positive conditional equations*:

$$t_0 = s_0 \Leftarrow t_1 = s_1, ..., t_n = s_n.$$

EXAMPLE. A specification of gcd on natural numbers with 0 and successor S, using conditional equations:

$0 < 0 = 0$	$S(x) - S(y) = x - y$	$gcd(x, y) = gcd(x - y, y) \Leftarrow y < x = S(0)$
$0 < S(x) = S(0)$	$0 - x = 0$	$gcd(x, y) = gcd(x, y - x) \Leftarrow x < y = S(0)$
$S(x) < 0 = 0$	$x - 0 = x$	$gcd(x, x) = x$
$S(x) < S(y) = x < y$		

Then the transition from conditional equations to conditional rewrite rules can be made by just orienting the equations in the left-hand sides. This gives rise to so-called *semi-equational* systems. In Bergstra & Klop [1986], one finds also some other types of CTRSs, here listed in Definition 2.7; they are derived from the semi-equational ones, according to different choices that can be made in the implementation of the equational conditions. The terminology we use is taken from Dershowitz, Okada & Sivakumar [1988]; as a matter of fact, they have a more extended classification. CTRSs that do not correspond to any of the special categories are sometimes called *generalized* systems.

2.7. DEFINITION. We distinguish three special types of CTRS, with the format of the rewrite rules as displayed. (the sign $\downarrow$ in (ii) stands for joinability: $t \downarrow s$ iff $t \twoheadrightarrow u$ and $s \twoheadrightarrow u$ for some term u.)

(i) *semi-equational* systems

$$t_0 \to s_0 \Leftarrow t_1 = s_1, ..., t_n = s_n,$$

(ii) *join* systems

$$t_0 \to s_0 \Leftarrow t_1 \downarrow s_1, ..., t_n \downarrow s_n$$

(iii) *normal* systems

$$t_0 \to s_0 \Leftarrow t_1 \twoheadrightarrow n_1, ..., t_k \twoheadrightarrow n_k$$

(with n_1, ..., n_k ground normal forms with respect to the unconditional system R_u).

In each of these three cases the definition of $\to$ depends on conditions involving a reference to $\to$ (via =, $\downarrow$ or $\twoheadrightarrow$). The rewrite rules should be taken as constituting a positive inductive definition of $\to$; this is all right since the conditions are positive. In the case of generalized CTRSs one has to take care in formulating conditions involving $\to$, in order to ensure that $\to$ is well-defined (see Note 2.8).

Notice that the normal systems are a special case of the join systems, since, when s is a ground normal form, the conditions $t \twoheadrightarrow s$ and $t \downarrow s$ are equivalent.

2.8. NOTE. Incorporating negative conditions containing $\to$ in a generalized CTRS would disturb the inductive definition. A simple example already illustrates the point. Consider the generalized CTRS consisting of the single conditional rewrite rule:

$$a \to b \Leftarrow a \neq b.$$

Does $a \to b$ hold? If not, then yes by the conditional rule. If yes, then by which reduction rule?

For this reason the conditions of normal systems can not be put in the form $t \twoheadrightarrow_! s$: t reduces to s and s is a normal form *with respect to the relation* $\to$ *being defined*. Indeed, this type of condition would have a hidden negative part: t does *not* reduce. E.g. consider the problematic single rule CTRS:

$$a \to b \Leftarrow a \twoheadrightarrow_! b.$$

Allowing conditions of the form $t \twoheadrightarrow s$ without requiring s to be a normal form at all, is not very attractive. The conditions would in general be unstable, even if the reduction relation corresponding to the CTRS turns out to be confluent.

Obviously, the convertibility conditions $t_i = s_i$ (i = 1, ..., n) in a rewrite rule of a semi-equational CTRS are stable. So the first part of the following theorem from Bergstra & Klop [1986] is in fact a corollary of Theorem 2.6. The second part involves an induction on the definition of $\to$.

2.9. THEOREM. (i) *Orthogonal semi-equational CTRSs are confluent.*
(ii *Orthogonal normal CTRSs are confluent.*

2.10. EXAMPLE. Let CL-e* be the orthogonal, semi-equational CTRS obtained by extending Combinatory Logic with a 'test for convertibility':

r-e*: $\quad Dxy \to E \Leftarrow x = y.$

Then R is confluent.

Orthogonal normal CTRSs are used in the logic language K-LEAF, see Bosco e.a. [1987].

Orthogonal join CTRSs are in general not confluent. In Bergstra & Klop [1986] a counterexample is given in the CTRS R_0 in Table 2.1.

R_0: $C(x) \to E \Leftarrow x \downarrow C(x)$ $B \to C(B)$.	R_1: $C(x) \to E \Leftarrow x = C(x)$ $B \to C(B)$

Table 2.1

Indeed, in the diagram exhibited in Figure 2.1, we do not have $C(E) \downarrow E$, since this would require $C(E) \to E$, i.e. $C(E) \downarrow E$ again.

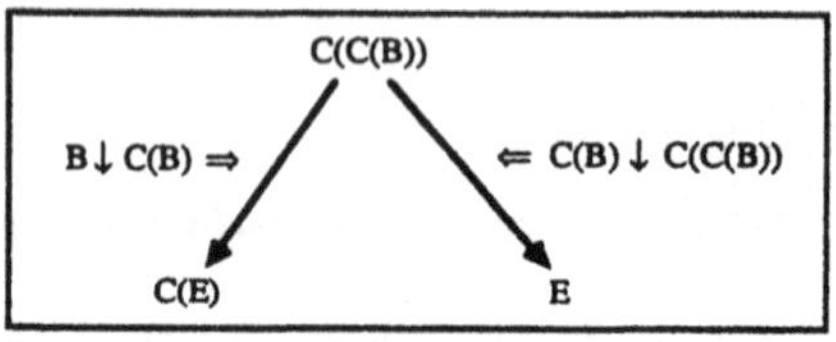

Figure 2.1

This counterexample exhibits an interesting phenomenon, or rather, makes a pitfall explicit. According to Theorem 2.9 above, the semi-equational CTRS R_1 in Table 2.1 is confluent. Hence its convertibility =, coincides with the joinability relation ↓, that is, x = C(x) iff x ↓ C(x). Yet the join CTRS obtained by replacing the condition x = C(x) by x ↓ C(x) is R_0 again, and hence *not* confluent.

Modularity

Modularity results for CTRSs have been studied in recent work of Middeldorp [1989, 1990]. We list some of his results in Theorem 2.11. Part (i) is a generalization of Toyama's theorem (Toyama [1987], stating that confluence is a modular property of TRSs, to CTRSs. As a matter of fact, the proof is by a non-trivial application of Toyama's theorem. Note that all positive results concerning CTRSs can be seen as generalizations of the corresponding ones for unconditional TRSs, since a TRS is just a CTRS without conditions. And of course the results on join CTRSs generalize to normal CTRSs.

2.11. THEOREM. (i) *CR is both a modular property of semi-equational and of join CTRSs.*
(ii) *UN is a modular property of semi-equational CTRSs.*
(iii) *Semi-completeness is a modular property of semi-equational and of join CTRSs.*

Decidability of reduction and of normal forms

Let us for the moment restrict our attention to systems with only finitely many rewrite rules. Then in the unconditional case one-step reduction will be an easily decidable relation. This needs no longer to be so in the case of CTRSs. E.g. consider the example of CL-e*, Combinatory Logic with test for convertibility of Example 2.10. It is not difficult to show that CL-e* is a conservative extension of CL. So deciding whether for pure CL terms s and t, the term Dst is a redex, amounts to deciding whether t and s are convertible in CL. This is an undecidable problem.

From this observation it does not yet follow that being a normal form is in general undecidable. As a matter of fact, CL-e* *has* a decidable set of normal forms, since, as it will turn out in the next section, this set coincides with the normal forms of the non-leftlinear system CL-e, Combinatory Logic extended with the rule r-e of Example 2.2 (see also Table 2.2 below). CL-e has decidable one-step reduction and also decidable normal forms.

All the same, there do exist both semi-equational and normal orthogonal CTRSs for which the set of normal forms is undecidable (and hence not even r.e., since the complement of the set of normal forms is r.e). The following example from Bergstra & Klop [1986] is a normal CTRS; it can easily turned into a semi-equational one..

Consider again Combinatory Logic; it is well-known (cf. Barendregt [1981]) that there is a *representation* $\underline{n}$, a ground CL-term in normal form, of each natural number n, together with a *computable coding* # from the set of ground CL-terms into natural numbers, and an '*enumerator*' E (also a ground CL-term in normal form) such that $E\underline{\#(M)} \twoheadrightarrow M$ for every ground CL-term M. Now let R be the normal CTRS obtained by extending CL with a new constant symbol F and the rule

$$Fx \to \underline{1} \Leftarrow Ex \twoheadrightarrow \underline{0}.$$

This is a conservative extension, and the reduction relation $\rightarrow$ of R satisfies $Fx \rightarrow \underline{1} \Leftrightarrow Ex \twoheadrightarrow \underline{0}$.

Now suppose R has decidable normal forms; then in particular the set $\{\underline{n} \mid F\underline{n} \rightarrow \underline{1}\}$ is decidable, and hence the set $\{\underline{n} \mid E\underline{n} \twoheadrightarrow \underline{0}\}$. However, then also the set

$$\mathsf{X} = \{M \text{ a ground CL-term} \mid M \twoheadrightarrow \underline{0}\}$$

is decidable; for, given M we can compute #(M) and decide whether $E(\underline{\#(M)}) \twoheadrightarrow \underline{0}$ or not. (By confluence for R it follows from $E(\underline{\#(M)}) \twoheadrightarrow \underline{0}$ and $E\underline{\#(M)} \twoheadrightarrow M$ that $M \twoheadrightarrow \underline{0}$.) But then we have a contradiction; from the theorem of Scott stating that *any non-empty proper subset of the set of ground CL-terms which is closed under convertibility in CL, must be undecidable* it follows that X is undecidable.

Proof theoretic applications of CTRSs

We now turn to a proof-theoretic application of CTRSs in the field of term rewriting itself. We describe a general method for proving the *unique normal forms property* (UN) for certain, non-confluent, non-left-linear TRSs. It proceeds by proving confluence for an associated left-linear CTRS, that originates from the original non-leftlinear TRS by—what might be called—'linearizing' the rules. This type of use of a linearized CTRS has originally been proposed by the second author and was applied in Klop [1980] and in Klop & de Vrijer [1989]. In these papers the method has been put to use in different situations, but in a rather ad hoc manner. A systematic presentation can be found in de Vrijer [1990], to which we refer for further details.

We will explain two new applications of the method of conditional linearization, taken from de Vrijer [1990]. First, it yields very easily that all TRSs that are non-ambiguous after linearization have unique normal forms (Theorem 2.16). A second new application is the case of Combinatory Logic plus Parallel Conditional. These two results can also be obtained via a theorem stated in Chew [1981], establishing UN for a somewhat wider class of non-leftlinear TRSs. However, the proof of Chew's theorem is very complicated and our new proofs are much simpler.

Let us first mention three specific non-leftlinear extensions of Combinatory Logic to which the method can be applied: CL-sp, CL-e and CL-pc. The usual rewrite rules of CL are: $Sxyz \rightarrow xz(yz)$, $Kxy \rightarrow x$, $Ix \rightarrow x$. Two interesting ways of extending Combinatory Logic are with Surjective Pairing (CL-sp) and with Parallel Conditional (CL-pc); see Table 2.2.

CL-sp	CL-pc	CL-e
CL + $D_1(Dxy) \rightarrow x$ $D_2(Dxy) \rightarrow y$ $D(D_1x)(D_2x) \rightarrow x$	CL + r-t: $CTxy \rightarrow x$ r-f: $CFxy \rightarrow y$ r-pc: $Czxx \rightarrow x$	CL + r-e: $Dxx \rightarrow x$

Table 2.2

CL-sp was the first non-leftlinear term rewriting system to be extensively studied, mostly in the related lambda calculus version (see e.g. Mann [1973], Barendregt [1974], Klop [1980]). In de Vrijer [1987, 1989] a linearized CTRS was used to prove that the surjective pairing rules are conservative. The system CL-e came up in the study of CL-sp for theoretical purposes.

Each of the three non-leftlinear rewriting systems of Table 2.2 lacks the Church-Rosser property (Klop [1980]). But nevertheless, each one can be shown to have unique normal forms, by the method of conditional linearization. Only the existing proof for the case of CL-sp is very complicated (see Klop & de Vrijer [1989]); the other cases are much simpler and will be covered here.

As a starting point consider the following simple observation concerning Abstract Reduction Systems (ARSs); recall that an ARS is just any set with a binary relation $\to$, considered as a reduction relation.

2.12. PROPOSITION. *Let* R_0 *and* R_1 *be two ARSs with the same set of objects, and with reduction relations* $\to_0$, $\to_1$ *and convertibility relations* $=_0$, $=_1$ *respectively. Let* NF_i *be the set of normal forms of* R_i $(i = 0, 1)$. *Then* R_0 *is UN if each of the following conditions hold:*

(i) $\to_1$ *extends* $\to_0$;
(ii) R_1 *is CR;*
(iii) NF_1 *contains* NF_0.

PROOF. Easy. □

The interest of Proposition 2.12 derives from its use in the method proving UN. E.g., in order to be able to use this proposition for establishing UN for CL-e, we 'break' the non-leftlinearity constraint in the rule r-e by replacing it with a conditional rule r-e* (the resulting system is CL-e* of Example 2.10).

r-e*: $\quad Dxy \to E \Leftarrow x = y.$

Then it turns out that Proposition 2.12 can be applied with respect to CL-e and CL-e*.

Note that the rule r-e* can be seen as resulting from r-e, written in the conditional format of Example 2.2(i), by just relaxing the condition $x \equiv y$ to $x = y$. More in general, we have the following definition.

2.13. DEFINITION. (i) If r is a rewrite rule $t \to s$, we say that $r' = t' \to s'$ is a *left-linear version* of r if there is a substitution σ: VAR $\to$ VAR such that $r'^{\sigma} = r$ and r' is left-linear.
(ii) If $r = t \to s$ is a rewrite rule, and $r' = t' \to s'$ is a left-linear version of r, such that $r = r'^{\sigma}$, then the *conditionalized left-linear version* or *linearization* of r (associated to r') is the conditional rewrite rule:

$$t' \to s' \quad \Leftarrow \bigwedge \{x_i = x_j \mid i > j,\ x_i^{\sigma} = x_j^{\sigma},\ x_i, x_j \in t'\}.$$

(In case r is already left-linear, it will coincide with its left-linear version r' and with the associated conditional rule.)

EXAMPLE. $Czxy \to y$ is a left-linear version of the non-leftlinear rule $Czxx \to x$, by the substitution σ with $\sigma(z) = z$, $\sigma(x) = x$, $\sigma(y) = x$. The associated conditional rule is $Czxy \to y \Leftarrow x = y$. The only other left-linear version of $Czxx \to x$ is $Czxy \to x$, with associated conditional rule $Czxy \to x \Leftarrow x = y$.

2.14. DEFINITION. (*Linearization*) (i) If R is a TRS, then a *linearization* of R is a semi-equational CTRS that consists of linearizations of the rules of R, for each rule of R at least one.
(ii) If R is a TRS, then R^L, the *full linearization* of R, is defined as the linearization of R that is obtained by including for each rule $r \in R$ *all* its conditionalized left-linear versions.

EXAMPLE. The system CL-e* is the result of linearizing the system CL-e. In fact, CL-e* = CL-e^L.

2.15. THEOREM. *If a linearization of a term rewriting system* R *is confluent, then* R *is UN.*

PROOF. We want to apply Proposition 3.1 with $R_0 = R$ and R_1 a confluent linearization R' of R; so we must check the clauses (i), (ii) and (iii) of 3.1 for R and R'.

(i) $\to_{R'}$ extends $\to_R$ since a linear rule is always a restriction of each of its linearizations.
(ii) R' is Church-Rosser holds by assumption.

As to (iii), we prove by induction on X the implication $X \in NF_R \Rightarrow X \in NF_{R'}$. Assume $X \in NF_R$. Then X can only be not an R'-normal form, if it contains a redex Y that is an instance of a linearization r^* of some non-leftlinear rule $r = t \to s$ of R. That is, $X \equiv C[Y]$ and for a left-linear version $r' = t' \to s'$ of r (such that $r = r'^{\sigma}$), we have $Y \equiv t'^{\tau}$; moreover the conditions of r^* must be satisfied, amounting to the implication $x_i^{\sigma} \equiv x_j^{\sigma} \Rightarrow x_i^{\tau} = x_j^{\tau}$, for all $x_i, x_j \in t'$. Since the x_i^{τ}'s are proper subterms of X, and hence R-normal forms, they are by the induction hypothesis also R'-normal forms. Hence, since R' has unique normal forms: $x_i^{\sigma} \equiv x_j^{\sigma} \Rightarrow x_i^{\tau} \equiv x_j^{\tau}$. But then Y would be also an R-redex, contradicting the assumption that $X \in NF_R$. □

Theorem 2.15 yields a general method to prove UN for non-leftlinear TRSs: try to prove CR for one of its linearizations. Whether the method will work in a particular case, and how difficult it is, depends on the CR problem that ensues.

We first show a quite general result, that is an immediate consequence of Theorem 2.15. Call a TRS *strongly non-ambiguous* if after replacing each non-leftlinear reduction rule by a left-linear version the resulting TRS is non-ambiguous.

2.16. THEOREM. *Any strongly non-ambiguous TRS has unique normal forms.*

PROOF. Let R be a strongly non-ambiguous TRS. Consider a linearization R' of R consisting of exactly one conditionalized left-linear version for each rule of R. Then R' will be an orthogonal semi-equational CTRS. Hence the result follows by Theorems 2.15 and 2.9(i). □

EXAMPLES. A non-leftlinear TRS to which Theorem 2.16 can be applied is the system CL-e. A non-ambiguous but not strongly non-ambiguous TRS that does not have unique normal forms is (Huet [1980]):

R: $F(x, x) \to A$, $F(x, G(x)) \to B$, $C \to G(C)$.

R is non-ambiguous; there are no critical pairs since x and G(x) cannot be unified. However, R is not strongly non-ambiguous, since {F(x, y) → A, F(x, G(y) → B} has a critical pair. The term F(C, C) has the two distinct normal forms A and B.

Unicity of normal forms for Combinatory Logic plus Parallel Conditional

Finally, we sketch how one can prove confluence for the full linearization $CL\text{-}pc^L$ of the ambiguous and non-leftlinear system CL-pc (Table 2.2). The rules of the full linearization $CL\text{-}pc^L$ are summed up in Table 2.3.

$CL\text{-}pc^L$		$CL\text{-}pc^{L-}$	
CL +		CL +	
r-t:	$CTxy \to x$	r-t:	$CTxy \to x$
r-f:	$CFxy \to y$	r-f:	$CFxy \to y$
$r\text{-}pc^1$:	$Czxy \to x \Leftarrow x = y$	$r\text{-}pc^{1-}$:	$Czxy \to x \Leftarrow x =_{CL\text{-}pc} y,\ z \neq_{CL\text{-}pc} F$
$r\text{-}pc^2$:	$Czxy \to y \Leftarrow x = y$	$r\text{-}pc^{2-}$:	$Czxy \to y \Leftarrow x =_{CL\text{-}pc} y,\ z =_{CL\text{-}pc} F$

Table 2.3

Solving the CR problem for $CL\text{-}pc^L$ may at first look not very promising, because of the vicious cases of overlap between the pairs of rules r-t / $r\text{-}pc^2$, r-f / $r\text{-}pc^1$ and $r\text{-}pc^1$ / $r\text{-}pc^2$. Now the idea is to add extra conditions in order to remove these overlaps. This will involve the use of negative conditions, however, and therefore, in order to avoid the difficulty indicated in Note 2.8, we have in the system $CL\text{-}pc^{L-}$ 'fixed' the conditions, making them refer to $=_{CL\text{-}pc}$, convertibility in CL-pc. In this way, the conditions in CL-pc^{L-} have a determinate meaning, independent of the inductive definition of conversion ($=_{CL\text{-}pcL-}$) they are part of. What we get is not a semi-equational, but a generalized CTRS, called CL-pc^{L-}.

It is now crucial that all this fiddling with the original reduction relation of CL-pc, did not change conversion. That is, the conversion relation relations of CL-pc^{L} and CL-pc^{L-} both coincide with $=_{CL\text{-}pc}$. Moreover, in order to prove CR for CL-pc^{L-} we need to know that T $\neq_{CL\text{-}pc}$ F; this will guarantee that there is indeed no overlap in CL-pc^{L-} between the rules r-t and r-pc^{2-}, etc. A model construction within the Graph Model Pω for CL can be used for this purpose.

2.17. PROPOSITION. (i) *The system* CL-pc^{L-} *is Church-Rosser.*
(ii) *The system* CL-pc^{L} *is Church-Rosser.*

PROOF. (i) Since $=_{CL\text{-}pcL-} = =_{CL\text{-}pc}$, the conditions of CL-pc^{L-} are stable. Moreover, between the rules of CL-pc^{L-} there are no harmful cases of overlap, due to the negative condition and since T $\neq_{CL\text{-}pc}$ F. Then proving CR is a routine matter (compare Theorem 2.6).
(ii) Assume t $=_{CL\text{-}pcL}$ s. Then t $=_{CL\text{-}pcL-}$ s. Hence by (i), the terms t and s must have a common reduct in CL-pc^{L-}. But now since $=_{CL\text{-}pcL} = =_{CL\text{-}pc}$, obviously $\rightarrow_{CL\text{-}pcL-} \subseteq \rightarrow_{CL\text{-}pcL}$: in CL-pc^{L} reduction is more liberal than in CL-pc^{L-}, two conditions have been lifted. So t and s have the same common reduct in CL-pc^{L}. □

2.18. COROLLARY. *The system* CL-pc *has unique normal forms.*

3. TERM REWRITING WITH BOUND VARIABLES

We will now introduce TRSs with as additional feature: *bound variables*. The well-known paradigm is, of course, λ-calculus. We want to exhibit a framework for term rewriting incorporating apart from the usual TRSs, also specimens like λ-calculus, various typed λ-calculi and extended λ-calculi, for instance combinations of some typed λ-calculus with an ordinary TRS. Recently, there has been quite some attention for such combinations. (See e.g. Breazu-Tannen [1988].)

For the sake of abbreviation, we will refer to TRSs (possibly) with bound variables as *Combinatory Reduction Systems* or CRSs for short. TRSs will form a subclass of the class of CRSs. CRSs were introduced in Klop [1980], where a study is made of especially orthogonal CRSs. We start with considering several examples.

3.1. EXAMPLES.

(i) **λ-calculus.** The only rewrite rule is the β-reduction rule:

$$(\lambda x.Z_1(x))Z_2 \rightarrow Z_1(Z_2),$$

presented in an informal notation; the formal notation would use a substitution operator [:=] and we would write $(\lambda x.M)N \rightarrow [x := N]M$. Still, this informal notation has a direct appeal, and in the sequel we will make it formal; this is essential for the syntax definition of CRSs.

(ii) **Polyadic λ-calculus.** Here we have n-ary λ-abstraction and reduction rules (β_n) for every $n \geq 1$:

$$(\beta_n) \quad (\lambda x_1x_2...x_n.\ Z_0(x_1, x_2, ..., x_n))Z_1Z_2...Z_n \rightarrow Z_0(Z_1, Z_2,..., Z_n).$$

(iii) **μ-calculus.** This is the well-known notation system designed to deal with recursively defined objects (processes, program statements, ...) with as basic rewrite or reduction rule:

$$\mu x.Z(x) \rightarrow Z(\mu x.Z(x))$$

We have used it in Section 1, to extend an ordinary TRS R with recursion, resulting in R_μ.

(iv) **Some rewrite rules in Proof Theory.**

$$\mathcal{P}(\mathcal{L}Z_0)(\lambda x.Z_1(x))(\lambda y.Z_2(y)) \rightarrow Z_1(Z_0)$$
$$\mathcal{P}(\mathcal{R}Z_0)(\lambda x.Z_1(x))(\lambda y.Z_2(y)) \rightarrow Z_2(Z_0)$$

The operational meaning of this pair of rewrite rules should be self-explaining: according to whether Z_0 is prefixed by $\mathcal{L}$ or $\mathcal{R}$ it is substituted in the left or the right part of the 'body' of the redex headed by $\mathcal{P}$, for all the free occurrences of x respectively y. The rules occur as *normalization procedures for proofs* in Natural Deduction (Prawitz [1971], p. 252), albeit not in the present linear notation. (For more explanation see Klop [1980].)

(v) **λ-calculus with δ-rules of Church.** This is an extension of λ-calculus with a constant δ and a possibly infinite set of rules of the form

$$\delta M_1 \ldots M_n \rightarrow N$$

where the M_i ($i = 1, \ldots, n$) and N are closed terms and the M_i are moreover in "βδ-normal form", i.e. contain no β-redex and no subterm as in the left-hand side of a δ-rule. To ensure non-ambiguity (defined below) there should moreover not be two left-hand sides of different δ-rules of the form $\delta M_1 \ldots M_n$ and $\delta M_1 \ldots M_m$, $m \geq n$. (So every left-hand side of a δ-rule is a normal form with respect to the other δ-rules.)

The preceding examples suggest that a general definition of what we will call *Combinatory Reduction Systems* (CRSs) may be profitable, in order to be able to derive properties like confluence at once for a whole class of such CRSs, rather than repeating similar proofs or using 'proof by hand-waving'. The account below follows Klop [1980]. The concept of a CRS was first suggested in Aczel [1978], where a confluence proof for a subclass of the orthogonal CRSs was given.

Term formation in a Combinatory Reduction System.

3.2. DEFINITION. The *alphabet* of a CRS consists of
(i) a set Var = $\{x_n \mid n \geq 0\}$ of *variables* (also written as x, y, z, ...);
(ii) a set Mvar of *metavariables* $\{Z_n^k \mid k, n \geq 0\}$; here k is the *arity* of Z_n^k;
(iii) a set of *constants* $\{Q_i \mid i \in I\}$ for some I; constants will be written also as $\mathcal{P}$, Q, λ, μ,... ;
(iv) improper symbols ( ) and [ ].

The arities k of the metavariables Z_n^k can always be read off from the term in which they occur—hence we will often suppress these superscripts. E.g. in $(\lambda x.Z_0(x))Z_1$ the Z_0 is unary and Z_1 is 0-ary.

Usually, TRSs are presented as *functional* TRSs (where all operators have a fixed arity); then we can consider *applicative* TRSs as a subclass (where the only non-constant operator is a binary operator 'application'; the paradigm example is CL, Combinatory Logic). The reverse way is also possible, once the notion of substructure is available (as will be the case later on); then functional TRSs can be seen as restricted versions (substructures) of the corresponding applicative TRSs where operators are 'varyadic', i.e. permit any number of arguments. So, the set-ups via the functional and the applicative 'format' are entirely equivalent. Below we will use the applicative format. For a syntax definition of CRSs in the functional format, see Kennaway [1988].

3.3. DEFINITION. The set Mter of *meta-terms* of a CRS with alphabet as in 3.2 is defined inductively as follows:

(i) constants and variables are meta-terms;

(ii) if t is a meta-term, x a variable, then ([x]t) is a meta-term, obtained by *abstraction*;

(iii) if t, s are meta-terms, then (ts) is a meta-term, obtained by *application*, provided t is not an abstraction meta-term ([x]t');

(iv) if $t_1, ..., t_k$ ($k \geq 0$) are meta-terms, then $Z_n^k(t_1, ..., t_k)$ is a meta-term (in particular the Z_n^0 are meta-terms).

Note that meta-variables Z_n^{k+1} are not meta-terms; they need arguments. Meta-terms in which no metavariable Z occurs, are *terms*. Ter is the set of terms.

3.4. NOTATION. (i) As in applicative TRSs such as CL, the convention of association to the left is adopted. The outermost pair of brackets is dropped.

(ii) An iterated abstraction meta-term $[x_1](...([x_{n-1}]([x_n]t))...)$ is written as $[x_1, ..., x_n]t$ or $[\mathbf{x}]t$ for $\mathbf{x} = x_1, ..., x_n$. A meta-term Q([x]t) will be written as Qx.t.

(iii) We will not be precise about the usual problems with renaming of variables, α-conversion etc. That is, this is treated like in λ-calculus when one is not concerned with implementations. Thus we will adopt the following conventions:

- All occurrences of abstractors $[x_i]$ in a meta-term are different; e.g. λxx.t is not legitimate, nor is λx.(tλx.t').
- Furthermore, terms differing only by a renaming of bound variables are considered syntactically equal. (The notion of 'bound' is as in λ-calculus: in [x]t the free occurrences of x in t (hence by (i) *all* occurrences) are bound by the abstractor [x].)

3.5. DEFINITION. A (meta-)term is *closed* if every variable occurrence is bound.

Rewriting in a Combinatory Reduction System.

3.6. DEFINITION. A *rewrite* (or *reduction*) *rule* in a CRS is a pair (t, s), written as t → s, where t, s are meta-terms such that:

(i) t has a constant as 'head' (i.e. leftmost) symbol;

(ii) t, s are closed meta-terms;

(iii) the metavariables Z_n^k that occur in s, also occur in t;

(iv) the metavariables Z_n^k in t occur only in the form $Z_n^k(x_1, ..., x_k)$ where the x_i ($i = 1, ..., k$) are variables (no meta-terms). Moreover, the x_i are pairwise distinct.

If, moreover, no metavariable Z_n^k occurs twice or more in t, the rewrite rule t → s is called *left-linear*.

In order to generate actual rewrite steps from the rewrite rules, we have to define substitution:

3.7. DEFINITION. (i) A *valuation* $\underline{\sigma}$: Mvar → Var* × Ter is a map such that

$$\underline{\sigma}(Z_n^k) = (x_1, ..., x_k)t,$$

i.e. σ assigns to a k-ary metavariable Z_n^k a term t *together with a list of* k *pairwise different variables* $x_1, ..., x_k$. It is not required that the x_i actually occur in t.

Furthermore we define:

$$((x_1, ..., x_k)t)(t_1, ..., t_k) = t[x_1 := t_1, ..., x_k := t_k]$$

where $[x_1 := t_1, ..., x_k := t_k]$ denotes simultaneous substitution of t_i for x_i ($i = 1, ..., k$).

(ii) A *substitution* σ corresponding to the valuation $\underline{\sigma}$ is a map from Mter to Ter as follows:

$\sigma(x) = x$ for $x \in$ Var, $\sigma(Q) = Q$ for constants Q;
$\sigma([x]t) = [x]\,\sigma(t)$;
$\sigma(ts) = (\sigma(t)\,\sigma(s))$
$\sigma(Z_n{}^k(t_1, ..., t_k) = \underline{\sigma}(Z_n{}^k)\,(\sigma(t_1), ..., \sigma(t_k))$.

(So if $\underline{\sigma}(Z_n{}^k) = (x_1, ..., x_k)t$, then $\sigma(Z_n{}^k(t_1, ..., t_k) = t[x_1{:=}\sigma(t_1), ..., x_k{:=}\sigma(t_k)]$.)

3.8. EXAMPLE. (i) Let $\sigma(Z_1{}^1) = (u)\ uy$. Then $\underline{\sigma}(Z_1{}^1(x)) = xy$.
(ii) Let $\sigma(Z_1{}^1) = (u)\ zy$. Then $\underline{\sigma}(Z_1{}^1(x)) = zy$.
(iii) Let $\sigma(Z^2) = (x,y)\ xyxz$, $\sigma(Z^1) = (z)\ xzy$, $\sigma(Z^0) = u$. Then

$$\underline{\sigma}(Z^2(Z^2(Z^0, Z^0), Z^1(Z^0))) = uuuz(xuy)(uuuz)z.$$

As in ordinary term rewriting, if $r = t \to s$ is a rewrite rule, then $\sigma(t)$ is an *r-redex*, and r-reduction (or r-rewrite) steps have the form $C[\sigma(t)] \to C[\sigma(s)]$ for some context $C[\ ]$, with the proviso that $\underline{\sigma}(Z) = (x_1, ..., x_k)p$ for some p if $Z(x_1, ..., x_k)$ occurs in t. (The definition of 'context' is left to the reader.)

3.9. EXAMPLE. In this example we write t^σ instead of $\sigma(t)$. We reconstruct a step according to the β-reduction rule of λ-calculus

$$(\lambda x.\ Z(x))Z' \to Z(Z').$$

Let the valuation $Z^{\underline{\sigma}} = (x)\ yxx$, $Z'^{\underline{\sigma}} = ab$ be given. Then we have the reduction step

$$
\begin{aligned}
&((\lambda x.\ Z(x))Z')^\sigma \\
&\quad = (\lambda x.\ Z(x)^\sigma)Z'^{\underline{\sigma}} \\
&\quad = (\lambda x.\ Z^{\underline{\sigma}}\,(x^\sigma))Z'^{\underline{\sigma}} \\
&\quad = (\lambda x.((x)yxx)(x))(ab) \\
&\quad = (\lambda x.\ yxx)(ab) \to \\
&(Z(Z'))^\sigma \\
&\quad = Z^{\underline{\sigma}}\,(Z'^{\underline{\sigma}}) \\
&\quad = ((x)(yxx))(ab) \\
&\quad = y(ab)(ab).
\end{aligned}
$$

3.10. DEFINITION. The notion of "*non-ambiguity*" for a CRS containing rewrite rules $\{r_i = t_i \to s_i \mid i \in I\}$ is as for ordinary TRSs:

(i) Let the left-hand side t_i of r_i be in fact $t_i(Z_1(x_1), ..., Z_m(x_m))$ where all metavariables in t_i are displayed. Now if the r_i-redex $\underline{\sigma}(t_i(Z_1(x_1), ..., Z_m(x_m))$ contains an r_j-redex $(i \neq j)$, then this r_j-redex must be already contained in one of the $\underline{\sigma}(Z_p(x_p))$.

(ii) Likewise if the r_i-redex *properly* contains an r_i-redex.

Also as usual, a CRS is *orthogonal* if it is left-linear and non-ambiguous.

A large part of the theory for orthogonal TRSs carries over to orthogonal CRSs (see Klop [1980]). The main fact is:

3.11. THEOREM. *All orthogonal CRSs are confluent.* □

Hence normal forms are unique in orthogonal CRSs. Also Church's Theorem for λI-calculus generalizes to orthogonal CRSs. Church's Theorem states that for a term in λI-calculus the properties WN (Weak Normalization) and SN (Strong Normalization) coincide. (A term is WN if it has a normal form, and SN if all its reductions are terminating.) The same is true for orthogonal TRSs which are *non-erasing*; this means that in every reduction rule $t \to s$ both sides t, s contain the same variables. Here the definition of 'non-erasing' reduction rule for CRSs generalizes from that for TRSs as follows: A rule $t \to s$ is non-erasing if all metavariables Z occurring in t, have an occurrence in s *which is not in the scope of a metavariable* (i.e. not occurring in an argument of a metavariable). Without this proviso, which for TRSs is vacuously fulfilled since there all metavariables in the rewrite rules are 0-ary, also rules like the β-reduction rule of λ-calculus $(\lambda x.Z(x))Z' \to Z(Z')$ would be non-erasing, which obviously is not the intention.

REMARK. It would be interesting to investigate which CRSs can be 'defined' (or 'interpreted', or 'implemented') in λ-calculus. First, a good notion of 'interpretation', of which there seem to be many variants, should be developed—for some proposals concerning TRSs see O'Donnell [1985].

Note that even for orthogonal CRSs which are in a very direct sense definable in λ-calculus (e.g. CL, Combinatory Logic, is 'directly definable' in λ-calculus in an obvious way), theorems like the Church-Rosser theorem (3.11) are not superfluous: if a reduction system R_1 can be interpreted in a "finer" reduction system R_2, the confluence of R_2 need not imply the confluence of R_1.

Substructures of Combinatory Reduction Systems.

Above, all CRSs had an *unrestricted term formation* by some inductive clauses. However, often one will be interested in CRSs where some restrictions on term formation are present. A typical example is λI-calculus, mentioned above, where the restriction is that in a subterm λx. t there must be at least one occurrence of x in t.

Other typical examples of restricted term formation arise when *types* are introduced, as in typed λ-calculus (λ^{τ}-calculus) or typed Combinatory Logic (CL^{τ}) (see Hindley & Seldin [1986]). In a simple way a type restriction occurs already when one considers *many-sorted* TRSs. This leads us to the following definition:

3.12. DEFINITION.
(i) Let $(R, \to_R)$ be a CRS as defined above. Let $\mathcal{T}$ be a subset of Ter(R), which is closed under $\to_R$. Then $(\mathcal{T}, \to_R|\mathcal{T})$, where $\to_R|\mathcal{T}$ is the restriction of $\to_R$ to $\mathcal{T}$, is a *substructure* of $(R, \to_R|\mathcal{T})$.
(ii) If $(R, \to_R)$ is orthogonal, so are its substructures.

We now declare that also a substructure of a CRS is a CRS. It is not hard to see (by a patient inspection of the proofs of the theorems mentioned above) that almost everything carries over to our new notion of CRS: for orthogonal CRSs we have confluence, the Parallel Moves Lemma (mentioned in Section 1), Finite Developments (see Klop [1980]), Church's Theorem for non-erasing CRSs, etc. The 'almost' refers to cases where *expansions* (i.e. inverse reductions $M \leftarrow \leftarrow \ldots$) are considered (since we did not require that substructures are closed under expansion). In Klop [1991] several more examples of well-known orthogonal CRSs can be found, such as polymorphic λ-calculus and PCF (Plotkin [1977]).

Acknowledgement.

Discussions with Fer-Jan de Vries, who pointed out some shortcomings in an earlier version concerning Section 1, are gratefully acknowledged. We thank Aart Middeldorp and Vincent van Oostrom for carefully reading the manuscript and suggesting improvements.

References

ACZEL, P. (1978). *A general Church-Rosser theorem.* Preprint, Univ. of Manchester.

APT, K.R. (1990). *Logic Programming.* In: Formal models and semantics, Handbook of Theoretical Computer Science, Vol.B (J. van Leeuwen, editor), Elsevier - The MIT Press, Chapter 10, p.493-574.

BARENDREGT, H.P. (1974). *Pairing without conventional restraints.* Zeitschrift für Math. Logik und Grundl. der Math. 20, p. 289-306.

BARENDREGT, H.P. (1981). *The Lambda Calculus, its Syntax and Semantics.* 1st ed. North-Holland 1981, 2nd ed. North-Holland 1984.

BARENDREGT, H.P. (1989). *Lambda calculi with types.* in: Handbook of Logic in Computer Science (eds. S. Abramsky, D. Gabbay and T. Maibaum) Vol.1, Oxford University Press, to appear.

BARENDREGT, H.P. (1989). *Functional programming and lambda calculus.* In: Formal models and semantics, Handbook of Theoretical Computer Science, Vol.B (J. van Leeuwen, editor), Elsevier - The MIT Press, Chapter 7, p.321-364.

BARENDREGT, H.P., VAN EEKELEN, M.C.J.D., GLAUERT, J.R.W., KENNAWAY, J.R., PLASMEIJER, M.J. & SLEEP, M.R. (1987). *Term graph rewriting.* In: Proc. PARLE Conf., Springer LNCS 259, 141-158.

BERGSTRA, J.A. & KLOP, J.W. (1986). *Conditional rewrite rules: confluence and termination.* JCSS Vol.32, No.3, 1986, 323-362.

BÖHM, C. (ed.) (1975), *λ-Calculus and Computer Science Theory.* Springer Lecture Notes in Computer Science 37.

BOSCO, P.G., GIOVANETTI, E., LEVI, G., MOISO, C. & PALAMIDESSI, C. (1987). *A complete semantic characterization of K-LEAF, a logic language with partial functions.* Proc. 4th Symp. on Logic Programming, San Francisco (1987).

BREAZU-TANNEN, V. (1988). *Combining Algebra and Higher-Order Types.* In: Proc. 3rd Symp. on Logic in Computer Science, Edinburgh. 82-90.

CHEW, P. (1981). *Unique normal forms in term rewriting systems with repeated variables.* In: Proc. 13th Annual Symp. on the Theory of Computing, 7-18.

CHURCH, A. (1941). *The calculi of lambda conversion.* Annals of Mathematics Studies, Vol.6. Princeton University Press.

DERSHOWITZ, N. & JOUANNAUD, J.-P. (1990). *Rewrite systems.* In: Formal models and semantics, Handbook of Theoretical Computer Science, Vol.B (J. van Leeuwen, editor), Elsevier - The MIT Press, Chapter 6, p.243-320.

DERSHOWITZ, N. & KAPLAN, S. (1989). *Rewrite, rewrite, rewrite, rewrite, rewrite,* Principles of programming languages, Austin, Texas, pp. 250–259.

DERSHOWITZ, N., KAPLAN, S. & PLAISTED, D.A. (1989). *Infinite Normal Forms (plus corrigendum),* ICALP, pp. 249–262.

DERSHOWITZ, N. & OKADA, M. (1990). *A rationale for conditional equational programming.* Theor. Comp. Sci. 75 (1990) 111-138.

DERSHOWITZ, N., OKADA, M. & SIVAKUMAR, G. (1987). *Confluence of Conditional Rewrite Systems.* In: Proc. of the 1st International Workshop on Conditional Term Rewrite Systems, Orsay, Springer LNCS 308, 31-44.

DERSHOWITZ, N., OKADA, M. & SIVAKUMAR, G. (1988). *Canonical Conditional Rewrite Systems*. In: Proc. of 9th Conf. on Automated Deduction, Argonne, Springer LNCS 310, 538-549.

DERSHOWITZ, N. & PLAISTED, D.A. (1985). *Logic Programming cum Applicative Programming*. In: Proc. of the IEEE Symp. on Logic Programming, Boston, 54-66.

DERSHOWITZ, N. & PLAISTED, D.A. (1987). *Equational Programming*. In: Machine Intelligence 11 (eds. J.E. Hayes, D. Michie and J. Richards), Oxford University Press, 21-56.

EHRIG, H. & MAHR, B. (1985). *Fundamentals of Algebraic Specification 1. Equations and Initial Semantics*. Springer Verlag.

FARMER, W.M. & WATRO, R.J. (1989). *Redex capturing in term graph rewriting*, in *Computing with the Curry Chip* (eds. W.M. Farmer, J.D. Ramsdell and R.J. Watro), Report M89-59, MITRE.

GRÄTZER, G. (1979). *Universal Algebra. (Second Edition)*. Springer 1979.

HINDLEY, J.R. & SELDIN, J.P. (1986). *Introduction to Combinators and λ-Calculus*. London Mathematical Society Student Texts, Nr.1, Cambridge University Press 1986.

HUDAK, P. et al (1988). *Report on the Functional Programming Language Haskell*. Draft Proposed Standard, 1988.

HUET, G. (1980). *Confluent reductions: Abstract properties and applications to term rewriting systems*. Journal of the ACM 27, No.4, p. 797-821.

HUET, G. & LÉVY, J.-J. (1979). *Call-by-need computations in non-ambiguous linear term rewriting systems*. Rapport INRIA nr.359.

HUET, G. & LÉVY, J.-J. (1991). *Computations in orthogonal term rewriting systems*. In: *Computational Logic: Essays in Honour of Alan Robinson* (J.-L. Lassez & G. Plotkin, eds.), MIT Press, Cambridge, MA (to appear).

HUET, G. & OPPEN, D.C. (1980). *Equations and rewrite rules: A survey*. In: Formal Language Theory: Perspectives and Open Problems (ed. R. Book), Academic Press, 1980, 349-405.

JOUANNAUD, J.-P. & WALDMANN, B. (1986). *Reductive conditional Term Rewriting Systems*. In: Proc. of the 3rd IFIP Working Conf. on Formal Description of Programming Concepts, Ebberup, 223-244.

KAPLAN, S. (1984). *Conditional Rewrite Rules*. TCS 33(2,3), 1984.

KAPLAN, S. (1985). *Fair conditional term rewriting systems: Unification, termination and confluence*. Recent Trends in Data Type Specification (ed. H.-J. Kreowski), Informatik-Fachberichte 116, Springer-Verlag, Berlin, 1985.

KAPLAN, S. (1987). *Simplifying conditional term rewriting systems: Unification, termination and confluence*. J. of Symbolic Computation 4 (3), 1987, 295-334.

KENNAWAY, J.R., KLOP, J.W., SLEEP, M.R. & DE VRIES, F.J. (1990a). *Transfinite reductions in orthogonal term rewriting systems* (Full paper), report CS-R9041, CWI, Amsterdam.

KENNAWAY, J.R., KLOP, J.W., SLEEP, M.R. & DE VRIES, F.J. (1990b). *An infinitary Church-Rosser property for non-collapsing orthogonal term rewriting systems*, report CS-R9043, CWI, Amsterdam.

KENNAWAY, J.R., KLOP, J.W., SLEEP, M.R. & DE VRIES, F.J. (1991), *Transfinite reductions in orthogonal term rewriting systems*, to appear in Proc. RTA '91, Springer LNCS.

KENNAWAY, J.R. & SLEEP, M.R. (1989). *Neededness is hypernormalizing in regular combinatory reduction systems*. Preprint, School of Information Systems, Univ. of East Anglia, Norwich.

KLOP, J.W. (1987). *Term rewriting systems: a tutorial*. Bulletin of the EATCS 32, p. 143-182.

KLOP, J.W. (1980). *Combinatory Reduction Systems*. Mathematical Centre Tracts Nr.127, CWI, Amsterdam.

KLOP, J.W. (1991). *Term rewriting systems*, to appear in *Handbook of Logic in Computer Science*, Vol I (eds. S. Abramsky, D.Gabbay and T. Maibaum), Oxford University Press.

KLOP, J.W. & VRIJER, R.C. DE (1989). *Unique normal forms for lambda calculus with surjective pairing*. Information and Computation, Vol.80, No.2, 1989, 97-113.

MANN, C.R. (1973). *Connections between proof theory and category theory*. Dissertation, Oxford University.

MEINKE, K. & TUCKER, J.V. (1990). *Universal algebra*. In: Handbook of Logic in Computer Science (eds. S. Abramsky, D. Gabbay and T. Maibaum) Vol.1, Oxford University Press, to appear.

MIDDELDORP, A. (1989). *Confluence of the Disjoint Union of Conditional Term Rewriting Systems*. Report CS-R8944, CWI, Amsterdam (also in this volume).

MIDDELDORP, A. (1990). *Modular properties of term rewriting systems*. Ph.D. Thesis, Free University Amsterdam.

O'DONNELL, M.J. (1977). *Computing in systems described by equations*. Springer Lecture Notes in Computer Science 58.

O'DONNELL, M.J. (1985). *Equational logic as a programming language*. The MIT Press, Cambridge MA.

PLOTKIN, G.D. (1977). *LCF as a programming language*. TCS 5, 223-257.

PRAWITZ, D. (1971). *Ideas and results in proof theory*. In: Proc. 2nd Scandinavian Logic Symposium (ed. J.E. Fenstad), North-Holland, 235-307.

TURNER, D.A., (1985). *Miranda: a non-strict functional language with polymorphic types*. In J.-P. Jouannaud (ed.), Proc. ACM Conf. on Functional Programming Languages and Computer Architecture, LNCS, vol. 201, Springer-Verlag.

VRIJER, R.C. DE (1987). *Surjective pairing and strong normalization: two themes in lambda calculus*. Dissertation, University of Amsterdam.

VRIJER, R.C. DE (1989). *Extending the lambda calculus with surjective pairing is conservative*. In Proc. of the 4th Annual IEEE Symposium on Logic in Computer Science, Pacific Grove, p. 204-215.

VRIJER, R.C. DE (1990). *Unique normal forms for Combinatory Logic with Parallel Conditional, a case study in conditional rewriting*. Techn. Report Free University Amsterdam, 1990.

A proof system for conditional algebraic specifications

Emmanuel Kounalis and Michaël Rusinowitch
CRIN, BP239, 54506 Vandœuvre-les-Nancy, France &
INRIA Lorraine, BP 101, 54600 Villers-les-Nancy, France
e-mail: {kounalis,rusi}@loria.fr

Draft

Abstract

Algebraic specifications provide a formal basis for designing data-structures and reasoning about their properties. Sufficient-completeness and consistency are fundamental notions for building algebraic specifications in a modular way. We give in this paper effective methods for testing these properties for specifications with conditional axioms.

1 Introduction

For the description of large data structures and complex system of data-structures it is necessary to design specifications in a modular way which means that larger specifications are built from smaller ones. Algebraic specifications are widely accepted as a formal basis for designing structured specifications and reasoning about their properties. The foundations for a mathematical theory of algebraic specifications as well as the use of model-theoretical methods for proving their correctness were given by the ADJ group [6]. Berlin's group [4] has developed a method of stepwise extension of equational specifications and has shown how this method allows us to prove the correctness of a software system specification in parallel to its software design. The key idea is first to define a *base* specification which contains all those operations and equations that are necessary for the construction of data, and then to design an extension of it. The whole specification is correct if its semantics agrees with the one of the base specification.

The *sufficient-completeness* and *consistency* properties introduced by Guttag have been found useful in providing guidelines for designing specifications and proving their properties: a specification is correct with respect to a base specification, if and only if, it is sufficiently complete and consistent. Guttag has shown that both properties are undecidable. Subsequently, these properties have been investigated for restricted classes of specifications. Most efforts have been focused on building tools to check these properties for equational specifications [4, 19, 17, 7, 22, 10, 12, 21, 1, 8, 9, 13]. Based on our recent results on inductive theorem proving [16, 14] we show how to check sufficient-completeness and consistency of *conditional* algebraic specifications. These problems have also been investigated in recent work by Padawitz [20, 18]. However, it seems that his approach cannot be fully automatized.

The structure of this paper is as follows: in Section 2 we describe our approach on a simple example. In Section 3 we introduce some preliminary definitions. In Section 4 we show inductive reasoning techniques applies to the verification of sufficient completeness and we provide a simple procedure to check this property. In Section 5 we give two methods for checking consistency of algebraic specifications.

2 Overview of our approach: an example

Before discussing the technical details of the method we propose for checking sufficient completeness and consistency of conditional specifications, we first describe our proof system on a simple example, namely the specification of *odd* and *even* on positive integers.

Example 2.1 *Consider the base specification $SPEC0 = (F_0, A_0)$, where the set of function symbols is $F_0 = \{0, s, true, false\}$ and the set of axioms A_0 is empty. Let us now enrich our specification $SPEC0$ with new functions and axioms. Let F_1 be the set $\{+, odd, even\}$ and let A_1 be the following set of axioms which define these functions. The first important fact to notice is that they can be oriented to a terminating simplification system. The arrow $\rightarrow$ indicates how to apply a (conditional) equation for simplification:*

$$x + 0 \rightarrow x \tag{1}$$
$$x + s(y) \rightarrow s(x + y) \tag{2}$$
$$even(0) \rightarrow true \tag{3}$$
$$even(s(0)) \rightarrow false \tag{4}$$
$$even(s(s(x))) \rightarrow even(x) \tag{5}$$
$$odd(0) \rightarrow false \tag{6}$$
$$even(x) = true \Rightarrow odd(x) \rightarrow false \tag{7}$$
$$even(x) = false \Rightarrow odd(x) \rightarrow true \tag{8}$$

There are two main problems when verifying that the whole specification $SPEC = SPEC0 + SPEC1 = (\{F_0 \cup F_1\}, \{A_0 \cup A_1\})$ is correct *(see [4]) with respect to the base specification $SPEC0$.*

- *Is $SPEC$ sufficiently complete w.r.t. $SPEC0$ (which means that any ground term is equivalent modulo the axioms to a term built only with symbols from F_0) ?*
- *Is $SPEC$ consistent w.r.t. $SPEC0$ (which means that two different terms of the base specification are still different in the enriched specification) ?*

In this paper, we develop techniques for answering these two questions.

For checking the sufficient completeness of $SPEC$ w.r.t. $SPEC0$ we build what we call a complete pattern tree for each new function symbol $f \in F_1$. See Section 4. A pattern tree is a tree whose nodes are terms, whose root is $f(x_1, \ldots, x_n)$ and such that the successors of any internal node $t = f(t_1, \ldots, t_n)$ are obtained by replacing a variable in t by all possible terms $q(z_1, \ldots, z_k)$ where $q \in F_0$ and the z_i are new fresh variables not already in t. Here are the complete pattern trees for the symbols in F_1.

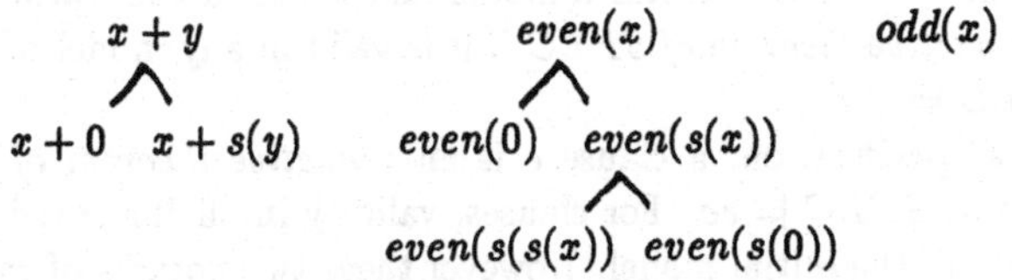

We can easily verify that the leaves $x + 0, x + s(y), even(0), even(s(s(x)), even(s(0)), odd(0)$ are all instances of the left-hand side of an unconditional (oriented) axiom of $SPEC$.

However, the term odd(s(x)) is not, although we can notice that all its ground instances are reducible by some conditional axiom. This can be proved formally by checking that $even(x) = true \vee even(x) = false$ *is valid for any ground term x built with function symbols in the base specification SPEC0. Now any ground term which contains operators in* F_1 *is reducible. Therefore, by theorem 4.1, SPEC is sufficiently with respect to SPEC0.*

To verify the consistency property of SPEC w.r.t. SPEC0, we apply the set of inference rules given in subsection5.1. By the Deduce rule, we obtain:

$$even(x) = true \wedge even(x) = false \Rightarrow true = false$$

which can be simplified into:

$$even(x) = true \wedge true = false \Rightarrow true = false$$

which is deleted by the Trivial rule. No new axiom can be derived and, therefore, the system is consistent.

3 Preliminaries

3.1 Terms and conditional equations

Let F be a signature of function symbols, and X a set of variables. We shall denote by $T(F,X)$ the set of terms built with F and X. The set of ground terms is denoted by $T(F)$. Given an object t, $t\sigma$ represents the image of t by the substitution σ. As usual, occurrences in terms are represented as sequences of integers. Let u be an occurrence, we use t/u for the subterm of t at u. More precisely, $t/\lambda = t$ where λ is the empty occurrence, and $g(t_1, \ldots, t_n)/i.u = t_i/u$. We also use $s = s[u \leftarrow t]$ to denote that s is a term whose subterm at occurrence u is t. For convenience, we sometimes express it by $s = s[t]$ if the particular occurrence is not important.

An equation is a pair of terms, written $s = t$. A conditional equation is an equation or an expression of one of the types: $e_1 \wedge \ldots \wedge e_n \Rightarrow e$ or $e_1 \wedge \ldots \wedge e_n \Rightarrow$ or $\Rightarrow$ where $e, e_1, \ldots e_n$ are equations, $e_1, \ldots e_n$ are *conditions*, a positive literal e in a conditional equation is a *conclusion* and $\Rightarrow$ is the empty clause. In this paper, axiomatizations are built from conditional equations and goals to be proved are clauses (i.e. disjunction of equational literals, since $=$ is the only predicate [1]).

3.2 Specifications

A specification $SPEC$ is given by a couple (F, A) where F is a signature and A is a set of conditional equations. We recall that a Herbrand model of $SPEC$ is a model of A whose domain is $T(F)$ (axioms for equality are implicitly assumed to be valid, too). The intersection of all Herbrand models of $SPEC$ is still a model and is called the *initial model* of $SPEC$. A formula $\mathcal{F}$ is a *deductive theorem of SPEC* if it is valid in any model of $SPEC$. This will be denoted by $SPEC \models$ F.

Let $SPEC$ be a specification, a clause e is an *inductive theorem of SPEC* iff for any ground substitution σ, $SPEC \models e\sigma$. For clauses, validity in all Herbrand models differs, in general, from validity in the initial model. However these two notions of validity coincide for unconditional equations:

[1] we shall identify a conditional equation and its corresponding representation as a clause

Proposition 3.1 (see Padawitz [20]) *A clause e is an* **inductive theorem** *iff it is valid in any Herbrand model of SPEC. An unconditional equation e is an inductive theorem iff it is valid in the initial model of SPEC.*

A base specification $SPEC0 = (F_0, A_0)$ of $SPEC$ is a specification such that $F_0 \subseteq F$ and $A_0 \subseteq A$. The specification $SPEC$ is called an *enrichment* of $SPEC0$. The elements of $T(F_0)$ are called *constructor terms.* The set Σ_0 denotes the substitutions which apply the variables of their domain to constructor terms.

Definition 3.1 *Let SPEC be an enrichment of a base specification SPEC0. A clause e is an* constructor based theorem *of SPEC iff for any substitution σ which applies the variables to ground terms of the base specification, $H \models e\sigma$.*

Let us introduce the classical properties which ensure that the enrichment $SPEC$ is correct w.r.t. the base specification $SPEC0$ which means that it does not change the equality relation on ground terms of the base specification.

Definition 3.2 *Let SPEC be an enrichment of SPEC0, then:*

- *SPEC is sufficiently complete w.r.t. SPEC0 if for any term t in $T(F)$ there exists some term s in $T(F_0)$ such that $SPEC \models s = t$.*
- *SPEC is consistent w.r.t. SPEC0 if for any terms $s, t \in T(F_0)$ $SPEC \models s = t \Rightarrow SPEC0 \models s = t$*

3.3 Rewriting

Given a binary relation $\rightarrow$, $\rightarrow^*$ denotes its reflexive transitive closure. Given two binary relations, R, S, RoS denotes their composition. A relation R is noetherian if there is no infinite sequence $t_1 \; R \; t_2 \; R \; \ldots$. In the following, we suppose we are given a *reduction ordering* $\succ$ on the set of terms, that is, a transitive irreflexive relation which is noetherian, monotonic ($s \succ t$ implies $w[s] \succ w[t]$) and stable ($s \succ t$ implies $s\sigma \succ t\sigma$). A reduction ordering can be extended to literals by comparing the multiset of their members with the multiset extension of $\succ$ (see [3]). Formulae are compared by using the multiset extension of this last ordering to the multiset of their atomic subformulas. Since there is no ambiguity, all these extensions will also be denoted by $\succ$. An equation $s = t$ will be written $s \rightarrow t$ if $s\sigma \succ t\sigma$ for all ground substitutions σ; in that case we say that the equation is *orientable.*

The idea of rewriting is to impose a direction when using equations in proofs. This direction is indicated by an arrow when it is independant from the instantiation: $l \rightarrow r$ means that, we can replace l by r in some context. When an instance of a conditional equation is orientable and has a valid conditional part, it can be applied as a rule. The conditions are checked by a recursive call to the theorem-prover. Termination of such calls is ensured by requiring the conditions to be smaller (w.r.t. the reduction ordering $\succ$) than the conclusion:

Definition 3.3 *Let H be a set of conditional equations. Let A be a term or a clause, and n an occurrence of A. Then $A[s\sigma] \rightarrow_H A[t\sigma]$ if σ is a substitution and there is a conditional equation $c \Rightarrow s = t$, in H such that $s\sigma \succ t\sigma$ and $H \models c\sigma$ and $(s = t)\sigma \succ c\sigma$.*

A term A is reducible w.r.t. $\rightarrow_H$, if there is a term B such that $A \rightarrow_H B$. Otherwise we say that A is *H-irreducible.* The system H will be qualified as *convergent* if

$$\forall a, b \in T(F) \;\; H \models a = b \text{ implies } \exists c \;\text{ such that } a {\rightarrow_H}^* c \;\text{ and } b {\rightarrow_H}^* c.$$

This is usually a definition of the Church-Rosser property. But, since the relation $\rightarrow_H$ is *always noetherian* (as a consequence of our definition), these two notions are equivalent here. One can easily see that it is equivalent to the property that every ground term possesses an unique irreducible form. Note also that the rewriting relation may be undecidable since we have to find solutions for c in a Horn theory, which is known to be semi-decidable. The relation $\rightarrow_H$ will be extended to sets of clauses in a natural way: by definition, $S \cup \{c\} \rightarrow_H S \cup \{d\}$ whenever $c \rightarrow_H d$.

Case reasoning is a very powerful technique which is the basis of many theorem proving strategies. It is also a most important rule in the context of inductive theorem proving where case splitting arises naturally from induction hypothesis.

Definition 3.4 *Let H be a set of conditional equations and $c \Rightarrow s = t$ a conditional equation in H. Let $A[n \leftarrow s\sigma]$ be a clause (where σ is a substitution). The case rewriting rule can be stated as follows*

$$S \cup \{A[n \leftarrow s\sigma]\} \rightharpoonup_H S \cup \{(c\sigma \vee A[n \leftarrow s\sigma]), (\neg c\sigma \vee A[n \leftarrow t\sigma])\}$$

if neither $c\sigma$ nor $\neg c\sigma$ is a subclause of $A[n \leftarrow s\sigma]$, n occurs in a maximal literal of A, $s\sigma \succ t\sigma$ and $(s = t)\sigma \succ c\sigma$ and $A[s\sigma] \succ c\sigma$. Let us denote the relation $\rightharpoonup_H \cup \rightarrow_H$ by $\hookrightarrow_H$.

4 How to check sufficient-completeness

Let $SPEC = (F_0 \cup F_1, A_0 \cup A_1)$ be an enrichment of the base specification $SPEC0 = (A_0, F_0)$. **In the following sections, we assume that A_0 is empty.** We shall assume in this section that the set of axioms A_1 of the specification can be oriented to constitute a reductive set of rules. This means that for every one of them, say $c \Rightarrow l = r$, and for any ground substitution ϕ, we have $l\phi \succ r\phi, c\phi$.

In this section, we propose a method of checking sufficient-completeness of conditional theories with empty base specification.

4.1 Pattern trees

The verification of sufficient-completeness needs some kind of inductive reasoning.

Definition 4.1 *Let $TS(SPEC0)$ be the set $\{g(x_1, \ldots, x_n); g \in F_0\}$. A* **pattern tree** *$T$ of $f \in F_1$ is a tree whose nodes are terms. The root is $f(x_1, \ldots, x_n)$. Every successor of a node s is obtained by replacing a variable of s by an element of $TS(SPEC0)$ whose variables have been renamed.*

Example 4.1 *Let $TS(SPEC0)$ be $\{0, s(x)\}$. Here is a pattern tree of even:*

$even(x)$

$even(0)$ $\quad$ $even(s(x_1))$

Note that the sets of ground instances of the leaves of a pattern tree for f constitute a partition of the set of ground instances of $f(x_1, \ldots, x_n)$.

Now, we assume that we are given a set of conditional equations defining a function f. Let us introduce two notions which allow to build a suitable pattern-tree for f. First, we must identify the nodes to be expanded and the variables to be replaced. The next definition tells us how to do it.

Definition 4.2 *A term t is* **extensible at position** *u w.r.t. $SPEC$ if t/u is a variable and there is a conditional rule $c \Rightarrow l \rightarrow r \in SPEC$ such that l/u is a function symbol or a variable which occurs more than once in l. A term t is* extensible *w.r.t. $SPEC$ if it is extensible at some position u.*

Definition 4.3 *A complete pattern tree of $f \in F_1$ is a pattern tree which is obtained by starting from the tree $T = f(x_1, ..., x_n)$, and iterating the following operations:*

- *Select non-deterministically a leaf t which is extensible at some position u*
- *For any c in $TS(SPEC0)$, rename c in c' with new variables and add $t[u \leftarrow c']$ as a son of t.*

Note that a finite complete pattern tree of f can always be derived.

Let us now show how complete pattern trees may be used to verify the sufficient completeness property. For that, we need the following definition.

Definition 4.4 *We say that t is* **pseudo-reducible** *by $SPEC$ if there exists a set of conditional equations $\{c_1 \Rightarrow l_1 \rightarrow r_1, \ldots, c_n \Rightarrow l_n \rightarrow r_n\}$ in $SPEC$ and a set of occurrences $\{u_1, ..., u_n\}$ such that $t/u_1 = l_1\sigma_1, \ldots, t/u_n = l_n\sigma_n$ and $C_1\sigma_1 \vee \ldots \vee C_n\sigma_n$ is a constructor based theorem of $SPEC$. Otherwise t is said to be* **pseudo-irreducible.**

Note immediately that if a term is pseudo-reducible, all its ground instances by constructor terms are reducible.

Theorem 4.1 *Assume that $SPEC$ is an enrichment of the base specification $SPEC0$, with an empty set of axioms. Let $A = \{T_f; f \in F_1\}$ be a set of complete pattern trees of the function symbols $f \in F_1$. If all the leaves of all the elements of A are pseudo-reducible then $SPEC$ is sufficiently complete with respect to $SPEC0$.*

Note that the main problem to get effective methods to derive complete pattern trees consist of proving (or disproving) constructor based theorems.

4.2 Proving and disproving constructor based theorems

In this section, we show how to prove and disprove constructor based theorems. The method is based on our general method of [16, 14]. The technique allows us to replace inductive reasoning by pure (algebraic) simplification.

4.3 Proving constructor based theorems

Our notion of induction uses a noetherian ordering on ground terms, which contains the conditional rewriting relation. **We can use as an inductive hypothesis any instance of the theorem we want to prove, as soon as this instance is smaller (w.r.t. $\succ$) than the one that is currently considered.** We propose here a rewriting relation which is sound, with regard to the use of induction hypothesis. This rewriting uses inductive hypothesis when attempting to satisfy the conditions of a conditional rule.

Definition 4.5 *Let $u = v$ be an equation and A, B two terms.*

- *Let H be a set of conditional equations. Let A be a term, and n an occurrence of A. We write $A[n \leftarrow s\sigma] \rightarrow_{H[u=v]} A[n \leftarrow t\sigma]$ if σ is a substitution and there is a conditional equation $a_1 = b_1 \wedge \cdots \wedge a_n = b_n \Rightarrow s = t$ in H, such that*
- $s\sigma \succ t\sigma$
- $\forall i \in \{1,\ldots,n\}\ \exists c\ \ a_i\sigma{\rightarrow_{H\cup\{u=v\}}}^* c$ *and* $b_i\sigma{\rightarrow_{H\cup\{u=v\}}}^* c$
- $(s = t)\sigma \succ (a_1 = b_1 \wedge \cdots \wedge a_n = b_n)\sigma$
- $A[s\sigma] \succ (a_1 = b_1 \wedge \cdots \wedge a_n = b_n)\sigma$

Let $SPEC$ be an enrichment of $SPEC0 = (F_0, A_0)$. We denote by $AS(SPEC)$ the set of minimal elements, for the subsumption ordering, of the leaves arguments of the complete pattern-trees for the elements of F_1. For instance, if F_1 only contains the symbol f whose complete pattern tree is the following:

$$f(x,y,z)$$
$$f(0,y,z) \quad f(s(x),y,z)$$
$$f(s(0),y,z) \quad f(s(s(x)),y,z)$$
$$f(s(s(x)),0,z) \quad f(s(s(x)),s(y),z)$$

then $AS(SPEC) = \{0, s(0), s(s(x))\}$. A test substitution for SPEC is a substitution ν such that for every x, y in the domain of ν, $\nu(x)$ is a renaming of an element of $AS(SPEC)$ and $\nu((x)$ and $\nu(y)$ admit no common variables.

The following theorem shows how to prove that equations are constructor based theorems.

Theorem 4.2 *Let $SPEC$ be an enrichment of $SPEC0 = (F_0, A_0)$ and and let $u = v$ (with $u \succ v$) be an equation. We suppose that the following hypothesis is verified: $\rightarrow_{A\cup\{u=v\}}$ is noetherian and we define $\rightsquigarrow$ as the reflexive closure of the relation:$(\rightarrow_A \cup \rightarrow_{A[u=v]})o(\rightarrow_{A\cup\{u=v\}})^*$. If, for all test substitutions ν of $SPEC$, there is a term α such that $u\nu \rightsquigarrow \alpha$ and $v\nu(\rightarrow_{A\cup\{u=v\}})^*\alpha$ then $u = v$ is a constructor based theorem of $SPEC$.*

Let us illustrate the previous theorem on a simple example:

Example 4.2 *Consider the following base specification $SPEC0 = (\{0, s, true\}, \emptyset)$ and let us now enrich $SPEC0$ by the new function symbol p with axioms A: $p(0) = true$ and $p(x) = true \Rightarrow p(s(x)) = true$. Let us check that this new specification is sufficiently complete: by computing its pattern tree we get*

$$p(x)$$
$$p(0) \quad p(s(z))$$

Therefore the leaves are $p(0), p(s(z))$ and $AS(SPEC) = \{0, s(z)\}$. We note that $p(0)$ is reducible. Consider now $p(s(z))$ which is an instance of the left-hand side of $p(x) = true \Rightarrow p(s(x)) = true$. To see that $p(s(z))$ is pseudo-reducible, we must show that $p(z) = true$ is a constructor-based theorem. Using the previous theorem, we have to verify $p(0) = true$ and $p(s(w)) = true$. This last equation is proved by simplification:

$$p(s(w)) \rightarrow_{A[p(x)=true]} true$$

It is straightforward to generalize the previous method to proving that clauses are constructor based theorems:

Theorem 4.3 *Let $SPEC$ be an enrichment of $SPEC0$ and C a clause. If, for all test substitutions ν, $\{C\nu\} \hookrightarrow_H^* \{p_1, p_2 \cdots, p_n\}$, and every clause p_j is either* a tautology *(contains two complementary literals or an instance of $x{=}x$) or* is subsumed by an axiom *or* contains an instance of C which is strictly smaller w.r.t. $\succ$ than $C\nu$, *then C is an constructor based theorem of H.*

In the example above, to show $odd(x)$ is pseudoreducible, we need to verify that $even(x) = true \vee even(x) = false$ is a constructor based theorem. To check it, we need to prove: $even(0) = true \vee even(0) = false$ and $even(s(0)) = true \vee even(s(0)) = false$ and $even(s(s(x))) = true \vee even(s(s(x))) = false$. The first two are trivial. The third one is simplified to $even(x) = true \vee even(x) = false$ by rule 4. This clause is subsumed by the initial goal.

4.4 Disproving constructor based theorems

We show now how to prove that conjectures are not constructor based theorems.

Definition 4.6 *Let $SPEC$ be an enrichment of $SPEC0$ and C a clause. Let A be the set of positive literals of $SPEC$. Then, a clause $\neg e_1 \vee \ldots \vee \neg e_m \vee g_1 = d_1 \vee \ldots \vee g_n = d_n$ is* **quasi-inconsistent** *with respect to $SPEC$ if there is a test substitution σ such that, for all $i \leq m$, $e_i\sigma$ is a constructor based theorem and for all $j \leq n$ at least one of the following is verified:*

- *$g_j\sigma \not\equiv d_j\sigma$ and both $g\sigma$ and $d\sigma$ are irreducible by A.*
- *$g_j\sigma \succ d_j\sigma$ and $g_j\sigma$ is irreducible by A.*
- *$g_j\sigma \prec d_j\sigma$ and $d_j\sigma$ is irreducible by A.*

The next result shows that, when the set of axioms is convergent, a quasi-inconsistent clause cannot be a constructor based theorem. This is proved by building a well-chosen ground instance of the clause, which is false in some Herbrand model of the axioms.

Theorem 4.4 *Let H be a convergent set of conditional equations. If C is quasi-inconsistent w.r.t. H then C is not a constructor based theorem of H.*

5 How to check consistency

In this section we propose two methods for verifying the consistency of algebraic specifications with respect to a base specification with an empty set of axioms. They both conclude to consistency when a ground convergent system can be derived from the initial specification. However the ways to obtain this property are different.

The first method relies on a complete set of inference rules for Horn logic, and corresponds to Knuth and Bendix algorithm in Huet-Hullot approach [7]. The second one uses inductive reasoning to test ground convergence.

5.1 The saturation method

The saturation technique is based on a set of inference rules which is refutationally complete for conditional equations. Some of these rules have been discussed in [15, 23]. We will show next how to apply them to get consistency of a set of axioms.

We assume in this subsection that $\prec$ is total on $T(F)$ and that for any t and for any proper subterm s of t we have $s \prec t$.

As the main deduction rule, we introduce *superposition*, which is a refinement of paramodulation: the only inferences which are allowed are those obtained by paramodulating maximal members of conclusions into maximal members of conclusions (*maximal* refers here to the ordering $\prec$). We formally define the set of superpositions which can be derived from the set of equations E. Define,

$$\begin{array}{rcl} SP(E) & = & \{(c \wedge q \Rightarrow l[v] = r)\sigma : \exists (c \Rightarrow l[u'] = r) \in E, \exists (q \Rightarrow u = v) \in E) \\ & & \text{such that } u' \text{ is not a variable and } u'\sigma \equiv u\sigma \text{ and } l\sigma \not\preceq .c\sigma, r\sigma \text{ and } u\sigma \not\preceq q\sigma, v\sigma\} \end{array}$$

There is also a refinement of paramodulation to deal with conditions. This is the conditional narrowing rule: paramodulation is performed from maximal members of conclusions into maximal conditions.

$$\begin{array}{rcl} CN(E) & = & \{(c \wedge q \wedge (l[v] = r) \Rightarrow e)\sigma : \exists (c \wedge (l[u'] = r) \Rightarrow e) \in E, \exists (q \Rightarrow u = v) \in E \\ & & \text{such that } u' \text{ is not a variable and } u'\sigma \equiv u\sigma \text{ and} \\ & & l\sigma \not\prec r\sigma, (l = r)\sigma \not\prec c\sigma, e\sigma \text{ and } u\sigma \not\preceq q\sigma, v\sigma\} \end{array}$$

The new rules *IDelete* and *IRefute* are specific to the framework of constructor based specifications. They take advantage of the fact that the operators of F_0 are free.

$$\begin{array}{llcll}
\textit{Deduce:} & E & \vdash & E \cup \{c\} & \text{if } c \in SP(E) \\
\textit{Narrow:} & E & \vdash & E \cup \{c\} & \text{if } c \in CN(E) \\
\textit{Reflect:} & E \cup \{c \wedge s = t \Rightarrow l = r\} & \vdash & E \cup \left\{ \begin{array}{c} c \wedge s = t \Rightarrow l = r \\ (c \Rightarrow l = r)\sigma \end{array} \right\} & \text{if } \begin{array}{l} s\sigma \equiv t\sigma \text{ and} \\ (s = t)\sigma \not\preceq c\sigma, (l = r \end{array} \\
\textit{Delete:} & E \cup \{c \Rightarrow s = s\} & \vdash & E & \\
\textit{Trivial:} & E \cup \{e \wedge c \Rightarrow e\} & \vdash & E & \\
\textit{Subsume:} & E \cup \{C, D\} & \vdash & E \cup \{C\} & \text{if } C\sigma \subseteq D \\
\textit{Simplify:} & E \cup \{c[s\sigma]\} & \vdash & E \cup \{c[t\sigma]\} & \text{if } \begin{array}{l} E \cup \{c[s\sigma]\} \models s\sigma = \\ \text{and } (s = t)\sigma \prec c[s\sigma \end{array} \\
\textit{S1:} & E \cup \{a = b \wedge c \Rightarrow e[a]\} & \vdash & E \cup \{a = b \wedge c \Rightarrow e[b]\} & \text{if } a \succ b \\
\textit{S2:} & E \cup \{a = b \wedge c[a] \Rightarrow e\} & \vdash & E \cup \{a = b \wedge c[b] \Rightarrow e\} & \text{if } a \succ b
\end{array}$$

IRefute: $E \cup \{c \Rightarrow l = r\} \ \vdash \ Disproof$ if $l \in T(F_0, X)$ and $\exists \sigma \in \Sigma_0$ such that $E \models c\sigma$ and $l\sigma \succ r\sigma, c\sigma$

IDelete: $E \cup \{c \wedge l = r \Rightarrow e\} \ \vdash \ E$ if $l, r \in T(F_0, X)$ and they are not unifiable

We shall suppose that the rules *Delete, Trivial, Subsume, Simplify, S1, S2,Idelete* have a higher priority than the others. The Knuth and Bendix procedure [11] has been designed to derive convergent systems from equational presentations. Huet and Hullot [7] have shown how to use this procedure to check the consistency of constructor based specifications. This procedure halts with success when all possible deductions are trivial. Applying our set of inference rules to conditional equations, we can easily adapt the Huet-Hullot result to conditional specifications.

Theorem 5.1 *We assume that SPEC is sufficiently complete. If the procedure does not generate Disproof, then SPEC is consistent with respect to the base specification* $(F_0, \emptyset)$.

Example 5.1 *Consider the following set of conditional equations which defines the minimum of a list of integers. We orient equations with the lrpo [2] with the following precedence on functions:*

$$min \succ \ < \ \succ cons \succ nil \succ succ \succ 0 \succ true \succ false$$

Here $F_0 = \{true, false, cons, nil, 0, succ\}$.

$$(0 < succ(x)) \rightarrow true \tag{9}$$
$$(x < 0) \rightarrow false \tag{10}$$
$$succ(x) < succ(y) \rightarrow x < y \tag{11}$$
$$min(cons(x, nil)) \rightarrow x \tag{12}$$
$$(x < min(l)) = true \Rightarrow min(cons(x, l)) \rightarrow x \tag{13}$$
$$(x < min(l)) = false \Rightarrow min(cons(x, l)) \rightarrow min(l) \tag{14}$$
$$min(l) = x \Rightarrow min(cons(x, l)) \rightarrow min(l) \tag{15}$$

The inductive completion procedure generates only two equations, namely:

$$(x < min(nil)) = false \Rightarrow x \ = \ min(nil)$$
$$(x < min(l)) = false \wedge (x < min(l)) = true \Rightarrow x \ = \ min(l)$$

The first one is non operational and the second one can be simplified by applying $S1$ *into the following equation:*

$$(x < min(l)) = false \wedge false = true \Rightarrow x \ = \ min(l)$$

which is deleted by IDelete. *We can conclude that the specification of min is consistent with the base specification of lists and booleans.*

5.2 The inductive convergence method

Let us now introduce a more powerful method. It is based on a new technique for deriving systems which are ground convergent. We do not assume anymore that $\prec$ is total on ground terms nor that for any t and for any proper subterm s of t we have $s \prec t$. However we shall assume that the set of axioms A_1 of the specification can be oriented as a reductive set of rules R.

We give now a criteria to detect equations that do not contribute to change the closure of the rewriting relation on ground terms. This test is related to the simplification relations introduced in 4.2.

Definition 5.1 *A conditional equation $e = c \Rightarrow l = r$ (with $l \succ r, c$) e is inductively convergent if for every test-substitution ϕ of $SPEC$ there is a term m such that:*

$$l\phi\ (\rightarrow_R)\ o\ (\rightarrow_{R\cup\{e\}})^*\ m \text{ and } r\phi\ (\rightarrow_{R\cup\{e\}})^*\ m$$

Now if all critical pairs derived from a given system R pass the previous test, then we can be sure that R is ground convergent:

Theorem 5.2 *If R is a reductive set of conditional equations, and if all the critical pairs of R are inductively convergent, then R is convergent. Moreover, if E is a set of equations which is inductively convergent w.r.t. a convergent system R, then $R \cup E$ is convergent.*

Example 5.2 *Consider the base specification $(\{0, s\}, \emptyset)$ and its enrichment by the new function $+$ and the set of axioms R which defines $+$:*

$$x + 0 \rightarrow x \tag{16}$$

$$x + s(y) \rightarrow s(x + y) \tag{17}$$

$$(x + y) + z \rightarrow x + (y + z) \tag{18}$$

A test-set is $\{0, s(x)\}$. The system is convergent by the previous theorem. Consider, for example, a critical pair between associativity and the second rule:

$$s(x + y) + z \rightarrow x + (s(y) + z) \tag{19}$$

and the following instance:

$$s(s(x) + s(y)) + s(z)\ (\rightarrow_R)^*\ s((s(x) + s(y)) + z)\ \rightarrow_{R\cup\{(x+y)+z\rightarrow x+(y+z)\}}\ s(s(x) + (s(y) + z))$$

$$s(x) + (s(s(y)) + s(z))\ (\rightarrow_R)\ s(s(x) + (s(y) + z))$$

A straightforward corollary is the following:

Theorem 5.3 *We assume that we are given a specification $SPEC$ which is an enrichment of $SPEC0 = (F_0, \emptyset)$ and such that every left-hand side of a rule contains a symbol which is not in F_0. If every critical pair is inductively convergent then $SPEC$ is consistent with respect to $SPEC0$.*

6 Conclusion

We have presented new methods for verifying the correctness (i.e. sufficient completeness and consistency) of conditional specifications. We feel that our results should also apply to parameterized specifications, following [5], for instance. We also plan to generalize the technique to the case where there are relations between the constructors.

References

[1] N. Dershowitz. Computing with rewrite systems. *Information and Control*, 65(2/3):122–157, 1985.

[2] N. Dershowitz. Termination of rewriting. *Journal of Symbolic Computation*, 3(1 & 2):69–116, 1987.

[3] N. Dershowitz and Z. Manna. Proving termination with multiset orderings. *Communications of the Association for Computing Machinery*, 22(8):465–476, 1979.

[4] H. Ehrig, H. Kreowsky, and P. Padawitz. Stepwise specifications and implementation of adt. In *Proceedings International Colloquium on Automata, Languages and Programming, Lecture Notes in Computer Science, volume 62*, 1978.

[5] H. Ganzinger. Ground term confluence in parametric conditional equational specifications. In F. Brandenburg, G. Vidal-Naquet, and M. Wirsing, editors, *Proceedings STACS 87*, pages 286–298. Springer-Verlag, 1987. Lecture Notes in Computer Science, volume 247.

[6] J.A. Goguen, J.W. Thatcher, and E.G. Wagner. An initial algebra approach to the specification, correctness and implementation of abstract data types. In Yeh R., editor, *Current Trends in Programming methodology IV: Data structuring*, pages 80–144. Prentice Hall, 1978.

[7] G. Huet and J-M. Hullot. Proofs by induction in equational theories with constructors. *Journal of Computer and System Sciences*, 25(2):239–266, October 1982. Preliminary version in Proceedings 21st Symposium on Foundations of Computer Science, IEEE, 1980.

[8] J.-P. Jouannaud and E. Kounalis. Proof by induction in equational theories without constructors. In *Proceedings 1st Symp. on Logic In Computer Science*, pages 358–366, Boston (USA), 1986.

[9] D. Kapur, P. Narendran, and H. Zhang. On sufficient completeness and related properties of term rewriting systems. *Acta Informatica*, 24:395–415, 1987.

[10] H. Kirchner. A general inductive completion algorithm and application to abstract data types. In R. Shostak, editor, *Proceedings 7th international Conference on Automated Deduction*, pages 282–302. Springer-Verlag, Lecture Notes in Computer Science, 1984.

[11] D.E. Knuth and P.B. Bendix. Simple word problems in universal algebras. In J. Leech, editor, *Computational Problems in Abstract Algebra*, pages 263–297. Pergamon Press, Oxford, 1970.

[12] E. Kounalis. Completeness in data type specifications. In B. Buchberger, editor, *Proceedings EUROCAL Conference*, pages 348–362, Linz (Austria), 1985. Springer-Verlag. Lecture Notes in Computer Science, volume 204.

[13] E. Kounalis. Pumping lemmas for tree languages generated by rewrite systems. In *Fifteenth International Symposium on Mathematical Foundations of Computer Science*, Banská Bystrica (Czechoslovakia), 1990. Springer-Verlag. Lecture Notes in Computer Science.

[14] E. Kounalis and Rusinowitch M. Mechanizing inductive reasoning. *EATCS Bulletin*, 41:216–226, 1990.

[15] E. Kounalis and M. Rusinowitch. On word problem in Horn logic. In J.-P. Jouannaud and S. Kaplan, editors, *Proceedings on the first international workshop on Conditional Term Rewriting*, pages 144–160. Springer-Verlag, 1987. Lecture Notes in Computer Science, volume 308. See also the extended version published in Journal of Symbolic Computation, number 1 & 2, 1991.

[16] E. Kounalis and M. Rusinowitch. Mechanizing inductive reasoning. In *Proceedings of the AAAI Conference*, pages 240–245, Boston, 1990. AAAI Press and the MIT Press.

[17] D.R. Musser. On proving inductive properties of abstract data types. In *Proceedings 7th ACM Symp. on Principles of Programming Languages*, pages 154–162. Association for Computing Machinery, 1980.

[18] Padawitz P. Reductive validity. In *Proceedings 2nd International Workshop on Conditional Rewriting Systems*, Montreal (Canada), 1990. S. Kaplan and M. Okada.

[19] P. Padawitz. New results on completeness and consistency of abstract data types. *Proceedings 9th Symposium on Mathematical Foundations of Computer Science, Springer Verlag, Lecture Notes in Computer Science*, 88:460–473, 1980.

[20] P. Padawitz. *Computing in Horn Clause Theories*. Springer-Verlag, 1988.

[21] D. Plaisted. Semantic confluence tests and completion methods. In *Journal Information and Control 65*, pages 182–215, 1985.

[22] J.-L. Rémy. Etude des systèmes de réécriture conditionnels et applications aux types abstraits algébriques. Thèse d'Etat de l'Institut National Polytechnique de Lorraine, Nancy (France), 1982.

[23] M. Rusinowitch. Démonstration automatique par des techniques de réécriture. Thèse d'Etat de l'Université de Nancy I, 1987. Also published by InterEditions, Collection Science Informatique, directed by G. Huet, 1989.

Conditional Rewriting Logic: Deduction, Models and Concurrency*

José Meseguer
SRI International, Menlo Park, CA 94025, and
Center for the Study of Language and Information
Stanford University, Stanford, CA 94305

Abstract

Term rewriting has so far been understood almost exclusively as a technique for equational reasoning. This paper proposes a much broader interpretation in which term rewriting can be used both as a semantic foundation and as a programming paradigm in areas such as reactive systems, concurrency theory and object-oriented programming that do not fit naturally within the functional world of equational logic. The interpretation proposed views conditional rewriting as a logic in its own right, with its own proof theory and with a very general model theory of wide applicability. The logic is sound and complete and admits initial models. Equational logic appears as a special refinement of the general framework; this supports a natural unification of the functional and concurrent programming paradigms with a purely declarative style. Throughout the paper, the concurrent nature of term rewriting is emphasized; the role of rewriting as a unified model of concurrency is also discussed.

1 Introduction

During the past few years conditional rewrite systems have experienced a vigorous development. This conference—a sequel to [36] and the second explicitly devoted to the topic—attests to the broad interest that rewriting in general and conditional rewriting in particular has attracted in automated deduction, functional and logic programming, parallel computation and type theory among other areas.

Conditional rewrite rules are particularly attractive because they are more general and expressive than ordinary rewrite rules. This is very useful for functional programming languages such as OBJ [15, 20] and for languages that combine functional and logic programming, an area that has undergone very energetic development (see [14, 22, 10] for some early developments and [8] for a recent study). New techniques such as conditional narrowing (see, e.g., [33, 8]) and conditional completion (see [37, 35, 9, 16]) support both automated deduction and computation in the conditional case.

In these recent developments, exactly as in the previous work on unconditional rewriting, a basic assumption taken of course for granted has been that term rewriting is a *technique for equational deduction*. Indeed, both in the applications to automated deduction that have been a constant source of stimulus for term rewriting techniques as well as in functional and logic programming applications, equational logic has entirely dominated our thinking.

A central aim of this paper is to suggest that, while the use of term rewriting within equational logic is of course very important, a much broader range of exciting applications—applications for which equational logic allows no room at all—fit amazingly well within a broader interpretation of rewriting as a logic on its own right. In fact, it is precisely by breaking away from the tutelage of equational logic and acquiring a life of its own as a logic with its own proof theory (see Section 2) and model theory (see Section 3) that rewriting logic can both overcome its previous restrictive

*Supported by Office of Naval Research Contracts N00014-90-C-0086, N00014-88-C-0618 and N00014-86-C-0450, and NSF Grant CCR-8707155.

image as an operational technique and provide a much needed semantic understanding for other areas of computer science such as concurrency theory and object-oriented programming.

Indeed, one of the important benefits which can be gained from understanding rewriting as a logic and concurrent systems as models in such a logic is a *unification of programming language paradigms* by which functional programming can be nicely integrated with concurrent programming and in particular with concurrent object-oriented programming. Such a unification is remarkably simple and precise because it can be attained by means of a *map of logics*

$$\mathit{RewritingLogic} \longrightarrow \mathit{EquationalLogic}$$

in the precise sense of [48], which provides a systematic relationship between the syntaxes, deductions and model theories of both logics. Functional modules receive their meaning in equational logic, whereas "system modules" are given semantics in rewriting logic. The Maude language [50, 49], which contains OBJ [15, 20] as its functional sublanguage, has been designed precisely for the purpose of providing this unification; although Maude as such is not discussed in this paper (see [50, 49] for Maude's semantics and examples) a few examples written in Maude are used to illustrate the ideas, and the logic on which it is based is studied in detail.

The literature on rewrite systems has for the most part formalized rewriting in terms of a sequential rewriting relation. However, the most natural understanding of rewriting is as an intrinsically concurrent activity; this opens up many possibilities for programming and for parallel architecture such as those proposed in [21] and in work by other authors (see e.g., [38]). The close connection between rewriting and concurrency—leading to the identification of concurrent computation with rewriting logic deduction—is emphasized throughout this paper, and is explored in more detail in Section 4 where rewriting is proposed as a unified model of concurrency from which many other existing models can be obtained as direct straightforward specializations.

1.1 Rewriting is Naturally Concurrent

The idea of concurrent rewriting is very simple. It is the idea of *equational simplification* that we are all familiar with from our secondary school days, *plus* the obvious remark that we can do many of those simplifications independently, i.e., in *parallel*. Consider for example the following functional modules in Maude, written with an OBJ3 like syntax:

```
fmod NAT is
  sort Nat .
  op 0 : -> Nat .
  op s_ : Nat -> Nat .
  op _+_ : Nat Nat -> Nat [comm] .
  vars N M : Nat .
  eq N + 0 = N .
  eq (s N) + (s M) = s s (N + M) .
endfm
```

```
fmod REVERSE is
  protecting NAT .
  subsorts Nat < Tree .
  op _^_ : Tree Tree -> Tree .
  op rev : Tree -> Tree .
  var N : Nat .
  vars T T' : Tree .
  eq rev(N) = N .
  eq rev(T ^ T') = rev(T') ^ rev(T) .
endfm
```

The first module defines the natural numbers in Peano notation, and the second defines a function to reverse a binary tree whose leaves are natural numbers. Each module begins with the keyword fmod followed by the module's name, and ends with the keyword endfm. A module contains sort and subsort declarations introduced by the keywords sort and subsorts stating the different sorts of data manipulated by the module and how those sorts are related. As OBJ3's, Maude's functional modules are based on a particularly flexible variant of equational logic, namely *order-sorted logic* [25], in which it is possible to declare one sort as a subsort of another; for example,

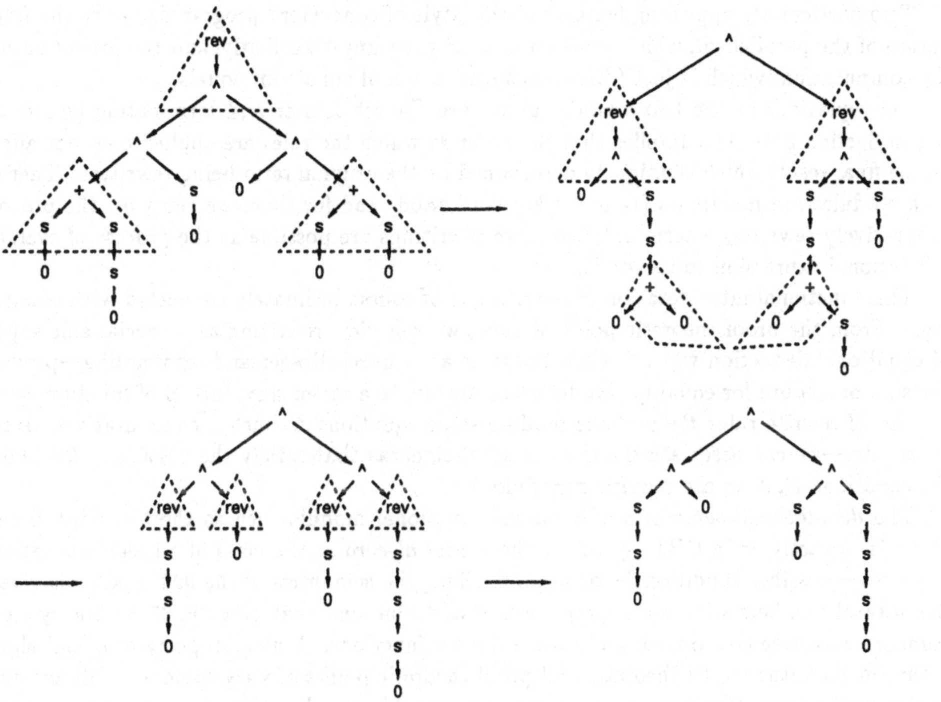

Figure 1: Concurrent rewriting of a tree of numbers.

the declaration Nat < Tree states that every natural number is a tree consisting of a single node. Each of the functions provided by the module, as well as the sorts of their arguments and the sort of their result, is introduced using the keyword op. The syntax is user-definable, and permits specifying function symbols in "prefix," (in the NAT example the function s_), "infix" (_+_) or any "mixfix" combination as well as standard parenthesized notation (rev). Variables to be used for defining equations are declared with their corresponding sorts, and then equations are given; such equations provide the actual "code" of the module. The statement protecting NAT imports the NAT module and asserts that the natural numbers are not modified in the sense that no new data of sort Nat is added and different numbers are not identified by the new equations declared in the module REVERSE.

To compute with such modules, one performs equational simplification by using the equations from left to right until no more simplifications are possible. Note that this can be done *concurrently*, i.e., applying several equations at once, as in the example of Figure 1, in which the places where the equations have been matched at each step are marked. Notice that the function symbol _+_ was declared to be commutative by the attribute[1] [comm]. This not only asserts that the equation N + M = M + N is satisfied in the intended semantics, but it also means that when doing simplification we are allowed to apply the rules for addition not just to *terms*—in a purely syntactic way—but to *equivalence classes* of terms *modulo* the commutativity equation. In the example of Figure 1, the rule

```
eq N + 0 = N .
```

is applied (modulo commutativity) with 0 both on the right *and* on the left.

[1]In Maude as in OBJ it is possible to declare several attributes of this kind for an operator, including also associativity and identity, and then do rewriting modulo such properties.

Two particularly appealing features of this style of concurrent programming are the *implicit* nature of the parallelism, which avoids having to program it explicitly, and the *logical* nature of the computation, which is just (directed) substitution of equals for equals.

The equations in the two modules above are Church-Rosser and terminating (more about this in Section 2.3). This implies that the *order* in which the rules are applied does not affect at all the final result which is uniquely determined by the original term being rewritten. Therefore, such modules are *functional* (thus the keyword `fmod`) and for them we can view the process of exhaustively rewriting a term until no more rewritings are possible as the process of evaluating a functional expression to its result.

This functional interpretation of rewriting is of course intimately connected with equational logic. From the proof theoretic point of view, we can view rewriting as a special efficient form of equational deduction which—when the rules are Church-Rosser and terminating—provides a decision procedure for equality. Model theoretically, to a given signature Σ of function symbols and set of rewrite rules R—perhaps modulo some equations E, such a commutativity, associativity, etc.—we can associate the class of all Σ-algebras that satisfy the equations R—and also the equations E, if we are rewriting modulo E.

The *denotational semantics* of a Maude functional module such as the two ones discussed above is—exactly as in OBJ—given by the *initial algebra* in the class of all algebras satisfying the given—possibly conditional—equations. Thus, the semantics of the `NAT` module is given by the natural numbers with zero, successor and addition, and that of `REVERSE` by binary trees of numbers with tree constructor and tree reversing functions. A nice property of initial algebras is that in them the model theoretic and proof theoretic points of view come into full agreement, in the precise sense that two ground terms are provably equal iff they denote identical elements in the initial algebra, and for Church-Rosser and terminating rules we can decide this equality efficiently by rewriting. All this goes back to the original ADJ proposal [27] and to Goguen's work on the relationship between initial algebras and rewriting [18], and generalizes nicely to order-sorted equational logic, on which OBJ modules and Maude's functional modules are based [25, 19, 40].

As long as we remain within the functional world, this view is indeed extremely satisfactory because of its simplicity and elegance. However, many natural examples exist for which the functional world of equational logic has no room.

1.2 Beyond Equational Logic

Consider for example the following Maude module, which adds a nondeterministic choice operator to the natural numbers:

```
mod NAT-CHOICE is
  extending NAT .
  op _?_ : Nat Nat -> Nat .
  vars N M : Nat .
  rl N ? M => N .
  rl N ? M => M .
endm
```

The intuitive *operational behavior* of this module is quite clear. Natural number addition remains unchanged and is computed using the two rules in the `NAT` module. Notice that any occurrence of the choice operator in an expression can be eliminated by choosing either of the arguments. In the end, we can reduce any ground expression to a natural number in Peano notation. The *mathematical semantics* of the module is much less clear. If we adopt an initial algebra semantics, it follows by the rules of equational deduction with the above two equations that `N = M`, i.e., everything collapses to one point and the module `NAT` is not protected at all. To indicate that this is not the semantics intended, the keyword mod (instead of the previous fmod) has been used, indicating that the module *is not functional*. Similarly, the rewrite rules are introduced by a a new keyword rl—instead of the usual eq—to suggest that they must be understood as "rules" and not as equations in the usual sense. Of course, at the operational level the equations introduced by the keyword eq in a functional module are also implemented as rewrite rules; the difference however lies in the *mathematical semantics* given to the module, which for modules like the one above should *not* be the initial algebra semantics.

But how can we make precise the intended meaning? Whatever it is, if we want to assert that NAT is not destroyed, as the behavior of the module shows and the statement extending NAT declares, it is clear that the module cannot be regarded as a theory in equational logic, and that the associated initial algebra does not at all provide an adequate semantics; therefore, both must be abandoned. My proposal is to seek a logic and a model theory that are the perfect match for this problem. For this solution to be in harmony with the old one, the new logic and the new model theory should in some sense *generalize* the old ones.

2 Rewriting Logic

This section defines the syntax and proof theory of rewriting logic, and identifies concurrent rewriting with deduction in such a logic. We first briefly recall some basic universal algebra needed in the exposition.

2.1 Basic Universal Algebra

For the sake of making the exposition simpler, I treat the *unsorted* case; the many-sorted and order-sorted cases can be given a similar treatment. Therefore, a set Σ of function symbols is a ranked alphabet $\Sigma = \{\Sigma_n \mid n \in \mathbb{N}\}$. A Σ-algebra is then a set A together with an assignment of a function $f_A : A^n \longrightarrow A$ for each $f \in \Sigma_n$ with $n \in \mathbb{N}$. As usual (see, e.g., [52]) I denote by T_Σ the Σ-algebra of ground Σ-terms, and by $T_\Sigma(X)$ the Σ-algebra of Σ-terms with variables in a set X. Similarly, given a set E of Σ-equations, $T_{\Sigma,E}$ denotes the Σ-algebra of equivalence classes of ground Σ-terms modulo the equations E (i.e., modulo provable equality using the equations E); in the same way, $T_{\Sigma,E}(X)$ denotes the Σ-algebra of equivalence classes of Σ-terms with variables in X modulo the equations E. We let $t =_E t'$ denote the congruence modulo E of two terms t, t', and $[t]_E$ or just $[t]$ denote the E-equivalence class of t.

Given a term $t \in T_\Sigma(\{x_1, \ldots, x_n\})$, and a sequence of terms $u_1, \ldots, u_n$, we denote by $t(u_1/x_1, \ldots, u_n/x_n)$ the term obtained from t by *simultaneously substituting* u_i for x_i, $i = 1, \ldots, n$. To simplify notation, we will often denote a sequence of objects $a_1, \ldots, a_n$ by $\overline{a}$, or, if we want to emphasize the length of the sequence, by $\overline{a}^n$; also, in many contexts we will find it convenient to identify a sequence $a_1, \ldots, a_n$ of length n and its associated n-tuple $(a_1, \ldots, a_n)$. With this notation, $t(u_1/x_1, \ldots, u_n/x_n)$ can be abbreviated to $t(\overline{u}/\overline{x})$.

2.2 Deduction in Rewrite Theories

The syntax of rewriting logic is given by signatures. A *signature* is a pair (Σ, E) with Σ a ranked alphabet of function symbols[2] and E a set of Σ-equations. Rewriting will operate on equivalence classes of terms modulo a given set of equations E. In this way, we free rewriting from the syntactic constraints of a term representation and gain a much greater flexibility, thanks to the "structural axioms" E, in deciding what counts as a *data structure*; for example, string rewriting is obtained by imposing an associativity axiom, and multiset rewriting by imposing associativity and commutativity. Of course, standard term rewriting is obtained as the particular case in

[2] As already mentioned, we could consider an *order-sorted* family Σ of function symbols; however, for the sake of a simpler exposition we treat the unsorted case.

which the set E of equations is empty. The idea of rewriting in equivalence classes is well known (see, e.g., [31, 7]).

Given a signature (Σ, E), the *sentences* that we consider are sequents of the form $[t]_E \longrightarrow [t']_E$ with t, t' Σ-terms, where t and t' may possibly involve some variables from the countably infinite set $X = \{x_1, \ldots, x_n, \ldots\}$; sentences of this form can be provable consequences of *rewrite theories*. The notion of rewrite theory presented below is very general and expressive. In the first place, as already mentioned, we allow the important extra generality of rewriting modulo "structural axioms" E. In addition, we allow *conditional* rules of a very general form, where the conditions need not require equalities to hold but only the existence of rewritings among pairs of terms in the condition, which increases the expressive power. Finally, we allow *labelling* of the rewrite rules; this is very natural for many applications, and customary for automata—viewed as labelled transition systems—and for Petri nets, which are both particular instances of our definition (see Section 4). The categorical semantics of Section 3 will further clarify why this last extra generality is natural and desirable.

Definition 1 A (*labelled*) *rewrite theory*[3] $\mathcal{R}$ is a 4-tuple $\mathcal{R} = (\Sigma, E, L, R)$ where Σ is a ranked alphabet of function symbols, E is a set of Σ-equations, L is a set called the set of *labels*, and R is a set of pairs $R \subseteq L \times (T_{\Sigma,E}(X)^2)^+$ whose first component is a label and whose second component is a nonempty sequence of pairs of E-equivalence classes of terms, with $X = \{x_1, \ldots, x_n, \ldots\}$ a countably infinite set of variables. Elements of R are called *rewrite rules*[4]. For a rewrite rule $(r, ([t], [t'])([u_1], [v_1]) \ldots ([u_k], [v_k]))$ we use the notation

$$r : [t] \longrightarrow [t'] \;\; \textit{if} \;\; [u_1] \longrightarrow [v_1] \wedge \ldots \wedge [u_k] \longrightarrow [v_k].$$

We call the part $[u_1] \longrightarrow [v_1] \wedge \ldots \wedge [u_k] \longrightarrow [v_k]$ the *condition* of the rule, and may abbreviate it with the letter C. To indicate that $\{x_1, \ldots, x_n\}$ is a set of variables occurring in either t, t', or C, we write[5] $r : [t(x_1, \ldots, x_n)] \longrightarrow [t'(x_1, \ldots, x_n)]$ *if* $C(x_1, \ldots, x_n)$, or in abbreviated notation $r : [t(\overline{x}^n)] \longrightarrow [t'(\overline{x}^n)]$ *if* $C(\overline{x}^n)$.

Rules of the form $(r, ([t], [t']))$, i.e., with an empty condition, are called *unconditional rewrite rules*, and we use for them the notation $r : [t] \longrightarrow [t']$. A rewrite theory where all the rules are unconditional is called an *unconditional rewrite theory*. □

Given a rewrite theory $\mathcal{R}$, we say that $\mathcal{R}$ *entails* a sequent $[t] \longrightarrow [t']$ and write

$$\mathcal{R} \vdash [t] \longrightarrow [t']$$

if and only if $[t] \longrightarrow [t']$ can be obtained by finite application of the following *rules of deduction*:

1. **Reflexivity.** For each $[t] \in T_{\Sigma,E}(X)$,

$$\overline{[t] \longrightarrow [t]}$$

2. **Congruence.** For each $f \in \Sigma_n$, $n \in \mathbb{N}$,

$$\frac{[t_1] \longrightarrow [t'_1] \quad \ldots \quad [t_n] \longrightarrow [t'_n]}{[f(t_1, \ldots, t_n)] \longrightarrow [f(t'_1, \ldots, t'_n)]}$$

[3] I consciously depart from the standard terminology, that would call $\mathcal{R}$ a *rewrite system*. The reason for this departure is very specific. I want to keep the term "rewrite system" for the *models* of such a theory, which will be defined in Section 3 and which really are systems with a dynamic behavior. Strictly speaking, $\mathcal{R}$ is not a system; it is only a static, linguistic, *presentation* of a class of systems—including the initial and free systems that most directly formalize our dynamic intuitions about rewriting.

[4] I.e., all rules are assumed *conditional* unless said otherwise.

[5] Note that, in general, the set $\{x_1, \ldots, x_n\}$ will depend on the representatives t, t', u_i, v_i chosen; therefore, we allow any possible such qualification with explicit variables.

3. **Replacement.** For each rewrite rule

$$r : [t(\overline{x})] \longrightarrow [t'(\overline{x})] \text{ if } [u_1(\overline{x})] \longrightarrow [v_1(\overline{x})] \wedge \ldots \wedge [u_k(\overline{x})] \longrightarrow [v_k(\overline{x})]$$

in R,

$$\frac{[w_1] \longrightarrow [w_1'] \quad \ldots \quad [w_n] \longrightarrow [w_n'] \qquad [u_1(\overline{w}/\overline{x})] \longrightarrow [v_1(\overline{w}/\overline{x})] \quad \ldots \quad [u_k(\overline{w}/\overline{x})] \longrightarrow [v_k(\overline{w}/\overline{x})]}{[t(\overline{w}/\overline{x})] \longrightarrow [t'(\overline{w'}/\overline{x})]}$$

4. **Transitivity.**

$$\frac{[t_1] \longrightarrow [t_2] \quad [t_2] \longrightarrow [t_3]}{[t_1] \longrightarrow [t_3]}$$

Note that for unconditional rules the rule of replacement specializes to the simpler:

5. **Unconditional Replacement.** For each $r : [t(x_1, \ldots, x_n)] \longrightarrow [t'(x_1, \ldots, x_n)]$ in R,

$$\frac{[w_1] \longrightarrow [w_1'] \ldots [w_n] \longrightarrow [w_n']}{[t(\overline{w}/\overline{x})] \longrightarrow [t'(\overline{w'}/\overline{x})]}$$

Equational logic (modulo a set of axioms E) is obtained from rewriting logic by adding the following rule:

6. **Symmetry.**

$$\frac{[t_1] \longrightarrow [t_2]}{[t_2] \longrightarrow [t_1]}$$

Because of this new rule, sequents derivable in equational logic are always *bidirectional*; therefore, in this case we can adopt the notation $[t] \leftrightarrow [t']$ throughout and call such bidirectional sequents *equations*.

For the moment we have only considered the rules of deduction for rewriting logic. Therefore, this logic might at first sight seem somewhat of an empty formal game. Such an impression would be mistaken. The true importance of this logic, and its radical novelty when compared with equational logic, will become apparent when we study its semantics in Section 3. However, a few remarks are in order at present. First, note that a sequent $[t] \longrightarrow [t']$ should not be read as "$[t]$ *equals* $[t']$," but as "$[t]$ *becomes* $[t']$." Therefore, rewriting logic is a logic of *becoming* or *change*, not a logic of equality in a static Platonic sense. Adding the symmetry rule is a *very strong* restriction, namely assuming that *all change is reversible*, thus bringing us into a timeless Platonic realm in which "before" and "after" have been identified. A second related observation is that $[t]$ should not be understood as a *term* in the usual first-order logic sense, but as a *proposition* —built up using the *logical connectives* in Σ—that asserts being in a certain *state* having a certain *structure*. The rules of rewriting logic are therefore rules to reason about *change in a concurrent system*. They allow us to draw valid conclusions about the evolution of the system from certain basic types of change known to be possible thanks to the rules R.

2.3 Concurrent Rewriting

We can now give a precise definition of concurrent rewriting. A nice consequence of having defined rewriting logic is that concurrent rewriting, rather than emerging as an operational notion, actually *coincides* with deduction in such a logic.

Definition 2 Given a rewrite theory $\mathcal{R} = (\Sigma, E, L, R)$, a (Σ, E)-sequent $[t] \longrightarrow [t']$ is called:

- a *0-step concurrent $\mathcal{R}$-rewrite* iff it can be derived from $\mathcal{R}$ by finite application of the rules 1 and 2 of rewriting deduction (in which case $[t]$ and $[t]'$ necessarily coincide);
- a *one-step concurrent $\mathcal{R}$-rewrite* iff it can be derived from $\mathcal{R}$ by finite application of the rules 1–3, with at least one application of rule 3; if rule 3 was applied exactly once, we then say that the sequent is a one-step *sequential* $\mathcal{R}$-rewrite;
- a *concurrent $\mathcal{R}$-rewrite* (or just a *rewrite*) iff it can be derived from $\mathcal{R}$ by finite application of the rules 1–4.

We call the rewrite theory $\mathcal{R}$ *sequential* if all one-step $\mathcal{R}$-rewrites are necessarily sequential. A sequential rewrite theory $\mathcal{R}$ is in addition called *deterministic* if for each $[t]$ there is at most one one-step (necessarily sequential) rewrite $[t] \longrightarrow [t']$. The notions of sequential and deterministic rewrite theory can be made relative to a given subset $S \subseteq T_{\Sigma,E}(X)$ by requiring that the corresponding property holds for each $[t']$ "reachable from S," i.e., for each $[t']$ such that for some $[t] \in S$ there is a concurrent $\mathcal{R}$-rewrite $[t] \longrightarrow [t']$. □

Traditionally, rewriting is defined by repeating one-step sequential rewrites. This makes the notion of a *sequence* of rewrites—an "operational" and even "sequential" notion indeed—the basic notion. By contrast, the basic view of rewriting stressed in this paper is that of a *logical deduction.* If we had followed the more traditional and operational view, we would have defined concurrent rewriting as a sequence of one-step concurrent rewrites. Implicit in such a view is the idea of an *observation* of "snapshots," but of course many different observations, all consistent with each other, may be possible. Section 3 will develop an algebraic theory of equivalence between such observations that can free us from an interleaving or snapshot view and bring us to the *paradiso* of "true concurrency." For the moment we have,

Lemma 3 For each concurrent $\mathcal{R}$-rewrite $[t] \longrightarrow [t']$, either $[t] = [t']$ or there is an $n \in \mathbb{N}$ and a chain of one-step (concurrent) rewrites

$$[t] \longrightarrow [t_1] \longrightarrow \ \ldots \ \longrightarrow [t_n] \longrightarrow [t'].$$

We call such a chain a *step sequence* for $[t] \longrightarrow [t']$. In addition, we can always choose all the steps to be sequential; we then call it an *interleaving* or *firing* sequence for $[t] \longrightarrow [t']$. □

We call a rewrite theory $\mathcal{R}$ *terminating* if there is no infinite chain of one-step rewrites (whether sequential or concurrent)

$$[t] \longrightarrow [t_1] \longrightarrow \ \ldots \ \longrightarrow [t_n] \longrightarrow \ \ldots$$

We say that $[t']$ is an *$\mathcal{R}$-normal form* of $[t]$ if $[t] \longrightarrow [t']$ is an $\mathcal{R}$-rewrite and there does not exist any one-step $\mathcal{R}$-rewrite of the form $[t'] \longrightarrow [t'']$. If each $[t]$ has at least one normal form, we call the theory $\mathcal{R}$ *weakly terminating*.

We say that a rewrite theory $\mathcal{R}$ is *Church-Rosser* or *confluent* if given any two concurrent rewrites $[t] \longrightarrow [t']$, $[t] \longrightarrow [t'']$, there is a $[t''']$ and concurrent rewrites $[t'] \longrightarrow [t''']$, $[t''] \longrightarrow [t''']$. Likewise, we call $\mathcal{R}$ *ground Church-Rosser* when the property is only asserted for equivalence classes of ground terms. This situation is shown in Figure 2. Note that, by Lemma 3, this notion coincides with the usual one defined in terms of sequential rewriting. As expected, we then have

Theorem 4 If $\mathcal{R}$ is Church-Rosser, then an equation $[t] \leftrightarrow [t']$ is provable from $\mathcal{R}$ by equational deduction, i.e., by using the rules 1–4 and rule 6, iff there is a $[t'']$ and $\mathcal{R}$-rewrites $[t] \longrightarrow [t'']$, $[t'] \longrightarrow [t'']$. In addition, if $\mathcal{R}$ is terminating, any $[t]$ has a unique normal form, denoted $can_{\mathcal{R}}[t]$; under these conditions, an equation $[t] \leftrightarrow [t']$ is provable from $\mathcal{R}$ by equational deduction iff $can_{\mathcal{R}}[t] = can_{\mathcal{R}}[t']$. □

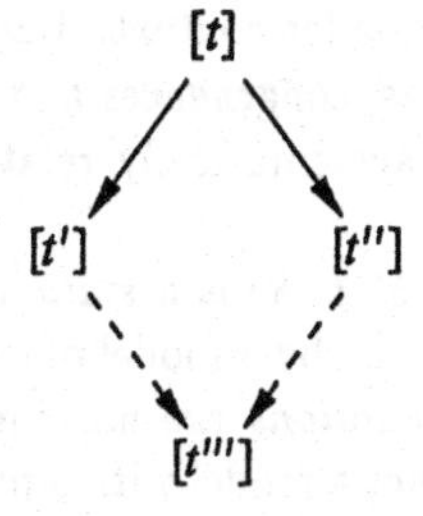

Figure 2: The Church-Rosser property.

The case of Church-Rosser rewrite rules provides a straightforward specialization of the notion of conditional rewrite rule that we have been considering to the more traditional notion in which the condition is a finite conjunction of equalities. The specialization is as follows. Consider a Church-Rosser rewrite theory $\mathcal{R}$ such that all the rules are of the form

$$(\dagger)\ r : [t(\overline{x})] \longrightarrow [t'(\overline{x})]\ \mathit{if}$$
$$[u_1(\overline{x})] \longrightarrow [y_1] \wedge [v_1(\overline{x})] \longrightarrow [y_1] \wedge \ldots \wedge [u_k(\overline{x})] \longrightarrow [y_k] \wedge [v_k(\overline{x})] \longrightarrow [y_k]$$

with the y_j and the x_i pairwise disjoint. Then, application of the replacement rule 3 is possible relative to instances $[w_1] \longrightarrow [w'_1] \quad \ldots \quad [w_n] \longrightarrow [w'_n]$ of the variables $x_1, \ldots, x_n$ if and only if we can find terms $p_1, \ldots, p_k$ such that

$$[u_1(\overline{w}/\overline{x})] \longrightarrow [p_1] \wedge [v_1(\overline{w}/\overline{x})] \longrightarrow [p_1] \wedge \ldots \wedge [u_k(\overline{w}/\overline{x})] \longrightarrow [p_k] \wedge [v_k(\overline{w}/\overline{x})] \longrightarrow [p_k]$$

and by Theorem 4 this holds iff

$$[u_1(\overline{w}/\overline{x})] \leftrightarrow [v_1(\overline{w}/\overline{x})] \wedge \ldots \wedge [u_k(\overline{w}/\overline{x})] \leftrightarrow [v_k(\overline{w}/\overline{x})].$$

Therefore, for a Church-Rosser rewrite theory $\mathcal{R}$ with rules of the form ($\dagger$) above, we can express the rules with the alternative notation

$$r : [t(\overline{x})] \longrightarrow [t'(\overline{x})]\ \ \mathit{if}\ \ [u_1(\overline{x})] \leftrightarrow [v_1(\overline{x})] \wedge \ldots \wedge [u_k(\overline{x})] \leftrightarrow [v_k(\overline{x})]$$

which is fully justified, since checking the condition is in this case equivalent to checking provable equality from $\mathcal{R}$ under the rules of equational logic. If, in addition, the theory $\mathcal{R}$ is terminating, we can implement application of the replacement rule by reducing the two sides of each condition to canonical form, i.e., by checking the identities

$$can_{\mathcal{R}}[u_1(\overline{w}/\overline{x})] = can_{\mathcal{R}}[v_1(\overline{w}/\overline{x})] \wedge \ldots \wedge can_{\mathcal{R}}[u_k(\overline{w}/\overline{x})] = can_{\mathcal{R}}[v_k(\overline{w}/\overline{x})]$$

which is of course the nicest possible situation from an implementation point of view. Careful examination of the above argument (and some of the reasoning that goes into the proof of Theorem 4) shows that all we have really used is something much weaker than requiring $\mathcal{R}$ to be Church-Rosser. All we need to require is that the Church-Rosser property holds for all terms p that can be proved equal by the rules of equational logic (using $\mathcal{R}$) to a substitution instance of a term $[u_j(\overline{x})]$ or a term $[v_j(\overline{x})]$ appearing in a condition of a rule ($\dagger$) in $\mathcal{R}$.

3 Models

This section presents a new model theory for rewriting logic. Many of the nice properties enjoyed by algebras—the classical models of equational logic—including initiality, soundness and completeness hold also in this much broader context. Besides, algebras and also other models such as ordered and preordered algebras, congruences and groupoids appear as particular cases of the general model theory, and are systematically related to each other and to the general models by means of adjoint functors.

As such, a rewrite theory $\mathcal{R} = (\Sigma, E, L, R)$ is a *static description* of what a system can do. The *meaning* of the theory should be given by a model of its actual *behavior*. Since our approach has emphasized that *concurrent computations* are nothing other than *deductions* in rewriting logic, it is now possible for us to construct a model with a precise algebraic structure which is in a sense the most informative model possible; in it, behavior exactly corresponds to deduction. The procedure for building such a model is straightforward. As in the Curry-Howard correspondence, which this approach greatly generalizes, we can just read the rules 1–4 of rewriting logic as *rules of generation*. As already hinted at in Section 2.3, the question of when are two proofs (and therefore two computations) equal still remains. This is answered by postulating a set of equations that are natural and express intuitive identities between computations. What they allow us to get at—by abstracting away from particular "snapshot" descriptions—is a precise algebraic formulation of the notion of "true concurrency." The model obtained in this way is just the initial (or free when we allow variables) model of a very rich class of models that is further explored in Section 3.2.

Given a rewrite theory $\mathcal{R} = (\Sigma, E, L, R)$, the model that we are seeking is a category $\mathcal{T}_{\mathcal{R}}(X)$ whose objects are the equivalence classes of terms $[t] \in T_{\Sigma,E}(X)$ and whose morphisms are equivalence classes of terms representing proofs in rewriting deduction, i.e., concurrent $\mathcal{R}$-rewrites. The rules for generating such terms, with the specification of their respective domain and codomain, are given below. Note that in the rest of this paper we always use "diagrammatic" notation for morphism composition, i.e., $\alpha;\beta$ always means the composition of α *followed by* β.

1. **Identities.** For each $[t] \in T_{\Sigma,E}(X)$,

$$\frac{}{[t] : [t] \longrightarrow [t]}$$

2. **Σ-structure.** For each $f \in \Sigma_n$, $n \in \mathbb{N}$,

$$\frac{\alpha_1 : [t_1] \longrightarrow [t'_1] \quad \ldots \quad \alpha_n : [t_n] \longrightarrow [t'_n]}{f(\alpha_1, \ldots, \alpha_n) : [f(t_1, \ldots, t_n)] \longrightarrow [f(t'_1, \ldots, t'_n)]}$$

3. **Replacement.** For each rewrite rule

$$r : [t(\overline{x})] \longrightarrow [t'(\overline{x})] \;\; if \;\; [u_1(\overline{x})] \longrightarrow [v_1(\overline{x})] \wedge \ldots \wedge [u_k(\overline{x})] \longrightarrow [v_k(\overline{x})]$$

in R,

$$\frac{\begin{array}{c}\alpha_1 : [w_1] \longrightarrow [w'_1] \quad \ldots \quad \alpha_n : [w_n] \longrightarrow [w'_n] \\ \beta_1 : [u_1(\overline{w}/\overline{x})] \longrightarrow [v_1(\overline{w}/\overline{x})] \quad \ldots \quad \beta_k : [u_k(\overline{w}/\overline{x})] \longrightarrow [v_k(\overline{w}/\overline{x})]\end{array}}{r(\overline{\alpha}^n, \overline{\beta}^k) : [t(\overline{w}/\overline{x})] \longrightarrow [t'(\overline{w'}/\overline{x})]}$$

4. **Composition.**

$$\frac{\alpha : [t_1] \longrightarrow [t_2] \qquad \beta : [t_2] \longrightarrow [t_3]}{\alpha;\beta : [t_1] \longrightarrow [t_3]}$$

For the case of equational logic we can also define a similar model as a category $\mathcal{T}_{\mathcal{R}}^{\leftrightarrow}(X)$ (actually a *groupoid*[6]) by using the rule of symmetry to generate additional terms:

[6] A category C is called a *groupoid* iff any morphism $f : A \longrightarrow B$ in C has an inverse morphism $f^{-1} : B \longrightarrow A$ such that $f; f^{-1} = 1_A$, and $f^{-1}; f = 1_B$.

5. **Inversion.**

$$\frac{\alpha : [t_1] \longrightarrow [t_2]}{\alpha^{-1} : [t_2] \longrightarrow [t_1]}$$

Convention and Warning. In the case when the same label r appears in two different rules of R, the "proof terms" $r(\overline{\alpha}, \overline{\beta})$ can be *ambiguous*, either because different sources or targets are assigned to the same term by different rules, or because of the even more ambiguous situation whereby the same source and target are assigned to the same proof term $r(\overline{\alpha}, \overline{\beta})$ by two different rules, which can happen when the set E of axioms is nonempty. We will always assume that such ambiguity problems *have been resolved* by disambiguating the label r in the proof terms $r(\overline{\alpha}, \overline{\beta})$. With this understanding, the simpler notation $r(\overline{\alpha}, \overline{\beta})$ is adopted to ease the exposition.

Each of the above rules of generation defines a different operation taking certain proof terms as arguments and returning a resulting proof term. In other words, proof terms form an algebraic structure $\mathcal{P}_{\mathcal{R}}(X)$ defined by the generation rules. Specifically, this structure is a graph[7] with set of nodes $T_{\Sigma,E}(X)$ and with set of arrows the set $\mathcal{P}_{\mathcal{R}}(X)$ of all proof terms generated by the above rules, with source and target maps

$$\partial_0, \partial_1 : \mathcal{P}_{\mathcal{R}}(X) \longrightarrow T_{\Sigma,E}(X)$$

as specified by the rules; in particular, rule 1 requires each node $[t]$ to be also an arrow $[t] : [t] \longrightarrow [t]$, i.e., it makes the graph reflexive[8]. In addition, by rule 2, the set $\mathcal{P}_{\mathcal{R}}(X)$ has a Σ-algebra structure such that the source and target maps ∂_0 and ∂_1 are Σ-homomorphisms; in other words, we have a Σ-algebra structure *on a graph* instead than just on a set. Rule 4 defines an operation on composable[9] pairs of arrows:

$$\_;\_ : \mathit{Composable}(\mathcal{P}_{\mathcal{R}}(X)) \longrightarrow \mathcal{P}_{\mathcal{R}}(X)$$

with $\partial_0(\alpha;\beta) = \partial_0(\alpha)$ and $\partial_1(\alpha;\beta) = \partial_1(\beta)$. Finally, rule 3 defines for each rewrite rule

$$r : [t(\overline{x})] \longrightarrow [t'(\overline{x})] \ \ \mathit{if} \ \ [u_1(\overline{x})] \longrightarrow [v_1(\overline{x})] \wedge \ldots \wedge [u_k(\overline{x})] \longrightarrow [v_k(\overline{x})]$$

in R an operation[10]

$$r(\_,\_) : \{(\overline{\alpha}^n, \overline{\beta}^k) \in \mathcal{P}_{\mathcal{R}}(X)^{n+k} \mid \partial_0^k(\overline{\beta}^k) = \overline{[u(\partial_0^n(\overline{\alpha}^n)/\overline{x})]}^k \wedge \partial_1^k(\overline{\beta}^k) = \overline{[v(\partial_0^n(\overline{\alpha}^n)/\overline{x})]}^k\} \longrightarrow \mathcal{P}_{\mathcal{R}}(X)$$

with source and target requirements as specified by rule 3. The structure of $\mathcal{P}_{\mathcal{R}}(X)$ just made explicit is an instance of the following general concept.

Definition 5 Given a rewrite theory $\mathcal{R} = (\Sigma, E, L, R)$, an *$\mathcal{R}$-presystem* is a reflexive graph $G = (\partial_0, \partial_1 : \mathit{Arrows} \longrightarrow \mathit{Nodes}, j)$ together with a Σ-algebra structure on G such that the Σ-algebra *Nodes* of nodes satisfies the equations E, with an operation

$$\_;\_ : \mathit{Composable}(G) \longrightarrow \mathit{Arrows}$$

with $\partial_0(a;b) = \partial_0(a)$ and $\partial_1(a;b) = \partial_1(b)$, and for each rewrite rule

$$r : [t(\overline{x})] \longrightarrow [t'(\overline{x})] \ \ \mathit{if} \ \ [u_1(\overline{x})] \longrightarrow [v_1(\overline{x})] \wedge \ldots \wedge [u_k(\overline{x})] \longrightarrow [v_k(\overline{x})]$$

[7] By a *graph* G we will always mean a directed graph, i.e., a structure $G = (\partial_0, \partial_1 : \mathit{Arrows} \longrightarrow \mathit{Nodes})$. A *graph homomorphism* $(f, g) : (\partial_0, \partial_1 : \mathit{Arrows} \longrightarrow \mathit{Nodes}) \longrightarrow (\partial_0', \partial_1' : \mathit{Arrows}' \longrightarrow \mathit{Nodes}')$ is then defined as a pair of functions $f : \mathit{Arrows} \longrightarrow \mathit{Arrows}'$, $g : \mathit{Nodes} \longrightarrow \mathit{Nodes}'$ such that $f;\partial_i' = \partial_i;g$, $i = 0,1$.

[8] A *reflexive graph* is a graph $G = (\partial_0, \partial_1 : \mathit{Arrows} \longrightarrow \mathit{Nodes})$ together with a function $j : \mathit{Nodes} \longrightarrow \mathit{Arrows}$ such that $j;\partial_0 = j;\partial_1 = 1_{\mathit{Nodes}}$.

[9] Given a graph $G = (\partial_0, \partial_1 : \mathit{Arrows} \longrightarrow \mathit{Nodes})$, the set of its *composable pairs of arrows* is the set $\mathit{Composable}(G) = \{(a, a') \in \mathit{Arrows}^2 \mid \partial_1(a) = \partial_0(a')\}$.

[10] Recall again that we implicitly assume that the operator $r(\_,\_)$ has been disambiguated if this is necessary in order to keep the proof terms unambiguous.

in R, an operation

$$r(\_,\_) : \{(\overline{a}^n, \overline{b}^k) \in Arrows^{n+k} \mid \partial_0^k(\overline{b}^k) = \overline{u_{Nodes}(\partial_0^n(\overline{a}^n))}^k \wedge \partial_1^k(\overline{b}^k) = \overline{v_{Nodes}(\partial_0^n(\overline{a}^n))}^k\} \longrightarrow Arrows$$

(where for $t(x_1, \ldots, x_n)$ a Σ-term, t_{Nodes} denotes the n-ary derived operation on nodes associated to the term t) and such that

$$r(\overline{a}^n, \overline{b}^k) : t_{Nodes}(\partial_0^n(\overline{a}^n)) \longrightarrow t'_{Nodes}(\partial_1^n(\overline{a}^n)).$$

Given two $\mathcal{R}$-presystems G and G', an $\mathcal{R}$-*prehomomorphism* is a graph homomorphism that preserves all the operations, i.e., the operations in Σ, the operation $\_;\_$ and the operations $r(\_,\_)$ for $r \in R$. This defines a category $\underline{\mathcal{R}\text{-}PreSys}$ in the obvious way. □

The definition of $\mathcal{R}$-presystem just given is "essentially algebraic." The style of universal algebra involved is slightly more subtle than standard universal algebra (because operations such as $\_;\_$ or the operations $r(\_,\_)$ for $r \in R$ are only defined in a *subset* of tuples of elements determined by some *equations*) but well known. It can be formalized by means of "essentially algebraic theories" [13, 60] or "sketches" [2].

Note that in the $\mathcal{P}_{\mathcal{R}}(X)$ construction the set X of variables is actually a parameter, and we need not assume X to be fixed and countable. The following proposition can be proved easily by induction on the depth of proof terms:

Proposition 6 For $\mathcal{R}$ a rewrite theory, given a set X and a function $F : X \longrightarrow Nodes$ to the set *Nodes* of nodes of an $\mathcal{R}$-presystem G, there exists a unique $\mathcal{R}$-prehomomorphism $F^\sharp$: $\mathcal{P}_{\mathcal{R}}(X) \longrightarrow G$ whose map of nodes, say $F^\sharp_{nodes}$, extends F, i.e., $F^\sharp_{nodes}([x]) = F(x)$ for each $x \in X$. □

In the language of adjoint functors [43], this proposition can be rephrased by saying that the forgetful functor

$$Nodes : \underline{\mathcal{R}\text{-}PreSys} \longrightarrow \underline{Set}$$

sending each $\mathcal{R}$-presystem to its set of nodes has a left adjoint.

The system $\mathcal{T}_{\mathcal{R}}(X)$ is defined as the quotient presystem of $\mathcal{P}_{\mathcal{R}}(X)$ modulo the following equations (in the expressions appearing in the equations, when compositions of morphisms are involved, we always implicitly assume that the corresponding domains and codomains match):

1. **Category.**
   (a) *Associativity.* For all α, β, γ, $\quad (\alpha;\beta);\gamma = \alpha;(\beta;\gamma)$
   (b) *Identities.* For each $\alpha : [t] \longrightarrow [t']$, $\quad \alpha;[t'] = \alpha$ *and* $[t];\alpha = \alpha$
2. **Functoriality of the Σ-algebraic structure.** For each $f \in \Sigma_n$, $n \in \mathbb{N}$,
   (a) *Preservation of composition.* For all $\alpha_1, \ldots, \alpha_n, \beta_1, \ldots, \beta_n$,
   $$f(\alpha_1;\beta_1, \ldots, \alpha_n;\beta_n) = f(\alpha_1, \ldots, \alpha_n); f(\beta_1, \ldots, \beta_n)$$
   (b) *Preservation of identities.* $\quad f([t_1], \ldots, [t_n]) = [f(t_1, \ldots, t_n)]$
3. **Axioms in E.** For $t(x_1, \ldots, x_n) = t'(x_1, \ldots, x_n)$ an axiom in E, for all $\alpha_1, \ldots, \alpha_n$,
   $$t(\alpha_1, \ldots, \alpha_n) = t'(\alpha_1, \ldots, \alpha_n)$$
4. **Decomposition.** For each rewrite rule
   $$r : [t(\overline{x})] \longrightarrow [t'(\overline{x})] \text{ if } [u_1(\overline{x})] \longrightarrow [v_1(\overline{x})] \wedge \ldots \wedge [u_k(\overline{x})] \longrightarrow [v_k(\overline{x})]$$

in R,

$$\frac{\alpha_1 : [w_1] \longrightarrow [w_1'] \quad \ldots \quad \alpha_n : [w_n] \longrightarrow [w_n'] \qquad \beta_1 : [u_1(\overline{w}/\overline{x})] \longrightarrow [v_1(\overline{w}/\overline{x})] \quad \ldots \quad \beta_k : [u_k(\overline{w}/\overline{x})] \longrightarrow [v_k(\overline{w}/\overline{x})]}{r(\overline{\alpha}^n, \overline{\beta}^k) = r(\overline{[w]}^n, \overline{\beta}^k); t'(\overline{\alpha}^n)}$$

5. **Exchange.** For each rewrite rule

$$r : [t(\overline{x})] \longrightarrow [t'(\overline{x})] \;\; \mathit{if} \;\; [u_1(\overline{x})] \longrightarrow [v_1(\overline{x})] \wedge \ldots \wedge [u_k(\overline{x})] \longrightarrow [v_k(\overline{x})]$$

in R,

$$\frac{\begin{array}{c}\alpha_1 : [w_1] \longrightarrow [w_1'] \quad \ldots \quad \alpha_n : [w_n] \longrightarrow [w_n'] \\ \beta_1 : [u_1(\overline{w}/\overline{x})] \longrightarrow [v_1(\overline{w}/\overline{x})] \quad \ldots \quad \beta_k : [u_k(\overline{w}/\overline{x})] \longrightarrow [v_k(\overline{w}/\overline{x})] \\ \beta_1' : [u_1(\overline{w'}/\overline{x})] \longrightarrow [v_1(\overline{w'}/\overline{x})] \quad \ldots \quad \beta_k' : [u_k(\overline{w'}/\overline{x})] \longrightarrow [v_k(\overline{w'}/\overline{x})] \\ \beta_1; v_1(\overline{\alpha}) = u_1(\overline{\alpha}); \beta_1' \quad \ldots \quad \beta_k; v_k(\overline{\alpha}) = u_k(\overline{\alpha}); \beta_k'\end{array}}{r(\overline{[w]}^n, \overline{\beta}^k); t'(\overline{\alpha}) = t(\overline{\alpha}); r(\overline{[w']}^n, \overline{\beta'}^k)}$$

Similarly, the groupoid $\mathcal{T}_{\mathcal{R}}^{\leftrightarrow}(X)$ is obtained by identifying the terms generated by rules 1–5 modulo the above equations plus the additional:

6. **Inverse.** For any $\alpha : [t] \longrightarrow [t']$ in $\mathcal{T}_{\mathcal{R}}^{\leftrightarrow}(X)$, $\quad \alpha; \alpha^{-1} = [t]$ *and* $\alpha^{-1}; \alpha = [t']$

Note again that the set X of variables is actually a parameter of these constructions, and we need not assume X to be fixed and countable. In particular, for $X = \emptyset$ we adopt the notations $\mathcal{T}_{\mathcal{R}}$ and $\mathcal{T}_{\mathcal{R}}^{\leftrightarrow}$, respectively.

Some comments can help clarify the intended meaning behind the above equations. The equations in 1 make $\mathcal{T}_{\mathcal{R}}(X)$ into a category, the equations in 2 give it a Σ-algebra structure *as a category*, i.e., each $f \in \Sigma_n$ determines a functor $f : \mathcal{T}_{\mathcal{R}}(X)^n \longrightarrow \mathcal{T}_{\mathcal{R}}(X)$ and the equations in 3 force such Σ-algebra structure to satisfy the equations E. Note that it follows easily (by induction) from the functoriality of the basic operations and from the satisfaction of the equations E that each $[t(x_1, \ldots, x_n)]$ defines a functor $[t(x_1, \ldots, x_n)] : \mathcal{T}_{\mathcal{R}}(X)^n \longrightarrow \mathcal{T}_{\mathcal{R}}(X)$ in the obvious way; this is just what the algebraic notion of a *derived operation* means in this context.

The decomposition law states that any rewriting of the form $r(\overline{\alpha}, \overline{\beta})$—which represents the *simultaneous* rewriting of the term at the top using rule r (once proofs $\overline{\beta}$ for the conditions have been provided) *and* "below," i.e., in the subterms matched by the rule—is equivalent to the sequential composition $r(\overline{[w]}, \overline{\beta}); t'(\overline{\alpha})$ corresponding to first rewriting on top with r and then below on the matched subterms.

The intuitive meaning of the exchange law is that—under certain conditions—rewriting at the top by means of a rule r and rewriting "below," i.e., in the subterms matched by the rule, are independent processes and therefore can be done in any order. Since the exchange law is somewhat complicated by the conditions in the conditional rule, it is simpler to first make precise the meaning of this law for *unconditional* rules, for which it specializes to:

7. **Unconditional Exchange.** For each $r : [t(x_1, \ldots, x_n)] \longrightarrow [t'(x_1, \ldots, x_n)]$ in R,

$$\frac{\alpha_1 : [w_1] \longrightarrow [w_1'] \quad \ldots \quad \alpha_n : [w_n] \longrightarrow [w_n']}{r(\overline{\alpha}) = r(\overline{[w]}); t'(\overline{\alpha}) = t(\overline{\alpha}); r(\overline{[w']})}$$

Therefore, in the unconditional case rewritings on top using r and rewritings "below" can always be exchanged. Since $[t(x_1, \ldots, x_n)]$ and $[t'(x_1, \ldots, x_n)]$ can be regarded as functors $\mathcal{T}_{\mathcal{R}}(X)^n \longrightarrow \mathcal{T}_{\mathcal{R}}(X)$, the exchange law just asserts that r is a natural transformation [43], i.e.,

Lemma 7 For each $r : [t(x_1,\ldots,x_n)] \longrightarrow [t'(x_1,\ldots,x_n)]$ in R, the family of morphisms

$$\{r(\overline{[w]}) : [t(\overline{w}/\overline{x})] \longrightarrow [t'(\overline{w}/\overline{x})] \mid \overline{[w]} \in T_{\Sigma,E}(X)^n\}$$

is a natural transformation $r : [t(x_1,\ldots,x_n)] \Longrightarrow [t'(x_1,\ldots,x_n)]$ between the functors

$$[t(x_1,\ldots,x_n)], [t'(x_1,\ldots,x_n)] : \mathcal{T}_{\mathcal{R}}(X)^n \longrightarrow \mathcal{T}_{\mathcal{R}}(X).$$

□

In the conditional case, the exchange law is subject to certain requirements. Such requirements essentially mean that the proofs of the conditions "before" and "after" rewriting below with the α's should be equivalent, i.e., that we have in essence provided the same proof for each of the conditions. This is important, because for a non Church-Rosser rewrite theory it may even be *impossible* to prove the conditions *after* the term has been rewritten below by the α's, and therefore exchanging rewritings on top and "below" would in this case be illegal and ill defined.

Now that the intuitive meaning of the exchange law has been explained, we can make precise its mathematical meaning. For this, we need the notion of a *subequalizer* [41]; this notion generalizes that of an equalizer of two functors, by requiring a natural transformation instead of just an identity of functors.

Definition 8 Given functors $F, G : \mathcal{A} \longrightarrow \mathcal{B}$, the *subequalizer* of F and G is a category $Subeq(F, G)$ together with a functor $J : Subeq(F,G) \longrightarrow \mathcal{A}$ and a natural transformation[11] $\alpha : J{*}F \Longrightarrow J{*}G$ with the following universal property: Given a functor $H : \mathcal{C} \longrightarrow \mathcal{A}$ and a natural transformation $\beta : H * F \Longrightarrow H * G$, there exists a unique functor $(H,\beta) : \mathcal{C} \longrightarrow Subeq(F,G)$ such that

- $(H,\beta) * J = H$
- $(H,\beta) * \alpha = \beta$

Similarly, given a family of pairs of functors $\{F_i, G_i : \mathcal{A} \longrightarrow \mathcal{B}_i \mid i \in I\}$, the (simultaneous) *subequalizer* of this family is a category $Subeq((F_i,G_i)_{i\in I})$ together with a functor $J : Subeq((F_i,G_i)_{i\in I}) \longrightarrow \mathcal{A}$ and a family of natural transformations $\{\alpha_i : J * F_i \Longrightarrow J * G_i \mid i \in I\}$ with the following universal property: Given a functor $H : \mathcal{C} \longrightarrow \mathcal{A}$ and a family of natural transformations $\{\beta_i : H * F_i \Longrightarrow H * G_i \mid i \in I\}$, there exists a unique functor $(H, \{\beta_i\}_{i\in I}) : \mathcal{C} \longrightarrow Subeq((F_i,G_i)_{i\in I})$ such that

- $(H, \{\beta_i\}_{i\in I}) * J = H$
- $(H, \{\beta_i\}_{i\in I}) * \alpha_i = \beta_i \quad (i \in I)$

□

The construction of $Subeq((F_i,G_i)_{i\in I})$ is quite simple. Its objects are pairs $(A, \{b_i\}_{i\in I})$ with A an object in $\mathcal{A}$ and $b_i : F_i(A) \longrightarrow G_i(A)$ a morphism in $\mathcal{B}_i$. Morphisms

$$a : (A, \{b_i\}_{i\in I}) \longrightarrow (A', \{b'_i\}_{i\in I})$$

are morphisms $a : A \longrightarrow A'$ in $\mathcal{A}$ such that for each $i \in I$, $b_i; G_i(a) = F_i(a); b'_i$. The functor J is just projection into the first component. The natural transformation α_j is defined by

$$\alpha_j(A, \{b_i\}_{i\in I}) = b_j.$$

[11] Note that we use diagrammatic order for the composition $J * F$ of two functors. More generally, we will also use diagrammatic order for the *horizontal*, $\alpha * \beta$, and *vertical*, $\gamma;\delta$, composition of natural transformations (see [43]).

For our present purposes, the subequalizers of interest are of the form

$$Subeq(([u_j(\overline{x})],[v_j(\overline{x})])_{1\leq j\leq k})$$

associated to rewrite rules[12]

$$r:[t(\overline{x})]\longrightarrow[t'(\overline{x})]\ \ if\ \ [u_1(\overline{x})]\longrightarrow[v_1(\overline{x})]\wedge\ldots\wedge[u_k(\overline{x})]\longrightarrow[v_k(\overline{x})]$$

in R. An object of such a subequalizer is a pair $(\overline{[w]}^n,\overline{[\beta]}^k)$ with $\overline{[w]}\in T_{\Sigma,E}(X)^n$ and with

$$[\beta_1]:[u_1(\overline{w}/\overline{x})]\longrightarrow[v_1(\overline{w}/\overline{x})]\quad\ldots\quad[\beta_k]:[u_k(\overline{w}/\overline{x})]\longrightarrow[v_k(\overline{w}/\overline{x})]$$

equivalence classes of proof terms modulo the equations 1-5, i.e., morphisms of $\mathcal{T}_{\mathcal{R}}(X)$. Therefore, the conditions

$$\beta_1;v_1(\overline{\alpha})=u_1(\overline{\alpha});\beta_1'\ \ \ldots\ \ \beta_k;v_k(\overline{\alpha})=u_k(\overline{\alpha});\beta_k'$$

in the exchange law just state that $\overline{\alpha}^n:(\overline{[w]}^n,\overline{[\beta]}^k)\longrightarrow(\overline{[w']}^n,\overline{[\beta']}^k)$ is a morphism in

$$Subeq(([u_j(\overline{x})],[v_j(\overline{x})])_{1\leq j\leq k})$$

and the equation in the conclusion just states that r is a natural transformation, i.e.,

Lemma 9 For each rewrite rule

$$r:[t(\overline{x})]\longrightarrow[t'(\overline{x})]\ \ if\ \ [u_1(\overline{x})]\longrightarrow[v_1(\overline{x})]\wedge\ldots\wedge[u_k(\overline{x})]\longrightarrow[v_k(\overline{x})]$$

in R, the family of morphisms

$$\{r(\overline{[w]}^n,\overline{\beta}^k):[t(\overline{w}/\overline{x})]\longrightarrow[t'(\overline{w}/\overline{x})]\mid(\overline{[w]}^n,\overline{[\beta]}^k)\in Subeq(([u_j(\overline{x})],[v_j(\overline{x})])_{1\leq j\leq k})\}$$

is a natural transformation

$$r:J*[t(x_1,\ldots,x_n)]\Longrightarrow J*[t'(x_1,\ldots,x_n)]$$

where $J:Subeq(([u_j(\overline{x})],[v_j(\overline{x})])_{1\leq j\leq k})\longrightarrow\mathcal{T}_{\mathcal{R}}(X)^n$ is the subequalizer functor. □

The $\mathcal{T}_{\mathcal{R}}$ construction just explained is very general and yields other well known constructions as particular cases. For example, for $\mathcal{R}$ a labelled transition system, $\mathcal{T}_{\mathcal{R}}$ is the well known *path category* associated to the transition system $\mathcal{R}$. For the case of Petri nets (see Section 4.2), the $\mathcal{T}_{\mathcal{R}}$ construction specializes to the construction of the monoidal category $\mathcal{T}[\mathcal{R}]$ associated to a Petri net $\mathcal{R}$ which I defined in joint work with Ugo Montanari [54, 53]. The following result holds for the $\mathcal{T}_{\mathcal{R}}$ construction in general:

Lemma 10 For each $[\alpha]:[t]\longrightarrow[t']$ in $\mathcal{T}_{\mathcal{R}}(X)$, either $[t]=[t']$ and $[\alpha]=[[t]]$, or there is an $n\in\mathbb{N}$ and a chain of morphisms $[\alpha_i]$, $0\leq i\leq n$ whose terms α_i describe one-step (concurrent) rewrites

$$[t]\xrightarrow{\alpha_0}[t_1]\xrightarrow{\alpha_1}\ \ldots\ \xrightarrow{\alpha_{n-1}}[t_n]\xrightarrow{\alpha_n}[t']$$

such that $[\alpha]=[\alpha_0;\ldots;\alpha_n]$. In addition, we can always choose all the α_i corresponding to sequential rewrites. □

The system $\mathcal{T}_{\mathcal{R}}(X)$ is just one among many *models* that can be assigned to the rewrite theory $\mathcal{R}$. The general notion of model, of which we shall later on discuss many examples, is called an $\mathcal{R}$-system and is defined as follows:

[12] If the condition of such a rule is abbreviated to C, then we will use the notation $Subeq(C)$ for the corresponding subequalizer.

Definition 11 Given a rewrite theory $\mathcal{R} = (\Sigma, E, L, R)$, an $\mathcal{R}$*-system* $\mathcal{S}$ is a category $\mathcal{S}$ together with:

- a (Σ, E)-algebra structure, i.e., for each $f \in \Sigma_n$, $n \in \mathbb{N}$, a functor $f_\mathcal{S} : \mathcal{S}^n \longrightarrow \mathcal{S}$, in such a way that the equations E are satisfied, i.e., for any $t(x_1, \ldots, x_n) = t'(x_1, \ldots, x_n)$ in E we have an identity of functors $t_\mathcal{S} = t'_\mathcal{S}$, where the functor $t_\mathcal{S}$ is defined inductively from the functors $f_\mathcal{S}$ in the obvious way.

- for each rewrite rule

$$r : [t(\overline{x})] \longrightarrow [t'(\overline{x})] \;\; \textit{if} \;\; [u_1(\overline{x})] \longrightarrow [v_1(\overline{x})] \wedge \ldots \wedge [u_k(\overline{x})] \longrightarrow [v_k(\overline{x})]$$

  in R, a natural transformation

$$r : J_\mathcal{S} * t_\mathcal{S} \Longrightarrow J_\mathcal{S} * t'_\mathcal{S}$$

  where $J_\mathcal{S} : \mathit{Subeq}((u_{j\mathcal{S}}, v_{j\mathcal{S}})_{1 \leq j \leq k}) \longrightarrow \mathcal{S}^n$ is the subequalizer functor.

An $\mathcal{R}$*-homomorphism* $F : \mathcal{S} \longrightarrow \mathcal{S}'$ between two $\mathcal{R}$-systems is then a functor $F : \mathcal{S} \longrightarrow \mathcal{S}'$ such that it is a Σ-algebra homomorphism—i.e., $f_\mathcal{S} * F = F^n * f_{\mathcal{S}'}$ for each f in Σ_n, $n \in \mathbb{N}$—and such that "F preserves R," i.e., for each rewrite rule $r : [t(\overline{x})] \longrightarrow [t'(\overline{x})]$ *if* C in R we have the identity of natural transformations:

$$r_\mathcal{S} * F = F^\bullet * r_{\mathcal{S}'}$$

where $F^\bullet : \mathit{Subeq}(C_\mathcal{S}) \longrightarrow \mathit{Subeq}(C_{\mathcal{S}'})$ is the unique functor induced by the universal property of $\mathit{Subeq}(C_{\mathcal{S}'})$ by the composition functor

$$\mathit{Subeq}(C_\mathcal{S}) \xrightarrow{J_\mathcal{S}} \mathcal{S}^n \xrightarrow{F^n} \mathcal{S}'^n$$

and the natural transformations $\alpha_j * F$ $(1 \leq j \leq k)$ where if C has k conditions $[u_j] \longrightarrow [v_j]$, $1 \leq j \leq k$, α_j is the jth natural transformation $\alpha_j : J_\mathcal{S} * u_{j\mathcal{S}} \Longrightarrow J_\mathcal{S} * v_{j\mathcal{S}}$ associated to the subequalizer $\mathit{Subeq}(C_\mathcal{S})$. Note that for this to make sense we have used the identities

$$u_{j\mathcal{S}} * F = F^n * u_{j\mathcal{S}'} \qquad\qquad v_{j\mathcal{S}} * F = F^n * v_{j\mathcal{S}'}$$

which hold for derived operations because F is a Σ-homomorphism and that allow us to express $\alpha_j * F$ as a natural transformation

$$\alpha_j * F : J_\mathcal{S} * F^n * u_{j\mathcal{S}'} \Longrightarrow J_\mathcal{S} * F^n * v_{j\mathcal{S}'}.$$

Despite the somewhat complicated definition of $F^\bullet$, its behavior on objects is quite simple; it is given by the equation

$$F^\bullet(\overline{C}^n, \overline{c}^k) = (F^n(\overline{C}^n), F^k(\overline{c}^k)).$$

This defines a category $\underline{\mathcal{R}\text{-}\mathit{Sys}}$ in the obvious way.

This category has the additional property that the homsets $\underline{\mathcal{R}\text{-}\mathit{Sys}}(\mathcal{S}, \mathcal{S}')$ are themselves categories with morphisms, called *modifications*, given by natural transformations $\delta : F \Longrightarrow G$ between $\mathcal{R}$-homomorphisms $F, G : \mathcal{S} \longrightarrow \mathcal{S}'$ satisfying the identities

$$\delta^n * f_{\mathcal{S}'} = f_\mathcal{S} * \delta$$

for each $f \in \Sigma_n$, $n \in \mathbb{N}$. This category structure actually makes $\underline{\mathcal{R}\text{-}\mathit{Sys}}$ into a 2-category [43, 39].

An $\mathcal{R}$*-groupoid* is an $\mathcal{R}$-system $\mathcal{S}$ whose category structure is actually a groupoid. This defines a full subcategory $\underline{\mathcal{R}\text{-}\mathit{Grpd}} \subseteq \underline{\mathcal{R}\text{-}\mathit{Sys}}$. □

As further confirmation that these definitions of $\mathcal{R}$-system and $\mathcal{R}$-homomorphism are very natural we leave for the reader to check the following

Lemma 12 The full subcategory of $\underline{\mathcal{R}\text{-}PreSys}$ determined by those presystems that satisfy the equations[13] 1–5 is isomorphic to the category $\underline{\mathcal{R}\text{-}Sys}$. □

3.1 Initiality, Soundness and Completeness

Lemma 12 shows that the notion of $\mathcal{R}$-system is also "essentially algebraic" and can be specified by adding equations of the type 1–5 to the essentially algebraic theory defining $\mathcal{R}$-presystems. As usual in universal algebra, whenever we have a full subcategory defined by a collection of equations, that subcategory is *reflective* inside the bigger one, i.e., its inclusion has a left adjoint called a *reflection* (see, e.g., [60] or [2] Thm. 4.4.1 for an even more general result implying this property). In our case, this means that the full subcategory inclusion

$$\underline{\mathcal{R}\text{-}Sys} \hookrightarrow \underline{\mathcal{R}\text{-}PreSys}$$

has a left adjoint. In fact, for presystems of the form $\mathcal{P}_{\mathcal{R}}(X)$ it is quite easy to see that the quotient $\mathcal{R}$-prehomomorphism $Q_X : \mathcal{P}_{\mathcal{R}}(X) \longrightarrow \mathcal{T}_{\mathcal{R}}(X)$ associated to the congruence defining $\mathcal{T}_{\mathcal{R}}(X)$ is in fact the reflection map or "counit" of such an adjunction. Indeed, given an $\mathcal{R}$-system $\mathcal{S}$ and an $\mathcal{R}$-prehomomorphism $F : \mathcal{P}_{\mathcal{R}}(X) \longrightarrow \mathcal{S}$, since $\mathcal{S}$, being an $\mathcal{R}$-system, satisfies equations just like those used in the definition of $\mathcal{T}_{\mathcal{R}}(X)$, there is a containment of the congruences associated to Q_X and F as follows:

$$Ker(Q_X) \subseteq Ker(F)$$

this defines a unique $\mathcal{R}$-prehomomorphism (therefore an $\mathcal{R}$-homomorphism) $F^{\dagger} : \mathcal{T}_{\mathcal{R}}(X) \longrightarrow \mathcal{S}$ such that $F = Q_X * F^{\dagger}$. The following important result becomes now a trivial consequence of this discussion.

Theorem 13 (*Initiality*) The $\mathcal{R}$-system $\mathcal{T}_{\mathcal{R}}$ is an initial object in the category $\underline{\mathcal{R}\text{-}Sys}$, and the $\mathcal{R}$-groupoid $\mathcal{T}_{\mathcal{R}}^{\leftrightarrow}$ is an initial object in the category $\underline{\mathcal{R}\text{-}Grpd}$. More generally, the system $\mathcal{T}_{\mathcal{R}}(X)$ has the following universal property: Given an $\mathcal{R}$-system $\mathcal{S}$, each function $F : X \longrightarrow Obj(\mathcal{S})$ extends uniquely to an *$\mathcal{R}$-homomorphism* $F^{\natural} : \mathcal{T}_{\mathcal{R}}(X) \longrightarrow \mathcal{S}$. The groupoid $\mathcal{T}_{\mathcal{R}}^{\leftrightarrow}(X)$ has the same universal property with respect to $\mathcal{R}$-groupoids. □

Since rewriting logic has a notion of *sentence*, namely a sequent $[t] \longrightarrow [t']$, we should consider what it means for an $\mathcal{R}$-system $\mathcal{S}$ to satisfy such a sequent. Intuitively, the sequent indicates the existence of an arrow in the category. However, since the terms t and t' may have variables, it is not a fixed arrow, but a "variable" one. Of course, the obvious candidate for the intuitive notion of a "variable arrow" is the concept of natural transformation. Lemmas 7 and 9 suggest that such an interpretation is indeed the right one.

Definition 14 A sequent $[t(x_1,\ldots,x_n)] \longrightarrow [t'(x_1,\ldots,x_n)]$ is *satisfied* by an $\mathcal{R}$-system $\mathcal{S}$ if there exists a natural transformation $\alpha : t_{\mathcal{S}} \Longrightarrow t'_{\mathcal{S}}$ between the functors $t_{\mathcal{S}}, t'_{\mathcal{S}} : \mathcal{S}^n \longrightarrow \mathcal{S}$. We use the notation

$$\mathcal{S} \models [t(x_1,\ldots,x_n)] \longrightarrow [t'(x_1,\ldots,x_n)]$$

to denote the satisfaction relation. Similarly,

$$\mathcal{R} \models [t(x_1,\ldots,x_n)] \longrightarrow [t'(x_1,\ldots,x_n)]$$

states that the sequent is satisfied by all $\mathcal{R}$-systems. □

[13]More precisely, the slightly generalized equations obtained from 1–5 by replacing nodes $[t] \in T_{\Sigma,E}(X)$ by nodes C in the set of nodes of an $\mathcal{R}$-presystem.

A detailed proof of the following theorem can be found in [51].

Theorem 15 (*Soundness and Completeness*) For $\mathcal{R}$ a rewrite theory,

$$\mathcal{R} \models [t(x_1,\ldots,x_n)] \longrightarrow [t'(x_1,\ldots,x_n)]$$

iff

$$\mathcal{R} \vdash [t(x_1,\ldots,x_n)] \longrightarrow [t'(x_1,\ldots,x_n)]$$

iff there is a morphism

$$[t(x_1,\ldots,x_n)] \longrightarrow [t'(x_1,\ldots,x_n)]$$

in $\mathcal{T}_{\mathcal{R}}(\{x_1,\ldots,x_n\})$. □

3.2 Equationally Defined Classes of Models

Since $\mathcal{R}$-systems are an "essentially algebraic" concept, we can consider classes Θ of $\mathcal{R}$-systems defined by the satisfaction of additional equations. Such classes give rise to full subcategory inclusions $\Theta \hookrightarrow \underline{\mathcal{R}\text{-}Sys}$, and by general universal algebra results about essentially algebraic theories (see, e.g., [2, 60]) such inclusions are *reflective* [43], i.e., for each $\mathcal{R}$-system $\mathcal{S}$ there is an $\mathcal{R}$-system $R_\Theta(\mathcal{S}) \in \Theta$ and an $\mathcal{R}$-homomorphism $\rho_\Theta(\mathcal{S}) : \mathcal{S} \longrightarrow R_\Theta(\mathcal{S})$ such that for any $\mathcal{R}$-homomorphism $F : \mathcal{S} \longrightarrow \mathcal{D}$ with $\mathcal{D} \in \Theta$ there is a unique $\mathcal{R}$-homomorphism $F^\diamond : R_\Theta(\mathcal{S}) \longrightarrow \mathcal{D}$ such that $F = \rho_\Theta(\mathcal{S}); F^\diamond$. The full subcategory $\underline{\mathcal{R}\text{-}Grpd} \subseteq \underline{\mathcal{R}\text{-}Sys}$ is also reflective, but it is not equationally definable. This situation generalizes that of the inclusion of the category of groups into the category of monoids. What we have in this case is an inclusion that is a *forgetful functor* from a category of algebras with additional operations (in this case the inversion operation). However, for any equationally definable (full) subcategory $\Theta \subseteq \underline{\mathcal{R}\text{-}Sys}$, defined by a collection of equations H, the intersection $\Theta \cap \underline{\mathcal{R}\text{-}Grpd}$ has a very simple description, since it is just the full subcategory of $\underline{\mathcal{R}\text{-}Grpd}$ definable by the equations H.

Therefore, we can consider subcategories of $\underline{\mathcal{R}\text{-}Sys}$ or of $\underline{\mathcal{R}\text{-}Grpd}$ that are defined by certain equations and be guaranteed that they have initial and free objects, that they are closed by subobjects and products, etc. Consider for example the following conditional equations:

$$\forall f,g \in Arrows,\ f = g \ \ if \ \ \partial_0(f) = \partial_0(g) \wedge \partial_1(f) = \partial_1(g)$$

$$\forall f,g \in Arrows,\ f = g \ \ if \ \ \partial_0(f) = \partial_1(g) \wedge \partial_1(f) = \partial_0(g).$$

where $\partial_0(f)$ and $\partial_1(f)$ denote the source and target of an arrow f respectively. The first equation forces a category to be a preorder, and the addition of the second requires this preorder to be a poset. By imposing the first one, or by imposing both, we get full subcategories

$$\underline{\mathcal{R}\text{-}Pos} \subseteq \underline{\mathcal{R}\text{-}Preord} \subseteq \underline{\mathcal{R}\text{-}Sys}.$$

A routine inspection of $\underline{\mathcal{R}\text{-}Preord}$ for $\mathcal{R} = (\Sigma, E, L, R)$ reveals that its objects are preordered Σ-algebras $(A, \leq)$ (i.e., preordered sets with a Σ-algebra structure such that all the operations in Σ are monotonic) that satisfy the equations E and such that for each rewrite rule

$$r : [t(\overline{x})] \longrightarrow [t'(\overline{x})] \ \ if \ \ [u_1(\overline{x})] \longrightarrow [v_1(\overline{x})] \wedge \ldots \wedge [u_k(\overline{x})] \longrightarrow [v_k(\overline{x})]$$

in R and for each $\overline{a} \in A^n$ such that $u_{jA}(\overline{a}) \geq v_{jA}(\overline{a})$ for $1 \leq j \leq k$, we have, $t_A(\overline{a}) \geq t'_A(\overline{a})$. The poset case is entirely analogous, except that the relation $\leq$ is a partial order instead of being a preorder. The reflection functor associated to the inclusion $\underline{\mathcal{R}\text{-}Preord} \subseteq \underline{\mathcal{R}\text{-}Sys}$, sends $\mathcal{T}_{\mathcal{R}}(X)$ to the familiar $\mathcal{R}$-*rewriting relation*[14] $\rightarrow_{\mathcal{R}(X)}$ on E-equivalence classes of terms with variables in

[14] It is perhaps more suggestive to call $\rightarrow_{\mathcal{R}(X)}$ the *reachability relation* of the system $\mathcal{T}_{\mathcal{R}}(X)$.

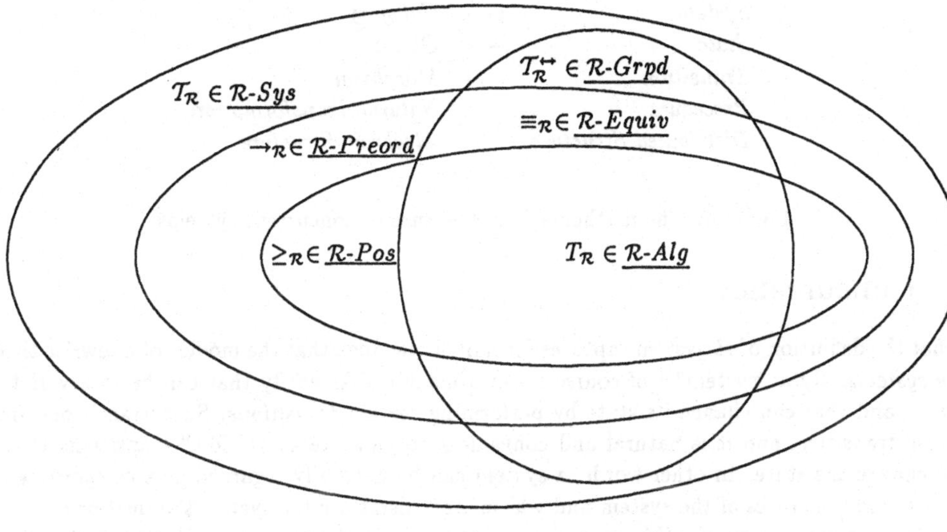

Figure 3: Subcategories of $\underline{\mathcal{R}\text{-}Sys}$ and their initial objects.

X. It is easy to show that rewriting logic remains complete when we restrict the models to be preorders [51]. Similarly, the reflection associated to the inclusion $\underline{\mathcal{R}\text{-}Pos} \subseteq \underline{\mathcal{R}\text{-}Sys}$ maps $\mathcal{T}_\mathcal{R}(X)$ to the partial order $\geq_{\mathcal{R}(X)}$ obtained from the preorder $\rightarrow_{\mathcal{R}(X)}$ by identifying any two $[t], [t']$ such that $[t] \rightarrow_{\mathcal{R}(X)} [t']$ and $[t'] \rightarrow_{\mathcal{R}(X)} [t]$. Again, rewriting logic remains complete for poset models [51].

Intersecting $\underline{\mathcal{R}\text{-}Pos}$ and $\underline{\mathcal{R}\text{-}Preord}$ with the category $\underline{\mathcal{R}\text{-}Grpd}$ we get two subcategories definable by the first equation or by both, but now in the context of $\underline{\mathcal{R}\text{-}Grpd}$. Combining the notions of a groupoid and a preorder we get exactly the notion of an *equivalence relation* and therefore a subcategory $\underline{\mathcal{R}\text{-}Equiv}$ whose initial object is the usual congruence $\equiv_\mathcal{R}$ on (E-equivalence classes of) ground terms modulo provable equality generated by the rules in $\mathcal{R}$ when regarded as equations. A poset that is also a groupoid yields a *discrete category* whose only arrows are identities, i.e., a set. It is therefore easy to see that the subcategory obtained by intersecting $\underline{\mathcal{R}\text{-}Pos}$ with $\underline{\mathcal{R}\text{-}Grpd}$ is just the familiar category $\underline{\mathcal{R}\text{-}Alg}$ of ordinary Σ-algebras that satisfy the equations $E \cup \mathit{unlabel}(R)$, where the *unlabel* function removes the labels from the rules and turns the sequent signs " $\longrightarrow$ " into equality signs. Similarly, the reflection functor into $\underline{\mathcal{R}\text{-}Alg}$ maps $\mathcal{T}_\mathcal{R}(X)$ to $T_\mathcal{R}(X)$, the free Σ-algebra on X satisfying the equations $E \cup \mathit{unlabel}(R)$. Figure 3 summarizes the relationships among all these categories.

Although any of the subcategories of $\underline{\mathcal{R}\text{-}Grpd}$ discussed above provides a suitable class of models for equational logic, the classical notion of model is of course provided by the algebras in $\underline{\mathcal{R}\text{-}Alg}$. The inclusion functor $\underline{\mathcal{R}\text{-}Alg} \hookrightarrow \underline{\mathcal{R}\text{-}Sys}$ is a key ingredient of the map of logics

$$\mathit{RewritingLogic} \longrightarrow \mathit{EquationalLogic}$$

mentioned in the introduction (note that, on models, the map goes backwards!). The two other ingredients are the assignment to each rewrite theory $\mathcal{R} = (\Sigma, E, L, R)$ of the equational theory[15] $(\Sigma, E, \mathit{unlabel}(R))$, and the assignment to each sequent $[t] \longrightarrow [t']$ of the equation $[t] = [t']$. Without entering into the technical details, we remark that these data determine a map of logics in the sense of [48], which is implicitly used to define the different semantics given in Maude to functional, system, and object-oriented modules (see [50, 49]).

[15]As for rewrite theories, we allow equational theories in which the equations are specified *modulo E*; this increases the expressive power of equational theories.

System	$\longleftrightarrow$	*Category*
State	$\longleftrightarrow$	*Object*
Transition	$\longleftrightarrow$	*Morphism*
Procedure	$\longleftrightarrow$	*Natural Transformation*
Distributed Structure	$\longleftrightarrow$	*Algebraic Structure*

Figure 4: The mathematical structure of concurrent systems.

4 Concurrency

What the definition of $\mathcal{R}$-system captures formally is the idea that the models of a rewrite theory *are systems*. By a "system" I of course mean a machine-like entity that can be in a variety of *states*, and that can change its state by performing certain *transitions*. Such transitions are of course transitive, and it is natural and convenient to view states as "idle" transitions that do not change the state. In other words, a system can be naturally regarded as a *category*, whose objects are the states of the system and whose morphisms are the system's transitions.

For *sequential* systems, this is in a sense the end of the story (see Section 4.1). As the examples discussed later in this section will make clear, what makes a system *concurrent* is precisely the existence of an additional *algebraic structure*. Ugo Montanari and I first observed this fact for the particular case of Petri nets for which the algebraic structure is precisely that of a commutative monoid [54, 53]. However, this observation holds in full generality for *any algebraic structure whatever*. What the algebraic structure captures is twofold. Firstly, *the states themselves are distributed according to such a structure*; for Petri nets (see Section 4.2) the distribution takes the form of a *multiset* that we can visualize with tokens and places; for a functional program involving just syntactic rewriting, the distribution takes the form of a *labelled tree structure* which can be spatially distributed in such a way that many transitions (i.e., rewrites) can happen concurrently in a way analogous to the concurrent firing of transitions in a Petri net. Secondly, *concurrent transitions are themselves distributed according to the same algebraic structure*; this is what the notion of $\mathcal{R}$-system captures, and is for example manifested in the concurrent firing of Petri nets and, more generally, in any type of concurrent rewriting.

The expressive power of rewrite theories to specify concurrent transition systems is greatly increased by the possibility of having not only transitions, but also *parameterized transitions*, i.e., *procedures*. This is what rewrite rules—with variables—provide. The family of states to which the procedure applies is given by those states where a component of the (distributed) state is a substitution instance of the lefthand side of the rule in question and—in addition—a proof of the corresponding substitution instance of the rule's condition can be obtained. The rewrite rule is then a *procedure*[16] which transforms the state *locally*, by replacing such a substitution instance by the corresponding substitution instance of the righthand side. The fact that this can take place concurrently with other transitions "below" is precisely what the concept of a *natural transformation* formalizes. The table of Figure 4 summarizes our present discussion; the most crucial correspondence listed in that figure is the identification of a system's distributed structure with its algebraic structure. The best way to exhibit that identification is by example. First, we put this in evidence by discussing labelled transition systems, where both concurrency and algebraic structure clearly shine by their absence; then we discuss truly concurrent systems such as Petri nets and concurrent object-oriented systems; finally, we briefly explain how concurrent rewriting is a very general model of concurrency, from which many other models can be obtained

[16] Its *actual parameters* are precisely given by a substitution.

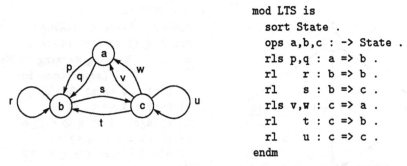

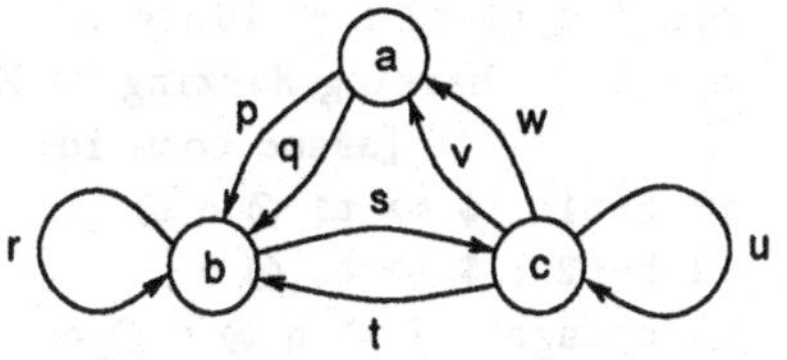


```
mod LTS is
  sort State .
  ops a,b,c : -> State .
  rls p,q : a => b .
  rl    r : b => b .
  rl    s : b => c .
  rls v,w : c => a .
  rl    t : c => b .
  rl    u : c => c .
endm
```

Figure 5: A labelled transition system and its code in Maude.

by specialization. Although the examples discussed are enough to illustrate the main ideas, space limitations preclude a more detailed and comprehensive discussion, for which we refer the reader to [51], and (for object-oriented matters) to [49].

4.1 Labelled Transition Systems

This is the particularly simple case of rewrite theories $\mathcal{R} = (\Sigma, E, L, R)$ such that $E = \emptyset$, $\Sigma = \Sigma_0$, i.e., Σ only involves constants, and all the rules in $\mathcal{R}$ are of the form $r : a \longrightarrow b$ for a, b constants. For example, the transition system of Figure 5 corresponds to the rewrite theory of the system module LTS in the same figure. Since Σ contains only constants and the rules have no variables, the rules 1–4 of rewriting logic specialize to very simple rules. The rule of congruence becomes a trivial subcase of reflexivity, and the rule of replacement just yields each rule $r : a \longrightarrow b$ in $\mathcal{R}$ as its own consequence. Thus, we just have reflexivity and transitivity with the rules $r : a \longrightarrow b$ in $\mathcal{R}$ as basic axioms. Therefore, $\mathcal{T}_\mathcal{R}$ is just the *free category*—also called the *path category*—generated by the labelled transition system when regarded as a graph. More generally, *any* $\mathcal{R}$-system with $\mathcal{R}$ a labelled transition system is just a *category* C together with the assignment of an object of C to each constant in Σ and a morphism in C for each rule in R in a way consistent with the assignment of objects. In other words, such systems are just *sequential systems*, and their sequentiality is precisely due to the *absence of any operations* other than constants. In fact, labelled transition systems are intrinsically *sequential* as rewrite theories, in the precise sense of Definition 2. However, since several transitions are in general possible from a given state, they exhibit a form of *nondeterminism*.

4.2 Petri Nets

This is one of the most basic models of concurrency. It has the great advantage of exhibiting concurrency *directly*, not through the indirect mediation of interleavings. Its relationship to concurrent rewriting can be expressed very simply. It is just the particular case of unconditional rewrite theories $\mathcal{N} = (\Sigma, E, L, R)$ with $\Sigma_0 = \Delta \uplus \{\lambda\}$, $\Sigma_2 = \{\otimes\}$, with all the other Σ_n empty, with $E = ACI$—where ACI are the axioms of *associativity* and *commutativity* for $\otimes$ and *identity* λ for $\otimes$—and with all terms in the rules R ground terms. Consider for example the Petri net in Figure 6, which represents a machine to buy subway tickets. With a dollar we can buy a ticket $t1$ by pushing the button $b-t1$ and get two quarters back; if we push $b-t2$ instead, we get a longer distance ticket $t2$ and one quarter back. Similar buttons allow purchasing the tickets with quarters. Finally, with one dollar we can get four quarters by pushing *change*. The corresponding order-sorted rewrite theory is that of the TICKET system module in the same figure.

A key point about this module is that the operator $\otimes$—corresponding to *multiset union* of markings on the net—provides the system's commutative monoid algebraic structure (this is

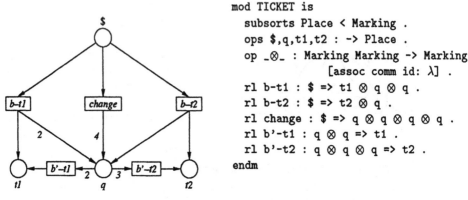

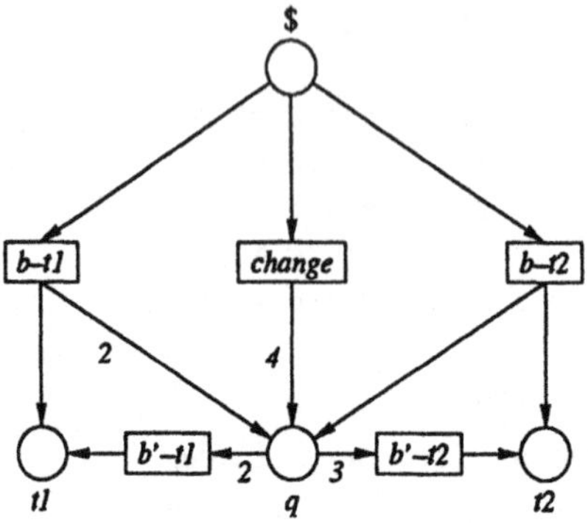


```
mod TICKET is
  subsorts Place < Marking .
  ops $,q,t1,t2 : -> Place .
  op _⊗_ : Marking Marking -> Marking
             [assoc comm id: λ] .
  rl b-t1 : $ => t1 ⊗ q ⊗ q .
  rl b-t2 : $ => t2 ⊗ q .
  rl change : $ => q ⊗ q ⊗ q ⊗ q .
  rl b'-t1 : q ⊗ q => t1 .
  rl b'-t2 : q ⊗ q ⊗ q => t2 .
endm
```

Figure 6: A Petri net and its code in Maude.

the "Petri nets as monoids" view [54, 53]) which is at the same time its concurrent structure. I.e., concurrent computations are possible in the net precisely due to the distributed nature of a marking as a multiset whose elements are put together by the binary union operator. By contrast, the state of a labelled transition system is *atomic*, and lacking any algebraic structure does, precisely for that reason, make any concurrency impossible. Of course, the commutative monoid structure is also enjoyed by the computations themselves. In this example, ACI-rewriting captures exactly the concurrent computations of the Petri net. Suppose, for example, that we begin in a state with four quarters and two dollars. Then, by first concurrently pushing the buttons b'–$t1$ and $b-t2$, and then concurrently pushing the buttons b'–$t2$ and $b-t2$ we end up with a ticket for the shorter distance, three tickets for the longer distance and a quarter, as shown in the two steps of concurrent ACI-rewriting below:

$$\texttt{q⊗q⊗q⊗q⊗\$⊗\$} \longrightarrow \texttt{q⊗q⊗t1⊗t2⊗q⊗\$} \longrightarrow \texttt{t2⊗t1⊗t2⊗t2⊗q}.$$

The rules of deduction specialize as follows. The congruence rule applies just to $\otimes$ and yields instances of reflexivity for the constants. Since the rewrite rules have no variables, the replacement rule yields each of the rewrite rules as axioms. Interpreting $\otimes$ as conjunction in linear logic [17], this specialization yields sound and complete rules of deduction for the linear theory having each of the rewrite rule sequents as axioms; i.e., rewriting logic specializes in this case to *conjunctive linear logic*. The models of rewrite theories $\mathcal{N}$ of this kind are categories with a commutative monoid structure in which we have chosen certain objects—the "places"—and certain morphisms—the "transitions." The initial system $\mathcal{T}_{\mathcal{N}}$ is exactly the category $\mathcal{T}[\mathcal{N}]$ that Ugo Montanari and I associated to a Petri net as its semantics in [54, 53]. Narciso Martí-Oliet and I later studied the connection of this model with models for linear logic in [45, 44] and obtained in this way a systematic triangular correspondence between Petri nets, linear logic and categories which is a particular instance of the more general triangular correspondence between concurrent systems, rewriting logic and categories developed in this paper.

4.3 Concurrent Object-Oriented Programming

This is also a very important case exhibiting true concurrency precisely due to its implicit algebraic structure. So far, most approaches in this area have been quite operational. Rewriting logic provides as a special case a nice logical theory of object-oriented concurrency that is entirely declarative; see [49] for an account of such a theory. Here we discuss the most basic ideas about objects and the evolution of an object-oriented system by concurrent rewriting of its configuration, which is made up of a collection of objects and messages.

An *object* can be represented as a term

$$\langle O : C \mid a_1 : v_1, \ldots, a_n : v_n \rangle$$

where O is the object's name, C is its class, the a_i's are the names of the object's *attributes*, and the v_i's are their corresponding *values*, which typically are required to be in a sort appropriate for their corresponding attribute. The *configuration* is the distributed state of the concurrent object-oriented system and is represented as a multiset of objects and messages according to the following syntax:

```
subsorts Object Message < Configuration .
op __ : Configuration Configuration -> Configuration [assoc comm id: ∅] .
```

where the operator __ is associative and commutative with identity $\emptyset$ and plays a role entirely similar to that played by the operator $\otimes$ for Petri nets. The system evolves by concurrent rewriting (modulo ACI) of the configuration by means of rewrite rules specific to each particular system, whose lefthand and righthand sides may in general involve patterns for several objects and messages. For example, objects in a class *Accnt* of bank accounts, each having a *bal*(ance) attribute, may receive messages for crediting or debiting the account and evolve according to the rules:

$$\begin{array}{ll} credit(B,M)\ \langle B : Accnt \mid bal : N \rangle & \longrightarrow \langle B : Accnt \mid bal : N + M \rangle \\ debit(B,M)\ \langle B : Accnt \mid bal : N \rangle & \longrightarrow \langle B : Accnt \mid bal : N - M \rangle. \end{array}$$

Again, the commutative monoid structure of the configuration makes it concurrent so that many rewrite rules can be applied at the same time. In addition, the values in the attributes of an object can also be computed by means of rewrite rules, and this adds yet another level of concurrency to a concurrent object-oriented system of this kind.

Concurrent object-oriented systems can be defined in Maude by means of *object-oriented module definitions* of the form omod $\mathcal{O}$ endom which provide special syntax taking advantage of the structural properties common to all such systems [49]. However, *object-oriented modules are entirely reducible to system modules*, i.e., we can systematically translate an object-oriented module omod $\mathcal{O}$ endom into a corresponding system module mod $\mathcal{O}^b$ endm whose semantics is the object-oriented module's intended semantics. Maude's functional, system and object-oriented modules and their initial model semantics are discussed in [49] and [50]. In comparison with the FOOPS language that Joseph Goguen and I developed in [23] and that has provided very valuable experience in the design of Maude, both FOOPS and Maude contain OBJ as their functional sublanguage, and both have declarative style object-oriented modules; also, the idea of transforming objects by rewrite rules goes back to [24], although the use of ACI to treat concurrency was not contemplated in that work. However, the semantic frameworks of FOOPS and Maude are quite different, and this leads to different styles of programming with objects and to different approaches to concurrency.

4.4 Unifying models of Concurrency

Space limitations preclude a detailed discussion of other models of concurrency to which concurrent rewriting specializes (see [51]). However, we can summarize such specializations using Figure 7, where CR stands for concurrent rewriting, the arrows indicate specializations, and the subscripts $\emptyset$, AI, and ACI stand for syntactic rewriting, rewriting modulo associativity and identity, and ACI-rewriting respectively. *Functional programming* (in particular Maude's functional modules) corresponds to the case of *confluent*[17] rules, and includes the λ-calculus

[17] Although not reflected in the picture, rules confluent *modulo* equations E are also functional.

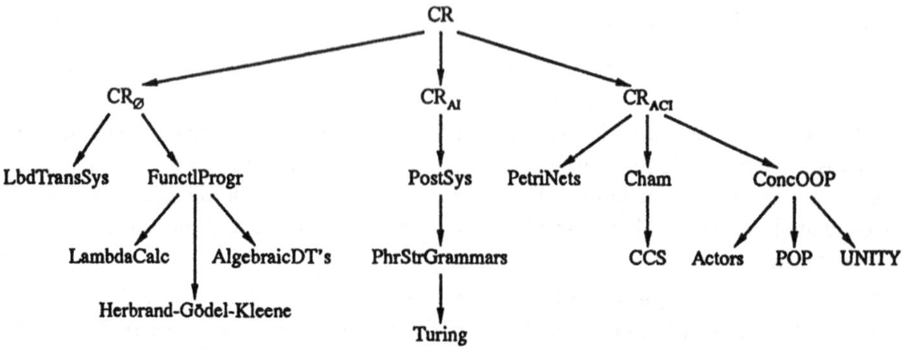

Figure 7: Specializations of concurrent rewriting.

(in combinator form) and the Herbrand-Gödel-Kleene theory of recursive functions. Rewriting modulo AI yields Post systems and related grammar formalisms, including Turing machines. Rewriting modulo ACI includes Berry and Boudol's *chemical abstract machine* [3] (which itself specializes to CCS [55]), as well as actors [1] and Unity's model of computation [5] which can both be regarded as special cases of concurrent object-oriented programming with rewrite rules; a third special case is Engelfriet et al.'s POPs and POTs higher level Petri nets [11, 12].

5 Related Work

There is a wealth of work on term rewriting, concurrency, Petri nets, linear logic, equational logic and the theory of abstract data types relevant to this paper; a good part of the most closely related work has already been mentioned in the body of the paper. However, now that the main ideas of this work have been explained, I would like to discuss some papers by other authors that bear some relationship to the logic and model theory of rewriting presented here.

An important area of common interest is the algebraic theory of *ordered* and *continuous* algebras. For the unconditional case, Bloom [4] studied the logic of (unconditional) inequalities and proved very nice results for classes of ordered algebras, including a Birkhoff theorem, using ordered algebraic theories [4]. The study of classes of *continuous algebras* definable by inequalities was advanced by a good number of authors including Iréne Guessarian [28], Bruno Courcelle and Maurice Nivat [6]. My own work in this area includes a very general completeness theorem for the logic of continuous algebras [46] and two Birkhoff theorems for classes of continuous algebras definable by continuous theories [46, 47].

Closely connected with the theory of ordered algebras is the algebraic approach to nondeterminism based on monotonic algebras whose elements are subsets of a given set. In the context of domain theory we have the Plotkin, Hoare and Smith powerdomains. In more direct connection with term rewriting and abstract data types, there is work by a number of authors including the ideas of O'Donnell [57], as well as work by Hussmann [32], Jayaraman and Plaisted [34], Hesselink [29] and others. The very nice theory of *unified algebras* developed by Peter Mosses [56] also deals with nondeterminism by means of subsets. The relationships of all these approaches to the work presented here, in particular within the equationally defined subclass of ordered and preordered models should be further clarified.

Within the area of term rewriting, the work of Huet is close in spirit to the approach taken here. In the lecture notes [30], Huet defined a category of rewritings for *regular* rewrite theories extending previous joint work with J.-J. Lévy and earlier ideas by Lévy [42]. However, only the categorical structure of those rewritings was emphasized. In [30], Huet also briefly discussed a nonconditional version of rewriting logic. Using ideas about residuals originating in the Huet-Lévy work, E. Stark developed a categorical model of transition systems exhibiting concurrency, and applied that model to obtain results on dataflow nets.

There is also important work on applications of 2-categories to rewriting, including work by Rydeheard and Stell [61], who for the unconditional case with $E = \emptyset$ constructed a 2-category of term rewritings, and Seely [62], who used 2-categories to treat reduction in lambda calculi. There is also important unpublished work of A. Pitts [58] brought to my attention by Axel Poigné, who has applied 2-category ideas to the semantics of fixpoints in denotational semantics, both at the level of functions and at the level of solving domain equations; in a sense, his work generalizes to the category-enriched setting previous work on continuous and rational theories. Independently of my own work, Pitts has also developed in [58] a logic of fixpoints, whose fixpoint-free fragment is very closely related to the unconditional fragment of the rewriting logic presented here. J. Power [59] has recently studied normalization of what we here call "proof expressions" by means of pasting techniques in 2-categories. Most of these approaches have focused on proof-theoretic results, using 2-categories as theories; by contrast, the main interest of this work is in the model theory of rewriting logic and its relationship to the classical theory of algebraic data types that this model theory generalizes.

Although this paper has avoided using the heavier machinery of 2-categories, they are very useful, and Appendix A of [51] develops both the proof theory and the model theory of a 2-categorical approach to rewriting that connects with the body of work already mentioned, extends already known results to the more complex conditional case, and provides a new perspective from which to view the results and ideas of this paper. An important fruit of the 2-categorical ideas presented in Appendix A of [51] is a theory of morphisms between rewrite theories—developed in Appendix B of the same paper—that generalizes the implementation morphisms of Petri nets introduced in [54] and permits treating parameterized rewrite theories.

6 Acknowledgements

I specially thank Prof. Joseph Goguen for our long term collaboration on the OBJ language and its multiparadigm extensions, the concurrent rewriting model of computation and its implementation on the RRM architecture, all of which have directly influenced this work; he has also provided many positive suggestions for improving a previous version of this paper. I also thank all my fellow members of the OBJ and RRM teams, past and present, and in particular Dr. Claude Kirchner, who worked with us on the first paper describing the concurrent rewriting model, Dr. Sany Leinwand, who suggested improvements to the exposition, and Mr. Timothy Winkler, who deserves special thanks for his many very good comments about the technical content as well as for his invaluable assistance with the pictures. I specially thank Prof. Ugo Montanari for our collaboration on the semantics of Petri nets; the algebraic ideas that we developed in that work have been a source of inspiration for the more general ideas presented here. I also thank Prof. Pierpaolo Degano, with whom I have also worked on further developments of the Petri net ideas. Mr. Narciso Martí-Oliet deserves special thanks for our collaboration on the semantics of linear logic and its relationship to Petri nets which is another source of inspiration for this work; he also provided very many helpful comments and suggestions for improving the exposition. I finally wish to thank Prof. Pierre-Louis Curien and Mr. Ross Casley, who suggested useful references, Dr. Mark Moriconi, who suggested improvements to the exposition, Dr. Axel Poigné who made valuable suggestions and brought to my attention related literature, Profs. Brian Mayoh and Peter Mosses, who both provided valuable detailed comments to a previous version of the paper, and Prof. Robin Milner, who provided valuable suggestions and encouraged this work in personal conversation during the FGCS'88 conference in Tokyo.

References

[1] G. Agha. *Actors.* MIT Press, 1986.

[2] M. Barr and C. Wells. *Toposes, Triples and Theories.* Springer-Verlag, 1985.

[3] Gérard Berry and Gérard Boudol. The Chemical Abstract Machine. In *Proc. POPL'90*, pages 81–94. ACM, 1990.

[4] Steven Bloom. Varieties of ordered algebras. *Journal of Computer and System Sciences*, 13:200–212, 1976.

[5] K. Many Chandy and Jayadev Misra. *Parallel Program Design: A Foundation.* Addison-Wesley, 1988.

[6] Bruno Courcelle and Maurice Nivat. The algebraic semantics of recursive program schemes. In *Proceedings, Mathematical Foundations of Computer Science.* Springer-Verlag, 1978. Lecture Notes in Computer Science, Volume 64.

[7] N. Dershowitz and J.-P. Jouannaud. Rewrite systems. In *Handbook of Theoretical Computer Science, Vol. B.* North-Holland, 1990.

[8] N. Dershowitz and M. Okada. A rationale for conditional equational programming. *Theoretical Computer Science*, 75:111–138, 1990.

[9] N. Dershowitz, M. Okada, and G. Sivakumar. Confluence of conditional rewrite systems. In Jean-Pierre Jouannaud and Stephane Kaplan, editors, *Proceedings, Conference on Conditional Term Rewriting, Orsay, France, July 8-10, 1987*, pages 31–44. Springer-Verlag, Lecture Notes in Computer Science No. 308, 1988.

[10] Nachum Dershowitz and David A. Plaisted. Equational programming. In J. Richards, editor, *Machine Intelligence 11: The logic and acquisition of knowledge*, pages 21–56. Oxford University Press, 1988.

[11] J. Engelfriet. Net-based description of parallel object-based systems, or POTs and POPs. Technical report, Noordwijkerhout FOOL Workshop, May 1990.

[12] J. Engelfriet, G. Leih, and G. Rozenberg. Parallel object-based systems and Petri nets, I and II. Technical Report 90-04-5, Dept. of Computer Science, University of Leiden, February 1990.

[13] P. Freyd. Aspects of topoi. *Bull. Austral. Math. Soc.*, 7:1–76, 1972.

[14] Laurent Fribourg. Oriented equational clauses as a programming language. *Journal of Logic Programming*, 1(2):179–210, 1984.

[15] Kokichi Futatsugi, Joseph Goguen, Jean-Pierre Jouannaud, and José Meseguer. Principles of OBJ2. In Brian Reid, editor, *Proceedings of 12th ACM Symposium on Principles of Programming Languages*, pages 52–66. ACM, 1985.

[16] H. Ganzinger. A completion procedure for conditional equations. In Jean-Pierre Jouannaud and Stephane Kaplan, editors, *Proceedings, Conference on Conditional Term Rewriting, Orsay, France, July 8-10, 1987*, pages 62–83. Springer-Verlag, Lecture Notes in Computer Science No. 308, 1988. Final version will appear in J. Symb. Comp.

[17] Jean-Yves Girard. Towards a geometry of interaction. In J.W. Gray and A. Scedrov, editors, *Proc. AMS Summer Research Conference on Categories in Computer Science and Logic, Boulder, Colorado, June 1987*, pages 69–108. American Mathematical Society, 1989.

[18] Joseph Goguen. How to prove algebraic inductive hypotheses without induction: with applications to the correctness of data type representations. In Wolfgang Bibel and Robert Kowalski, editors, *Proceedings, Fifth Conference on Automated Deduction*, pages 356–373. Springer-Verlag, 1980. Lecture Notes in Computer Science, Volume 87.

[19] Joseph Goguen, Jean-Pierre Jouannaud, and José Meseguer. Operational semantics of order-sorted algebra. In W. Brauer, editor, *Proceedings, 1985 International Conference on Automata, Languages and Programming.* Springer-Verlag, 1985. Lecture Notes in Computer Science, Volume 194.

[20] Joseph Goguen, Claude Kirchner, Hélène Kirchner, Aristide Mégrelis, José Meseguer, and Timothy Winkler. An introduction to OBJ3. In Jean-Pierre Jouannaud and Stephane Kaplan, editors, *Proceedings, Conference on Conditional Term Rewriting, Orsay, France, July 8-10, 1987*, pages 258–263. Springer-Verlag, Lecture Notes in Computer Science No. 308, 1988.

[21] Joseph Goguen, Claude Kirchner, and José Meseguer. Concurrent term rewriting as a model of

computation. In R. Keller and J. Fasel, editors, *Proc. Workshop on Graph Reduction, Santa Fe, New Mexico*, pages 53–93. Springer LNCS 279, 1987.

[22] Joseph Goguen and José Meseguer. Equality, types, modules and generics for logic programming. In S.-A. Tärnlund, editor, *Proc. 2nd Intl. Logic Programming Conf., Uppsala, July 2-6, 1984*, pages 115–125. Uppsala University, 1984.

[23] Joseph Goguen and José Meseguer. Unifying functional, object-oriented and relational programming with logical semantics. In Bruce Shriver and Peter Wegner, editors, *Research Directions in Object-Oriented Programming*, pages 417–477. MIT Press, 1987. Preliminary version in *SIGPLAN Notices*, Volume 21, Number 10, pages 153-162, October 1986; also, Technical Report CSLI-87-93, Center for the Study of Language and Information, Stanford University, March 1987.

[24] Joseph Goguen and José Meseguer. Software for the rewrite rule machine. In *Proceedings of the International Conference on Fifth Generation Computer Systems, Tokyo, Japan*, pages 628–637. ICOT, 1988.

[25] Joseph Goguen and José Meseguer. Order-sorted algebra I: Equational deduction for multiple inheritance, overloading, exceptions and partial operations. Technical Report SRI-CSL-89-10, SRI International, Computer Science Lab, July 1989. Given as lecture at Seminar on Types, Carnegie-Mellon University, June 1983. Submitted for publication.

[26] Joseph Goguen, José Meseguer, Sany Leinwand, Timothy Winkler, and Hitoshi Aida. The rewrite rule machine. Technical Report SRI-CSL-89-6, SRI International, Computer Science Lab, March 1989.

[27] Joseph Goguen, James Thatcher, and Eric Wagner. An initial algebra approach to the specification, correctness and implementation of abstract data types. Technical Report RC 6487, IBM T. J. Watson Research Center, October 1976. Appears in *Current Trends in Programming Methodology, IV*, Raymond Yeh, Ed., Prentice-Hall, 1978, pages 80-149.

[28] Irene Guessarian. *Algebraic Semantics*. Springer-Verlag, 1981. Lecture Notes in Computer Science, Volume 99.

[29] W.H. Hesselink. A mathematical approach to nondeterminism in data types. *ACM Trans. Prog. Lang. and Sys.*, 10:87–117, 1988.

[30] G. Huet. *Formal Structures for Computation and Deduction*. INRIA, 1986.

[31] Gerard Huet. Confluent reductions: Abstract properties and applications to term rewriting systems. *Journal of the Association for Computing Machinery*, 27:797–821, 1980. Preliminary version in 18th Symposium on Mathematical Foundations of Computer Science, 1977.

[32] H. Hussmann. Nondeterministic algebraic specifications and nonconfluent term rewriting. Computer Science Dept., TU Munich.

[33] H. Hussmann. Unification in conditional equational theories. In *Proc. EUROCAL 2*, pages 543–553, 1985.

[34] B. Jayaraman and D. Plaisted. Functional programming with sets. In Kahn, editor, *Proceedings, IFIP Conference on Functional Programming Languages and Computer Architecture*, pages 194–210. Springer-Verlag, Portland, September 1987. Lecture Notes in Computer Science, Volume 274.

[35] J.-P. Jouannaud and B. Waldmann. Reductive conditional term rewriting systems. In *Proceedings of Third IFIP Working Conference on Formal Description of Programming Concepts*. 1986. Ebberup, Denmark.

[36] Jean-Pierre Jouannaud and Stephane Kaplan, editors. *Proceedings, Conference on Conditional Term Rewriting, Orsay, France, July 8-10, 1987*. Springer-Verlag, Lecture Notes in Computer Science No. 308, 1988.

[37] S. Kaplan and J.-L. Rémy. Completion algorithms for conditional rewriting systems. In Maurice Nivat and Hassan Aït-Kaci, editors, *Resolution of Equations in Algebraic Structures*, volume 2, pages 141–170. Academic Press, 1989.

[38] R. Keller and J. Fasel, editors. *Proc. Workshop on graph reduction, Santa Fe, New Mexico*. Springer LNCS 279, 1987.

[39] G.M. Kelly and R. Street. Review of the elements of 2-categories. In G.M. Kelly, editor, *Category Seminar, Sydney 1972/73*. Springer Lecture Notes in Mathematics No. 420, 1974.

[40] Claude Kirchner, Hélène Kirchner, and José Meseguer. Operational semantics of OBJ3. In T. Lepistö

and A. Salomaa, editors, *Proceedings, 15th Intl. Coll. on Automata, Languages and Programming, Tampere, Finland, July 11-15, 1988*, pages 287–301. Springer-Verlag, Lecture Notes in Computer Science No. 317, 1988.

[41] J. Lambek. Subequalizers. *Canadian Math. Bull.*, 13:337–349, 1970.

[42] J.-J. Lévy. Optimal reductions in the lambda calculus. In *To H.B. Curry: Essays on Combinatory Logic, Lambda Calculus and Formalism*. Academic Press, 1980.

[43] Saunders MacLane. *Categories for the working mathematician*. Springer, 1971.

[44] Narciso Martí-Oliet and José Meseguer. An algebraic axiomatization of linear logic models. Technical Report SRI-CSL-89-11, SRI International, Computer Science Lab, December 1989. To appear in G.M. Reed, A.W. Roscoe and R. Wachter (eds.), *Proceedings of the Oxford Symposium on Topology in Computer Science*, Oxford University Press, 1990.

[45] Narciso Martí-Oliet and José Meseguer. From Petri nets to linear logic. In D.H. Pitt et al., editor, *Category Theory and Computer Science*, pages 313–340. Springer Lecture Notes in Computer Science, Vol. 389, 1989. Full version to appear in *Mathematical Structures in Computer Science.*

[46] J. Meseguer. Varieties of chain-complete algebras. *Journal of Pure and Applied Algebra*, 19:347–383, 1980.

[47] J. Meseguer. A Birkhoff-like theorem for algebraic classes of interpretations of program schemes. In *Formalization of Programming Concepts*, pages 152–168. Springer-Verlag, 1981. Lecture Notes in Computer Science, Volume 107.

[48] José Meseguer. General logics. In H.-D. Ebbinghaus et al., editor, *Logic Colloquium'87*, pages 275–329. North-Holland, 1989.

[49] José Meseguer. A logical theory of concurrent objects. In *ECOOP-OOPSLA'90 Conference on Object-Oriented Programming, Ottawa, Canada, October 1990*, pages 101–115. ACM, 1990.

[50] José Meseguer. Rewriting as a unified model of concurrency. In *Proceedings of the Concur'90 Conference, Amsterdam, August 1990*, pages 384–400. Springer LNCS Vol. 458, 1990.

[51] José Meseguer. Rewriting as a unified model of concurrency. Technical Report SRI-CSL-90-02, SRI International, Computer Science Laboratory, February 1990. Submitted for publication; revised June 1990.

[52] José Meseguer and Joseph Goguen. Initiality, induction and computability. In Maurice Nivat and John Reynolds, editors, *Algebraic Methods in Semantics*, pages 459–541. Cambridge University Press, 1985.

[53] José Meseguer and Ugo Montanari. Petri nets are monoids: A new algebraic foundation for net theory. In *Proc. LICS'88*, pages 155–164. IEEE, 1988.

[54] José Meseguer and Ugo Montanari. Petri nets are monoids. *Info and Co*, 88:105–155, 1990. Appeared as SRI Tech Report SRI-CSL-88-3, January 1988.

[55] Robin Milner. *Communication and Concurrency.* Prentice Hall, 1989.

[56] Peter Mosses. Unified algebras and institutions. Technical Report DAIMI PB-274, Computer Science Department, Aarhus University, 1989.

[57] Michael J. O'Donnell. Survey of the equational logic programming project. In Maurice Nivat and Hassan Aït-Kaci, editors, *Resolution of Equations in Algebraic Structures.* MCC Corporation, 1987. Preliminary Proceedings.

[58] A. Pitts. An elementary calculus of approximations. Unpublished manuscript, University of Sussex, December 1987.

[59] A.J. Power. An abstract formulation of rewrite systems. In D.H. Pitt et al., editor, *Category Theory and Computer Science*, pages 300–312. Springer Lecture Notes in Computer Science, Vol. 389, 1989.

[60] H. Reichel. *Initial Computability, Algebraic Specifications, and Partial Algebras.* Oxford University Press, 1987.

[61] D.E. Rydeheard and J.G. Stell. Foundations of equational deduction: A categorical treatment of equational proofs and unification algorithms. In *Proceedings of the Summer Conference on Category Theory and Computer Science, Edinburgh, Sept. 1987.* Springer LNCS 283, 1987.

[62] R. Seely. Modelling computations: a 2-categorical framework. In D. Gries, editor, *Second IEEE Symposium on Logic in Computer Science*, pages 65–71. IEEE Computer Society, June 1987.

Equivalences of Rewrite Programs

Chilukuri K. Mohan
School of Computer and Information Science
Syracuse University, N.Y. 13244-4100, U.S.A.

ABSTRACT: In this paper, we formulate syntactic, operational, and semantic equivalence concepts for rewrite programs in various formalisms, and study their properties and inter-relations.

1 Introduction

For standard notation and terminology, see [3]. Syntactic identity among terms p, q is denoted by "$p \equiv q$". We denote by "$M[s]$" that s is a subterm of M. "$M[s/t]$" denotes M with one or more occurences of subterm s in it replaced by occurences of another term t. "$p \rightarrow_R q$" denotes that there exists a reduction (in one step) from the term p to the term q using the rewrite program R. Programs written in the following (and other) conditional and priority rewriting formalisms are jointly referred to as "rewrite programs".

- In Equational-CTRS (E-CTRS) the antecedent of each rule is a conjunction of equations [7, 6].
- In Equational-Inequational-CTRS (EI-CTRS), the antecedent of each rule is a conjunction of equations and inequations (negated equality literals, *e.g.*, $p \neq q$) [11, 8].
- In Priority rewriting systems (abbreviated "P-x" where x is any rewrite formalism, "PRS" for P-TRS), a "priority" partial ordering is imposed on the rules, determining which of many competing rules may be used to reduce a given term [1, 9, 13].

2 Syntactic Equivalence

To be considered syntactic, an equivalence relation among rewrite programs must be determined without actually executing any of the rewrite rules; to be useful, it must be a sufficient condition for most notions of operational equivalence. The latter requirement rules out a relation like *"have the same number of variables"* which is syntactic but is not useful from the viewpoint of actual rewriting mechanisms.

Definition 1 : A rewrite rule $\lambda_j \rightarrow \rho_j$ subsumes another rule $\lambda_i \rightarrow \rho_i$ iff $\rho_i \equiv \lambda_i[\lambda_j\sigma/\rho_j\sigma]$ for some substitution σ. A TRS P_2 subsumes another TRS P_1 iff each rule in P_1 is subsumed by some rule in P_2.

Example 1 The rule $f(x) \rightarrow x$ subsumes $g(f(a)) \rightarrow g(a)$.
The TRS $\{f(x) \rightarrow x\}$ subsumes $\{g(f(a)) \rightarrow g(a),\ g(f(b)) \rightarrow g(b)\}$.

A similar notion of subsumption is defined (in the full paper) for prioritized and conditional rewrite rules and systems, taking into account the antecedents of rules and the priority ordering among rules.

Definition 2 Two rewrite programs are subsumption-equivalent iff they subsume each other.

3 Operational Equivalence

In this section, we discuss various notions of operational equivalences among rewrite programs. Their mutual relationships and relative strength are summarized in the conclusions (section 5).

Definition 3 Let P_1 and P_2 be two rewrite programs defining the one-step rewrite relations $\rightarrow_{P_1}$ and $\rightarrow_{P_2}$ respectively. P_1 and P_2 are said to be $\rightarrow$-equivalent iff $\rightarrow_{P_1}$ and $\rightarrow_{P_2}$ are the same relation.

In most cases, subsumption-equivalence implies $\rightarrow$-equivalence. This result is not straightforward for conditional and prioritized rewriting systems, and needs a complicated proof, since subsumption does not imply inclusion of rewrites. In other words, it is possible that P_2 subsumes P_1 and $p \rightarrow_{P_1} q$, but not $p \rightarrow_{P_2} q$.

Example 2 Let P_1 be the EI-CTRS $\{x = y \Rightarrow f(x,y) \rightarrow b,\ x \neq y \Rightarrow f(x,y) \rightarrow a\}$. Let P_2 be $P_1 \cup \{c \rightarrow d\}$. Clearly, P_2 subsumes P_1, since $P_1 \subset P_2$. But we have $f(c,d) \rightarrow_{P_1} a$, in contrast to $f(c,d) \rightarrow_{P_2} b$.

Theorem 1 : Two TRS's are $\rightarrow$-equivalent iff they are subsumption-equivalent. Hence $\rightarrow$-equivalence of TRS's is decidable.

This result does not hold for conditional and prioritized rewrite systems. With some CTRS's, the attempt to perform even one step of reduction does not always terminate. In such cases, we say the rewrite relation itself is not decidable, and there are two possible ways of defining $\rightarrow$-equivalence, which differ in whether they consider only "finite success" cases in judging equivalence. For instance, different answers are possible when asked whether $\{p \rightarrow q\}$ and $\{p \rightarrow q,\ \ m = n \Rightarrow m \rightarrow n\}$ are $\rightarrow$-equivalent. The approach we adopt does consider the two above programs to be $\rightarrow$-equivalent, since the two rewrite relations are identical. But the opposite approach is also feasible, since m is known to be irreducible using the first program but the second program loops on this input without positively answering that m is irreducible. A variant of the latter approach is used later in formalizing the notion of normal-form equivalence.

Theorem 2 *(a)* There is an E-CTRS R such that $\rightarrow_R$ is undecidable.
Hence: *(b)* Checking $\rightarrow$-equivalence of E-CTRS's is not decidable.

However, a property generalizing the noetherian property assures decidability of the $\rightarrow_R$-relation for E-CTRS's. Essentially, if each rule in an E-CTRS satisfies the property that all terms in the antecedent and *rhs* are smaller than the *lhs* in some well-founded stable monotonic term ordering, termination of every rewrite sequence is assured, implying that evaluation of instances of the antecedent always terminates [7].

Theorem 3 *(a)* There is a PRS R such that $\rightarrow_R$ is undecidable.
Hence: *(b)* Checking $\rightarrow$-equivalence of PRS's is not decidable.

Definition 4 If P is a rewrite program, then $\rightarrow_P^*$ is the reflexive transitive closure of $\rightarrow_P$. Two programs P_1 and P_2 are $\rightarrow^*$-equivalent iff $\rightarrow_{P_1}^*$ and $\rightarrow_{P_2}^*$ are identical relations.

Theorem 4 Checking $\rightarrow^*$-equivalence of TRS's is semi-decidable, but not decidable.

The root of the problem may be seen in the undecidability of termination of rewriting sequences. In fact, if two TRS's satisfy the 'noetherian' property, so that every rewrite sequence is of finite length, then their $\rightarrow^*$-equivalence becomes decidable.

Lemma 1 If R_1 is a TRS, and R_2 is a noetherian TRS, then it is decidable whether: $\forall p,q.(p \rightarrow^*_{R_1} q) \Rightarrow (p \rightarrow^*_{R_2} q)$.

Theorem 5 Checking $\rightarrow^*$-equivalence of noetherian TRS's is decidable.

Definition 5 "$p \downarrow_R q$" denotes that there exist converging reduction sequences from terms p and q using the rewrite program R. Two rewrite programs R_1 and R_2 are convergence-equivalent (or $\downarrow$-equivalent) <u>iff</u> for every pair of terms p and q, we have $[p \downarrow_{R_1} q$ iff $p \downarrow_{R_2} q]$.

Note that (unlike the $\rightarrow^*$ relation) convergence is not transitive in general; *e.g.*, we have $p \downarrow_R q$ and $q \downarrow_R r$ but not $p \downarrow_R r$ where R is the TRS $\{p \rightarrow c,\ q \rightarrow c,\ q \rightarrow d,\ r \rightarrow d\}$.

Theorem 6 $\rightarrow$-equivalence of rewrite programs implies $\rightarrow^*$-equivalence, and $\rightarrow^*$-equivalence implies $\downarrow$-equivalence (but not conversely).

For every rewrite program P, we define a "normal-form" relation $\mathcal{N}_P$ (or $\triangleright_P$), such that $\mathcal{N}_P(p,q)$ (or $p \triangleright_P q$) whenever $p \rightarrow^*_P q$ and q has decidably no reduct. Note that the latter condition also implies that every attempt to reduce q terminates with failure (cf., "finite failure" in logic programming); this is particularly relevant to conditional rewriting systems where it is possible that the evaluation of the antecedent of a rule may not terminate. A term may have multiple normal forms, when the program is not confluent. In general, some rewrite sequences may not terminate, and some terms may not have normal forms. When rewriting is decidable <u>and</u> all rewriting sequences are finite, $\triangleright_P$ is decidable and every term has a finite set of normal forms.

Definition 6 : Two rewrite programs P_1, P_2 are <u>n-equivalent</u> iff they compute the same normal-form relation, *i.e.*, $\mathcal{N}_{P_1} \equiv \mathcal{N}_{P_2}$, *i.e.*, $\forall p,q.p \triangleright_{P_1} q$ iff $p \triangleright_{P_2} q$.

For example, $\{a \rightarrow a;\ a \rightarrow b\}$ is n-equivalent to $\{a \rightarrow b\}$, but not $\{a \rightarrow a\}$. If two programs are confluent and all rewriting sequences finitely terminate, then all terms have unique normal forms, hence checking normal-form equivalence is even easier; it is then adequate to pursue any one R_2-rewrite sequence from each *lhs* of each rule in R_1, instead of having to explore all possible rewrite sequences.

The following examples illustrate that $\downarrow$-equivalence does not imply n-equivalence (even when the programs are confluent noetherian TRS's), nor does the converse hold in general.

Example 3 $\{p \rightarrow q,\ \ q \rightarrow r\}$ and $\{p \rightarrow r,\ \ r \rightarrow q\}$ are $\downarrow$-equivalent, but not n-equivalent; the normal forms of p, using the two programs, are r and q respectively.

Example 4 $\{p \rightarrow t,\ \ q \rightarrow r,\ \ r \rightarrow q,\ \ p \rightarrow q\}$ and $\{p \rightarrow t,\ \ q \rightarrow r,\ \ r \rightarrow q\}$ are n-equivalent but not $\downarrow$-equivalent, since p and t are the only terms which have a normal form t in both programs, although we have $p \downarrow q$ using the first program but not the second.

Theorem 7 : <u>If</u> R_1, R_2 are n-equivalent rewrite programs, every rewrite sequence using R_1, R_2 finitely terminates, and $\rightarrow_{R_1}, \rightarrow_{R_2}$ are decidable relations, <u>then</u> R_1, R_2 are $\downarrow$-equivalent.

Theorem 8 : <u>If</u> two rewrite programs R_1 and R_2 are $\rightarrow^*$-equivalent, and if the rewrite relations $\rightarrow_{R_1}, \rightarrow_{R_2}$ are decidable, and if every rewrite sequence (using R_1, R_2) finitely terminates, <u>then</u> R_1, R_2 are n-equivalent.

The converse does not hold, as shown by the following example:

Example 5 $\{a \rightarrow c,\ b \rightarrow c\}$ and $\{a \rightarrow b,\ b \rightarrow c\}$ are n-equivalent, since rewriting sequences from a, b, c all converge to the normal form c using both rewrite programs; but the programs are not $\rightarrow^*$-equivalent, since a cannot rewrite in any number of steps to b using the former rewrite program, unlike the latter.

The following examples illustrate the need for the decidability and termination conditions required in the above theorem.

Example 6 $\{p \rightarrow q\}$ and $\{p \rightarrow q,\ q \rightarrow q\}$ are $\rightarrow^*$-equivalent, but p has the normal form q in only the former program, because of non-terminating reduction sequences in the latter program.

Example 7 $\{p \rightarrow q\}$ and the E-CTRS $\{p \rightarrow q,\ [q = r \Rightarrow q \rightarrow r]\}$ are $\rightarrow^*$-equivalent, but p has the normal form q in only the former program, because the rewrite relation of the latter program is not decidable; an attempt to reduce q does not terminate, by the operational mechanism of conditional rewriting.

In theorem 8 above, the termination requirement is not needed if two rewrite programs are $\rightarrow$-equivalent. However, the decidability restriction is still needed, illustrated by example 7 above. The normal form of q is q itself using the first program; but q does not have a normal form using the second program, although the two programs are $\rightarrow$-equivalent.

Theorem 9 If R_1 and R_2 are $\rightarrow$-equivalent rewrite programs such that $\rightarrow_{R_1}$ and $\rightarrow_{R_2}$ are decidable, then R_1 and R_2 are n-equivalent.

Corollary 1 If R_1 and R_2 are $\rightarrow$-equivalent TRS's, then R_1 and R_2 are n-equivalent.

The following example disproves the converse:

Example 8 $\{a \rightarrow a,\ a \rightarrow b\}$ and $\{a \rightarrow b\}$ are n-equivalent, but not $\rightarrow$-equivalent, since a cannot rewrite in one step to a using the latter rewrite program.

4 Semantic Equivalence

The equational formulas denoted by prioritized systems and CTRS's with negation are complex [11, 12, 13]. In general, conditional and prioritized rewrite specifications assume an underlying set of implicit "default" inequations (Ψ). For example, in using the EI-CTRS $\{a \neq b \Rightarrow c \rightarrow d\}$ to reduce the term c to d, there is an implicit assumption $a \neq b$, which is not a consequence of the equational theory of $\{a \neq b \Rightarrow c = d\}$. The equational theory of $\mathcal{E}(R) \cup \Psi$ describes the meaning of such a CTRS, where $\mathcal{E}(R)$ is obtained by treating the (conditional/prioritized) rewrite rules as conditional equations, and Ψ is some set of formulas containing negated equality literals.

Definition 7 Two TRS's or E-CTRS's R, R' are <u>semantically equivalent</u> (or "=-equivalent") iff $\forall p, \forall q.\ [\mathcal{E}(R) \models p = q] \Leftrightarrow [\mathcal{E}(R') \models p = q]$. Two conditional and/or prioritized rewriting systems R, R' are <u>semantically equivalent with respect to Ψ</u> (or "$=_\Psi$-equivalent") iff $\forall p, \forall q.\ [\mathcal{E}(R) \cup \Psi \models p = q] \Leftrightarrow [\mathcal{E}(R') \cup \Psi \models p = q]$, where Ψ is a set of formulas.

We now relate this notion of semantic equivalence to the notions of operational equivalence. "$\leftrightarrow_R^*$" denotes the reflexive symmetric transitive closure of $\rightarrow_R$. From Birkhoff's result on the completeness of equational inference rules [2], we have:

$$\forall p, q. \mathcal{E}(R) \models p = q \Leftrightarrow p \leftrightarrow_R^* q, \quad \text{for any TRS } R. \qquad (1)$$

Hence, if two TRS's are $\rightarrow^*$-equivalent, then they are semantically equivalent; but the converse does not hold. We have also proved the following stronger result:

Theorem 10 If two TRS's are $\downarrow$-equivalent, then they are semantically equivalent.

Corollary 2 If two TRS's are noetherian and n-equivalent, then they are semantically equivalent.

However, non-noetherian n-equivalent TRS's need not be semantically equivalent, as shown by the following example.

Example 9 The TRS's R_1 defined as $\{a \rightarrow b, \;\; c \rightarrow c, \;\; d \rightarrow d\}$ and R_2 defined by $\{a \rightarrow b, \;\; c \rightarrow d, \;\; d \rightarrow c\}$ are n-equivalent, but not semantically equivalent. $c = d$ is a logical consequence of $\mathcal{E}(R_2)$ but not $\mathcal{E}(R_1)$.

The above results pertain to TRS's, and it is natural to ask whether similar results can also be obtained for more general rewrite formalisms. The crucial step in the proofs of the above theorems is equation 1. Similar results would also follow for the more complicated conditional and prioritized rewriting systems, whenever a property like equation (1) is applicable.

Theorem 11 Let R_1, R_2 be $\downarrow$-equivalent rewrite programs in some formalism with semantics parameterized by some set of formulas Ψ, each satisfying the property that $\forall p, q. \mathcal{E}(R_i) \cup \Psi \models p = q \Leftrightarrow p \leftrightarrow^*_{R_i} q$. Then R_1 and R_2 are semantically equivalent with respect to Ψ.

When a conditional rewrite program is non-confluent, the analog of equation 1 and the above theorem may not hold, since equations in the antecedents of conditional rules are verified by checking convergence of the reduction sequences issuing from the equated terms.

Example 10 If R is the E-CTRS $\{a \rightarrow b, \;\; a \rightarrow c, \;\; b = c \Rightarrow s \rightarrow t\}$, then $\mathcal{E}(R)$ is the set of formulas $\{a = b, \;\; a = c, \;\; b = c \Rightarrow s = t\}$. We clearly have $\mathcal{E}(R) \models s = t$, but not $s \leftrightarrow^*_R t$, since b and c do not have converging reduction sequences.

This may be considered to be a fault of the conditional rewriting formulation; however convergence is the natural rewriting mechanism to check equality of terms, and such a definition of conditional rewriting has become the accepted standard technique. Confluence of E-CTRS's guarantees the analogue of equation 1, as shown in [7]; hence the following result (with empty Ψ).

Corollary 3 Confluent and $\downarrow$-equivalent E-CTRS's are semantically equivalent.

On the other hand, if the conditional rewriting mechanism for E-CTRS's is operationally defined such that each equation $m = n$ in the antecedent of a conditional rule in R_i is to be verified by checking whether $m \leftrightarrow^*_{R_i} n$, and this mechanism is assumed in recursively defining $\rightarrow_{R_i}$, then we do have the property that $\forall p, q. \mathcal{E}(R_i) \models p = q \Leftrightarrow p \leftrightarrow^*_{R_i} q$ (shown in [7]); hence the above theorem can again be applied (with empty Ψ) without the confluence criterion.

5 Conclusions

We have formulated syntactic, operational and semantic notions of equivalence of rewrite programs. We have studied their decidability, and the relations between these notions for different rewrite formalisms. The operational notions of equivalence can be used to compare rewrite formalisms.

The following diagram illustrates the relations between various notions of equivalence; we have proved these results, and also given counterexamples to illustrate that most of the implications cannot be reversed, *i.e.*, there are no missing non-redundant edges in the graph. Attached to some edges are conditions ("d": decidable, "t": terminating) that must be satisfied for the implication to hold.

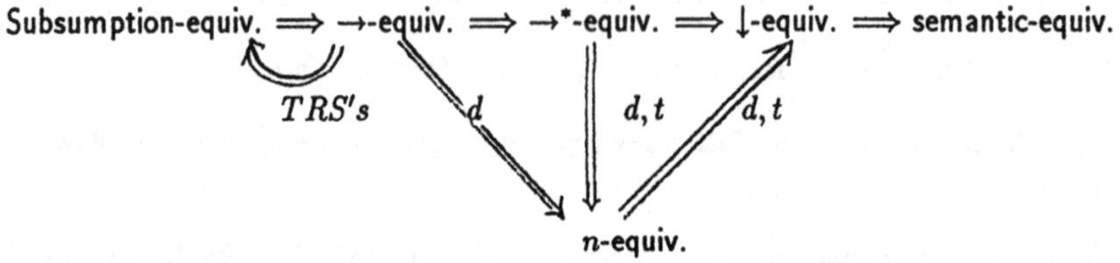

Some other equivalence relations among rewrite programs are also interesting. We may ask whether two programs are equivalent on inputs which satisfy specific restrictions, or whether they behave identically on terms from a common sub-language. In some situations, when the primary interest is in manipulating ground terms, it makes sense to compare programs with respect to ground rewrite sequences alone. Sometimes, we need to compare structurally similar programs defined over different languages. These aspects of equivalence may be conjoined to the other operational notions of equivalence defined earlier.

References

[1] J.C.M.Baeten, J.A.Bergstra, J.W.Klop, *Term Rewriting Systems with Priorities,* Proc. 2nd Conf. Rewriting Techniques and Applications, Springer-Verlag LNCS 256, 1987.

[2] G.Birkhoff, *On the Structure of Abstract Algebras,* Proc. Cambridge Philosophical Society 31, pp.433-454, 1935.

[3] N.Dershowitz, J.-P.Jouannaud, *Rewrite Systems,* (to appear) in Handbook of Theoretical Computer Science, 1990.

[4] J.-P.Jouannaud, B.Waldmann, *Reductive Conditional Term Rewriting Systems,* Proc. Third TC2 Working Conf. on the Formal Description of Programming Concepts, Ebberup (Denmark), Aug. 1986.

[5] S. Kaplan, *Simplifying Conditional Term Rewriting Systems: Unification, Termination and Confluence,* Rapport de Recherche no. 194, Universite de Paris-Sud, Nov. 1984.

[6] S.Kaplan, *Positive/Negative Conditional Rewriting,* Proc. First Int'l. Workshop on Conditional Term Rewriting Systems, Paris, Springer-Verlag LNCS 308, 1987.

[7] A.Laville, *Comparison of Priority Rules in Pattern Matching and Term Rewriting,* Technical Report, INRIA, Le Chesnay (France), 1988.

[8] M.J.Maher, *Equivalences of Logic Programs,* in Deductive Databases and Logic Programming, (ed.) J.Minker, Morgan Kaufmann Publishers, 1988.

[9] C.K.Mohan, M.K.Srivas, *Conditional Specifications with Inequational Assumptions,* Proc. First Int'l. Workshop on Conditional Term Rewriting Systems, Paris, Springer-Verlag LNCS 308, 1987.

[10] C.K.Mohan, M.K.Srivas, *Negation with Logical Variables in Conditional Rewriting,* Proc. 3rd Conf. on Rewriting Techniques and Applications, Chapel Hill (N.C.), 1989.

[11] C.K.Mohan, *Priority Rewriting: Semantics, Confluence and Conditionals,* Proc. 3rd Conf. on Rewriting Techniques and Applications, Chapel Hill (N.C.), 1989.

Chapter 2

Infinite Terms, Non-Terminating Systems, and Termination

On Finite Representations of Infinite Sequences of Terms

Hong Chen[1], Jieh Hsiang[1]
Department of Computer Science
SUNY at Stony Brook
Stony Brook, NY 11794

Hwa-Chung Kong
Department of Computer Science
National Taiwan University
Taipei, Taiwan, R.O.C.

Abstract

In this paper we introduce a notion of recurrence-terms for finitely representing infinite sequences of terms. A recurrence-term utilizes the structural similarities among terms and expresses them explicitly using recurrence relations. Its formalism is natural and simple, and based on which algebraic operations such as unification, matching, and reductions can be defined. Recurrence-rewrite rules, defined respectively, also yield finite representation of certain divergent term rewriting systems.

Recurrence-rules do not only play a passive role in detecting divergence, they can also be incorporated as part of the completion process. In addition to giving the formalism, we present methods of inferring recurrence-terms from finite sets of regular terms, and a matching algorithm between a recurrence-term and a regular term. Recurrence-term rewriting systems are also defined, and we prove the equivalence between a recurrence-system and the (infinite) term rewriting system it schematizes, as well as the preservation of desirable properties such as termination and confluence.

1 Introduction

In computer science one frequently encounters infinite sequences of objects during computations. These infinite sequences often lead to non-termination of computation. For example, given a finite set of equations, an infinite set of rules may be generated during Knuth-Bendix completion process, and causes the completion process to run forever.

The papers [Her 89] and [Mop 89] investigated certain patterns of rules which may lead to generating infinite sequences of rules during the Knuth-Bendix completion process. If such patterns are detected, equations are re-oriented and the Knuth-Bendix completion is executed again. However, re-orientation cannot always be done and they did not suggest other useful solutions beyond that.

Formulating an infinite sequence of rules using a meta-representation was first studied in [Kir 87], where infinitely many rules are expressed via a meta-rule with meta-variables ranging over infinite sets of terms. This approach has several problems: First, no systematic way of deriving meta-rules was given. The derivation can only be done through user observation. Second, the meta-rule method shifted the difficulty of representing an infinite sequence of rules to representing an infinite set of terms on which the meta-variable is defined. This merely moved the problem of finitely representing infinite objects from one level to another, while no attempt was made to represent this infinite set finitely. In [HeK 90], it was suggested to interpret meta-rules by order-sorted rewriting; however, as pointed out in [Com 89], sorts are regular-tree languages and are insufficient to express the domains of meta-variables.

A method based on primitive recursive auxiliary operators was proposed in [LaJ 89]. The expressive power of this method is quite limited. It can only represent a sequence in which the number of variables in the terms does not increase with the sequence. This is hardly the case with divergent term rewriting systems.

[1]This work was partially supported by NSF grants INT-8715231, CCR-8805734, and CCR-8901322

The idea of using recurrence relation to schematize an infinite sequence of terms was first proposed by Kong[Kon 89] who called it the *hyper-term method*. Kong's method, however, is incomplete and contains several incorrect definitions and conclusions. In this paper, we rename it *recurrence-terms* and define its semantics differently. We show how recurrence-terms can be deduced automatically from a (finite or infinite) sequence of terms. Comparing to other methods, recurrence-terms allow us to express the infinite sequence of terms explicitly and, consequently, enable us to define algebraic operations on them naturally. We emphasize that a recurrence-term is an abstraction of an infinite sequence or a set of *finite* terms. It should not be confused with infinite terms such as used in Prolog II[Col 82].

The paper is organized as follows: Some preliminary definitions are given in Section 2. In section 3, we define the semantics of recurrence-terms. Mechanical ways of constructing recurrence-terms are presented in Section 4. An algorithm for matching a recurrence-term with a first order term is described in Section 5, with completeness, correctness, and termination proven. In section 6 we define recurrence-rewriting systems and show that they preserve various properties, such as confluence and termination, of the term rewriting systems which they schematize.

2 Definitions

Let $\mathcal{V}$ denote a countable set of variables and $\mathcal{F}$ denote a finite set of function symbols. $\mathcal{T}(\mathcal{F},\mathcal{V})$ is the set of terms freely generated from $\mathcal{F}$ and $\mathcal{V}$. A *substitution* is a function mapping $\delta : \mathcal{V} \to \mathcal{T}(\mathcal{F},\mathcal{V})$ with $\delta(x) \neq x$ for only a finite subset (denoted by $Dom(\delta)$) of $\mathcal{V}$, and $Dom(\delta) \cap \mathcal{V}(\{\delta(x) : x \in Dom(\delta)\}) = \emptyset$. The result of applying δ to a term t, denoted by $t\delta$, is the result of replacing every variable x of t by $\delta(x)$. Without loss of generality, we denote a substitution δ by a finite set $\{\delta(x)/x : x \in Dom(\delta)\}$. A substitution δ is a *ground substitution* if for all $x \in Dom(\delta)$, $\delta(x)$ is ground. A substitution δ is an *empty substitution* if $Dom(\delta) = \emptyset$. Throughout the paper, we reserve ϕ as the empty substitution. A substitution δ is a *variable renaming substitution* if it is a one to one function from $\mathcal{V}$ to $\mathcal{V}$.

The composition $\sigma \circ \delta$ of two substitutions $\sigma : \mathcal{V} \to \mathcal{T}(\mathcal{F},\mathcal{V})$, $\delta : \mathcal{V} \to \mathcal{T}(\mathcal{F},\mathcal{V})$, is the substitution λ such that

$$\lambda(x) = \begin{cases} \delta(x) & \text{if } x \in Dom(\delta) - Dom(\sigma) \\ \delta(\sigma(x)) & \text{if } x \in Dom(\sigma) \\ x & \text{otherwise} \end{cases}$$

A term t has a set of position $P(t)$ defined as:

1. $\epsilon \in P(t)$
2. $i.p_i \in P(t)$ if $t = f(t_1, ..., t_i, ..., t_n)$ and $p_i \in P(t_i)$, where . is concatenation.

We use t/p to denote the subterm at position p of t and $t[p \leftarrow s]$ to denote the term after replacing t/p by s.

A binary relation $\trianglelefteq^h$ on $\mathcal{T}(\mathcal{F})$ is a *homeomorphic embedding relation*[Kru 72] $\trianglelefteq^h$ if:

$$s = f(s_1, ..., s_n) \trianglelefteq^h g(t_1, ..., t_m) = t$$

if and only if

- $f = g$ and $s_i \trianglelefteq^h t_{j_i}, \forall i, 1 \leq i \leq n$, where $1 \leq j_1 < j_2 < ... < j_n \leq m$, or
- $s \trianglelefteq^h t_j$ for some $j, 1 \leq j \leq m$.

Example : $f(a,b) \trianglelefteq^h g(f(g(a),b))$, $f(a,b) \trianglelefteq^h f(a,b)$, $f(a,b) \not\trianglelefteq^h f(g(b),a)$

The extension of $\trianglelefteq^h$ to $\mathcal{T}(\mathcal{F},\mathcal{V})$, called *homeomorphic variable embedding relation* $\trianglelefteq^{hv}$, is defined as:

$$s \trianglelefteq^{hv} t \text{ if and only if } s \trianglelefteq^h t$$

when the variables in $\mathcal{V}(s)$ and $\mathcal{V}(t)$ are assumed to be new constants, and $\trianglelefteq^h$ is defined on $\mathcal{T}(\mathcal{F} \cup \mathcal{V}(s) \cup \mathcal{V}(t))$.

Example: $f(x) \trianglelefteq^{hv} f(f(x))$, but $f(x) \not\trianglelefteq^{hv} f(f(y))$.

If σ is a variable renaming substitution such that $s\sigma \trianglelefteq^{hv} t$, then we call σ an *hv-substitution from s to t*. A binary relation $\trianglelefteq^{vr}$ on $\mathcal{T}(\mathcal{F},\mathcal{V})$ is a *variable renaming embedding* relation if it satisfies: $s \trianglelefteq^{vr} t$ if and only if there exists an *hv*-substitution σ of s and t such that $s\sigma \trianglelefteq^{hv} t$.

Example: $f(x_1, y_1) \trianglelefteq^{vr} f(g(x), g(y))$, but $f(x,y) \not\trianglelefteq^{vr} f(g(x), g(x))$ because $f(x,y)$ has more distinguished variables than $f(g(x), g(x))$, and no variable renaming substitution from $\mathcal{V}(f(x,y))$ to $\mathcal{V}(f(g(x), g(x)))$ exists.

Proposition 1 *Given s and t, $s \trianglelefteq^h t$, $s \trianglelefteq^{hv} t$ and $s \trianglelefteq^{vr} t$ are decidable.*

3 Recurrence-terms

In Knuth-Bendix completion, rules in a divergent sequence of rules usually enjoys a certain repetitive pattern along the sequence. For instance, completing a one rule system $\{f(g(f(x))) \to g(f(x))\}$ generates a divergent sequence of rules $\{f(g^i(f(x))) \to g^i(f(x))\}_{i=1}^{\infty}$ where where each $i+1^{st}$ rule has an extra $g(\_)$, a recurring pattern, comparing to the ith rule. The goal of recurrence-terms is to extract those recurring patterns in an infinite sequence and schematizes the sequence finitely.

3.1 Syntax

Let $\mathcal{F}$ be a finite set of function symbols, and Φ be a special symbol of arity 3. Let $\mathcal{V}_\Phi$ be a set of variables over natural numbers, which we call *degree variables.* Normally we use the lowercase letters to denote variables and the uppercase ones to denote degree variables. For convenience, we still refer to the first order terms in $\mathcal{T}(\mathcal{F},\mathcal{V})$ as terms. The set of *recurrence-terms* is defined as the inductive closure of the following:

1. Every variable in $\mathcal{V}$ is a recurrence-term.
2. If $f \in \mathcal{F}$ is of arity n and $t_1, ..., t_n$ are recurrence-terms, then $f(t_1, ..., t_n)$ is a recurrence-term.
3. If h is a term, p is a non-root position of h, N is a degree variable or a natural number, and l is a recurrence-term, then $\Phi(h[p \leftarrow \diamond], N, l)$ is a recurrence-term, where $\diamond$ is a special symbol serving as a place holder.

We call $\Phi(h[p \leftarrow \diamond], N, l)$, also abbreviated as $\Phi_N(h[p \leftarrow \diamond], l)$, a *generator*, or an N-generator when N is emphasized. The first argument $h[p \leftarrow \diamond]$ is called an *N-pattern*. All the variables occurring in the N-patterns of a recurrence-term, denoted as $\mathcal{V}_{H,N}$, are local with respect to the degree variable N and share the same scope. For example, $h(\Phi_N(f(x,\diamond), \Phi_M(f(x,\diamond), x)), \Phi_N(f(x,\diamond), x))$ is the same as $h(\Phi_N(f(x_1,\diamond), \Phi_M(f(x_2,\diamond), x)), \Phi_N(f(x_1,\diamond), x))$.

3.2 Semantics

The degree variables can be instantiated into natural numbers. If all of the degree variables of a recurrence-term are instantiated, it is said to be *fully instantiated*, otherwise it is *partially instantiated.* We use $H[N_1 \leftarrow n_1 ..., N_m \leftarrow n_m]$ to denote the instantiation of degree variables N_1, ..., N_m for a recurrence-term H. A *proper recurrence-term* is a recurrence-term with at least one uninstantiated degree variable. Intuitively, a proper recurrence-term is an abstraction of an infinite sequence of terms, where each term in the sequence can be obtained by fully instantiating its degree variables and unfolding it by repeated *level unfolding.*

We define level unfolding informally in this section. The formal (procedural) definition is given in the Appendix.

In the case where a recurrence-term H constains at most one N-generator for each particular degree variable N, one *level unfolding* of a generator term can be defined inductively:

$$\Phi(h[p\leftarrow\diamond], n+1, l) \stackrel{def}{\equiv} h\sigma[p\leftarrow\Phi(h[p\leftarrow\diamond], n, l)]$$
$$\Phi(h[p\leftarrow\diamond], 0, l) \stackrel{def}{\equiv} l$$

where σ renames the variables in $h[p\leftarrow\diamond]$ into new variables. As we can see, the N-head $h[p\leftarrow\diamond]$ specifies the repeated pattern governed by degree variable N.

Example: In the following, x_1 and x_2 are new variables introduced to rename x.

$$\begin{aligned} g(\Phi(f(x,\diamond),2,a)) &= g(f(x_1,\Phi(f(x,\diamond),1,a))) \\ &= g(f(x_1,f(x_2,\Phi(f(x,\diamond),0,a)))) \\ &= g(f(x_1,f(x_2,a))) \end{aligned}$$

In general, a recurrence-term H may contain several N-generators for a specific N, and the first arguments of those N-generators may share common variables. To ensure that the shared variables in N-patterns are renamed consistently, all the N-generators must be unfolded simultaneously.

An *unfolding renaming* for a set V of variables is a variable renaming substitution from V to a set of new variables which do not appear anywhere in the context of discussion. Let N be a degree variable, applying one *level unfolding* of N-generators in a recurrence-term H is to replace each N-generator $\Phi(h[p\leftarrow\diamond], n, l)$ in H by $h\sigma[p\leftarrow\Phi(h[p\leftarrow\diamond], n-1, l')]$ if $n > 0$, where σ is an unfolding renaming for $\mathcal{V}_{H,N}$, and l' is obtained from l by recursively applying the same level unfolding (with the same σ) to l. If l does not contain N-generators, l' is simply l. If $n = 0$, then each $\Phi(h[p\leftarrow\diamond], n, l)$ of H is recursively replaced by l'.

Example: In the following, the unfolding renamings for $\{x\}$ are $\sigma_1 = \{x\leftarrow x_1\}$ and $\sigma_2 = \{x\leftarrow x_2\}$ respectively.

$$\begin{aligned} h(\Phi(f(x,\diamond),N,\Phi(g(x,\diamond),N,b)))[N\leftarrow 2] &= h(\Phi(f(x,\diamond),2,\Phi(g(x,\diamond),2,b))) \\ &= h(f(x_1,\Phi(f(x,\diamond),1,g(x_1,\Phi(g(x,\diamond),1,b))))) \\ &= h(f(x_1,f(x_2,\Phi(f(x,\diamond),0,g(x_1,g(x_2,\Phi(g(x,\diamond),0,b))))))) \\ &= h(f(x_1,f(x_2,g(x_1,g(x_2,b))))) \end{aligned}$$

Thus, the recurrence-term $h(\Phi(f(x,\diamond),N,\Phi(g(x,\diamond),N,b)))$ schematizes an infinite sequence of terms: $h(b)$, $h(f(x_1,g(x_1,b)))$, $h(f(x_1,f(x_2,g(x_1,g(x_2,b)))))$, ...

Example: Generators with distinct degree variables are to be unfolded independently. A recurrence-term $f(\Phi_N(g(\diamond),a),\Phi_M(h(\diamond),b))$ schematizes an infinte sequence of terms $f(a,b)$, $f(g(a),b)$, $f(a,h(b))$, $f(g(a),h(b))$,$\cdots$.

In general, unfolding an instantiated recurrence-term $H[N_1\leftarrow n_1,...,N_m\leftarrow n_m]$ is to apply repeated level unfoldings to all its instantiated generators until all instantiated generators dispear. We also denote the recurence-term so obtained by $H[N_1\leftarrow n_1,...,N_m\leftarrow n_m]$. For example, $f(\Phi(f(\diamond),N,a))[N\leftarrow 2]$ denotes $f(f(f(a)))$.

A fully instantiated recurrence-term $H[N_1\leftarrow n_1...,N_m\leftarrow n_m]$ becomes a first order term. We use $\Omega(H)$ to denote the set of terms $\{H[N_1\leftarrow n_1,...,N_m\leftarrow n_m] \;:\; H[N\leftarrow n_1,...,N_m\leftarrow n_m]$ is fully instantiated and $n_1,...,n_m \in \mathcal{N}\}$. Each element of $\Omega(H)$ is called a *constituent* of H.

Since we do not explicitly define what are the fresh variables during each level unfolding, unfolding an instantiated recurrence-term may result in different recurrence-terms which are unique up to variable renaming. In this paper, we do not differentiate them.

We sometimes denote $h[p \leftarrow t]$ by $h[t]$ and $h[\diamond]$ by h whenever the omission of the position p and $\diamond$ does not cause confusion.

4 Generating Recurrence-terms from a Sequence of Terms

4.1 Construction of Recurrence-terms

As mentioned before, we use recurrence-terms to schematize infinite sequences of terms. It is therefore necessary to have some way of generating recurrence-terms from a set of terms. In this section we present a method of automatically constructing recurrence-terms from two given terms. The intuition behind this is to pick two consecutive terms in a sequence, find an embedding relation and guess the recurring pattern. Note that using our construction, several different recurrence-terms may be generated from two terms, and some of them may not correctly schematize the intended sequence. Therefore a verification process is needed to ensure that the recurrence-term is indeed what one wants. An easy way to check is to match the recurrence-term with the ensuing terms in the sequence. If there is any which does not match with the recurrence-term, then the recurrence-term can be discarded. More elaborate verification methods have been designed, but they will not be discussed in this paper.

A *recurrence term generator* **cgen** generates a recurrence-term from two terms s and t in a nondeterministic way. The operation $cgen(s,t,N)$ is equal to

- t if $s = t$, *or*
- $f(cgen(s_1,t_1,N),...,cgen(s_m,t_m,N))$ if $s = f(s_1,\ldots,s_m)$ and $t = f(t_1,\ldots,t_m)$, *or*
- $\Phi(t[p \leftarrow \diamond], N, cgen(s, t/p, N))$ if $s \trianglelefteq^{hv} t/p$.

Sometimes no recurrence-term can be generated from two terms. This can be characterized by the following proposition:

Proposition 2 *If every $f \in \mathcal{F}$ has a unique arity, then $s \trianglelefteq^{hv} t$ if and only if a recurrence-term can be generated by* $cgen(s,t,N)$.

Proof: By simulation on the construction of recurrence-terms and definition of homeomorphic embedding relation.□

Remarks: If we allow a functor f to have different arities, then Proposition 2 may no longer hold. For example, if f has arities of 2 and more, then $f(x,b) \triangleleft^{hv} f(g(x),a,g(b))$, but $cgen(f(x,b), f(g(x),a,g(b)),N)$ can not generate any recurrence-terms.

Proposition 3 *There are only finitely many hv-substitutions σ from s to t such that $s\sigma \trianglelefteq^{hv} t$, and their number is bounded by $\frac{n_t!}{(n_t-n_s)!}$, where n_s and n_t are the number of distinct variables of s and t respectively.*

Proof: By definition of $\trianglelefteq^{hv}$, $s\sigma \trianglelefteq^{hv} t$, implies that σ has to be a one-to-one mapping from $\mathcal{V}(s)$ to $\mathcal{V}(t)$, and n_s must be less than or equal to n_t. Since the total number of one-to-one mappings from $\mathcal{V}(s)$ to $\mathcal{V}(t)$ is $\frac{n_t!}{(n_t-n_s)!}$, there can be only finitely many hv-substitutions which satisfy the requirement.□

Let $Gen(s,t,N)$ be the set of all the recurrence-terms generated by $cgen(s,t,N)$. Since there are only finitely many ways to embed s in t, it follows that

Proposition 4 *$Gen(s,t,N)$ is finite.*

The set of all the possible recurrence-terms constructed from two terms s and t are

$$Gen_{all}(s,t) = \bigcup_i \{Gen(s\sigma_i, t, N) : s\sigma_i \trianglelefteq^{hv} t\}.$$

Proposition 5 *$Gen_{all}(s,t,N)$ is finite.*

Proof: A direct consequence of proposition 3 and 4.□

If we consider $\rightarrow$ also as a functor with the lowest precedence, then the above method can be easily extended to handle recurrence rules as well:

$$cgen(l_1 \rightarrow r_1, l_2 \rightarrow r_2, N) = cgen(l_1, l_2, N) \rightarrow cgen(r_1, r_2, N)$$

Example: Given two terms $s = f(x,y), t = f(x_1, f(x_2, y_1))$. There exist three different *hv*-substitutions from s to t: $\sigma = \{x \leftarrow x_1, y \leftarrow x_2\}$, $\sigma = \{x \leftarrow x_1, y \leftarrow y_1\}$, and $\sigma = \{x \leftarrow x_2, y \leftarrow y_1\}$ such that $s\sigma \trianglelefteq^{hv} t$.

substitution σ	generated recurrence-term	sequences schematized
$\sigma = \{x \leftarrow x_1, y \leftarrow x_2\}$	$f(x_1, \Phi(f(\diamond, y_1), N, x_2))$	$f(x_1, f(f(...f(x_2, y_k)...), y_2))$
$\sigma = \{x \leftarrow x_1, y \leftarrow y_1\}$	$f(x_1, \Phi(f(x_2, \diamond), N, y_1))$	$f(x_1, f(x_3, f(x_4, f(...f(x_k, y_1)...)$
$\sigma = \{x \leftarrow x_2, y \leftarrow y_1\}$	$\Phi(f(x_1, \diamond), N, f(x_2, y_1))$	$f(x_3, f(x_4, f(...f(x_k, f(x_2, y_1))...)$

4.2 Comparisons and Examples

As mentioned briefly in the introduction, [Her 89a, Her 89b, HeP 89, MoP 89] were aimed at identifying patterns of rules, such as crossed pairs, which may cause divergence in Knuth-Bendix completion. But they did not suggest concrete methods in dealing with such divergence other than re-orienting the rules.

Kirchner[Kir 87, Kir 89] and Lange and Jantke[LaJ 89] took another approach aiming at finitely schematizing divergent sequences. The problems with Kirchner's meta-rule solution are that the meta-rules can only be constructed heuristically and that the difficulty of schematizing infinite many terms (in the domains of the meta-variables) is still unsolved. Thus the meta-rules cannot be involved in the completion process. Lang and Jantke's primitive recursive auxiliary operator method requires that the number of variables of the terms does not change among the terms in the sequence. Thus, it cannot schematize a sequence such as

$$\begin{array}{rcl} f(x_1, x_1) & \rightarrow & x_1, \\ f(x_1, f(x_2, f(x_1, x_2))) & \rightarrow & f(x_1, x_2), \\ f(x_1, f(x_2, f(x_3, f(x_1, f(x_2, x_3))))) & \rightarrow & f(x_1, f(x_2, x_3)), \\ & \vdots & \end{array}$$

In the following, we discuss two examples from [Kir 89]. The first example can be dealt with by an order-sorted solution, while the second can not. We first show how they were presented and solved using meta-rules as done by Kirchner, then we show how recurrence-rules can be constructed automatically from the given sequences of rules by simply utilizing the syntactic structural similarities of the terms in the sequence.

Example 1:[Kir 89] Associativity and idempotency of f.

$$f(f(x,y),z) \rightarrow f(x, f(y,z))$$
$$f(x,x) \rightarrow x$$

Knuth-Bendix completion will generate two infinite sequences of rules:

$$f(x_1, f(x_1, z)) \to f(x_1, z),\ f(x_1, f(x_2, f(x_1, f(x_2, z)))) \to f(x_1, f(x_2, z)), \ldots$$

$$f(x_1, x_1) \to x_1,\ f(x_1, f(x_2, f(x_1, x_2))) \to f(x_1, x_2), \ldots.$$

Kirchner's meta-rule approach first tries to find an equivalence relation $\sim_o$ defined in the theory, then under $\sim_o$ transforms the sequence of terms into a meta-term with meta-variables. Each meta-variable ranges over a domain of infinitely many terms, and each instantiation represents a term in the domain. The obtained *meta-rule* $G_l \to G_r$ satisfies the condition that for each rule $l_k \to r_k$ in the sequence, there is a corresponding substitution θ_k such that $G_l\theta_k \sim_o l_k$ and $G_r\theta_k \sim_o r_k$. In this particular example, $f(f(x, y), z) = f(x, f(y, z))$ is chosen to define $\sim_o$ and the meta-rule is

$$f(X, f(X, z)) \to f(X)$$

for the first divergent sequence. The domain of the meta-variable X is

$$\{x_1, f(x_1, x_2), f(x_1, f(x_2, x_3)), \ldots\}.$$

The difficulties in Kirchner's method lie in the selection of equivalence relation where there may be many candidates, the derivation of the meta-rules, and the representation of the infinite domain of meta-variables. In [Ker 89], no systematic way of defining $\sim_o$ or meta-rule was given. Even though an infinite sequence of rules may be schematized by a meta-rule, the infinite domain of meta-variables has to be dealt with separately. Hermann and Kirchner[HeK 90] propose to interpret above meta-rule system by order-sorted rewriting where the sort of X is $\{x_1, f(x_1, x_2), f(x_1, f(x_2, x_3)), \ldots\}$.

Now we show how the same example is done in the recurrence-term framework. A recurrence-rule can be constructed from the first two rules of the first sequence as follows:

$$\begin{aligned} &cgen(f(x_2, f(x_2, z)) \to f(x_2, z),\ f(x_1, f(x_2, f(x_1, f(x_2, z)))) \to f(x_1, f(x_2, z)), N) \\ &\quad = cgen(f(x_2, f(x_2, z)), f(x_1, f(x_2, f(x_1, f(x_2, z)))), N) \to cgen(f(x_2, z), f(x_1, f(x_2, z)), N). \end{aligned}$$

The left hand side of $\to$ can be further evaluated into

$$\begin{aligned} &= cgen(f(x_2, f(x_2, z)), f(x_1, f(x_2, f(x_1, f(x_2, z)))), N) \\ &= \Phi(f(x_1, \diamond), N, cgen(f(x_2, f(x_2, z)), f(x_2, f(x_1, f(x_2, z))), N)) \\ &= \Phi(f(x_1, \diamond), N, f(x_2, cgen(f(x_2, z), f(x_1, f(x_2, z)), N))) \\ &= \Phi(f(x_1, \diamond), N, f(x_2, \Phi(f(x_1, \diamond), N, f(x_2, z))). \end{aligned}$$

The right hand side of $\to$ is evaluated into

$$\begin{aligned} &= cgen(f(x_2, z), f(x_1, f(x_2, z)), N) \\ &= \Phi(f(x_1, \diamond), N, cgen(f(x_2, z), f(x_2, z), N)) \\ &= \Phi(f(x_1, \diamond), N, f(x_2, z)). \end{aligned}$$

It is not difficult to see that the generated recurrence-rule

$$\Phi(f(x_1, \diamond), N, f(x_2, \Phi(f(x_1, \diamond), N, f(x_2, z)))) \to \Phi(f(x_1, \diamond), N, f(x_2, z))$$

does indeed schematize the first divergent sequence of rules of this example.

Similarly, a recurrence-rule can also be constructed to schematize the second divergent sequence of rules of this example

$$\Phi(f(x_1, \diamond), N, f(x_2, \Phi(f(x_1, \diamond), N, x_2))) \to \Phi(f(x_1, \diamond), N, x_2).$$

A divergent sequence of terms or rules may have different recurrence-term or recurrence-rule schematizations. For instance, the first divergent sequence of this example can also be schematized by

$$f(x_1, \Phi(f(x_2, \diamond), N, f(x_1, \Phi(f(x_2, \diamond), N, z)))) \rightarrow f(x_1, \Phi(f(x_2, \diamond), N, z)).$$

Example 2: [Kir 89] Associativity and endomorphism

$$\lambda_1 : (x * y) * z \rightarrow x * (y * z)$$
$$\lambda_2 : f(x) * f(y) \rightarrow f(x * y)$$

Completing λ_1 and λ_2 generates the infinite sequence

$$\{f^i(x * y) * z \rightarrow f^i(x) * (f^i(y) * z)\}_{i=0}^{\infty}$$

where f^i denotes i applications of the functor f.

If one chooses the equivalence relation $\sim_o$ to be λ_2, then the above infinite sequence can be expressed by the meta-rule

$$(X * Y) * z \rightarrow X * (Y * z)$$

where the pair (X, Y) ranges over

$$\{(x_1, x_2), (f(x_1), f(x_2)), ..., (f^i(x_1), f^i(x_2)), ...\}.$$

However, in this case, the elements of X and Y are paring in a synchronized manner and the domain of (X, Y) are not a regular tree language, therefore, sorts of order-sorted rewriting are not expressive enough here. Using the recurrence-term approach, we can generate the following recurrence-rule which schematizes the infinite sequence:

$$\begin{aligned}
&cgen((x * y) * z \rightarrow x * (y * z), f(x * y) * z \rightarrow f(x) * (f(y) * z), N)\\
&= cgen((x * y) * z, f(x * y) * z, N) \rightarrow cgen(x * (y * z), f(x) * (f(y) * z), N)\\
&= cgen(x * y, f(x * y), N) * cgen(z, z, N) \rightarrow cgen(x, f(x), N) * cgen(y * z, f(y) * z, N)\\
&= \Phi(f(\diamond), N, x * y) * z \rightarrow \Phi(f(\diamond), N, x) * cgen(y, f(y), N) * cgen(z, z, N)\\
&= \Phi(f(\diamond), N, x * y) * z \rightarrow \Phi(f(\diamond), N, x) * \Phi(f(\diamond), N, y) * z.
\end{aligned}$$

4.3 The Expressive Power of Recurrence-Terms

Recurrence-terms can be used to schematize most of the examples of divergent rewriting systems given in [HeP 86, MoP 89, Kir 89]. However, there are some divergent systems which recurrence-terms cannot schematize. This is mainly because the recurrence-terms we defined do not allow proper recurrence-terms to appear in the first argument of a generator. Furthermore, the second argument of a generator has to be a degree variable. Intuitively, our recurrence-terms are linear with respect to the degree variables. Therefore, they can only schematize those divergent sequences of terms whose sizes grow linearly. For example, the following divergent rewriting systems generated during Knuth-Bendix completion can not be schematizeed by recurrence-terms:

Example: Type(Λ, λ_0) divergent sequence in [MoP 89]

$$\begin{aligned}
\text{EQ1}: \;& \gamma : f(g(f(x))) \rightarrow h(f(x))\\
& \lambda_0 : f(g(f(x))) \rightarrow h(f(x))\\
\text{DS1}: \;& (\lambda_0, \lambda_0) \Rightarrow \lambda_1 : f(g(h(f(x)))) \rightarrow h^2(f(x))\\
& (\lambda_1, \lambda_1) \Rightarrow \lambda_2 : f(g(h^3(f(x)))) \rightarrow h^4(f(x))\\
& (\lambda_2, \lambda_2) \Rightarrow \lambda_3 : f(g(h^7(f(x)))) \rightarrow h^8(f(x))\\
& (\lambda_3, \lambda_3) \Rightarrow \lambda_4 : f(g(h^{15}(f(x)))) \rightarrow h^{16}(f(x))\\
& \vdots
\end{aligned}$$

Example : [Her 89]

EQ2 : $\gamma : (x \otimes h(y)) \oplus y \rightarrow (x \oplus y) \otimes y$
$\lambda_0 : (x \oslash y) \otimes y \rightarrow x$
DS2 : $(...(x \oslash h^{n+1}(y)) \oplus h^n(y)) \oplus ... \oplus h(y)) \oplus y) \otimes y \rightarrow (...(x \oplus h^n(y)) \oplus ... \oplus h(y)) \oplus y$

The terms in the divergent sequence DS1 "grow" both horizontally and vertically, and as a result the terms grow exponentially in size. The terms in the divergent sequence DS2 also grow exponentially.

Our definition of recurrence-terms can be extended to express the above sequences. For instance, if we allow the second argument of a generator to be a function, then the first divergent sequence (DS1) can be schematized by

$$f(g(\Phi(h(\diamond), 2^N - 1, f(x)))) \rightarrow \Phi(h(\diamond), 2^N, f(x))$$

If we allow the first argument of a generator to be a proper recurrence-term, then the second divergent sequence (DS2) can be schematized by

$$\Phi_N(\diamond \oplus \Phi_N(h(\diamond), y), x \oslash h(\Phi_N(h(\diamond), y))) \otimes y \rightarrow \Phi_N(\diamond \oplus \Phi_N(h(\diamond), y), x)$$

Obviously, we can also define a generator to be of the form $\Phi(h[p_1 \leftarrow \diamond, ..., p_n \leftarrow \diamond], f(N), l)$ where f is a monotoniclly increasing function from $\mathcal{N}$ to $\mathcal{N}$. Regardless of how unfolding and unfolding renaming are defined for this kind of representation, we can effectively decide whether a constituent of such a recurrence-term matches a first order term since the size of terms schematized by an N-generator grows monotoniclly with respect to N. However, decision procedures for some algebraic opeartions, such as unification, are more difficult to obtain in this case than in the case adopted in this paper.

5 Recurrence Matching

In this section we present an algorithm for matching a term with a recurrence-term. A recurrence-term H *matches* a term s if there exists a substitution θ such that $s = t\theta$ for some constituent t of H.

A term s may be matched, at the same position, with several constituents of a recurrence-term H. A naive way of matching is to enumerate all constituents of H up to the size of s, and try to match each one of them with s. These constituents can be obtained effectively since the size of terms in a sequence schematized by any N-generator increases monotonically. However, this method is not efficient since the matchings for the common prefixes of constituents of H have to be done repeatedly. In the following, we will give a more efficient matching algorithm such that no repeated computation is done.

During each level unfolding for N-generators, either the N-generators are unfolded with an unfolding renaming substitution σ or N is instantiated to 0. We use *unfolding operators* $N \leftarrow \sigma$ to denote the former and $N \leftarrow 0$ to denote the latter. An *unfolding sequence es* is a sequence of unfolding opeartors sequence $[e_1, ..., e_n]$ such that for all $0 \leq i, j \leq n$ and $i \neq j$, *Range*$(\sigma_i) \cap$ *Range*$(\sigma_j) =$ whenever e_i is $N \leftarrow \sigma_i$ and e_j is $M \leftarrow \sigma_j$ or $N \leftarrow \sigma_j$.

Thus, several level unfoldings can be imitated by an *unfolding sequence.* We call (σ, es) an *unfolding substitution* where σ is a substitution and es is an unfolding sequence. The composition of two unfolding substitutions (σ, es) and (δ, es') is an unfolding subsititution (λ, es'') such that

$$\lambda = \sigma \circ \delta$$
$$es'' = es \bullet es'$$

where • stands for concatenation.

We use $H|e$ to denote the application of level unfolding e to H, and $H\|es$ to denote applying a sequence of level unfoldings es to H. We use $H[x \leftarrow s]$ to denote the result of replacing all occurrences of x in H by s. Let M be a set of pairs $s_i \hookleftarrow H_i$ where s_i is a term and H_i is a recurrence-term. We use $M[x \leftarrow s], M|e, M\|es$ to denote the results of applying $[x \leftarrow s], e$ and es to every H_i of $s_i \hookleftarrow H_i \in M$.

Remarks: Sometimes $Dom(\sigma)$ of an operator $N \leftharpoonup \sigma$ may not cover all the variables of N-patterns in the recurrence-term to which it applies. For instance, applying $N \leftharpoonup \{x \leftarrow x_1\}$ to $\Phi_N(f(x, y, \diamond), a)$ has to consider the renaming for y also. Since the soly purpose of unfolding renaming is to rename the variables into unused ones, without losing generality, we can imagine that $N \leftharpoonup \{x \leftarrow x_1\}$ is the same as its extension $N \leftharpoonup \{x \leftarrow x_1, y \leftarrow y_{new}\}$ where y_{new} is a *fresh* variable.

Applying an unfolding sequence (σ, es) to a recurrence-term H obtains $(H\|es)\sigma$. In the following, we frequently use Θ, Υ, Γ to denote unfolding substitutions.

Proposition 6 *A term t is a constituent of a recurrence-term H if and only if there is an unfolding sequence es such that t is equal to H‖es upon variable renaming.*

Proof: Followed directly from the definition of constituents of recurrence-terms.□.

We now define the matching algorithm via a set of inference rules. Each rule will be of the form

$$\frac{M, \Theta}{M', \Theta'} \text{ or } \frac{M, \Theta}{\Box}$$

where M is a set of pairs $s_i \hookleftarrow H_i$ to be matched, and Θ is an unfolding substitution. The symbol □ denotes failure, and M' and Θ' are the results after applying an inference rule. The symbol ϵ denotes empty unfolding substitution.

Recurrence-Matching Inference Rules

M1(Replacement):

$$\frac{\{s \hookleftarrow x\} \cup M,\ \Theta}{M[x \leftarrow s],\ \Theta \circ (\{x \leftarrow s\}, [])}$$

M2(Failure1):

$$\frac{\{x \hookleftarrow g(t_1, ..., t_m)\} \cup M, \Theta}{\Box}$$

M3(Decomposition):

$$\frac{\{f(s_1, ..., s_n) \hookleftarrow f(t_1, ..., t_n)\} \cup M,\ \Theta}{\{s_1 \hookleftarrow t_1, ..., s_n \hookleftarrow t_n)\} \cup M,\ \Theta}$$

M4(Failure2):

$$\frac{\{f(s_1, ..., s_n) \hookleftarrow g(t_1, ..., t_m)\} \cup M,\ \Theta}{\Box} \text{ if } f \neq g$$

M5(Instantation):

$$\frac{\{(s, \Phi_N(h, l))\} \cup M,\ \Theta}{\{s \hookleftarrow l[N \leftharpoonup 0])\} \cup M[N \leftharpoonup 0],\ \Theta \circ (\phi, [N \leftharpoonup 0])}$$

M6(Unfolding):

$$\frac{\{s \hookleftarrow \Phi_N(h[u \leftarrow \diamond], l)\} \cup M,\ \Theta}{\{s \hookleftarrow h\theta[u \leftarrow \Phi_N(h, l')]\} \cup M',\ \Theta \circ (\phi, [N \leftharpoonup \theta])} \ \textit{where} \ \begin{array}{l} \theta \text{ is an unfolding renaming for } N\text{-patterns} \\ \text{in } \Phi(h[u \leftarrow \diamond], N, l) \text{ and } M, \\ l' = l|N \leftharpoonup \theta, \quad M' = M|N \leftharpoonup \theta \end{array}$$

One step of application of an inference rule is also denoted as $(M, \Theta) \vdash_{MI} (M', \Theta')$. We use $\vdash^*_{MI}$ to denote the transitive and reflexive closure of $\vdash_{MI}$. Thus to match s by H, we start with the *initial configuration* $(\{s \hookleftarrow H\}, \epsilon)$ and try to obtain $(\emptyset, \Theta)$. We start from an example:

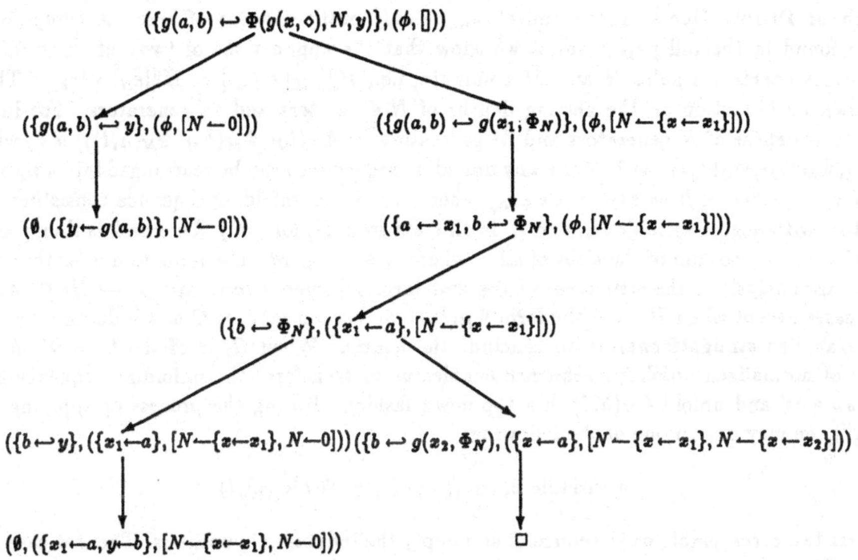

An unfolding substitution Θ is said to be *successful* with respect to (M, Θ_0) if $(M, \Theta_0) \vdash^*_{MI} (\emptyset, \Theta_0 \circ \Theta)$. Let us use $S_{match}(s, H)$ to denote the set of all the successful unfolding substitutions of $(\{s \hookleftarrow H\}, \epsilon)$. We now show that the above set of inference rules are sound, complete and terminating.

Lemma 1 *If* $(M, \Theta_0) \vdash^*_{MI} (\emptyset, \Theta_0 \circ (\sigma, es))$, *then for each* $s_i \hookleftarrow H_i \in M$,

$$H_i \| es \text{ is a constituent of } H_i \text{ and } s_i = (H_i \| es)\sigma.$$

Proof: By induction on the number of steps of matching derivations.

Corollary 1 (Soundness) *If* $(\{s \hookleftarrow H\}, \epsilon) \vdash^*_{MI} (\emptyset, (\sigma, es))$, *then* $H \| es$ *is a constituent of* H *and* $(H \| es)\sigma = s$.

Theorem 1 *The applications of the set of inference rules M1 through M6 terminate.*

Proof: To prove that repeated applications of inference rules M1 through M6 terminate, it is sufficient to show that the complexity measure $(|S|, |D_H|)$ associated with each configuration (M, Θ) decreases according to the (well-founded) lexicographic ordering on numbers. The measure $|S|$ is the summation of the sizes of all the terms s_i and $|D_H|$ is the number of degree variables in all the H_i where $s_i \hookleftarrow H_i \in M$.

Inference rules M1 and M3 strictly reduce the measure $|S|$, and therefore reduce the complexity measure of (M, Θ). (since the variable x cannot appear in the first argument of any term/recurrence-term pair, replacing x by s does not increase the size of $|S|$.) Rules M2 and M4 reduce everything to the failure configuaration □ and the algorithm terminates obviously. The measure $|S|$ remains the same when applying M5 or M6, but M5 reduces $|D_H|$. Even the application of M6 does not change the complexity measure; however, the only succeeding rules after M6 are among M2 through M4, which strictly reduce $|S|$, hence reduce the complexity measure of (M, Θ). Then the theorem follows. □.

Lemma 2 *Given a configuration* (M, Θ_0), *if there is an unfolding substitution* Γ *such that for all* $s_i \hookleftarrow H_i \in M$, $H_i\Gamma = s_i$, *then there exists a derivation* $(M, \Theta_0) \vdash^*_{MI} (\emptyset, \Theta_0 \circ \Upsilon)$ *such that* $H_i\Upsilon = s_i$.

Sketch of Proof: Due to space limitation, we only outline the proof here. A complete proof can be found in the full paper. First we show that the applications of two unfolding operators for distinct degree variables N and M commute, i.e., $H\|[e_M] \bullet [e_N] = H\|[e_N] \bullet [e_M]$. This can be shown by induction on the nesting depths of N-generators and M-generators. For instance, the nesting depths of N-generators and M-generators for $\Phi_N(h_N, \Phi_M(h_M, \Phi_N(h, l)))$ is 3, while for $\Phi_N(h_N, \Phi_M(h_M, \Phi_Q(h, l)))$ is 2. Thus, any unfolding sequence es can be rearranged into a *normalized unfolding sequence* such as $es_{N_1} \bullet ... \bullet es_{N_k}$ where es_{N_i} is an unfolding sequence consisting of only operators with degree variable N_i, and N_i is different from N_j for $i \neq j$. Lemma 2 is then proved by induction on C, the sum of the sizes of all s_i where $s_i \hookleftarrow H_i \in M$. The induction is further divided into a case analysis of the structures of the first term/recurrence-term pair $s_1 \hookleftarrow H_1 \in M$. In all other cases except when H_1 is of the form $\Phi_N(h, l)$, derivation reduces C and induction hypothesis can be applied straightforwardly to conclude the lemma. When H_1 is of the form $\Phi_N(h, l)$, the notion of normalized unfolding sequence enables us to transform the unfolding sequence es of Γ into $es_N \bullet es'$ and unfold $\Phi_N(h, l)$ in a top-down fashion. During the process of applying es_N to $\Phi_N(h, l)$, we may obtain one of the following:

$$\text{a variable } x,\ g(t_1, ..., t_m),\ \text{or } \Phi(h[\diamond], Q, l).$$

The first two cases enable us to reduce C and apply the induction hypothesis. The last case cannot occur infinitely many times since there are only finitely many degree variables in M. Eventually we can apply induction hypothesis and conclude the lemma.□.

Theorem 2 (Completeness) *Let t be a constituent of a recurrence-term H such that t matches s, then $(\{s \hookleftarrow H\}, \epsilon) \vdash^*_{MI} (\emptyset, (\sigma, es))$ for some (σ, es) such that $s = (H\|es)\sigma$ and t is equal to $H\|es$ upon variable renaming.*

Proof: If t is a constituent of H such that t matches a term s, then there must exist an unfolding sequence $\tilde{es}$ and a substitution θ such that $t = H\|\tilde{es}$ and $s = t\theta$. If we start with the initial configuration $(\{s \hookleftarrow H\}, \epsilon)$ and by Proposition 2, the theorem follows. □

Corollary 2 *$S_{match}(s, H)$ has only a finite number of successful unfolding substitutions.*

In the next section, we will use the results obtained so far to define a recurrence-rewriting system R_H which schematizes an (infinite) rewriting systems R, and show that R_H is terminating and confluent if and only if R is.

6 Recurrence Rewriting System

A term rewriting system (TRS) R is a set of oriented equations $\{l_i \rightarrow r_i\}$ and each oriented equation is called a *rewrite rule.* A term s *rewrites* to another term t by R, denoted as $s \rightarrow_R t$, if there exists a nonvariable subterm of s at position p and a rewrite rule $l \rightarrow r$ in R such that $s/p = l\sigma$ for some substitution σ and $t = s[p \leftarrow r\sigma]$. We write $s \rightarrow t$ whenever R is obvious, and $s \rightarrow_{l \rightarrow r} t$ whenever the rewrite rule $l \rightarrow r$ is emphasized. We use $\rightarrow^*$ to denote the transitive-reflexive closure of R.

A TRS R is *terminating* if there does not exist an infinite sequence of rewritings $s_1 \rightarrow_R s_2 \rightarrow_R s_3 \rightarrow_R \ldots$. R is *confluent* if for every term s, $s \rightarrow^* t_1$ and $s \rightarrow^* t_2$ imply that there exists a term t such that $t_1 \rightarrow^* t$ and $t_2 \rightarrow^* t$.

Similarly, a *recurrence-term rewriting system*(RTRS) R_H is a set of oriented rules on recurrence-terms $\{L_i \Rightarrow R_i\}$, and each oriented rule is called a *recurrence-rule* which may finitely schematize an infinite set of rewrite rules. For example, a recurrence-rule $f(\Phi_N(g(\diamond), f(a))) \Rightarrow \Phi_N(g(\diamond), f(a))$

schematizes an infinite set of rules: $\{f(f(a)) \to f(a),\ f(g(f(a))) \to g(f(a)),\ f(g(g(f(a)))) \to g(g(f(a))), ,...\}$.

A term s *recurrence-rewrites* to a term t by a RTRS R_H, denoted as $s \mapsto_{R_H} t$, if there exists a recurrence-rule $H_l \Rightarrow H_r$ in R_H and a nonvariable position p in s such that

1. $\Theta \in S_{match}(s/p, H_l)$, and
2. $t = s[p \leftarrow H_r\Theta]$.

We may write $\mapsto_{R_H}$ as $\mapsto$ when the context R_H is obvious. Sometimes we write $s \mapsto_{H_l \Rightarrow H_r} t$ to emphasize that s recurrence-rewrites to t via $H_l \Rightarrow H_r$. A term s is *recurrence-reducible* if s recurrence-rewrites to some term t. A term can recurrence-rewrite to several different terms by a recurrence-rule at the same position of s.

Example : $s = f(g(a, g(a, b))), R_H = \{\lambda : \Phi_N(g(x, \diamond), g(y, z)) \Rightarrow \Phi_N(h(x, \diamond), a)\}$ and $g(a, g(a, b))$ is a nonvariable subterm of s at position 1.

$$S_{match}(g(a, g(a, b)), \Phi_N(g(x, \diamond), g(y, z))) = \{ \ (\{y \leftarrow a, z \leftarrow g(a, b)\}, [N \leftarrow 0]), \ (\{x_1 \leftarrow a, y \leftarrow a, z \leftarrow b\}, [N \leftarrow \{x \leftarrow x_1\}, N \leftarrow 0]) \ \}$$

By the first rule $g(y, z) \to a$ in the sequence, $f(g(a, g(a, b))) \mapsto_\lambda f(a)$. By the second rule $g(x, g(y, z)) \to h(x, a)$, $f(g(a, g(a, b))) \mapsto_\lambda f(h(a, a))$.

A term s is in *recurrence normal form* in R_H if and only if s is not recurrence-reducible by any recurrence-rule of R_H. A term s has a recurrence normal form in R_H if and only if there is a normal form t such that $s \mapsto^* t$, where $\mapsto^*$ denotes the transitive-reflexive closure of R_H.

A RTRS R_H is said to be *recurrence-terminating* if there does not exist an infinite rewrite sequence $s \mapsto t_1 \mapsto t_2 \mapsto t_3 \mapsto \ldots$. A RTRS R_H is *recurrence-confluent* if and only if $s \mapsto^* s_1, s \mapsto^* s_2$ implies $s_1 \mapsto^* t, s_2 \mapsto^* t$ for some term t.

Lemma 3 *If a recurrence-rule $\lambda : H_l \Rightarrow H_r$ schematizes an infinite sequence of rules $\{l_1 \to r_1, l_2 \to r_2, \ldots\}$, then $s \to_{l_i \to r_i} t$ for some i iff $s \mapsto_\lambda t$.*

Proof: $\Longrightarrow$: If $s \to_{l_i \to r_i} t$, that is, there exists a nonvariable subterm s/p of s and a substitution θ such that $s/p = l_i\theta$ and $t = s[p \leftarrow r_i\theta]$. Since $l_i \to r_i$ is a constituent of $H_l \Rightarrow H_r$, there exists some complete unwinding sequence es such that $l_i = H_l\|es$ and $r_i = H_r\|es$. By the completeness of recurrence-matching and the definition of recurrence-rewrting, we have $s \mapsto_\lambda t$.
$\Longleftarrow$: Obvious from the definition of recurrence-rewriting and the soundness of recurrence-matching.□.

In the following, we assume that a RTRS R_H schematizes a TRS R.

Lemma 4 *$s \to_R^* t$ iff $s \mapsto_{R_H}^* t$.*

Proof: Since a 'normal' rule is a recurrence-rule, and by lemma 3, a term s is rewritten to a term t if and only if s is recurrence-rewritten to t, the theorem follows.□.

Theorem 3 *R is terminating if and only if R_H is recurrence-terminating.*

Proof: Suppose R is terminating. If R_H is not recurrence-terminating, then there exists an infinite sequence of recurrence rewritings $s \mapsto s_1 \mapsto s_2 \mapsto s_3 \mapsto \ldots$. By lemma 4, there is an infinite sequence of normal rewritings: $s \to s_1 \to s_2 \to s_3 \to \ldots$, which is a contradiction.
The other side of the proof is the same.□.

Theorem 4 *R is confluent if and only if R_H is recurrence-confluent.*

Proof: Direct consequence of lemma 4. □.

7 Conclusion

In this paper we presented a method of finitely schematizing infinite sequences of terms as recurrence-terms. We showed how it can be used to finitely schematize divergent term rewriting systems. Our method has the advantage that recurrence-terms can be generated automatically and algebraic operations can be defined on them. We also presented a recurrence-matching algorithm as well as proved the equivalence between the recurrence-term rewriting system and the "regular" one which it schematizes.

The applications of our notion of recurrence-terms are not limited to the term rewriting systems and the completion process. They can be utilized in the places where infinite sets or sequences of terms occur, e.g, to schematize infinite queries in deductive databases, to schematize an infinite, complete set of most general unifiers in E-unification, and to be used as primitive terms in logic programming [ChH 90].

References

[ChH 90] H. Chen, J. Hsiang, *Logic Programming with Recurrence Domains*, in preparation.

[Com 89] H. Comon, *Inductive Proofs by Specifications Transformation*, RTA 89, LNCS 355, Page 76-91, Springer-Berlag, April 1989.

[Der 82] N. Dershowitz, Ordering for Term Rewriting Systems, Theoretical Computer Science, 17, 1982, Page 279-301

[HeK 90] M. Hermann, H. Hirchner, *Meta-rule Synthesis from Crossed Rewrite Systems*, CTRS 90, June 11-14, 1990, Montreal, Canada.

[Her 89] M. Hermann, *Chain Properties of Rule Closures*, Lecture Notes in CS, Stacs 89, 349, Page 339-347.

[HeP 86] M. Hermann, I. Privara, *On Nontermination of Knuth-Bendix Algorithm*, Automata, Languages and Programming, 13th International Colloquium, Rennes, France, July 1986, Ed G.Goos and J. Hartmanis, Page 146-156.

[Kir 87] H. Kirchner, *Schematization of Infinite Sets of Rewrite Rules, Application to the Divergence of Completion Processes*, Lecture Notes in CS, Rewriting Techniques and Applications, Bordequs, France, May 1987, Page 180-191.

[Kir 89] H. Kirchner, *Schematization of Infinite Sets of Rewrite Rules, Application to the Divergence of Completion Processes*, Theoretical Computer Science 67 (1989) 303-332.

[Kon 89] H. C. Kong, *The Embedding Property and Its Application to Divergent Term Rewriting Systems*, M.S. thesis, Dept of Computer Science, National Taiwan University, June 1989.

[Kru 72] J.B. Kruskal, *The Theory of Well-quasi-ordering: A Frequently Discovered Concept*, J. Combinatorial Theory Ser.A 13(3), pp. 297-305, November 1972.

[LaJ 89] S. Lange, K. Jantke, *Towards A Learning Calculus for Solving Divergence in Knuth-Bendix Completion*, Communications of the Algorithm Learning Group, Nov. 1989

[MiR 85] P. Mishra, U. Reddy, *Declaration-free Type Checking*, POPL 1985, Page 7-21.

[MoP 89] R. Mong, P. Purdom, *Divergence in the Completion of Rewriting Systems*, Memo.

8 Appendix

As we mentioned before, a recurrence-terms may contain several N-generators for a specific N and renamings for the shared variables in the N-heads have to synchronized. To accomplish such "synchronization", we introduce a function $sync(H, N, \sigma_{in})$ to perform one level unfolding of N-generators. The substitution σ_{in} records the variables which have been instantiated during current level unfolding. If during the same level unfolding, we need to rename a variable in $Dom(\sigma_{in})$ again, we use its binding in σ_{in} instead of getting a new one. A tuple (H', σ_{out}) is returned where H' is the recurrence-term obtained from H after one level unfolding, and σ_{out} records the variable renamings. Since we do not need to synchronize different level unfoldings, we set $\sigma_{in} = \phi$ (ϕ denotes empty substitution) at the beginning of each level unfolding. For instance, in $sync(\Phi(f(x,y,\diamond), N, a), N, \{x \leftarrow x_1\})$, x has already been instantiated, then one step of level unfolding of $\Phi(f(x,y,\diamond), N, a)$ will only introduce a new variable for y and returns the tuple

$$(f(x_1, y_1, \Phi(f(x,y,\diamond), N, a)), \{x \leftarrow x_1, y \leftarrow y_1\})$$

$$sync(H, N, \sigma_{in}) = \begin{cases} (H, \sigma_{in}) & \text{if } H \text{ is a variable } x \text{ or a constant.} \\ (f(s'_1, \ldots, s'_n), \sigma_{out}) & \text{if } H \text{ is } f(s_1, \ldots, s_n), \text{ and} \\ & (s'_1, \sigma_1) = sync(s_1, N, \sigma_{in}), \\ & (s'_2, \sigma_2) = sync(s_2, N, \sigma_1), \\ & \vdots \\ & (s'_n, \sigma_{out}) = sync(s_n, N, \sigma_{n-1}). \\ (h\sigma'[\Phi(h, N, l')], \sigma_{out}) & \text{if } H = \Phi(h, N, l), \sigma' \text{ renames variables of } \mathcal{V}(h) - Dom(\sigma_{in}), \\ & \sigma'' = \sigma_{in} \circ \sigma', (l', \sigma_{out}) = sync(l, N, \sigma''). \\ (\Phi(h, M, l'), \sigma_{out}) & \text{if } H = \Phi(h, M, l), \text{ and } (l', \sigma_{out}) = sync(l, N, \sigma_{in}). \end{cases}$$

Now we give the full operational semantics of recurrence-terms.

$$H[N_1 \leftarrow n_1, \ldots, N_m \leftarrow n_m] \stackrel{def}{\equiv} ev_{all}(H, [(N_1, n_1), \ldots, (N_m, n_m)])$$

$$ev_{all}(H, L) = \begin{cases} H & : L = [] \\ ev_{all}(ev(H, N_1, n_1), L_{rest}) & : L = [(N_1, n_1) | L_{rest}] \end{cases}$$

$$ev(H, N, n) = \begin{cases} H[N \leftarrow 0] & : n = 0 \\ ev(H', N, n-1) & : n \geq 1 \text{ and } (H', \sigma_{out}) = sync(H, N, \phi) \end{cases}$$

Example :

$$\begin{aligned} & f(\Phi(f(x,\diamond), N, \Phi(g(\diamond), M, a)), \Phi(f(x,\diamond), N, b))[N \leftarrow 1, M \leftarrow 2] \\ & = ev_{all}(f(\Phi(f(x,\diamond), N, \Phi(g(\diamond), M, a)), \Phi(f(x,\diamond), N, b)), [(N,1), (M,2)]) \\ & = ev_{all}(ev(f(\Phi(f(x,\diamond), N, \Phi(g(\diamond), M, a)), \Phi(f(x,\diamond), N, b)), N, 1), [(M,2)]) \\ & \vdots \\ & = ev_{all}(ev(f(f(x_1, \Phi(g(\diamond), M, a)), f(x_1, b)), [(M,2)]) \\ & = ev_{all}(ev(f(f(x_1, \Phi(g(\diamond), M, a)), f(x_1, b)), M, 2), []) \\ & \vdots \\ & = ev_{all}(f(f(x_1, g(g(a))), f(x_1, b)), []) \\ & = f(f(x_1, g(g(a))), f(x_1, b)) \end{aligned}$$

Infinite Terms and Infinite Rewritings

Yiyun Chen[*†]
Michael J. O'Donnell[*]
The University of Chicago

Abstract

We discuss here the properties of term rewriting systems that allow infinite derivations leading to infinite normal forms. We extend work of Dershowitz and Kaplan to preserve more information about infinite derivations in the resulting normal forms, and to give finite representations of cyclic infinite terms. We develop algebraic semantics of the new system that are very similar to those of the old.

1 Introduction

Term Rewriting Systems(TRSs) are sets of rules given as pairs $L \to R$ of term schemata. These rules define nondeterministic computations by repeatedly replacing left-hand side instances in a given term by corresponding right-hand instances, as long as possible. The final result of rewriting is called a *normal form*. Term rewriting systems can be used as programming languages and in theorem provers. Key properties for term rewriting systems are *termination* and *confluence*, which together ensure that a unique normal form always exists. [HU80].

Dershowitz and Kaplan [DK89] investigated analogous properties of systems that allow infinite rewriting sequences leading to infinite normal forms. Such systems are not terminating in the classical sense, so they defined *ω-termination* and *ω-confluence*, which together imply the existence of a possibly infinite unique normal form.

We extend the Dershowitz-Kaplan work in two ways. First, we study finite representations of infinite terms. For example, the rule set

$$\mathsf{R} = \{a \to s(a)\}$$

yields the infinite normal form $s(s(s(\ldots s(a))))$, which might be represented by $s^\omega(a)$, but with the rule set

$$\mathsf{S} = \{a \to f(a,b), b \to g(b,a)\}$$

the normal form of a is the complicated infinite term shown in Figure 1, which is more difficult to describe finitely.

Second, the Dershowitz-Kaplan form of infinite terms displays infinitely many symbols, but each one is a finite distance from the root of the term. For example, with R above, they give the normal form $s(s(\ldots))$ denoted as s^ω, dropping the final a which gets pushed off to infinity. So, they cannot distinguish the normal forms of a and b under the rule set

$$\mathsf{T} = \{a \to s(a), b \to s(b)\},$$

[*]The research of these authors was supported in part by NSF grant CMPS 8805503.
[†]Yiyun Chen is supported by a grant from the People's Republic of China.

even though these are logically distinct, in the sense that it does not follow from $a = s(a)$ and $b = s(b)$ that $a = b$.

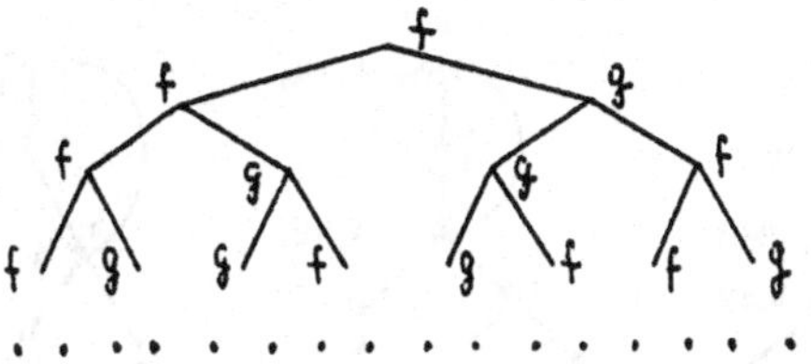

Figure 1

In Section 2, we define a finite representation of infinite cyclic terms, which keeps more information than the conventional representation. In Section 3, we define reductions on infinite cyclic terms with this representation. In Section 4, we discuss the new concepts of ω-termination and ω-confluence that are appropriate to this representation. Finally, in Section 5 we develop algebraic semantics of infinite rewriting which are almost the same as those in [DK89].

2 Term Graphs - Finite Representations of Infinite Cyclic Terms

For any term t, we use its *set of occurrences* $O(t) \subseteq N_+^*$ to represent the positions of its subterms [HU80]. Terms can be represented by trees. An occurrence also represents a node location in a tree if we consider that the occurrence describes a path from the root of that tree to a node in the tree. A term t over an alphabet Σ may be formally viewed as a function $f_t : O(t) \to \Sigma$, mapping each location in its term tree to the symbol at that location.

For an infinite term, we can imagine that it can be rolled into a finite graph and use loops in the graph to represent infinite cyclic behavior in the term. So the rolling is possible only when the infinity is caused by infinite cycles. We use term graphs uniformly to represent finite terms and infinite cyclic terms. A term and its graph representation roughly relate to each other as a real number relates to its continued fraction expansion. If a term has a finite representation, it is analogous to a rational number.

First we extend a term tree to a *term graph* by making every edge in the tree has direction from the father node to the son node and then adding $n (n \geq 0)$ *back edges* on the tree to compose loops. Back edges are used to indicate which parts of the graph can be unrolled. Every back edge is out of a nonleaf node (*loop exit*) to one of its ancestor node (*loop entry*) and the loop is from the entry to the exit along the tree edges and then back to the entry by the back edge. All these edges composing the loop are called the *loop edges*. The locations of two loops must satisfy the following restrictions. Assuming one loop has entry u_1 and exit v_1 and the other one has entry u_2 and exit v_2 respectively, then

1. If $u_1 \prec u_2$ then $v_1 \preceq u_2$ or $v_1 = v_2$ (such as G_1 and G_2 in Figure 2).

2. If $u_1 = u_2$ then neither $v_1 \prec v_2$ nor $v_2 \prec v_1$ (such as G_3 and G_4 in Figure 2).

Each tree edge and back edge is labeled by a positive integer d, called the *direction* that is used to indicate where the dth son node of a node is. For every nonleaf node, the directions of all the edges out of it (*out-edge*) satisfy the following restrictions.

1. If d $(d > 1)$ is a direction of one of its out-edges, then $d - 1$ is also a direction of one of its out-edges.

2. If the node is a nonexit node, all directions of its out-edges are different.

3. If the node is an exit node, each loop edge out of it must have the same direction as some nonloop edge out of it has, and all loop edges out of it must have different directions.

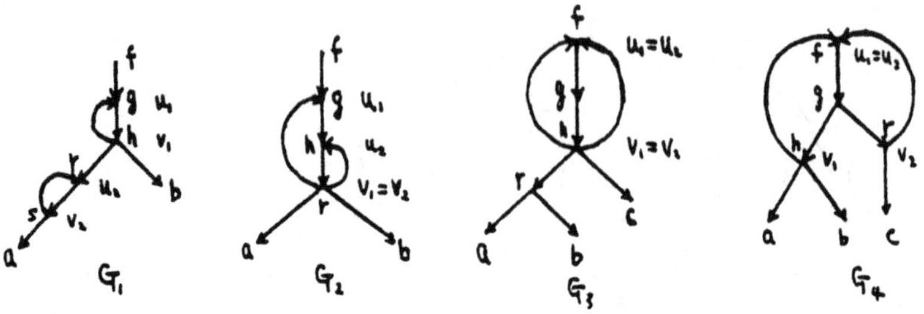

Figure 2

Two loops are *connected* if they share a node, or if they are both connected to the same loop. For example, the three loops in Figure 4 are connected one another. Clearly, in a connected component of loops, the entry of one loop is a prefix of the entries of all other loops.

Why do we call these graphs term graphs? Because they can be unrolled into an infinite number of finite term trees. For example, the term graph in Figure 3 can be unrolled to $f(s(a)), f(s(s(a))), f(s(s(s(a)))), \ldots$, these terms depend on how many cycles we take while unrolling the loop. The term graph in Figure 4 can be unrolled to the finite term in Figure 5. An infinite term is just a visualization of the limit of an infinite set of finite terms.

Figure 3

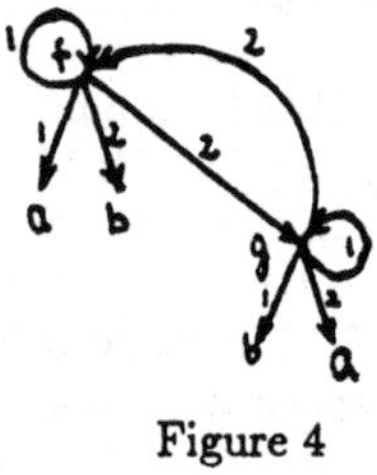

Figure 4

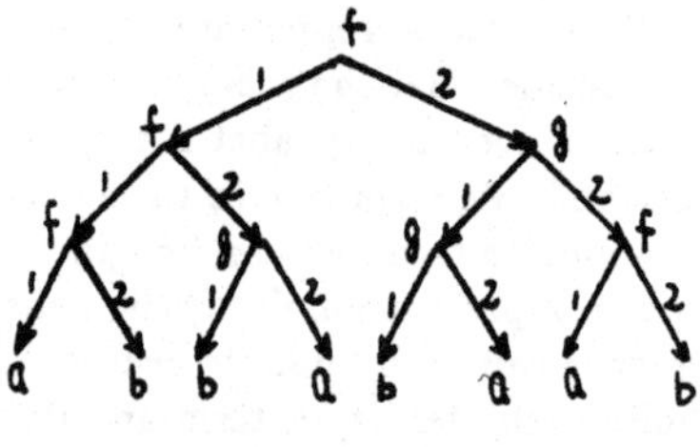

Figure 5

Now we define the unrolling process for a term graph G exactly. First we describe the result term graph for unrolling a group of connected loops. If one loop is unrolled, all other loops that connect this loop have to be unrolled also. Let $(u_1, v_1), \ldots, (u_m, v_m)$ be the loop entries and exits of a maximal set C of connected loops in G. The result of unrolling this set of loops $n(n > 0)$ times is G', defined below.

1. If C is the only connected component of loops in G, then G' is a tree, and it is easy to define G' directly as a function mapping occurrences to symbols. In a term graph, an occurrence describes a path from the root to some node. The only problem is the nondeterminism at loop exits, where there may be a loop edge and a nonloop edge with the same direction. The principle of solving the nondeterminism is that when a path reaches these exit nodes at $v_1, \ldots, v_m$ for the kth time and $k < n$, then it goes along a loop edge and when $k = n$ it exits the loops. Since there is only one connected component of loops, after leaving these loops, there will be no more looping, and no more nondeterminism. So, G' is defined by the function that maps an occurrence u to the symbol at the node in G reached by following the directions in u, starting at the root, and resolving nondeterminism by the rule above.

2. If there are other connected components of loops in G besides C, and C contains the root of G, then we may simply treat each subgraph rooted at another component as if it were a

new unique symbol of arity zero. Unroll as in case 1, then replace each of the new symbols with a copy of the subgraph that it represents.

3. If C does not contain the root of G, then replace the subgraph rooted at C by the unrolled version of the subgraph, as described in case 2.

Now we can describe the process of unrolling a graph G to a finite term.

1. If G has no loops, then nothing needs to be done, the result is the same as G.
2. If G has loops, then using the method above, we can reduce the number of loops in a term graph, and finally produce a finite tree.

For example, Figure 5 is the unrolled term graph of Figure 4 with $n = 3$. In Figure 6, as another example, the term graph G_1 can be unrolled into G_2 and also into G_3, each of G_2 and G_3 can be unrolled into G_4.

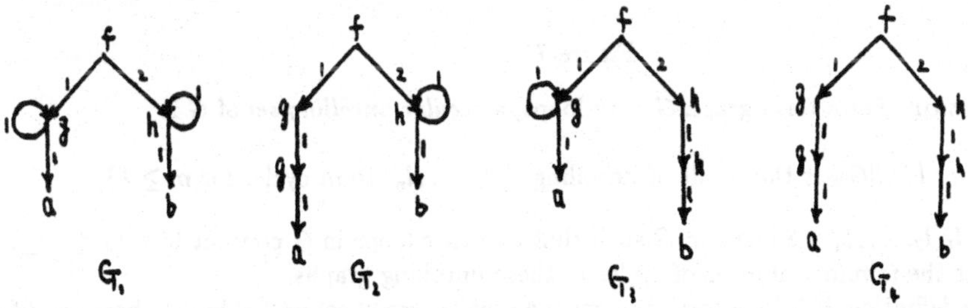

Figure 6

We can also define the unrolling process that unrolls more than one maximal group of connected loops to n cycles. In Figure 6, G_1 can be unrolled to G_4 directly.

Why do we need to define the unrolling process in terms of whether loops are connected or not? The term graph G_1 in Figure 7 can be unrolled to the graph G_2 if the above unrolling method is used. The finite term can be produced from d by using the rule set:

$$\mathsf{R}_1 = \{d \rightarrow h(a), a \rightarrow f(g(a,b)), b \rightarrow g(a,b)\}.$$

If the smaller loop in G_1 of Figure 7 is unrolled first and then so is the larger one (Of course, there is another method in reverse order), the graph G_3 can be gotten and the finite term can be produced from $h(a)$ by using the rule set:

$$\mathsf{R}_2 = \{h(x) \rightarrow h(f(g(x,b))), g(x,b) \rightarrow g(a, g(x,b))\}.$$

This unrolling method will not be considered in this paper because in such a TRS there is always some terms that cannot be represented by term graphs. To increase the cycles of the outer loop, there are some rules making $t_1 \xrightarrow{*} t_2$ to do so, such as the first rule in R_2; to increase the cycles of the inner loop, there are also some rules making $t_3 \xrightarrow{*} t_4$ to do so, and t_3 and t_4 have the same head symbols, after these same head symbols there must be some variables in t_3 in order to match all subterms produced from a subterm of t_2 such as the second rule in R_2. Then there must be an infinite term in the system that cannot be represented by such a term graph. For example, in the TRS

$$\mathsf{R}'_2 = \{a \rightarrow f(g(a)), g(x) \rightarrow g(h(x))\},$$

$a \xrightarrow{*} fga \xrightarrow{*} fghfga \xrightarrow{*} fgh^2fghfga \xrightarrow{*} fgh^3fgh^2fghfga \rightarrow \ldots$ (All parentheses are dropped.). The infinite term cannot be represented by a term graph.

A term graph with loops can be unrolled to infinite set of other term graphs. We need to define regular unrolling sets as follows.

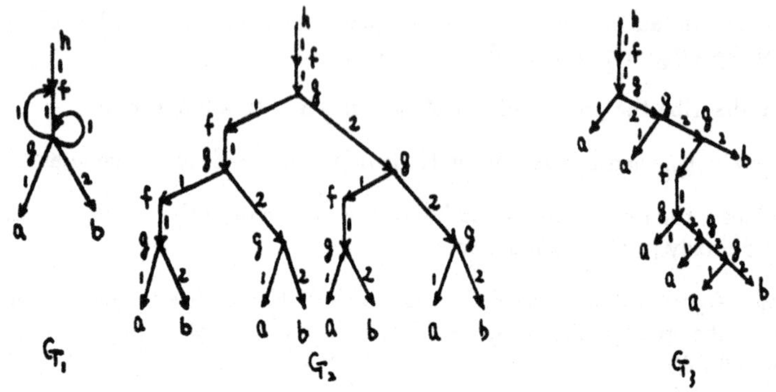

Figure 7

Definition For a term graph G with loops, a *regular unrolling set* of G is

$$\{G_m | G_m \text{ is the result of unrolling } l_1, l_2, \ldots, l_n \text{ to } m \text{ cycles for } m \geq k\},$$

where $l_1, l_2, \ldots, l_n$ are loops in G such that no other loops in G connect to l_i $(1 \leq i \leq l)$ and k indicates the minimal number of cycles in these unrolling graphs.

By the definition, if G has more than one maximal group of connected loops, there are different kinds of regular unrolling sets that depend on the loops chosen to unroll. In a regular unrolling set, all graphs have loops in them or all graphs have no loops in them.

Clearly, different graphs may be unrolled to the same set of term graphs. We need to define simplest term graphs. First we define a quasi-order on term graphs.

Definition Term graphs $G_1 \preceq G_2$ if there are regular unrolling sets S_{G_1} and S_{G_2} for G_1 and G_2 respectively such that $S_{G_2} \subseteq S_{G_1}$.

For example, in Figure 8. $G_1 \preceq G_2$.

Figure 8

If both $G_1 \preceq G_2$ and $G_2 \preceq G_1$, we say $G_1 \simeq G_2$, this is an equivalence relation. In Figure 9, $G_1 \simeq G_2 \simeq G_3$. If $G_1 \preceq G_2$ but $(G_1 \not\simeq G_2)$, we say that $G_1 \prec G_2$. In Figure 8, $G_1 \prec G_2$.

Definition In a set of equivalent term graphs, we call a graph in the set a *simplest term graph* if it has the least nodes.

In Figure 9, G_3 is a simplest term graph.

We use simplest term graphs to represent both finite term and infinite cyclic terms. Finite terms are represented by graphs without loops, infinite terms are represented by graphs with loops and regular unrolling sets describe how the infinity can be produced. From now on we do not distinguish term graphs from terms. For some simple infinite terms we still use the conventional representation of [DK89] to avoid drawing a lot of term graphs.

Figure 9

3 Reduction on Term Graphs

After getting uniform finite representation for finite terms and infinite cyclic terms, we can consider reductions on term graphs. Because a term graph is the graphical representation of a term, the reduction on a graph corresponds to reduction on the corresponding terms. But there are some differences for graphs with loops.

Let $t_1 \underset{u}{\longrightarrow} t_2$ mean that $t_1 \rightarrow t_2$ and the reduction is at the occurrence u in t_1. Then we define parallel reductions.

Definition For any TRS, if

$$t = t_0 \underset{u_1}{\longrightarrow} t_1 \underset{u_2}{\longrightarrow} \cdots \underset{u_n}{\longrightarrow} t_n = t',$$

where $t_i \in T$, and u_j and u_k are disjoint for any $1 \leq j \neq k \leq n$. We say that t reduces in *parallel* to t' [OD77] [HU80], which we write $t \twoheadrightarrow\!\!\!\!\!\!\!+\ t'$.

Now we first extend the relation $\rightarrow$ on T to T^G. (T^G is the term set of the alphabet Σ of a TRS, including infinite terms that can be represented by term graphs.) We need to define reduction from an infinite cyclic term to an infinite cyclic term.

Definition Term graphs t_1 and t_2 have the relation $t_1 \rightarrow t_2$ if and only if there exist regular unrolling sets S_{t_1} and S_{t_2} for t_1 and t_2 respectively, such that for every term $s_1 \in S_{t_1}$ there is a term $s_2 \in S_{t_2}$ such that $s_1 \twoheadrightarrow\!\!\!\!\!\!\!+\ s_2$, and for every $s_2' \in S_{t_2}$, there is a term $s_1' \in S_{t_1}$ such that $s_1' \twoheadrightarrow\!\!\!\!\!\!\!+\ s_2'$.

Because connected loops must be unrolled together, we use parallel reduction in the definition. For example, $f^\omega(a) \rightarrow f^\omega(a)$ by the rule $a \rightarrow f(a)$. For the TRS

$$\mathsf{R} = \{b \rightarrow f(a, a), a \rightarrow b\},$$

in Figure 10 term G_1 can be reduced to terms G_2 and G_3.

The definition above allows an infinite term t_1 to reduce to a finite term t_2, if every term $s_1 \in S_{t_1}$ reduces to t_2. This happens only when there is a rule where some variable appears in its left hand side but does not appear in its right hand side. The reverse, a finite term reduced to an infinite term by one step of reduction, is impossible.

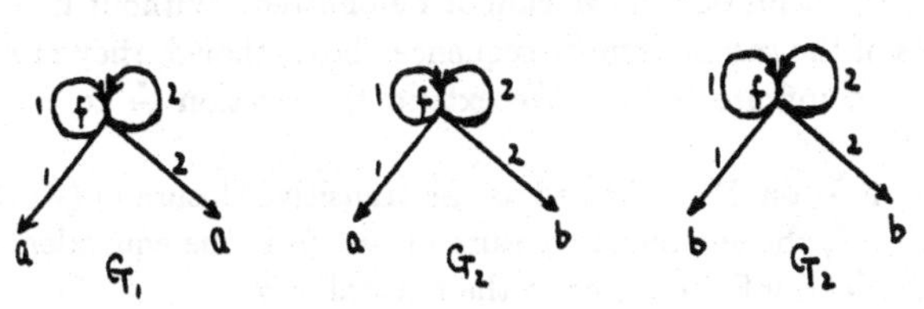

Figure 10

Next we define infinite reductions.

Definition For a TRS with reduction relation $\rightarrow$ over T^G, the relation $\stackrel{\omega}{\rightarrow}$ over T^G is defined by $s \stackrel{\omega}{\rightarrow} t$ (or $t \stackrel{\omega}{\rightarrow} s$) for s and t in T^G, if and only if

1. there exists a chain $s = s_0 \rightarrow s_1 \rightarrow \ldots \rightarrow s_n \rightarrow \ldots$ (or $\ldots \rightarrow s_n \rightarrow \ldots \rightarrow s_1 \rightarrow s_0 = s$).

2. there exists a regular unrolling set S_t for t, such that $\{s_n | n \geq 0\} \supseteq S_t$.

3. t is a minimal ($\prec$) term graph satisfying condition 2.

The idea behind the definition is that from a finite term to an infinite cyclic term by infinite reductions, for example, the reductions behave periodically after some term and an infinite number of terms after that can be produced by unrolling a term graph.

Note that this definition includes all kinds of infinite reductions: a finite term to an infinite term, the reverse, and an infinite term to an infinite term.

For example, in the TRS

$$\mathsf{R} = \{a \rightarrow f(a), a \rightarrow c, f(c) \rightarrow c\},$$

$a \stackrel{\omega}{\rightarrow} f^\omega(a) \rightarrow f^\omega(c) \stackrel{\omega}{\rightarrow} c$.

In the TRS

$$\mathsf{R} = \{a \rightarrow g(a), f(x) \rightarrow f(f(x))\},$$

$f(a) \stackrel{\omega}{\rightarrow} f^\omega(a), f^\omega(a) \stackrel{\omega}{\rightarrow} f^\omega(g^\omega(a))$ and $f(a) \stackrel{\omega}{\rightarrow} f^\omega(g^\omega(a))$. The first reduction and the third one depend on the infinite sequence from $f(a)$. So, a term may reduce to different terms by $\stackrel{\omega}{\rightarrow}$.

In the TRS

$$\mathsf{R} = \{f(b,a) \rightarrow g(f(c,a)), f(c,a) \rightarrow g(f(b,a))\},$$

then $f(b,a) \stackrel{\omega}{\rightarrow} g^\omega(f(b,a)), f(b,a) \stackrel{\omega}{\rightarrow} g^\omega(f(c,a))$ and they are gotten from the same infinite reduction sequence. Furthermore, $g^\omega(f(b,a)) \rightarrow g^\omega(f(c,a))$ and $g^\omega(f(c,a)) \rightarrow g^\omega(f(b,a))$.

Why do we not define the relation $\stackrel{\omega}{\rightarrow}$ by $\{s_n | n \geq 0\} \subseteq S_t$ instead of the condition 2 above? If it is used, in the TRS

$$\mathsf{R} = \{a \rightarrow f(b), a \rightarrow b, b \rightarrow f(f(b))\},$$

both sequences $a \rightarrow f(b) \rightarrow f(f(f(b))) \rightarrow \ldots$ and $a \rightarrow b \rightarrow f(f(b)) \rightarrow \ldots$ will generate the cyclic term G_1 in Figure 11. So, two disjoint sequences have the same limit, and the two inequivalent terms $f(b)$ and $f(f(b))$ in the original TRS yield the same infinite result.

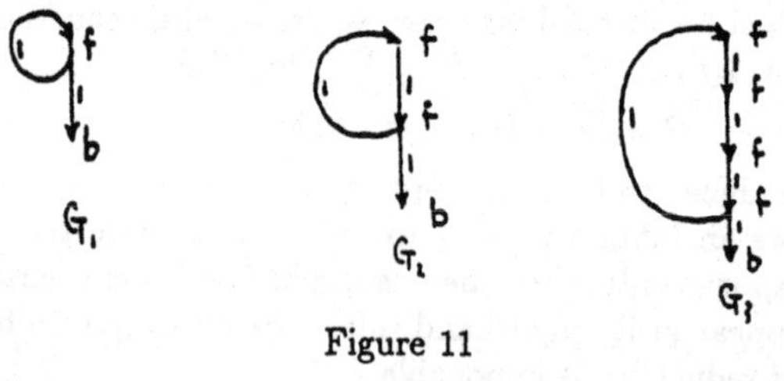

Figure 11

The third condition in the definition above cannot be omitted. Without it, both G_2 and G_3 in Figure 11 become limits of the second infinite sequence above, though they are not equivalent.

Also we can define $\stackrel{*}{\rightarrow}$ on infinite terms. We extend the relation $\stackrel{*}{\rightarrow}$ to include $\stackrel{\omega}{\rightarrow}$ by the following definition.

Definition The relation $\stackrel{*\omega}{\rightarrow}$ on T^G is defined as the transitive closure of $(\stackrel{\omega}{\rightarrow} \cup \rightarrow)$.

Definition $\leftrightarrow = \rightarrow \cup \rightarrow^{-1}$, the symmetric closure of $\rightarrow$. $\stackrel{*}{\leftrightarrow}$ is the equivalence generated by the relation $\rightarrow$. $\stackrel{*\omega}{\leftrightarrow}$ extends $\stackrel{*}{\leftrightarrow}$ to infinite terms in the natural way.

4 ω–Termination and ω–Confluence

Before discussing ω–termination and ω–confluence we need to define infinite normal forms. In this section we only discuss the TRSs where all infinite terms that can be produced from finite terms are able to be represented by term graphs.

Definition An *ω-normal form* for a TRS is a term t such that, either

1. t is finite and minimal with respect to $\rightarrow$, i.e. $\neg(\exists t_1\ t \rightarrow t_1)$; or

2. t is infinite and if $t \rightarrow t_1$ or $t \stackrel{\omega}{\rightarrow} t_1$ then $t_1 = t$.

An ω–normal form of a term s in T^G is a term t in T^G, such that $s \stackrel{*\omega}{\rightarrow} t$ and t is an ω–normal form.

In the TRS

$$\mathsf{R} = \{a \rightarrow f(a), a \rightarrow c, f(c) \rightarrow c\},$$

neither $f^\omega(a)$ nor $f^\omega(c)$ is an ω–normal form because $f^\omega(a) \rightarrow f^\omega(c)$ and $f^\omega(c) \stackrel{\omega}{\rightarrow} c$ though $f^\omega(c) \rightarrow f^\omega(c)$.

The definitions of ω–normal form here and in [DK89] are essentially the same, but the definitions of $\stackrel{\omega}{\rightarrow}$ are different. The advantage of our definitions can be seen in the following example. In the TRS

$$\mathsf{R} = \{a \rightarrow s(a), b \rightarrow s(b), h(x,x) \rightarrow c\},$$

there is $h(a,b) \stackrel{*\omega}{\rightarrow} h(s^\omega(a), s^\omega(b))$, but no $h(s^\omega(a), s^\omega(b)) \stackrel{*\omega}{\rightarrow} c$. [DK89] drops a and b of the infinite term, so it may be reduced to c in their system.

Now we weaken the termination requirement, to guarantee that every derivation leads to an ω–normal form.

Definition For an infinite reduction sequence $t_0 \rightarrow \ldots \rightarrow t_n \ldots$, if there is an ω–normal form t such that for all n, $t_n \stackrel{\omega}{\rightarrow}\stackrel{\omega}{\rightarrow} \ldots \stackrel{\omega}{\rightarrow} t$ by finite number of $\stackrel{\omega}{\rightarrow}$, we say that the infinite sequence *leads to an ω-normal form.*

Definition A TRS is *ω-terminating* if every infinite sequence $t_0 \rightarrow \ldots \rightarrow t_n \rightarrow \ldots$ leads to an ω–normal form.

Clearly, in an ω–terminating TRS, for every loop in a term, one step and at most one step of reduction can increase one cycle of the loop.

Definition A TRS is *ω-confluent* if whenever $t_1 \stackrel{*\omega}{\rightarrow} t_2$ and $t_1 \stackrel{*\omega}{\rightarrow} t_3$, then there exists a term t_4 such that $t_2 \stackrel{*\omega}{\rightarrow} t_4$ and $t_3 \stackrel{*\omega}{\rightarrow} t_4$.

Before proving some theorems about ω–confluence and ω–termination, we need some lemmas.

Lemma 1 If $t_1 \stackrel{\omega}{\rightarrow} t$ and $t_2 \stackrel{\omega}{\rightarrow} t$ and t_1, t_2 are finite terms and t is a finite or infinite one, then there is a finite term t' such that $t_1 \stackrel{*}{\rightarrow} t'$ and $t_2 \stackrel{*}{\rightarrow} t'$.

Proof Direct from the definition of infinite terms.

Lemma 2 If $t_1 \stackrel{*\omega}{\rightarrow} t$ and $t_2 \stackrel{*\omega}{\rightarrow} t$ and t_1, t_2 are finite terms and t is infinite, then there is a finite term t' such that $t_1 \stackrel{*}{\rightarrow} t'$ and $t_2 \stackrel{*}{\rightarrow} t'$.

Proof By the lemma above, using induction on the number of $\stackrel{\omega}{\rightarrow}$ steps.

Lemma 3 For an infinite term t, if there are a regular unrolling set of finite terms S_t and a finite term s such that $t_m \stackrel{*}{\rightarrow} s$ for every term $t_m \in S_t$, then $t \stackrel{*\omega}{\rightarrow} s$ with at most one $\stackrel{\omega}{\rightarrow}$ step in this $\stackrel{*\omega}{\rightarrow}$ sequence.

Proof There are only two cases:

1. There are some rules with some variables appearing only in their left-hand sides and they are used to make the unrolled cycles of all loops disappear. The conclusion is direct from the definition of one step of reduction from an infinite term to a finite term.

2. There are some rules that can be used to decrease the numbers of cycles of all loops. The conclusion is also obvious from the definition of $\stackrel{\omega}{\rightarrow}$ from an infinite term to an finite term.

Lemma 4 In a confluent TRS, if $t_1 \xrightarrow{\omega} t_2$ and $t_1 \xrightarrow{\omega} t_3$, then there is a term t_4 such that $t_2 \xrightarrow{*\omega} t_4$ and $t_3 \xrightarrow{*\omega} t_4$ with at most one $\xrightarrow{\omega}$ step in each of these $\xrightarrow{*\omega}$ sequences.

Proof Without losing generality, assuming t_1 has no loops that are not affected by these reductions to t_2 and t_3.

1. If t_1 is finite, then t_2 and t_3 are infinite. If the loops in t_2 and t_3 are the same, then they can reduce to a common form by finite number of reduction. Otherwise, there exist regular unrolling sets S_{t_2} and S_{t_3} of finite terms for t_2 and t_3 respectively. By confluence, every $t_2' \in S_{t_2}$ and every $t_3' \in S_{t_3}$ must be joinable to a finite term. All these cases can be joinable to the same finite term by using rules to decrease the numbers of cycles of all loops or using some rules with some variable only appearing in their left-hand sides. Then the conclusion is direct from Lemma 3.

2. If t_1 is infinite, then t_2 and t_3 are finite. The conclusion direct from the definition of $\xrightarrow{\omega}$ from an infinite term to a finite term.

Lemma 5 In a confluent TRS, if $t_1 \xrightarrow{*\omega} t_2$ and $t_1 \xrightarrow{*\omega} t_3$ with at most one $\xrightarrow{\omega}$ step in each of these $\xrightarrow{*\omega}$ sequences, then there is a term t_4 such that $t_2 \xrightarrow{*\omega} t_4$ and $t_3 \xrightarrow{*\omega} t_4$ with at most one $\xrightarrow{\omega}$ step in each of these $\xrightarrow{*\omega}$ sequences.

Proof the proof is based on the different cases of reduction sequences $t_1 \xrightarrow{*\omega} t_2$ and $t_1 \xrightarrow{*\omega} t_3$.

1. Both these sequences are $\xrightarrow{*}$. Direct from confluence.

2. Both them are $\xrightarrow{\omega}$. Direct from Lemma 4.

3. One of them is $\xrightarrow{*}$, the other is $\xrightarrow{\omega}$, the proof is analogous to that of Lemma 4.

4. There are $\xrightarrow{*}$ before or behind of $\xrightarrow{\omega}$. Easy to prove from case 2 and case 3.

Theorem 1 A TRS is confluent if and only if it is ω–confluent.

Proof First we prove ω–confluence implies confluence. For all finite terms $t_1 \xrightarrow{*} t_2$ and $t_1 \xrightarrow{*} t_3$, suppose there is no finite term t_4 such that $t_2 \xrightarrow{*} t_4$ and $t_3 \xrightarrow{*} t_4$, there must exist an infinite term t_4', such that $t_2 \xrightarrow{*\omega} t_4'$ and $t_3 \xrightarrow{*\omega} t_4'$. Then by Lemma 2 the finite term t_4 must exist.

Then we prove the converse. By Lemma 5, this goes by induction on the number of $\xrightarrow{\omega}$ steps in $t_1 \xrightarrow{*\omega} t_2$ and $t_1 \xrightarrow{*\omega} t_3$.

Theorem 2 If a TRS is left-linear, ω–terminating and locally confluent, then it is confluent and ω–confluent.

Proof By theorem 1, we only need to prove it is confluent and only consider the case that it is not terminating.

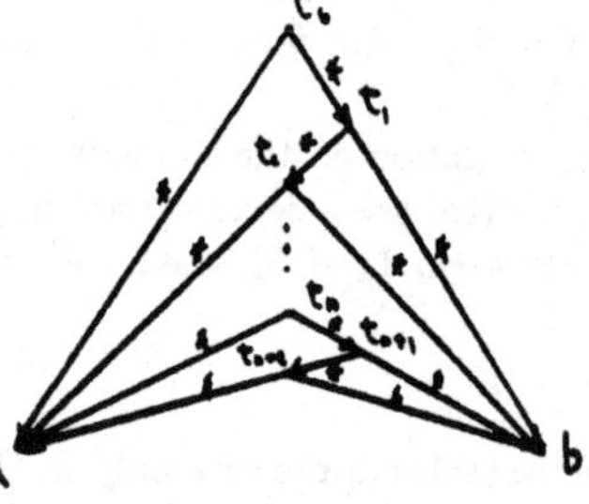

Figure 12

Suppose it is not confluent. From [CO90a], we know that there is an infinite critical pair dependence chain $t_0 \xrightarrow{*} t_1 \ldots \xrightarrow{*} t_n \xrightarrow{*} \ldots$, as in Figure 12 with $a \neq b$, and suppose the ω–normal form of the sequence is t. For all i, $t_i \xrightarrow{*} a$ and $t_i \xrightarrow{*} b$. By Lemma 3, $t \xrightarrow{*} a$ and $t \xrightarrow{*} b$. t cannot equal both of a and b, so t is not an ω–normal form, contradicting ω–terminating. Thus the TRS is confluent and therefore ω–confluent.

Theorem 3 If a TRS is ω–confluent, then the following extended "Church–Rosser" property holds: $t_1 \overset{*\omega}{\leftrightarrow} t_2$ iff there exists an t_3 such that $t_1 \overset{*\omega}{\rightarrow} t_3$ and $t_2 \overset{*\omega}{\rightarrow} t_3$.

Proof Obvious.

5 Algebraic Semantics

We have mentioned that the substantial difference between this paper and [DK89] is that we keep more information about an infinite cyclic term but [DK89] drop all information appearing below a loop. This difference does not trouble us in discussing algebraic semantics and getting the same conclusion as [DKP89] did. In this section, we consider algebraic aspects of infinite terms with continuous models. We prove the same theorems for our ω–terminating systems as [DKP89] did for their systems.

First we give some definitions about term set in a TRS R.

1. T_R is the set of all finite ground terms.
2. $T_{R,G}$ is the set of all infinite ground terms that can be produced by infinite reductions from a finite term t_1, i.e., $t_1 \overset{*\omega}{\rightarrow} t$.
3. $T_R^G = T_R \bigcup T_{R,G}$.

Note that T_R^G does not include all possible infinite terms, but only these infinite terms that can be produced by infinite reductions from a finite term. Compared with [DKP89], we include more infinite terms. We also have sets $T_R(X), T_{R,G}(X)$ and $T_R^G(X)$ corresponding to $T_R, T_{R,G}$ and T_R^G augmented by variables in the set X.

Before discussing algebraic semantics, we first prove a lemma. For an infinite term t, let $t_{(k)}$ denote the finite term that is produced by unrolling all the loops in t to k cycles.

Lemma 6 In an ω–terminating TRS, for each infinite term $t \in T_R^G(X)$, there exists k such that for all $m \geq k$, $t_{(m)} \overset{*}{\rightarrow} t_{(m+1)}$.

Proof Obvious.

Definition Given a signature Σ, a *continuous Σ–algebra* consists of:

1. a partially ordered universe $(M, \preceq)$, such that each nonempty increasing sequence has a *least upper bound (lub)* in M, and
2. for each $f \in \Sigma$, a continuous function $f^M : M^{arity(f)} \rightarrow M$, called the interpretation of f.

Now we discuss term algebra for ω–terminating TRSs. The partial order $\preceq$ on $T_R^G(X)$ is $\overset{*\omega}{\rightarrow}$. We prove a lemma first.

Lemma 7 In an ω–terminating TRS, for each infinite term t, there is an infinite sequence of finite terms $t_1, t_2, \ldots, t_m, \ldots$ with $t_m \preceq t_{m+1} (m > 0)$ where t is the least upper bound of this sequence.

Proof By the above Lemma 6, we know there are k and an infinite sequence $t_{(k)}, t_{(k+1)}, t_{(k+2)}, \ldots$ with $t_{(k+i)} \preceq t_{(k+i+1)}$ $(i \geq 0)$. Obviously, t is an upper bound of the sequence. For any upper bound $t'(t' \neq t)$:

1. Clearly, $t \overset{*\omega}{\rightarrow} t'$.
2. By ω–termination, there are no two upper bounds that can be reduced to each other.

So, t is the least upper bound of the sequence.

Given a continuous Σ–algebra M, for an ω–terminating TRS R, any assignment $\sigma : X \rightarrow M$ extends to a morphism $\sigma : T_R^G(X) \rightarrow M$, as follows:

1. If $t = x$ for some $x \in X$, then $\sigma(t) = \sigma(x)$.

2. If $t = f(t_1, \ldots, t_n) \in T_R(X)$, then $\sigma(t) = f^M(\sigma(t_1), \ldots, \sigma(t_n))$.

3. If t is infinite, by Lemma 7 there is a finite term t_m such that $t_m \xrightarrow{\omega} t$. We define $\sigma(t) = \sigma(lub\ \{t_m\}) = lub\ \{\sigma(t_m)\}$ for sequence $\{t_m\}(\{t_m\}$ represents the term sequence in reduction sequence $t_m \to t_{m+1} \to \ldots)$. $\sigma(t)$ must satisfy two conditions:

   (a) If $t = f(t_1, \ldots, t_n)$ and $t_i (1 \le i \le n) \in T_R^G(X)$, then $\sigma(lub\ \{t_m\}) = f^M(\sigma(t_1), \ldots, \sigma(t_n))$.

   (b) If $u_m \xrightarrow{\omega} w \xleftarrow{\omega} u'_m$ for some w, then $\sigma(lub\ \{u_m\}) = \sigma(lub\ \{u'_m\})$.

   The two conditions ensure that for each infinite term, different approaches of evaluations of the morphism must get the same value.

Definition Given an ω–terminating TRS R over $T_R^G(X)$, an R–model is a continuous Σ–algebra M that satisfies that for any rule $l \to r$ in R, assignment $\sigma : X \to M$ and context c in $T_R^G(X)$, the inequality $\sigma(c[l]) \preceq \sigma(c[r])$ holds.

Lemma 8 If R is an ω–terminating TRS, then the universes $(T_R^G, \xrightarrow{*\omega})$ and $(T_R^G(X), \xrightarrow{*\omega})$ are R–models. We express them as $M_{T_R^G}$ and $M_{T_R^G(X)}$.

Proof It's not difficult to check that these two universes are continuous Σ–algebras and satisfy the conditions for R–model.

Theorem 6 If a TRS R is ω–terminating, then $M_{T_R^G}$ is initial in the class Mod^ω of all R–models and $M_{T_R^G(X)}$ is free over X in Mod^ω.

Proof For any $M \in Mod^\omega$, we define a morphism $\phi : M_{T_R^G} \to M$ the same as the evaluation of the interpretation $f^M(f \in \Sigma)$. It is well known that ϕ is unique for finite terms and it can be proved that $\phi(lub\{t_m\}) = lub\{\phi(t_m)\}$. Suppose there is another morphism ϕ'. For infinite term t,

$$\phi'(t) = \phi'(lub\{t_m\}) = lub\{\phi'(t_m)\} = lub\{\phi(t_m)\} = \phi(lub\{t_m\}) = \phi(t).$$

So, ϕ is unique and continuous.

Similarly, $M_{T_R^G(X)}$ is free over X in Mod^ω.

Definition Given $t, t' \in T_R^G(X)$ and $M \in Mod^\omega$, we say that M *satisfies the inequality* $t \preceq t'$ if for every assignment $\sigma : X \to M$ we have $\sigma(t) \preceq \sigma(t')$. In this case, we write $M \models t \preceq t'$. If $M \models t \preceq t'$ for every $M \in Mod^\omega$, we say that Mod^ω satisfies $t \preceq t'$ and write $Mod^\omega \models t \preceq t'$.

Theorem 7 If a TRS R is ω–terminating and $t, t' \in T_R^G(X)$, then $Mod^\omega \models t \preceq t'$ if and only if $t \xrightarrow{*\omega} t'$.

Proof Suppose that $Mod^\omega \models t \preceq t'$, then in particular, $T_R^G(X) \models t \preceq t'$. It means $t \xrightarrow{*\omega} t'$. Conversely, suppose that $t \xrightarrow{*\omega} t'$. Let $M \in Mod^\omega$ and $\sigma : X \to M$. $t \xrightarrow{*\omega} t'$ means $t = t_0 \to$ (or $\xrightarrow{\omega}$) $\ldots \to$ (or $\xrightarrow{\omega}$) $t_n = t'$. For any reduction $t_{i-1} \to t_i$ or $t_{i-1} \xrightarrow{\omega} t_i$, we have $\sigma(t_{i-1}) \preceq \sigma(t_1)$ and then $\sigma(t) \preceq \sigma(t')$.

Definition The class Eq^ω is a subclass of Mod^ω. For each M of the class, $\sigma(c[l]) = \sigma(c[r])$ for any rule $l \to r$ in R, assignment $\sigma : X \to M$ and context c in $T_R^G(X)$.

theorem 8 If R is an ω–terminating and ω–confluent TRS, then the universe NF^ω composed by all ω–normal forms of the system is initial in Eq^ω.

Proof Analogous to above proofs.

Corollary If R is an ω–terminating and ω–confluent TRS and $t, t_1 \in T_R^G$, then $Eq^\omega \models t \preceq t_1$ & $t_1 \preceq t$ if and only if $t \xrightarrow{*\omega} w \xleftarrow{*\omega} t_1$ for some w.

6 Conclusion

We extend term rewritings from terminating systems to nonterminating systems and study infinite reduction sequences with some special stability. In this paper we discuss infinite cyclic terms, i.e., loops appear in infinite reduction sequences. In [CO90b], we discuss another useful stability which is called head boundedness, i.e., in an infinite reduction sequence the head symbol will never change after some finite number of standard reduction steps.

One thing should be mentioned here. To make the definition of term graph understood easily, we did not use a more complex definition. So, for the system

$$R = \{a \rightarrow f(a,b,c), b \rightarrow g(b,c,a), c \rightarrow h(c,a,b)\},$$

we cannot draw term graphs for some infinite terms of this system. But it is easy to extend the definition to cover this kind of systems and all conclusions we have gotten in this paper still hold.

References

[CO90a] Yiyun Chen, Michael J. O'Donnell. *Testing Confluence of Nonterminating Rewriting Systems.* In this Proceedings.

[CO90b] Yiyun Chen, Michael J. O'Donnell. *Nonterminating Rewritings with Head Boundedness.* Technical Report, The University of Chicago, 1990.

[DK89] Nachum Dershowitz and Stéphane Kaplan. *Rewrite, rewrite, rewrite, rewrite, rewrite ...* In 16th Annual Symposium on Principles of Programming Languages, pp. 250–259, SIGPLAN and SIGACT, 1989.

[DKP89] Nachum Dershowitz, Stéphane Kaplan and David A. Plaisted. *Infinite Normal Forms.* Volume 372 of Lecture Notes in Computer Science, Springer–Verlag, pp. 249–262, 1989.

[HU80] G. Huet. *Confluent reductions:Abstract properties and applications to term rewriting systems.* J. of the Association for Computing Machinery 27(4), pp. 797–821, 1980.

[OD77] M. J. O'Donnell. *Computing in systems described by equations.* Volume 58 of Lecture Notes in Computer Science, Springer–Verlag, 1977.

Testing Confluence of Nonterminating Rewriting Systems

Yiyun Chen*†
Michael J. O'Donnell*
The University of Chicago

Abstract
A key problem in the implementation of equational logic programming is how to check that an equational program is confluent. The current version of the equation compiler allows no critical pairs, or overlaps, between left-hand sides of rules. Normal approaches to proving confluence in the presence of critical pairs require that all reductions terminate, but many of the most useful techniques in equational logic programming involve lazy evaluation of terms with nonterminating reductions. In this paper, we give a theorem, showing how confluence can be determined only by critical pairs and their relations in left-linear term rewriting system.

1 Introduction

The *equational logic programming* project started in 1975. O'Donnell initiated a small project to explore the consequences of implementing equational logic programming with no semantic compromises. In essence, he produced a programming language, which uses a set of equations, applied as *rewrite rules* as its computational mechanism. A rewrite rule can be thought of as the natural mechanism used by people to evaluate. In the language, a set of equations, when treated as (from left to right) rewrite rules, are *regular systems* of rules [KL80], i.e., they have no overlaps between left-hand sides, and the *left-linear*. O'Donnell [OD77] and also Rosen, Berry Lévy proved that regular systems have the *confluence*, or *Church-Rosser property*, which is sufficient to guarantee the completeness of term rewriting as an implementation of equational logic programming.

Unfortunately, although the regular systems of equations are theoretically much more general than LISP programs, typical examples of concepts that do not code nicely into LISP are also troublesome to express with regular sets of equations. It appears, however, that a really desirable equational logic programming language should allow benign overlaps that do not destroy the Church-Rosser property. There are certainly many overlaps which do not destroy the Church-Rosser property. For examples, $(i+j)+k = i+(j+k)$ overlaps itself in $((1+2)+3)+4$ and $i*(j+k) = i*j+i*k$ overlaps with $i+0=i$ in $2*(5+0)$, but these overlaps do not destroy the Church-Rosser property.

Normal approaches to proving confluence in the presence of critical pairs require that all reductions terminate, but many of the most useful techniques in equational logic programming involve lazy evaluation of terms with nonterminating reductions. So, a new method to test confluence needs to develop. In this paper a new theorem about confluence is given which shows that a left-linear term rewriting system is confluent if and only if there is no infinite dependence chains of critical pairs with unjoinable dependences. Then a method of using the theorem to check confluence is explained. Section 2 gives some preliminaries for term rewriting systems. Section 3 proves the main theorem of this paper. Section 4 describes how to use the theorem in practice.

*The research of these authors was supported in part by NSF grant CMPS 8805503.
†Yiyun Chen is supported by a grant from the People's Republic of China.

2 Preliminaries

We only mention some important definitions and properties of Term Rewriting Systems (TRSs) which are concerned with the proofs of next section, other definitions and properties about TRSs come from [HU80]. In this paper we only consider finite left-linear TRSs (LTRSs).

Now we recall the notion of occurrence of a subterm in a term[HU80]. Let N_+^* be the set of sequences of positive integers, Λ is the empty sequence in N_+^*, and . the concatenation operation on sequences. We shall call the members of N_+^* *occurrences* and denote them u, v, w. We define the *prefix ordering* $\preceq$ in N_+^* by $u \preceq v \Leftrightarrow \exists w\ v = u.w$; in this case we define $v/u = w$. Occurrences u and v are said to be *disjoint*, denoted by $u|v$, iff $u \not\preceq v$ and $v \not\preceq u$. We let $u \prec v$ iff $u \preceq v$ and $u \neq v$.

For any term M, we define its *set of occurrences* $\mathsf{O}(M) \subseteq N_+^*$ and the *subterm of M at u*, M/u is a term, for $u \in \mathsf{O}(M)$, as follow.

1. If $M = x$, x is a variable, then $\mathsf{O}(M) = \{\Lambda\}$ and $M/\Lambda = M$.

2. If $M = F(M_1 \ldots M_n)$, then $\mathsf{O}(M) = \{\Lambda\} \cup \{iu | i \leq n, u \in \mathsf{O}(M_i)\}$, $M/\Lambda = M$, and $M/iu = M_i/u$.

We say that u is *an occurrence of M/u in M*.

Because an occurrence is a sequence of positive integers, we also use *lexicographical ordering* on the set of occurrences.

Definition A *redex-pattern* is (informally) the left-hand side of a rule in which the variables are discarded [KM87].

According to the Lemma 3.1 in [HU80], there are three cases of the relation between two redexes in a term. Suppose the occurrences of two redexes are u_1 and u_2 and the corresponding rules and substitutions are $l_1 \to r_1, l_2 \to r_2$ and σ_1, σ_2 respectively.(V is the variable set of the TRS.)

1. *Disjoint redexes* $u_1|u_2$.

2. *Nested redexes* $u_1 \prec u_2$ and there is a $v, v = u_2/u_1$ and $\sigma_1(l_1)/v = \sigma_2(l_2)$, furthermore, $v = v_1.v_2, l_1/v_1 = x \in \mathsf{V}$ and $\sigma_2(l_2) = \sigma_1(x)/v_2$.

3. *Overlapping redexes* $u_1 \preceq u_2$ and there is a $v, v = u_2/u_1$ and $\sigma_1(l_1)/v = \sigma_2(l_2)$, furthermore, $l_1/v \notin \mathsf{V}$ and $\sigma_2(l_2) = \sigma_1(l_1/v)$.

Note that, in nonleft-linear TRSs, a term has a redex-pattern at somewhere does not mean there must be a redex at the same position, but in LTRSs it does. So, in a LTRS, if two redex-patterns overlap in a term, there are two correspondent redexes in the term and they also overlap, but it's not true for nonleft-linear TRSs.

The following definition of critical pair is also out of [HU80].

Definition Let $l_1 \to r_1, l_2 \to r_2 \in \mathsf{R}$ and $u \in \mathsf{O}(l_1)$ such that $M = l_1/u$, $M \notin \mathsf{V}$ and M and l_2 are unifiable. Let N be a most general instance of M and l_2 such that $\mathsf{V}(N) \cap \mathsf{V}(l_1) = \phi$ ($\mathsf{V}(M)$ means the set of variables of term M). There are σ_1 and σ_2 such that $\sigma_1(M) = N$ & $\sigma_2(l_2) = N$. We say that the superposition of $l_2 \to r_2$ on $l_1 \to r_1$ in u determines the *critical pair* (P, Q) defined by

$$P = \sigma_1(l_1)[u \leftarrow \sigma_2(r_2)], \quad Q = \sigma_1(r_1).$$

Term $\sigma_1(l_1)$ is called the *normal generating term* of (P, Q), because $\sigma_1(l_1) \to P$ and $\sigma_1(l_1) \to Q$. These two reductions lead two branches out of $\sigma_1(l_1)$, they are called *critical pair branches.*

In a locally confluent TRS, for each critical pair (P, Q) and its normal generating term S, there is a *common form* R such that $P \xrightarrow{*} R$ & $Q \xrightarrow{*} R$ as in Figure 1(a). We call such a reduction figure *locally confluent figure.*

By stability and compatibility [HU80], Figure 1(b) holds. In this figure, M is called an *abnormal generating term* of (P, Q), M_3 is also called a common form and the two branches out of

M are also called critical pair branches. In the following sections, we simply describe the situation as *M with critical pair branches.*

Both normal generating terms and abnormal generating terms are called generating terms.

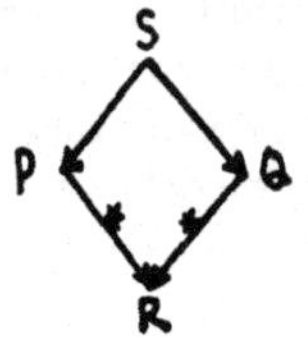

Figure 1(a)

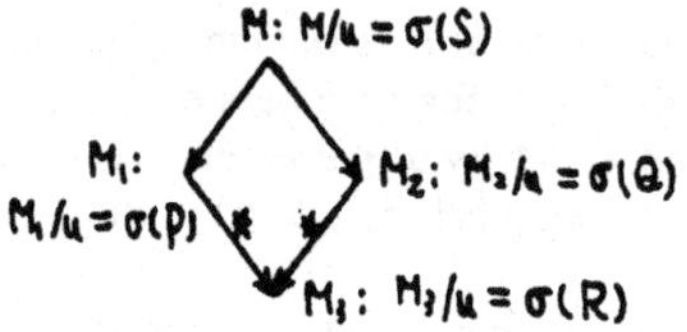

Figure 1(b)

3 The Confluence Theorem

The main theorem of this paper is given in this section, which shows how confluence of a LTRS can be determined only by critical pairs and their relations.

Definition If R is a common form for P and Q, we say that P and Q are *joinable*. If there is not any other common form R_1 for P and Q such that $R_1 \xrightarrow{*} R$, but not $R \xrightarrow{*} R_1$, R is called a *minimal common form* for P and Q.

Note that, minimal common form may not be unique for a pair of terms.

Definition In a locally confluent TRS, for normal generating terms S_1 with critical pair (P_1, Q_1) and S_2 with critical pair (P_2, Q_2), if for some minimal common form R_1 of (P_1, Q_1), every path (restricted to a certain reduction strategy) from Q_1 (or P_1) to R_1 contains a term M such that there are substitutions σ_1, σ_2 and position u in M with $\sigma_1(M/u) = \sigma_2(S_2)$ $(M/u \notin V)$, and in Figure 2(a), the first step from M' to R_1' is a critical pair branch concerned with (P_2, Q_2) and furthermore, if there is no critical pair (P_3, Q_3) such that (P_1, Q_1) and (P_3, Q_3) have such relation and so do (P_3, Q_3) and (P_2, Q_2), we say that critical pair (P_1, Q_1) is *dependent on* critical pair (P_2, Q_2) *at first level* and this relation is expressed as $(P_1, Q_1) \prec_1 (P_2, Q_2)$.

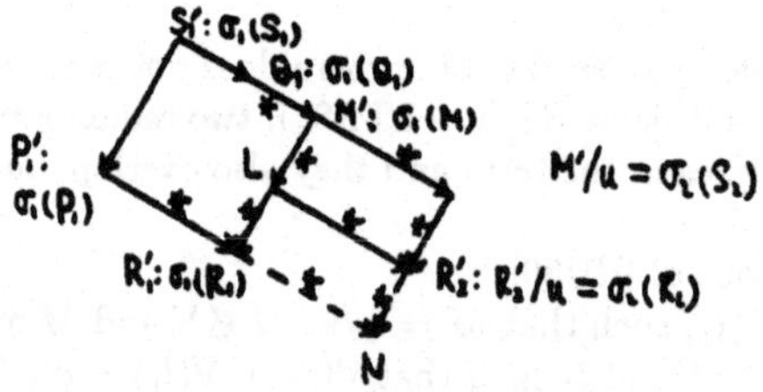

Figure 2(a)

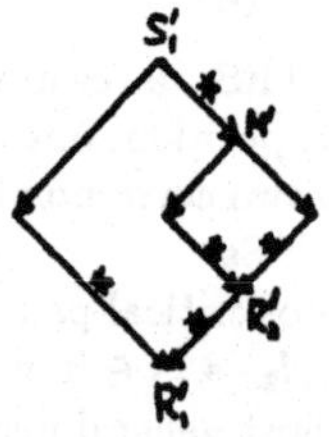

Figure 2(b)

There are two cases of the dependence $(P_1, Q_1) \prec_1 (P_2, Q_2)$ in Figure 2(a).

1. There is a common form N for R_1' and R_2'. We say this dependence at first level is *joinable*.

2. Otherwise, we say this dependence at first level is *unjoinable*.

If $N = R_1'$ and sequence $L \xrightarrow{*} R_1'$ is the same as $L \xrightarrow{*} R_2' \xrightarrow{*} N$, Figure 2(a) can be simplified to Figure 2(b).

In order to understand the definition well, more examples are given. In Figure 3(a), $(P_1, Q_1) \prec_1 (P_2, Q_2)$ and $(P_1, Q_1) \prec_1 (P_3, Q_3)$. In Figure 3(b), $(P_1, Q_1) \prec_1 (P_2, Q_2)$ and $(P_2, Q_2) \prec_1 (P_3, Q_3)$, but $(P_1, Q_1) \not\prec_1 (P_3, Q_3)$. In Figure 3(c), though S_3 happens to appear at all paths, (P_1, Q_1) (P_3, Q_3) does not hold.

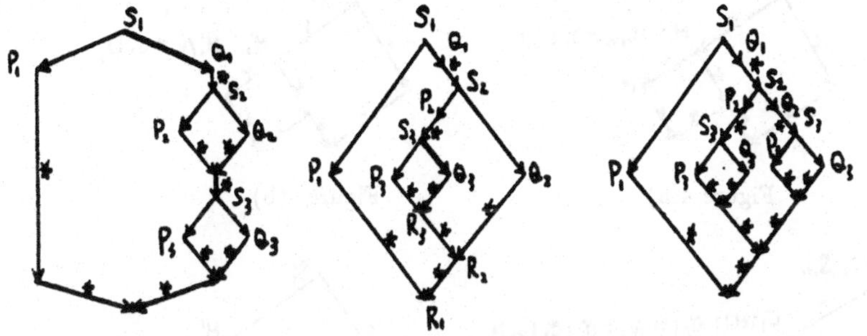

Figure 3(a) Figure 3(b) Figure 3(c)

If $(P_1, Q_1) \prec_1 (P_2, Q_2)$ and $(P_2, Q_2) \prec_1 (P_3, Q_3)$ as in Figure 4(a) and 4(b) and $\sigma_2(S_2)$ and $\sigma_1'(S_2)$ are unifiable, then there are substitutions σ_3, σ_4 such that $\sigma_3(M_2/v) = \sigma_4(S_3)$ & $u \preceq v$ and Figure 4(c) holds. We say that critical pair (P_1, Q_1) is *dependent on* critical pair (P_3, Q_3) *at second level.* This relation is expressed as $(P_1, Q_1) \prec_2 (P_3, Q_3)$.

1. If $(P_1, Q_1) \prec_1 (P_2, Q_2)$ is joinable, then as in Figure 4(c), there is a common form R_1' for R_1 and R_2. If there is also a common form N for R_1' and R_2', then we say this is a joinable dependence at second level. (In this case, R_2 and R_2' are also joinable).
2. Otherwise, $(P_1, Q_1) \prec_1 (P_2, Q_2)$ is unjoinable or there is no such a common form N, then we say this is an unjoinable dependence at second level.

Clearly, if $(P_1, Q_1) \prec_2 (P_3, Q_3)$ then $(P_1, Q_1) \prec_1 (P_2, Q_2)$ and $(P_2, Q_2) \prec_1 (P_3, Q_3)$ holds. But the reverse may not be true. The case in Figure 4(d) can be defined analogously.

In this way, relation $\prec_n$, joinable dependence at level n and unjoinable dependence at level n for any number n can be defined. They can also be extended by stability and compatibility.

If $(P_1, Q_1) \prec_n (P_{n+1}, Q_{n+1})$, there is a *dependence chain* $(P_1, Q_1) \prec_1 (P_2, Q_2), \ldots, (P_n, Q_n) \prec_1 (P_{n+1}, Q_{n+1})$, but, as above, the reverse may not be true.

We define a new relation $\prec$ as the union of $\prec_n$ for all n.

Definition In a locally confluent TRS, critical pair (P_1, Q_1) is *dependent on* critical pair (P_2, Q_2) if there is an m such that $(P_1, Q_1) \prec_m (P_2, Q_2)$. This relation is expressed as $(P_1, Q_1) \prec (P_2, Q_2)$. If for any possible m, $(P_1, Q_1) \prec_m (P_2, Q_2)$ is joinable, then $(P_1, Q_1) \prec (P_2, Q_2)$ is called joinable, otherwise, it is unjoinable.

It is not difficult to prove that the transitive closure of relation $\prec_1$ is a superset of $\prec$. The reason is the same as above, $(P_1, Q_1) \prec_1 \ldots \prec_1 (P_n, Q_n)$ does not imply $(P_1, Q_1) \prec (P_n, Q_n)$.

If there is a dependence chain $(P_1, Q_1) \prec_1 \ldots \prec_1 (P_{n+1}, Q_{n+1})$ and there are $i, j (1 \leq i < j \leq n+1)$ such that $(P_i, Q_i) = (P_j, Q_j)$, we say that there is a *dependence loop* in the chain.

Obviously, if there is an unjoinable dependence in a locally confluent TRS, it is not confluent.

Now we need to define standard reduction sequence. Before giving the definition, we first show a lemma.

Definition In a LTRS, if a term M_1 has two nonoverlapping redex-patterns at u_1 and u_2 (So, there are two redexes at u_1 and u_2), and in $M_1 \rightarrow M_2 \rightarrow M_3$ the first step reduces the redex-pattern at u_1 and the second one reduces the redex-pattern at u_2, then if another reduction sequence $M_1 \xrightarrow{*} M_3$ can be found such that the redex-pattern at u_2 in M_1 is reduced first, we say that these two successive reductions are *reorderable*.

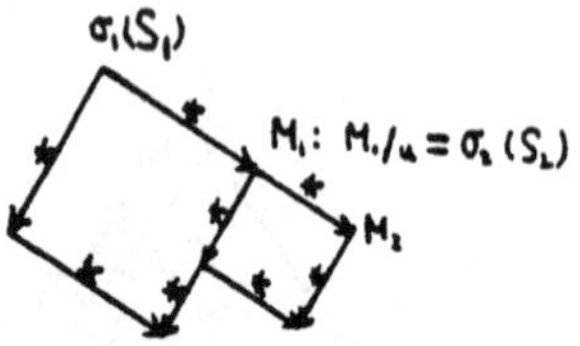

Figure 4(a)

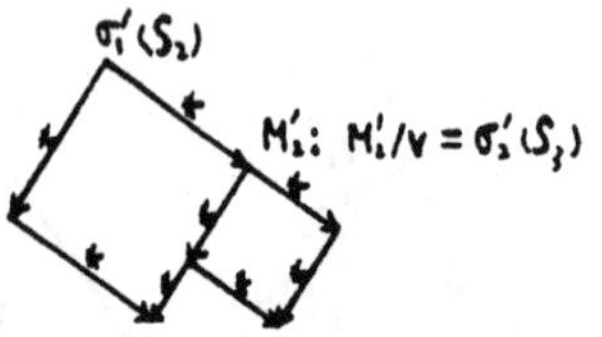

Figure 4(b)

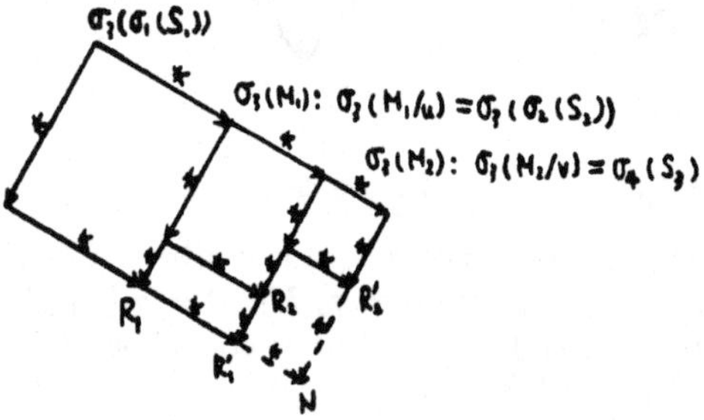

Figure 4(c)

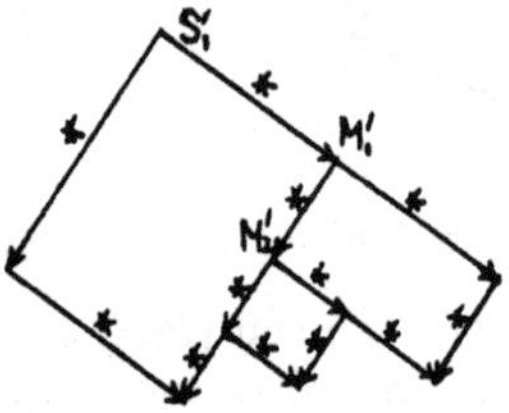

Figure 4(d)

Lemma 1 In the definition above, if u_2 is lexicographically less than u_1, $M_1 \to M_2 \to M_3$ is always reorderable.

Proof The proof is analogous to the proof of Lemma 3.1 in[HU80].

For example, if a rule set is

$$\mathsf{R} = \{f(x) \to g(x,x),\ a \to b\},$$

$f(h(a)) \to f(h(b)) \to g(h(b), h(b))$ can be reordered to

$f(h(a)) \to g(h(a), h(a)) \to g(h(b), h(a)) \to g(h(b), h(b))$.

If a rule set is

$$\mathsf{R} = \{f(x) \to c,\ a \to b\},$$

$f(a) \to f(b) \to c$ can be reordered to $f(a) \to c$.

The lemma does not apply to nonleft-linear systems. For example, in system

$$\mathsf{R} = \{f(x,x) \to a, g(x) \to f(x, g(x)), b \to g(b)\},$$

$f(b, g(b)) \to f(g(b), g(b)) \to a$ is not reorderable, though there are redex-patterns at 1 and Λ in $f(b, g(b))$ and they are reduced at those two reductions.

In this paper only this kind of reorderable reductions are considered.

Definition In a LTRS, a reduction sequence $M \xrightarrow{*} N$ is standard if there are no reorderable successive reductions.

Note that in a standard sequence, at any reduction steps, current leftmost and outermost redex is reduced or else it will be killed later or will not be reduced in the sequence.

Lemma 2 In a LTRS, any reduction sequence $M \xrightarrow{*} N$ can be converted into a unique standard sequence by reordering reorderable successive reductions.

Proof Trivial.

Lemma 3 In a locally confluent LTRS without unjoinable dependence of critical pairs, if there are standard sequences $M \xrightarrow{*} M_n$ and $M \to N$, and M is a generating term, and the first step $M \to M_1$ of $M \xrightarrow{*} M_n$ and $M \to N$ are a critical pair branches of the generating term, then there is an N_n such that $M_n \xrightarrow{*} N_n$ and $N \xrightarrow{*} N_n$, as in Figure 5.

Proof If the length of $M \xrightarrow{*} M_n$ is one, by local confluence, N_n exists. Assume that the length is more than one. By local confluence and no unjoinable dependence of critical pairs at any level, a suitable reduction figure as Figure 6 can be gotten, where all sequences are standard. First we consider the figure is finite. $M \to M_1 \xrightarrow{*} M_n$ in Figure 5 and some sequence $M \to M_1 \xrightarrow{*} N_i'(0 < i \leq n)$ in Figure 6 must agree with some initial steps. Suppose they agree as long as possible and reach M_u and $M_v(M_u = M_v)$ respectively. If $M_u = M_n$ or $M_v = N_i'$, it is easy to show that N_n exists. Suppose they reduce the redexes at positions u and v respectively at next step. The redexes at u and v are not overlapping redexes, otherwise the two sequences can agree at least one more step by going along one branch. There are two cases for the relative position of u and v.

1. u is lexicographically less than v. The redex at u in M_v must remain in N_i'. Add the same reduction as the first step of $M_u \xrightarrow{*} M_n$ to N_i', then restandardize all sequences in Figure 6 and get a new figure that we also express by Figure 6. $M \to M_1 \xrightarrow{*} M_n$ and $M \to M_1 \xrightarrow{*} N_i'$ agree at least one more step.

2. u is lexicographically greater than v. The redex at v in M_u must remain in M_n. Add the same reduction as the first step of $M_v \xrightarrow{*} N_i'$ to M_n, restandardize the sequence $M \xrightarrow{*} M_n$ with the new step, then get $M \to M_u' \xrightarrow{*} M_n^1$. $M \to M_u' \xrightarrow{*} M_n^1$ and $M \to M_1 \xrightarrow{*} N_i'$ agree at least one more step.

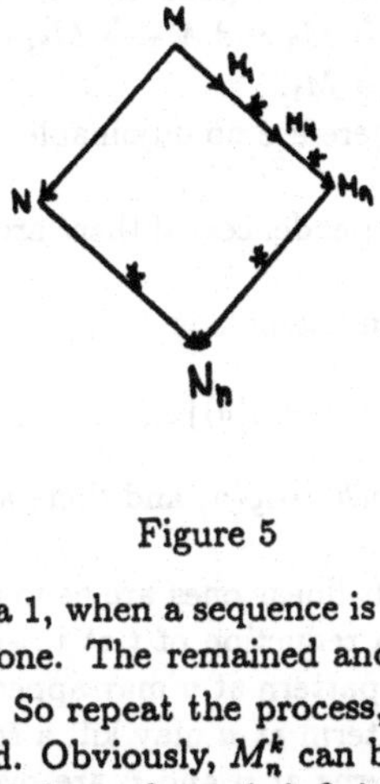

Figure 5

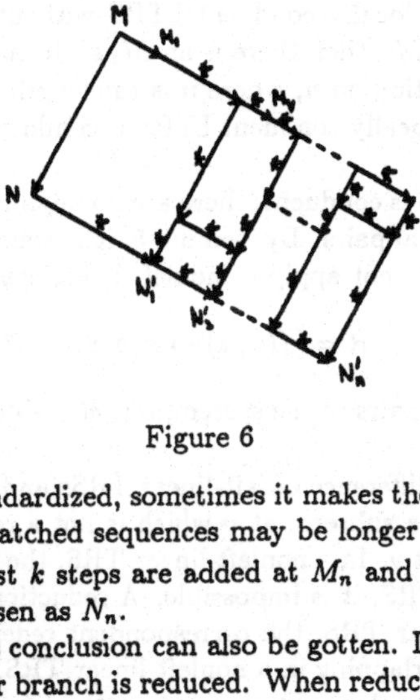

Figure 6

By Lemma 1, when a sequence is restandardized, sometimes it makes the new sequence longer than the old one. The remained and unmatched sequences may be longer than before, but this case is finite. So repeat the process, at last k steps are added at M_n and $M \xrightarrow{*} M_n \xrightarrow{k} M_n^k$ and $M_n^k = N_i'$ hold. Obviously, M_n^k can be chosen as N_n.

If the dependence figure is infinite, this conclusion can also be gotten. In $M \xrightarrow{*} M_n$, there are finite generating terms where a critical pair branch is reduced. When reductions are added at the end of $M \xrightarrow{*} M_n$, generating terms are not increased in the sequence. So, at last $M \xrightarrow{*} M_n$ can be agreed into a left branch somewhere in Figure 6.

Lemma 4 In a locally confluent LTRS without unjoinable dependence of critical pairs, if $M \xrightarrow{*} M_n$ and $M \to N$, then there is an N_n such that $M_n \xrightarrow{*} N_n$ and $N \xrightarrow{*} N_n$.

Proof First $M \xrightarrow{*} M_n$ is standardized to $M \xrightarrow[std]{*} M_n$ (*std* means standard). If it can be proved that there is an N_n suited for $M \xrightarrow[std]{*} M_n$, then the N_n is also suited for $M \xrightarrow{*} M_n$, too. This lemma is proved by induction on the length n of $M \xrightarrow[std]{*} M_n$.

1. If $n = 0$, choose $N_n = N$.

2. If $n = 1$, by local confluence, N_n exists.

3. If $n \geq 2$, there are five cases in terms of the first step $M \to M_1$ of $M \xrightarrow[std]{*} M_n$. Suppose $M \to M_1$ and $M \to N$ reduce redexes at positions u_1 and u_2 respectively.

(a) Same redexes: $u_1 = u_2$ and same rules. Let $N_n = M_n$.

(b) Disjoint redexes: $u_1|u_2$. There is N_1 such that $N \rightarrow N_1$ and $M_1 \rightarrow N_1$. The remained $M_1 \xrightarrow[std]{*} M_n$ is short than $M \xrightarrow[std]{*} M_n$. By hypothesis, N_n exists.

(c) Nested redexes: $u_2 \prec u_1$. By the proof of Lemma 1, there is N_1 such that $N \xrightarrow{*} N_1$ and $M_1 \rightarrow N_1$. The remained $M_1 \xrightarrow[std]{*} M_n$ is shorter than $M \xrightarrow[std]{*} M_n$. By hypothesis, N_n exists.

(d) Nested redexes: $u_1 \prec u_2$. By the proof of Lemma 1, there is N_1 such that $N \rightarrow N_1$ and $M_1 \xrightarrow{*} N_1$. The remained $M_1 \xrightarrow[std]{*} M_n$ is short than $M \xrightarrow[std]{*} M_n$. Reduction $M \rightarrow M_1$ produces n disjoint redexes at $v_1, \ldots, v_n$, all these positions lexicographically greater than u_1, and $M_1 \xrightarrow{*} N_1$ just reduce these redexes at $v_1, \ldots, v_n$. For any subterm T_i of M_1 at $v_i (1 \leq i \leq n)$, in the sequence $M_1 \xrightarrow[std]{*} M_n$, the reductions on T_i is no more than the sequence $M_1 \xrightarrow[std]{*} M_n$ itself. By hypothesis, for any subterm T_i of M_1 at $v_i (1 \leq i \leq n)$, we have the conclusion. Obviously, N_n exists.

(e) Overlapping redexes. By Lemma 3.

By 1,2 and 3, the proof is completed.

Lemma 5 In a locally confluent LTRS without unjoinable dependence of critical pairs, if $M \xrightarrow{*} M_1$ and $M \xrightarrow{*} M_2$, then there is an M_3 such that $M_1 \xrightarrow{*} M_3$ and $M_2 \xrightarrow{*} M_3$, i.e., confluence.

Proof By induction on n, where n is the length of $M \xrightarrow{*} M_1$.

Theorem 1 A locally confluent LTRS is confluent iff there are no unjoinable dependences of critical pairs.

Proof If a LTRS is confluent, there are no unjoinable dependences. If there are no unjoinable dependences of critical pairs. By Lemma 5, it is confluent.

This theorem does not apply to nonleft-linear system, an example is

$$\mathsf{R} = \{f(x,x) \rightarrow a, g(x) \rightarrow f(x, g(x)), b \rightarrow g(b)\}.$$

There are no critical pairs in the system, but $g(b) \xrightarrow{*} a$ and $g(b) \xrightarrow{*} g(a)$ and there are no common forms for a and $g(a)$.

The substantial difference of left-linear TRSs and nonleft-linear ones are as follows.

In a term t with a subterm at v which is not a redex, a reduction of t at u with $v \prec u$ may create a new redex at v. In a nonleft-linear TRS, the redex-pattern at v may appear in t already, but in a left-linear TRS, it is impossible. A reduction of a term at u may kill a redex at v with $v \prec u$. In a left-linear TRS, the correspondent redex-patterns at v and u are overlapping, but they may not be overlapping in a nonleft-linear TRS.

The theorem above does not give us a method to determine if there are no unjoinable dependences in a locally confluent LTRS. We need to develop another theorem.

Lemma 6 In a locally confluent LTRS, if M is a generating term with critical pair branches as in Figure 7, and M_1 and M_2 cannot be joinable, then there is an M' such that the relation in Figure 7(a) or 7(b) or 7(c) holds, M_1' and/or M_2' are generating terms with critical pair branches, M_1 and M' (and/or M_2 and M') cannot be joinable, and the critical pair concerned with M is dependent on the critical pair concerned with M_1' (and/or M_2') at first level.

Proof Using the same techniques in the proof of Lemma 4.

Theorem 2 In a locally confluent LTRS, if there is an unjoinable dependence of critical pairs, then there must be an infinite dependence chain with $\prec_k$ existing for any number k and in the chain, there is an m such that $\prec_n$ is unjoinable for all $n > m$,

Proof If there is an unjoinable dependence, then, as in Figure 8, M_1 and N_1 are not able to be joinable. The same techniques in proving Lemma 4 can be used to make L be a generating term with critical pair branches. The critical pair concerned with M is dependent on the critical pair concerned with L. By Lemma 6, there must be an infinite dependence chain and some $\prec_n$ is unjoinable.

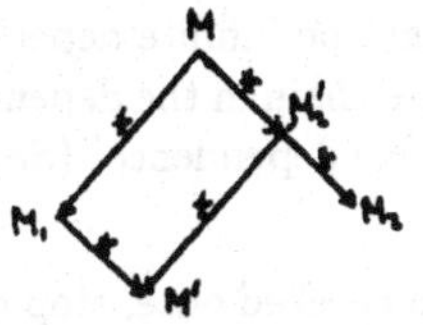

Figure 7(a)

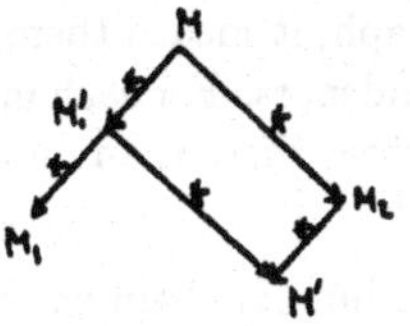

Figure 7(b)

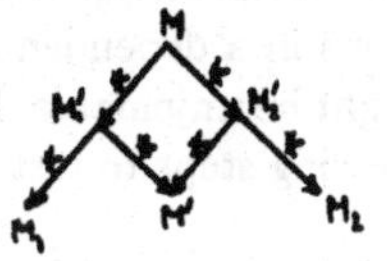

Figure 7(c)

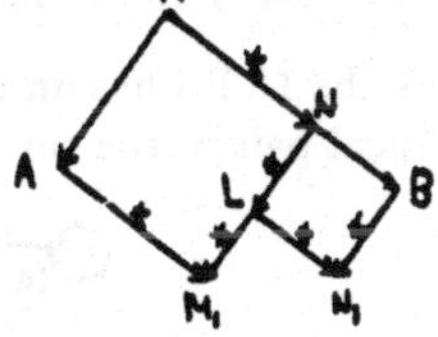

Figure 8

4 Examining Confluence

By the theorems in last section, for a locally confluent LTRS, if there are no dependence loops in it, then there are no infinite dependence chains. So in order to determine confluence, it only needs to examine whether dependence loops will lead to an unjoinable dependence.

Now we first discuss how to get all dependence relations of critical pairs in a locally confluent LTRS. For a normal generating term with a critical pair, there may be more than one minimal common form, as in Figure 9. In Figure 9(a), minimal common forms R_1 and R_2 cannot reduce to each other. In Figure 9(b), minimal common forms R_1 and R_2 can reduce to each other. So, for a critical pair there may be more than one locally confluent figures with different minimal common forms.

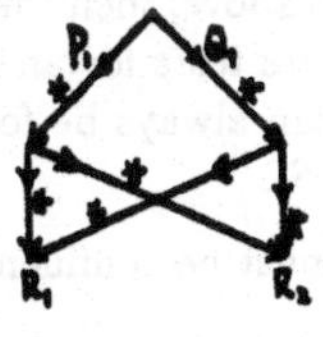

Figure 9(a)

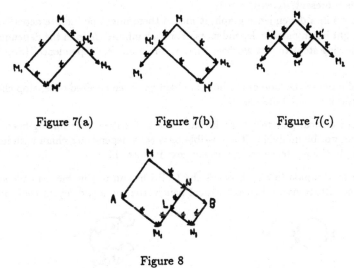

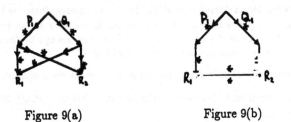

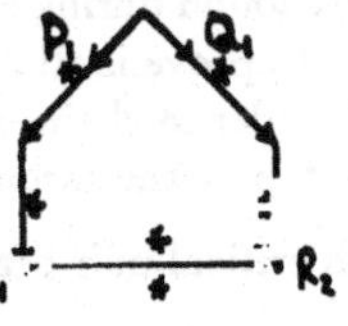

Figure 9(b)

For every normal generating term with critical pair (P, Q), check locally confluent figure(s), then all critical pairs, which (P, Q) depends on and appear in the figure(s), can be gotten. These critical pairs include all (P_i, Q_i) with $(P, Q) \prec_1 (P_i, Q_i)$. So, from all locally confluent figures of a LTRS, a superset of relation $\prec_1$ can be gotten.

It can be proved that for the case of Figure 9(a), if only one locally confluent figure is checked, we can also get a superset of relation $\prec_1$. It can also be proved that if common forms are used instead of minimal common forms in local reduction figures, the same conclusion holds though the superset of $\prec_1$ may not the same as above .

For example, let the rule set of a LTRS is expressed in Figure 10(a). Local confluence figures are in Figure 10(c) (The middle figure of Figure 10(c) is a short notation of three locally confluent figures.) and dependence graph is in Figure 10(b) that is produced from Figure 10(c). In this graph, some edges represent $\prec_n$ with $n > 1$.

If there are loops in a dependence graph, it means there might be infinite dependence chains and then there might be unjoinable dependences. For each infinite chain in the dependence graph, we repeat the following steps to test whether there is an unjoinable dependence. (See Figure 4(c) and 4(d)).

- If all critical pairs in the loop of the infinite chain has been checked once, stop checking this chain and check next infinite chain.
- If the current critical pair to be checked and the last critical pair having been checked of the chain can not be unifiable (It is possible because a dependence chain with length $n+1$ does not imply the existence of $\prec_n$), check next infinite chain.
- If there are no common forms, it means the LTRS has an unjoinable dependence and it is not confluent. Otherwise, take next critical pair in the chain and repeat these steps.

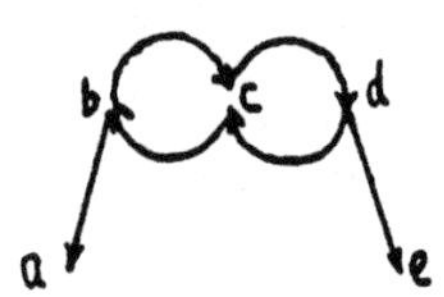

Figure 10(a)

Figure 10(b)

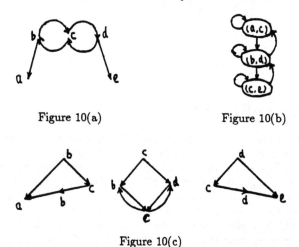

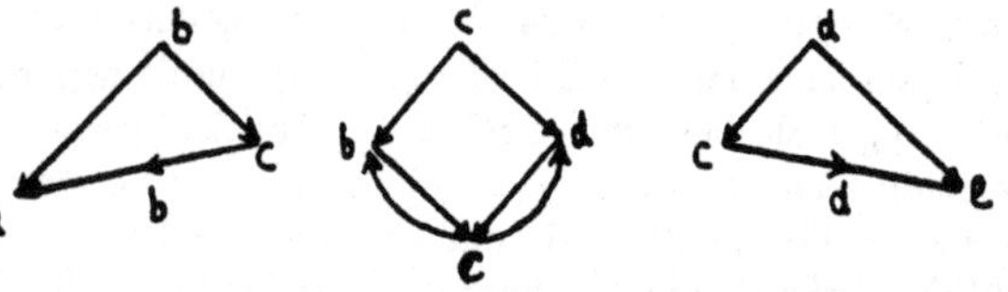

Figure 10(c)

Theorem 3 In a locally confluent LTRS, for every infinite chain of critical pairs, if no unjoinable dependences are found during executing the process above, then the LTRS is confluent.

Proof What we need to prove is that why we can use finite tests for an infinite dependence chain. For an infinite dependence chain, if a common form can always be found, at last Figure 11 can be gotten and we stop at the second time of checking S_2.

- If there is no L with critical pair branches, R_1 and R_2 must be confluent.
- If L is a generating term and there is no infinite chain through (P, Q), R_1 and R_2 are also confluent.
- Otherwise, there is an infinite chain through (P, Q). It can be proved whether there is a common form for R_1 and R_2 depends on the checking of another infinite dependence chain. For the case in Figure 4(d), the same conclusion holds.

For example, for the LTRS in Figure 10, it is not confluent because there is an infinite dependence chain with an unjoinable dependence as in Figure 12. Another simple example with unjoinable dependence is

$$\mathsf{R} = \{f(x,y) \to g(x,y), g(x,y) \to f(x,y), f(h(x),y) \to a, g(x,h(y)) \to b\}.$$

Examining confluence for an arbitrary LTRS can be concluded as three steps

1. Examining local confluence and get locally confluent figure(s) for every critical pair.
2. Constructing dependence graph of critical pairs.
3. Checking whether there are unjoinable dependences of critical pairs.

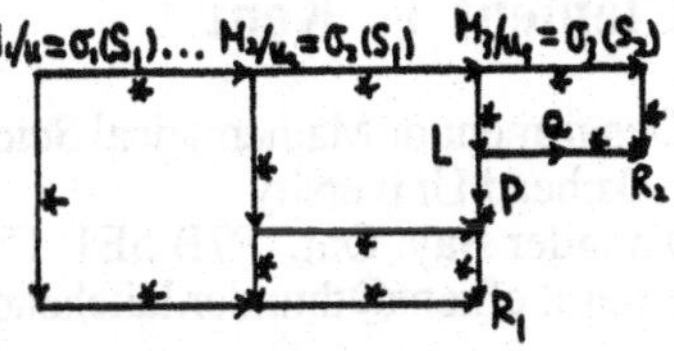

Figure 11

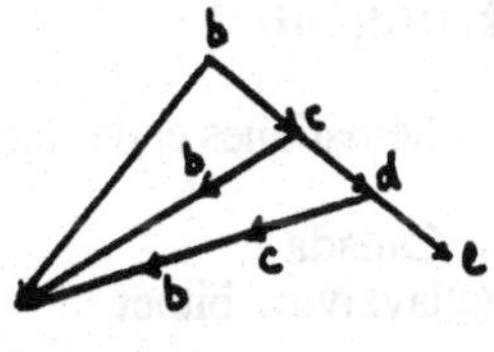

Figure 12

It does not mean confluence is decidable. The problem is that the procedure of looking for a common form for a pair of terms may not terminate because of infinite reductions. So testing confluence is still troubled by the problem of terminating of a system. But the problem here is not so severe as determining termination for a whole TRS.

According to our experience, in practical use, there are not many reduction steps for a critical pair of terms to get to a common form. We can use breadth search to find a common form and the depth in breadth search for a common form can be limited.

Another problem of the method is that we need to determine whether there are more than one minimal common forms (or common forms) for a critical pair and they can reduce to each other. It might also be solved by breadth search.

The theorems proved in this paper is potentially useful, but the method of determining confluence presented in this section needs to be improved.

References

[HU80] G. Huet. *Confluent reductions:Abstract properties and applications to term rewriting systems.* J. of the Association for Computing Machinery 27(4), pp. 797–821, 1980.

[KL80] J. W. Klop. *Combinatory reduction systems.* PhD thesis, Mathematisch Centrum, Amsterdam, 1980.

[KM87] J. W. Klop, A. Middeldorp. *Strongly Sequential Term Rewriting Systems.* Report CS–R8730, Center for Mathematics and Computer Science, Amsterdam.

[OD77] M. J. O'Donnell. *Computing in systems described by equations.* Volume 58 of Lecture Notes in Computer Science, Springer-Verlag, 1977.

[OD85] M. J. O'Donnell. *Equational Logic as a Programming Language.* MIT Press, Cambridge, MA, 1985.

A SURVEY OF ORDINAL INTERPRETATIONS OF TYPE ε_0 FOR TERMINATION OF REWRITING SYSTEMS

Bernard R. Hodgson †

Département de mathématiques et de statistique
Université Laval
Québec G1K 7P4 Canada
e-mail: bhodgson@lavalvm1.bitnet

Clement F. Kent †

Department of Mathematical Sciences
Lakehead University
Thunder Bay, Ont. P7B 5E1 Canada
e-mail: cfkent@thunder.lakeheadu.ca

Abstract

This paper discusses termination of term-rewriting systems from the point of view of formal provability. We examine certain extensions of the formal theory of elementary arithmetic in which termination proofs can take place and compare their relative strengths.

1. Introduction

The use of well-founded orderings as a method for proving termination of general calculations goes back to Floyd [Fl67]. When applied to computations with a term-rewriting system ([De85]), this technique, as pointed out by Manna & Ness [MN70], rests on finding an *interpretation function* τ mapping the terms of the system into a certain *well-founded set* $(W, \prec)$ and such that, throughout the rewriting process, the value of τ in W is monotonically decreasing with respect to the order $\prec$ (i.e. such that

(*) $\quad \tau(r) \prec \tau(l)$ for each rewrite rule $l \to r$ of the system).

Since well-foundedness prevents infinite descending sequences in W, rewriting must thus terminate. (Detailed expositions of this and other termination proof techniques for rewriting can be found in the surveys Dershowitz [De87] and Dershowitz & Jouannaud [DJxx].)

The purpose of this paper is to discuss termination of rewriting systems from the point of view of *formal provability*. Our concern is thus with formal theories, K, in which the termination proofs can be carried out, i.e. so that

- $\vdash_K$ "$(W, \prec)$ is a well-founded set"
- •• for any rule $l \to r$ of the system, $\vdash_K$ "$\tau(r) \prec \tau(l)$",

where $\vdash_K$ means provability in the theory K and " · " indicates an appropriate description of the given property in the formal language of K. It is natural to take K to be at least as strong as elementary (Peano) arithmetic, PA.

---

† Authors supported by the Natural Sciences and Engineering Research Council (grants A4519 and A5603).

Using ideas introduced by Routledge [Ro53], it can be shown that for any terminating rewrite system over a recursive set T of terms, there is a recursive interpretation function τ mapping T in a non-standard primitive recursive well-ordering $(\mathbb{N}, \prec)$ of type ω for which (*) is satisfied. If we choose those standard orderings having the properties of the Cantor Normal Forms (such restriction avoids pathological orders), then no new formal termination proofs result from suborderings of type α, $\omega \leq \alpha < \varepsilon_0$. The ordinal ε_0 is the so-called "proof-theoretic ordinal" of PA — it is well known that the standard ordering of type ε_0 cannot be proved a well-order in PA.

Consequently, the basic well-founded set for our formal proofs of termination of rewriting systems will be the (standard) well-ordering $(\Omega,<)$ of type ε_0 obtained by encoding ordinals below ε_0 in Cantor Normal Forms(1). Our object is to find a ("weak") well-order assertion, WO, for Ω such that the formal theory $K = PA \cup \{WO\}$ can prove the finiteness of all rewriting sequences of the form

$$\tau(t) > \tau(t\!\to) > \tau(t\!\to^{(2)}) > \cdots > \tau(t\!\to^{(n)}) > \cdots$$

(where $t\!\to^{(n)}$ denotes the term obtained after n rewrites of the term t). A reasonable restriction on the interpretation funtion, τ, is that it be a "combinatorial" function, i.e one whose totality is provable in PA.

The rest of this note is organized as follows. We first briefly recall two basic termination techniques based on well-founded orderings: *polynomial interpretations* and *multiset orderings*. We then show how these techniques can lead to another approach in which the functions τ take values into Ω and finally present some general properties of such *ordinal interpretations*.

2. Polynomial interpretations

One particularly successful instance of a well-founded approach to termination is the *polynomial interpretation* method introduced by Lankford [La75]. With each *n*-ary function symbol f of the system is associated a *n*-place polynomial $\tau(f)$ with integer coefficients. These coefficients must be chosen so that $\tau(f)$ is monotonic and takes only non-negative values, as is the case, for example, when they are all positive. Each of the rewrite rule must then be shown to be reducing under τ in the sense of (*).

---

(1) For instance, given the ordinal $\alpha = \omega^{\omega^2 \cdot 2} \cdot 3 + \omega^{\omega \cdot 4} \cdot 2 + \omega$, we can chose any "scale" $z > 4$ (4 is the largest finite ordinal in the CNF for α) and code α by the integer $z^{z^2 \cdot 2} \cdot 3 + z^{z \cdot 4} \cdot 2 + z$.

Various examples of polynomial interpretations can be found in the literature. For instance, Dershowitz & Manna [DM79] consider a rewriting system for symbolic differentiation with respect to x given by the rules

$$Dx \to 1 \qquad Dy \to 0 \qquad D(\alpha \pm \beta) \to D\alpha \pm D\beta \qquad D(-\alpha) \to -D\alpha$$

plus rules for differentiation of $\alpha{\cdot}\beta$, α/β, $\ln \alpha$ and α^β (exponentiation). Each of these nine rewrite rules is reducing under the following polynomial interpretation function τ:

$$\begin{array}{ll} \tau(u) = 4 & \text{where } u \text{ is an atom,} \\ \tau(\alpha \perp \beta) = \tau(\alpha) + \tau(\beta) & \text{where } \perp \text{ is } +, -, \cdot, / \text{ or } \wedge, \\ \tau(\Diamond\alpha) = \tau(\alpha) + 1 & \text{where } \Diamond \text{ is } - \text{ or } \ln, \\ \tau(D\alpha) = \tau(\alpha)^2. & \end{array}$$

3. Multiset and nested multiset orderings

Well-foundedness is also used in the *multiset ordering* approach introduced by Dershowitz & Manna [DM79]. This technique rests on the fact that the set of all finite multisets(2) with elements taken from a well-founded set is itself well-founded, with multiset ordering appropriately defined: for a base set of order type α, this multiset ordering is of order type ω^α. Dershowitz & Manna also consider the notion of a nested multiset(3) and give a definition of a *nested multiset ordering* such that when the base set is of order type less than ε_0, then the nested multiset ordering is of order type ε_0. To each occurence of an element from the base set in a given nested multiset is associated a *depth* corresponding to the number of braces surrounding this occurence and, intuitively, one can think of each level of nesting depth as corresponding to a level of ordinal exponentiation:

$$(**) \qquad \{\cdots\} \quad \Leftrightarrow \quad \omega^{(\cdots)}$$

The rewriting system for symbolic differentiation is proved in [DM79] to be terminating also from the standpoint of multiset ordering and nested multiset ordering. For instance, they introduce the interpretation function τ defined by (***) and taking values into the set of nested multisets over $\mathbb{N}$, where $\perp$ and $\Diamond$ are as above and $|\alpha|$ denotes the length of the term α. When translating this last approach in terms of ordinals below ε_0, using (**), one thus gets a termination proof with index set Ω and interpretation function τ as given by (****), where $\oplus$ denotes "natural" (commuting) ordinal addition.

---

(2) $\{2,2,2,4,0,1,1\}$ is a multiset over $\mathbb{N}$, in which multiple occurences of identical elements are allowed.

(3) $\{\{3,3\},0,\{\{0\},2,1\}\}$ is a nested multiset over $\mathbb{N}$.

$$(***)\quad \begin{aligned} \tau(u) &= \{\ \} \\ \tau(\alpha \perp \beta) &= \tau(\alpha) \cup \tau(\beta) \\ \tau(\Diamond\alpha) &= \tau(\alpha) \\ \tau(D\alpha) &= \{\{|\alpha|\} \cup \tau(\alpha)\} \end{aligned} \qquad (****)\quad \begin{aligned} \tau(u) &= 1 \\ \tau(\alpha \perp \beta) &= \tau(\alpha) \oplus \tau(\beta) \\ \tau(\Diamond\alpha) &= \tau(\alpha) \\ \tau(D\alpha) &= \omega^{\omega^{|\alpha|}} \oplus \tau(\alpha) \end{aligned}$$

4. Ordinal interpretations and Goodstein reduction process

The use of *ordinal interpretations* for termination proofs is present in the work of Kirby & Paris [KP82]. They consider a combinatorial game, the "Battle of Hercules and the Hydra", and show how the tree corresponding to a given Hydra can be coded as an ordinal below ε_0. One verifies that the successive trees generated during the battle give rise to a strictly decreasing sequence of ordinals, so that the game must terminate. However it is shown in [KP82] that this fact cannot be proved in PA, as transfinite (ε_0-)induction is required. The rules for this battle can be presented in the form of a rewriting system, as done in [DJxx]. It thus follows that this system cannot be proved terminating over standard orderings smaller than the well-ordering $(\Omega,<)$ introduced in Section 1.

Kirby & Paris [KP82] also proved similar independence results for another combinatorial principle, the "Goodstein reduction scheme", which can be defined as follows (see [KH89]). Given an ordinal α and an integer $z \geq 2$, the *scale-*z *predecessor of* α, $P_z(\alpha)$, is defined to be the largest ordinal β having a notation of scale z and such that $\beta < \alpha$. We set $P_z(0) = 0$. For α such that $0 < \alpha < \varepsilon_0$ and for the scale sequence $z = z_0 \leq z_1 \leq z_2 \leq \cdots \leq z_k \leq \cdots$, the *Goodstein sequence from* α *and* z is defined as the decreasing sequence of ordinals

$$\alpha > P_z(\alpha) > P_{z_1}P_z(\alpha) > \cdots > P_{z_k}\cdots P_{z_1}P_z(\alpha) > \cdots.$$

For $z_{k+1} = z_k + 1$, we get *standard* Goodstein sequences, studied by Goodstein in [Go44].

Going back to the rewriting system for differentiation, let us consider the sequence of decreasing ordinals generated by translation of the nested multiset approach, as given by (****) above. If one wants to simulate this interpretation process within the context of Goodstein reductions, the scale increase, from one stage to the next, must be such that for each differentiation rule $l \to r$, a notation for the ordinal corresponding to $\tau(l)$ must insure a notation for the ordinal of $\tau(r)$. For the mentioned assignment, the standard scale increase $z_{k+1} = z_k + 1$ is not sufficient for such a notation always to exist, as some of the rules (for instance the rule for differentiation of exponentiation) sometimes require greater scale growth. These considerations have led the present authors to consider in [KH89] Goodstein reduction schemes governed by various types of scale growth. In particular, they were interested in *slow* Goodstein sequences, given by the relation $z_{k+1} = \rho(z_k)$ for a function ρ such that $\rho(x) > x+1$.

5. Equivalence of standard and slow Goodstein reductions

Let us use the notation $gl_z(\alpha)$ [resp. $gl_{\rho,z}(\alpha)$] to represent the *Goodstein length function* giving the number of applications of the predecessor operator needed to reach 0 in Goodstein standard [resp. slow ρ-governed] sequence from α and z. The termination of these reduction processes is thus equivalent to the totality of the length functions $gl_z(\alpha)$ and $gl_{\rho,z}(\alpha)$. This leads us to introduce the following two statements, called *Goodstein's Axioms*:

$$GA:\ \forall\alpha\forall z\exists x\,(gl_z(\alpha) = x) \qquad \text{and} \qquad GA_\rho:\ \forall\alpha\forall z\exists x\,(gl_{\rho,z}(\alpha) = x)$$

for any fixed ρ as above. The main result of [KH89] is that, under the hypothesis that ρ is a provably-total recursive function of PA (satisfying some mild growth condition), GA and GA_ρ are equivalent over PA, so that both statements can be seen as having the same power for formal proofs of termination for rewriting systems.

The independence results of Kirby & Paris [KP82] mentioned above give indications about the limitations of formal proofs of termination for rewriting systems. The intuition behind these independence results is that the growth of the Goodstein length functions is so enormous that it exceeds the "working capacity" of PA. This can be seen by relating the Goodstein length functions to hierarchies of number-theoretic functions known, by previous work of Wainer, to have connection with provable-computability within PA (see [BW87] for a precise statement and a recent proof of Wainer's result). Cichon [Ci83] has made explicit the connections between Goodstein (standard) reduction process and the *Hardy Hierarchy* of functions, showing that the value of the scale upon reaching 0 from α and z is given by $H_\alpha(z) - 1$, where H_α is the αth Hardy function. (Since we then have $gl_z(\alpha) = H_\alpha(z) - z$, the independence of GA from PA follows from the fact that H_{ε_0} is not provably recursive in PA, by the work of Wainer.) Similar results can be given in the case of a slow reduction by considering so-called *inner iteration hierarchies* (see [CW83] for the definition of these hierarchies).

6. Descending sequences on ε_0

Goodstein sequences can be used to dominate some more general types of descending sequences of ordinals below ε_0. We consider *primitive* [resp. *provably-total*] *recursive sequences* $\{f(n)\}$ in the ordinal notations less than ε_0, in which the function f is primitive [resp. provably-total] recursive. Denoting by PRWO(ε_0) [resp. PTRWO(ε_0)] the statement: "All primitive [resp. provably-total] recursive descending sequences on ε_0 are finite," then the equivalence of GA, PRWO(ε_0) and PTRWO(ε_0) can be proved by extending arguments in [KH89].

Since proofs of termination of rewriting systems which use combinatorial interpretation assignments in ordinals less than ε_0 follow from PTRWO(ε_0), then they can be given in PA with the single additional axiom, GA, rather than an axiom scheme of induction. While rewriting systems like the symbolic differentiation algorithm of [DM79] can be proved terminating, under a polynomial interpretation, using only ω-induction, the next step, to ε_0-induction, seems to be a flexible tool going beyond ω, especially when ordinal assignments are interpreted in the context of a combinatorial principle such as GA.

References

[BW87] W. Buchholz and S.S. Wainer, "Provably computable functions and the fast growing hierarchy." In: S.G. Simpson, ed., *Logic and Combinatorics*. (Proceedings of the AMS-IMS-SIAM Joint Summer Research Conference, August 1985), *Contemporary Mathematics* 65 (1987) 179-198.

[Ci83] E.A. Cichon, "A short proof of two recently discovered independence results using recursion theoretic methods." *Proceedings of the American Mathematical Society* 87 (1983) 704-706.

[CW83] E.A. Cichon and S.S. Wainer, "The slow-growing and the Grzegorczyk hierarchies." *Journal of Symbolic Logic* 48 (1983) 399-408.

[De85] N. Dershowitz, "Computing with rewrite systems." *Information and Control* 65 (1985) 122-157.

[De87] N. Dershowitz, "Termination of rewriting." *Journal of Symbolic Computation* 3 (1987) 69-115. (Corrigendum: 4 (1987) 409-410)

[DJxx] N. Dershowitz and J.-P. Jouannaud, "Rewriting systems." In: J. van Leeuwen, ed., *Handbook of Theoretical Computer Science*. North-Holland, to appear.

[DM79] N. Dershowitz and Z. Manna, "Proving termination with multiset ordering." *Communications of the Association for Computing Machinery* 22 (1979) 465-476.

[Fl67] R.W. Floyd, "Assigning meanings to programs." *Mathematical Aspects of Computer Science*, Proceedings of Symposia in Applied Mathematics, vol. 19, American Mathematical Society, 1967, pp. 19-32.

[Go44] R.L. Goodstein, "On the restricted ordinal theorem." *Journal of Symbolic Logic* 9 (1944) 33-41.

[KH89] C.F. Kent and B.R. Hodgson, "Extensions of arithmetic for proving termination of computations." *Journal of Symbolic Logic* 54 (1989) 779-794.

[KP82] L. Kirby and J. Paris, "Accessible independence results for Peano arithmetic." *Bulletin of the London Mathematical Society* 14 (1982) 285-293.

[La75] D.S. Lankford, *Canonical algebraic simplification in computational logic*. Report ATP-25 (Automatic Theorem Proving Project), University of Texas, Austin, TX, 1975.

[MN70] Z. Manna et S. Ness, "On the termination of Markov algorithms." *Proceedings of the Third Hawaii International Conference on System Science*, 1970, pp. 789-792.

[Ro53] N.A. Routledge, "Ordinal recursion." *Proceedings of the Cambridge Philosophical Society* 49 (1953) 175-182.

Meta-rule Synthesis from Crossed Rewrite Systems

Hélène KIRCHNER *and* Miki HERMANN
CRIN and INRIA-Lorraine
Campus Scientifique, BP 239,
54506 Vandœuvre-lès-Nancy, France

{hkirchne, hermann}@loria.crin.fr

Abstract

Infinite sets of rewrite rules may be generated by completion of term rewriting systems. To cope with this problem, detection of divergence and automatic generation of meta-rules from syntactic conditions of divergence are proposed in this paper. We show that in a reasonably large class of divergent systems, equational rewriting is enough to simulate rewriting with meta-rules, but the full power of typed rewriting and conditional rewriting is needed for some divergence problems.

1 Introduction

Confluent and terminating term rewriting systems present an important tool for deciding the word problem in equational theories. They are commonly produced by the Knuth-Bendix completion procedure [BDH86, Hue81,KB70]. The goal of a completion process is to construct a terminating and confluent term rewriting system R from a given set of equalities E.

The practical interest of the completion processes is limited by the possibility of generating infinite sets of rewrite rules. Moreover, the uniqueness of the result of the completion procedure [Hue81,JK86], given a fixed ordering for orienting equalities, implies that it cannot be expected to find another completion strategy for which the completion terminates. However, changing the ordering can have an influence on the behavior of the completion procedure and, in some cases, it can even be possible to avoid its divergence, for example by means described in [Her88].

The problem of generating an infinite set of rewrite rules has been attacked using two different approaches. On one hand, sufficient conditions for predicting the divergence have been given in [Her89]. They are based on the notion of *forward or backward crossed rewrite systems* which is recalled in Section 2.

On the other hand, the notion of *schematization using meta-rules* has been proposed in [Kir89]. Given an infinite set of rules, the problem is to find a finite set of schemas, called meta-rules, where special variables, called meta-variables, may have infinite sets of possible values. A formal notion of schematization was proposed in [Kir89], but discovering the schemas was yet a matter of heuristics. Meta-rules are in general a special kind of conditional rewrite rules, in which conditions express that meta-variables are constrained to belong to a recursively defined domain. In this sense, rewriting with meta-rules is an approach close to rewriting with membership conditions [Toy87]. In [Kir89], it was shown that, under weak assumptions on the domains of meta-variables, rewriting with meta-rules can be interpreted by order-sorted rewriting. This is summarized in Section 3.

In this paper, we bring together these results to deduce, from the syntactic conditions of divergence, an automatic generation method for meta-rules. We show that, in a reasonably large class of divergent systems, equational rewriting is enough to simulate rewriting with meta-rules, but the full power of order-sorted rules may be necessary even in very simple cases. Moreover, for some divergent problems, it is necessary to introduce conditional meta-rules. This is the subject of Section 4. Complete proofs and several examples of divergent systems with their meta-rules synthesized by our method can be found in [KH90].

2 Crossed term rewrite systems

Several examples of divergent systems can be found in [Her88], together with suggestions to avoid their divergence. In all these examples, the divergence phenomenon is due to the existence of a crossed rewrite system, composed of a rule and of an overlap closure. This section introduces these notions, studied in [Her89, Her90].

Let us first introduce some notations consistent with those in [DJ89,DJ90]. Classically $\mathcal{V}, \mathcal{W}\ldots$ denote variable sets, $\mathcal{T}(\mathcal{F},\mathcal{V})$ denotes the free $\mathcal{F}$-algebra generated by the set of variables $\mathcal{V}$, whose elements are called terms and are denoted by $t, t' \ldots$. $\mathcal{V}ar(t)$ denotes the set of variables of the term t. $\mathcal{P}os(t)$ is the set of *positions* of the term t. The subset of non-variable positions of t is denoted by $\mathcal{FP}os(t)$. The subterm of t at some position $a \in \mathcal{P}os(t)$ is denoted by $t|_a$. Let $s[t]_a$ denote a new term obtained from the term s by replacing its subterm $s|_a$ by t.

For a substitution φ, the sets $\mathcal{D}om(\varphi)$, $\mathcal{R}an(\varphi)$, and $\mathcal{V}ar(\varphi)$ denote respectively the domain, range, and all variables of φ, i.e. $\mathcal{V}ar(\varphi) = \mathcal{D}om(\varphi) \cup \mathcal{R}an(\varphi)$. The substitution σ on a term t is a *substitution in own variables* of t if it does not introduce new variables, i.e. $\mathcal{V}ar(t\sigma) \subseteq \mathcal{V}ar(t)$, and does not contain variable renamings.

Two substitutions φ and ψ are **coherent** (denoted by $\varphi \perp \psi$) if $\mathcal{D}om(\varphi) \cap \mathcal{V}ar(\psi) = \emptyset$ or $\mathcal{V}ar(\varphi) \cap \mathcal{D}om(\psi) = \emptyset$.

The following constructions come from [Der87,GKM83].

Definition 2.1 *Let R be an arbitrary set of rules. The set of* **overlap closures** $OC(R)$ *is inductively defined as follows:*

1. *Every rule $s \rightarrow t$ from R is an overlap closure $s \mapsto\!\!\!\rightarrow t$.*
2. *Let $s_1 \mapsto\!\!\!\rightarrow t_1$, $s_2 \mapsto\!\!\!\rightarrow t_2$ be two overlap closures. If $t_1|_a\sigma = s_2\sigma$ holds for a most general unifier σ and position $a \in \mathcal{FP}os(t_1)$, then $s_1\sigma \mapsto\!\!\!\rightarrow t_1\sigma[t_2\sigma]_a$ is an overlap closure.*
3. *Let $s_1 \mapsto\!\!\!\rightarrow t_1$, $s_2 \mapsto\!\!\!\rightarrow t_2$ be two overlap closures. If $t_1\sigma = s_2|_a\sigma$ holds for a most general unifier σ and position $a \in \mathcal{FP}os(s_2)$, then $s_2\sigma[s_1\sigma]_a \mapsto\!\!\!\rightarrow t_2\sigma$ is an overlap closure.*

An overlap closure $s \mapsto\!\!\!\rightarrow s$ with the same terms on both sides is called **reflexive**.

Overlap closures with specific structure (described as closure chains [Her90]) represent the clue for generating crossed rewrite systems [Her89].

Definition 2.2 *The rewrite rule $s_1 \rightarrow t_1$ and the nonreflexive overlap closure $s_2 \mapsto\!\!\!\rightarrow t_2$ (with supposed disjoint variables) form a* **forward crossed rewrite system** *if there are substitutions σ_2, φ_1, φ_2 in own variables of s_2, an idempotent substitution σ_1, and positions $a \in \mathcal{FP}os(s_1)$, $b \in \mathcal{FP}os(t_2)$ such that*

1. $s_1|_a\sigma_1 = s_2\sigma_2$
2. $t_2|_b\varphi_1 = s_2\varphi_2$
3. $\varphi_1 \perp (\varphi_2 \cup \sigma_2)$

Definition 2.3 *Let $s_1 \rightarrow t_1$ and $s_2 \mapsto\!\!\!\rightarrow t_2$ form a forward crossed rewrite system S. Assume that each nontrivial critical pair $\langle s\sigma[t'\sigma]_c, t\sigma\rangle$ computed by the completion procedure from S and an ordering $\succ$ satisfies $s\sigma[t'\sigma]_c \succ t\sigma$.*

Let

$$\begin{aligned} u_1 \rightarrow v_1 &= s_1\sigma_1[t_2\sigma_2]_a \rightarrow t_1\sigma_1 \\ u_{n+1} \rightarrow v_{n+1} &= u_n\omega_n[t_2\omega_n]_{ab^n} \rightarrow v_n\omega_n \end{aligned}$$

define a sequence of rewrite rules $u_n \rightarrow v_n$, where ω_n is the superposition unifier of $u_n|_{ab^n}$ and s_2. The rewrite rules $u_n \rightarrow v_n$ form an infinite **forward iterated family**. *Denote the forward iterated family, generated by a forward crossed system S wrt the positions a and b, as $\mathcal{I}^{a,b}_{FC}(S)$.*

Each unifier ω_n is a function of the substitutions σ_2, φ_1, and φ_2. The iterative method to compute the unifiers ω_n is given in [Her89].

Example 2.4 A natural example of a divergent rewrite system is *Associativity & Endomorphism.*

$$\begin{aligned} (x * y) * z &\rightarrow x * (y * z) \\ f(x) * f(y) &\rightarrow f(x * y) \end{aligned}$$

If we try to complete this system choosing the operator precedence $* > f$ with the left-to-right status of $*$ in the recursive path ordering, we get a divergent process. The completion procedure generates the infinite family of rules

$$f^n(x * y) * z \rightarrow f^n(x) * (f^n(y) * z)$$

The overlap closure is $f(x) * f(y) \rightarrowtail f(x * y)$.

The following definitions describe another divergence pattern for the class of rewrite systems with backward oriented critical pairs.

Definition 2.5 *The overlap closure $s_1 \rightarrowtail t_1$ and the (nonreflexive) rewrite rule $s_2 \rightarrow t_2$ (with supposed disjoint variables) form a* **backward crossed rewrite system** *if t_1 is not a variable and if there are substitutions σ_1, φ_1, φ_2 in own variables of s_1, an idempotent substitution σ_2, and position $b \in \mathcal{FP}os(s_1)$ such that*

1. $s_1|_b\sigma_1 = s_2\sigma_2$
2. $t_1\varphi_1 = s_1|_b\varphi_2$
3. $\varphi_1 \perp (\varphi_2 \cup \sigma_1)$

Definition 2.6 *Let $s_1 \rightarrowtail t_1$ and $s_2 \rightarrow t_2$ form a backward crossed rewrite system S. Assume that each nontrivial critical pair $\langle s\sigma[t'\sigma]_c, t\sigma\rangle$ computed by the completion procedure from S and an ordering $\succ$ satisfies $t\sigma \succ s\sigma[t'\sigma]_c$.*

Let

$$\begin{aligned} u_1 \rightarrow v_1 &= t_1\sigma_1 \rightarrow s_1\sigma_1[t_2\sigma_2]_b \\ u_{n+1} \rightarrow v_{n+1} &= t_1\omega_n \rightarrow s_1\omega_n[v_n\omega_n]_b \end{aligned}$$

define a sequence of rewrite rules $u_n \rightarrow v_n$, where ω_n is the superposition unifier of $s_1|_b$ and u_n. The rewrite rules $u_n \rightarrow v_n$ form an infinite **backward iterated family**. *Denote the backward iterated family, generated by a backward crossed system S wrt the position b, as $\mathcal{I}^b_{BC}(S)$.*

Each unifier ω_n is a function of σ_1, φ_1, and φ_2, similarly to the previous case. The iterative method to compute the unifiers ω_n is given in [Her89], as previously.

Example 2.7 Consider the rewrite system

$$\begin{aligned} (x + f(y)) * y &\rightarrow (x * y) + y \\ (x - y) + y &\rightarrow x \end{aligned}$$

where f is a function that satisfies further the identity $f(x) * x = x$. If we try to complete this rewrite system, choosing the operator precedence $* > +$ in the recursive path ordering, the process diverges while generating the infinite family of rules

$$((((x - f^{n+1}(y)) * f^n(y)) * \ldots * f(y)) * y) + y \rightarrow ((x * f^n(y))) * \ldots * f(y)) * y$$

The overlap closure is $(x + f(y)) * y \rightarrowtail (x * y) + y$.

Let R be a term rewriting system and $\succ$ a reduction ordering. R is *divergent* in the ordering $\succ$ if the completion does not terminate (i.e. generates an infinite set of rules) on inputs R and $\succ$. R is *weakly divergent* in the ordering $\succ$ if the completion without interreduction of rules does not terminate on inputs R and $\succ$. The following result is proved in [Her89].

Theorem 2.8 *If R contains a forward crossed rewrite system S, then R^∞, produced by a completion procedure without interreduction, contains the infinite forward iterated family $\mathcal{I}^{a,b}_{FC}(S)$, hence R is weakly divergent.*

If R contains a backward crossed rewrite system S, then R^∞, produced by a completion procedure without interreduction, contains the infinite backward iterated family $\mathcal{I}^b_{BC}(S)$, hence R is weakly divergent.

The extension to standard divergence (originated from completion with interreduction) is described in [Her89] as well.

3 Finite description of infinite rewrite systems

Schematization is a way to finitely describe infinite families of rules. The following example gives an intuitive idea of the notions introduced hereafter.

Example 3.9 (Example 2.4 continued).
Let us consider the equivalence $\sim$ generated by the rewrite system:

$$\Gamma = \{f(x) * f(y) \to f(x * y)\}.$$

The sequence of generated rules ($l_k \to r_k$) can be generalized modulo $\sim$ by

$$(X * Y) * Z \quad \to \quad X * (Y * Z)$$

using substitutions that associate with the meta-variable vector $\langle X, Y, Z\rangle$ any value in the set of term vectors

$$\{\langle x_1, x_2, z\rangle, \langle f(x_1), f(x_2), z\rangle, \langle f(f(x_1)), f(f(x_2)), z\rangle, \ldots, \langle f^n(x_1), f^n(x_2), z\rangle \ldots\}$$

The meta-rule here is the associativity rule:

$$\forall X, Y, Z,\ (X * Y) * Z \to X * (Y * Z).$$

3.1 Schematization

In order to schematize infinite sets of rewrite rules in a finite way, generic variables, called meta-variables, are introduced. A set of possible instantiations is associated with the meta-variables.

Definition 3.10 *A variable schematization $(\mathcal{X}, \Psi)$ is given by a set $\mathcal{X}$ of symbols called* **meta-variables,** *together with a set Ψ of mappings ψ defined from $\mathcal{X}$ to $\mathcal{T}(\mathcal{F}, \mathcal{W})$ with $\mathcal{W} \cap \mathcal{V} = \emptyset$, and called* **principal instantiations** *of $\mathcal{X}$.*

Note that in this definition, a principal instantiation of a meta-variable may collapse to a single variable. This allows us to consider, for instance in Example 3.9, a meta-variable Z whose principal instantiation is simply a variable z and to avoid introducing two kinds of variables in the set of variable schemas.

In the following, we denote variables in $\mathcal{V}$ by lower case letters $x, y, z \ldots$, and meta-variables by upper case letters $X, Y, Z \ldots$.

The set of principal instantiations of a meta-variable vector $\mathcal{X} = \{\langle X_1, \ldots, X_n\rangle\}$ will be described hereafter by the set $P = \{\langle \psi(X_1), \ldots, \psi(X_n)\rangle \mid \psi \in \Psi\}$.

Definition 3.11 *A schematization $\mathcal{S} = (\mathcal{X}, \Psi, MR, \sim)$ is defined by a set of meta-variables $\mathcal{X}$, a set of principal instantiations Ψ of $\mathcal{X}$, a finite set MR of rewrite rules in $\mathcal{T}(\mathcal{F}, \mathcal{W})$, called* **meta-rules,** *and a* **schematization congruence** $\sim$ *generated by a confluent and terminating rewrite system* Γ.

To reproduce the infinite set of rules from a meta-rule, the meta-variables have to be instantiated with principal instantiations. This notion of principal instantiation expresses the fact that a meta-rule is a generalization of all the rules in the infinite family of rules. The notion of schematization congruence deals with the application of the equivalence relation pointed out in Example 3.9.

A schematization $\mathcal{S} = (\mathcal{X}, \Psi, MR, \sim)$ represents the infinite family of rules

$$\{G\psi{\downarrow_\Gamma} \to D\psi{\downarrow_\Gamma} \mid \psi \in \Psi, G \to D \in MR\}.$$

This family is obtained by applying to each meta-rule $G \to D$, all the possible principal instantiations $\psi \in \Psi$ of $\mathcal{X}$, followed by computing the Γ-normal forms (denoted by ${\downarrow_\Gamma}$) of the previously instantiated terms. The redexes for such family have the form $(G\psi{\downarrow_\Gamma})\gamma$ and thus are equivalent modulo $\sim$ to $G\psi\gamma$.

3.2 Rewriting with a schematization

We are now interested in defining an adequate notion of rewriting in a given schematization, and in finding properties that must be satisfied by this rewriting relation.

Rewriting with a schematization takes into account the schematization congruence and principal instantiations.

Definition 3.12 *The* **rewriting relation associated with the schematization** $\mathcal{S} = (\mathcal{X}, \Psi, MR, \sim)$ *is defined on* $\mathcal{T}(\mathcal{F}, \mathcal{V})$ *by:*

$$t \Longrightarrow_{\mathcal{S}}^{u,G \to D} t'$$

if

- $(G \to D) \in MR$,
- *there exist* $\psi \in \Psi$ *and* γ *from* W *to* $\mathcal{T}(\mathcal{F}, \mathcal{V})$ *satisfying:* $G\psi\gamma \sim t|_u$.
- $t' = t[D\psi\gamma]_u$.

We write $\Longrightarrow$ instead of $\Longrightarrow_{\mathcal{S}}$ if the schematization is clear from the context. Let $\overset{*}{\Longleftrightarrow}$ denote the reflexive, symmetric and transitive closure of $(\Longrightarrow \cup \sim)$.

In this rewriting relation, meta-variables are constrained to belong to some domain.

Definition 3.13 *Given a schematization* $\mathcal{S} = (\mathcal{X}, \Psi, MR, \sim)$ *the* **domain of a meta-variable** $X \in \mathcal{X}$ *is the set*

$$\mathcal{D}om(X) = \{t \sim X\psi\gamma \mid \psi \in \Psi,\ \gamma : W \mapsto \mathcal{T}(\mathcal{F}, \mathcal{V})\}.$$

Definition 3.14 *Let* $|\mathcal{T}|$ *denote the carrier of the free algebra* $\mathcal{T}(\mathcal{F}, \mathcal{V})$. *Whenever* $\mathcal{D}om(X) = |\mathcal{T}|$ *the meta-variable* X *is said to be* **unconstrained.**

An interpretation of the rewriting relation $\Longrightarrow$ in an order-sorted framework [GKK90,GJM85] is developed in [Kir89]. Informally, a meta-variable X with the domain $\mathcal{D}om(X)$ is identified with a sorted variable $X : \mathcal{D}om(X)$. The set of sorts is the set of meta-variables domains, ordered by set inclusion $\subseteq$, and there exists a greatest sort, namely $|\mathcal{T}|$. With each operator $f \in F$ with arity n is associated a rank declaration $f : |\mathcal{T}|^n \to |\mathcal{T}|$. Meta-rules then become order-sorted rules.

It should be emphasized that any order-sorted substitution is not always of the form $\psi\gamma$, required in Definition 3.12, as shown in the following example:

Example 3.15 Let consider the following set of principal instantiations of a meta-variable vector $\langle X, Y \rangle$:

$$\Psi = \{\psi_n = [X \mapsto s^n(0), Y \mapsto s^{n+1}(0)] \mid n \geq 0\}.$$

Then the domains of X and Y are respectively: $\mathcal{D}om(X) = \{s^p(0) \mid p \geq 0\}$ and $\mathcal{D}om(X) = \{s^p(0) \mid p \geq 1\}$. The order-sorted substitution $\sigma = [X \mapsto s^3(0), Y \mapsto s^2(0)]$ does not factorize through a principal instantiation.

A schematization such that any order-sorted substitution factorizes through a principal instantiation is called $\mathcal{D}om$-based. The relation $\Longrightarrow$ then coincides with order-sorted rewriting with MR modulo a set of equalities Γ.

A $\mathcal{D}om$-based schematization where all meta-variables are unconstrained is called an **unconstrained schematization.** In the case of an unconstrained schematization, the relation $\Longrightarrow$ coincides with rewriting modulo a set of equalities Γ.

Example 3.16 (Example 2.4 continued)
Although there seems to be some correlation between the meta-variables in this example, the proposed schematization is $\mathcal{D}om$-based. Consider the $\mathcal{D}om$-instantiation ξ defined as

$$[X \mapsto f^m(t_1), Y \mapsto f^n(t_2), Z \mapsto t_3]$$

with for instance $m > n$. The instantiation ξ can be decomposed using the principal instantiation ψ defined as

$$[X \mapsto f^n(x_1), Y \mapsto f^n(x_2), Z \mapsto z]$$

and the substitution γ

$$[x_1 \mapsto f^{m-n}(t_1), x_2 \mapsto t_2, z \mapsto t_3].$$

Moreover the schematization is unconstrained since any term $t \in |\mathcal{T}|$ can be expressed at least as an instance of x_1 or x_2.

3.3 Deciding equality with a schematization

In practice, we want to use a set of meta-rules MR to decide the equality generated by R^∞, that is the reflexive symmetric and transitive closure of the rewriting relation $\longrightarrow_{R^\infty}$, denoted $\overset{*}{\longleftrightarrow}_{R^\infty}$. For this purpose, we introduce two properties of a schematization, namely *soundness* and *completeness* with respect to R^∞. Soundness means that all deductions performed with meta-rules are equational consequences of R^∞.

Definition 3.17 *A schematization* $\mathcal{S} = (\mathcal{X}, \Psi, MR, \sim)$ *is* **sound** *with respect to an infinite rewrite rule system* R^∞ *if for any terms* t *and* t', $t \overset{*}{\Longleftrightarrow}_{MR} t'$ *implies* $t \overset{*}{\longleftrightarrow}_{R^\infty} t'$.

The soundness of a schematization is ensured if all the realizations of a meta-rule are equational consequences of the infinite set of rewrite rules R^∞.

Proposition 3.18 *A schematization* $\mathcal{S} = (\mathcal{X}, \Psi, MR, \sim)$ *of* R^∞ *is sound if*

1. $\sim$ *is included in the congruence* $\overset{*}{\longleftrightarrow}_{R^\infty}$
2. *for any meta-rule* $(G \to D) \in MR$ *and for any principal instantiation* ψ *of* $\mathcal{X}$,

$$G\psi \overset{*}{\longleftrightarrow}_{R^\infty} D\psi.$$

It is important to require a finite set of meta-rules in the schematization, since otherwise the infinite set R^∞ could be considered as a sound schematization for itself with the syntactic equality as schematization congruence.

A very simple sufficient condition to ensure the condition (2) in Proposition 3.18 will be used in what follows for a meta-rule $G \to D$. If the substitution φ that renames each meta-variable X into a variable $x \in \mathcal{W}$ is a principal instantiation, condition (2) is equivalent to: $G\varphi \overset{*}{\longleftrightarrow}_{R^\infty} D\varphi$. A more general sufficient condition can be found in [Kir89].

Soundness is not enough if the set of meta-rules is designed to decide the equality in the quotient algebra $\mathcal{T}(\mathcal{F}, \mathcal{V})/R^\infty$. The schematization is actually supposed to satisfy the following completeness property, which expresses the fact that rewriting with a schematization provides a complete method for proving equational theorems in $\mathcal{T}(\mathcal{F}, \mathcal{V})$. Composition of relations is denoted by $\circ$.

Definition 3.19 *A schematization* $\mathcal{S} = (\mathcal{X}, \Psi, MR, \sim)$ *is* **complete** *with respect to an infinite rewrite rule system* R^∞ *if for any terms* $t, t' \in \mathcal{T}(\mathcal{F}, \mathcal{V})$, $t \overset{*}{\longleftrightarrow}_{R^\infty} t'$ *implies* $t \overset{*}{\Longrightarrow} \circ \sim \circ \overset{*}{\Longleftarrow} t'$.

Whenever $\Longrightarrow_{\mathcal{S}}$ is order-sorted rewriting modulo $\sim$, the completeness proof can then be decomposed in two parts:

Proposition 3.20 *A schematization* $\mathcal{S} = (\mathcal{X}, \Psi, MR, \sim)$ *is complete with respect to an infinite rewrite rule system* R^∞ *if*

1. *for any terms* t *and* t' *of* $\mathcal{T}(\mathcal{F}, \mathcal{V})$, $t \overset{*}{\longleftrightarrow}_{R^\infty} t'$ *implies* $t \overset{*}{\Longleftrightarrow}_{MR} t'$,
2. *the set of meta-rules* MR *is convergent modulo* $\sim$.

The first condition will be ensured in our meta-rule inference from crossed rewrite systems presented in Section 4. On the contrary, the set of meta-rules given in a schematization does not always satisfy the Church-Rosser property and a completion procedure is needed. Techniques presented in [Kir89] provide a completion procedure to transform a sound schematization into a sound and complete one. They rely on order-sorted completion modulo $\sim$ [GKK90] in the case of $\mathcal{D}om$-based schematization, and on completion modulo $\sim$ [Bac87,JK86,PS81] for unconstrained schematization.

4 Automatic inference of meta-rules

In this section, we propose an automatic method for constructing sound schematizations, developed in two stages. The first one presents a schematization of a single iterated family generated by a crossed system. Of course, the infinite system R^∞ can contain more iterated families with no mutual inclusion. Therefore the next stage presents a schematization of the whole infinite system R^∞, based on the schematization synthesis of all iterated families. The technique is illustrated by a reasonably complex example that contains an infinite set of crossed systems.

4.1 Single iterated family

Remind that crossed rewrite systems generate forward or backward iterated families. From the description of these two kinds of families, a meta-rule can be automatically inferred in each case, provided the overlap closure is itself a convergent rewrite system. Actually, the meta-rule is the rule itself with specialized variables that become meta-variables and the overlap closure generates the schematization congruence.

4.1.1 Forward crossed rewrite systems

Proposition 4.21 *Let $\mathcal{I}_{FC}^{a,b}(C) = \{u_n \to v_n\}$ form an infinite forward iterated family built from the forward crossed rewrite system $C = \{s_1 \to t_1, s_2 \mapsto\!\!\to t_2\}$ as in Definition 2.3, where $\{s_2 \mapsto\!\!\to t_2\}$ presents itself a convergent system. Let consider the following schematization:*

- *the schematization congruence $\sim$ is generated by $s_2 \to t_2$,*
- *the (unconstrained) meta-rule $S_1 \to T_1$ is obtained from $s_1 \to t_1$ by upgrading each variable $x \in \mathcal{V}ar(s_1)$ to an unconstrained meta-variable X. Let $MR = \{S_1 \to T_1\}$.*

Then $\forall n, u_n \overset{*}{\Longleftrightarrow}_S v_n$. Moreover, this schematization is sound with respect to $\mathcal{I}_{FC}^{a,b}(C)$.

Example 4.22 (Example 2.4 continued).
Let us explain the Proposition 4.21 on an example. The rewrite system

$$\begin{array}{rcl}(x * y) * z & \to & x * (y * z) \\ f(x) * f(y) & \to & f(x * y)\end{array}$$

generates under completion the infinite set of rewrite rules of the form

$$f^n(x * y) * z \quad \to \quad f^n(x) * (f^n(y) * z)$$

The variables x, y, and z from the *Associativity* rule become unconstrained meta-variables X, Y and Z. The principal instantiations of the meta-variable vector $\langle X, Y, Z\rangle$ are given by the set

$$P \;=\; \{\langle f^n(x), f^n(y), z\rangle \mid n \geq 0\}.$$

The meta-rule is

$$(X * Y) * Z \to X * (Y * Z)$$

and the schematization congruence $\sim$ is generated by $f(x) * f(y) \to f(x * y)$.

4.1.2 Backward crossed rewrite systems

Proposition 4.23 *Let $\mathcal{I}_{BC}^{b}(C) = \{u_n \to v_n\}$ form an infinite backward iterated family built from the backward crossed rewrite system $C = \{s_1 \mapsto\!\!\to t_1, s_2 \to t_2\}$ as in Definition 2.6, where $\{s_1 \mapsto\!\!\to t_1\}$ presents itself a convergent system. Let consider the following schematization:*

- *the schematization congruence $\sim$ is generated by $s_1 \to t_1$,*
- *the meta-rule $S_2 \to T_2$ is obtained from $s_2 \to t_2$ by upgrading each variable $x \in \mathcal{V}ar(s_2)$ to an unconstrained meta-variable X. Let $MR = \{S_2 \to T_2\}$.*

Then $\forall n,\ u_n \overset{*}{\Longleftrightarrow}_S v_n$. Moreover, this schematization is sound with respect to $\mathcal{I}_{BC}^{b}(C)$.

Example 4.24 (Example 2.7 continued).
Let us illustrate the last theorem on a previous example. The rewrite system

$$\begin{array}{rcl}(x + f(y)) * y & \to & (x * y) + y \\ (x - y) + y & \to & x\end{array}$$

generates under completion the infinite set of rewrite rules of the form

$$((((x - f^{n+1}(y)) * f^n(y)) * \ldots * f(y)) * y) + y \quad \to \quad ((x * f^n(y))) * \ldots * f(y)) * y$$

The variables x and y in the second rule become unconstrained meta-variables X and Y. The principal instantiations of the meta-variable vector $\langle X, Y\rangle$ are given by

$$P = \{\langle x, f^n(y)\rangle \mid n \geq 0\}.$$

The meta-rule is

$$(X - Y) + Y \to X$$

and the schematization congruence $\sim$ is generated by $(x + f(y)) * y \to (x * y) + y$.

The reader can check that

$$((((x - f^{n+1}(y)) * f^n(y)) * \ldots * f(y)) * y) + y \sim ((((x - f^n(y)) + f^n(y)) * f^n(y)) \ldots * f(y)) * y.$$

4.1.3 One-rule systems

Simple divergent systems consisting of only one rule fail to get a proper schematization with the previous method. The problem is that the schematization congruence $\sim$ as well as the meta-rule are generated by the same object. Actually this case needs the full power of order-sorted rewriting.

Example 4.25 The rule

$$f(g(f(x))) \to g(f(x))$$

generates the infinite family $\{f(g^i(f(x))) \to g^i(f(x))\}$. This family of rules is represented by the meta-rule $f(g(X)) \to g(X)$.

The domain of X is instances of the set of terms $P = \{f(x), g(f(x)), g(g(f(x))), \ldots\}$.

Proposition 4.26 *Let $\mathcal{I}_{FC}^{a,b}(C) = \{u_n \to v_n\}$ form an infinite forward iterated family built from the forward crossed rewrite system $C = \{s \to t, s \mapsto t\}$ as in Definition 2.3. Assume in addition that $s|_a = t|_b, \sigma_2 = \varphi_2 = \emptyset$. Let us consider the following schematization:*

- *the schematization congruence $\sim$ is the syntactic equality,*
- *the meta-rule is $s[X]_a \to t[X]_b$ where X is a meta-variable whose set of principal instantiations is given by $P = \{t|_b, t, C[t], \ldots, C^i[t], \ldots\}$ with $C[\cdot]$ being the context $t[\cdot]_b$. Let $MR = \{s[X]_a \to t[X]_b\}$.*

Then $\forall n$, $u_n \overset{*}{\Longleftrightarrow}_S v_n$. Moreover, this schematization is sound with respect to $\mathcal{I}_{FC}^{a,b}(C)$.

Another kind of difficulty arises when the schematization introduces different meta-variables with a correlation between them. This can be illustrated by a simple example of a one-rule backward crossed rewrite system.

Example 4.27 The rule

$$f(g(f(x))) \to f(h(x))$$

generates the infinite family $\{f(h^i(g(f(x)))) \to f(h^{i+1}(x))\}$. The family of left-hand sides of these rules is schematized by the term $f(X)$, while the family of right-hand sides is schematized by the term $f(h(Y))$. The principal instantiations of the meta-variable vector $\langle X, Y\rangle$ are given by the set

$$P = \{\langle g(f(x)), x\rangle, \langle h(g(f(x))), h(x)\rangle, \ldots, \langle h^i(g(f(x))), h^i(x)\rangle, \ldots\}$$

A schematization is provided by a conditional meta-rule $(X, Y) \in P^* \mid f(X) \to f(h(Y))$, where P^* is the set of instances of P.

Sufficient conditions that lead to this kind of conditional schematization are expressed in the following proposition.

Proposition 4.28 *Let $\mathcal{I}_{BC}^{b}(C) = \{u_n \to v_n\}$ form an infinite backward iterated family built from the backward crossed rewrite system $C = \{s \to t, s \mapsto t\}$ as in Definition 2.6. Assume in addition that $\sigma_2 = \varphi_1 = \emptyset$ and $Var(t) \cap Dom(\sigma_1) = \{x\}$. This family of rules has the following schematization:*

- *the schematization congruence $\sim$ is the syntactic equality,*

- *the conditional meta-rule is* $H \mid G = D$ *where*
  - G *is obtained from* $s|_b$ *by upgrading* $x \in \mathcal{D}om(\sigma_1)$ *to a meta-variable* X,
  - D *is is obtained from* t *by upgrading* $x \in \mathcal{V}ar(t)$ *to a meta-variable* Y,
  - $\langle X,Y\rangle$ *is a meta-variable vector whose set of principal instantiations is given by* $P = \{\langle x\sigma_1, x\rangle, \langle x\varphi_2\sigma_1, x\varphi_2\rangle, \ldots, \langle x\varphi_2^i\sigma_1, x\varphi_2^i\rangle, \ldots\}$.
  - H *is* $\langle X,Y\rangle \in P^*$, *the set of instances of* P.

Then $\forall n,\ u_n \overset{*}{\Longleftrightarrow}_S v_n$. *Moreover, this schematization is sound with respect to* $\mathcal{I}^b_{BC}(C)$.

4.2 Class of iterated families

A more complex example is now considered to illustrate the general case where different iterated families are present together. Each divergent rewrite system R can be split into two disjoint parts, the subsystem R_{diverg} whose all rules participate in the divergent process, and the subsystem R_{fix} whose all rules stay apart from divergence. It is fairly clear from the previous considerations that the rules of R_{fix} are raised to meta-rules only by upgrading each standard variable x to the (unconstrained) meta-variable X with the value x as the set of principal instantiations. For this reason, we will further assume that $R = R_{diverg}$ without loss of generality.

It is obvious that a divergent system need not contain only one iterated family of rules.

Example 4.29 Let us consider the following set of rewrite rules that axiomatizes *the signed binary trees theory* [KK82], presented as an extract from a group specification:

$$R = \left\{ \begin{array}{rcl} i(i(x)) & \rightarrow & x, \\ i(x*y) & \rightarrow & i(y)*i(x), \\ (y*x)*i(x) & \rightarrow & y, \\ i(x)*(x*y) & \rightarrow & y \end{array} \right\}$$

From these rules, the completion procedure generates two infinite families of rules:

$$F_1 = \left\{ \begin{array}{rcl} i(x_1)*(x_1*y) & \rightarrow & y, \\ (i(x_2)*i(x_1))*((x_1*x_2)*y) & \rightarrow & y, \\ (i(x_2)*x_1)*((i(x_1)*x_2)*y) & \rightarrow & y, \\ (x_2*i(x_1))*((x_1*i(x_2))*y) & \rightarrow & y, \ldots \end{array} \right\}$$

$$F_2 = \left\{ \begin{array}{rcl} (y*x_1)*i(x_1) & \rightarrow & y, \\ (y*(x_1*x_2))*(i(x_2)*i(x_1)) & \rightarrow & y, \\ (y*(i(x_1)*x_2))*(i(x_2)*x_1) & \rightarrow & y, \\ (y*(x_1*i(x_2)))*(x_2*i(x_1)) & \rightarrow & y, \ldots \end{array} \right\}$$

It can be noticed that the subset of rules

$$\Gamma = \left\{ \begin{array}{rcl} i(i(x)) & \rightarrow & x, \\ i(x*y) & \rightarrow & i(y)*i(x) \end{array} \right\}$$

is confluent and terminating. The equivalence generated by Γ is denoted $\sim$. Then the sequence of rules $(l_k \rightarrow r_k)$ of the first family can be generalized modulo $\sim$ by

$$i(X)*(X*Y) \rightarrow Y$$

where principal instantiations of the meta-variable vector $\langle X,Y\rangle$ are given by the set

$$\{\langle x_1, y\rangle, \langle x_1*x_2, y\rangle, \langle i(x_1)*x_2, y\rangle, \langle x_1*i(x_2), y\rangle, \langle i(x_1)*i(x_2), y\rangle, \langle x_1*(x_2*x_3), y\rangle \ldots\}$$

The meta-rules associated respectively with each family of rules are

$$\begin{array}{rcl} \forall X,Y,\ i(X)*(X*Y) & \rightarrow & Y, \\ \forall X,Y,\ (Y*X)*i(X) & \rightarrow & Y. \end{array}$$

The previous example contains two infinite families of rules, but an infinite number of iterated families due to the fact that the set of overlap closures $OC(\Gamma)$ contains an infinite set of *saturated crossed systems.*

Definition 4.30 *Let R be a rewrite system and $S \subseteq R$ a crossed subsystem. The system S is called* **saturated** *in R if there is no other crossed system $S' \subseteq R$, such that $\mathcal{I}(S) \subseteq \mathcal{I}(S')$. Denote further by $SatCr(R)$ the class of* **all saturated crossed systems** *in R.*

Note: If we do not distinguish between forward and backward iterated families (we assume one of them), we denote them $\mathcal{I}(S)$ for a crossed system S.

We now introduce definitions in order to classify rules into families with a generator set of saturated crossed systems. Denote *clos*(S) the overlap closure, and *start*(S) the starting rule of the crossed system S. Denote further their extensions to a class C of crossed systems

$$Start(C) = \bigcup_{S \in C} start(S)$$

the set of all starting rules, and

$$Clos(C) = \bigcup_{S \in C} clos(S)$$

the set of all overlap closures present in the class C. The set of starting rules of all crossed systems in R is then, according to this notation, $Start(SatCr(R))$.

Example 4.31 (Example 4.29 cont.)
The set of all starting rules $Start(SatCr(R))$ is $\{(y * x) * i(x) \to y, i(x) * (x * y) \to y\}$.

Note that if R is a finite rewrite system, then the set $Start(SatCr(R))$ is finite.

Now two rules are in the same family if they can be generated from a same rule p with a saturated crossed system S. To each family is associated the set of saturated crossed systems with the same starting rule p.

Definition 4.32 *Denote*

$$Fam(p, R) = \{p\} \cup \{q \in \mathcal{I}(S) \mid S \in SatCr(R), start(S) = p\}$$

the **p-based** *(infinite)* **family** *of rules in R^∞ and*

$$Gen(p, R) = \{S \in SatCr(R) \mid start(S) = p\}$$

the **p-based generator** *in R^∞. Denote further*

$$Fam^*(R) = \{Fam(p, R) \mid p \in Start(SatCr(R))\}$$

the class of **all infinite families** *of rules in R^∞ and*

$$Gen^*(R) = \{Gen(p, R) \mid p \in Start(SatCr(R))\}$$

the class of **all generators** *in R^∞.*

Example 4.33 (Example 4.29 cont.)
The class $Fam^*(R)$ contains two families: $Fam(i(x) * (x * y)) \to y, R) = F_1$ and $Fam((y * x) * i(x) \to y, R) = F_2$, according to the set $Start(SatCr(R))$ that contains two rules.
The class $Gen^*(R)$ contains two generators, $Gen((y * x) * i(x) \to y, R)$ and $Gen(i(x) * (x * y) \to y, R)$. Both generators in $Gen^*(R)$ are infinite.

In Example 4.29, an infinite set of saturated crossed systems corresponds to each family. The problem is then to describe the schematization congruence $\sim$ using a finite set of axioms. The difficulty can be solved thanks to the existence of a finite convergent term rewriting system Γ.

Example 4.34 (Example 4.29 cont.)
The equivalence relation generated by the convergent system Γ contains the equivalence relations generated by $Clos(SatCr(R))$ as well as by $Clos(Gen((y*x)*i(x) \to y, R))$ and $Clos(Gen(i(x)*(x*y) \to y, R))$.

Note however that Γ contains the rewrite rule $i(i(x)) \to x$ that does not belong to any of these closure sets.

Theorem 4.35 *Let R be a divergent rewrite system. If there exists a finite convergent system Γ, such that for each generator $\mathcal{G} \in Gen^*(R)$,*

$$\begin{array}{rcl} \overset{*}{\longleftrightarrow}_{Clos(\mathcal{G})} & \subseteq & \overset{*}{\longleftrightarrow}_{\Gamma} \\ \overset{*}{\longleftrightarrow}_{\Gamma - Clos(\mathcal{G})} & \subseteq & \overset{*}{\longleftrightarrow}_{R} \end{array}$$

then R^∞ has the following schematization:

- *the schematization congruence $\sim$ is generated by Γ,*
- *the (unconstrained) meta-rules MR are obtained from $Start(SatCr(R))$ by upgrading each variable x to an unconstrained meta-variable X.*

This schematization is sound with respect to R^∞. Moreover, if MR is convergent modulo $\sim$ then the schematization is complete with respect to R^∞.

Note that whenever each generator $\mathcal{G} \in Gen^*(R)$ is finite, then Γ can be chosen as $Clos(\mathcal{G})$, as in the case of single iterated families of Section 4.1, provided Γ is convergent.

5 Conclusion

The results presented here show how divergence can be detected by a syntactical check, and then handled by a schematization process. In a reasonable number of cases, this schematization is provided by a rewriting relation modulo an equivalence generated by the overlap closure. This provides an effective solution for a reasonably large class of divergence problems. In addition, it is a first step towards an "intelligent" completion system that detects divergence and proposes an equivalent and finite class rewrite system for computing and proving properties in the considered equational theory.

In other cases, the schematization is provided by an order-sorted rewrite rule or a conditional rule. Simple cases of divergent systems consisting of only one rule fall into this scope.

Last but not least, let us mention other complementary research directions for handling the divergence problem: generalizing the notion of schematization using functional meta-variables [Gra88], introducing new function symbols in a conservative extension of the initial theory [JT88,Lan89], generating inductive consequences of the equational theory [DP89,JT88].

Acknowledgments

We thank anonymous referees for their constructive remarks.

References

[Bac87] L. Bachmair. *Proof methods for equational theories.* PhD thesis, University of Illinois, Urbana-Champaign, 1987. Revised version, August 1988.

[BDH86] L. Bachmair, N. Dershowitz, and J. Hsiang. Orderings for equational proofs. In *Proceedings Symp. Logic in Computer Science*, pages 346–357, IEEE, 1986.

[Der87] N. Dershowitz. Termination of rewriting. *Journal of Symbolic Computation*, 3(1 & 2):69–116, 1987.

[DJ89] N. Dershowitz and J.-P. Jouannaud. Notations for rewriting. October 1989. Unpublished note.

[DJ90] N. Dershowitz and J.-P. Jouannaud. *Handbook of Theoretical Computer Science*, chapter 15: Rewrite systems. Volume B, North-Holland, 1990. Also as: Research report 478, LRI.

[DP89] N. Dershowitz and E. Pinchover. Inductive synthesis of equational programs. 1989. Unpublished article.

[GJM85] J.A. Goguen, J.-P. Jouannaud, and J. Meseguer. Operational semantics for order-sorted algebra. In W. Brauer, editor, *Proceeding of the 12th International Colloquium on Automata, Languages and Programming*, pages 221–231, Springer-Verlag, Nafplion (Greece), 1985. Lecture Notes in Computer Science, volume 194.

[GKK90] I. Gnaedig, C. Kirchner, and H. Kirchner. Equational completion in order-sorted algebras. *Theoretical Computer Science*, 72, 1990.

[GKM83] J.V. Guttag, D. Kapur, and D.R. Musser. On proving uniform termination and restricted termination of rewrite systems. *SIAM Journal on Computation*, 12(1):189–214, February 1983.

[Gra88] B. Gramlich. *Unification of Term Schemes - Theory and Applications.* Technical Report SR-88.18, SEKI, University of Kaiserslautern, RFA, 1988.

[Her88] M. Hermann. *Vademecum of divergent term rewriting systems.* Research report 88–R–082, Centre de Recherche en Informatique de Nancy, 1988. Presented at *Term Rewriting Workshop, Bristol (UK).*

[Her89] M. Hermann. *Crossed term rewriting systems.* Research report 89-R-003, Centre de Recherche en Informatique de Nancy, 1989.

[Her90] M. Hermann. Chain properties of rule closures. *Formal Aspects of Computing*, 2(3):207–225, 1990.

[Hue81] G. Huet. A complete proof of correctness of the Knuth and Bendix completion algorithm. *Journal of Computer Systems and Sciences*, 23:11–21, 1981.

[JK86] J.-P. Jouannaud and H. Kirchner. Completion of a set of rules modulo a set of equations. *SIAM Journal of Computing*, 15(4):1155–1194, 1986. Preliminary version in Proceedings 11th ACM Symposium on Principles of Programming Languages, Salt Lake City, 1984.

[JT88] K.P. Jantke and M. Thomas. *Inductive inference for solving divergence in Knuth-Bendix completion.* Research report 88/R6, University of Glasgow, Department of Computer Science, Glasgow, UK, September 1988.

[KB70] D.E. Knuth and P.B. Bendix. Simple word problems in universal algebras. In J. Leech, editor, *Computational Problems in Abstract Algebra*, pages 263–297, Pergamon Press, Oxford, 1970.

[KH90] H. Kirchner and M. Hermann. *Meta-rule synthesis from crossed rewrite systems.* Research report 90-R-143, Centre de Recherche en Informatique de Nancy, 1990.

[Kir89] H. Kirchner. Schematization of infinite sets of rewrite rules generated by divergent completion processes. *Theoretical Computer Science*, 67(2-3):303–332, October 1989.

[KK82] C. Kirchner and H. Kirchner. Résolution d'équations dans les algèbres libres et les variétés équationnelles d'algèbres. 1982. Thèse de troisième cycle, Université de Nancy I.

[Lan89] S. Lange. Towards a set of inference rules for solving divergence in Knuth-Bendix completion. In K.P. Jantke, editor, *Proceedings of the International Workshop on Analogical and Inductive Inference, Reinhardsbrunn Castle (GDR)*, pages 304–316, Springer-Verlag, October 1989. Lecture Notes in Computer Science (in Artificial Inteligence), volume 397.

[PS81] G. Peterson and M. Stickel. Complete sets of reductions for some equational theories. *Journal of the Association for Computing Machinery*, 28:233–264, 1981.

[Toy87] Y. Toyama. Confluent term rewriting systems with membership conditions. In S. Kaplan and J.-P. Jouannaud, editors, *Proceedings of the First International Workshop on Conditional Term Rewriting Systems, Orsay (France)*, pages 228–241, Springer-Verlag, July 1987. Lecture Notes in Computer Science, volume 308.

Chapter 3

Extension of Knuth-Bendix Completion

An Application of Automated Equational Reasoning to Many-valued Logic

Siva Anantharaman
LIFO, Dépt. Math-Info.
Université d'Orléans
45067 ORLEANS Cedex 02 (Fr.)
e-mail: siva@univ-orleans.fr

Maria Paola Bonacina*
Department of Computer Science
SUNY at Stony Brook
Stony Brook NY 11794-4400 USA
e-mail: bonacina@sbcs.sunysb.edu

Abstract

In this paper we present a new set of axioms of an algebaric nature, for the many-valued logic of Lukasiewicz. These axioms are similar to those given by J.Hsiang for the Boolean Algebra. The equivalence of our set of axioms with those given by Lukasiewicz himself is proved *mechanically*, by resorting to the Automatic UKB-based Equational Theorem Prover *SBR3*. These new axioms may be helpful for further equational reasoning in such logics, or for interpreting the 'equality symbol' of linear logic. This paper is organized as follows. Section 1 presents briefly the many-valued logic. Section 2 indicates the proof-steps leading to our new set of axioms. Section 3 presents the equational prover SBR3.

1 An Introduction to the Lukasiewicz Logic

Many-valued propositional logic was first introduced by Jan Lukasiewicz in the 1920's. All the following results about early work on many-valued logic are reported in [TL-56].

The original definition of many-valued logic is purely semantical. No axioms and no inference rules are given. Lukasiewicz defines first a model and then the logic is defined as the set of all sentences in propositional calculus which are true on that model. More precisely, the n-valued logic L_n is defined as the set of all sentences satisfied by the structure

$$\mathcal{L}_n = < \{\frac{k}{n-1} | 0 \leq k \leq n-1\}, g, f >$$

where $A_n = \{\frac{k}{n-1} | 0 \leq k \leq n-1\}$ is the domain, $g : A_n \rightarrow A_n$ is the unary function $g(x) = 1 - x$ and $f : A_n \times A_n \rightarrow A_n$ is the binary function $f(x,y) = min(1 - x + y, 1)$.

L_1 is the set of all legal propositional sentences, L_2 is classical two-valued propositional logic with model

$$\mathcal{L}_2 = < \{0,1\}, g, f >$$

where the functions g and f are classical negation and implication. L_3 is three-valued logic, the first one introduced by Lukasiewicz.

*Research supported in part by NSF grants INT-8715231, CCR-8805734 and CCR-8901322. Also supported by Dottorato di Ricerca in Informatica, Universitá degli Studi di Milano.

As n increases, the domain A_n grows:

$$A_1 = \{1\}, \quad A_2 = \{0,1\}, \quad A_3 = \{0, \tfrac{1}{2}, 1\},$$
$$A_4 = \{0, \tfrac{1}{3}, \tfrac{2}{3}, 1\}, \quad A_5 = \{0, \tfrac{1}{4}, \tfrac{2}{4}, \tfrac{3}{4}, 1\} \ldots$$

The limit of this sequence is the set Q_0 of all rational numbers in the interval $[0,1]$, which is the domain of the many valued logic

$$\mathcal{L}_{\aleph_0} =< \{\frac{k}{l} | 0 \leq k \leq l\}, g, f >$$

As n increases, the set L_n shrinks and $L_{\aleph_0}$ is the smallest such set, i.e. the intersection of all the L_n. It has been later proved by Lindenbaum that the domain $\mathcal{L}_{\aleph_0}$ can be any arbitrary set of numbers $\{x | 0 \leq x \leq 1\}$ closed with respect to g and f.

Lukasiewicz conjectured that the following is an axiomatization for $L_{\aleph_0}$:

(1) $p \Rightarrow (q \Rightarrow p)$

(2) $(p \Rightarrow q) \Rightarrow ((q \Rightarrow r) \Rightarrow (p \Rightarrow r))$

(3) $((p \Rightarrow q) \Rightarrow q \Rightarrow (q \Rightarrow p) \Rightarrow p$

(4) $((p \Rightarrow q) \Rightarrow (q \Rightarrow p)) \Rightarrow (q \Rightarrow p)$

(5) $(not(p) \Rightarrow not(q)) \Rightarrow (q \Rightarrow p)$

where *not* and $\Rightarrow$ are interpreted as g and f in the model $\mathcal{L}_{\aleph_0}$. We will write *not*, $\Rightarrow$ and *true* rather than g, f and 1 whenever we are working on the axiomatization rather than on the model. The fourth axiom was subsequently proved dependent on the others.

Several classes of algebras related to Lukasiewicz logics have been given an equational axiomatization. The following is the equational axiomatization for the *Wajsberg algebras* ([FRT-84]) :

(1) $true \Rightarrow x == x$

(2) $(x \Rightarrow y) \Rightarrow ((y \Rightarrow z) \Rightarrow (x \Rightarrow z)) == true$

(3) $((x \Rightarrow y) \Rightarrow y == (y \Rightarrow x) \Rightarrow x$

(4) $(not(x) \Rightarrow not(y)) \Rightarrow (y \Rightarrow x) == true$

We will denote by $\mathcal{W}$ this equational presentation of Wajsberg algebras.

A certain connective, that we will call *weakor*, is then defined as

$$x \vee y == (x \Rightarrow y) \Rightarrow y$$

The problem known as the *fifth Lukasiewicz conjecture* (and solved in the late-20's) is that of proving :

$$(x \Rightarrow y) \vee (y \Rightarrow x) == true$$

If the operator $\Rightarrow$ in the definition of the operator $\vee$ is interpreted as $min(1-x+y,1)$, then $x \vee y$ gets interpreted as $max(x,y)$:

$$min(1 - min(1 - x + y) + y, 1) = \begin{cases} min(1 - 1 + y, 1) = y \ if \ y \geq x \\ min(1 - 1 + x - y + y, 1) = x \ if \ x \geq y \end{cases}$$

It follows that the theorem $(x \Rightarrow y) \vee (y \Rightarrow x) == 1$ is interpreted as $max(min(1 - x + y, 1), min(1 - y + x, 1)) == 1$, which is intuitively true, since the above evaluates to $max(1 - x + y, 1) = 1$ if $x \geq y$ and to $max(1 - y + x, 1) = 1$ if $y \geq x$.

1.1 Boolean-Like Connectives in the Lukasiewicz Logic

To compare this many-valued logic with classical Boolean logic, we may define two auxiliary connectives '*or*', and '*and*', respectively explicitly and implicitly, as below :

$$or(x,y) = not(x) \Rightarrow y$$

$$and(x,y) \Rightarrow z = (x \Rightarrow y) \Rightarrow z$$

It can be shown that *both are associative-commutative.* For instance, a quick mechanical proof has been obtained by the equational theorem-prover *SBR3* (see Section 3), along the following lines.

Lemma 1. Prove that $not(x) \Rightarrow not(y) == y \Rightarrow x$, in the theory $\mathcal{W}$.

We also get among others, the following important result as a by-product of the proof of Lemma 1 : $not(not(x)) == x$

Lemma 2. Prove that the above defined connective *and* is commutative; that is $x \Rightarrow (y \Rightarrow z) == y \Rightarrow (x \Rightarrow z)$, with $\mathcal{W}$, and the lemma deduced in Step 1, as input.

Lemma 3 . Prove then that the connective *or* is associative-commutative from $\mathcal{W}$ and Lemmas 1, 2 as input.

This proof produces the by-product : $and(x,y) = not(or(not(x), not(y)))$, so *and* is AC too.

In the model $\mathcal{L}_{\aleph_0}$, described in Section 1, our new conective *or* is interpreted as $min(1-1+x+y,1) = min(x+y,1)$. Thus $\vee$ and *or* are different connectives, except when the domain of interpretation is $\{0,1\}$, i.e. when the logic is 2-valued.

Note that since $not(not(x)) == x$, we may write henceforth $x \Rightarrow y = or(not(x), y)$. Thus 'everything' can be expressed in terms of AC-operators in many-valued logic. In fact, a very quick (mechanical) rewrite-proof for the above 'fifth-conjecture' of Lukasiewicz can be derived by expressing it in terms of *or* and using the AC-UKB of *SBR3* (Section 3).

2 A new set of axioms for Wajsberg Algebras

The new axiomatic structure presented below, involves two binary functions, denoted $+, *$, and a unary *not*, related by the following equations :

$$((1+x)*y)+1)*y == ((1+y)*x)+1)*x$$

$$not(x) == x+1$$

$$x+0 == x$$

$$x+x == 0$$

$$x*1 == x$$

$$x*0 == 0$$

$$(1+x)*x == 0$$

$$x+(1+y) == (x+1)+y$$

And it is stipulated that '+' is commutative, while '*' is associative-commutative.

(The difference between the specifications of J. Hsiang for Boolean algebra, and these, resides in the fact that our '+' is only commutative, whereas in the Boolean case it is AC. There is no distributivity axiom between our '*'and '+'; and '*' is *not* idempotent, in general). Here are the 'translation tables' between the earlier Wajsberg specifications and the current ones.

i) If we start with $\mathcal{W}$ and the added connectives *or* and *and*, we may define :

$$x * y = and(x, y) == not(or(not(x), not(y))), \text{ and}$$

$$x + y == or(and(x, not(y)), and(not(x), y))$$

Then all the current equations are equational theorems; besides, we have :

$$or(x, y) = 1 + ((1 + x) * (1 + y))$$

ii) Inversely, if we start with our current equations as axioms, then by defining :

$$(x \Rightarrow y) = 1 + (x * (1 + y)) \ ,$$

we get back all the equations of $\mathcal{W}$ as theorems.

This axiomatization for the many-valued logic is perhaps more suited for automated equational reasoning in such logics. For instance a mechanical proof for the fifth Lukasiewicz conjecture is derived more quickly this way. Finally, the binary '+' appears as a candiadte for 'interpreting the equality symbol' of linear logic.

3 The Equational Theorem-Prover SBR3

SBR3 is the latest version of a long line of term rewriting based *automatic* theorem provers for equational logic. The first of which, *Reve*, was written by P. Lescanne ([LE-83]), in CLU. The entire *Reve2.n* family, written and maintained at the MIT, is mostly concerned with completing a set of equations into a canonical system. But several of the interesting equational theorems invole rewrite-systems containing many unorientable equations, and do not admit finite canonical systems under all known orderings. So, for reasoning automatically in such systems, the Unfailing Knuth-Bendix Approach was developed ([Bac-87], [Hr-87]).

In 1986, Mzali and Hsiang built a new system *SbReve1*, based on *Reve2.4* and incorporating UKB. *SbReve1* was overhauled into *SbReve2* by the first-named author at the Université d'Orléans ([AHM-89]), to incorporate UKB modulo commutativity, and associativity-commutativity axioms. Sophisticated search and deletion mechanisms have also been incorporated into its current version, labelled *SBR3*.

3.1 The Inference Mechanisms of SBR3

SBR3 takes as inputs an equational theory E and an equation $s = t$ and tries to prove that $s = t$ is a theorem of E, in the refutational way. That is, it replaces all (free) variables in $s = t$ by new Skolem constants and tries to find a contradiction to $E \cup \{\bar{s} \neq \bar{t}\}$) (where the 'bar' designates skolemization).

For applying rewrite-methods, a reduction-ordering for ordering the terms has to be indicated too. For classical UKB, the ordering should be a *cso* = *complete simplification ordering*, i.e., a simplification ordering which is total on ground terms. However this restriction has been relaxed in *SBR3*, since such a constraint is in general incompatible with axiomatic rewriting, for arbitrarily chosen orderings; and also because *the refutational completeness of UKB can be established even for non-csos* ([AA-90]).

As a matter of fact, the lack of totality on ground terms may help in getting a shorter proof (Cf. Order-Saturation strategy below). Especially for the C/AC-UKB procedures ([AM-88]), which are the backbones of *SBR3*.

Another feature in *SBR3* is the implementation of inference rules for the *cancellation* axioms. These help us replace many big equations by smaller equations through complete sets of inference rules for cancellation [HRS-87].

Other automatic inference mechanisms of *SBR3* include *functional subsumption*, and simplification.

3.2 The Search Mechanisms of SBR3

The simplification-first search strategy coupled with cancellation controls the growth of the number and size of rules to some extent, but cleverer means are quite often needed.

The first problem to tackle is one of finding a shorter path to a solution. If UKB guarantees the existence of a proof through simplification and superposition (should there be one), it cannot guarantee a *short* proof. *SBR3* provides a facility for working on a reasonable number of (equivalent) inequalities, to get shorter proofs. This is the *ordered-saturation strategy.* This strategy, and the Cancellation-Laws, have been indispensable for proving some of the more difficult problems which we have successfully experimented ([AH-90]), thru' UKB-modulo axioms.

Eliminating 'redundant' critical pairs is another big problem in automated reasoning, which is especially serious in AC-rewriting. Most of the critical pair criteria in the literature, designed to meet this purpose, try not to destroy the confluence property of *any two* given terms. In the refutational approach however, we are only interested in the confluence of the two terms of the targeted theorem. Therefore a critical pair can be deleted or suspended as long as it does not destroy the confluence of the intended terms.

For achieving this, a notion called *measure* is employed in *SBR3*. A measure is defined syntactically on the structure of terms, such as the number of occurrences of a specific operator. See [AA-90] for precise definitions. Suitably well-defined measures can estimate the likelihood of whether a critical pair contributes to an eventual proof of the intended theorem. Such measures can also be used to order the set of rules, to help us decide optimally which next equation to choose. (Such strategies are called *filtration-sorted strategies*). Certain measures even allow us to delete critical pairs on a sound basis (loc. cit.). Several such strategies are implemented in *SBR3*. And they have played a decisive role in our experiments so far, including the above application to many-valued logic.

4 Conclusion : Perspectives for SBR3

The approach adopted in *SBR3* for valid-theorem-proving in equational theories, has recently been 'extended' to give an inference system for proving valid deductive-theorems in Horn theories ([AA-91]); it corresponds to the 'unit-strategy' approach of [De-90], and has been proved 'deductively complete' even for non-csos. Thus it is applicable without difficulty, even when the 'equality' symbol in the clauses is interpreted as *syntactic equality modulo a set of equational axioms.*

Finally owing to the evolution on the materiel side, *SBR3* (currently available in CLU, for SUN-3/VAX) is being rewritten in C/C++, (hopefully) incorporating all the above described features. A first working version for Horn theories (in C/C++, for all SUNs) must be available by the fall of 1991.

References

[AA-90] S.Anantharaman, N. Andrianarivelo, Heuristic Criteria in Refutational Theorem Proving, *Research-Report, LRI-Orsay(Fr.), no. 541, 1990, Extended-Abstract in Proc. DISCO'90, LNCS. no. 429*

[AA-91] S.Anantharaman, N. Andrianarivelo, A Complete Semi-Decision Procedure for Horn Theories, *Research Report, LIFO - LRI(Fr.), (to appear in Feb.-March 1991)*

[AH-90] S.Anantharaman, J.Hsiang, Automated Proofs of the Moufang Identities in Alternative Rings, *Vol. 1, J. Automated Reasoning, 1990*

[AHM-89] S.Anantharaman, J.Hsiang, J.Mzali, SbReve2: A term rewriting laboratory with (AC-)Unfailing Completion, *RTA, Springer-Verlag LNCS Vol. 355, pp533–537, 1989*

[AM-88] S.Anantharaman, J.Mzali, Unfailing Completion Modulo a set of Equations, *Research Report, no. 470, LRI-Orsay (Fr.), 1989*

[BDP-87] L.Bachmair, N.Dershowitz, D.Plaisted, Completion without failure, *Proc. Coll. on Resolution of Equations in Algebraic Structures, Lakeway, Texas, 1987*

[Der-90] N. Dershowitz, A Maximal-Literal Unit Strategy for Horn Clauses, *Proc. of the 2nd CTRS, Montreal, 1990*

[FRT-84] J.M.Font, A.J.Rodriguez, A.Torrens, Wajsberg algebras, *Stochastica, Vol. 8, No. 1, pp5-31, 1984*

[HR-87] J. Hsiang, M. Rusinowitch, On word problems in equational theories, *14th ICALP, Springer-Verlag LNCS Vol. 267, pp54–71, 1987*

[HRS-87] J. Hsiang, M. Rusinowitch, K. Sakai, Complete set of inference rules for the cancellation laws, *IJCAI 87, Milan, Italy, 1987*

[Le-83] P. Lescanne, Computer Experiments with the REVE term rewriting system generator, *10th POPL, pp99-108, 1983*

[Mu] D.Mundici, Personal communication.

[TL-56] A.Tarski, J.Lukasiewicz, Investigations into the sentential calculus, *Chapter 4th in A.Tarski, Logic, Semantics, Meta-mathematics, pp38-56, Clarendon Press, Oxford, 1956*

Completion of First-Order Clauses with Equality by Strict Superposition*

Extended Abstract†

Leo Bachmair

Department of Computer Science
SUNY at Stony Brook
Stony Brook, New York 11794, U.S.A.
leo@sbcs.sunysb.edu

Harald Ganzinger‡

FB Informatik
Universität Dortmund
D-4600 Dortmund 50, F.R.G.
hg@informatik.uni-dortmund.de

We have previously shown that strict superposition together with merging paramodulation is refutationally complete for first-order clauses with equality. This paper improves these results by considering a more powerful framework for simplification and elimination of clauses. The framework gives general criteria under which simplification and elimination do not destroy the refutation completeness of the superposition calculus. One application is a proof of the refutation completeness for alternative superposition strategies with arbitrary selection functions for negative literals. With these powerful simplification mechanisms it is often possible to compute the closure of nontrivial sets of clauses under superposition in a finite number of steps. Refutation or solving of *goals* for such closed or *complete* sets of clauses is simpler than for arbitrary sets of clauses. The results in this paper contain as special cases or generalize many known results about about ordered Knuth-Bendix-like completion of equations, of Horn clauses, of Horn clauses over built-in Booleans, about completion of first-order clauses by clausal rewriting, and inductive theorem proving for Horn clauses.

1 Introduction

A variety of techniques have been suggested for theorem proving in theories with equations. The use of equations $s \approx t$ as directed rewrite rules $s \Rightarrow t$ can be very effective, as the various applications of the Knuth-Bendix completion method (Knuth and Bendix 1970) indicate. The completion procedure can be applied to theories described by (universally quantified) equations. Its two main components are *superposition*, which is a

***The research described in this paper was supported in part by the National Science Foundation under grant CCR-8901322, by the ESPRIT project PROSPECTRA (ref. no. 390), and by a travel grant from Deutsche Forschungsgemeinschaft.**

†A full version of this paper containing all the proofs of the theorems is available as Research Report No. 372, FB Informatik, Universität Dortmund, 1990.

‡Part of this work was done while the second author was on leave at SUNY at Stony Brook.

restricted form of paramodulation (Robinson and Wos 1969), and *simplification by rewriting*, of which demodulation (Wos et al. 1967) is essentially a special case. Extensions of completion to a refutationally complete proof method (for purely equational theories) have been described by Lankford 1975, Hsiang and Rusinowitch 1987, and Bachmair, Dershowitz and Plaisted 1989.

Similar techniques have also been applied to first-order clauses. Consider, for instance, paramodulation for variable-free formulas:

$$\frac{\Gamma \rightarrow \Delta, s \approx t \quad \Lambda \rightarrow \Pi, u[s] \approx v}{\Gamma, \Lambda \rightarrow \Delta, \Pi, u[t] \approx v}$$

and suppose that $\succ$ is an ordering which is total on variable-free terms and formulas. We say that the paramodulation inference is *ordered* (with respect to $\succ$) if and only if (i) $s \succ t$; (ii) $s \approx t$ is maximal with respect to $\Gamma \cup \Delta$; and (iii) $u[s] \approx v$ is maximal with respect to $\Lambda \cup \Pi$. An ordered paramodulation inference is said to be a *superposition inference*, if (iv) $u[s] \succ v$. The superposition is called *strict*, if in addition (v) s does not occur in Γ. By a *weak superposition* inference we mean a paramodulation inference for which conditions (i), (iii), and (iv)—but not necessarily (ii)—are satisfied.

Hsiang and Rusinowitch (1989) have proved that ordered paramodulation is refutationally complete, whereas Rusinowitch (1988) has established the refutation completeness of weak superposition. We have recently shown (Bachmair and Ganzinger 1990) that superposition (and hence strict superposition) is not complete if deletion of tautologies is permitted.

For example, consider the set of clauses

$$\begin{array}{rcl} c \approx d & \rightarrow & \\ & \rightarrow & b \approx d \\ a \approx d & \rightarrow & a \approx c \\ & \rightarrow & a \approx b, a \approx d \end{array}$$

where a, b, c, and d are constants. If $\succ$ is an ordering in which $a \succ b \succ c \succ d$, then the only clause that can be obtained from the above clauses by superposition is $a \approx d \rightarrow b \approx c, a \approx d$, which is a tautology. In other words, even though the above set of clauses is unsatisfiable, with superposition a contradiction (i.e., the empty clause) can only be derived by paramodulating into a tautology. (Ordered paramodulation, on the other hand, need never be applied to tautologies.)

However, several moderate enrichments of the superposition calculus are refutationally complete and compatible with various simplification and deletion mechanisms, including tautology deletion (Bachmair and Ganzinger 1990).

In this paper, we improve the results of Bachmair and Ganzinger (1990) by presenting a much stronger notion of redundancy of formulas and inferences which preserves the refutation completeness of the calculus. The notion of redundancy is basis for powerful techniques of simplification and deletion of clauses. One further application is a proof of refutation completeness for alternative superposition strategies with arbitrary selection functions for negative literals. An abstract scheme for theorem proving procedures with simplification and elimination of clauses is described and proved complete.

Finally we investigate some ways of how to extend the fundamental idea of Knuth-Bendix completion to first-order clauses. A completed system of clauses is simply one which is closed under the inferences of the strict superposition calculus up to redundancy as defined within our framework. Under certain additional assumptions, the refutation or solving of goals with respect to a given closed set of clauses is much more efficient that in the general case. The results presented here for the case of first-order clauses include and generalize results about ordered completion of equations (Bachmair, Dershowitz and Plaisted 1989), completion of Horn clauses (Kounalis and Rusinowitch 1988, Ganzinger 1987b), and ground completion of Horn clauses over built-in Booleans (Rémy and Zhang 1985, Ganzinger 1987a, Nieuwenhuis and Orejas 1991).

2 Equality Herbrand Interpretations

We formulate our inference rules in an equational framework and define clauses in terms of multisets.

A *multiset* over a set X is a function M from X to the set of natural numbers N. Intuitively, $M(x)$ specifies the number of occurrences of x in M. We say that x is an *element* of M if and only if $M(x) > 0$. A multiset M is a *submultiset* of M' (written $M \subseteq M'$) if and only if $M'(x) \geq M(x)$, for all x. *Union* and *intersection* of multisets are defined in the usual way by the identities $M_1 \cup M_2(x) = M_1(x) + M_2(x)$ and $M_1 \cap M_2(x) = \min(M_1(x), M_2(x))$. If M is a multiset and S a set, we write $M \subseteq S$ to indicate that every element of (the multiset) M is an element of (the set) S, and use $M \cap S$ to denote the *set* $\{x \in S : M(x) \geq 1\}$.

For simplicity, we often use a set-like notation to describe multisets. For example, $\{x, x, x\}$ denotes the multiset M such that $M(x) = 3$ and $M(y) = 0$, for $y \neq x$.

An *equation* is a multiset $\{s, t\}$, where s and t are (first-order) *terms* built from given function symbols and variables. We write $s \approx t$ to denote the equation $\{s, t\}$. By a *ground* expression we mean an expression containing no variables.

A *clause* is a pair of multisets of equations, written $\Gamma \rightarrow \Delta$. The multiset Γ is called the *antecedent*; the multiset Δ, the *succedent*. We usually write Γ_1, Γ_2 instead of $\Gamma_1 \cup \Gamma_2$; Γ, A or A, Γ instead of $\Gamma \cup \{A\}$; and $A_1, \ldots, A_m \rightarrow B_1, \ldots, B_n$ instead of $\{A_1, \ldots, A_m\} \rightarrow \{B_1, \ldots, B_n\}$. A clause $A_1, \ldots, A_m \rightarrow B_1, \ldots, B_n$ may be regarded as representing an implication $A_1 \wedge \cdots \wedge A_m \supset B_1 \vee \cdots \vee B_m$. The empty clause $\rightarrow$ indicates a contradiction. Clauses of the form $\Gamma, s \approx t \rightarrow s \approx t, \Delta$ or $\Gamma \rightarrow \Delta, t \approx t$ are called *tautologies*.

We write $u[s]$ to indicate that s is a subterm of u and (ambiguously) denote by $u[t]$ the result of replacing a particular occurrence of s by t. By $t\sigma$ we denote the result of applying the substitution σ to t, and call $t\sigma$ an *instance* of t. If $t\sigma$ is ground, we speak of a *ground instance*. We shall also consider instances of multisets of equations and of clauses. For example, the multiset $\{a \approx b, a \approx b\}$ is an instance of $\{x \approx b, a \approx y\}$. The composition $\tau\rho$ of two substitutions τ and ρ is defined by: $x\tau\rho = (x\tau)\rho$, for all variables x.

An *equivalence* is a reflexive, transitive, symmetric binary relation. A *congruence* is an equivalence $\sim$ on terms, such that $s \sim t$ implies $u[s] \sim u[t]$, for all terms u, s, and t. If E is a set of ground equations, we denote by E^* the smallest congruence $\sim$ such that $s \sim t$ whenever $s \approx t \in E$. By an *(equality Herbrand) interpretation* we mean a congruence on ground terms.

An interpretation I is said to *satisfy* a ground clause $\Gamma \rightarrow \Delta$ if and only if either $\Gamma \not\subseteq I$ or else $\Delta \cap I \neq \emptyset$. We also say that a ground clause C is *true in* I, if I satisfies C; and that C is *false in* I, otherwise. An interpretation I is said to *satisfy* a (non-ground) clause $\Gamma \rightarrow \Delta$ if and only if it satisfies all ground instances $\Gamma\sigma \rightarrow \Delta\sigma$. For instance, a tautology is satisfied by any interpretation. A clause which is satisfied by no interpretation is called *(equality) unsatisfiable*. The empty clause is unsatisfiable by definition. If I satisfies all clauses of a set N, we say that I is a *(equality Herbrand) model* of N. A set N of clauses is *unsatisfiable* if it has no model. We say that a set of clauses N *implies* a clause C (and write $N \models C$) if and only if, for each ground substitution σ, $C\sigma$ is satisfied by any model of N.

3 Canonical Rewrite Systems

We shall use canonical rewrite systems to describe and reason about equality interpretations.

A binary relation $\Rightarrow$ on terms is called a *rewrite relation* if and only if $s \Rightarrow t$ implies $u[s\sigma] \Rightarrow u[t\sigma]$, for all terms s, t and u, and substitutions σ. A transitive, well-founded rewrite relation is called a *reduction ordering*. By $\Leftrightarrow$ we denote the symmetric closure of $\Rightarrow$; by $\Rightarrow^*$ the transitive, reflexive closure; and by $\Leftrightarrow^*$ the symmetric, transitive, reflexive closure. Furthermore, we write $s \Downarrow t$ to indicate that there exists a term v such that $s \Rightarrow^* v$ and $t \Rightarrow^* v$. A rewrite relation $\Rightarrow$ is said to be *Church-Rosser* if and only if the two relations $\Leftrightarrow^*$ and $\Downarrow$ are the same.

We say that a set of equations E is a *rewrite system* with respect to an ordering $\succ$ if and only if for all equations $s \approx t$ in E, either $s \succ t$ or $t \succ s$. Equations in a rewrite system are also called *(rewrite) rules*. (We adopt the convention to write rules in the form $s \approx t$, if and only if $s \succ t$.) We speak of a ground rewrite system E if all equations in E are ground. If E is a rewrite system with respect to $\succ$, we denote by $\Rightarrow_{E\succ}$ (or simply $\Rightarrow_E$) the smallest rewrite relation for which $s \Rightarrow_E t$ whenever $s \approx t \in E$ and $s \succ t$. A term s is said to be *reducible* by E if there is a term t such that $s \Rightarrow_E t$; and *irreducible*, otherwise. For instance, if $s \Downarrow_E t$ and $s \succ t$, then the term s is reducible by E. A rewrite system E is said to be *canonical* if and only if the rewrite relation $\Rightarrow_E$ is well-founded and Church-Rosser.

Lemma 3.1 *Let E be a ground rewrite system with respect to some reduction ordering $\succ$. If there are no two distinct equations $s \approx t$ and $u \approx v$ in E, such that $s \succ t$ and $u \succ v$ and s is a subterm of u, then E is canonical.*

Any ordering $\succ$ on a set S can be extended to an ordering $\succ_{mul}$ on multisets over S as follows: $M \succ_{mul} N$ if and only if (i) $M \neq N$ and (ii) whenever $N(x) > M(x)$ then $M(y) > N(y)$, for some y such that $y \succ x$. If $\succ$ is a total [well-founded] ordering, so is $\succ_{mul}$.

Thus, if $\succ$ is an ordering on terms, the corresponding multiset ordering $\succ_{mul}$ is an ordering on equations. We say that an equation A is *maximal* with respect to a multiset of equations Γ if and only if there is no equation B in Γ such that $B \succ_{mul} A$. We say that A is *strictly maximal* with respect to Γ if and only if there is no equation B in Γ such that $B \succeq_{mul} A$. We say that a term s is *maximal* in a multiset of equations Γ if and only if Γ contains an equation $s \approx t$, but no equation $u \approx v$ such that $u \succ s$. Similarly, s is called *maximal* in a clause $\Gamma \rightarrow \Delta$ if and only if s is maximal in $\Gamma \cup \Delta$.

Term orderings can also be extended to orderings on clauses. For that purpose we shall distinguish between occurrences of an equation in the antecedent and succendent, respectively, of a clause. More precisely, a *negative equation* is a multiset $\{s, t, \top\}$, written $s \approx_\top t$, and a *positive equation* is a multiset $\{s, t, \bot\}$, written $s \approx_\bot t$, where $\top$ and $\bot$ are new symbols with $\succ$ extended by $\top \succ \bot$, and $s \succ \top$, for any ground term s.[1] Then, occurrences of equations are ordered by $\succ^l$ which is defined as the multiset extension $\succ_{mul}$ of $\succ$.

If C is a clause

$$s_1 \approx t_1, \ldots, s_m \approx t_m \rightarrow u_1 \approx v_1, \ldots, u_n \approx v_n,$$

let γ^C be the multiset

$$\{s_1 \approx_\top t_1, \ldots, s_m \approx_\top t_m, u_1 \approx_\bot v_1, \ldots, u_n \approx_\bot v_n\},$$

We define the ordering $\succ^c$ on clauses by: $C \succ^c C'$ if and only if $\gamma^C \succ^l_{mul} \gamma^{C'}$. As can be seen easily, $\succ^c$ is total [well-founded] on ground clauses, if $\succ$ is total [well-founded] on ground terms.

The ordering $\succ^c$ assigns more weight to occurrences of an equation in the antecedent than to occurrences in the succedent. On the other hand, if $s \approx t \succ_{mul} u \approx v$ then $s \approx_p t \succ^l u \approx_q t$, for any two "signs" p and q in $\{\top, \bot\}$. Note also that, in the ground case, $C \succ^c D$ whenever the maximal term in C is greater (with respect to $\succ$) than the maximal term in D. Only if the maximal terms of two ground clauses coincide, the smaller sides of equations for the maximal term and the signs of these literals have an impact on $\succ^c$.

4 Ordered Inference Rules

Let $\succ$ be a reduction ordering. We say that a clause $\Gamma \rightarrow \Delta, s \approx t$ is *reductive* (for $s \approx t$) if and only if (i) $t \not\succeq s$, (ii) $s \approx t$ is strictly maximal with respect to $\Gamma \cup \Delta$, and (iii) s does not occur in Γ.

The following inference rules are defined with respect to $\succ$.

Ordered equality resolution: $\dfrac{\Lambda, u \approx v \rightarrow \Pi}{\Lambda\sigma \rightarrow \Pi\sigma}$

where σ is a most general unifier of u and v, and $u\sigma \approx v\sigma$ is maximal with respect to $\Lambda\sigma$ and strictly maximal with respect to $\Pi\sigma$.

This inference rule encodes the reflexivity of equality.

Ordered factoring: $\dfrac{\Gamma \rightarrow \Delta, A, B}{\Gamma\sigma \rightarrow \Delta\sigma, A\sigma}$

where σ is a most general unifier of A and B, and $A\sigma$ is maximal with respect to $\Gamma\sigma \cup \Delta\sigma$.

We do not include a corresponding inference rule for factoring in the succedent, as such a rule is not necessary in order to obtain a refutationally complete inference system.

[1] Note that $\top$ and $\bot$ are *not* considered as additional constants in the vocabulary of non-logical symbols.

Strict Superposition, left:
$$\frac{\Gamma \rightarrow \Delta, s \approx t \quad u[s'] \approx v, \Lambda \rightarrow \Pi}{u[t]\sigma \approx v\sigma, \Gamma\sigma, \Lambda\sigma \rightarrow \Delta\sigma, \Pi\sigma}$$

where (i) σ is a most general unifier of s and s', (ii) the clause $\Gamma\sigma \rightarrow \Delta\sigma, s\sigma \approx t\sigma$ is reductive for $s\sigma \approx t\sigma$, (iii) $v\sigma \not\succeq u\sigma$ and $u\sigma \approx v\sigma$ is maximal with respect to $\Lambda\sigma$ and strictly maximal with respect to $\Pi\sigma$, and (iv) s' is not a variable.

Strict Superposition, right:
$$\frac{\Gamma \rightarrow \Delta, s \approx t \quad \Lambda \rightarrow u[s'] \approx v, \Pi}{\Gamma\sigma, \Lambda\sigma \rightarrow u[t]\sigma \approx v\sigma, \Delta\sigma, \Pi\sigma}$$

where (i) σ is a most general unifier of s and s', (ii) the clause $\Gamma\sigma \rightarrow \Delta\sigma, s\sigma \approx t\sigma$ is reductive for $s\sigma \approx t\sigma$, (iii) $v\sigma \not\succeq u\sigma$ and $u\sigma \approx v\sigma$ is strictly maximal with respect to $\Lambda\sigma \cup \Pi\sigma$, and (iv) s' is not a variable.

Strict superposition is a sharpened version of superposition. Observe the difference, in condition (iii), between the two superposition rules: when superposing on an equation $u \approx v$ in the antecedent of a clause, we require $u\sigma \approx v\sigma$ to be maximal, but not necessarily strictly maximal. This difference is due to the fact that ordered factoring is applied only to equations in the succedent of a clause. Indeed, if ordered factoring is also applied to antecedents, it suffices to superpose on strictly maximal equations.

Strict superposition is not refutationally complete, but has to be combined with an additional inference rule.

Merging Paramodulation:
$$\frac{\Gamma \rightarrow \Delta, s \approx t \quad \Lambda \rightarrow u \approx v[s'], u' \approx v', \Pi}{\Gamma\sigma, \Lambda\sigma \rightarrow u\sigma \approx v[t]\sigma, u\sigma \approx v'\sigma, \Delta\sigma, \Pi\sigma}$$

where (i) σ is the composition $\tau\rho$ of a most general unifier τ of s and s', and a most general unifier ρ of $u\tau$ and $u'\tau$, (ii) the clause $\Gamma\sigma \rightarrow \Delta\sigma, s\sigma \approx t\sigma$ is reductive for $s\sigma \approx t\sigma$, (iii) the clause $\Lambda\sigma \rightarrow \Pi\sigma, u\sigma \approx v\sigma$ is reductive for $u\sigma \approx v\sigma$, (iv) $u\tau \succ v\tau$ and $v'\sigma \not\succeq v\sigma$, and (v) s' is not a variable.

Merging paramodulation is designed in such a way that its repeated application to ground clauses (in conjunction with ordered factoring) has the effect of merging certain atoms in the succedent with contain the maximal term.

By $\mathcal{M}$ we denote the inference system consisting of all of the above inference rules, with the additional restrictions: (a) the premises of an inference rule must not share any variables (if necessary, the variables in one premise are renamed); and (b) no premise or conclusion must be a tautology.

When describing certain other strategies of paramodulation, the maximality constraints for $u\sigma \approx v\sigma$ in the inference rules for equality resolution and strict left-superposition are dropped. These weaker inferences will be called *selective equality resolution* and *selective paramodulation*, respectively, and the weaker inference system, with the restrictions (a) and (b) as above, will be denoted by $\mathcal{S}$. The term "selective" is meant to indicate that *arbitrary*, not necessarily maximal negative literals can be chosen for paramodulation or resolution within $\mathcal{M}$.

5 Refutation Completeness

In this section we shall explain the construction of an equality interpretation from a given set of clauses.

5.1 Construction of Equality Interpretations

Let N be a set of clauses and $\succ$ be reduction ordering which is total on ground terms. We construct an interpretation I by means of a canonical rewrite system R as follows.

First, we use induction on $\succ^c$ to define sets of equations E_C, R_C, and I_C, for all ground instances C of clauses of N. Let C be such a ground instance and suppose that $E_{C'}$, $R_{C'}$, and $I_{C'}$ have been defined for all ground instances C' of N for which $C \succ^c C'$. Then we define

$$R_C = \bigcup_{C \succ^c C'} E_{C'} \text{ and } I_C = R_C^*.$$

Moreover

$$E_C = \{s \approx t\}$$

if C is a clause $\Gamma \rightarrow s \approx t, \Delta$ such that (i) C is reductive for $s \approx t$, (ii) s is irreducible by R_C, (iii) $\Gamma \subseteq I_C$, and (iv) $\Delta \cap I_C = \emptyset$. In that case, we also say that C *generates* the equation (or rule) $s \approx t$. In all other cases, $E_C = \emptyset$.

Finally, we define I to be the equality interpretation R^*, where $R = \bigcup_C E_C$ is the set of all equations generated by ground instances of clauses of N. (It is evident that R is a rewrite system with respect to $\succ$.)

The following properties of the rewrite system R and the interpretation I will be used repeatedly.

Lemma 5.1 *Any of the rewrite systems R_C and R is canonical.*

Lemma 5.2 *Let C be a ground instance of some clause of N with maximal term s and let u and v be terms with $s \succeq u$ and $s \succeq v$. Furthermore suppose that either $u \neq s \neq v$ or else s is reducible by R_C. Then $u \approx v \in I$ if and only if $u \Downarrow_{R_C} v$.*

Lemma 5.3 *Suppose $C\sigma$ is a ground instance of a clause C in N and $x\sigma$ is reducible by R, for some variable x occurring in C. Then there is a ground instance $C\tau$ of C, such that (i) $C\sigma \succ^c C\tau$ and (ii) $C\tau$ is true in I if and only if $C\sigma$ is true in I.*

Lemma 5.4 *Suppose $C\sigma$ is a ground instance of a clause C in N and $x\sigma$ is reducible by R, for some variable x occurring in an equation $u \approx v$ in the antecedent of C and $u\sigma \neq v\sigma$. Then there is a ground instance $C\tau$ of C, such that (i) $C\sigma \succ^c C\tau$ and (ii) $C\tau$ is false in $I_{C\sigma}$ whenever $C\sigma$ is false in $I_{C\sigma}$.*

The preceding lemmas ensure that the following Lifting Lemma can be applied to certain ground instances of clauses.

Lemma 5.5 (Lifting Lemma) *Let C be a clause $\Gamma \rightarrow \Delta, s \approx t$ and D a clause $\Lambda, A[u] \rightarrow \Pi$ or $\Lambda \rightarrow A[u], \Pi$, such that u is not a variable and C and D have no variables in common. Suppose $s\sigma = u\sigma$ and C' is a ground clause*

$$\Gamma\sigma, \Lambda\sigma, A[t]\sigma \rightarrow \Delta\sigma, \Pi\sigma$$

or

$$\Gamma\sigma, \Lambda\sigma \rightarrow A[t]\sigma, \Delta\sigma, \Pi\sigma$$

obtained from $C\sigma$ and $D\sigma$ by strict superposition, selective paramodulation, or merging paramodulation. Then C' is a ground instance of a clause that is obtained from C and D by strict superposition, selective paramodulation, or merging paramodulation.

Note that (ordered) factoring is not required for this version of the Lifting Lemma, since we have defined clauses in terms of *multisets* (rather than sets) of atoms. Also note that a merging paramodulation on the ground level may correspond to a strict superposition inference on the non-ground level.

5.2 Redundancy and Saturation

We want to prove that the construction of the interpretation I yields a model for N, provided N is consistent and closed under sufficiently many applications of the inference rules of the superposition calculus $\mathcal{M}$. Although $\mathcal{M}$ greatly constrains paramodulation and factoring inferences, the closure of a set of clauses under $\mathcal{M}$ may still be too large, as deduced clauses are often redundant in that they do not contribute to the construction of a contradiction. Inferences from such clauses need not be computed. We will define a suitable notion of redundancy which will place severe additional, more global restrictions on the search space for refutation proofs, so that in many cases the closure of a set of clauses up to redundancy will be finite.

Let N be a set of clauses and let I be the interpretation constructed from N. Then we define three sets N^{mf}, N^{f} and N^{gen} of clauses as follows:

(i) N^{mf} is the set consisting of the smallest ground instance of N that is false in I, provided such an instance exists. Otherwise, $N^{\mathrm{mf}} = \emptyset$.

(ii) A ground instance C of a clause in N is in N^{f} if and only if C is false in I_C and each ground instance of N which is smaller than C with respect to $\succ^c$ is satisfied by I_C.

(iii) N^{gen} is the subset of clauses in N^{f} which generate a rule in R.

A ground instance $C\sigma$ of a clause C in N is called *essential* (in N) if and only if $C\sigma$ is in N^{mf} or in N^{f}. Otherwise, $C\sigma$ is called *redundant* (in N). A clause in N is called essential (in N) if any one of its ground instances is essential in N. Otherwise, the clause is called redundant in N.[2]

A ground inference from instances of N is said to be *essential* (in N) if and only if it belongs to one of the following classes:

(a) Ordered factoring inferences whose premise is essential.

(b) Equality resolution inferences whose premise is contained in N^{mf}.

(c) Strict superposition inferences whose first premise is contained in N^{gen} and whose second premise is contained in N^{mf}.

(d) Merging paramodulation inferences whose premises are both contained in N^{gen}.

Otherwise, the inference is called *redundant* or *inessential* (in N).

Let us call an inference

$$\frac{C_1\tau_1 \ldots C_n\tau_n}{C_0\tau_0}$$

in $\mathcal{M}$ a *ground instance* of some other inference

$$\frac{C_1 \ldots C_n}{C_0}$$

[2] In Bachmair and Ganzinger (1991) we have used, for technical simplicity, a simpler notion of redundancy. In the latter paper a clause C is called redundant if it is satisfied both by I and I_C. Evidently, clauses which are satisfied both by I and I_C are redundant with respect to the present definition, so that the present notion of redundancy is (strictly) more general.

in $\mathcal{M}$, if the $C_i\tau_i$ are ground instances of the C_i. Note that not every combination of substitution instances of the C_i will define a legal inference in $\mathcal{M}$ and, hence, a ground instance of the inference from the C_i. An inference is called *essential* (in N) if and only if any one of its ground instances is essential.

A set of clauses N is said to be *saturated* (with respect to $\mathcal{M}$) if and only if every clause C that is the conclusion of an essential ground instance of an inference in $\mathcal{M}$ from premises in N is a ground instance of N.

Lemma 5.6 *Suppose that N is saturated and let I be the interpretation constructed from N. Furthermore, let C be a clause generating an equation $s \approx t$ and let all instances of N with a maximal term smaller than s be satisfied by I. Then C is in N^{gen}.*

Lemma 5.7 *Suppose that N is saturated and let I be the interpretation constructed from N. If a clause $C = \Gamma \rightarrow s \approx t, \Delta$ in N^{gen} generates $s \approx t$, then* (i) $\Gamma \subseteq I$, (ii) *if $\Delta = s \approx w, \Delta'$, then t is irreducible by R and $t \succ w$, and* (iii) $\Delta \cap I = \emptyset$.

Theorem 5.8 *A saturated set of clauses N is consistent if and only if it does not contain the empty clause. If N is saturated and consistent, the interpretation I constructed from the ground instances of N is a model of N.*

Lemma 5.9 *A set of clauses N is saturated if and only if any $\mathcal{M}$-inference from N is redundant (in N).*

6 Modular Redundancy Criteria

The notion of essential inferences is too general to be useful in a theorem prover. An inference which is redundant at one point in the theorem proving process may become essential at a later point. Moreover we need criteria for when clauses can be deleted without loosing refutation completeness.

Let N be a set of clauses and C be a ground clause. Let us call C *composite* (in N), if there exist ground instances $C_1, \ldots, C_k$ of N such that $C_1, \ldots, C_k \models C$ and $C \succ^c C_j$, for $1 \leq j \leq k$. A clause is called *composite* (in N) if all its ground instances are *composite* in N.[3]

A ground inference

$$\frac{B}{C} \quad \text{or} \quad \frac{B' \; B}{C},$$

respectively, in S is called *composite* in N, if

(i) any of the premises B or B' is composite in N, or

(ii) there exist ground instances $C_1, \ldots, C_k$ of N such that $C_1, \ldots, C_k \models C$ and, for all j with $1 \leq j \leq k$, $B \succ^c C_j$ or $C \succeq^c C_j$. An inference is called *composite* (in N) if all its ground instances are composite in N.

---

[3] A more refined way of comparing substitution instances of clauses in N might be based on pairs of clauses and substitutions and involve the subsumption ordering on clauses, cf. Ganzinger and Bachmair (1990). For the present paper we have decided not to overly complicate matters technically and simply compare the substituted clauses. To include subsumption as a simplification technique, one should (slightly ambiguously) identify any instance $C\sigma$ of a clause C in some given set of clauses N with the pair (D, τ) such that $C\sigma = D\tau$ and D a minimal (with respect to the subsumption ordering) clause in N. Clauses which are subsumed by other clauses are then useless and can be deleted.

Lemma 6.1 *Composite clauses are redundant.*

Lemma 6.2 *An inference is composite in N whenever its conclusion is either an instance of N or else is composite in N.*

Lemma 6.3 *If π is a composite (in N) inference by selective equality resolution with premise B, then B, and hence π, is redundant in N.*

Lemma 6.4 *Let π be a composite (in N) inference by selective paramodulation or by strict right-superposition of a ground clause B' on a ground clause B. If B' is in N^{gen}, then B, and hence π, is redundant in N.*

Lemma 6.5 *Composite inferences are redundant.*

Lemma 6.6 *(i) If $N \subseteq N'$, then any inference or clause which is composite in N is also composite in N'.*

(ii) If $N \subseteq N'$ and all clauses in $N' \setminus N$ are composite in N', then any inference or clause which is composite in N' is also composite in N.

The last lemma shows that compositeness is stable under enrichments and under deletion of *composite* clauses, making the concept of compositeness a useful basis for simplification and deletion in a theorem prover.

7 Theorem Proving with Simplification and Deletion

In this section we will consider the problem of how to construct a saturated set of clauses from an initially given set N of clauses.

7.1 Theorem Proving Derivations

Theorem proving can be viewed as computing derivations using the following two inference rules on sets of clauses N:

Deduction: $$\frac{N}{N \cup \{C\}} \quad \text{if } N \models C$$

Deletion: $$\frac{N \cup \{C\}}{N} \quad \text{if } C \text{ is composite in } N$$

Deduction (DED) adds formulas which follow from already derived formulas; deletion (DEL) eliminates composite clauses. An addition of a clause D to N which triggers a subsequent deletion of a clause C can be viewed as a simplification. If D is needed to prove the compositeness of C, it will be smaller than C with respect to $\succ^c$. Hence, we have a derived inference

Simplification: $$DEL\frac{DED\dfrac{N \cup \{C\}}{N \cup \{C, D\}}}{N \cup \{D\}}$$

Note that simplification may require the deduction of (logically sound) clauses other than those that can be obtained by $\mathcal{S}$.

A finite or countably infinite sequence $N_0, N_1, N_2, \ldots$ of sets of clauses is called a *theorem proving derivation* if N_{i+1} results from N_i by deduction or deletion.

Let now $\mathcal{R}$ denote any subset of inference rules from $\mathcal{S}$. We call a theorem proving derivation *fair* with respect to $\mathcal{R}$ if for any ground instance π of an inference in $\mathcal{R}$ from *persisting formulas* in $N_\infty = \bigcup_j \bigcap_{k \geq j} N_k$ there exists an index $j \geq 0$ such that π is composite in N_j.

Lemma 7.1 *A derivation $N_0, N_1, N_2, \ldots$ is fair (with respect to $\mathcal{R}$), if for any inference in $\mathcal{R}$ with conclusion C and premises in N_∞ there exists an index j, such that $C \in N_j$.*

In practice, if an inference is scheduled for step j, one will first try to prove the compositeness of the inference and only if this fails add the conclusion.

We call a set of clauses N *complete* (with respect to $\mathcal{R}$) if all inferences in $\mathcal{R}$ from N are composite in N. (For certain systems $\mathcal{R}$, completeness will imply saturation, cf. below.)

Lemma 7.2 *Let $N_0, N_1, N_2, \ldots$ be a fair theorem proving derivation with respect to $\mathcal{R}$.*
(i) Any clause C in $(\bigcup_j N_j) \setminus N_\infty$ is composite in N_∞.
(ii) N_∞ is complete with respect to $\mathcal{R}$.

7.2 Maximal-Literal Strategies

Theorem proving with respect to $\mathcal{M}$ is refutationally complete.

Theorem 7.3 *Let $N_0, N_1, N_2, \ldots$ be a fair theorem proving derivation with respect to $\mathcal{M}$. Then N_∞ is saturated (with respect to $\mathcal{M}$), and N_0 is inconsistent if and only if the empty clause is in N_k, for some $k \geq 0$.*

7.3 Literal Selection Strategies

We now consider theorem proving with respect to arbitrary selection functions for negative literals. Formally *selection functions* may be described by sets S of pairs (C, Λ) of clauses C and non-empty subsets Λ of equations of the antecedent of C, i.e., C is of form $\Lambda, \Gamma \rightarrow \Delta$. Observe that only negative literals can be selected. For a clause C one or more negative literals may be selected by S. Selection functions are in general partial, so that for some clauses no selection is made at all. By $\mathcal{S}_S$, where S is a selection function, we denote the inference system consisting of the following rules:

- selective equality resolution, restricted to negative literals $u \approx v$ in clauses C, such that (i) (C, Λ) is in S, (ii) $u \approx v$ is in Λ, and (iii) $u\sigma \approx v\sigma$, with σ the mgu of u and v, is maximal in $\Lambda\sigma$;

- selective paramodulation on negative literals $u \approx v$ in clauses C such that (i) (C, Λ) is in S, (ii) $u \approx v$ is in Λ, (iii) $u\sigma \approx v\sigma$, with σ the mgu of the superposition, is maximal in $\Lambda\sigma$, and (iv) the clause which is paramodulated into $u \approx v$ does not occur in S;

- all rules in $\mathcal{M}$, with the further restriction that no clause C where (C, Λ) is in S, for some Λ, occurs as a premise in any of the inferences.

The difference between $\mathcal{M}$ and $\mathcal{S}_S$ is that whenever the selection function selects some subset Λ of negative literals for a clause, all corresponding selective inferences onto the maximal literals in Λ are computed, even if the clause has some other literal which is strictly greater than Λ. Clauses for which no literal is selected are subject to the general maximal-literal strategy, where the maximality constraint refers to all literals of the clause.

One of the main advantages in selecting literals for superposition explicitly seems to be the fact that in $\mathcal{S}_S$ a clause for which a literal is selected needs not be paramodulated into any other clause, even if some ground instance is reductive for some equation.

Special cases of $\mathcal{S}_S$ are $\mathcal{M}$ (for S the empty set) and *positive strategies* (in which case S selects some *part* of antecedent of any clause with a non-empty antecedent). In positive strategies only clauses with an empty antecedent are paramodulated into other clauses. In the particular case of Horn clauses, and selection functions S which select the *complete* antecedent of any clause with a non-empty antecedent, the strategy $\mathcal{S}_S$ has been called *maximal-literal unit strategy* by Dershowitz (1991), who gives a simple completeness proof based on proof transformations.

A selection function may be called *unitary*, if whenever (C, Λ) is a selection in S, Λ consists of exactly one literal. A completion procedure (albeit not an unfailing one) on the basis of $\mathcal{S}_S$ for Horn clauses and with arbitrary unitary selection functions S has been proved complete in (Ganzinger 1987a).

In the following let S be an arbitrary selection function.

Theorem 7.4 *Let $N_0, N_1, N_2, \ldots$ be a fair theorem proving derivation with respect to $\mathcal{S}_S$, and let I be the interpretation constructed from N_∞.*

(i) If (D, Λ) is in S, then no ground instance of D generates a rule in I.

(ii) N_∞ is saturated (with respect to $\mathcal{M}$).

(iii) N_0 is inconsistent if and only if the empty clause is in N_k, for some $k \geq 0$.

The theorem shows in particular the refutation completeness of $\mathcal{S}_S$ for arbitrary selection functions S.

7.4 Simplification and Deletion Techniques

If N is a set of clauses and C is a clause, we will in the following write $[N] \models C$ if and only if $C\sigma$ is satisfied by any model of $N\sigma$, for any ground substitution σ.

Contextual reductive rewriting. If C is a clause and N is a set of clauses, let in the following N_C denote the set of all substitution instances $D\sigma$ of clauses D in N such that $C \succ^c D\sigma$.

Let $C = \Gamma, A[u\rho] \rightarrow \Delta$ (or $C = \Gamma \rightarrow \Delta, A[u\rho]$) be a clause in N. Suppose that there exists a clause $D = \Lambda \rightarrow u \approx v, \Pi$ in N, such that (i) $u \succ v$, (ii) $D\rho \in N_C$, (iii) $[N_C] \models \Gamma \rightarrow s\rho \approx t\rho$, for any $s \approx t$ in Λ, and (iv) $[N_C] \models s\rho \approx t\rho \rightarrow \Delta$, for any $s \approx t$ in Π. Then C can be simplified to $C' = \Gamma, A[v\rho] \rightarrow \Delta$ (or $C' = \Gamma \rightarrow \Delta, A[v\rho]$). In other words, if C' is added to the set of clauses then C can be deleted afterwards.

Replacing C by C' is a generalization of simplification by reductive conditional rewriting in contexts as described by Ganzinger (1987a) for a completion procedure for conditional equations. The contextual aspect is that $u\rho$ and $v\rho$ have to be equal only for those substitutions that make Γ true and Δ false. In practice, proofs of $[N_C] \models \Gamma \rightarrow s\rho \approx t\rho$, for any $s \approx t$ in Λ, may be conducted by reductive conditional rewriting with rule instances in N_C, and using the (skolemized) antecedent Γ as additional rewrite rules. Reductive conditional rewrite rules D in N do not have an occurrence of the maximal term in the antecedent. If they match a proper subterm in C, the corresponding substitution instances belong to N_C.

The ordering $\succ^c$ as defined above places a too hard restriction on rewriting occurrences of the maximal term $u\rho$ of C in a positive literal of C. In this case $v\rho$ must be smaller than the smaller term t in $A[u]\rho = u\rho \approx t$ as otherwise $D\rho$ would not be in N_C. A slightly more complex ordering on clause-substitution pairs exists which would allow to rewrite the root of $u\rho$, regardless of the size of $v\rho$, if $\Pi = \emptyset$, $D\rho$ is reductive for $u\rho \approx v\rho$ and if u is a proper prefix of $u\rho$.

Elimination of redundant atoms. Let $C = \Gamma, u \approx v \rightarrow \Delta$ be a clause in N. If there exists a clause $D = \Gamma \rightarrow u \approx v, \Delta$ such that $[N_C] \models D$, then C can be simplified to $C' = \Gamma \rightarrow \Delta$. A particular case is the elimination of multiple occurrences of atoms in the antecedent. If $C = \Gamma, u \approx v, u \approx v \rightarrow \Delta$, then the clause $\Gamma, u \approx v \rightarrow u \approx v, \Delta$ is a tautology and hence trivially implied by N_C. In a similar way, redundant atoms in the succedent can be eliminated.

Case analysis. The main idea of case analysis is to split a clause $C = \Gamma \rightarrow \Delta$ into a set of n clauses $C_i = A_i, \Gamma \rightarrow \Delta$, provided $[N_C] \models \Gamma \rightarrow A_1, \ldots, A_n, \Delta$, and checking as to whether each of the cases C_i of C can be simplified by clauses smaller than C to some logically equivalent D_i which is smaller than C in the deletion ordering. If this succeeds, C becomes redundant. Contextual rewriting (Rémy and Zhang 1985, Navarro 1987) or splitting of clauses (Ganzinger 1987b) are particular instances of case analysis techniques.

8 Refutation of Goals in the Strict Superposition Calculus

In this section we consider situations in which a fair theorem proving derivation in $\mathcal{S}_S$, starting with some finite set of clauses N_0, terminates after finitely many, say k, steps without encountering an inconsistency. In this case $N_\infty = N_k$ and N_k is a finite and complete set of clauses. Because of the powerful concept of simplification and elimination which we have introduced before there is reason to believe that termination of completion is not unusual in practice.

As an example of a complete (with respect to $\mathcal{M}$) set of clauses consider the theory of a total order p.

$$
\begin{aligned}
&\rightarrow p(x,x) \\
&\rightarrow p(x,y), p(y,x) \\
p(x,y), p(y,z) &\rightarrow p(x,z) \\
p(x,y), p(y,x) &\rightarrow x \approx y
\end{aligned}
$$

The proof of completeness is rather tedious and proceeds by case analysis on the inequalities with respect to $\succ$ between the terms which can be substituted for the variables. We assume that $p(s,t) \succ p(u,v)$ if and only if either $s \succ u$, or else $s = u$ and $t \succ v$. For example, consider the inference by ordered resolution

$$\frac{p(x,y), p(y,z) \rightarrow p(x,z) \quad p(x,z), p(z,x) \rightarrow x \approx z}{p(x,y), p(y,z), p(z,x) \rightarrow x \approx z}$$

If any two of the variables are instantiated to the same term, the inference is composite as one of the premises would be composite. If the variables denote pairwise distinct terms, then the inference is ordered if $x \succ z \succ y$. Observe that, for ground terms x, y and z,

$$\left[\begin{array}{lcl} p(y,z), p(z,x) & \rightarrow & p(y,x), \\ p(x,y), p(y,x) & \rightarrow & x \approx y, \\ p(z,x), p(x,y) & \rightarrow & p(z,y), \\ p(y,z), p(z,y) & \rightarrow & y \approx z \end{array}\right] \models p(x,y), p(y,z), p(z,x) \rightarrow x \approx z .$$

It is easy to verify that all the premises of the latter implication are smaller with respect to $\succ^c$ than the second premise $p(x,z), p(z,x) \rightarrow x \approx z$ of the inference. Hence the inference is composite. Note that this case analysis on the ordering between variables—a technique that has also been described by Martin and Nipkow (1989) in the context of unfailing ordered completion of equations—is independent of the signature, that is, of any additional function or predicate symbols that might exist besides p.

Finite, complete and consistent sets N of clauses will be called *programs* from now on. A formula

$$\neg G = \exists \vec{x}\, (A_1 \wedge \ldots \wedge A_n \wedge \neg B_1 \wedge \ldots \wedge \neg B_k)$$

is a logical consequence of a program N if and only if the set of clauses $N \cup \{G\}$ is inconsistent. The clause $G = A_1, \ldots, A_n \rightarrow B_1, \ldots, B_k$ is called a *goal* in such a case.

The search for a refutation of $N \cup \{G\}$ may be simpler and more efficient for several reasons than the search for a refutation in general.

- Since N is complete, inferences between clauses in N are composite and remain composite at any step of a theorem proving process starting with $N \cup \{G\}$.

- Under certain additional assumptions, it may be sufficient to compute only paramodulation inferences

$$\frac{C \quad D}{B}$$

  where C is a clause in N and D is not it N. Such inferences are called *linear* in N, or *N-linear.* For example, SLD-resolution and conditional narrowing are linear refutation proof strategies which are complete for certain classes of Horn clause programs.

- Clauses for which a negative literal has been selected for superposition during completion need in many cases not be considered for the refutation of a goal.

- Theorem proving procedures usually employ some backtracking mechanism. We speak of a *don't care nondeterministic* inference system if backtracking is not needed, in the sense that a refutation can be constructed regardless of the order in which inference rules are applied. Rewriting by confluent term rewriting systems as used in certain Knuth-Bendix-like completion procedures is "don't care nondeterministic" in this sense.

In certain cases it is decidable whether a goal G is refutable by a program N (or equivalently, whether $\neg G$ is a logical consequence of N). This holds in particular if the inferences in any refutation proof tree for any goal (in some particular class of goals) are strictly decreasing, meaning that the conclusion is smaller than one of its premises in some well-founded ordering. An example is the decidability of the validity problem in an equational theory generated by a finite canonical rewrite system.

The completeness of linear superposition for refutation of goals may depend on the syntactic properties of a given program. For that purpose we shall assume that a set of predicate symbols is given in addition to the set of function symbols. Thus we also consider expressions $P(t_1, \ldots, t_n)$, where P is some predicate symbol and $t_1, \ldots, t_n$ are terms. Predicate symbols are not allowed in proper subterms of any term.

In this context, we consider equations $s \approx t$, where s and t are (non-predicate) terms, and equations $P(t_1, \ldots, t_n) \approx \mathrm{tt}$, where tt is a distinguished unary predicate symbol that is taken to be minimal in the given reduction ordering $\succ$. For simplicity, we abbreviate $P(t_1, \ldots, t_n) \approx \mathrm{tt}$ by $P(t_1, \ldots, t_n)$. We refer to $P(t_1, \ldots, t_n)$ as a *non-equational atomic formula.*

Clauses of the form $\Gamma, P(t_1, \ldots, t_n) \approx \mathrm{tt} \rightarrow \Delta$ or $\Gamma \rightarrow \Delta, P(t_1, \ldots, t_n) \approx \mathrm{tt}$ in which $P(t_1, \ldots, t_n) \approx \mathrm{tt}$ is a maximal equation can evidently not be part of an equality resolution. Furthermore, the strict right superposition of a clause $\Gamma \rightarrow \Delta, P(s_1, \ldots, s_n) \approx \mathrm{tt}$ on a clause $\Lambda \rightarrow \Pi, P(t_1, \ldots, t_n) \approx \mathrm{tt}$ results in a tautology $\Gamma\sigma, \Lambda\sigma \rightarrow \Delta\sigma, \Pi\sigma, \mathrm{tt} \approx \mathrm{tt}$ and is therefore unnecessary.

Strict left superposition or selective paramodulation inferences of the form

$$\frac{\Gamma \rightarrow \Delta, P(s_1, \ldots, s_n) \approx \mathrm{tt} \quad P(t_1, \ldots, t_n) \approx \mathrm{tt}, \Lambda \rightarrow \Pi}{\Gamma\sigma, \Lambda\sigma, \mathrm{tt} \approx \mathrm{tt} \rightarrow \Delta\sigma, \Pi\sigma}$$

on the other hand, are needed in general. Note that the trivial equation $\mathrm{tt} \approx \mathrm{tt}$ in the antecedent can be eliminated by equality resolution. Thus we obtain a derived inference rule, called *ordered resolution*:

$$\frac{\Gamma \rightarrow \Delta, P(s_1, \ldots, s_n) \quad P(t_1, \ldots, t_n), \Lambda \rightarrow \Pi}{\Gamma\sigma, \Lambda\sigma \rightarrow \Delta\sigma, \Pi\sigma}$$

with the usual restrictions imposed on strict left superposition. The selective variant without the maximality constraint for the selected negative literal in the second clause will be called *selective resolution*. From now on the terms "strict left superposition" and "(selective) paramodulation" are reserved for the cases in which proper equations are used for superposition (on arbitrary atoms).

A *quasi-Horn clause* is either a clause $\Gamma \rightarrow \Delta$ or a clause $\Gamma \rightarrow \Delta, s \approx t$, where Δ contains only non-equational atoms and $s\sigma \approx t\sigma$ is strictly maximal in $\Gamma'\sigma \cup \Delta\sigma$, for all ground substitutions σ, where Γ' is the subset of non-equational atoms in Γ.

Quasi-Horn clause programs correspond to what are sometimes called Horn clause specifications over "built-in booleans." In quasi-Horn programs, predicates are defined by clauses $\Gamma \rightarrow \Delta$ with no equations in the succedent, whereas functions are defined by clauses ("conditional equations") $\Gamma \rightarrow \Delta, s \approx t$, where Δ contains only non-equational atoms. The predicates in the non-equational parts of a function definition $\Gamma \rightarrow \Delta, s \approx t$ have to be simpler with respect to the ordering $\succ$ than the equation $s \approx t$, which generalizes the idea of a hierarchical specification over built-in Booleans.

Lemma 8.1 *Let C be a quasi-Horn clause.*

(i) If C contains an equation in the succedent, it cannot be premise of an ordered factoring or a selective resolution inference.

(ii) If D is a clause which has no equation in its succedent, and if B is deduced from C and D by (one application of) strict superposition or selective paramodulation, then B has no equation in the succedent.

Theorem 8.2 *Let $N \cup M$ be a set of quasi-Horn clauses which is complete with respect to S_S, where, for any clause in M, S selects a subset of negative* equational *literals for superposition. Moreover, let G_i be $n \geq 1$ clauses which do not contain an equation in their succedent. Let $N \cup M \cup \{G_1, \ldots, G_n\} = N_0, N_1, \ldots$ be a theorem proving derivation in which no deduction step (DED) adds a clause with an equation in the succedent. Then any inference from premises in N_∞ by strict superposition or selective paramodulation which is not N-linear is composite in N_0.*

The theorem states that the refutation, by quasi-Horn programs, of goals without equations in the succedent is linear with respect to the equality part of the logic. Selective resolution which covers the non-functional aspects of the program, is still nonlinear. Fortunately, selective resolution is a rather restricted form of resolution. Clauses for which some set of negative literals has been selected for superposition during the completion process that led to $N \cup M$, these are the clauses in M, might be called *nonoperational*. They are not used at all for the refutation of the goal.

Corollary 8.3 *Let $N \cup M$ be a set of Horn clauses which is complete with respect to S_S, where, for any clause in M, S selects a subset of (arbirtary) negative literals for superposition. Moreover, let G be a clause with an empty succedent and let $N \cup M \cup \{G\} = N_0, N_1, \ldots$ be a theorem proving derivation in which any deduction step (DED) adds a clause with an empty succedent. Then N_∞ is complete with respect to S_S if and only if it is complete with respect to N-linear selective paramodulation, N-linear selective resolution and selective equality resolution.*

The corollary says in particular that certain ordered generalizations of conditional narrowing is complete for refuting goals which are conjunctions of equations in complete sets of conditional equations. Note that there are no restrictions about variables. The slightly weaker result obtained by Bertling and Ganzinger (1989) did not admit conditional rewrite rules with variables as their left side.

The above result does not say anything about *goal solving*, that is, finding complete sets of substitutions for which the goal becomes true. In the case of Horn clauses, all irreducible substitutions which solve the goal can be enumerated by ordered conditional narrowing. This does not hold for quasi-Horn clauses, in general, as shown by the following example:

$$
\begin{array}{rcl}
 & \rightarrow & p, q \\
p & \rightarrow & a \approx b \\
q & \rightarrow & c \approx d \\
 & \rightarrow & f(x,x,y,z) \approx z \\
 & \rightarrow & f(x,y,z,z) \approx z
\end{array}
$$

where $a \succ b \succ c \succ d \succ p \succ q$. p and q are predicates without arguments. This set of quasi-Horn clauses is complete, however the solution $\{x \mapsto h(a), y \mapsto h(b), u \mapsto h(c), v \mapsto h(d)\}$ for $f(x,y,u,v) \approx v$ cannot be generated from the given axioms and the goal

$$f(x,y,u,v) \approx v \rightarrow$$

by any sequence of paramodulation inferences if the functional reflexive axioms are not available. The difficulty arises from disjunctions of equations which, as in the example above, can easily be specified via propositional variables.

If we restrict attention to ground goals, then goal refutation may be decidable. A clause $C = \Gamma \rightarrow \Delta$ is called *universally reductive* if either Δ is empty, or else $\Delta = A, \Delta'$ and C is universally reductive for A. By that we mean that (i) any variable of C occurs in A, and (ii) $C\sigma$ is reductive for $A\sigma$, for any ground substitution σ. In addition, if A is an equation $s \approx t$, we require that $s\sigma \succ t\sigma$ and that all the variables of A occur in s.

Theorem 8.4 *Let $N \cup M$ be a finite set of quasi-Horn clauses which is complete with respect to $\mathcal{S}_S$, where S selects some set of negative equational literals for any clause in M. Moreover let all clauses in N be universally reductive and let G_i be $n \geq 1$ ground clauses which do not contain an equation in their succedent. Then $N \cup M \models \neg(G_1 \wedge \ldots \wedge G_n)$ is decidable by N-linear strict superposition, N-linear selective paramodulation, ordered factoring, selective equality resolution, and (non-linear) selective resolution.*

We conclude this section with a few remarks about "don't care nondeterminism." The most frequent case of "don't care nondeterminism" is when a superposition inference actually is a simplification by contextual rewriting

$$\frac{\Gamma \rightarrow s \approx t, \Delta \quad \Lambda, A[s\sigma] \rightarrow \Pi}{\Gamma\sigma, \Lambda, A[t\sigma] \rightarrow \Delta\sigma, \Pi}$$

such that $\Gamma\sigma$ is true and $\Delta\sigma$ is false in context $\Lambda \rightarrow \Pi$, as described formally in Section 7. Here $\Lambda, A[s\sigma] \rightarrow \Pi$ typically is a formula of goal type, i.e. without an occurrence of an equation in Π. Then, after having added the new goal $\Gamma\sigma, \Lambda, A[t\sigma] \rightarrow \Delta\sigma, \Pi$, the old goal $\Lambda, A[s\sigma] \rightarrow \Pi$ can be deleted as it has become composite. In other words, no further inference on it is required. In the case of N being a Horn clause program this implies the ground confluence of reductive, recursive rewriting.

Theorem 8.5 *Let $N \cup M$ be a set of Horn clauses which is complete with respect to $\mathcal{S}_S$, where S selects some subset of the negative literals for any clause in M. Let N_R be a set of universally reductive substitution instances of clauses in N such that the sets of reductive ground instances of the clauses in N and N_R, respectively, coincide. Then the initial algebras of N_R, N and $N \cup M$ concide, and recursive conditional rewriting with N_R is ground confluent.*

Note that if $N \cup M$ is also complete with respect to an extension of the given signature (and term ordering) by infinitely many new constants, the last result means in fact confluence on general terms. In general, the completeness of $N \cup M$ need not be preserved under such an extension as the compositeness of an inference or clause may depend on the signature.

The theorem also opens up new and general ways of completion-based inductive theorem proving for Horn clauses, avoiding the problems with the undecidability of inductive reducibility in the Horn clause case. (The clauses in M are inductive theorems of the clauses in N.) This is another interesting application of explicit selection strategies for negative literals. With an appropriate coding, the selection of literals can correspond directly to the selection of induction variables in more traditional induction techniques. A more elaborate treatment of these ideas goes beyond the scope of the present paper.

Acknowledgements. M. Haberstrau and R. Nieuwenhuis have found two serious errors in the preliminary version of the paper that was distributed at the workshop. We are very grateful for their careful reading and for the questions they have asked. We are also grateful to U. Waldmann for his valuable comments on the paper.

References

[1] L. Bachmair, N. Dershowitz, and D. Plaisted, 1989. Completion without failure. In H. Ait-Kaci and M. Nivat, editors, *Resolution of Equations in Algebraic Structures, vol. 2*, pp. 1–30. Academic Press.

[2] H. Bertling and H. Ganzinger, 1989. Completion-time optimization of rewrite-time goal solving. In *Proc. 3rd Int Conf. Rewriting Techniques and Applications*, Lect. Notes in Comput. Sci., vol. 355, Berlin, Springer-Verlag.

[3] L. Bachmair and H. Ganzinger, 1990. On restrictions of ordered paramodulation with simplification. In *Proc. 10th Int. Conf. on Automated Deduction*, Lect. Notes in Comput. Sci., vol. 449, pp. 427–441, Berlin, Springer-Verlag.

[4] L. Bachmair and H. Ganzinger, 1991. Perfect Model Semantics for Logic Programs with Equality. Submitted for publication.

[5] N. Dershowitz, 1991. A Maximal-Literal Unit Strategy for Horn Clauses. In *Proc. Second Int. Workshop on Conditional and Typed Rewriting Systems*, Lect. Notes in Comput. Sci., vol. to appear, Berlin, Springer-Verlag.

[6] H. Ganzinger, 1987. A completion procedure for conditional equations. In S. Kaplan and J.-P. Jouannaud, editors, *Conditional Term Rewriting Systems*, Lect. Notes in Comput. Sci., vol. 308, pp. 62–83, Berlin, Springer-Verlag. To appear in *J. Symbolic Computation.*

[7] H. Ganzinger, 1987. Ground term confluence in parametric conditional equational sepcifications. In *STACS'87*, Lect. Notes in Comput. Sci., vol. 247, pp. 286–298, Berlin, Springer-Verlag.

[8] J. Hsiang and M. Rusinowitch, 1987. On word problems in equational theories. In T. Ottmann, editor, *Proc. 14th ICALP*, Lect. Notes in Comput. Sci., vol. 267, pp. 54–71, Berlin, Springer-Verlag.

[9] J. Hsiang and M. Rusinowitch, 1989. Proving refutational completeness of theorem proving strategies: The transfinite semantic Tree method. Submitted for publication, 1989.

[10] D. Knuth and P. Bendix, 1970. Simple word problems in universal algebras. In J. Leech, editor, *Computational Problems in Abstract Algebra*, pp. 263–297. Pergamon Press, Oxford.

[11] E. KOUNALIS AND M. RUSINOWITCH, 1988. On word problems in Horn theories. In E. Lusk and R. Overbeek, editors, *Proc. 9th Int. Conf. on Automated Deduction*, Lect. Notes in Comput. Sci., vol. 310, pp. 527–537, Berlin, Springer-Verlag.

[12] D. LANKFORD, 1975. Canonical inference. Technical Report ATP-32, Dept. of Mathematics and Computer Science, University of Texas, Austin.

[13] U. MARTIN AND T. NIPKOW, 1989. Ordered Rewriting and Confluence. Technical Report, Univ. of Cambridge, Cambridge, U.K.

[14] R. NIEUWENHUIS AND F. OREJAS, 1991. Clausal Rewriting. In *Proc. Second Int. Workshop on Conditional and Typed Rewriting Systems*, Lect. Notes in Comput. Sci., vol. to appear, Berlin, Springer-Verlag.

[15] M. RUSINOWITCH, 1988. Theorem proving with resolution and superposition: An extension of the Knuth and Bendix procedure as a complete set of inference rules. Submitted for publication, 1988.

[16] G.A. ROBINSON AND L. T. WOS, 1969. Paramodulation and theorem proving in first order theories with equality. In B. Meltzer and D. Michie, editors, *Machine Intelligence 4*, pp. 133–150. American Elsevier, New York.

[17] L. T. WOS, G. A. ROBINSON, D. F. CARSON, AND L. SHALLA, 1967. The concept of demodulation in theorem proving. *Journal of the ACM*, Vol. 14, pp. 698–709.

[18] H.T. ZHANG AND J-L. RÉMY, 1985. Contextual Rewriting. In J.-P. Jouannaud, editor, *Rewriting Techniques and Applications*, Lect. Notes in Comput. Sci., vol. 202, pp. 46–62, Berlin, Springer-Verlag.

Knuth-Bendix Completion of Horn Clause Programs for Restricted Linear Resolution and Paramodulation *

Hubert Bertling
Fachbereich Informatik, Universität Dortmund
D-4600 Dortmund 50, Germany
uucp, bitnet: hubert@ls5.informatik.uni-dortmund.de

The aim of Knuth-Bendix completion is to transform an input specification into another, equivalent specification such that a given set of inference rules becomes complete for the transformed specification. In the classical case, the specifications consist of unconditional equations, and the inference rule is term rewriting.

Here we investigate Knuth-Bendix completion methods for full Horn clause specifications and introduce a class of inference systems defined by an abstract concept of restrictions imposed on linear resolution and paramodulation.

The usual technical ingredients of Knuth-Bendix completion methods, e.g. critical pairs, a corresponding critical pair lemma, simplification and elimination methods and critical pair criteria are generalized with respect to the concept of restrictions. We show that for any concrete restriction, our completion methods produce specifications for which linear resolution and paramodulation under the given restriction are refutationally complete. Conventional completion methods are instances of completion with respect to restrictions.

1 Introduction

In [Gan88] Ganzinger explored that the Knuth-Bendix completion process can be guided in a way that certain conditional equations, declared as "should become non-operational", become indeed superfluous in the final canonical term rewrite system. Here completion serves not only as a mechanism for generating canonical term rewrite systems. It is also a mechanism by which an initial set of equations can be made complete with respect to certain (additional) restrictions: certain equations in a complete set don't have to be applied. Consider the set

$$
\begin{array}{ll}
f(a) \approx g(a) \Leftarrow & (1) \\
h(a) \approx a \Leftarrow & (2) \\
f(x) \approx g(x) \Leftarrow h(x) \approx a & (3)
\end{array}
$$

*This work is partially supported by the ESPRIT-project PROSPECTRA, ref#390.

with a recursive path ordering induced by the operator precedence $f > g$, $f > h$, $f > a$. All three equations are reductive term rewrite rules. As equation (1) is an instantiation of (the head of) equation (3), equation (1) can be eliminated without changing the theory. This is what a conventional completion procedure would do in this case. Following the idea of Ganzinger we can alternatively declare that "equation (3) should become non-operational". In this case his completion procedure terminates for the example stating that the system is complete, and that equation (3) is operationally superfluous. If Ganzinger's completion procedure starts under the same assumptions with equation (2) and (3) then it generates equation (1) by paramodulating (2) into the condition of (3). Declaring that a conditional equation "should become non-operational" changes the needed paramodulations from and into that equation.

This paper generalizes the idea of restrictions like "should become non-operational" and investigates the relationship between restrictions and needed resolvents and paramodulants during completion. The key concepts are *Knuth-Bendix restrictions, critical pairs with respect to Knuth-Bendix restrictions* and a corresponding *critical pair lemma.* This work is an extension of ideas which have previously been reported in [BG89]. It extracts and summarizes results of the thesis [Ber90].

We assume the reader being familiar with standard notions and concepts in the fields of term rewriting, cf. for instance [DJ89] and Horn clause logic, cf. for instance [Llo87]. For convenience let us recall some of them. A *Horn specification* $SPEC = (\Sigma, S)$ consists of a signature Σ and of a set S of Horn clauses with respect to this signature.

A *signature* is a triple $\Sigma = (P, F, X)$ consisting (i) of a family of disjoint sets of predicate symbols $P = \bigcup_{m\geq 0} P_m$, with the understanding that P_i contains all predicate symbols with i arguments and, (ii) of a family of disjoint sets of function symbols $F = \bigcup_{n\geq 0} F_n$, with the understanding that F_i contains all function symbols with i arguments and, (iii) of a countably infinite set of variable symbols X. We assume that the sets P, F and X are disjoint. Members of F_0 are called *constants.* P_2 is assumed to contain the distinguished symbol $\approx$ for equality. We write $s \mathrel{\dot\approx} t$ in cases where we are not interested whether the given equation is $s \approx t$ or $t \approx s$.

Terms with respect to Σ are expressions built from function symbols in F and variable symbols in X. Variable-free terms are said to be *ground.* We assume that F_0 is non-empty, i.e. that there exists at least one constant and therefore at least one ground term. *Atomic formulas* or *atoms* with respect to Σ are expressions of the form $p(t_1, \ldots, t_n)$ where $p \in P_n$ and t_i for $i = 1, \ldots, n$ are terms with respect to Σ. We also regard the special symbol $\Diamond \notin \Sigma$ as an atom. It symbolizes the contradiction.

We use the terms *occurrences, substitutions, unifiers, most general unifiers* etc. with their usual meanings. The set of variables occurring in E is denoted by $var(E)$. We write $E[s]$ to indicate that s is a subterm of the expression E and denote by $E[t]$ the result of replacing a particular occurrence of s by t.

Finite multisets of atoms with respect to Σ are called *goals.* The letters Γ, Δ denote goals, the letters A, B atoms. Elements and subsets of a goal Γ are also called *subgoals.* We sometimes write A, Γ to denote $\{A\} \cup \Gamma$ and Γ_1, Γ_2 to denote $\Gamma_1 \cup \Gamma_2$.

A *Horn clause* with respect to Σ is a formula of the form $B_0 \Leftarrow B_1, \ldots, B_n$ where $B_0, B_1, \ldots, B_n$ are atoms with respect to Σ and $B_1, \ldots, B_n$ are different from $\Diamond$. If B_0 is different from $\Diamond$ then $B_0 \Leftarrow B_1, \ldots, B_n$ is called a *definite* Horn clause, otherwise it is called a *goal clause.* B_0 is called the *head* and $B_1, \ldots, B_n$ the *body* of the Horn clause. A *unit clause* is a clause of the form $B_0 \Leftarrow \emptyset$ i.e. a clause with an empty body. We also write B_0 instead of $B_0 \Leftarrow \emptyset$. A goal clause $\Diamond \Leftarrow B_1, \ldots, B_n$ is a shorthand notation for $\neg(B_1 \wedge \ldots \wedge B_n)$ where $\neg$ stands for the negation. It is always assumed that variables in clauses are universally quantified. The clause $\Diamond \Leftarrow \emptyset$ is called the *empty clause*.

Let $SPEC = (\Sigma, S)$ be a Horn specification. The set of all ground atoms with respect to Σ different from $\Diamond$ is also known as the *Herbrand universe*. Any non-empty subset of the Herbrand universe is called an *interpretation*. An interpretation I *satisfies* a ground clause $A \Leftarrow \Gamma$, if $\Gamma \not\subseteq I$ or $A \in I$. An interpretation *satisfies* a (possibly non-ground) clause C, if and only if it satisfies all ground instances $C\sigma$.

A clause C is a *logical consequence of* S, written $S \models C$, if every interpretation I that satisfies S also satisfies C. We write $S \models \sigma \approx \sigma'$, whenever σ and σ' are substitutions with equal domains and $S \models x\sigma \approx x\sigma'$ for all variables in the domain of σ. We assume the standard properties of equality expressed by the set EAX of the following clauses:

$$\begin{array}{rcl} x \approx x & \Leftarrow & \\ y \approx x & \Leftarrow & x \approx y \\ x \approx z & \Leftarrow & x \approx y,\ y \approx z \\ f(\ldots x \ldots) \approx f(\ldots y \ldots) & \Leftarrow & x \approx y \\ p(\ldots y \ldots) & \Leftarrow & x \approx y,\ p(\ldots x \ldots) \end{array}$$

for all $f \in F$, and for all $p \in P$.

A Horn specification $SPEC = (\Sigma, S)$ is called *satisfiable*, if and only if there is an interpretation I satisfying all clauses in $S \cup EAX$. The *theory* $\mathcal{TH}(S)$ of a specification $SPEC = (\Sigma, S)$ consists of all (not only ground) Σ-atoms A with $S \cup EAX \models A$. Elements and finite subsets of $\mathcal{TH}(S)$ are called *S-theorems*. A substitution σ is a *correct answer substitution* in S for a goal Γ iff $\Gamma\sigma$ is an S-theorem.

2 Knuth-Bendix Restrictions

Probably the most popular inference rules on Horn clauses are linear resolution and paramodulation. We have to distinguish between a *goal-solving* and a *goal-verifying* variant of linear resolution and paramodulation.

Linear goal-solving resolution:

$$B,\ \Gamma_2 \xrightarrow[A \Leftarrow \Gamma_1,\ \sigma]{}_S \Gamma_1\sigma,\ \Gamma_2\sigma$$

if (i) either $A \Leftarrow \Gamma_1$ is $x \approx x$ or non-equational, (ii) $\sigma = mgu(A, B)$ and (iii) $A \Leftarrow \Gamma_1 \in S$. The triple $(A \Leftarrow \Gamma_1,\ \sigma|_{var(A \Leftarrow \Gamma_1)},\ \bot)$ is called the *application* of this inference step.

Linear goal-solving paramodulation:

$$A[u],\ \Gamma_2 \xrightarrow[s \dot{\approx} t \Leftarrow \Gamma_1,\ \sigma]{}_S A\sigma[t\sigma],\ \Gamma_1\sigma,\ \Gamma_2\sigma$$

if (i) $\sigma = mgu(s, u)$ and (ii) $s \dot{\approx} t \Leftarrow \Gamma_1 \in S$. Dependent on whether $s \dot{\approx} t$ stands for $s \approx t$ or $t \approx s$ the triple $(s \approx t \Leftarrow \Gamma_1,\ \sigma|_{var(s \dot{\approx} t \Leftarrow \Gamma_1)},\ lr)$ or $(t \approx s \Leftarrow \Gamma_1,\ \sigma|_{var(s \dot{\approx} t \Leftarrow \Gamma_1)},\ rl)$ is called the *application* of this inference step.

In contrast to goal-solving resolution and paramodulation verifying resolution and paramodulation do not instantiate goals.

Linear goal-verifying resolution:

$$B,\ \Gamma_2 \xrightarrow[A \Leftarrow \Gamma_1,\ \sigma]{}_S \Gamma_1\sigma,\ \Gamma_2$$

if (*i*) either $A \Leftarrow \Gamma_1$ is $x \approx x$ or non-equational, (*ii*) σ is a substitution such that $A\sigma = B$ and (*iii*) $A \Leftarrow \Gamma_1 \in S$. The triple $(A \Leftarrow \Gamma_1,\ \sigma,\ \bot)$ is called the *application* of this inference step.

Linear goal-verifying paramodulation:

$$A[u],\ \Gamma_2 \xrightarrow[s \dot{\approx} t \Leftarrow \Gamma_1,\ \sigma]{}_S A[t\sigma],\ \Gamma_1\sigma,\ \Gamma_2$$

if (*i*) σ is a substitution such that $s\sigma = u$ and (*ii*) $s \approx t \Leftarrow \Gamma_1 \in S$. Dependent on whether $s \dot{\approx} t$ stands for $s \approx t$ or $t \approx s$ the triple $(s \approx t \Leftarrow \Gamma_1,\ \sigma,\ lr)$ or $(t \approx s \Leftarrow \Gamma_1,\ \sigma,\ rl)$ is called the *application* of this inference step.

We implicitly assume that an applied clause does not share any variable with the clause to which it is applied (if necessary, variables are renamed). Note that we admit the application of goal clauses in resolution inference steps, i.e. A may be $\Diamond$.

A *proof* is any sequence of applications of the above inference rules. We are interested in proofs of the form

$$P : \Gamma \dashrightarrow^*_S \emptyset \quad \text{and} \quad Q : \Gamma \longrightarrow^*_S \emptyset$$

which are often called *refutations*.

Both variants of resolution and paramodulation are sound and complete provided that $x \approx x$ and the functional reflexive axioms $f(x_1, \ldots, x_n) \approx f(x_1, \ldots, x_n)$, $f \in F$ are in S. Completeness means: (*i*) For each correct answer substitution δ of Γ in S there exists a goal-solving refutation P with computed answer substitution σ and a substitution γ such that $\delta = \sigma\gamma$. (As usual, the ***computed answer substitution*** of a refutation is obtained by restricting the composition of all mgus occurring in the refutation to variables of the original goal.) (*ii*) For every S-theorem Γ there exists a verifying refutation Q. These soundness and completeness results can be studied in numerous textbooks, for instance [Llo87], [Pad88] or [Höl89].

Proofs are called ***ground***, if only ground instances of clauses are applied and the occurring goals are ground. Obviously, ground proofs are verifying proofs.

Knuth-Bendix restrictions, as described below, have to be reductive w.r.t. reduction orderings. ***Reduction orderings*** are partial strict orderings $>$ on expressions, i.e. atoms or terms, which are stable under substitution (i.e. $E_1 > E_2$ implies that $E_1\sigma > E_2\sigma$), monotonic (i.e. $s > t$ implies that $K[s] > K[t]$) and Noetherian (i.e. there are no infinite chains $E_1 > E_2 > \ldots$). We assume the following additional properties: (*i*) $A[s] > (s \dot{\approx} t)$, whenever $s \geq t$ and $A[s]$ is not an equation; (*ii*) $(u[s] \dot{\approx} v) > (s \dot{\approx} t)$, whenever $s \geq t$ and $s \neq u[s]$ and (*iii*) $E > \Diamond$, for all $E \neq \Diamond$. Conditions (*i*) and (*ii*) are needed to ensure that reductive verifying paramodulation steps reduce the size of goals. Because of (*iii*) goal clauses are not reductive. In the following we assume familiarity with ***multiset orderings*** [DM79]. By $\gg$ we denote the multiset extension of $>$.

Definition 2.1 *Let SPEC* $= (\Sigma,\ S)$ *be a specification. A SPEC-***restriction** $\mathcal{R}$ *is a pair of functions* $(\mathcal{SR},\ \mathcal{VR})$*, both of type*

$$\text{Reduction Orderings} \longrightarrow 2^{\text{Applications}}$$

where Applications $= (\Sigma\text{-Clauses} \times \Sigma\text{-Substitutions} \times \{lr,\ rl,\ \bot\})$.

$\mathcal{SR}$ and $\mathcal{VR}$ define for any given reduction ordering $>$ subsets of admissible applications of linear resolution and paramodulation. Instead of $\mathcal{R}(>)$, $\mathcal{SR}(>)$ and $\mathcal{VR}(>)$ we write $\mathcal{R}^>$, $\mathcal{SR}^>$ and $\mathcal{VR}^>$, respectively. We shall ambiguously use the term *restriction* for both $\mathcal{R}$ and $\mathcal{R}^>$.

A goal-solving inference step is called $\mathcal{R}^>$*-relevant*, if its application is in $\mathcal{SR}^>$, and a goal-verifying inference step is called $\mathcal{R}^>$*-relevant*, if its application is in $\mathcal{VR}^>$. A proof is called $\mathcal{R}^>$*-relevant*, if all its inference steps are $\mathcal{R}^>$-relevant. A clause $A \Leftarrow \Gamma$ that is either non-equational or $x \approx x$ is called *universally ground* $\mathcal{R}^>$*-relevant for a substitution* σ, if $(A \Leftarrow \Gamma,\ \sigma\delta,\ \bot) \in \mathcal{VR}^>$ for all ground substitutions δ. An equation $s \approx t \Leftarrow \Gamma$ is called *universally ground* $\mathcal{R}^>$*-relevant for a substitution* σ, if either $(s \approx t \Leftarrow \Gamma,\ \sigma\delta,\ lr) \in \mathcal{VR}^>$ or $(s \approx t \Leftarrow \Gamma,\ \sigma\delta,\ rl) \in \mathcal{VR}^>$ for all ground substitutions δ. A clause is called *universally ground* $\mathcal{R}^>$*-relevant*, if it is universally ground $\mathcal{R}^>$-relevant for *id*, the identity substitution.

A clause $B_0 \Leftarrow B_1,\ \ldots,\ B_n$ that is either non-equational or $x \approx x$ is called *reductive* w.r.t. $>$, if $B_0 > B_i$, $1 \leq i \leq n$. An equation $s \approx t \Leftarrow B_1,\ \ldots,\ B_n$ is called *lr-reductive* w.r.t. $>$, if $\{s \approx x\} > B_i$, $1 \leq i \leq n$, and $s > t$. The same equation is called *rl-reductive* w.r.t. $>$, if $\{t \approx x\} > B_i$, $1 \leq i \leq n$, and $t > s$. In both cases $x \notin var(s) \cup var(t) \cup \bigcup_{i=1}^{n} var(B_i)$ is assumed. Applications are said to be *reductive*, if their clauses are reductive under their specified substitution in their specified direction.

A *SPEC*-restriction $\mathcal{R} = (\mathcal{SR},\ \mathcal{VR})$ is called *reductive*, if for all reduction orderings $>$, the applications in $\mathcal{VR}^>$ are reductive w.r.t. $>$.

Lemma 2.2 *If* $\mathcal{R} = (\mathcal{SR},\ \mathcal{VR})$ *is a reductive restriction and* $>$ *a reduction ordering as specified above, then every verifying* $\mathcal{R}^>$*-relevant inference step*

$$\Gamma_1 \xrightarrow{\mathcal{R}^>} \Gamma_2$$

implies that $\Gamma_1 \gg \Gamma_2$*, and therefore this relation is Noetherian.*

Simpler substitutions should not make the application of clauses "illegal" w.r.t. $\mathcal{R}^>$. For any two substitutions δ, δ' with equal domain we define $\delta \geq \delta'$ iff $x\delta \geq x\delta'$ for all variables in the domain of δ and δ'. A *SPEC*-restriction $\mathcal{R} = (\mathcal{SR},\ \mathcal{VR})$ is called *closed*, if for all reduction orderings $>$, $(s \approx t \Leftarrow \Gamma,\ \delta,\ lr) \in \mathcal{VR}^>$ implies that $(s \approx t \Leftarrow \Gamma,\ \delta',\ lr) \in \mathcal{VR}^>$, whenever $S \cup EAX \models \delta \approx \delta'$, $\delta \geq \delta'$ and $s\delta = s\delta'$ and analogously for the rl- and the non-equational case.

A *SPEC*-restriction $\mathcal{R} = (\mathcal{SR},\ \mathcal{VR})$ is called *liftable*, if $(A \Leftarrow \Gamma,\ \sigma\delta,\ \_\,) \in \mathcal{VR}^>$ implies that $(A \Leftarrow \Gamma,\ \sigma,\ \_\,) \in \mathcal{SR}^>$ for all reduction orderings $>$.

Definition 2.3 *Let SPEC* $= (\Sigma,\ S)$ *be a specification. A SPEC-restriction* $\mathcal{R}$ *is called a* **Knuth-Bendix restriction,** *if* $\mathcal{R}$ *is reductive, closed and liftable.*

Standard rewriting and narrowing are an example of a Knuth-Bendix restriction. Rewriting can be expressed as the restriction on verifying paramodulation to apply reductive equations only and narrowing as the restriction of goal-solving paramodulation to apply reductive equations only. Similarly, rewriting with orientable instances of equations and paramodulation with maximal literals or terms is a further example of a Knuth-Bendix restriction. These restrictions can be combined with restrictions which exclude the application of certain clauses (cf. section 1). More examples of Knuth-Bendix restrictions can be found in [Ber90]. By ***restricted linear resolution and paramodulation*** we mean linear goal-solving resolution and paramodulation with respect to some Knuth-Bendix restriction $\mathcal{R}^>$.

3 Complete Specifications w.r.t. KB-Restrictions

From now on we assume signatures being extended by Skolem constants K, where K is a set of symbols such that $F \cap K = \emptyset$. Furthermore, if not stated otherwise let in the following $\mathcal{R}$ always denote a Knuth-Bendix restriction.

Definition 3.1 *A specification S is called* **ground $\mathcal{R}^>$-complete**, *if S is satisfiable and for all ground S-theorems Γ there exists an $\mathcal{R}^>$-relevant refutation*

$$Q: \ \Gamma \xrightarrow{\mathcal{R}^>}{}^*_S \emptyset$$

or if S is unsatisfiable and $\Diamond \Leftarrow \emptyset \in S$.

A term t_1 is called $\mathcal{R}^>$*-reducible*, if there exists a t_2, $t_1 > t_2$ and a verifying $\mathcal{R}^>$-relevant refutation of $t_1 \approx t_2$. A term is called $\mathcal{R}^>$-irreducible, if it is not $\mathcal{R}^>$-reducible. The $\mathcal{R}^>$-reducibility is extended to substitutions as usual.

Theorem 3.2 *Let S be satisfiable and ground $\mathcal{R}^>$-complete and Γ be a goal. For each correct ground answer substitution θ of Γ which is $\mathcal{R}^>$-irreducible there exists an $\mathcal{R}^>$-relevant refutation*

$$P: \ \Gamma \stackrel{\mathcal{R}^>}{- - \rightarrow}{}^*_S \emptyset$$

with a computed answer substitution σ and a substitution γ such that $\theta = \sigma\gamma$. Moreover, P does neither paramodulate into variable positions of goals nor apply functional reflexive axioms.

Proof: Since S is ground $\mathcal{R}^>$-complete there exists for every ground theorem an $\mathcal{R}^>$-relevant ground refutation. Moreover, because of the closedness of $\mathcal{R}$ there exists an $\mathcal{R}^>$-relevant ground refutation which applies clauses using $\mathcal{R}^>$-irreducible substitutions only. Such a ground refutation can be lifted to a (goal-solving) refutation. Because of the liftability of $\mathcal{R}$ the lifted refutation is a refutation with respect to $\mathcal{R}^>$. □

Corollary 3.3 *Let S be satisfiable and ground $\mathcal{R}^>$-complete. For each S-theorem Γ there exists an $\mathcal{R}^>$-relevant refutation*

$$P: \ \Gamma \stackrel{\mathcal{R}^>}{- - \rightarrow}{}^*_S \emptyset$$

with the identity as computed answer substitution. P does neither paramodulate into variable positions of goals nor apply functional reflexive axioms.

4 Completion w.r.t. KB-Restrictions

The objective of completion is to transform a given specification S into an equivalent S' which is ground $\mathcal{R}^>$-complete.

We are now going to formulate our notion of critical pairs. As Ganzinger has pointed out in [Gan88] it is sufficient to select one literal in the condition of an equation declared as "should become non-operational" and to compute paramodulants only into this literal in order to make the equation superfluous in the final canonical system. This idea is incorporated into our notion of critical pairs.

For this purpose let us assume that for every non-unit clause $A \Leftarrow \Gamma$ (goal clauses included) a unique literal in the condition can be selected once and for all.

(Non-linear) $\mathcal{R}^>$-resolution/paramodulation

$$\mathcal{R}^>\textit{-resolution}: \frac{A_1 \Leftarrow \Gamma_1 \quad A_2 \Leftarrow B,\ \Gamma_2}{A_2\sigma \Leftarrow \Gamma_1\sigma,\ \Gamma_2\sigma}$$

if (i) $A_1 \Leftarrow \Gamma_1$ is $x \approx x$ or non-equational, (ii) $\sigma = mgu(A_1, B)$, (iii) the application $(A_1 \Leftarrow \Gamma_1,\ \sigma|_{var(A_1 \Leftarrow \Gamma_1)},\ \bot)$ is in $\mathcal{SR}^>$, (iv) $A_2 \Leftarrow B,\ \Gamma_2$ is not universally ground $\mathcal{R}^>$-relevant for $\sigma|_{var(A_2 \Leftarrow B,\ \Gamma_2)}$ and (v) if $A_2 \Leftarrow B,\ \Gamma_2$ is a clause with selected condition literal then B is the selected literal of $A_2 \Leftarrow B,\ \Gamma_2$ or else $D\sigma \not> B\sigma$ for all $D \in \Gamma_2$.

$$\mathcal{R}^>\textit{-paramodulation into the condition}: \frac{s \dot{\approx} t \Leftarrow \Gamma_1 \quad A_2 \Leftarrow A[u],\ \Gamma_2}{A_2\sigma \Leftarrow A\sigma[t\sigma],\ \Gamma_1\sigma,\ \Gamma_2\sigma}$$

if (i) $\sigma = mgu(u, s)$, (ii) u is not a variable, (iii) $(s \approx t \Leftarrow \Gamma_1,\ \sigma|_{var(s \dot{\approx} t \Leftarrow \Gamma_1)},\ lr) \in \mathcal{SR}^>$, (analogously $(t \approx s \Leftarrow \Gamma_1,\ \sigma|_{var(s \dot{\approx} t \Leftarrow \Gamma_1)},\ rl) \in \mathcal{SR}^>$ if the applied clause is $t \approx s \Leftarrow \Gamma_1$), (iv) $A_2 \Leftarrow A[u],\ \Gamma_2$ is not universally ground $\mathcal{R}^>$-relevant for $\sigma|_{var(A_2 \Leftarrow A[u],\ \Gamma_2)}$, (v) if $A_2 \Leftarrow A[u],\ \Gamma_2$ is a clause with selected condition literal then $A[u]$ is the selected literal of $A_2 \Leftarrow A[u],\ \Gamma_2$ or else $D\sigma \not> A\sigma[u\sigma]$ for all $D \in \Gamma_2$ and (vi) if $A[u] = (l[u] \dot{\approx} r)$ then $r\sigma \not\geq l\sigma[u\sigma]$.

$$\mathcal{R}^>\textit{-paramodulation into the head}: \frac{s \dot{\approx} t \Leftarrow \Gamma_1 \quad A[u] \Leftarrow \Gamma_2}{A\sigma[t\sigma] \Leftarrow \Gamma_1\sigma,\ \Gamma_2\sigma}$$

if (i) $\sigma = mgu(u, s)$, (ii) u is not a variable, (iii) $(s \approx t \Leftarrow \Gamma_1,\ \sigma|_{var(s \dot{\approx} t \Leftarrow \Gamma_1)},\ lr) \in \mathcal{SR}^>$, (analogously $(t \approx s \Leftarrow \Gamma_1,\ \sigma|_{var(s \dot{\approx} t \Leftarrow \Gamma_1)},\ rl) \in \mathcal{SR}^>$ if the applied clause is $t \approx s \Leftarrow \Gamma_1$), (iv) if $A[u] \Leftarrow \Gamma_2$ is not an equation then $(A[u] \Leftarrow \Gamma_2,\ \sigma|_{var(A[u] \Leftarrow \Gamma_2)},\ \bot) \in \mathcal{SR}^>$ or else, if $A[u] = (l[u] \approx r)$ then $(l[u] \approx r \Leftarrow \Gamma_2,\ \sigma|_{var(l[u] \approx r \Leftarrow \Gamma_2)},\ lr) \in \mathcal{SR}^>$ or else, if $A[u] = (l \approx r[u])$ then $(l \approx r[u] \Leftarrow \Gamma_2,\ \sigma|_{var(l \approx r[u] \Leftarrow \Gamma_2)},\ rl) \in \mathcal{SR}^>$.

Note that $\mathcal{R}^>$-resolution/paramodulation refers to both components $(\mathcal{SR}^>, \mathcal{VR}^>)$ of a Knuth-Bendix restriction $\mathcal{R}^>$. $\mathcal{VR}^>$ is implicitly referred to when requiring that the clause to which the inference step applies must not be universally ground $\mathcal{R}^>$-relevant.

Definition 4.1 *Let $\mathcal{R}^>$ be a Knuth-Bendix restriction and C_1 and C_2 be two Horn clauses. A clause C_3 is called an $\mathcal{R}^>$-***critical pair** *from C_1 into C_2 if C_3 is an $\mathcal{R}^>$-resolvent or an $\mathcal{R}^>$-paramodulant from C_1 into C_2. By $\mathcal{R}^>$-$CP(S)$ we denote the set of $\mathcal{R}^>$-critical pairs in S.*

Like the conventional notions of critical pairs $\mathcal{R}^{>}$-$CP(S)$ is finite, whenever S is.

The following critical pair lemma is based on the idea of Noetherian proof transformations introduced by [BDH86] for the unconditional equational case and by Ganzinger for the conditional equational case. For this purpose a Noetherian ordering on linear verifications is defined. We define an ordering which is quite similar to the orderings used by Bachmair in [Bac87] for standard and unfailing completion.

As usual the idea is to associate to every verifying inference step a complexity. The complexity of a whole verification is the multiset of the complexities of its inference steps.

$$Cpl(\ A_0\delta,\ \Gamma_1 \xrightarrow[A_0 \Leftarrow \Gamma_2,\ \delta]{} \Gamma_2\delta,\ \Gamma_1\ )$$

$$= \begin{cases} (\{A_0\delta\},\ A_0\delta,\ A_0,\ \Gamma_2\delta) & \text{if } (A_0 \Leftarrow \Gamma_2,\ \delta,\ \bot) \in \mathcal{VR}^{>} \\ (\{A_0\delta,\ \bot\} \cup \Gamma_2\delta,\ \bot,\ \bot,\ \bot), & \text{otherwise} \end{cases}$$

$$Cpl(\ A[l\delta],\ \Gamma_1 \xrightarrow[l \approx r \Leftarrow \Gamma_2,\ \delta]{} A[r\delta],\ \Gamma_2\delta,\ \Gamma_1\ )$$

$$= Cpl(\ A[r\delta],\ \Gamma_1 \xrightarrow[l \approx r \Leftarrow \Gamma_2,\ \delta]{} A[l\delta],\ \Gamma_2\delta,\ \Gamma_1\ )$$

$$= \begin{cases} (\{A[l\delta]\},\ l\delta,\ l,\ \{A[r\delta]\} \cup \Gamma_2\delta) & \text{if } (l \approx r \Leftarrow \Gamma_2,\ \delta,\ lr) \in \mathcal{VR}^{>} \\ (\{A[l\delta],\ A[r\delta]\} \cup \Gamma_2\delta,\ \bot,\ \bot,\ \bot), & \text{otherwise} \end{cases}$$

Analogously the rl-case. These 4-tuples are compared from left-to-right: $\gg$ for the first component, the proper subterm ordering for the second component, the proper subsumption ordering for the third component and $\gg$ for the fourth component. Let $>^{Cpl}$ denote this ordering on 4-tuples. We assume that $\bot$ is minimal w.r.t. all of these orderings. Now we define $Q >_{\mathcal{P}} Q'$ *iff* $Cpl(Q) \gg^{Cpl} Cpl(Q')$ for proofs Q and Q'. We say that $>_{\mathcal{P}}$ is *the proof ordering derived from* $>$.

Lemma 4.2 *For any reduction ordering $>$ the ordering $>_{\mathcal{P}}$ is Noetherian.*

Lemma 4.3 (Critical Pair Lemma for Horn Clauses) *Let $\mathcal{R}^{>}$ be a Knuth-Bendix restriction such that $>_{\mathcal{P}}$ is the proof ordering derived from $>$. Then for all specifications S and for every ground verification*

$$Q \ = \ (\Gamma_1 \longrightarrow_S \Gamma_2 \xrightarrow{\mathcal{R}^{>}}{}^{*}_{S} \emptyset)$$

such that the first step is not $\mathcal{R}^{>}$-relevant and applies either a non-unit clause or is inversely $\mathcal{R}^{>}$-relevant, there exists an alternative verification

$$Q' \ = \ (\Gamma_1 \longrightarrow^{*}_{S \,\cup\, \mathcal{R}^{>}\text{-}CP(S)} \emptyset)$$

with $Q >_{\mathcal{P}} Q'$.

An inference step is said to be *inversely $\mathcal{R}^{>}$-relevant*, if it is a paramodulation step whose inverse application is in $\mathcal{VR}^{>}$. The *inverse* of an application $(s \approx t \Leftarrow \Gamma,\ \sigma,\ lr)$ is the application $(s \approx t \Leftarrow \Gamma,\ \sigma,\ rl)$ and vice versa. This is similar to a conventional peak situation, where the first step applies a rewrite rule in the "wrong" direction.

Proof: The proof of this lemma consists of a set of Noetherian transformation rules on verifications. Like in the conventional critical pair lemmata non-overlapping, variable-overlapping and proper overlapping situations have to be distinguished. Different from the conventional case here also overlaps into the condition of Horn clauses may occur.

Let Q be a verification whose first step is not $\mathcal{R}^>$-relevant. The step generates new subgoals. We always select one of the generated subgoals for further consideration. The selection depends on whether the clause applied in the first step is a clause with selected condition literal or not. If it is a clause with selected literal then we consider the corresponding literal in the generated goal. Otherwise we select a literal in the generated goal that is maximal w.r.t. $>$.

W.o.l.o.g. let us assume that the second step of Q is a resolution or paramodulation step into the selected literal. The ideas for transformation proposed below remain the same when the inference steps into the selected literals occur later in Q but need a more complicated presentation.

Due to the lack of space we consider only two examples from the whole case analysis. In both examples the first inference step of Q is neither $\mathcal{R}^>$-relevant nor inversely $\mathcal{R}^>$-relevant. In the first example there is a proof Q of the form

$$
Q: \quad
\begin{array}{ll}
 & A[s\delta] \\
\xrightarrow[s\approx t\Leftarrow B_1,\,\ldots,\,B_n,\,\delta]{} & A[t\delta],\ B_1\delta,\ \ldots,\ B_j[x\delta, x\delta],\ \ldots,\ B_n\delta \\
\xrightarrow[l\approx r\Leftarrow\Gamma,\,\gamma]{\mathcal{R}^>} & A[t\delta],\ B_1\delta,\ \ldots,\ B_j[x\delta, x\delta'],\ \ldots,\ B_n\delta,\ \Gamma\gamma \\
\xrightarrow{\mathcal{R}^>}{}^{*} & \emptyset
\end{array}
$$

We assume that there is a paramodulation step into $B_j\delta$ at or below a variable position in B_j. Let us w.o.l.o.g. assume that the domain of δ is only one variable, say x, and let us denote by $B_j[x\delta, x\delta]$ that there are possibly several occurrences of x in B_j. A simpler alternative for Q is

$$
Q': \quad
\begin{array}{ll}
 & A[s\delta] \\
\xrightarrow[l\approx r\Leftarrow\Gamma,\,\gamma]{\mathcal{R}^>}{}^{*} & A[s\delta'],\ (\Gamma\gamma)^m \\
\xrightarrow[s\approx t\Leftarrow B_1,\,\ldots,B_n,\,\delta']{} & A[t\delta'],\ B_1\delta',\ \ldots,\ B_j[x\delta', x\delta'],\ldots,\ B_n\delta',\ (\Gamma\gamma)^m \\
\xrightarrow[l\approx r\Leftarrow\Gamma,\,\gamma]{\leftarrow,\ \mathcal{R}^>}{}^{*} & A[t\delta],\ B_1\delta,\ \ldots,\ B_j[x\delta, x\delta'],\ \ldots,\ B_n\delta,\ (\Gamma\gamma)^{m+m'} \\
\xrightarrow{\mathcal{R}^>}{}^{*} & \emptyset
\end{array}
$$

The first steps of Q' paramodulate into the different occurrences of $x\delta$ in $s\delta$ (assume m occurrences of $x\delta$ in $s\delta$). They apply always $l \approx r \Leftarrow \Gamma$ under substitution γ thereby producing m-times the subgoal $\Gamma\gamma$. Then the clause from the first step of Q is applied but now with the reduced substitution δ'. After this step in the generated subgoals all reductions to δ' are undone by reversing the reducing applications of $l \approx r \Leftarrow \Gamma$ (assume $m' + 1$ occurrences of $x\delta'$ in $A[t\delta'],\ B_1\delta',\ \ldots,\ B_j[x\delta', x\delta'],\ldots,\ B_n\delta'$). These reverse applications are inversely $\mathcal{R}^>$-relevant because the original application is $\mathcal{R}^>$-relevant. $Q >_{\mathcal{P}} Q'$ is true because all inference steps in Q' are dominated by the first step of Q.

The second example is a proper overlap case. Consider the case that there is a resolution step into $B_j\delta$. Then Q is of the form

$$Q: \quad \begin{array}{ll} & A[s\delta] \\ \xrightarrow[s\approx t\Leftarrow B_1,\ \ldots,\ B_n,\ \delta]{} & A[t\delta],\ B_1\delta,\ \ldots,\ B_j\delta,\ \ldots,\ B_n\delta \\ \xrightarrow[L\Leftarrow\Gamma,\ \gamma]{\mathcal{R}^>} & A[t\delta],\ B_1\delta,\ \ldots,\ B_{j-1}\delta,\ B_{j+1}\delta,\ \ldots,\ B_n\delta,\ \Gamma\gamma \\ \xrightarrow{\mathcal{R}^>}{}^{*} & \emptyset \end{array}$$

A simpler alternative is

$$Q': \quad \begin{array}{ll} & A[s\delta] \\ \xrightarrow[s\sigma\approx t\sigma\Leftarrow B_1\sigma,\ \ldots,\ B_{j-1}\sigma,\ B_{j+1}\sigma,\ldots,B_n\sigma,\ \Gamma\sigma,\ \kappa]{} & A[t\delta],\ B_1\delta,\ \ldots,\ B_{j-1}\delta, \\ & B_{j+1}\delta,\ \ldots,\ B_n\delta,\ \Gamma\gamma \\ \xrightarrow{\mathcal{R}^>}{}^{*} & \emptyset \end{array}$$

where we show that $s\sigma \approx t\sigma \Leftarrow B_1\sigma,\ \ldots,\ B_{j-1}\sigma,\ B_{j+1}\sigma,\ldots,B_n\sigma,\ \Gamma\sigma$ is an $\mathcal{R}^>$-critical pair obtained by the resolution inference

$$\frac{L \Leftarrow \Gamma \qquad s \approx t \Leftarrow B_1,\ \ldots, B_j,\ \ldots,\ B_n}{s\sigma \approx t\sigma \Leftarrow B_1\sigma,\ \ldots,\ B_{j-1}\sigma,\ B_{j+1}\sigma,\ \ldots,\ B_n\sigma,\ \Gamma\sigma}$$

Conditions (i) and (ii) are fulfilled because the resolution rule does not apply equations and because there exists always a most general unifier $\sigma = mgu(L,\ B_j)$. Therefore there exists a substitution κ with $(L \Leftarrow \Gamma)\sigma\kappa = (L \Leftarrow \Gamma)\gamma$. Because of the liftability of Knuth-Bendix restrictions $(L \Leftarrow \Gamma,\ \sigma|_{var(L\Leftarrow\Gamma)},\ \bot) \in \mathcal{SR}^>$ is true and therefore condition (iii). $s \approx t \Leftarrow B_1,\ \ldots,\ B_j,\ \ldots,\ B_n$ is not universally ground $\mathcal{R}^>$-relevant for $\sigma|_{var(s\approx t\Leftarrow B_1,\ \ldots,\ B_j,\ \ldots,\ B_n)}$ because otherwise the first step of Q cannot be neither $\mathcal{R}^>$-relevant nor inversely $\mathcal{R}^>$-relevant. This implies condition (iv). If $s \approx t \Leftarrow B_1,\ \ldots,\ B_n$ is a clause with selected condition literal B_j then $B_j\delta$ is chosen for consideration by the assumptions for this case. Otherwise, if there is some literal $D \in \{B_1,\ \ldots, B_n\}$ with $D\sigma > B_j\sigma$ then also $D\delta > B_j\delta$ must hold true which contradicts our assumptions for this case. Therefore also (v) is fulfilled. Since $\{B_j\delta\} \gg \Gamma\gamma$ is true we can conclude that $Q >_{\mathcal{P}} Q'$. □

Abstract Completion with respect to $\mathcal{R}^>$ As usual we present our completion method as a set of inferences rules on specifications. To make things easier we separate the correctness consideration of the completion method from the correctness consideration of specific simplification and elimination methods. Therefore we have an abstract view of completion.

Abstract Addition : $\dfrac{S}{S \cup \{C\}}$ if $C \in \mathcal{R}^>\text{-}CP(S)$

Abstract Elimination : $\dfrac{S \cup \{C\}}{S}$ if S $\mathcal{R}^>$-subsumes C

Abstract Simplification : $\dfrac{S \cup \{C_1\}}{S \cup \{C_2\}}$ if $S \cup \{C_1\} \models C_2$ and $S \cup \{C_2\}$ $\mathcal{R}^>$-subsumes C_1.

Any concrete simplification and elimination rule is considered to be *correct*, if it fulfills the $\mathcal{R}^>$-subsumption property of its abstract counterpart:

Definition 4.4 *A set S of clauses* **$\mathcal{R}^>$-subsumes** *a clause $A \Leftarrow \Gamma$ that is either $x \approx x$ or non-equational, if (i) $S \cup EAX \models (A \Leftarrow \Gamma)$ and (ii) if for arbitrary sets of clauses AnySet the following is true:*

(a) if $S \cup AnySet$ is satisfiable, then there exists for all ground verifications of the form

$$Q = (A\delta \xrightarrow[A \Leftarrow \Gamma,\ \delta]{} \Gamma\delta \xrightarrow{\mathcal{R}^>}{}^{*}_{S \cup AnySet} \emptyset)$$

a ground verification

$$Q' = (A\delta \longrightarrow^{*}_{S \cup AnySet} \emptyset)$$

such that $Q \geq_{\mathcal{P}} Q'$.

(b) if $S \cup AnySet$ is unsatisfiable then there exists for all ground verifications of the form

$$Q = (\Diamond \longrightarrow^{*}_{S \cup AnySet} \Gamma_1 \xrightarrow[A \Leftarrow \Gamma,\ \delta]{} \Gamma_2 \xrightarrow{\mathcal{R}^>}{}^{*}_{S \cup AnySet} \emptyset)$$

a ground verification

$$Q' = (\Diamond \longrightarrow^{*}_{S \cup AnySet} \emptyset)$$

with $Q \geq_{\mathcal{P}} Q'$. Analogously the definition for equations.

Note condition (a) implies condition (b). But condition (a) alone is a too strong requirement for allowing an elimination as shown below. Intuitively, if a specification is unsatisfiable, then the objective of completion is to ensure the existence of the empty clause $\Diamond \Leftarrow \emptyset$ but not the existence of $\mathcal{R}^>$-relevant refutations for all theorems (everything is a logical consequence in an unsatisfiable set).

$$\textit{Elimination}: \quad \frac{S' \cup \{\Diamond \Leftarrow tt \approx ff,\ 0 \approx s(0) \Leftarrow tt \approx ff\}}{S' \cup \{\Diamond \Leftarrow tt \approx ff\}}$$

Let $S = S' \cup \{\Diamond \Leftarrow tt \approx ff\}$. The elimination is correct, because (i) $S \models 0 \approx s(0) \Leftarrow tt \approx ff$ and (ii), if there exists a refutation

$$A[0] \xrightarrow[0 \approx s(0) \Leftarrow tt \approx ff,\ id]{} A[s(0)],\ tt \approx ff \xrightarrow{\mathcal{R}^>}{}^{*}_{S \cup AnySet} \emptyset$$

then $S \cup AnySet$ is unsatisfiable since $tt \approx ff$ and $tt \not\approx ff$ cannot be true at the same time. For all ground verifications of the form

$$Q = (\Diamond \longrightarrow^{*}_{S \cup AnySet} \Gamma_1 \xrightarrow[0 \approx s(0) \Leftarrow tt \approx ff]{} \Gamma_2,\ tt \approx ff \xrightarrow{\mathcal{R}^>}{}^{*}_{S \cup AnySet} \emptyset)$$

there exists

$$Q' = (\Diamond \xrightarrow[\Diamond \Leftarrow tt \approx ff]{}{}_{S \cup AnySet}\ tt \approx ff \xrightarrow{\mathcal{R}^>}{}^{*}_{S \cup AnySet} \emptyset)$$

with $Q \geq_{\mathcal{P}} Q'$.

In contrast to the non-feasibility condition of Kaplan in [Kap85] and Jouannaud and Waldmann in [JW86] $\mathcal{R}^{>}$-subsumption is monotonic with respect to arbitrary extensions of specifications.

Lemma 4.5 *If a set S of Horn clauses $\mathcal{R}^{>}$-subsumes a clause C, then also every extension $S \cup S'$ $\mathcal{R}^{>}$-subsumes C.*

Proof: (i) $S \cup EAX \models C$ implies that $S \cup S' \cup EAX \models C$. (ii) The second condition of the $\mathcal{R}^{>}$-subsumption property is fulfilled because $S \cup S' \cup AnySet$ can be considered as $S \cup AnySet'$ for some $AnySet'$. □

A *completion system* CS is a set consisting of correct addition, elimination and simplification rules. A *CS-derivation for* S is a sequence S_0, S_1, ... with $S_0 = S$ so that all subsequent sets of program clauses are obtained by applications of inference rules in CS.

Definition 4.6 *A CS-derivation is called* **fair**, *if (i) for all $C \in \mathcal{R}^{>}\text{-}CP(S_\infty)$ there is a step i at which $C \in S_i$, and (ii) for all unit clauses $C \in \bigcup_{i \geq 0} S_i$ which are not universally ground $\mathcal{R}^{>}$-relevant there exists a step k at which C will be simplified or eliminated.*

Theorem 4.7 *Let $\mathcal{R}^{>}$ be a Knuth-Bendix restriction, CS be a completion system and, S_0, S_1, ... be a fair CS-derivation. Then S_∞, the limit of the sequence S_0, S_1, ..., is ground $\mathcal{R}^{>}$-complete.*

Proof: non-$\mathcal{R}^{>}$-relevant applications of unit clauses, not covered by the critical pair lemma, are reflected by point (ii) of the fairness definition. □

5 Conclusions

The consideration of different Knuth-Bendix completion methods known for the unconditional equational case, together with the ideas of restricting the application of equations by Ganzinger in [Gan88] and later on in [BG89] led us to a formulation of KB-completion which is parameterized by its objective, namely the completeness of resolution and paramodulation with respect to certain restrictions.

The key concept of the approach are *Knuth-Bendix restrictions.* Standard completion concepts, like *critical pairs* and a corresponding *critical pair lemma* have been reformulated with respect to KB-restrictions. As a consequence prior separately discussed completion methods like standard completion or unfailing completion of Bachmair cf. [BDH86], [Bac87], or conditional equational completion as has been proposed by [Gan88] can be considered as special instantiations of completion with respect to KB-restrictions. Let us emphasize that the proof of our critical pair lemma is an alternative proof of the completeness of inference systems similar to that proposed by Kounalis and Rusinowitch in [KR88]. They introduce a factor rule which is in contrast to the full first-order predicate logic not really needed for Horn clause logic. In their context all unit clauses are universally ground-relevant, because they assume a total reduction ordering on ground expressions.

We did not consider concrete simplification and elimination methods. Instead we have formulated an abstract correctness condition and were so being able to separate the correctness consideration of the completion method from that of concrete simplification and elimination methods.

Concerning concrete simplification methods one has to take care of conditions which are always fulfilled for the standard case but not for arbitrary KB-restrictions. For example, a simplified clause must preserve the $\mathcal{R}^>$-relevance: $(s \approx t \Leftarrow \Gamma \cup \{C\},\ \sigma,\ lr) \in \mathcal{VR}^>$ implies that $(s \approx t \Leftarrow \Gamma \cup \{D\},\ \sigma,\ lr) \in \mathcal{VR}^>$, if the literal C in the first clause is simplified to the literal D in the second clause. Also *critical pair cirteria* can be adapted to KB-completion w.r.t. KB-restrictions. This has been done in [Ber90].

Different from [Kap85], [JW86] or [Gan88] we admit goal clauses in specifications. Aside from the fact that negative assertions are interesting when formulating consistency conditions they are interesting for simplification and elimination as can be seen in the shown example.

References

[Bac87] L. Bachmair. *Proof Methods for Equational Theories.* PhD thesis, University of Illinois at Urbana-Champaign, USA, 1987.

[BDH86] L. Bachmair, N. Dershowitz, and J. Hsiang. Orderings for equational proofs. In *Proc. of the 1st IEEE Symposium on Logic in Computer Science, Boston, MA*, pp. 346–357, 1986.

[Ber90] H. Bertling. *Knuth-Bendix Completion of Horn Clause Programs for Restricted Linear Resolution and Paramodulation.* PhD thesis, FB Informatik, Universität Dortmund, 1990.

[BG89] H. Bertling and H. Ganzinger. Completion-Time Optimization of Rewrite-Time Goal Solving. In *Proc. of the 3rd International Conference on Rewriting Techniques and Applications, Chapel Hill*, LNCS 355, pp. 45–58, Berlin-Heidelberg-New York, April 1989. Springer-Verlag.

[DJ89] N. Dershowitz and J.P. Jounannaud. Rewrite Systems. In *Handbook of Theoretical Computer Science, Volume B, Chapter 15.* North-Holland, 1989.

[DM79] N. Dershowitz and Z. Manna. Proving termination with multiset orderings. Commun. ACM 22:465-476, 1979.

[Gan88] H. Ganzinger. A completion procedure for conditional equations. In *Proc. of the 1st International Workshop on Conditional Term Rewriting, Orsay*, volume 308 of *LNCS*, pp. 62–83, Berlin-Heidelberg-New York, 1988. Springer-Verlag.

[Höl89] S. Hölldobler. *Foundations of Equational Logic Programming.* Springer-Verlag, Berlin-Heidelberg-New York, 1989.

[JW86] J.P. Jouannaud and B. Waldmann. Reductive conditional term rewriting systems. In *Proc. of the 3rd TC2 Working Conference on the Formal Description of Programming Concepts*, Ebberup, Denmark, August 25–28 1986.

[Kap85] S. Kaplan. Fair Conditional Term Rewriting Systems: Unification, Termination and Confluence, 1985.

[KR88] E. Kounalis and M. Rusinowitch. On word problems in Horn logic. In N. Dershowitz and S. Kaplan, editors, *Proc. of the 1st International Workshop on Conditional Term Rewriting, Orsay*, LNCS 308, pp. 144–160, Berlin-Heidelberg-New York, 1988. Springer-Verlag.

[Llo87] J. W. Lloyd. *Foundations of Logic Programming.* Springer-Verlag, Berlin-Heidelberg-New York, 1987.

[Pad88] P. Padawitz. *Computing in Horn Clause Theories.* Springer-Verlag, Berlin-Heidelberg-New York, 1988.

Proof by Consistency in Conditional Equational Theories

Eddy Bevers†, Johan Lewi

Department of Computer Science, K.U.Leuven
Celestijnenlaan 200A, B-3001 Leuven (Heverlee), BELGIUM
E-mail: eddy@cs.kuleuven.ac.be

Abstract.

In this paper we deal with the problem of proving inductive theorems in conditional equational theories. We propose a proof by consistency method that can be employed when the theory is representable as a ground Church-Rosser conditional equational system. The method has a *linear* proof strategy and is shown to be sound and refutational complete, i.e. it refutes any conditional equation which is not an inductive theorem. Moreover it can handle rewrite rules as well as (unorientable) equations and therefore it will not fail when an unorientable equation comes up (as was the case in the earliest proof by consistency (*inductionless induction*) methods). The method extends the work on unconditional equational theories of [Bachmair 1988].

1. Introduction

Over the last years, formal specification methods have gained a lot of interest in computer science. Many mathematical based formalisms to describe programs and properties of programs have been developed. With these formalisms come deduction rules for deducing and proving new properties of the program. One of these formalisms, the algebraic specification method, is frequently employed to describe abstract data types. The characteristic properties of the data type's objects and functions are described by means of equations between first order terms. The deduction mechanism used to derive or prove new equations from a given specification is known as *equational logic*. Thus derived equations are valid in all models which satisfy the given specification (i.e. in the *equational theory*). When specifying abstract data types we are most of the time only interested in one standard model, namely the *initial algebra*. Proving equations in the *initial model* needs besides equational logic also some kind of *induction* e.g. on the structure of terms. Properties which hold in the initial model are therefore also called *inductive properties*.

Most of the equational logic theorem proving tools are based on the Knuth-Bendix completion procedure, which tries to transform a given set of equations into a convergent term rewriting system, in which equations have been turned into *one-way* rewriting rules and in which each term has a unique irreducible (*normal*) form. Two terms are proved to be equal in the equational theory if they have the same normal form.

In the early 1980's a proof method for inductive properties was developed, employing this Knuth-Bendix completion procedure instead of the more traditional induction (it was therefore originally called the *inductionless induction* method). Basically this method is a *proof by consistency method*: the equation to be proved is added to the given specification after which the completion procedure tries to derive a convergent rewriting system without generating an *inconsistency*. When this succeeds, the original equation is valid in the initial model. If an inconsistency comes up, the equation is not valid. During the years, several variants of the original method (as described in [Musser 1980, Goguen 1980]), have been developed: [Huet & Hullot 1982] refined the method for theories with fully free *constructors*, [Paul 1984] for theories with constructors with certain relations between them, and [Jouannaud & Kounalis 1985] and [Kapur et al. 1986] for theories with no constructors (i.e. all theories). Besides the area of application, these methods essentially differ only in the definition of the *consistency check*. [Kapur & Musser 1987] describes a general theory of proof by consistency.

Important drawbacks of these methods are divergence and failure. Divergence has been tackled in [Fribourg 1986], where a *linear* proof by consistency method was introduced by drastically limiting the

† Research supported by the "Instituut tot Aanmoediging van het Wetenschappelijk onderzoek in Nijverheid en Landbouw - I.W.O.N.L."

number of equations (critical pairs) to be computed. Further optimizations, [Küchlin 1989] and [Göbel 1987], restrict the number of equations to be computed even more. [Bachmair 1988] solves the failure problem by defining an *unfailing* proof by consistency method that is, in addition, linear and *refutational complete* (in the sense that any equation which is not an inductive theorem will be refuted).

A more powerful and general formalism for specifying abstract data types is by means of *conditional equations*, i.e. equations (between terms) which have a premiss, being a conjunction of equations (between terms). This leads to *conditional algebraic specifications*. More and more research has been going on in this field, especially on the topic of *conditional term rewriting* and completion procedures for conditional theories [Bergstra & Klop 1982, Kaplan 1984, Zhang & Rémy 1985, Jouannaud & Waldmann 1986, Ganzinger 1987a, Kaplan 1987, Bousdira & Rémy 1987, Dershowitz et al. 1987, Ganzinger 1987b, Okada 1987, Dershowitz et al. 1988]. Very recently also an extension to conditional theories of Fribourg's linear proof by consistency method, for proving conditional theorems in the initial model, was published [Fribourg 1989]. The main limitation of this method is again the fact that it is not unfailing.

In this paper we propose a proof by consistency method for proving conditional theorems in the initial model of conditional equational theories (i.e. inductive theorems), and which is linear (in the sense of [Fribourg 1986]), unfailing and refutational complete. The method is essentially based on Bachmair's proof by consistency for equational theories [Bachmair 1988] and the approach is similar to his (therefore supporting Bachmair's claim that the approach is *well-suited for studying extensions of the method (e.g. to conditional equations)*). The main difference (besides the extension to the conditional case) lies in the fact that we distinguishes between (unorientable) equations and (*reductive*) rewrite rules, making the method more powerful. The rewrite relation employed, is inspired by the contextual rewriting relation of [Ganzinger 1987b].

In section 2 we introduce some preliminary definitions and notations. Section 3 briefly explains how inductive proofs and proofs by consistency are related. In the main section, section 4, we define the inference rules for our proof by consistency method for conditional equational theories. This results in a linear and unfailing proof by consistency procedure from which we can prove the soundness and the refutational completeness. The theory is illustrated by a small (but non trivial) example in section 5. Finally, section 6 contains a short conclusion.

2. Preliminary Definitions and Notations

We assume the reader is familiar with the basic concepts of equational theories and term rewriting systems, such as signatures, terms, equations, rewrite rules, initial algebra etc. (for a thorough overview see e.g. [Huet & Oppen 1980]. Below we introduce the essential notions and notations, used in this paper.

Given a set of function symbols $\mathbb{F}$ and a set of variables X. $T(\mathbb{F}, X)$ will denote the set of terms built on $\mathbb{F}$ and X. The set of ground terms (i.e. terms without variables) is written $GT(\mathbb{F})$. A term is built up of a number of subterms. We write $t[\,r\,]$ to indicate that t contains r as a subterm (if $t[\,r\,] \neq r$, r is a proper subterm of $t[\,r\,]$). Every subterm of a term t is located at a certain *position*. The term t/p stands for the subterm of t at position p. The term $t[\,p \leftarrow r\,]$ is the result of replacing the subterm of t at position p, by the term r. A substitution σ is a mapping from a subset of X to $T(\mathbb{F}, X)$. The application of a substitution σ to a term t is written as σt. σt is an *instance* of t. A ground substitution maps variables into ground terms. We call σ a (most general) unifier of the conjunction of equations $\bigwedge_{i=1}^{n} u_i{=\!=}v_i$ iff it is a (most general) unifier of $(u_1, \ldots, u_n)$ and $(v_1, \ldots, v_n)$. The *rewrite* relations we will consider in this paper, are binary relations on terms generated by conditional equational systems. As usual $\rightarrow^*$ stands for the transitive and reflexive closure of the rewrite relation $\rightarrow$, while $\rightarrow^+$ is the transitive closure and $\leftrightarrow^*$ the reflexive, transitive and symmetric closure of $\rightarrow$. A term s is irreducible for $\rightarrow$ if no term t exists such that $s \rightarrow t$. If $s \rightarrow^* t$ and t is irreducible, we call t the *normal form* of s (for $\rightarrow$). Two terms s and t are *joinable* under $\rightarrow$ (written as $s \downarrow t$), iff for some term u: $s \rightarrow^* u$ and $t \rightarrow^* u$. We say that $\rightarrow$ is *terminating* if there is no infinite chain $t_1 \rightarrow t_2 \rightarrow \ldots$. A partial ordering $>$ on terms is an irreflexive and transitive binary relation on terms. By $s \not< \not> t$ we mean that neither $s>t$ nor $t>s$. An ordering is well-founded if there is no infinite descending sequence of terms $t_1 > t_2 > \ldots$. A *reduction ordering* is a well-founded partial ordering on terms that is *monotonic w.r.t. the term structure* (i.e. $s \rightarrow t$ implies $u[s] \rightarrow u[t]$) and *monotonic w.r.t. instantiation* (i.e. $s \rightarrow t$ implies $\sigma s \rightarrow \sigma t$). A *ground reduction ordering* is a reduction ordering which is total on ground terms. The *specialization ordering* $\rhd$ is defined as follows: $s \rhd t$ iff some subterm of s is an instance of t but not vice versa. $\rhd$ is well-founded.

A conditional equation is a formula of the form $(u_1 = v_1 \wedge ... \wedge u_n = v_n \Rightarrow s = t)$, where $u_1, v_1, ..., u_n, v_n, s, t \in T(\mathbb{F}, X)$. A set E of conditional equations is called a *conditional equational system*, or short: CES. If $c \Rightarrow s = t$ (where c stands for a conjunction of equations of the form $u_i = v_i$) is valid in all algebra's of the *variety* of E, then it is a *(conditional) equational consequence* of E. This will be written as $\langle E \vdash_{eq} c \Rightarrow s = t \rangle$, or when c is empty: $s =_E t$. We will say a conditional equation $c \Rightarrow s = t$ is an *inductive consequence* of E, if it is valid in the initial model. This will be notated by $\langle E \vdash_{ind} c \Rightarrow s = t \rangle$. It can be easily shown that $\langle E \vdash_{ind} c \Rightarrow s = t \rangle$ iff for all ground substitutions σ: $\langle E \vdash_{eq} \sigma c \Rightarrow \sigma s = \sigma t \rangle$.

Given a reduction ordering $>$. We will call a conditional equation $c \Rightarrow l = r$ *reductive* (w.r.t. $>$) if $l > r$ and $\{l\} >_M M(c)$, where $>_M$ stands for the multiset extension of $>$, and $M(c)$ for the multiset of terms occurring in the equations of c. This will be written as: $\langle c \Rightarrow l \to_> r \rangle$, or just $\langle c \Rightarrow l \to r \rangle$ if the reduction ordering is clear from the context. A reductive equation will also be called a *rewrite rule* or just *rule*. If the equation is not reductive (w.r.t. $>$), we notate this as: $\langle c \Rightarrow l \leftrightarrow_> r \rangle$, or just $\langle c \Rightarrow l \leftrightarrow r \rangle$. We also say the equation is *unorientable*. If all equations of a CES E are reductive (w.r.t. $>$), we call E *reductive*.

Next we introduce the notion of term rewriting (w.r.t. $>$) with conditional equations:

Definition 2.1. Given a conditional equational system E. A term s rewrites to a term t, under E (notation of this rewrite relation: $\to_E$) iff there exists an equation $c \Rightarrow l = r$ in E, such that:

(1) s/p matches l with matching substitution σ (i.e. $s/p = \sigma l$) for a position p in s;
(2) $\sigma c \Rightarrow \sigma l \to_> \sigma r$;
(3) (let c be $\bigwedge_{i=1}^{n} u_i = v_i$) for all i ($1 \leqslant i \leqslant n$): $\sigma u_i \downarrow_E \sigma v_i$;
(4) $t = s[p \leftarrow \sigma r]$. □

The matching substitution σ is assumed to substitute only for variables in l (i.e. for all variables x, not in l: $\sigma x = x$). In this way $\to_E$ is terminating and basic notions such as one-step reduction (is $s \to_E t$?), finite reduction (is $s \to_E^* t$?), joinability (is $s \downarrow_E t$?) and reducibility (is s reducible?) all are decidable. However this implies also that rewriting by conditional equations with extra variables in the premisses, is impossible by our rewrite relation.

Remark that this rewrite relation is more general than the one used with *simplifying* [Kaplan 1987], *reductive* [Jouannaud & Waldmann 1986] or *decreasing* [Dershowitz et al. 1988] CTRS, since we allow rewriting with *non-reductive* equations, as long as the equation instantiated with the matching substitution, is reductive. Notice also that, if no equation of E has any premises, our definition is analogous to the definition of rewriting with equations given in [Hsiang & Rusinowitch 1986].

The (ground) Church-Rosser property of the rewriting relation is defined as usual:

Definition 2.2. $\to_E$ is *(ground) Church-Rosser* iff for all (ground) terms s and t, $s =_E t$ implies $s \downarrow_E t$. □

The CES E is called *(ground) Church-Rosser* iff the associated rewrite relation $\to_E$ is (ground) Church-Rosser. Unlike in unconditional equational systems, ground convergence (i.e. ground confluence and termination) and the ground Church-Rosser property of $\to_E$ are not equivalent, even if the reduction ordering $>$ used in $\to_E$, is total on ground terms. In order to obtain a (ground) Church-Rosser system, some kind of completion with superposition on maximal literals is needed (e.g. [Dershowitz 90] or [Bachmair & Ganzinger 90]). This was brought to our attention by H. Ganzinger, for which we are grateful.

We also define a rewrite relation for reducing terms that appear in a certain *context* (e.g. when reducing s in a conditional equation $c \Rightarrow s = t$, s appears in *context* c). The context can be used to infer condition (3) of definition 2.1 (joinability of the instantiated premises), for checking applicability of a certain equation. This generalization of the rewrite relation $\to_E$, will be called the *contextual rewrite relation*. The concept is based on the notion of contextual rewriting in [Ganzinger 1987b], but is not identical the same.

Definition 2.3. Given a conditional equational system E and a reduction ordering $>$. Let c denote the conjunction $\bigwedge_i u_i = v_i$. Let $\bar{t}$ stand for the skolemized version of the term t and $\bar{c}$ is $\bigwedge_i \overline{u_i} = \overline{v_i}$. Then the term s *contextually rewrites under E and context* c to the term t, notated as $s \to_{E,c} t$ iff one of the following two holds:

1. $\bar{s} =_{\bar{c}} \bar{t}$ and $s>t$. We call this a *C-rewrite* (C from *Condition*).
2. There is an equation $\bigwedge_i u_{ij} == v_{ij} \Rightarrow l_j == r_j$ in E such that:

   (1) a subterm of s (say at position p) matches l_j with matching substitution σ;
   (2) $\bigwedge_i \sigma u_{ij} == \sigma v_{ij} \Rightarrow \sigma l_j \rightarrow_> \sigma r_j$;
   (3) $\sigma u_{ij} \downarrow_{E,c} \sigma v_{ij}$ which stands for: $\sigma u_{ij} \rightarrow^*_{E,c} u'_{ij}$ and $\sigma v_{ij} \rightarrow^*_{E,c} v'_{ij}$ and $\overline{u'_{ij}} =_{\bar{c}} \overline{v'_{ij}}$;
   (4) $t = s[p \leftarrow \sigma r_j]$.

   We call this an *E-rewrite*. If in addition: $\bigwedge_i u_{ij} == v_{ij} \Rightarrow l_j \rightarrow_> r_j$, we call it an *R-rewrite*. □

Again the matching substitution σ is assumed to substitute only for variables in l. Then it can be easily shown that the contextual rewriting relation $\rightarrow_{E,c}$ is terminating and that the basic properties (one step reduction, finite reduction, joinability and reducibility) are all decidable (making use of the fact that $=_{\bar{c}}$ is decidable, since $\bar{c}$ is a finite set of ground equations, cfr. [Ackermann 1962]).

3. Inductive Theorems and Consistency

The aim of this paper is the development of a proof procedure for inductive theorems of the form $c \Rightarrow s==t$ in a conditional equational system: given a CES E, we want to show that $E \vdash_{ind} c \Rightarrow s==t$. This is equivalent with proving that for all ground substitutions σ: $E \vdash_{eq} \sigma c \Rightarrow \sigma s==\sigma t$.

If we can transform the CES E into an equivalent CES R_0 (i.e. $=_E$ and $=_{R_0}$ are the same) that is ground Church-Rosser, then we have a decision procedure for $=_E$ on ground terms: indeed, $=_{R_0}$ and $\downarrow_{R_0}$ are the same on ground terms, and thus also $=_E$ and $\downarrow_{R_0}$. Therefore: $\langle E \vdash_{ind} c \Rightarrow s==t \rangle$ iff $\langle \bigwedge_{i=1}^{n} \sigma u_i \downarrow_{R_0} \sigma v_i \Rightarrow \sigma s \downarrow_{R_0} \sigma t$, for all ground substitutions $\sigma \rangle$. If the latter condition is fulfilled $\langle \bigwedge_{i=1}^{n} u_i==v_i \Rightarrow s==t \rangle$ is called *consistent* with R_0. A set of conditional equations C is consistent with R_0, iff every equation of C is consistent with R_0. A conditional equation $c \Rightarrow s==t$ is inconsistent with R_0 if there exists a ground substitution σ such that $R_0 \vdash_{eq} \sigma c$ and $\sigma s \not\downarrow_{R_0} \sigma t$.

Now clearly $\bigwedge_{i=1}^{n} u_i==v_i \Rightarrow s==t$ is consistent with R_0 iff all equational consequences of $\bigwedge_{i=1}^{n} u_i==v_i \Rightarrow s==t$ and R_0 are consistent with R_0. Hence if we can find an equational consequence $d \Rightarrow p==q$ and a ground substitution τ for which we can prove that $R_0 \vdash_{eq} \tau d$ (i.e. for all equations $w_i==z_i$ in d: $\tau w_i \downarrow_{R_0} \tau z_i$) and $\tau p \not\downarrow_{R_0} \tau q$ (i.e. τp and τq are not joinable w.r.t. R_0), then $\bigwedge_{i=1}^{n} u_i==v_i \Rightarrow s==t$ is not consistent with R_0, and therefore not an inductive theorem of E. Such an equation $d \Rightarrow p==q$ will be called *provably inconsistent* (w.r.t. R_0). A set of conditional equations is provably inconsistent if one of its elements is.

The whole idea of a *proof by consistency* (as we will see in the next section) is based on generating (certain) equational consequences and checking that each of them is not provably inconsistent, by means of a so called *consistency check*. The consistency check we will define, is in fact a generalization of the inductive reducibility test of [Jouannaud & Kounalis 1986] and the consistency check in [Bachmair 1988] to conditional equations.

Definition 3.1. [Jouannaud & Kounalis 1986], [Bachmair 1988]
A term s is *inductively reducible* (w.r.t. R_0) iff for all ground substitutions σ: σs is reducible by R_0.
An equation $s==t$ is *inductively reducible* (w.r.t. R_0) iff for all ground substitutions σ for which σs and σt are distinct: σs or σt is reducible by R_0. □

Both notions of inductive reducibility are decidable for finite unconditional rewrite systems ([Plaisted 1985], [Bachmair 1988]), but not for conditional rewrite systems in general [Kaplan & Choquer EATCS Bulletin 28]. Henceforth we will assume that for the CES R_0 we will work with, inductive reducibility is decidable.

The consistency check *CC* is then defined as follows:

Definition 3.2. Given a ground Church-Rosser CES R_0, and a conditional equation $c \Rightarrow s==t$.
Then CC($c \Rightarrow s==t$) (w.r.t. R_0) is false iff c is unifiable with most general unifier σ and one of the

following cases is fulfilled:

— $\sigma s == \sigma t$ is not inductively reducible with respect to R_0;
— $\sigma s > \sigma t$ (resp. $\sigma t > \sigma s$) and σs (resp. σt) is not inductively reducible with respect to R_0;

Otherwise it is true. □

This consistency check is decidable in systems where inductive reducibility is decidable. We have:

Theorem 3.1. If CC($c \Rightarrow s == t$) is false than $c \Rightarrow s == t$ is inconsistent with R_0.

Proof

Since σ is the mgu of c, trivially for all ground substitutions τ: $R_0 \vdash_{eq} \tau\sigma c$. Furthermore, (1) if $\sigma s == \sigma t$ is not inductively reducible (w.r.t. R_0), then there exists a ground substitution τ such that $\tau\sigma s$ and $\tau\sigma t$ are distinct and irreducible under R_0, proving inconsistency of $c \Rightarrow s == t$. Also (2) if $\sigma s > \sigma t$ (resp. $\sigma t > \sigma s$) and σs (resp. σt) is not inductively reducible w.r.t. R_0, then there exists a ground substitution τ such that $\tau\sigma s$ (resp. $\tau\sigma t$) is irreducible under R_0, and since $\tau\sigma s > \tau\sigma t$ (resp. $\tau\sigma t > \tau\sigma s$) $\tau\sigma s$ and $\tau\sigma t$ do not rewrite to the same term, proving inconsistency of $c \Rightarrow s == t$. □

In both cases (1) and (2), $\langle \tau\sigma c \Rightarrow \tau\sigma s == \tau\sigma t \rangle$ is called a *direct* inconsistency witness for $c \Rightarrow s == t$. An inconsistency witness is *indirect* if it is not direct. Consequently, CC($c \Rightarrow s == t$) is false iff there exists a direct inconsistency witness for $c \Rightarrow s == t$.

4. Proof by Consistency in Conditional Equational Systems

As stated earlier, a *proof by consistency* essentially tries do derive (by *conditional equational reasoning*) from the given set of *axioms* R_0 and the set of *conjectures* C to prove, an equation which is provable inconsistent. This can be done by generating all possible equational consequences of R_0 and C, and checking for each whether the consistency check returns false. In theory this could be a potential method for *disproving* certain conjectures, but certainly not for proving they are inductively valid. Indeed in most cases the number of equational consequences is infinite. Fortunately not all possible equational consequences need to be computed and checked, to have *refutational completeness* of the method, i.e. such that every inconsistent set of conjectures can always be transformed into a provable inconsistent one. Indeed, it will be shown that, whenever the derivation (of equational consequences) has the so called *fairness*-property and the set of conjectures is inconsistent, then necessarily, an equation will be generated for which the consistency check returns false. In many cases even, only a finite number of equations has to be computed and checked for consistency. Consequently if the consistency check returns true for all these computed consequences, the original conjectures are all inductive theorems.

The main idea behind our approach is *proof simplification*, or more precisely *inconsistency witness simplification*: Given a set of conjectures C that is inconsistent, but for which the consistency check returns true (i.e. C is not provably inconsistent), then there must exist an indirect inconsistency witness. Now our proof by consistency inference system will be constructed such that during a *fair* inference derivation, for any indirect inconsistency witness a smaller (in some well-founded partial ordering) inconsistency witness will be deduced. Since the ordering is well-founded, this will eventually lead to a direct inconsistency witness, which will be detected by the consistency check (yielding disproof). And hence we have refutational completeness.

4.1 A Partial Ordering on Inconsistency Witnesses

As suggested above, we need a well-founded partial ordering (call it $>_{GP}$) on inconsistency witnesses.

Definition 4.1.a. Let $>$ be the ground reduction ordering containing $\rightarrow_{R_0}$.
The *complexity* c of the inconsistency witness $\sigma c \Rightarrow \sigma s == \sigma t$ is defined as:

$$
\begin{aligned}
c(\,\sigma c \Rightarrow \sigma s == \sigma t\,) &= (\,\{\sigma s\}, M(\sigma c), \{s\}, \sigma t\,) \text{ if } c \Rightarrow s \rightarrow_{>} t \\
&= (\,\{\sigma t\}, M(\sigma c), \{t\}, \sigma s\,) \text{ if } c \Rightarrow t \rightarrow_{>} s \\
&= (\,\{\sigma s, \sigma t\} \cup M(\sigma c), M(\sigma c), \{s, t\}, -\,) \text{ otherwise (if } c \Rightarrow s \leftrightarrow_{>} t)
\end{aligned}
$$

Let P_1 and P_2 be two inconsistency witnesses. Then $P_1 >_{GP} P_2$ iff $c(\,P_1\,) >_c c(\,P_2\,)$, where $>_c$ stands for the lexicographic combination of twice the multiset ordering $>_M$ induced by $>$ (for the first two elements of the tuple), the multiset ordering $\rhd_M$ induced by the specialization ordering $\rhd$, and $>$. □

Note that $>_{GP}$ is well-founded, since $>_M$, $\rhd_M$ and $>$ are all well-founded and therefore also $>_c$.

We also need a well-founded ordering $>_P$ on conditional equations, satisfying the following (*compatibility with* $>_{GP}$) property: $\langle c_1 \Rightarrow s_1{==}t_1\rangle >_P \langle c_2 \Rightarrow s_2{==}t_2\rangle$ implies $\langle \sigma c_1 \Rightarrow \sigma s_1{==}\sigma t_1\rangle >_{GP} \langle \sigma c_2 \Rightarrow \sigma s_2{==}\sigma t_2\rangle$ for all ground substitutions σ. (Notice $>_{GP}$ is used in a somewhat extended context, since $\sigma c_1 \Rightarrow \sigma s_1{==}\sigma t_1$ and/or $\sigma c_2 \Rightarrow \sigma s_2{==}\sigma t_2$ are not necessarily inconsistency witnesses). This ordering will be employed in some of the inference rules in order to obtain *inconsistency witness simplifying* inference derivations. Obviously, the following definition for $>_P$ fulfills the stated requirements:

Definition 4.1.b. The *complexity* c of the conditional equation $c \Rightarrow s{==}t$ is defined as:

$$\begin{aligned} c(\, c \Rightarrow s{==}t \,) &= (\, \{s\}, M(c), \{s\}, t \,) \text{ if } c \Rightarrow s \to_> t \\ &= (\, \{t\}, M(c), \{t\}, s \,) \text{ if } c \Rightarrow t \to_> s \\ &= (\, \{s, t\} \cup M(c), M(c), \{s, t\}, - \,) \text{ otherwise (if } c \Rightarrow s \leftrightarrow_> t) \end{aligned}$$

Let e_1 and e_2 be two conditional equations. Then $e_1 >_P e_2$ iff $c(\, e_1 \,) >_c c(\, e_2 \,)$. □

4.2 Inference Rules for Proof by Consistency

Let R_0 be a ground Church-Rosser CES for the CES E, in which we want to do inductive proofs. Furthermore let $>$ be the ground reduction ordering containing $\to_{R_0}$. By C we denote the set of *conjectures*, i.e. the conditional equations which we want to prove in I(E). The subset of reductive (w.r.t. $>$) conjectures of C is denoted by CR. L stands for the set of *lemma's*, which can be used for simplification of the conjectures. The inference system *PCC* for *Proof by Consistency in Conditional Equational Systems* then contains the following inference rules:

Deduction: $\dfrac{(L, C)}{(L, C \cup \{c \Rightarrow s{==}t\})}$ if $(C \cup R_0) \vdash_{eq} c \Rightarrow s{==}t$

Induction: $\dfrac{(L, C)}{(L \cup \{c \Rightarrow s{==}t\}, C)}$ if $R_0 \vdash_{ind} c \Rightarrow s{==}t$

Deletion:

- of a trivial equation: $\dfrac{(L, C \cup \{c \Rightarrow s{==}t\})}{(L, C)}$ if $R_0 \vdash_{ind} c \Rightarrow s{==}t$

- of a trivial condition: $\dfrac{(L, C \cup \{c \wedge u{==}v \Rightarrow s{==}t\})}{(L, C \cup \{c \Rightarrow s{==}t\})}$ if $R_0 \vdash_{ind} u{==}v$

Simplification:

- of an equation: $\dfrac{(L, C \cup \{c \Rightarrow s \leftrightarrow t\})}{(L, C \cup \{c \Rightarrow u{==}t\})}$ if $s \to_{CR \cup R_0 \cup L, c} u$

- of the rhs of a rule: $\dfrac{(L, C \cup \{c \Rightarrow s \to t\})}{(L, C \cup \{c \Rightarrow s \to u\})}$ if $t \to_{CR \cup R_0 \cup L, c} u$

- of the lhs of a rule: $\dfrac{(L, C \cup \{c \Rightarrow s \to t\})}{(L, C \cup \{c \Rightarrow u \to t\})}$

if $s \to_{CR \cup R_0 \cup L, c} u$ *by a C-rewrite* or *an E-rewrite with an equation of* $R_0 \cup L$ or *an R-rewrite with a rule* $(c' \Rightarrow l \to r)$ *of CR and matching substitution* ρ, *and* $\langle c \Rightarrow s{==}t\rangle >_P \langle \rho c' \Rightarrow \rho l{==}\rho r\rangle$

- of a condition: $\dfrac{(L, C \cup \{c \wedge u{==}v \Rightarrow s{==}t\})}{(L, C \cup \{c \wedge w{==}v \Rightarrow s{==}t\})}$ if $u \to_{CR \cup R_0 \cup L, c} w$

Subsumption by an equation: $\dfrac{(L, C \cup \{c_1 \Rightarrow s \leftrightarrow t,\ c_2 \Rightarrow w[\rho s]{==}w[\rho t]\})}{(L, C \cup \{c_1 \Rightarrow s \leftrightarrow t\})}$

if (*let* c_1 *be* $\bigwedge_i u_{1i}{==}v_{1i}$) $\forall i$: $\rho u_{1i} \downarrow_{CR \cup R_0 \cup L, c_2} \rho v_{1i}$ and $\langle c_2 \Rightarrow w[\rho s]{==}w[\rho t]\rangle >_P \langle \rho c_1 \Rightarrow \rho s{==}\rho t\rangle$

The deduction rule is the heart of the inference system. Indeed by generating equational consequences from $C \cup R_0$ we are able to derive a provable inconsistent set from an inconsistent set C. The other inference rules (induction, deletion, simplification and subsumption) are not essential, but are needed for efficiency reasons. The idea behind the simplification and subsumption rules, for instance, is to simplify the conjectures in C, so that they possibly turn into trivialities (which can be deleted, reducing C), or that they become equal to other equations in C (also reducing C). In either case less equational consequences have to be computed by the deduction rule. By the induction rule, useful lemma's can be added to L, useful in the sense that they help the simplification and subsumption process in reducing the set of conjectures C.

These inference rules are heavily inspired by those in [Bachmair 1988]. However, besides the fact that they work on and with conditional equations (instead of *unconditional* equations), there are some other notable differences. The most important difference is that our approach distinguishes between reductive conjectures and others. This makes simplification of unorientable equations and right hand sides of rules with these reductive conjectures possible, without any additional constraint; [Bachmair 1988] however requires that in this case the redex (i.e. the term to simplify) must be greater in the specialization ordering than the left hand side of the conjecture, used to simplify.

We will write $(L, C) \vdash_{PCC} (L', C')$, if (L', C') is obtained from (L, C) by application of an inference rule of the system PCC. A sequence $(L_0, C_0), (L_1, C_1), \ldots$ is called a *PCC-inference derivation* if $(L_{i-1}, C_{i-1}) \vdash_{PCC} (L_i, C_i)$, for all $i>0$. The *limit* of a PCC-inference derivation is the set C^∞, which stands for the set of all *persisting conjectures*, i.e. $\bigcup_{i\geq 0} \bigcap_{j\geq i} C_j$.

4.3 Soundness and Refutational Completeness

The inference system PCC is sound in the following sense:

Theorem 4.1. Let R_0 be a ground Church-Rosser system. If (L', C') can be derived from (L, C), by using inference rules of the system PCC, then C is consistent with R_0 iff C' is. □

The proof of this theorem can be found in [Bevers & Lewi 1990].

A *proof by consistency procedure for conditional equational systems* is any procedure that accepts as inputs a ground reduction ordering $>$, a ground Church-Rosser CES R_0, a set of inductive lemma's L_0 and a set of conjectures C_0, and generates a derivation starting from (L_0, C_0) using applications of the inference rules of PCC as the only elementary computation steps. We say a proof by consistency procedure for CES is *refutational complete* iff from any inconsistent set of conjectures C_0 it generates a derivation for which some set C_i is provably inconsistent, i.e. every conditional equation which is not an inductive theorem will eventually be disproved.

We will prove that a sufficient condition for refutational completeness is that any derivation generated by the proof by consistency procedure, must be *fair*. Such a procedure will be called a *fair* proof by consistency procedure. *Fairness* is defined as follows:

Definition 4.2. A derivation $(L_0, C_0), (L_1, C_1), \ldots$ in PCC is *fair* iff for every conditional equation $c \Rightarrow s{=}{=}t$ in C^∞ for which $CC(c \Rightarrow s{=}{=}t)$ is true, and for every ground substitution σ for which $R_0 \not\vdash_{eq} \sigma c \Rightarrow \sigma s{=}{=}\sigma t$, there exists a conditional equation $d \Rightarrow u{=}{=}v$ in $\bigcup_i C_i$ and a ground substitution τ, such that $R_0 \not\vdash_{eq} \tau d \Rightarrow \tau u{=}{=}\tau v$ and $\langle \sigma c \Rightarrow \sigma s{=}{=}\sigma t \rangle >_{GP} \langle \tau d \Rightarrow \tau u{=}{=}\tau v \rangle$. □

So a PCC-derivation is fair if for every equation in the limit C^∞ for which an indirect inconsistency witness exists, another equation has been derived for which a smaller (direct or indirect) inconsistency witness exists. Of course this definition does not tell you *how* to build a fair proof by consistency procedure (it is not *constructive*). Therefore in the next section we will see more practical conditions and an outline of a practical fair proof by consistency procedure. Nevertheless this definition is very useful, not only because of its generality, but also because it simplifies the proof of the next important theorem:

Theorem 4.2. Every fair proof by consistency procedure for conditional equations is refutational complete.

Proof

The idea of the proof is the following: suppose there exists a conjecture $\langle c \Rightarrow s{=}{=}t \rangle$ in any C_j (during the fair inference derivation) for which an inconsistency witness $\langle\sigma c \Rightarrow \sigma s{=\!=}\sigma t\rangle$ exists. If the inconsistency witness is direct then $CC(c \Rightarrow s{=\!=}t)$ returns false and the theorem is proved. Otherwise if it is indirect ($CC(c \Rightarrow s{=\!=}t)$ returns true) then we will prove that there will always exist a conditional equation $\langle d \Rightarrow u{=\!=}v\rangle \in \bigcup_i C_i$ that has an inconsistency witness $\langle\tau d \Rightarrow \tau u{=\!=}\tau v\rangle$ which is smaller (in the well-founded ordering $>_{GP}$) than $\langle\sigma c \Rightarrow \sigma s{=\!=}\sigma t\rangle$ (I). Now we can repeat this whole reasoning on $d \Rightarrow u{=\!=}v$, i.e. if $CC(d \Rightarrow u{=\!=}v)$ returns false then the theorem is proved, and if it returns true we can always find another conditional equation ($\in \bigcup_i C_i$) with an inconsistency witness smaller than $\langle \tau d \Rightarrow \tau u {=\!=} \tau v \rangle$, for which we can repeat the same reasoning again. However, since $>_{GP}$ is well-founded, this descendent chain of indirect inconsistency witnesses cannot be infinite. Hence there must exist a conditional equation in $\bigcup_i C_i$ for which the consistency check returns false, proving refutational completeness.

Now we can prove (I) by case analysis on $\langle c \Rightarrow s{=\!=}t\rangle$:

1. $c \Rightarrow s{=\!=}t$ is persistent, i.e. $(c \Rightarrow s{=\!=}t) \in C^\infty$. The proof is trivial since $CC(\, c \Rightarrow s{=\!=}t\,)$ is true and by to the fairness of the inference derivation.
2. $c \Rightarrow s{=\!=}t$ is not persistent. Then it must have been deleted by deletion, by simplification, or by subsumption. In all three cases the further proof is fairly easy but too elaborated to put here. The interested reader is referred to [Bevers & Lewi 1990].

□

Corollary 4.1. Given a set of conjectures C_0 and a PCC-derivation starting from C_0. If this derivation is fair and no C_i in it is provably inconsistent, then C_0 is consistent. □

4.4 A Linear and Unfailing Proof by Consistency Procedure

We will now show that, in order to obtain fairness, only so called *contextual critical pairs* and/or *superposition instances* (cfr. [Ganzinger 1987b]) of rules in R_0 on conditional equations in C have to be computed. Contextual critical pairs are in fact the extension to the conditional case of the well known *critical pairs*, while superposition instances can be seen as the results of critical overlappings on the premisses of a conditional equation. They are defined as follows:

Definition 4.3. Given two conditional equations $d \Rightarrow l{=\!=}r$ and $c \Rightarrow s{=\!=}t$, which have no variables in common. Assume that p is a position in s, such that s/p and l unify with the most general unifier ρ. Then:

$$\rho c \wedge \rho d \Rightarrow \rho s[p \leftarrow \rho r] {=\!=} \rho t$$

is a *contextual critical pair* of $\langle d \Rightarrow l{=\!=}r\rangle$ on $\langle c \Rightarrow s{=\!=}t\rangle$ (at the position p).

Assume that q is a position in $u{=\!=}v$, such that $(u{=\!=}v)/q$ and l unify with the most general unifier τ (i.e. $\tau(u{=\!=}v)/q = \tau l$). Then:

$$(\,\tau d \wedge \tau c \wedge \tau(u{=\!=}v)[q \leftarrow \tau r]\,) \Rightarrow \tau s {=\!=} \tau t$$

is a *superposition instance* of $\langle d \Rightarrow l{=\!=}r\rangle$ on $\langle c \wedge u{=\!=}v \Rightarrow s{=\!=}t\rangle$ (at the position q of $u{=\!=}v$). □

In the case of proof by consistency in unconditional theories, [Fribourg 1986] restricts the number of equational consequences to be computed, by only considering critical pairs at so called *complete positions*. This notion was extended in [Küchlin 1989] to *inductively complete sets of positions* of a term. [Bachmair 1988] proved that this kind of proof by consistency procedure was fair and therefore also refutational complete. We now adapt this notion to the conditional case and speak about *inductively reducible set of positions* of a conditional equation:

Definition 4.4. A set P of non-variable positions of a conditional term $c \Rightarrow t$ is called an *inductively reducible set of positions* for $R_0' \subseteq R_0$ of $c \Rightarrow t$ iff, for every (R_0-)normalized ground substitution σ, if $\sigma c \Rightarrow \sigma t$ is reducible by R_0, then there is a position $p \in P$, such that $\sigma c \Rightarrow \sigma t$ is reducible by R_0' at position p.

A set P of non-variable positions of a conditional equation $\langle c \Rightarrow s{=\!=}t\rangle$ is called an *inductively reducible set of positions* for $R_0' \subseteq R_0$ of $\langle c \Rightarrow s{=\!=}t\rangle$ iff one of the following three cases is met:

— $s>t$ and P is an inductively reducible set of positions for R_0' of $c \Rightarrow s$;
— $t>s$ and P is an inductively reducible set of positions for R_0' of $c \Rightarrow t$;

— $s \not> t$ and $t \not> s$ and for every R_0-normalized ground substitution σ, if $\langle \sigma c \Rightarrow \sigma s == \sigma t \rangle$ is reducible by R_0, then there is a position $p \in P$, such that $\langle \sigma c \Rightarrow \sigma s == \sigma t \rangle$ is reducible by R'_0 at position p.

□

We still have one more improvement, which is not in Bachmair's proof by consistency, and for which we need the concept of ground joinability w.r.t. the contextual rewriting relation:

Definition 4.5. $c \Rightarrow s == t$ (contextual critical pair or superposition instance) is ground joinable w.r.t. $\rightarrow_{CRUR_0UL,c}$ iff for all ground substitutions σ for which $R_0 \vdash_{eq} \sigma c$: $\langle \sigma s \downarrow_{CRUR_0UL,\sigma c} \sigma t \rangle$. □

Now we will prove we do not need to consider contextual critical pairs $d \Rightarrow u == v$ which are ground joinable w.r.t. $\rightarrow_{CRUR_0UL,d}$, nor superposition instances $d \Rightarrow u == v$ which are ground joinable w.r.t. $\rightarrow_{R_0UL,d}$. Of course ground joinability is not decidable. However if we have a (non-complete) ground joinability test at our disposal (such as for instance the one proposed in [Göbel 1987], for unconditional rewrite systems), then we can employ it to rule out superfluous contextual critical pairs and/or superposition instances (i.e. pairs or instances which can be proved to be ground joinable, using the test). Remark that in the unconditional case, ground joinability is a special case of the (more general) *ground subconnectedness* concept in [Küchlin 1989].

For all conjectures $\langle c_i \Rightarrow s_i == t_i \rangle$ of C, let P_i be an inductively reducible set of positions for $R_{0_i} \subseteq R_0$ of $\langle c_i \Rightarrow s_i == t_i \rangle$. By $CCP(R_{0_i}, P_i, c_i \Rightarrow s_i == t_i)$ we denote the set of all contextual critical pairs $\langle d \Rightarrow u == v \rangle$, for which we cannot prove ground joinability w.r.t. $\rightarrow_{CRUR_0UL,d}$, of the rules in R_{0_i} on $\langle c_i \Rightarrow s_i == t_i \rangle$, at all positions p (in $s_i == t_i$) that belong to P_i. Similarly by $SI(R_{0_i}, P_i, c_i \Rightarrow s_i == t_i)$ we denote the set of all superposition instances $\langle d \Rightarrow u == v \rangle$, for which we cannot prove ground joinability w.r.t. $\rightarrow_{R_0UL,d}$, of the rules in R_{0_i} on $\langle c_i \Rightarrow s_i == t_i \rangle$, at all positions p (in c) that belong to P_i. Furthermore let $CCP \oplus SI(R_0, C)_{IRP}$ stand for the set:

$$\bigcup_{\langle c_i \Rightarrow s_i == t_i \rangle \in C} CCP(R_{0_i}, P_i, c_i \Rightarrow s_i == t_i) \oplus \bigcup_{\langle c_i \Rightarrow s_i == t_i \rangle \in C} SI(R_{0_i}, P_i, c_i \Rightarrow s_i == t_i).$$

The next theorem proves that only $CCP \oplus SI(R_0, C)_{IRP}$ (in the terminology of [Bachmair 1988], this is called a *coverset* of C) has to be computed to obtain fair derivations:

Theorem 4.3. The derivation $(L_0, C_0), (L_1, C_1), \ldots$ in PCC is *fair* if every conditional equation of $CCP \oplus SI(R_0, C^\infty)_{IRP}$ is contained in the set of all derived conjectures $\bigcup_k C_k$. □

The proof of this theorem can, again, be found in [Bevers & Lewi 1990].

As a consequence of theorems 4.2 and 4.3 we are able to build a refutational complete proof by consistency procedure for CES, by seeing to it that $CCP \oplus SI(R_0, C^\infty)_{IRP}$ (i.e. the coverset of C^∞) will be contained in $\bigcup_k C_k$. Below you find an outline of a possible refutational complete proof by consistency procedure for CES:

1. Initially $C := C_0$ and all conjectures in C unmarked;
2. Repeat until all conjectures in C are marked:
   - 2.1. check if for some conjecture in C the consistency check CC returns false; if so stop with *DISPROOF*;
   - 2.2. take an unmarked conjecture $\langle c \Rightarrow s == t \rangle$ in C (without ignoring any conjecture of C indefinitely long) and mark $\langle c \Rightarrow s == t \rangle$;
   - 2.3. choose a set $R'_0 \subseteq R_0$ and an inductively reducible set of positions P for R'_0 of $\langle c \Rightarrow s == t \rangle$;
   - 2.4. compute the coverset of $\langle c \Rightarrow s == t \rangle$:
     Coverset $:= CCP(R'_0, P, c \Rightarrow s == t) \cup SI(R'_0, P, c \Rightarrow s == t)$;
   - 2.5. $C := C \cup$ *Coverset*;
   - 2.6. apply the other inference rules (i.e. all except D) to (L, C);

Fig. 4.1.

If the original set of conjectures C_0 was inconsistent, this procedure will derive a C which is provably inconsistent, and therefore stop with *DISPROOF*. However if C_0 is consistent, then the procedure may either terminate, or run forever. If it terminates (without *DISPROOF*), then all conjectures in the resulting C (which is in the terminology of the inference system PCC equal to C^∞) are marked, which means all coversets of all equations in C are computed and added. And thus $CCP \oplus SI(R_0, C^\infty)_{IRP}$ is contained in the set of all derived equations $\bigcup_k C_k$. Consequently we have a fair derivation, without generating a provable inconsistent equation, so by corollary 4.1, it is proved that C_0 is consistent and all deduced conjectures of $\bigcup_k C_k$ are inductive theorems of R_0. It is also possible that a fair PCC-procedure runs forever for a (consistent) set C_0 and deduces more and more conjectures without ever obtaining a provably inconsistent set. The only solution in our framework to this problem, consists in the addition and use of appropriate inductive lemma's, to simplify (some of) the conjectures.

The proof procedure of Fig. 4.1 is called *linear* because only contextual critical pairs and/or superposition instances of R_0 on C have to be computed and not of C on R_0 or C on C. Furthermore it is *unfailing* in the sense that it does not fail whenever a certain generated equation cannot be oriented in one way or the other (in our terminology: the equation is not reductive). Therefore we can speak about a *linear, unfailing, refutational complete proof by consistency procedure for conditional equational theories.*

5. An Example

It is time to illustrate the theory with a small example. The following equations specify ordered lists of natural numbers:

(e1)		$0 \leqslant n == true;$
(e2)		$succ(m) \leqslant 0 == false;$
(e3)		$succ(m) \leqslant succ(n) == m \leqslant n;$
(e4)		$insert(nil, m) == cons(m, nil);$
(e5)	$m \leqslant n == true \Rightarrow$	$insert(cons(n, l), m) == cons(m, cons(n, l));$
(e6)	$m \leqslant n == false \Rightarrow$	$insert(cons(n, l), m) == cons(n, insert(l, m));$
(e7)		$ordered(nil) == true;$
(e8)		$ordered(cons(m, nil)) == true;$
(e9)	$m \leqslant n == true \Rightarrow$	$ordered(cons(m, cons(n, l))) == ordered(cons(n, l));$
(e10)	$m \leqslant n == false \Rightarrow$	$ordered(cons(m, cons(n, l))) == false;$

Fig. 5.1.

The operation *insert* is used to build up a list of natural numbers which is ordered according to $\leqslant$. *ordered* checks whether its argument is ordered or not. So the following conditional equation should be an inductive theorem of this specification:

(c1) $ordered(l) == true \Rightarrow ordered(insert(l, m)) \rightarrow_> true;$

R_0 consists of the equations (e1) - (e10). All these equations are reductive w.r.t. a simple ground reduction ordering > like for instance the recursive path ordering [Kamin & Levy 1980, Dershowitz 1982]. Furthermore it can be easily proved that R_0 is ground Church-Rosser; and thus we can use it for our proof by consistency method. We will prove that (c1) holds in I(R_0) by following the sketched proof procedure of Fig. 5.1. Initially C consists of the unmarked equation (c1) (marked equations will be denoted by placing a ✱ in front of it). The consistency check on (c1) returns true since the premiss of (c1) is not unifiable. We then mark (c1) and compute its coverset. The set P containing the position 1 of the left hand side of the conclusion of (c1) is an inductively reducible set of positions for R_0 of (c1). Superposing (e4) (e5) and (e6) on (c1) yields the following contextual critical pairs:

(ccp1) $ordered(nil) == true \Rightarrow ordered(cons(m_1, nil)) \rightarrow_> true;$
(ccp2) $m_1 \leqslant n_1 == true \wedge ordered(cons(n_1, l_1)) == true \Rightarrow ordered(cons(m_1, cons(n_1, l_1))) \rightarrow_> true;$
(ccp3) $m_1 \leqslant n_1 == false \wedge ordered(cons(n_1, l_1)) == true \Rightarrow ordered(cons(n_1, insert(l_1, m_1))) \rightarrow_> true;$

These critical pairs are all added to C. We then apply the other inference rules. (ccp1) can be simplified by applying the inference rule **SLR**; indeed, $ordered(cons(m_1, nil)) \rightarrow_{\{(e8)\}} true$ and the conclusion reduces to true=true and so the resulting equation may be deleted by **DTE**. Also (ccp2) can be simplified by applying inference rule **SLR**: $ordered(cons(m_1, cons(n_1, l_1))) \rightarrow_{\{(e9)\},\langle m_1 \leqslant n_1 == true\rangle} ordered(cons(n_1, l_1)) \rightarrow_{\emptyset,\langle ordered(cons(n_1,l_1)) == true\rangle} true$, and therefore (ccp2) reduces to $m_1 \leqslant n_1 == true \wedge ordered(cons(n_1, l_1)) == true \Rightarrow true == true$, which is obviously deleted by **DTE**. Finally, after one iteration of the loop in the proof procedure of Fig. 5.1, we obtain the following set C:

∗ (c1) $ordered(l) == true \Rightarrow ordered(insert(l, m)) \rightarrow_{>} true;$
(c2) $m_1 \leq n_1 == false \wedge ordered(cons(n_1, l_1)) == true \Rightarrow ordered(cons(n_1, insert(l_1, m_1))) \rightarrow_{>} true;$

We have still one unmarked equation, so we start a second iteration of the loop. Clearly CC((c2)) is true, since the premiss is not unifiable. So we compute its coverset, with the inductively reducible set of positions (for R_0) containing one position, namely the position of $ordered(cons(n_1, l_1))$. Thus we obtain the following superposition instances:

(e8) on (c2): (si1) $m_1 \leq n_1 == false \wedge true == true \Rightarrow ordered(cons(n_1, insert(nil, m_1))) \rightarrow_{>} true;$
(e9) on (c2): (si2) $m_1 \leq n_1 == false \wedge n_1 \leq n_2 == true \wedge ordered(cons(n_2, l_2)) == true \Rightarrow$
$ordered(cons(n_1, insert(cons(n_2, l_2), m_1))) \rightarrow_{>} true;$
(e10) on (c2): (si3) $m_1 \leq n_1 == false \wedge n_1 \leq n_2 == false \wedge false == true \Rightarrow ordered(cons(n_1, insert(cons(n_2, l_2), m_1))) \rightarrow_{>} true;$

Again after simplification and then deletion of the trivialities we obtain the following set C:

∗ (c1) $ordered(l) == true \Rightarrow ordered(insert(l, m)) \rightarrow_{>} true;$
∗ (c2) $m_1 \leq n_1 == false \wedge ordered(cons(n_1, l_1)) == true \Rightarrow ordered(cons(n_1, insert(l_1, m_1))) \rightarrow_{>} true;$
(c3) $m_1 \leq n_1 == false \Rightarrow ordered(cons(n_1, cons(m_1, nil))) \rightarrow_{>} true;$

We now have one unmarked equation (c3). The consistency check returns true. Let us take (c3), mark it and compute its coverset, with the inductively reducible set of positions (for R_0) containing the position of the left hand side of the conclusion. We have as contextual critical pairs:

(e9) on (c3): (ccp4) $m_1 \leq n_1 == false \wedge n_1 \leq m_1 == true \Rightarrow ordered(cons(m_1, nil)) \leftrightarrow_{>} true;$
(e10) on (c3): (ccp5) $m_1 \leq n_1 == false \wedge n_1 \leq m_1 == false \Rightarrow false \leftrightarrow_{>} true;$

After simplification and deletion C becomes:

∗ (c1) $ordered(l) == true \Rightarrow ordered(insert(l, m)) \rightarrow_{>} true;$
∗ (c2) $m_1 \leq n_1 == false \wedge ordered(cons(n_1, l_1)) == true \Rightarrow ordered(cons(n_1, insert(l_1, m_1))) \rightarrow_{>} true;$
∗ (c3) $m_1 \leq n_1 == false \Rightarrow ordered(cons(n_1, cons(m_1, nil))) \rightarrow_{>} true;$
(c4) $m_1 \leq n_1 == false \wedge n_1 \leq m_1 == false \Rightarrow false \leftrightarrow_{>} true;$

Clearly all conjectures of C are not provably inconsistent. We now take (c4), mark it and compute its coverset. The inductively reducible set of positions (for R_0) contains the position of $m_1 \leq n_1$ in the premiss. The superposition instances are then:

(e1) on (c4): (si4) $true == false \wedge n_1 \leq 0 == false \Rightarrow false \leftrightarrow_{>} true;$
(e2) on (c4): (si5) $false == false \wedge 0 \leq succ(m_2) == false \Rightarrow false \leftrightarrow_{>} true;$
(e3) on (c4): (si6) $m_2 \leq n_2 == false \wedge succ(n_2) \leq succ(m_2) == false \Rightarrow false \leftrightarrow_{>} true;$

Simplification and deletion gives as set C:

∗ (c1) $ordered(l) == true \Rightarrow ordered(insert(l, m)) \rightarrow_{>} true;$
∗ (c2) $m_1 \leq n_1 == false \wedge ordered(cons(n_1, l_1)) == true \Rightarrow ordered(cons(n_1, insert(l_1, m_1))) \rightarrow_{>} true;$
∗ (c3) $m_1 \leq n_1 == false \Rightarrow ordered(cons(n_1, cons(m_1, nil))) \rightarrow_{>} true;$
∗ (c4) $m_1 \leq n_1 == false \wedge n_1 \leq m_1 == false \Rightarrow false \leftrightarrow_{>} true;$

Since all conjectures in C are marked, the proof procedure stops. We have a fair derivation without encountering a provably inconsistent equation, therefore all conjectures in C (and a fortiori (c1)) are inductive theorems of the specification of fig. 5.1.

6. Conclusion

We have presented an extension of Bachmair's proof by consistency method, to conditional equational theories. The method follows a linear superposition strategy and is unfailing, in the sense that it can handle conditional (reductive) rewrite rules as well as unorientable (non-reductive) conditional equations. Furthermore we proved its refutational completeness. It can be employed for all CES in which inductive reducibility is decidable, for instance in systems where inductive reducibility and inductive matchability coincide.

We believe this proof by consistency procedure method is very well suited as a firm base for an automated theorem prover for (conditional) algebraic specifications. Furthermore extensions can be formulated (and proved to be correct) very easily in the framework of this paper. We think for instance of incorporating equational rewriting, or also employing all conjectures (not only the reductive ones) for simplification. Yet another powerful extension is to allow for *generalization* of terms by variables, such as in the traditional induction method of [Boyer & Moore 1979]. How this can be done and the possible implications with respect to soundness and refutational completeness, still have to be studied very carefully.

References

Ackermann, W. (1962). *Solvable cases of the decision problem*. North-Holland.

Bachmair, L. (1988). Proof by Consistency in Equational Theories. *Logic in Computer Science*, Edinburgh 1988.

Bachmair, L., Ganzinger, H. (1990). Completion of first-order clauses with equality by strict superposition. *2nd CTRS*, Logic and Formal Method Lab, Dept. of Computer Science, Concordia University, Montreal.

Bergstra, J., Klop, J.W. (1982). *Conditional rewrite rules: Confluence and termination*. Report IW198/82, Mathematisch Centrum, Amsterdam.

Bevers, E., Lewi, J. (1990). *Proof by Consistency in Conditional Equational Theories*. Report CW 102, Department of Computer Science, K.U.Leuven.

Bousdira, W., Rémy, J.L. (1987). Hierarchical contextual rewriting with several levels. *Proc. 1st CTRS*, LNCS 308, 15-30.

Dershowitz, N. (1982). Orderings for term-rewriting systems. *J. Theoretical Computer Science*, Vol 17, No 3, 279-301.

Dershowitz, N., Okada, M., Sivakumar, G. (1987). Confluence of Conditional Rewrite Systems. *1st CTRS*, LNCS 308, 31-44.

Dershowitz, N., Okada, M., Sivakumar, G. (1988). Canonical Conditional Rewrite Systems. *Proc. 9th CADE*, LNCS 310, 538-549.

Dershowitz, N. (1990). A Maximal-Literal Unit Strategy for Horn Clauses. *2nd CTRS*, Logic and Formal Method Lab, Dept. of Computer Science, Concordia University, Montreal.

Fribourg, L. (1986). A strong restriction of the inductive completion procedure. *ICALP '86*, LNCS 226, 105-115.

Fribourg, L. (1989). A strong restriction of the inductive completion procedure. *Journal of Symbolic Computation*, 8, 253-276.

Ganzinger, H. (1987a). Ground term confluence in parametric conditional equational specifications. *Proc. STACS 1987*, LNCS 247.

Ganzinger, H. (1987b). A Completion Procedure for Conditional Equations. *1st CTRS*, LNCS 308, 62-83.

Göbel, R. (1987). Ground Confluence. *Proc. Rewriting Techniques and Applications*, Bordeaux, LNCS 256, 156-167.

Goguen, J.A., (1980). How to Prove Algebraic Inductive Hypotheses Without Induction, with Applications to the Correctness of Data Type Implementation, *Proc. 5th CADE*, LNCS 87, 356-373

Gramlich, B. (1989). *Inductive Theorem Proving Using Refined Unfailing Completion Techniques*. SEKI Report SR-89-14, Universität Kaiserslautern.

Hsiang, J., Rusinowitch, M. (1986). *On word problems in equational theories*. Tech. Rep. 86/29, SUNY at Stony Brook.

Huet, G., Oppen, D. (1980). Equations and rewrite rules: A survey. *Formal Language Theory: Perspectives and Open Problems*, Academic Press, New York, 1980, 349-405.

Huet, G., Hullot, J. M. (1982). Proofs by induction in equational theories with constructors. *21st IEEE symposium on Foundations of Computer Science*, 96-107.

Jouannaud, J.P., Kounalis, E. (1985). *Proofs by induction in equational theories without constructors*. CRIN 85-R-042, Nancy.

Jouannaud, J.P., Waldmann, B. (1986). Reductive Conditional term rewriting systems. *Proc. 3rd IFIP Working Conference on Formal Description of Programming Concepts*, Ebberup, Denmark, Aug. 1986, North-Holland.

Kamin, S., Levy, J.-J. (1980). *Two Generalisations of the Recursive Path Ordering*, Unpublished note, Dept. of Computer Science, University of Illinois, USA.

Kaplan, S. (1984). Conditional Rewrite Rules. *Journal of Theoretical Computer Science*, 33, 175-193.

Kaplan, S. (1987). Simplifying Conditional Term Rewriting Systems: Unification, Termination and Confluence. *Journal of Symbolic Computation*, 4, 95-334.

Kapur, D., Narendran, P., Zhang, H. (1986). Proof by induction using test sets. *Proc. 8th CADE*, LNCS 230, Springer New York, 99-117.

Kapur, D., Musser, D.R. (1987). Proof by Consistency. *Artificial Intelligence*, 31, 125-157.

Küchlin, W. (1989). Inductive completion by ground proof transformation. *Rewriting Techniques*, volume 2 of *Resolution of Equations in Algebraic Structures*, Aït-Kaci, H., Nivat, M. (eds.), Academic Press.

Musser, D. R. (1980). On proving inductive properties of abstract data types. *Proceedings 7th Symposium on Principles of Programming Languages*, ACM SIGPLAN, 154-162.

Okada, M. (1987). A Logical Analysis on Theory of Conditional Rewriting. *1st CTRS*, LNCS 308, 179-196.

Paul, E. (1984), Proof by induction in equational theories with relations between constructors. *Proceedings 9th Colloquium on trees in Algebra and Programming*, Bordeaux, 211-225.

Plaisted, D.A. (1985). Semantic confluence tests and completion methods. *Inf. Control* 65:182-215.

Zhang, H., Rémy, J.L. (1985). Contextual rewriting. *Rewriting Techniques and Applications*, LNCS 202, 46-62.

Completion procedures as Semidecision procedures *

Maria Paola Bonacina Jieh Hsiang
Department of Computer Science
SUNY at Stony Brook
Stony Brook, NY 11794-4400, USA
{bonacina,hsiang}@sbcs.sunysb.edu

Abstract

In this paper we give a new abstract framework for the study of Knuth-Bendix type completion procedures, which are regarded as *semidecision procedures* for theorem proving.

First, we extend the classical proof orderings approach started in [6] in such a way that proofs of different theorems can also be compared. This is necessary for the application of proof orderings to theorem proving derivations. We use proof orderings to uniformly define all the fundamental concepts in terms of *proof reduction.*

A completion procedure is given by a set of *inference rules* and a *search plan.* The inference rules determine what can be derived from given data. The search plan chooses at each step of the derivation which inference rule to apply to which data. Each inference step either reduces the proof of a theorem or deletes a *redundant* sentence. Our definition of *redundancy* is based on the assumed proof ordering. We have shown in [16] that our definition subsume those given in [50, 13].

We prove that if the inference rules are *refutationally complete* and the search plan is *fair*, a completion procedure is a semidecision procedure for theorem proving. The key part of this result is the notion of *fairness.* Our definition of fairness is the first definition of fairness for completion procedures which addresses the theorem proving problem. It is new in three ways: it is *target oriented,* that is it keeps the theorem to be proved into consideration, it is explicitly stated as a property of the search plan and it is defined in terms of proof reduction, so that expansion inferences and contraction inferences are treated uniformly. According to this definition of fairness, it is not necessary to consider all critical pairs in a derivation for the derivation to be fair. This is because not all critical pairs are necessary to prove a given theorem. Considering all critical pairs is an unnecessary source of inefficiency in a theorem proving derivation.

We also show that the process of diproving inductive theorems by the so called *inductionless induction* method is a semidecision process. Finally, we present according to our framework, some equational completion procedures based on Unfailing Knuth-Bendix completion.

*Research supported in part by grants CCR-8805734, INT-8715231 and CCR-8901322, funded by the National Science Foundation. The first author is also supported by Dottorato di ricerca in Informatica, Universitá degli Studi di Milano, Italy.

1 Introduction

The Knuth-Bendix completion procedure [49] computes a possibly infinite confluent rewrite system equivalent to a given set of equations [38]. If a set of equations E and an equation $s \simeq t$ are given, it semidecides whether $s \simeq t$ is a theorem of E, as first remarked in [51, 39]. These results hold if the procedure does not fail on an unoriented equation. Unoriented equations can be handled by adopting the Unfailing Knuth-Bendix method [35, 11], which gives a ground confluent set of equations.

Many completion procedures, related to Knuth-Bendix to different extents, have been designed. They include procedures for equational theories with special sets of axioms [54, 41, 7], Horn logic with equality [50, 27], first order logic [31, 32, 46, 8], first order logic with equality [33, 34, 36, 37, 57, 60, 12, 13], inductive theorem proving in equational and Horn theories [40, 29, 50] and logic programming [20, 21, 22, 15]. Surveys have been given in [24, 25].

A *completion procedure* is composed of *inference rules* and a *search plan*. The inference rules determine what can be derived from given data. The search plan chooses at each step of the derivation which inference rule to apply to which data and therefore it determines the unique derivation that the procedure computes from a given input.

The interpretation of Knuth-Bendix completion as generator of confluent systems is by far the most well known one, whereas theorem proving is basically regarded as a side effect of the generation of a confluent system. This view of completion is not acceptable from the theorem proving perspective, because a procedure which is guaranteed to eventually generate a confluent system cannot be efficient as theorem prover. In this work we reverse the traditional way of presenting completion procedures: we present them as semidecision procedures with the generation of confluent systems as a special side effect.

The interpretation of completion as semidecision procedure appeared first in [39]. Huet proved that if the search plan is fair, the limit of an unfailing Knuth-Bendix derivation is a confluent rewrite system and, as a consequence, if a theorem $s \simeq t$ is given to the procedure, it semidecides the validity of $s \simeq t$. The same result was obtained in a more general framework in [6].

We decouple the interpretation of completion as semidecision procedure from the interpretation of completion as generator of confluent systems. We prove that if the inference rules are *refutationally complete* and the search plan is *fair*, a completion procedure is a *semidecision procedure*. *Refutational completeness* means that for all unsatisfiable inputs, there exist successful derivations by the inference rules of the strategy. *Fairness* means that whenever successful derivations exist, the search plan guarantees that the computed derivation is successful, that is all the inference steps which are necessary to prove the goal are eventually done. In particular, all the critical pairs which are necessary to prove the goal are eventually considered. We give a new definition of fairness to capture this concept.

This notion of fairness is the key difference between completion for theorem proving and completion for the generation of confluent systems. In Huet's landmark paper [39] and in all the following work on completion [6, 9, 57, 13], fairness of a derivation consists in eventually considering all critical pairs. We call this property *uniform fairness* in order to distinguish it from fairness for theorem proving. Uniform fairness is necessary for the limit of a derivation to be confluent, but it is not necessary for theorem proving, because not all the critical pairs are necessary to prove a given theorem. All the definitions of fairness of completion procedures appeared so far in the literature [39, 9, 57, 13] require uniform fairness, because they do not separate theorem proving from the generation of a confluent system.

Therefore, we have proved Huet's classical result from weaker, strictly theorem proving oriented hypotheses, which do not imply any confluence property of the limit of the derivation, since such properties are not necessary for theorem proving. In our view, a completion procedure is first a theorem proving procedure, which also has the property of eventually generating confluent sets if it is uniformly fair.

The so called *inductionless induction* method is covered by the semidecision concept as well: completion for inductionless induction [40] is a *semidecision procedure for disproving inductive theorems.*

We conclude our work by presenting some completion procedures for equational logic: we show that the basic *Unfailing Knuth-Bendix procedure* [35, 11] and some of its extensions, such as the *AC-UKB procedure* [54, 41, 7, 1] with *Cancellation laws* [36], the *S-strategy* [35] and the *Inequality Ordered Saturation strategy* [3] fit nicely in our framework. To our knowledge, this is the first presentation of these extensions of the UKB procedure as sets of inference rules.

Our entire approach to completion procedures is coherently based on a notion of *proof reduction.* In theorem proving, one wants to reduce one single proof, the proof of the target theorem: the derivation halts successfully if the target has been reduced to some trivially true theorem, such as $s \simeq s$, whose proof is *empty.* In traditional completion, one wants to reduce all proofs: for instance, in Knuth-Bendix completion all equational proofs have to be reduced to rewrite proofs. In this proof reduction framework, fairness guarantees that the proof of the target is eventually reduced, whereas uniform fairness guarantees that all proofs are eventually reduced.

In order to formalize all concepts in terms of proof reduction, we need a notion of well founded *proof ordering.* Our starting point is the proof orderings approach originally given in [6, 9]. However, proof orderings as in [6] do not apply to a theorem proving derivation, because they allow to compare only two proofs of the same theorem. In theorem proving, the target is modified by inference steps applying to the theorem itself. Therefore, we give a new notion of proof ordering, where proofs of different theorems can be compared.

In this paper we concentrate on theorem proving. In [16], we compare our study of fairness with previous definitions of fairness in [9, 57, 13]. In [17], we complete our framework with a full generalization of Huet's theorem in [39] and of its extensions in [50, 13] for uniformly fair derivations.

In the following we assume that the reader is familiar with basic concepts and notations about theorem proving, term rewriting systems, completion procedures and orderings. We refer to [25, 26] for basic definitions and notations.

2 Completion procedures for theorem proving

2.1 Proof orderings for theorem proving

Given a finite set of sentences S, we denote by $Th(S)$ the *theory* of S, $Th(S) = \{\varphi | S \models \varphi\}$, and we say that S is a *presentation* of the theory $Th(S)$. The input for a theorem proving procedure is a pair $(S; \varphi)$, where S is a presentation of a theory and φ the *target.* A *theorem proving problem* is to decide whether $\varphi \in Th(S)$ and a *theorem proving derivation* is a sequence of deductions

$$(S_0; \varphi_0) \vdash (S_1; \varphi_1) \vdash \ldots \vdash (S_i; \varphi_i) \vdash \ldots,$$

where at each step the problem of deciding $\varphi_i \in Th(S_i)$ reduces to the problem of deciding $\varphi_{i+1} \in Th(S_{i+1})$. A step $(S_i; \varphi_i) \vdash (S_{i+1}; \varphi_i)$, where the presentation is modified, is a *forward reasoning* step. A step $(S_i; \varphi_i) \vdash (S_i; \varphi_{i+1})$, where the target is modified, is a *backward reasoning* step, which derives a new goal from the current one. Informally, the derivation halts successfully at stage k if

$\varphi_k \in Th(S_k)$ is trivially true and therefore it can be asserted that $\varphi_0 \in Th(S_0)$. In this section we introduce a notion of *proof ordering*, which allows us to describe a theorem proving derivation as a *proof reduction* process.

We denote proofs by capital Greek letters: $\Upsilon(S,\varphi)$ denotes a proof of φ from axioms in S. Proofs are often represented as trees whose nodes are labeled by sentences: the tree associated to $\Upsilon(S,\varphi)$ has φ as label of the root, elements in S as labels of the leaves and a node ψ has children $\psi_1 \ldots \psi_n$ if ψ is derived from $\psi_1 \ldots \psi_n$ by a step in $\Upsilon(S,\varphi)$. In equational logic, such a proof can also be represented as a chain [6]

$$s_1 \leftrightarrow_{l_1 \simeq r_1} s_2 \leftrightarrow_{l_2 \simeq r_2} \ldots \leftrightarrow_{l_{n-1} \simeq r_{n-1}} s_n,$$

where $s_1 \leftrightarrow_{l_1 \simeq r_1} s_2$ means that the equality of s_1 and s_2 is established by the equation $l_1 \simeq r_1$ because s_1 and s_2 are $c[l_1\sigma]$ and $c[r_1\sigma]$ for some context c and substitution σ. We write $s \rightarrow_{l \simeq r} t$ if $s \succ t$ is known a priori.

An ordering on proofs is defined in general starting from some ordering on the data involved in the proofs. We recall that a *simplification ordering* on terms is a monotonic and stable ordering, i.e. $s \succ t$ implies $c[s\sigma] \succ c[t\sigma]$ for all contexts c and substitutions σ, with the property that a term is greater than any of its subterms. A simplification ordering is well founded. A *complete simplification ordering* is also total on the set of ground terms. Some well known simplification orderings are the recursive path ordering [19], the lexicographic path ordering [45] and the Knuth-Bendix ordering [49]. Such orderings are surveyed in [23]. Our first basic assumption is to have a complete simplification ordering $\succ$ on terms. We prefer to have a simplification ordering, even if a well founded, monotonic and stable ordering total on ground terms is sufficient. Given a complete simplification ordering $\succ$ on terms, it is possible to define complete simplification orderings on equations, clauses and sets of clauses based on $\succ$, as shown for instance in [37].

A *proof ordering* is a monotonic, stable and well founded ordering on proofs [6]. As an example we give the following proof ordering from [27]:

Example 2.1 *A proof ordering to compare two ground equational proofs $\Upsilon(E, s \simeq t) = s \leftrightarrow^*_E t$ and $\Upsilon'(E', s \simeq t) = s \leftrightarrow^*_{E'} t$, can be defined as follows. We associate to a ground equational step $s \leftrightarrow_{l \simeq r} t$ the triple (s,l,t), if $s \succ t$. We compare these triples by the lexicographic combination $>^e$ of the complete simplification ordering $\succ$, the strict encompassment ordering $\rhd$ and again the ordering $\succ$. The encompassment ordering $\trianglerighteq$ is the composition of the subterm ordering and the subsumption ordering: $t \trianglerighteq s$ if $t|u = s\sigma$ for some position u and substitution σ; $t \rhd s$ if $t \trianglerighteq s$ and $s \neq t$ [25]. Then we compare two proofs $\Upsilon(E, s \simeq t)$ and $\Upsilon'(E', s \simeq t)$ by the multiset extension $>^e_{mul}$ of $>^e$.*

The proof orderings defined in [6] allow us to compare only two proofs $\Upsilon(S,\varphi)$ and $\Upsilon'(S',\varphi)$ of the same theorem φ in different presentations S and S' of a theory. This notion of proof ordering is not suitable for theorem proving, because in a theorem proving derivation

$$(S_0;\varphi_0) \vdash (S_1;\varphi_1) \vdash \ldots \vdash (S_i;\varphi_i) \vdash \ldots$$

both the presentation and the target are transformed. In order to compare the proof of φ_i in S_i and the proof of φ_{i+1} in S_{i+1}, we need a proof ordering such that two proofs $\Upsilon(S,\varphi)$ and $\Upsilon'(S',\varphi')$ may be comparable. Proof orderings with this property do exist and can actually be obtained quite easily. For instance the proof ordering of the previous example can be transformed into a proof ordering for proofs of different theorems as follows:

Example 2.2 *We can compare any two ground equational proofs $\Upsilon(E, s \simeq t) = s \leftrightarrow^*_E t$ and $\Upsilon'(E', s' \simeq t') = s' \leftrightarrow^*_{E'} t'$ by comparing the pairs $(\{s,t\}, s \leftrightarrow^*_E t)$ and $(\{s',t'\}, s' \leftrightarrow^*_{E'} t')$ by the lexico-*

graphic combination $>_u$ *of the multiset extension* $\succ_{mul}$ *of the ordering* $\succ$ *on terms and the multiset extension* $>^e{}_{mul}$ *of* $>^e$.

Henceforth a *proof ordering* is a monotonic, stable, well founded ordering on proofs. The minimum proof is the *empty proof.* We denote by *true* the theorem whose proof is empty. For instance in equational logic, *true* is a trivial equality $s \simeq s$. Given a pair $(S;\varphi)$, we can select a minimal proof among all proofs of φ from S:

Definition 2.1 *Given a proof ordering* $>_p$, *we denote by* $\Pi(S,\varphi)$ *a minimal proof of* φ *from* S *with respect to* $>_p$, *i.e. a proof such that for all proofs* $\Upsilon(S,\varphi)$ *of* φ *from* S, $\Upsilon(S,\varphi) \not<_p \Pi(S,\varphi)$.

Having introduced this notion of proof ordering, we can regard a theorem proving derivation

$$(S_0;\varphi_0) \vdash (S_1;\varphi_1) \vdash \ldots \vdash (S_i;\varphi_i) \vdash \ldots,$$

as a process of reducing $\Pi(S_0,\varphi_0)$ to the empty proof and φ_0 to *true*. At each step $\Pi(S_i,\varphi_i)$ is replaced by $\Pi(S_{i+1},\varphi_{i+1})$ and the derivation halts successfully at stage k if $\Pi(S_k,\varphi_k)$ is empty and φ_k is *true*.

Our generalization of the classical notion of proof orderings is more significant than it may seem at a first glance. Proof orderings were introduced in [6] to prove correctness of the Knuth-Bendix completion procedure as a procedure which generates possibly infinite, confluent term rewriting systems. A derivation by Knuth-Bendix completion in that context is a process of transforming a presentation

$$S_0 \vdash S_1 \vdash \ldots \vdash S_i \vdash \ldots.$$

In other words, it is a purely forward derivation. Since the purpose of such a derivation is to transform a presentation, it is sufficient to be able to compare $\Pi(S_i,\varphi)$ and $\Pi(S_{i+1},\varphi)$ for any theorem φ in the theory.

This is not the case in theorem proving, since the purpose of a derivation is to prove a specific theorem. Theorem proving requires *backward reasoning*, since a theorem proving problem includes a target. Furthermore, backward reasoning is necessary to obtain a *target-oriented* and therefore presumably efficient procedure. The classical proof orderings approach does not apply to theorem proving because it does not provide for backward reasoning. On the other hand, our proof orderings approach applies to both theorem proving and traditional completion.

2.2 Inference rules and search plans

Since completion procedures are theorem proving strategies with special properties, we start by introducing some basic concepts about theorem proving strategies.

A *theorem proving strategy* is a pair $\mathcal{P} = < I;\Sigma >$, where I is a set of *inference rules* and Σ is a *search plan.* Inference rules in I decide what consequences can be deduced from the available data and Σ decides which inference rule and which data to choose next. We discuss first the inference rules and next the search plan. The general form of an inference rule f is:

$$f\colon \frac{S}{S'}$$

where S and S' are sets of sentences. The rule says that given S, the set S' can be inferred. We distinguish between *expansion* inference rules and *contraction* inference rules, as they are called in [27]. An expansion inference rule expands a given set S into a new set S' by deriving new sentences from sentences in S:

$f\colon \frac{S}{S'}$ where $S \subset S'$.

A contraction inference rule contracts a given set S into a new set S' by either deleting some sentences in S or replacing them by others:

$f\colon \frac{S}{S'}$ where $S \not\subseteq S'$.

Different schemes for inference rules, called *deduction* and *deletion*, are given in [13]. We further distinguish between inference rules which transform the presentation and inference rules which transform the target. We assume that targets are clauses and therefore can be regarded as sets of literals:

- *Presentation inference rules:*
  - *Expansion inference rules:* $f\colon \frac{(S;\varphi)}{(S';\varphi)}$ where $S \subset S'$.
  - *Contraction inference rules:* $f\colon \frac{(S;\varphi)}{(S';\varphi)}$ where $S \not\subseteq S'$.
- *Target inference rules:*
  - *Expansion inference rules:* $f\colon \frac{(S;\varphi)}{(S;\varphi')}$ where $\varphi \subset \varphi'$.
  - *Contraction inference rules:* $f\colon \frac{(S;\varphi)}{(S;\varphi')}$ where $\varphi \not\subseteq \varphi'$.

Example 2.3 Deduction *of a critical pair in Unfailing Knuth-Bendix completion is an expansion inference rule on the presentation, since it adds to the given set a new equation:*

$$\frac{(E \cup \{p \simeq q, l \simeq r\}; \hat{s} \simeq \hat{t})}{(E \cup \{p \simeq q, l \simeq r, p[r]_u\sigma \simeq q\sigma\}; \hat{s} \simeq \hat{t})} \quad \begin{array}{l} p|u \notin X \\ p\sigma \not\preceq q\sigma, p[r]_u\sigma \end{array} \quad (p|u)\sigma = l\sigma$$

where X is the set of variables, σ is a most general unifier and $\succ$ is the assumed complete simplification ordering on terms. Simplification *of the target is a contraction inference rule:*

$$\frac{(E \cup \{l \simeq r\}; \hat{s} \simeq \hat{t})}{(E \cup \{l \simeq r\}; \hat{s}[r\sigma]_u \simeq \hat{t})} \quad \begin{array}{l} \hat{s}|u = l\sigma \\ \hat{s} \succ \hat{s}[r\sigma]_u \end{array}$$

The inference rules are required to be sound. A presentation inference rule is *sound* if $Th(S') \subseteq Th(S)$, since an application of a presentation rule involves the presentation only. A target inference rule is *sound* if $Th(S \cup \{\varphi'\}) \subseteq Th(S \cup \{\varphi\})$.

Deduction by presentation rules is deduction of consequences from the axioms or *forward reasoning*. A target rule applies to both the presentation and the target to infer a new target: deduction by target rules is *backward reasoning.* The *Knuth-Bendix procedure* computing critical pairs and simplifying rewrite rules to obtain a canonical system performs exclusively forward reasoning. If a theorem to be proved is given as target, the procedure also performs some backward reasoning, when it simplifies the target.

Finally, a search plan Σ decides which inference rule should be applied to what data at any given step during a derivation. It may set a *precedence* on the inference rules and a *well founded ordering* on data and proceed accordingly:

Example 2.4 *A* Simplification-first *search plan [35] for Unfailing Knuth-Bendix completion is a search plan where Simplification has priority over Deduction. Therefore Deduction is considered only if Simplification does not apply to any equation. Equations can be sorted by the multiset extension $\succ_{mul}$ of the ordering on terms, or by size, or by age such as in a first-in first-out plan.*

2.3 Completion procedures

A completion procedure has then three components $< I_p, I_t; \Sigma >$, where I_p is the set of presentation inference rules, I_t is the set of target inference rules and Σ is the search plan.

A derivation by a completion procedure is a process of *proof reduction*. A target inference step modifies the target and therefore it affects the proof of the target. We require that the proof of the target is reduced:

Definition 2.2 *A target inference step $(S;\varphi) \vdash (S;\varphi')$ is* proof-reducing *if $\Pi(S,\varphi) \geq_p \Pi(S,\varphi')$. It is* strictly proof-reducing *if $\Pi(S,\varphi) >_p \Pi(S,\varphi')$.*

Example 2.5 *Simplification of the target as given in Example 2.3 is strictly proof-reducing. We assume the proof ordering $>_u$ introduced in Example 2.2. We have $\{\hat{s},\hat{t}\} \succ_{mul} \{\hat{s}',\hat{t}'\}$, since $\hat{s} \succ \hat{s}'$ and $\hat{t} = \hat{t}'$, assuming $\hat{s}$ is simplified to $\hat{s}'$. Therefore $\Pi(E, \hat{s} \simeq \hat{t}) >_u \Pi(E, \hat{s}' \simeq \hat{t}')$.*

For a presentation inference step we allow more flexibility:

Definition 2.3 *Given two pairs $(S;\varphi)$ and $(S';\varphi')$, the relation $(S;\varphi) \triangleright_{p,\mathcal{T}} (S';\varphi')$ holds if*

1. *either $\Pi(S,\varphi) >_p \Pi(S',\varphi')$*
2. *or*
   (a) *$\Pi(S,\varphi) = \Pi(S',\varphi')$,*
   (b) *$\forall\psi \in \mathcal{T}$, $\Pi(S,\psi) \geq_p \Pi(S',\psi)$ and*
   (c) *$\exists\psi \in \mathcal{T}$ such that $\Pi(S,\psi) >_p \Pi(S',\psi)$.*

Definition 2.4 *A presentation inference step $(S;\varphi) \vdash (S';\varphi)$ is* proof-reducing *on $\mathcal{T}$ if $(S;\varphi) \triangleright_{p,\mathcal{T}} (S';\varphi)$ holds. It is* strictly proof-reducing *if $\Pi(S,\varphi) >_p \Pi(S',\varphi)$.*

The condition $(S_i;\varphi_i) \triangleright_{p,\mathcal{T}} (S_{i+1};\varphi_{i+1})$ says that either the step reduces the proof of the target, or it reduces the proof of at least one theorem in $\mathcal{T}$, while it does not increase the proof of any theorem in $\mathcal{T}$. A step which reduces the proof of the target is proof-reducing, regardless of its effects on other theorems. On the other hand, an inference step on the presentation may not immediately decrease the proof of the target and still be necessary to decrease it eventually. Such a step is proof-reducing too, if it does not increase any proof and strictly decreases at least one.

Example 2.6 *Deduction of a critical pair as given in Example 2.3 is proof-reducing. We assume the proof ordering $>_u$ introduced in Example 2.2. Given two equations $l \simeq r$ and $p \simeq q$, a* critical overlap *of $l \simeq r$ and $p \simeq q$ is any proof $s \leftarrow_{l \simeq r} v \rightarrow_{p \simeq q} t$, where v is $c[p\tau]$ for some context c and substitution τ and $(p|u)\tau = l\tau$ for some non variable subterm $p|u$ of p. The Deduction rule applied to $l \simeq r$ and $p \simeq q$ generates the critical pair $p[r]_u\sigma \simeq q\sigma$, where σ is the mgu of $p|u$ and l and therefore $\tau = \sigma\rho$ for some substitution ρ. Such a Deduction step affects a minimal proof by replacing any occurrence of the critical overlap $s \leftarrow_{l \simeq r} v \rightarrow_{p \simeq q} t$ by the equational step $s \leftrightarrow_{p[r]_u\sigma \simeq q\sigma} t$, justified by the critical pair. We have $\{(v,l,s),(v,p,t)\} >^e_{mul} \{(s,p[r]_u\sigma,t)\}$ or $\{(v,l,s),(v,p,t)\} >^e_{mul} \{(t,q\sigma,s)\}$, depending on whether $s \succ t$ or $t \succ s$, since $v \succ s,t$. Therefore $\Pi(E,\psi) >_u \Pi(E \cup \{p[r]_u\sigma \simeq q\sigma\},\psi)$ if the minimal proof of ψ in E contains a critical overlap between $l \simeq r$ and $p \simeq q$, $\Pi(E,\psi) = \Pi(E \cup \{p[r]_u\sigma \simeq q\sigma\},\psi)$ otherwise.*

This notion of proof reduction applies to presentation inference steps which are either expansion steps or contraction steps which replace some sentences by others. A contraction step which deletes sentences without adding any cannot reduce any minimal proof. In order to characterize these steps, we introduce a notion of *redundancy*:

Definition 2.5 *A sentence φ is* redundant *in S on domain $\mathcal{T}$ if $\forall\psi \in \mathcal{T}$, $\Pi(S,\psi) = \Pi(S\cup\{\varphi\},\psi)$.*

A sentence is redundant in a presentation if adding it to the presentation does not affect any minimal proof.

Example 2.7 *An inference rule of Unfailing Knuth-Bendix completion, which deletes an equation without adding any is* Functional subsumption*:*

$$\frac{(E\cup\{p\simeq q, l\simeq r\}; \hat{s}\simeq\hat{t})}{(E\cup\{l\simeq r\}; \hat{s}\simeq\hat{t})}\ (p\simeq q)\blacktriangleright(l\simeq r)$$

where $(p\simeq q)\blacktriangleright(l\simeq r)$ means that $p = c[l\sigma]$ and $q = c[r\sigma]$ for some context c and substitution σ, where either c is not empty or σ is not a renaming of variables. An equation $p\simeq q$ subsumed by $l\simeq r$ is redundant according to the proof ordering $>^e{}_{mul}$ and therefore to the proof ordering $>_u$ as defined in Example 2.2. No minimal proof contains a step $s\leftrightarrow_{p\simeq q} t$ since the step $s\leftrightarrow_{l\simeq r} t$ is smaller: either $\{(s,p,t)\}>^e{}_{mul}\{(s,l,t)\}$ or $\{(t,q,s)\}>^e{}_{mul}\{(t,r,s)\}$, depending on whether $s\succ t$ or $t\succ s$, since $p\blacktriangleright l$ and $q\blacktriangleright r$.

A notion of redundant clauses was introduced in [57] and in [13], where the term "redundant" was first used. We show in [16] that redundant clauses according to [57] and [13] are also redundant in our sense.

Definition 2.6 *An inference step $(S;\varphi)\vdash(S';\varphi')$ is* reducing *on $\mathcal{T}$ if either it is proof-reducing on $\mathcal{T}$ or it deletes a sentence which is redundant in S on domain $\mathcal{T}$.*

Definition 2.7 *An inference rule f is* reducing *if all the inference steps $(S;\varphi)\vdash_f(S';\varphi')$ where f is applied are reducing.*

We have finally all the elements to define a completion procedure:

Definition 2.8 *A theorem proving strategy $\mathcal{C} =< I_p, I_t; \Sigma >$ is a* completion procedure *on domain $\mathcal{T}$ if for all pairs $(S_0;\varphi_0)$, where S_0 is a presentation of a theory and $\varphi_0\in\mathcal{T}$, the derivation*

$(S_0;\varphi_0)\vdash_{\mathcal{C}}(S_1;\varphi_1)\vdash_{\mathcal{C}}\ldots\vdash_{\mathcal{C}}(S_i;\varphi_i)\vdash_{\mathcal{C}}\ldots$

has the following properties:

- *monotonicity: $\forall i\geq 0$, $Th(S_{i+1})\subseteq Th(S_i)$,*
- *relevance: $\forall i\geq 0$, $\varphi_{i+1}\in Th(S_{i+1})$ if and only if $\varphi_i\in Th(S_i)$ and*
- *reduction: $\forall i\geq 0$, the step $(S_i;\varphi_i)\vdash_{\mathcal{C}}(S_{i+1};\varphi_{i+1})$ is reducing on $\mathcal{T}$.*

The *domain $\mathcal{T}$* is the set of sentences where the inference rules of the completion procedure are reducing. For instance, for the *Knuth-Bendix completion procedure $\mathcal{T}$* is the set of all equations. For the *Unfailing Knuth-Bendix procedure*, $\mathcal{T}$ is the set of all ground equations.

The *monotonicity* and *relevance* properties establish the soundness of the presentation and the target inference rules respectively. Monotonicity ensures that a presentation inference step does not create new elements which are not true in the theory, while relevance ensures that a target inference step replaces the target by a new target in such a way that proving the latter is equivalent to proving the former. For instance, a simplification step which reduces a target φ to φ' satisfies the relevance requirement because if φ' is true, φ is true as well. An interesting expansion inference rule for the target, called *Ordered saturation* will be described in detail in Section 3.3.

Reduction is the property which characterizes completion procedures. Clearly, if all the inference rules of a procedure are reducing, the procedure has the reduction property. We shall see in the second part that the inference rules of the known equational completion procedures are reducing. Most inference rules are reducing because they are suitably restricted by the complete simplification ordering $\succ$ on terms. A complete simplification ordering on data turns out to be a key element in characterizing a theorem proving strategy as a completion procedure.

2.4 Completion procedures as semidecision procedures

Given an input pair $(S_0;\varphi_0)$, a completion procedure works by reducing the proof $\Pi(S_0,\varphi_0)$. If the proof of the target is minimal, the process halts. Since the empty proof is smaller than any other proof, the computation halts at stage k if $\Pi(S_k,\varphi_k)$ is empty and φ_k is *true*.

A procedure is *complete* if, whenever φ_0 is a theorem of S_0, the derivation from $(S_0;\varphi_0)$ reduces φ_0 to *true* and halts. Completeness involves both the inference rules and the search plan. First, it requires that whenever $\varphi_0 \in Th(S_0)$, there exist successful derivations by the inference rules of the procedure. Second, it requires that whenever successful derivations exist, the search plan guarantees that the computed derivation is successful. We call these two properties *refutational completeness* of the inference rules and *fairness* of the search plan respectively.

In order to describe them, we introduce a structure called *I-tree*. Given a theorem proving problem $(S_0;\varphi_0)$ and a set of inference rules I, the application of I to $(S_0;\varphi_0)$ defines a tree, the *I-tree rooted at* $(S_0;\varphi_0)$. The nodes of the tree are labeled by pairs $(S;\varphi)$. The root is labeled by the input pair $(S_0;\varphi_0)$. A node $(S;\varphi)$ has a child $(S';\varphi')$ if $(S';\varphi')$ can be derived from $(S;\varphi)$ in one step by an inference rule in I. The I-tree rooted at $(S_0;\varphi_0)$ represents all the possible derivations by the inference rules in I starting from $(S_0;\varphi_0)$.

A set I of inference rules is *refutationally complete* if whenever $\varphi_0 \in Th(S_0)$, the I-tree rooted at $(S_0;\varphi_0)$ contains successful nodes, nodes of the form $(S;true)$. More precisely, we define completeness as follows:

Definition 2.9 *A set* $I = I_p \cup I_t$ *of inference rules is* refutationally complete *if whenever* $\varphi \in Th(S)$ *and* $\Pi(S,\varphi)$ *is not minimal, there exist derivations*

$$(S;\varphi)\vdash_I(S_1;\varphi_1)\vdash_I\ldots\vdash_I(S';\varphi')$$

such that $\Pi(S,\varphi) >_p \Pi(S',\varphi')$.

A set of inference rules is refutationally complete if it can reduce the proof of the target whenever it is not minimal. Since a proof ordering is well founded, it follows that if $\varphi \in Th(S)$, the I-tree rooted at $(S;\varphi)$ contains successful nodes. The advantage of giving the definition of completeness in terms of proof reduction is that the problem of proving completeness of I is reduced to the problem of exhibiting a suitable proof ordering.

Given a completion procedure $\mathcal{C} =< I_p, I_t; \Sigma >$, $I = I_p \cup I_t$, the I-tree rooted at $(S_0;\varphi_0)$ represents the entire search space that the procedure can potentially derive from the input $(S_0;\varphi_0)$. The search plan Σ selects a path in the I-tree: the derivation from input $(S_0;\varphi_0)$ controlled by Σ is the path selected by Σ in the I-tree rooted at $(S_0;\varphi_0)$. Once both a set of inference rules and a search plan are given, the derivation from $(S_0;\varphi_0)$ is unique. A pair $(S_i;\varphi_i)$ reached at stage i of the derivation is a *visited node* in the I-tree. Each visited node $(S_i; \varphi_i)$ has generally many children, but the search plan selects only one of them to be $(S_{i+1}; \varphi_{i+1})$. A search plan Σ is *fair* if whenever the I-tree rooted at $(S_0; \varphi_0)$ contains successful nodes, the derivation controlled by Σ starting at $(S_0; \varphi_0)$ is guaranteed to reach a successful node. Similar to completeness, we define fairness in terms of proof reduction:

Definition 2.10 *A derivation*

$$(S_0;\varphi_0)\vdash_C(S_1;\varphi_1)\vdash_C\ldots\vdash_C(S_i;\varphi_i)\vdash_C\ldots$$

controlled by a search plan Σ is fair *if and only if for all $i \geq 0$, if there exists a path*

$$(S_i;\varphi_i)\vdash_I\ldots\vdash_I(S';\varphi')$$

in the I-tree rooted at $(S_0;\varphi_0)$ such that $\Pi(S_i,\varphi_i) >_p \Pi(S',\varphi')$, then there exists an $(S_j;\varphi_j)$ for some $j > i$, such that $\Pi(S',\varphi') \geq_p \Pi(S_j,\varphi_j)$. A search plan Σ is fair *if all the derivations controlled by Σ are fair.*

If the inference rules allow to reduce the proof of the target at $(S_i; \varphi_i)$, a fair search plan guarantees that the proof of the target will be indeed reduced at a later stage $(S_j; \varphi_j)$.

If the inference rules are complete and the search plan is fair, a completion procedure on domain $\mathcal{T}$ is a *semidecision procedure* for $Th(S) \cap \mathcal{T}$ for all presentations S:

Theorem 2.1 *Let $C =< I_p, I_t; \Sigma >$ be a completion procedure. If the set $I = I_p \cup I_t$ of inference rules is refutationally complete and the search plan Σ is fair, then for all derivations*

$$(S_0;\varphi_0)\vdash_C(S_1;\varphi_1)\vdash_C\ldots\vdash_C(S_i;\varphi_i)\vdash_C\ldots,$$

where $\varphi_0 \in Th(S_0)$, $\forall i \geq 0$, if $\Pi(S_i,\varphi_i)$ is not minimal, then there exists an (S_j,φ_j), for some $j > i$, such that $\Pi(S_i,\varphi_i) >_p \Pi(S_j,\varphi_j)$.

Proof: if $\Pi(S_i, \varphi_i)$ is not minimal, then by completeness of the inference rules, there exists a path $(S_i;\varphi_i)\vdash_I\ldots\vdash_I(S';\varphi')$ such that $\Pi(S_i,\varphi_i) >_p \Pi(S',\varphi')$. By fairness of the search plan, there exists an $(S_j;\varphi_j)$, for some $j > i$, such that $\Pi(S_i;\varphi_i) >_p \Pi(S',\varphi') \geq_p \Pi(S_j,\varphi_j)$. □

Corollary 2.1 *If a completion procedure C on domain $\mathcal{T}$ has refutationally complete inference rules and fair search plan, then for all inputs $(S_0;\varphi_0)$, if $\varphi_0 \in Th(S_0)$ then*

- *the derivation $(S_0;\varphi_0)\vdash_C(S_1;\varphi_1)\vdash_C\ldots\vdash_C(S_i;\varphi_i)\vdash_C\ldots$ halts at stage k for some $k \geq 0$ and*
- *$\varphi_k = true$.*

Proof: if $\varphi_0 \in Th(S_0)$, the derivation halts at some stage k by Theorem 2.1 and the well foundedness of $>_p$. Therefore, the proof $\Pi(S_k, \varphi_k)$ is minimal. Since we assume a proof ordering such that any two proofs can be compared, the only minimal proof is the empty proof and φ_k is *true*. □

In the following, we often write that a completion procedure is *complete* as a short hand for a completion procedure with complete inference rules and fair search plan.

2.5 Completion procedures as generators of decision procedures

In this paper we regard a completion procedure as a theorem proving procedure. In [17] we extend our framework to include completion procedures as *generators of decision procedures*.

If the search plan of a completion procedure satisfies a stronger fairness property, which we call *uniform fairness*, the procedure generates a possibly infinite *saturated set*. Uniform fairness is the fairness property which has been required so far for completion procedures [39, 9, 57, 13]. It basically consists in eventually considering all the inference steps. Saturated set is a generalization of confluent system: a set is saturated if no non-trivial consequences can be added [50, 13]. In [17], we define both uniform fairness and saturated set in terms of our notion of redundancy and we show that our definitions are equivalent to those given in [50, 13].

If a presentation is saturated, the derivations from that presentation are *linear input* derivations [18], that is derivations where each inference step applies to the goal to be proved. If linear input derivations from a saturated set are guaranteed to be well founded, a saturated set is a *decision procedure* and the completion procedure is a *generator of decision procedures.*

The well foundedness of the derivations is implied by additional requirements, which depend on the logic. In equational logic, a derivation $s \rightarrow^* \circ \leftarrow^* t$ made only of well founded simplification steps by a confluent rewrite system is a well founded linear input derivation and a confluent system is a decision procedure. Sufficient conditions for well-foundedness of derivations for ground targets in Horn logic with equality are also known [50, 13].

In [17], we give a full generalization of the classical results in [39], which covers also the extensions to Horn logic with equality in [50, 13].

Very few theories have a finite saturated presentation and even fewer satisfy the additional requirement for a saturated presentation to be a decision procedure. Therefore, the interpretation of completion as semidecision procedure which we have developed here is more useful in practice.

3 Completion procedures in equational logic

In the second part of this work we give a new presentation of some Knuth-Bendix type completion procedures for equational logic, in the framework developed so far.

3.1 Unfailing Knuth-Bendix completion

The *Unfailing Knuth-Bendix procedure* [35, 11] is a semidecision procedure for equational theories. A presentation is a set of equations E_0 and a theorem is an equational theorem $\forall \bar{x} s_0 \simeq t_0$. A derivation by UKB has the form

$$(E_0; \hat{s}_0 \simeq \hat{t}_0) \vdash_{UKB} (E_1; \hat{s}_1 \simeq \hat{t}_1) \vdash_{UKB} \ldots (E_i; \hat{s}_i \simeq \hat{t}_i) \vdash_{UKB} \ldots$$

where we denote by $\hat{s}_0 \simeq \hat{t}_0$ an equality which contains only universally quantified variables and therefore can be regarded as a ground equality. A derivation halts at stage k if $\hat{s}_k$ and $\hat{t}_k$ are identical. We assume that $\succ$ is a complete simplification ordering such that $\forall s, true \prec s$ and we extend the encompassment ordering to equations: $(p \simeq q) \trianglerighteq (l \simeq r)$ if $p|u = l\sigma$ and $q|u = r\sigma$, $(p \simeq q) \triangleright (l \simeq r)$ if $(p \simeq q) \trianglerighteq (l \simeq r)$ but $(p \simeq q) \neq (l \simeq r)$. At each step of the completion process the pair $(E_{i+1}; \hat{s}_{i+1} \simeq \hat{t}_{i+1})$ is derived from the pair $(E_i; \hat{s}_i \simeq \hat{t}_i)$ by applying one of the following inference rules:

- *Presentation inference rules*:
  - *Simplification*:

$$\frac{(E \cup \{p \simeq q, l \simeq r\}; \hat{s} \simeq \hat{t})}{(E \cup \{p[r\sigma]_u \simeq q, l \simeq r\}; \hat{s} \simeq \hat{t})} \quad \begin{matrix} p|u = l\sigma \\ p \triangleright l \vee q \succ p[r\sigma]_u \end{matrix} \qquad p \succ p[r\sigma]_u$$

  - *Deduction*:

$$\frac{(E \cup \{p \simeq q, l \simeq r\}; \hat{s} \simeq \hat{t})}{(E \cup \{p \simeq q, l \simeq r, p[r]_u\sigma \simeq q\sigma\}; \hat{s} \simeq \hat{t})} \quad \begin{array}{ll} p|u \notin X & (p|u)\sigma = l\sigma \\ p\sigma \not\preceq q\sigma, p[r]_u\sigma & \end{array}$$

- *Deletion:*

$$\frac{(E \cup \{l \simeq l\}; \hat{s} \simeq \hat{t})}{(E; \hat{s} \simeq \hat{t})}$$

- *Functional subsumption:*

$$\frac{(E \cup \{p \simeq q, l \simeq r\}; \hat{s} \simeq \hat{t})}{(E \cup \{l \simeq r\}; \hat{s} \simeq \hat{t})} \quad (p \simeq q) \blacktriangleright (l \simeq r)$$

- *Target inference rules:*

  - *Simplification:*

$$\frac{(E \cup \{l \simeq r\}; \hat{s} \simeq \hat{t})}{(E \cup \{l \simeq r\}; \hat{s}[r\sigma]_u \simeq \hat{t})} \quad \begin{array}{l} \hat{s}|u = l\sigma \\ \hat{s} \succ \hat{s}[r\sigma]_u \end{array}$$

  - *Deletion:*

$$\frac{(E; \hat{s} \simeq \hat{s}\})}{(E; true)}$$

The main inference rule of UKB is *Simplification.* A simplification step consists in applying an equation in E_i to simplify either another equation in E_i or the goal $\hat{s}_i \neq \hat{t}_i$. The step is performed only if $p \succ p[r\sigma]_u$, that is a term is replaced by a smaller term. The condition $p \blacktriangleright l \vee q \succ p[r\sigma]_u$ is explained as follows. If $q \succ p$, that is simplification applies to the smaller side of $p \simeq q$, the condition $q \succ p[r\sigma]_u$ is trivially satisfied and no other restriction is needed. If $q \not\succ p$, simplification applies to the greater side of $p \simeq q$ or $p \simeq q$ is not ordered. If $p \blacktriangleright l$, either p is a proper instance of l or l matches a proper subterm of p. Otherwise, $p \doteq l$, that is p and l are equal up to variables renaming, but $q \succ p[r\sigma]_u$, that is the newly generated term $p[r\sigma]_u$ is smaller than both sides of the simplified equation $p \simeq q$. We are going to see in a few paragraphs (Lemma 3.1) how these conditions ensures that Simplification is proof-reducing.

The *Deduction* inference rule is the only expansion rule of UKB. It works by checking for a *superposition* between two equations $p \simeq q$ and $l \simeq r$. Two given equations superpose if there exists a non variable subterm, say $p|u$, which unifies with mgu σ with a side l of the other equation $l \simeq r$. This means that the term $p\sigma$ is equal to both $q\sigma$ and $p[r]_u\sigma$. The new equation $p[r]_u\sigma \simeq q\sigma$ is called a *critical pair.* A critical pair is generated only if $p\sigma \not\preceq q\sigma, p[r]_u\sigma$, that is the two equations are applied according to the simplification ordering. The original definitions of Simplification and Deduction given in [35] have slightly different conditions. Simplification requires that $l\sigma \succ r\sigma$ and superposition requires that $p\sigma \not\preceq q\sigma$ and $l\sigma \not\preceq r\sigma$. We adopt here the conditions given in [27], because they put weaker requirements on simplification and stronger requirements on superposition than the original ones. However, these conditions may be more expensive to compute, since they require to perform both substitution application and term replacement.

The *Functional subsumption* inference rule deletes an equation $p \simeq q$ because it is subsumed by another equation $l \simeq r$, that is $p = c[l\sigma]$ and $q = c[r\sigma]$ for some context c and substitution σ, where either c is not empty or σ is not a renaming of variables.

Simplification is the most important among the above inference rules, because it reduces dramatically the number and the size of the generated equations. A search plan for UKB should give to Simplification the highest priority among all the inference rules, so that the target and the presentation are always kept fully simplified. A search plan with this property is called *Simplification-first* [35]. If Simplification is not applied, the Deduction inference rule rapidly saturates the memory space with equations, making impossible to reach a proof in reasonable time.

In order to characterize the UKB procedure as a completion procedure, we define a proof ordering $>_{UKB}$ to compare the proofs $\Pi(E_i, \hat{s}_i \simeq \hat{t}_i)$. We use the ordering $>_u$ introduced in Example 2.2.

We recall that we write and equational proof step $s \leftrightarrow_{l \simeq r} t$ meaning that s and t are $c[l\sigma]$ and $c[r\sigma]$ for some context c and substitution σ. We write $s \rightarrow_{l \simeq r} t$ if $s \succ t$ is known a priori. Then $\Pi(E_i, \hat{s}_i \simeq \hat{t}_i) >_{UKB} \Pi(E_j, \hat{s}_j \simeq \hat{t}_j)$ holds if and only if $(\{\hat{s}_i, \hat{t}_i\}, \hat{s}_i \leftrightarrow^*_{E_i} \hat{t}_i) >_u (\{\hat{s}_j, \hat{t}_j\}, \hat{s}_j \leftrightarrow^*_{E_j} \hat{t}_j)$ holds.

Lemma 3.1 *The presentation inference rules of the UKB procedure are reducing.*

Proof: we show that Deduction and Simplification are proof-reducing, Deletion and Functional subsumption delete redundant equations:

- the proof for Deduction was given in Example 2.6.
- A Simplification step where an equation $p \simeq q$ is simplified to $p[r\sigma]_u \simeq q$ by an equation $l \simeq r$, affects a minimal proof by replacing a step $s \leftrightarrow_{p \simeq q} t$ by two steps $s \rightarrow_{l \simeq r} v \leftrightarrow_{p[r\sigma]_u \simeq q} t$.
  - If $t \succ s$, we have $\{(t, q, s)\} >^e_{mul} \{(s, l, v), (t, q, v)\}$ since $t \succ s$ and $s \succ v$.
  - If $s \succ t$,
    - if $p \blacktriangleright l$, we have
      - if $t \succ v$, $\{(s, p, t)\} >^e_{mul} \{(s, l, v), (t, q, v)\}$ since $p \blacktriangleright l$ and $s \succ t$,
      - if $v \succ t$, $\{(s, p, t)\} >^e_{mul} \{(s, l, v), (v, q, t)\}$ since $p \blacktriangleright l$ and $s \succ v$;
    - if $p \doteq l$ and $q \succ p[r\sigma]_u$, $t \succ v$ follows from $q \succ p[r\sigma]_u$ by stability and monotonicity of $\succ$ and we have $\{(s, p, t)\} >^e_{mul} \{(s, l, v), (t, q, v)\}$ since $t \succ v$ and $s \succ t$.
- A trivial equation $l \simeq l$ is redundant: no minimal proof contains a step $s \leftrightarrow_{l \simeq l} s$ since the subproof given by the single term s is smaller: $\{(s, l, s)\} >^e_{mul} \{\epsilon\}$, where the empty triple ϵ is the proof complexity of s.
- the proof for Functional subsumption was given in Example 2.7. □

Lemma 3.2 *The target inference rules of the UKB procedure are strictly proof-reducing.*

Proof: the proof for Simplification was given already in Example 2.5. For a Deletion step we have $\{\hat{s}_i, \hat{t}_i\} \succ_{mul} \{true\}$, since *true* is smaller than any term. Therefore $\Pi(E_i, \hat{s}_i \simeq \hat{t}_i) >_{UKB} \Pi(E_i, \hat{s}_{i+1} \simeq \hat{t}_{i+1})$. □

We can then show that UKB is a completion procedure:

Theorem 3.1 *The Unfailing Knuth-Bendix procedure is a completion procedure on the domain $\mathcal{T}$ of all ground equalities.*

Proof: for all equational presentations E_0 and for all ground targets $\hat{s}_0 \simeq \hat{t}_0$ the derivation

$$(E_0; \hat{s}_0 \simeq \hat{t}_0) \vdash_{UKB} (E_1; \hat{s}_1 \simeq \hat{t}_1) \vdash_{UKB} \ldots (E_i; \hat{s}_i \simeq \hat{t}_i) \vdash_{UKB} \ldots$$

has the monotonicity, relevance and reduction properties. Monotonicity and relevance follow by soundness of the inference rules, which is proved among others in [39, 6, 9]. Reduction follows from Lemma 3.1 and Lemma 3.2. □

If a *fair* search plan is provided, the UKB procedure is a semidecision procedure for equational theories:

Theorem 3.2 (Hsiang and Rusinowitch 1987) [35], (Bachmair, Dershowitz and Plaisted 1989) [11] *An equation $\forall \bar{x} s \simeq t$ is a theorem of an equational theory E if and only if the Unfailing Knuth-Bendix procedure derives* true *from $(E; \hat{s} \simeq \hat{t})$.*

3.2 Extensions: AC-UKB and cancellation laws

Many equational problems involve associative and commutative (AC) operators. An AC function f satisfies the equations

$f(f(x,y),z) \simeq f(x,f(y,z))$ *(associativity)* and
$f(x,y) \simeq f(y,x)$ *(commutativity)*.

Handling associativity and commutativity as any other equation turns out to be very inefficient, since commutativity may generate a very high number of equations through the Deduction inference rule. Also, many instances of commutativity may not be ordered by the chosen simplification ordering, so that simplification does not apply as often as it is desirable to reduce the size and the number of the equations.

The efficiency of the UKB strategy can be greatly improved if associativity and commutativity are not given in the input, but built in the inference rules. The UKB procedure with associativity and commutativity built in the inference rules is called *AC-UKB* [1]. The basic idea is to replace syntactic identity by equality *modulo AC*. Let AC be a set of associativity and commutativity axioms. Two terms s and t are equal modulo AC, if $s \simeq t$ is a theorem of AC, which we write $s =_{AC} t$. The inference rules of the UKB procedure are modified in such a way that any two terms which are equal modulo AC are regarded as identical.

The first modification is to require that the complete simplification ordering on terms $\succ$ is in some sense "compatible" with replacing identity by equality modulo AC. More precisely, this "compatibility" requirement is a *commutation* property. Given two relations R and S, we say that R *commutes* over S if $S \circ R \subseteq R \circ S$, where $\circ$ is composition of relations. The complete simplification ordering $\succ$ is required to commute over $=_{AC}$: this means that for any two terms s and t, if there is a third term r such that $s =_{AC} r$ and $r \succ t$, there is also a term r' such that $s \succ r'$ and $r' =_{AC} t$. Secondly, matching and unification are replaced by AC-matching and AC-unification. A term s matches a term t *modulo AC* if there is a substitution σ such that $s\sigma =_{AC} t$. Similarly, two terms s and t unify *modulo AC* if there is a substitution σ such that $s\sigma =_{AC} t\sigma$. Finally, the strict encompassment ordering $\blacktriangleright$ is replaced by the ordering $\blacktriangleright_{AC}$, that is $s \blacktriangleright_{AC} t$ if and only if $s \blacktriangleright r$ and $r =_{AC} t$ for some term r.

The set of inference rules of the UKB procedure is therefore modified as follows:

- *Presentation inference rules:*
  - *Simplification:*
  $$\frac{(E \cup \{p \simeq q, l \simeq r\}; \hat{s} \simeq \hat{t})}{(E \cup \{p[r\sigma]_u \simeq q, l \simeq r\}; \hat{s} \simeq \hat{t})} \quad \begin{array}{l} p|u =_{AC} l\sigma \qquad p \succ p[r\sigma]_u \\ p \blacktriangleright_{AC} l \vee q \succ p[r\sigma]_u \end{array}$$
  - *Deduction:*
  $$\frac{(E \cup \{p \simeq q, l \simeq r\}; \hat{s} \simeq \hat{t})}{(E \cup \{p \simeq q, l \simeq r, p[r]_u\sigma \simeq q\sigma\}; \hat{s} \simeq \hat{t})} \quad \begin{array}{l} p|u \notin X \qquad (p|u)\sigma =_{AC} l\sigma \\ p\sigma \not\preceq q\sigma, p[r]_u\sigma \end{array}$$
  - *Extension:*
  $$\frac{(E \cup \{f(p,q) \simeq r\}; \hat{s} \simeq \hat{t})}{(E \cup \{f(p,q) \simeq r, f(p,q,z) \simeq f(r,z)\}; \hat{s} \simeq \hat{t})} \quad \begin{array}{l} f \text{ is } AC \\ f(p,q) \not\preceq r \end{array}$$
  - *Deletion:*
  $$\frac{(E \cup \{l \simeq l\}; \hat{s} \simeq \hat{t})}{(E; \hat{s} \simeq \hat{t})}$$
  - *Functional subsumption:*
  $$\frac{(E \cup \{p \simeq q, l \simeq r\}; \hat{s} \simeq \hat{t})}{(E \cup \{l \simeq r\}; \hat{s} \simeq \hat{t})} \quad (p \simeq q) \blacktriangleright_{AC} (l \simeq r)$$

- *Target inference rules:*
  - *Simplification:*
  $$\frac{(E \cup \{l \simeq r\}; \hat{s} \simeq \hat{t})}{(E \cup \{l \simeq r\}; \hat{s}[r\sigma]_u \simeq \hat{t})} \quad \begin{array}{l} \hat{s}|u =_{AC} l\sigma \\ \hat{s} \succ \hat{s}[r\sigma]_u \end{array}$$
  - *Deletion:*
  $$\frac{(E; \hat{s} \simeq \hat{s})}{(E; true)}$$

This set of inference rules is obtained from the set of inference rules of the UKB procedure by replacing identity by equality modulo AC as explained above and by adding a new inference rule, called *Extension.* The *Extension* inference rule is a specialized version of the *Deduction* inference rule, designed to compute superpositions of equations in E onto associativity axioms. Namely, if $f(p,q) \simeq r$ is an equation in E, f is AC and $f(p,q) \not\preceq r$, the equation $f(p,q) \simeq r$ trivially superposes onto the associativity axiom $f(f(x,y),z) \simeq f(x, f(y,z))$, yielding the critical pair $f(p, f(q,z)) \simeq f(r,z)$, which we write in *flattened* form as $f(p,q,z) \simeq f(r,z)$. These critical pairs are called *extended rules.* Computing the extended rules is sufficien to ensure completeness of the AC-UKB procedure: no other critical pairs between E and AC need to be computed [54].

The extension of UKB to AC-UKB is feasible because algorithms for AC-matching and AC-unification are available. An algorithm for AC-unification, its application in a completion procedure and the extended rules first appeared in [58, 59, 54]. The correctness of the AC-unification algorithm was proved in [28]. General theoretical frameworks for working with equations modulo a set of axioms A are given in [41] and in [7]. These results are surveyed in [25] and more specifically for unification problems in [44].

The UKB or AC-UKB procedure can be further improved by building in the inference rules for the *cancellation laws.* A function F is *right cancellable* if it satisfies the *right cancellation law*

$$\forall x, y, z \;\; f(x,y) = f(z,y) \;\supset\; x = z$$

The *left cancellation law* is defined symmetrically. Cancellation laws may reduce considerably the size of the equations. They are implemented as inference rules as follows [36]:

Cancellation 1:

$$\frac{(E \cup \{f(p,u) \simeq f(q,v)\}; \hat{s} \simeq \hat{t})}{(E \cup \{f(p,u) \simeq f(q,v), p\sigma \simeq q\sigma\}; \hat{s} \simeq \hat{t})} \quad u\sigma = v\sigma$$

Cancellation 2:

$$\frac{(E \cup \{f(d_1,d_2) \simeq y\}; \hat{s} \simeq \hat{t})}{(E \cup \{f(d_1,d_2) \simeq y, d_1\sigma \simeq x\}; \hat{s} \simeq \hat{t})} \quad \begin{array}{ll} y \in V(d_1) & \sigma = \{y \mapsto f(x,d_2)\} \\ y \notin V(d_2) & x \textit{ is a new variable} \end{array}$$

Cancellation 3:

$$\frac{(E \cup \{f(p_1,q_1) \simeq r_1, f(p_2,q_2) \simeq r_2\}; \hat{s} \simeq \hat{t})}{(E \cup \{f(p_1,q_1) \simeq r_1, f(p_2,q_2) \simeq r_2, p_1\sigma \simeq p_2\sigma\}; \hat{s} \simeq \hat{t})} \quad \begin{array}{l} q_1\sigma = q_2\sigma \\ r_1\sigma = r_2\sigma \end{array}$$

where the function f is right cancellable. In *Cancellation 2*, if the substitution $\sigma = \{y \mapsto f(x,d_2)\}$ is applied to the given equation, it becomes $f(d_1\sigma, d_2) \simeq f(x,d_2)$, since y does not occur in d_2. The cancellation law reduces this equation to $d_1\sigma \simeq x$.

In order to prove that the UKB procedure with the cancellation inference rules is a completion procedure, we need to prove that the Cancellation inference rules are proof-reducing. We adopt the proof ordering $>_{UKBC}$ defined as follows: a ground equational step $s \simeq t$ justified by an equation

$l \simeq r$ has complexity measure $(s, l\sigma, l, t)$, if s is $c[l\sigma]$, t is $c[r\sigma]$ and $s \succ t$. Complexity measures are compared by the lexicographic combination $>^{ec}$ of the orderings $\succ$, $\triangleright$, $\triangleright$ and $\succ$. Proofs are compared by the lexicographic combination $>_{uc}$ of the multiset extensions $\succ_{mul}$ and $>^{ec}{}_{mul}$: $\Pi(E_i, \hat{s}_i \simeq \hat{t}_i) >_{UKBC} \Pi(E_j, \hat{s}_j \simeq \hat{t}_j)$ if and only if $(\{\hat{s}_i, \hat{t}_i\}, \hat{s}_i \leftrightarrow^*_{E_i} \hat{t}_i) >_{uc} (\{\hat{s}_j, \hat{t}_j\}, \hat{s}_j \leftrightarrow^*_{E_j} \hat{t}_j)$. The proof of Lemma 3.1 is unaffected if $>_{UKBC}$ replaces $>_{UKB}$.

Lemma 3.3 *The* Cancellation *inference rules are proof-reducing.*

Proof: we show that if $(E_i; \hat{s}_i \simeq \hat{t}_i) \vdash_{UKB} (E_{i+1}; \hat{s}_i \simeq \hat{t}_i)$ is a Cancellation step, then if $\Pi(E_i, \hat{s} \simeq \hat{t}) \neq \Pi(E_{i+1}, \hat{s} \simeq \hat{t})$, that is the inference step affects the proof of $\hat{s} \simeq \hat{t}$, $\Pi(E_i, \hat{s} \simeq \hat{t}) >^{ec}{}_{mul} \Pi(E_{i+1}, \hat{s} \simeq \hat{t})$ holds.

- An application of the rule Cancellation 1 to an equation $f(p, u) \simeq f(q, v)$ affects any minimal proof in E_i which contains a step $s \leftrightarrow t$ such that $s = c[f(p, u)\tau]$, $t = c[f(q, v)\tau]$ and $\tau \geq \sigma$, where $\geq$ is the subsumption ordering and σ is the mgu such that $u\sigma = v\sigma$ of the application of Cancellation 1. The step $s \leftrightarrow_{f(p,u) \simeq f(q,v)} t$ has complexity $(s, f(p, u)\tau, f(p, u), t)$, if $s \succ t$. In the minimal proofs in E_{i+1} the step $s \leftrightarrow_{f(p,u) \simeq f(q,v)} t$ is replaced by a step $s \leftrightarrow_{p\sigma \simeq q\sigma} t$ justified by the new equation $p\sigma \simeq q\sigma$ generated by the application of Cancellation 1. The step $s \leftrightarrow_{p\sigma \simeq q\sigma} t$ has complexity $(s, p\tau, p\sigma, t)$. Since $f(p, u)\tau \triangleright p\tau$, $\{(s, f(p, u)\tau, f(p, u), t)\} >^{ec} \{(s, p\tau, p\sigma, t)\}$ follows. A symmetric argument applies if $t \succ s$.
- An application of the rule Cancellation 2 to an equation $f(d_1, d_2) \simeq y$ affects any minimal proof in E_i which contains a step $s \leftrightarrow t$ such that $s = c[f(d_1, d_2)\tau]$, $t = c[y\tau]$ and $\tau \geq \sigma$, where σ is $\{y \mapsto f(x, d_2)\}$. Since $y \in V(d_1)$, we have $f(d_1, d_2)\tau \succ y\tau$ by the subterm property and therefore $s \succ t$ by monotonicity, so that the step $s \leftrightarrow t$ has complexity $(s, f(d_1, d_2)\tau, f(d_1, d_2), t)$. In the minimal proofs in E_{i+1} the step $s \leftrightarrow t$ is replaced by a step $s \leftrightarrow_{d_1\sigma \simeq x} t$ justified by the new equation $d_1\sigma \simeq x$ generated by the application of Cancellation 2. The step $s \leftrightarrow_{d_1\sigma \simeq x} t$ has complexity $(s, d_1\tau, d_1\sigma, t)$. Since $f(d_1, d_2)\tau \triangleright d_1\tau$, $\{(s, f(d_1, d_2)\tau, f(d_1, d_2), t)\} >^{ec}{}_{mul} \{(s, d_1\tau, d_1\sigma, t)\}$ follows.
- An application of the rule Cancellation 3 to two equations $f(p_1, q_1) \simeq r_1$ and $f(p_2, q_2) \simeq r_2$ affects any minimal proof in E_i which contains a subproof $s \leftrightarrow u \leftrightarrow t$ such that $s = c[f(p_1, q_1)\tau]$, $u = c[r_1\tau]$, $t = c[f(p_2, q_2)\tau]$ and $\tau \geq \sigma$, where σ is the mgu such that $q_1\sigma = q_2\sigma$ and $r_1\sigma = r_2\sigma$ of the application of Cancellation 3. It follows that $q_1\tau = q_2\tau$ and $r_1\tau = r_2\tau$ too. The subproof $s \leftrightarrow u \leftrightarrow t$ is replaced in any minimal proof in E_{i+1} by a single step $s \leftrightarrow_{p_1\sigma \simeq p_2\sigma} t$ justified by the new equation $p_1\sigma \simeq p_2\sigma$ generated by the application of Cancellation 3.

  1. If $s \succ t \succ u$, the subproof $s \leftrightarrow u \leftrightarrow t$ has complexity $\{(s, f(p_1, q_1)\tau, f(p_1, q_1), u), (t, f(p_2, q_2)\tau, f(p_2, q_2), u)\}$ and the step $s \leftrightarrow_{p_1\sigma \simeq p_2\sigma} t$ has complexity $(s, p_1\tau, p_1\sigma, t)$. Since $f(p_1, q_1)\tau \triangleright p_1\tau$, the result follows. A symmetric argument applies if $t \succ s \succ u$.
  2. If $s \succ u \succ t$, the subproof $s \leftrightarrow u \leftrightarrow t$ has complexity $\{(s, f(p_1, q_1)\tau, f(p_1, q_1), u), (u, r_1\tau, r_1, t)\}$ and the step $s \leftrightarrow_{p_1\sigma \simeq p_2\sigma} t$ has complexity $(s, p_1\tau, p_1\sigma, t)$. Since $f(p_1, q_1)\tau \triangleright p_1\tau$, the result follows. A symmetric argument applies if $t \succ u \succ s$.
  3. If $u \succ s \succ t$, the subproof $s \leftrightarrow u \leftrightarrow t$ has complexity $\{(u, r_1\tau, r_1, s), (u, r_1\tau, r_1, t)\}$ and the step $s \leftrightarrow_{p_1\sigma \simeq p_2\sigma} t$ has complexity $(s, p_1\tau, p_1\sigma, t)$. Since $u \succ s$, the result trivially follows. A symmetric argument applies if $u \succ t \succ s$. □

The UKB or AC-UKB procedure enriched with the inference rules for cancellation is complete [36]. Most of the experimental results reported in [1, 2, 14, 3, 5] are obtained by AC-UKB with the inference rules for cancellation.

3.3 Efficiency of the Unfailing Knuth-Bendix procedure

The UKB procedure is complete, but it is not very efficient in general. The main source of inefficiency is the Deduction inference rule, that is the forward reasoning component of UKB. All the backward reasoning steps are Simplification steps, which are strictly proof-reducing. On the other hand, a Deduction step is guaranteed to reduce the proof of some theorem, but not necessarily the proof of the target. The UKB procedure is inefficient because it generates many critical pairs which do not help in proving the target.

Therefore, our goal is to reduce the number of critical pairs generated or equivalently to perform less forward reasoning and more backward reasoning.

For the forward reasoning part, a possible approach to the problem consists in designing search plans which generate first the critical pairs that are estimated to be likely to reduce the proof of the target. Since the effect of a critical pair on the target cannot be completely determined a priori, such a search plan is based on *heuristical criteria* that measure how useful a critical pair is expected to be with respect to the task of simplifying the goal. Some examples of these heuristics are given in [3, 4].

For the backward reasoning part, we observe that if the target $\hat{s}_i \simeq \hat{t}_i$ is fully simplified with respect to E_i, $\hat{s}_i \simeq \hat{t}_i$ is minimal in the ordering $\succ_{mul}$ among all the ground equations E-equivalent to the input target $s_0 \simeq t_0$, where $E = \bigcup_{0 \leq j \leq i} E_j$. If a Simplification-first plan is adopted, UKB maintains a minimal target. Therefore, it could seem that no improvement can be obtained on the target side. However, we shall see that this is not the case.

The notion of minimal target is relative to the assumed partially ordered set (poset) of targets. If we assume the poset of ground equalities ordered by $\succ_{mul}$, $\hat{s}_i \simeq \hat{t}_i$ is minimal among the ground equations E-equivalent to the input target $s_0 \simeq t_0$. The situation changes if we assume as poset of targets the poset of disjunctions of ground equalities ordered by an ordering $\succ'_{mul}$ defined as follows: $N_1 \succ'_{mul} N_2$ if $min(N_1) \succ_{mul} min(N_2)$, where N_1 and N_2 are disjunctions of ground equalities and $min(N)$ is the smallest equality in N according to $\succ_{mul}$. Since the equalities are ground and the simplification ordering is total on ground, there is a smallest element in a disjunction and this ordering is well defined. Furthermore, the poset of equalities is embedded[1] in the poset of disjunctions.

We show why the backward reasoning part of UKB is not guaranteed to compute a minimal target if the poset of disjunctions is assumed. Let $(E_i; \hat{s}_i \simeq \hat{t}_i)$ be the current stage in an UKB derivation and $l \simeq r$ be an un-orientable equation in E_i, such that $\hat{s}_i|u = l\sigma$ for some position u and substitution σ, but $\hat{s}_i \prec \hat{s}_i[r\sigma]_u$. In other words, l matches a subterm of $\hat{s}_i$ but Simplification does not apply because $\hat{s}_i$ would not be replaced by a smaller term. However, we assume that the target $\hat{s}_i[r\sigma]_u \simeq \hat{t}_i$ is generated nonetheless and that by simplification it reduces to an equation which is smaller than $\hat{s}_i \simeq \hat{t}_i$, that is $\hat{s}_i[r\sigma]_u \rightarrow^*_{E_i} \hat{s}'$, $\hat{t}_i \rightarrow^*_{E_i} \hat{t}'$ and $\{\hat{s}', \hat{t}'\} \prec_{mul} \{\hat{s}_i, \hat{t}_i\}$. If these conditions hold, we have that the disjunction $\hat{s}_i \simeq \hat{t}_i \vee \hat{s}' \simeq \hat{t}'$ is smaller than the disjunction given by $\hat{s}_i \simeq \hat{t}_i$ alone in the poset of disjunctions defined above. Therefore, if we assume the poset of disjunctions as posets of targets, it is not true that UKB maintains a minimal target.

The intuition behind the choice of considering disjunctions of equalities rather than equalities is that if we consider more than one target equality, we have a greater chance to find a short proof. In order to work on disjunctions of equalities, we need to add to the UKB procedure an expansion

[1] Given two posets $\mathcal{P}_1 = (D_1, >_1)$ and $\mathcal{P}_2 = (D_2, >_2)$, an *embedding* $h: \mathcal{P}_1 \rightarrow \mathcal{P}_2$ is an injective function $h: D_1 \rightarrow D_2$ which preserves the ordering: for all $x, y \in D_1$, $x >_1 y$ implies $h(x) >_2 h(y)$. The function which maps a ground equality into the disjunction given by the ground equality itself is clearly an embedding of the poset of ground equalities into the poset of disjunctions of ground equalities, since the smallest element in a disjunction given by a single equality is the equality itself.

inference rule, so that the target is eventually expanded into a disjunction of ground equalities. Such an expansion inference rule must satisfy the relevance requirement, so that proving the validity of any of the equalities in the disjunction is equivalent to prove the input target $s_0 \simeq t_0$. Also, the application of such rule must be restricted, in order to avoid the generation of a high number of target equalities, which may slow down the search for a solution.

This new inference rule is superposition of an un-orientable equation onto a target equality $\hat{s} \simeq \hat{t}$ to generate a new target equality. A newly generated target equality is first simplified as much as possible and then it is kept only if it is smaller than $\hat{s} \simeq \hat{t}$:

Ordered saturation:

$$\frac{(E \cup \{l \simeq r\}; N \cup \{\hat{s} \simeq \hat{t}\})}{(E \cup \{l \simeq r\}; N \cup \{\hat{s} \simeq \hat{t}, \hat{s}' \simeq \hat{t}'\})} \quad \begin{array}{lll} \hat{s}|u = l\sigma & \hat{s}[r\sigma]_u \rightarrow^*_E \hat{s}' & \hat{t} \rightarrow^*_E \hat{t}' \\ & \{\hat{s}', \hat{t}'\} \prec_{mul} \{\hat{s}, \hat{t}\} & \end{array}$$

Ordered saturation applies if $\hat{s} \prec \hat{s}[r\sigma]_u$, since if $\hat{s} \succ \hat{s}[r\sigma]_u$ holds, simplification would apply. The target equality $\hat{s}' \simeq \hat{t}'$ might have a shorter proof than the other target equalities. We do not know which one has the shortest proof. We keep all of them to broaden our chance of reaching the proof as soon as possible.

In addition, we need to modify the *Deletion* inference rule, since the computation halts successfully as soon as an equality in the disjunction is reduced to a trivial equality:

Deletion:

$$\frac{(E; N \cup \{\hat{s} \simeq \hat{s}\})}{(E; true)}$$

The procedure obtained by adding Ordered saturation to UKB and by modifying Deletion as above, is called the *Inequality Ordered-Saturation strategy* (IOS) [3]. A derivation by the IOS strategy has the form

$$(E_0; N_0) \vdash_{IOS} (E_1; N_1) \vdash_{IOS} \ldots \vdash_{IOS} (E_i; N_i) \vdash_{IOS} \ldots$$

where the set N_0 contains the initial goal $\hat{s}_0 \simeq \hat{t}_0$ and at stage i, N_i is the current set of target equalities. The derivation halts at stage k if N_k contains a target $\hat{s}_i \simeq \hat{t}_i$ such that $\hat{s}_i$ and $\hat{t}_i$ are identical and the clause in N_k reduces to *true*.

In order to show that the IOS strategy is a completion procedure, we assume a proof ordering $>_{IOS}$ defined as follows: $\Pi(E; N) >_{IOS} \Pi(E'; N')$ if and only if $\Pi(E; min(N)) >_{UKB} \Pi(E'; min(N'))$. In other words the proof of a disjunction is represented by the proof of the smallest target in the disjunction.

Lemma 3.4 *The* Ordered saturation *inference rule is proof-reducing.*

Proof: we show that if $(E_i; N_i) \vdash_{IOS} (E_i; N_{i+1})$ is an Ordered saturation step, then $\Pi(E_i, N_i) \geq_{IOS} \Pi(E_i, N_{i+1})$. Since $N_i \subset N_{i+1}$, $min(N_i) \succeq_{mul} min(N_{i+1})$ and the result follows. □

Theorem 3.3 *The Inequality Ordered-Saturation strategy is a completion procedure.*

Proof: it follows from Theorem 3.1 and Lemma 3.4. □

The IOS strategy has been implemented and observed to perform better than the UKB procedure [3]. In practice, few target equalities are kept, so that the overhead of handling them is negligible with respect to the advantage of keeping more than one target.

3.4 The S-strategy

The *S-strategy* [35] is an extension of the UKB procedure to the logic of equality and inequality. A presentation is a set of equations E_0 and a theorem φ is a sentence $\bar{Q}\bar{x}\ s_0 \simeq t_0 \vee \ldots \vee s_n \simeq t_n$, where $\bar{Q}\bar{x}$ is any sequence of quantifier-variable pairs. A theorem φ in this form is transformed into a target $N_0 = s_0 \simeq t_0 \vee \ldots \vee s_n \simeq t_n$, where all variables are implicitly existentially quantified, by replacing all the universally quantified variables by constants and by dropping the quantifiers. If φ is $\forall\bar{x}s_0 \simeq t_0$, N_0 is $\hat{s}_0 \simeq \hat{t}_0$ and the S-strategy reduces to the UKB procedure. A computation has the form

$$(E_0; N_0) \vdash_S (E_1; N_1) \vdash_S \ldots \vdash_S (E_i; N_i) \vdash_S \ldots$$

where $\forall i \geq 0$, E_i is a set of equalities and N_i is a disjunction of target equalities with existentially quantified variables. A derivation halts at stage k if N_k contains a target $s_i \simeq t_i$ whose sides are unifiable. The set of inference rules of UKB is modified as follows:

- *Presentation inference rules:*
  - *Simplification:*
  $$\frac{(E \cup \{p \simeq q, l \simeq r\}; N)}{(E \cup \{p[r\sigma]_u \simeq q, l \simeq r\}; N)} \quad \begin{matrix} p|u = l\sigma & p \succ p[r\sigma]_u \\ p \blacktriangleright l \vee q \succ p[r\sigma]_u & \end{matrix}$$
  - *Deduction:*
  $$\frac{(E \cup \{p \simeq q, l \simeq r\}; N)}{(E \cup \{p \simeq q, l \simeq r, p[r]_u\sigma \simeq q\sigma\}; N)} \quad \begin{matrix} p|u \notin X & (p|u)\sigma = l\sigma \\ p\sigma \not\preceq q\sigma, p[r]_u\sigma & \end{matrix}$$
  - *Deletion:*
  $$\frac{(E \cup \{l \simeq l\}; N)}{(E; N)}$$
  - *Functional subsumption:*
  $$\frac{(E \cup \{p \simeq q, l \simeq r\}; N)}{(E \cup \{l \simeq r\}; N)} \quad (p \simeq q) \blacktriangleright (l \simeq r)$$
- *Target inference rules:*
  - *Simplification:*
  $$\frac{(E \cup \{l \simeq r\}; N \cup \{s \simeq t\})}{(E \cup \{l \simeq r\}; N \cup \{s[r\sigma]_u \simeq t\})} \quad \begin{matrix} s|u = l\sigma \\ s \succ s[r\sigma]_u \end{matrix}$$
  - *Deduction:*
  $$\frac{(E \cup \{l \simeq r\}; N \cup \{s \simeq t\})}{(E \cup \{l \simeq r\}; N \cup \{s \simeq t, s[r]_u\sigma \simeq t\sigma\})} \quad \begin{matrix} s|u \notin X & (s|u)\sigma = l\sigma \\ s\sigma \not\preceq s[r]_u\sigma & \end{matrix}$$
  - *Deletion:*
  $$\frac{(E; N \cup \{s \simeq t\})}{(E; true)} \quad s\sigma = t\sigma$$

The *Deduction* inference rule applies to both equalities and inequalities. In the second case no ordering based condition applies to the inequality. The *Deletion* rule for the target is modified because the target contains variables: a contradiction is detected when the two sides of a target equality unify.

In order to characterize the S-strategy as a completion procedure, we define the following ordering: $\Pi(E_i; N_i) >_S \Pi(E_{i+1}; N_{i+1})$ if and only if $\Pi(\hat{E}_i; \hat{s}_i \simeq \hat{t}_i) >_{UKB} \Pi(\hat{E}_{i+1}; \hat{s}_{i+1} \simeq \hat{t}_{i+1})$, where $\forall i \geq 0$, $\hat{E}_i \cup \{\hat{s}_i \neq \hat{t}_i\}$ is the smallest finite unsatisfiable set of ground instances of clauses in $E_i \cup \neg N_i$. We show that this ordering is well defined. First we show how the pair $(\hat{E}_i; \hat{s}_i \simeq \hat{t}_i)$ is defined. N_i

is a theorem of E_i if and only if $E_i \cup \neg N_i$ is unsatisfiable, where N_i is a disjunction of equations $s_0 \simeq t_0 \vee \ldots \vee s_n \simeq t_n$ with existentially quantified variables and therefore $\neg N_i$ is a conjunction of inequalities $s_0 \neq t_0 \wedge \ldots \wedge s_n \neq t_n$ with universally quantified variables. By the Herbrand Theorem [18], the set $E_i \cup \neg N_i$ is unsatisfiable if and only if there is a finite subset of ground instances of the clauses in $E_i \cup \neg N_i$ which is unsatisfiable. Since $\neg N_i$ is a set of inequalities with universally quantified variables, an unsatisfiable ground instance of $E_i \cup \neg N_i$ needs to contain just one ground inequality: $\hat{E}_i \cup \{\hat{s}_i \neq \hat{t}_i\}$ is the smallest such set with respect to the ordering $\succ_{mul}$. Since $\succ$ is total on ground terms, there exists a smallest set.

The above definition of the ordering $>_S$ says that the complexity of the proof $\Pi(E_i; N_i)$ is measured by the complexity of the ground proof $\Pi(\hat{E}_i; \hat{s}_i \simeq \hat{t}_i)$ and that the impact of the inference steps on $\Pi(E_i; N_i)$ is measured by the impact of the inference steps on $\Pi(\hat{E}_i; \hat{s}_i \simeq \hat{t}_i)$. This approach is correct if to every inference step on $(E_i; N_i)$ corresponds an inference steps on $(\hat{E}_i; \hat{s}_i \simeq \hat{t}_i)$ and vice versa. In order to prove this, we need the following lemma, which rephrases for the S-strategy the *Paramodulation Lifting Lemma.* We recall that a ground substitution is E-irreducible if it does contain any pair $\{x \mapsto t\}$ such that t can be simplified by an equation in E.

Lemma 3.5 (Peterson 1983) [55], (Hsiang and Rusinowitch 1987) [37] *If σ is a ground, E-irreducible substitution, then for all inference rules f of S-strategy, if $(E\sigma; s\sigma \simeq t\sigma) \vdash_f (E'; s' \simeq t')$, then $(E; s \simeq t) \vdash_f (E''; s'' \simeq t'')$, where E' and $s' \simeq t'$ are instances of E'' and $s'' \simeq t''$ respectively.*

Lemma 3.6 $(E_i; N_i) \vdash_S (E_{i+1}; N_{i+1})$ *if and only if* $(\hat{E}_i; \hat{s}_i \simeq \hat{t}_i) \vdash_S (\hat{E}_{i+1}; \hat{s}_{i+1} \simeq \hat{t}_{i+1})$.

Proof:

$\Rightarrow$) An inference step on $(E_i; N_i)$ is trivially an inference step on $(\hat{E}_i; \hat{s}_i \simeq \hat{t}_i)$, since an inference step on non ground clauses is trivially an inference step on all their instances.

$\Leftarrow$) Since $(\hat{E}_i; \hat{s}_i \simeq \hat{t}_i)$ is minimal, $\hat{E}_i \subseteq E_i\sigma$ and $\hat{s}_i \simeq \hat{t}_i \in N_i\sigma$ for an E_i-irreducible substitution σ. Therefore, by Lemma 3.5, an inference step on $(\hat{E}_i; \hat{s}_i \simeq \hat{t}_i)$ is an inference step on $(E_i; N_i)$. □

We can finally state the following theorem:

Theorem 3.4 *The S-strategy is a completion procedure on the domain $\mathcal{T}$ of all ground equalities.*

Proof: monotonicity and relevance follow from the soundness of the inference rules. By the definition of the ordering $>_S$, the inference rules of S-strategy are proof-reducing if they are proof-reducing on ground proofs with respect to the ordering $>_{UKB}$. This follows from Lemma 3.1 and Lemma 3.2, since target Deduction is just target Simplification if the target is ground. □

If a *fair* search plan is provided, the S-strategy is is a semidecision procedure for theories in the logic of equality and inequality:

Theorem 3.5 (Hsiang and Rusinowitch 1987) [35] *A sentence $\bar{Q}\bar{x}\ s_0 \simeq t_0 \vee \ldots \vee s_n \simeq t_n$ is a theorem of an equational theory E if and only if the S-strategy derives* true *from* $(E; s_0 \simeq t_0 \vee \ldots \vee s_n \simeq t_n)$.

4 Semidecision procedures for disproving inductive theorems

The Knuth-Bendix completion procedure has been also applied to *disprove inductive theorem* in equational theories. This method has been called *inductionless induction*, *proof by consistency* or *proof by the lack of inconsistency* by several authors [53, 30, 40, 52, 42, 29, 47, 48, 10, 43]. Extensions of this method to Horn logic with equality are explored in [50].

First of all, we show that a completion procedure applied to disprove inductive theorems is a *semidecision procedure.* We denote by $G(S)$ the set of all ground terms on the signature of a presentation S and we use $Ran(\sigma)$ to represent the range of a substitution σ, so that a ground substitution is a substitution such that $Ran(\sigma) \subset G(S)$. A clause φ is an *inductive theorem* of S, written $S \models_{Ind} \varphi$, if and only if for all ground substitutions σ, $\varphi\sigma \in Th(S)$. We denote by $Ind(S)$ the set of all the inductive theorems of S, $Ind(S) = \{\varphi | \; S \models_{Ind} \varphi\}$, by $GTh(S)$ the set of all the ground theorems of S, $GTh(S) = \{\varphi | \; \varphi \in Th(S), \varphi \; ground\}$ and by $G(\varphi)$ the set of all the ground instances of φ, $G(\varphi) = \{\varphi\sigma | \; Ran(\sigma) \subset G(S)\}$.

The set $Ind(S)$ is not semidecidable. Even if we have a decision procedure for $G(\varphi) \cap GTh(S)$, we still cannot prove that φ is an inductive theorem, because the set $G(\varphi)$ is infinite. However, the complement problem, that is proving that φ is *not* an inductive theorem of S, is semidecidable in certain theories.

If $\varphi \notin Ind(S)$, then there is a ground instance $\varphi\sigma$ such that $\varphi\sigma \notin GTh(S)$. Therefore $GTh(S \cup \{\varphi\}) \neq GTh(S)$, since $\varphi\sigma \in GTh(S \cup \{\varphi\})$ for all ground instances $\varphi\sigma$. Thus, we can prove that φ is not an inductive theorem of S by proving the following target:

$$\Phi_0 = \exists\sigma \; Ran(\sigma) \subset G(S) \; \exists\psi \in S \cup \{\varphi\} \; such \; that \; \psi\sigma \in GTh(S \cup \{\varphi\}) - GTh(S).$$

If there exists an oracle $\mathcal{O}$ to decide such target, a completion procedure $C =< I_p, I_t; \Sigma; \mathcal{O} >$ equipped with the oracle $\mathcal{O}$ will be a semidecision procedure for disproving inductive theorems. A derivation has the form

$$(S \cup \{\varphi\}; \Phi_0) \vdash_C (S_1; \Phi_1) \vdash_C \ldots (S_i; \Phi_i) \vdash_C \ldots,$$

where at each step the target is

$$\Phi_i = \exists\sigma \; Ran(\sigma) \subset G(S) \; \exists\psi \in S_i \; such \; that \; \psi\sigma \in GTh(S_i) - GTh(S).$$

No inference step applies to the target: the procedure takes as input the presentation $S \cup \{\varphi\}$ given by the original presentation and the inductive conjecture and it proceeds by applying inference rules to the presentation until it obtains a presentation S_k such that the oracle applied to S_k answers positively and replaces Φ_k by *true*.

In the equational case, an oracle to decide Φ_i is available only under the assumption that the input set of equations E is ground confluent. Under this hypothesis, Φ_i is true if and only if there are two ground E-irreducible terms s and t such that $s_i\sigma \rightarrow^*_E s$, $t_i\sigma \rightarrow^*_E t$ and $s \simeq t \in GTh(E_i)$. Therefore, we can restrict our attention to ground E-irreducible terms.

A first oracle was given in [40] for equational presentations satisfying the *principle of definition:* the signature of E is given by the disjoint union of a set C of *constructors* and a set D of *defined symbols,* such that the set $T(C)$ of all ground constructor terms is *free* and all function symbols in D are *completely defined* on C, that is for all ground term $t \in T(F)$, there exists a unique ground constructor term $t' \in T(C)$ such that $t \leftrightarrow^*_E t'$.

A more general oracle was proposed in [42] for the Knuth-Bendix completion procedure and extended to the UKB procedure in [10]. This test is based on *ground reducibility:* a term t is *ground E-reducible* if for all ground substitutions σ, $t\sigma$ is E-reducible. Ground E-reducibility is decidable [56] only if E is a ground confluent rewrite system. Therefore, the test in [42, 10] applies only if the input presentation E is ground confluent and all its equations can be oriented into rewrite rules.

In order to characterize an inductive theorem proving strategy as a completion procedure, we define the proof of the target Φ_i as follows:

$$\Pi(S_i, \Phi_i) = \Pi(S_i, min\{\psi\sigma | \; \psi \in S_i, \psi\sigma \in GTh(S_i) - GTh(S)\}),$$

that is the proof of the target is the proof of the smallest ground instance of some clause in S_i which is a theorem in S_i but not in S.

In the equational case, a completion procedure which eventually generates a ground confluent set of equations, is able to reduce the proofs of all ground theorems and therefore the proof of the target. However, this is not necessary. Since the proof of the target is the proof of the smallest ground theorem which is not a theorem of the original presentation, we can restrict our attention to a smaller set of ground theorems:

Definition 4.1 (Fribourg 1986) [29] *Given a ground confluent presentation E, a ground substitution σ is* E-inductively complete *if for all ground substitutions ρ and for all terms t, there exists a ground substitution τ such that $t\rho \leftrightarrow_E^* t\sigma\tau$ and $t\rho \succ t\sigma\tau$.*

We denote by $\mathcal{IT}_E$ the set of ground instances which are E-inductively complete, that is $\mathcal{IT}_E = \{(s \simeq t)\sigma | \sigma \textit{ is } S\textit{-inductively complete}\}$. Since the proof of the target is the proof of the smallest ground theorem which is not a theorem of the original presentation, reducing the proofs of the theorems in $\mathcal{IT}_E$ is sufficient to guarantee that the proof of the target is reduced, as was first proved in [29] for the application of Knuth-Bendix completion to disprove inductive theorems in equational theories:

Theorem 4.1 (Fribourg 1986) [29] *A completion procedure $\mathcal{C} =< I_p, I_l; \Sigma; \mathcal{O} >$ on the domain $\mathcal{IT}_E$, with complete inference rules and fair search plan is a semidecision procedure for the complement of $Ind(E)$ for all equational presentations E, for which the oracle $\mathcal{O}$ is computable.*

As a consequence, the Deduction inference rule of UKB can be restricted in such a way that a superposition between $l \simeq r$ and $p \simeq q$ at position u in $p \simeq q$ is performed only if the mgu σ such that $l\sigma = (p|u)\sigma$ is E-inductively complete. The position u is called *completely superposable* in [29]. This result requires an algorithm to detect the completely superposable positions. An equivalent characterization is the following: a position u is completely superposable if the term $p|u$ is such that for all its ground instances $(p|u)\rho$ there is an equation $l \simeq r$ in E such that $(p|u)\rho = l\sigma$ and $l\sigma \succ r\sigma$. This shows that the problem of detecting completely superposable positions is basically an instance of the ground reducibility problem. However, if the presentation satisfies the principle of definition, a position u is completely superposable if $p|u$ is a term which has a defined symbol at the root and only constructor symbols and variables at the positions below the root. Therefore, the above theorem can be applied in practice to presentations satisfying the principle of definition.

5 Conclusions

We have given a new abstract framework for the study of Knuth-Bendix type completion procedures, which are regarded as *semidecision procedures* for theorem proving.

All the fundamental concepts are uniformly defined in terms of *proof reduction* with respect to a well founded proof ordering. In order to do this, we have given a new, more general notion of proof ordering, such that also proofs of different theorem can be compared.

A completion procedure is given by a set of *inference rules* and a *search plan*. We have emphasized the distinction between these two components throughout our work. This distinction is often overlooked in the literature, where most theorem proving strategies are presented by giving the set of inference rules only, whereas the search plan is what ultimately turns a set of inference rules into a procedure.

If the inference rules are *refutationally complete* and the search plan is *fair*, a completion procedure is a semidecision procedure for theorem proving. The key part of this result is the notion of *fairness.* Our definition of fairness is the first definition of fairness for completion procedures which addresses the theorem proving problem. It is new in three ways: it is *target oriented*, that is it keeps the theorem to be proved into consideration, it is explicitly stated as a property of the search plan and it is defined in terms of proof reduction, so that expansion inferences and contraction inferences are treated uniformly. We have also shown that the process of diproving inductive theorems by the so called *inductionless induction* method is a semidecision process.

In the second part of this work, we have presented some equational completion procedures based on Unfailing Knuth-Bendix completion, which include the AC-UKB procedure with Cancellation laws, the S-strategy and the Inequality Ordered Saturation strategy. These extensions of UKB had not been presented in a unified framework for completion before.

Directions for further research include the study of efficient, fair, but not uniformly fair, search plans and the full extension of this approach to completion procedures for Horn and first order logic with equality.

References

[1] S.Anantharaman and J.Mzali, Unfailing Completion modulo a set of equations, Technical Report, LRI, Université de Paris Sud, 1989.

[2] S.Anantharaman, J.Hsiang and J.Mzali, SbReve2: A Term Rewriting Laboratory with (AC)-Unfailing Completion, in N.Dershowitz (ed.), *Proceedings of the Third International Conference on Rewriting Techniques and Applications*, Chapel Hill, NC, USA, April 1989, Springer Verlag, Lecture Notes in Computer Science 355, 533–537, 1989.

[3] S.Anantharaman, J.Hsiang, Automated Proofs of the Mougang Identities in Alternative Rings, *Journal of Automated Reasoning*, Vol. 6, No. 1, 76–109, 1990.

[4] S.Anantharaman, N.Andrianarivelo, Heuristical Criteria in Refutational Theorem Proving, in *Proceedings of the Symposium on the Design and Implementation of Systems for Symbolic Computation*, 184–193, Capri, Italy, April 1990.

[5] S.Anantharaman, N.Andrianarivelo, M.P.Bonacina, J.Hsiang, SBR3: A Refutational Prover for Equational Theorems, to appear in *Proceedings of the Second International Workshop on Conditional and Typed Rewriting Systems*, Montreal, Canada, June 1990.

[6] L.Bachmair, N.Dershowitz, J.Hsiang, Orderings for Equational Proofs, in *Proceedings of the First Annual IEEE Symposium on Logic in Computer Science*, 346–357, Cambridge, MA, June 1986.

[7] L.Bachmair, N.Dershowitz, Completion for rewriting modulo a congruence, in P.Lescanne (ed.), *Proceedings of the Second International Conference on Rewriting Techniques and Applications*, Bordeaux, France, May 1987, Springer Verlag, Lecture Notes in Computer Science 256, 192–203, 1987.

[8] L.Bachmair, N.Dershowitz, Inference Rules for Rewrite-Based First-Order Theorem Proving, in *Proceedings of the Second Annual Symposium on Logic in Computer Science*, Ithaca, New York, June 1987.

[9] L.Bachmair, Proofs Methods for Equational Theories, Ph.D. thesis, Department of Computer Science, University of Illinois, Urbana, IL.,1987.

[10] L.Bachmair, Proof by consistency in equational theories, in *Proceedings of the Third Annual IEEE Symposium on Logic in Computer Science*, 228–233, Edinburgh, Scotland, July 1988.

[11] L.Bachmair, N.Dershowitz and D.A.Plaisted, Completion without failure, in H.Ait-Kaci, M.Nivat (eds.), *Resolution of Equations in Algebraic Structures*, Vol. II: Rewriting Techniques, 1–30, Academic Press, New York, 1989.

[12] L.Bachmair, H.Ganzinger, On Restrictions of Ordered Paramodulation with Simplification, in *Proceedings of the Tenth International Conference on Automated Deduction*, Kaiserslautern, Germany, July 1990.

[13] L.Bachmair, H.Ganzinger, Completion of First-Order Clauses with Equality by Strict Superposition, to appear in M.Okada, S.Kaplan (eds.), *Proceedings of the Second International Workshop on Conditional and Typed Rewriting Systems*, Montreal, Canada, June 1990.

[14] M.P.Bonacina, G.Sanna, KBlab: An Equational Theorem Prover for the Macintosh, in N.Dershowitz (ed.), *Proceedings of the Third International Conference on Rewriting Techniques and Applications*, Chapel Hill, NC, USA, April 1989, Springer Verlag, Lecture Notes in Computer Science 355, 548–550, 1989.

[15] M.P.Bonacina, J.Hsiang, Operational and Denotational Semantics of Rewrite Programs, to appear in *Proceedings of the North American Conference on Logic Programming*, Austin, TX, October 1990.

[16] M.P.Bonacina, J.Hsiang, On fairness of completion-based theorem proving strategies, Technical report, Department of Computer Science, SUNY at Stony Brook.

[17] M.P.Bonacina, J.Hsiang, The Knuth-Bendix-Huet theorem and its extensions, in preparation.

[18] C.L.Chang, R.C.Lee, *Symbolic Logic and Mechanical Theorem Proving*, Academic Press, New York, 1973.

[19] N.Dershowitz, Z.Manna, Proving termination with multisets orderings, *Communications of the ACM*, Vol. 22, No. 8, 465–476, August 1979.

[20] N.Dershowitz, N.A.Josephson, Logic Programming by Completion, in *Proceedings of the Second International Conference on Logic Programming*, 313–320, Uppsala, Sweden, 1984.

[21] N.Dershowitz, Computing with Rewrite Systems, *Information and Control*, Vol. 65, 122–157, 1985.

[22] N.Dershowitz, D.A.Plaisted, Logic Programming Cum Applicative Programming, in *Proceedings of the IEEE Symposium on Logic Programming*, 54–66, Boston, MA, 1985.

[23] N.Dershowitz, Termination of Rewriting, *Journal of Symbolic Computation*, Vol. 3, No. 1 & 2, 69–116, February/April 1987.

[24] N.Dershowitz, Completion and its Applications, in *Proceedings of Conference on Resolution of Equations in Algebraic Structures*, Lakeway, Texas, May 1987.

[25] N.Dershowitz, J.-P.Jouannaud, Rewrite Systems, Technical Report 478, LRI, Université de Paris Sud, April 1989 and Chapter 15 of Volume B of *Handbook of Theoretical Computer Science*, North-Holland, 1989.

[26] N.Dershowitz, J.-P.Jouannaud, Notations for Rewriting, Rapport de Recherche 478, LRI, Université de Paris Sud, January 1990.

[27] N.Dershowitz, A Maximal-Literal Unit Strategy for Horn Clauses, to appear in M.Okada, S.Kaplan (eds.), *Proceedings of the Second International Workshop on Conditional and Typed Rewriting Systems*, Montreal, Canada, June 1990.

[28] F.Fages, Associative-commutative unification, in R.Shostak (ed.), *Proceedings of the Seventh International Conference on Automated Deduction*, Napa Valley, CA, USA, 1984, Springer Verlag, Lecture Notes in Computer Science 170, 1984.

[29] L.Fribourg, A Strong Restriction to the Inductive Completion Procedure, in *Proceedins of the Thirteenth International Conference on Automata Languages and Programming*, Rennes, France, July 1986, Springer Verlag, Lecture Notes in Computer Science 226, 1986.

[30] J.A.Goguen, How to prove algebraic inductive hypotheses without induction, in W.Bibel and R.Kowalski (eds.), *Proceedings of the Fifth International Conference on Automated Deduction*, 356–373, Les Arcs, France, 1980, Springer Verlag, Lecture Notes in Computer Science 87, 1980.

[31] J.Hsiang, N.Dershowitz, Rewrite Methods for Clausal and Nonclausal Theorem Proving, in *Proceedings of the Tenth International Conference on Automata, Languages and Programming*, Barcelona, Spain, July 1983, Springer Verlag, Lecture Notes in Computer Science 154, 1983.

[32] J.Hsiang, Refutational Theorem Proving Using Term Rewriting Systems, *Artificial Intelligence*, Vol. 25, 255–300, 1985.

[33] J.Hsiang, M.Rusinowitch, A New Method for Establishing Refutational Completeness in Theorem Proving, in J.Siekmann (ed.), *Proceedings of the Eighth Conference on Automated Deduction*, Oxford, England, July 1986, Springer Verlag, Lecture Notes in Computer Science 230, 141–152, 1986.

[34] J.Hsiang, Rewrite Method for Theorem Proving in First Order Theories with Equality, *Journal of Symbolic Computation*, Vol. 3, 133–151, 1987.

[35] J.Hsiang, M.Rusinowitch, On word problems in equational theories, in Th.Ottman (ed.), *Proceedings of the Fourteenth International Conference on Automata, Languages and Programming*, Karlsruhe, West Germany, July 1987, Springer Verlag, Lecture Notes in Computer Science 267, 54–71, 1987.

[36] J.Hsiang, M.Rusinowitch and K. Sakai, Complete Inference Rules for the Cancellation Laws, in *Proceedings of the Tenth International Joint Conference on Artificial Intelligence*, Milano, Italy, August 1987, 990–992, 1987.

[37] J.Hsiang, M.Rusinowitch, Proving Refutational Completeness of Theorem Proving Strategies: the Transfinite Semantic Tree Method, to appear in *Journal of the ACM*, 1990.

[38] G.Huet, Confluent reductions: abstract properties and applications to term rewriting systems, *Journal of the ACM*, Vol. 27, 797–821, 1980.

[39] G.Huet, A Complete Proof of Correctness of the Knuth-Bendix Completion Algorithm, *Journal of Computer and System Sciences*, Vol. 23, 11–21, 1981.

[40] G.Huet, J.M.Hullot, Proofs by Induction in Equational Theories with Constructors, Journal of Computer and System Sciences, Vol. 25, 239–266, 1982.

[41] J.-P.Jouannaud, C.Kirchner, Completion of a set of rules modulo a set of equations, *SIAM Journal of Computing*, Vol. 15, 1155–1194, November 1986.

[42] J.-P.Jouannaud, E.Kounalis, Proofs by induction in equational theories without constructors, in *Proceedings of the First Annual IEEE Symposium on Logic in Computer Science*, 358–366, Cambridge, MA, June 1986.

[43] J.-P.Jouannaud, E.Kounalis, Automatic proofs by induction in equational theories without constructors, *Information and Computation*, 1989.

[44] J.-P.Jouannaud, C.Kirchner, Solving Equations in Abstract Algebras: A Rule-Based Survey of Unification, Rapport de Recherche, LRI, Université de Paris Sud, November 1989.

[45] S.Kamin, J.-J.Lévy, Two generalizations of the recursive path ordering, Unpublished note, Department of Computer Science, University of Illinois, Urbana, Illinois, February 1980.

[46] D.Kapur and P.Narendran, An equational approach to theorem proving in first order predicate calculus, in *Proceedings of the Ninth International Joint Conference on Artificial Intelligence*, 1146–1153, Los Angeles, CA, August 1985.

[47] D.Kapur and D.R.Musser, Proof by consistency, *Artificial Intelligence*, Vol. 31, No. 2, 125–157, February 1987.

[48] D.Kapur, P.Narendran and H.Zhang, Proof by induction using test sets, in J.Siekmann (ed.), *Proceedings of the Eighth Conference on Automated Deduction*, Oxford, England, July 1986, Springer Verlag, Lecture Notes in Computer Science 230, 99–117, 1986.

[49] D.E.Knuth, P.Bendix, Simple Word Problems in Universal Algebras, in J.Leech (ed.), *Proceedings of the Conference on Computational Problems in Abstract Algebras*, Oxford, England, 1967, Pergamon Press, Oxford, 263–298, 1970.

[50] E.Kounalis, M.Rusinowitch, On Word Problems in Horn Theories, in E.Lusk, R.Overbeek (eds.), *Proceedings of the Ninth International Conference on Automated Deduction*, 527–537, Argonne, Illinois, May 1988, Springer Verlag, Lecture Notes in Computer Science 310, 1988.

[51] D.S.Lankford, Canonical inference, Memo ATP-32, Automatic Theorem Proving Project, University of Texas, Austin, TX, May 1975.

[52] D.S.Lankford, A simple explanation of inductionless induction, Technical report MTP-14, Mathematics Department, Louisiana Technical University, Ruston, LA, 1981.

[53] D.Musser, On proving inductive properties of abstract data types, in *Proceedings of the Seventh ACM Symposium on Principles of Programming Languages*, 154–162, Las Vegas, Nevada, 1980.

[54] G.E.Peterson, M.E.Stickel, Complete sets of reductions for some equational theories, *Journal of the ACM*, Vol. 28, No. 2, 233–264, 1981.

[55] G.E.Peterson, A Technique for Establishing Completeness Results in Theorem proving with Equality, *SIAM Journal of Computing*, Vol. 12, No. 1, 82–100, 1983.

[56] D.A.Plaisted, Semantic confluence tests and completion methods, *Information and Control*, Vol. 65, 182–215, 1985.

[57] M.Rusinowitch, Theorem-proving with resolution and superposition: an extension of Knuth and Bendix procedure as a complete set of inference rules, Thèse d'Etat, Université de Nancy, 1987.

[58] M.E.Stickel, Unification Algorithms for Artificial Intelligence Languages, Ph.D. thesis, Carnegie Mellon University 1976.

[59] M.E.Stickel, A unification algorithm for associative-commutative functions, *Journal of the ACM*, Vol. 28, No. 3, 423–434, 1981.

[60] H.Zhang, D.Kapur, First Order Theorem Proving Using Conditional Rewrite RUles, in E.Lusk, R.Overbeek (eds.), *Proceedings of the Ninth International Conference on Automated Deduction*, 1–20, Argonne, Illinois, May 1988, Springer Verlag, Lecture Notes in Computer Science 310, 1988.

Errata corrige

Completion Procedures as Semidecision Procedures

(by M.P.Bonacina and J.Hsiang)

Page 226 Definition 4.1should read as follows:

Given a ground confluent presentation E, a set of substitutions H is *E-inductively complete* if for all ground substitutions ρ, there exist a substitution $\sigma \in H$ and a ground substitution τ such that $\rho \rightarrow^*_E \sigma\tau$.

And the following paragraph should start with:

We denote by H_E one such set and by $\mathcal{IT}_E$ the domain of all the ground equations which are instances of substitutions in H_E, that is $\mathcal{IT}_E = \{(l \simeq r)\sigma\tau | \sigma \in H_E, (l \simeq r)\sigma\tau \textit{ is ground}\}$. The proof of the target is the proof of the smallest ground theorem which is not a theorem of the original presentation. This smallest ground theorem is in $\mathcal{IT}_E$ and therefore reducing the proofs of the theorems in $\mathcal{IT}_E$ is sufficient to guarantee that the proof of the target is reduced,

Linear completion *

Hervé Devie

LRI, Université Paris-Sud, Bât 490
91405 ORSAY Cedex, France

Abstract

We give an example of a set of linear equational axioms such that no finite canonical rewrite system can be computed by ordered completion with a complete reduction ordering, although such a rewrite system does trivially exist. We then describe a set of inference rules for a completion procedure, called linear completion, that solves the problem when the axioms are linear.

1 Introduction

An equational theory is given by a set E_0 of equations between terms and will be denoted by $=_{E_0}$ in the sequel. We aim at solving the word problem for $=_{E_0}$, a problem known to be semi-decidable. An interesting semi-decision procedure is based on a method due to Knuth and Bendix [11], called *Knuth-Bendix completion*. It consists in computing a presentation of the theory $=_{E_0}$ by a new set E of equations equivalent to E_0 such that a proof of a theorem $s =_{E_0} t$ becomes constructive by rewriting s and t to their normal form with respect to the equations in E oriented into rules beforehand. Huet proposed and proved a completion procedure [9] based on eagerly orienting the equations into rules thanks to a reduction ordering on terms. But this method may fail when some equation cannot be oriented. A description of completion by inference rules was later used by Bachmair, Dershowitz and Hsiang [2]. In this framework, the completeness of the inference rules is proved by using meta level arguments on proofs being reduced when the inference rules are applied. This method was further developped by Bachmair [1], who also applied it to ordered completion, originally introduced by Hsiang and Rusinowitch [7] under the name of unfailing completion. In ordered completion, the ground instances of the equations in E are oriented, rather than the equations themselves.

Ordered completion enjoys an important property proved by Bachmair [1] : if a canonical rewrite system R exists for a given set of axioms E_0 which can be oriented by a reduction ordering total on ground terms, then ordered completion, when using a fair control, will compute it. In section 4, we exhibit an example of a set of axioms for which such a set R exists, but cannot be proved terminating by any reduction ordering total on ground terms. In section 5, we propose a new set of inference rules, linear completion,

*This work has been partially supported by GRECO de Programmation du CNRS

and prove its completeness by proof rewriting. In section 6, we study the particular case of linear presentations, for which linear completion has a very nice behaviour : first, if a canonical rewrite system R exists, then any deterministic version of linear completion, using a fair control and working with a reduction ordering containing R, will generate R ; second, linear completion is refutationally complete. In section 7, we point out the limits of linear completion.

2 Notations and preliminary definitions

Our definitions are consistant with Huet and Oppen [10] and Dershowitz and Jouannaud [4][5]. Notations are borrowed from the latter.

Given a finite graded set $\mathcal{F} = \cup_{n\geq 0}\mathcal{F}_n$, where $\mathcal{F}_n$ is the set of *function symbols* of arity n, and a denumerable set $\mathcal{X}$ of *variable symbols*, the *term algebra* $\mathcal{T}(\mathcal{F},\mathcal{X})$ is the set of elements, called *terms*, containing $\mathcal{X}$ and such that $f(t_1,\ldots,t_n)$ is a term if and only if $t_1,\ldots,t_n$ are terms and f is in $\mathcal{F}_n$.

Any term t in $\mathcal{T}(\mathcal{F},\mathcal{X})$ can be viewed as an application of its *domain* $\mathcal{P}(t)$ into $\mathcal{F}\cup\mathcal{X}$, where $\mathcal{P}(t)$ is the subset of N_+^*, the free monoid over N_+, such that we have $\Lambda \in \mathcal{P}(t)$ and $ip \in \mathcal{P}(f(t_1,\ldots,t_n))$ if and only if $1 \leq i \leq n$ and $p \in \mathcal{P}(t_i)$. Elements in $\mathcal{P}(t)$ are called *positions* (occurrences is also used in the litterature). If p and q are two positions such that neither is a left factor of the other, we say that they are *parallel* and write $p \parallel q$. We denote by $t|_p$ the subterm of t at position p, and by $t[s]_p$ the term obtained by replacement of $t|_p$ by s in t. Hence $t = t[t|_p]_p$. $\mathcal{GP}(t) = \{p \in \mathcal{P}(t) \,/ t|_p \notin \mathcal{X}\}$ denotes the subset of ground (also called non-variable in the litterature) positions of t.

A *substitution* is an homomorphism of $\mathcal{T}(\mathcal{F},\mathcal{X})$ considered as an $\mathcal{F}$-algebra. We denote by $t\sigma$ the application of the substitution σ to the term t. Given two substitutions σ and τ, we say that σ is *more general than* τ if $\tau = \sigma\theta$ for some substitution θ. Two terms s and t *unify* if there exists a substitution τ such that $s\tau = t\tau$. In this case, the most general substitution such that s and t unify is called the *most general unifier* (*mgu* for short) of s and t.

Given two strict orderings $>_1$ and $>_2$, $>_2$ is a *refinement* of $>_1$ (we write $>_1\subseteq>_2$) if $s >_2 t$ whenever $s >_1 t$. A *reduction ordering* $>$ on $\mathcal{T}(\mathcal{F},\mathcal{X})$ is a strict ordering which satisfies the properties of *well-foundedness* (every strictly decreasing chain is finite), of *monotonicity* ($u[s] > u[t]$ if $s > t$, where s, t and u are in $\mathcal{T}(\mathcal{F},\mathcal{X})$) and of *instanciation* ($s\sigma > t\sigma$ if $s > t$, where s, t are in $\mathcal{T}(\mathcal{F},\mathcal{X})$ and σ is a substitution). If in addition $>$ is total on ground terms, we call it a *complete reduction ordering*. A particular ordering on $\mathcal{T}(\mathcal{F},\mathcal{X})$ is the *encompassment ordering* $\trianglerighteq$: $l \trianglerighteq g$ if $l|_p = g\sigma$. The associated strict ordering, denoted by $\blacktriangleright$, is a well-founded ordering.

Actually, reduction ordering is too strong for completion, and can be weakened as follows :

Definition 1 *Let E be a set of axioms on $\mathcal{T}(\mathcal{F},\mathcal{X})$ and $\succ$ be a strict ordering on $\mathcal{T}(\mathcal{F},\mathcal{X})$.*
• *$\succ$ is said to be a* reduction ordering wrt *E if :*
i) $\succ$ is well-founded on $\mathcal{T}(\mathcal{F},\mathcal{X})$;
ii) $\succ$ has the monotonicity property wrt *E, i.e. : whenever $s =_E t$ and $s \succ t$, then $C[s] \succ C[t]$ for every context C;*

iii) $\succ$ *has the* instanciation property wrt E, *i.e. : whenever* $s =_E t$ *and* $s \succ t$, *then* $s\sigma \succ t\sigma$ *for every substitution* σ.
• $\succ$ *is said to be a* complete reduction ordering wrt E *if :*
i) $\succ$ *is a reduction ordering wrt* E;
ii) $\succ$ *is* total on ground terms wrt E, *i.e. : whenever* $s =_E t$ *and* $(s,t) \in \mathcal{T}(\mathcal{F}) \times \mathcal{T}(\mathcal{F})$, *then either* $s \succ t$, *or* $t \succ s$ *or* $s = t$.

Of course, a (respectively complete) reduction ordering is a (respectively complete) reduction ordering wrt E for any set E of axioms.

A *term rewriting system* is a set of rules written $\{l_i \to r_i\}_i$. A term t *rewrites to* t', denoted $t \to t'$, if $t = t[l_i\sigma]_p$ and $t' = t[r_i\sigma]_p$. The *reduction relation* is the reflexive transitive closure of the rewrite relation and is denoted by $\to^*$. The transitive closure is denoted by $\to^+$. A transitive and irreflexive rewrite relation is called a *rewrite ordering*. We denote by $t\!\downarrow$ a *normal form* of t, i.e. an irreducible term t' such that $t \to^* t'$. Two terms s and t are said to be *confluent* if they reduce to a same term v and *divergent* if they are obtained by two different reductions from a same term u. A term rewriting system R is *confluent* if every divergent pair is confluent, *terminating* if $\to^*$ is well-founded, *convergent* if it is both confluent and terminating. It is *inter-reduced* if, for each rule $l \to r$, r is irreducible under R and no term s less than l in the encompassment ordering $\blacktriangleright$ is reducible [4]. A term rewriting system is said to be *canonical* if it is convergent and inter-reduced.

The symbol $\approx$ will be used to denote equational axioms in order to avoid confusion with the syntactic equality (denoted by $=$). A term t in $\mathcal{T}(\mathcal{F},\mathcal{X})$ is *linear* if no variable of t occurs more than once in t. An equation $l \approx r$ is *linear* if both terms l and r are linear. A variable x is *linear* in a rule $l \to r$ if x occurs at most once in l and in r. Given a set of equations $E = \{l_i \to r_i\}_i$ and a rewrite ordering $\succ$ on $\mathcal{T}(\mathcal{F},\mathcal{X})$, the *ordered rewrite relation*, denoted by $\to_\succ$, is defined by $t \to_\succ t'$ if $t = t[l_i\sigma]_p$ and $t' = t[r_i\sigma]_p$ such that $l_i\sigma \succ r_i\sigma$.

Finally, we will denote terms by s,t,u,v,w,l,r,g,d, positions by p,q,o and substitutions by σ,τ.

3 Ordered Completion

We are given a set of equational axioms E_0 and a reduction ordering $>$. In the following, we do not distinguish between oriented and unoriented equations : consequently, we only use one generic set Ω for all equations. At the beginning, Ω is equal to E_0.

Here are the three inference rules (**OC**) of ordered completion :

Deduce : $\Omega \cup \{l \approx r\} \cup \{g \approx d\} \;\vdash\; \Omega \cup \{l \approx r\} \cup \{g \approx d\} \cup \{r\sigma \approx l\sigma[d\sigma]_p\}$
where $\sigma = mgu(l|_p, g)$, $r\sigma \not\geq l\sigma$ and $d\sigma \not\geq g\sigma$

Delete : $\Omega \cup \{l \approx l\} \;\vdash\; \Omega$

Simplify : $\Omega \cup \{l \approx r\} \cup \{g \approx d\} \;\vdash\; \Omega \cup \{u \approx r\} \cup \{g \approx d\}$
where $l \longrightarrow_{\{g\approx d\}\succ} u$ and ($l \blacktriangleright g$ if $l > r$)

Remarks :
• There is no rule for orientation. From a practical point of view, new equations are

implicitely oriented whenever possible.

• The conditions in **Deduce** are a little bit stronger than those in [1] or [4], which only impose $l\sigma[d\sigma]_p \not\succ l\sigma$ (instead of $d\sigma \not\succ r\sigma$). So our procedure will a priori compute more extended critical pairs. However, it is more consistent with the use of the ordered rewrite relation $\rightarrow_\succ$: only instances of equations are considered, not contexts.

A control must be added to this system of inference rules in order to get a deterministic procedure for ordered completion. Once a control is given, the sequence (or derivation) $(\Omega_i)_{i\geq 0}$ is totally defined with Ω_0 equal to E_0 and $\Omega_i \vdash \Omega_{i+1}$ (applying the chosen control). Let Ω_∞ be the inductive limit of this sequence, i.e. $\Omega_\infty = \bigcup_{i\geq 0} \bigcap_{j\geq i} \Omega_j$. If Ω_∞ is not a canonical rewrite system, we say that the procedure *fails.*

We denote by $OC(\Omega)$ the union of all sets of equations that can be obtained by applying at most one inference rule in **(OC)** from Ω. We require the control to be *fair* in the following sense [1] : $OC(\Omega_\infty)$ is a subset of $\bigcup_{i\geq 0} \Omega_i$.

4 An example of a theory E_0 for which no complete reduction ordering wrt E_0 is adapted

The problem we are interested in is to get sufficient conditions on the pair $(E_0, >)$ such that every fair control yields a canonical set of rules R if such one exists. A completion procedure having this property will be said to be *unfailing.* Bachmair [1] has obtained the following key result :

THEOREM 1 *Let R be a canonical system for E_0 and $>$ be a reduction ordering containing R. If $>$ can be extended to a complete reduction ordering, then every fair control on* **(OC)** *will generate from $(E_0, >)$ a derivation such that Ω_∞ and R are equal up to renaming.*

Actually, Bachmair's proof still holds when replacing 'reduction ordering' (respectively 'complete reduction ordering') by 'reduction ordering wrt E_0' (respectively 'complete reduction ordering wrt E_0'). In the following, when referring to Theorem 1, we shall consider this more general version. We want now to give an example of an equational theory E_0 for which ordered completion yields a finite canonical rewrite system, if and only if a reduction ordering wrt E_0 is used that cannot be extended to a complete reduction ordering wrt E_0. First, we prove an easy lemma about ordered completion :

LEMMA 1 *Let E_0 be a set of axioms and $\succ_1$, $\succ_2$ and $\succ$ be reduction orderings wrt E_0 such that $\succ_1 \subseteq \succ_2 \subseteq \succ$ and $\succ$ is a complete reduction ordering wrt E_0. If ordered completion using a fair control and starting with $(E_0, \succ_1)$ yields a canonical rewrite system R, then ordered completion using a fair control and starting with $(E_0, \succ_2)$ yields the same canonical rewrite system R up to renaming.*

Proof: Let R be the canonical system yielded with $(E_0, \succ_1)$. For each rule $l \rightarrow r$ in R, we have $l \succ_2 r$. Hence we get easily that R is a canonical rewrite system for $(E_0, \succ_2)$. From uniqueness of the canonical rewrite system for $(E_0, \succ_2)$ up to renaming [6], R is 'the' canonical rewrite system for $(E_0, \succ_2)$ up to renaming, and using Theorem 1, ordered completion using a fair control and starting with $(E_0, \succ_2)$ will compute it. □

Proposition 1 *Let E_0 be the set of axioms*

$$E_0 = \{fgix \approx gifgjx, f'g'jx \approx g'jf'g'ix, hgix \approx gix, h'g'jx \approx g'jx, fa \approx a, f'a \approx a, ga \approx a, g'a \approx a, ha \approx a, h'a \approx a, ia \approx a, ja \approx a\},$$

where a is a constant, x is a variable and all other letters denote unary function symbols. The following properties hold :
i) Orienting all equations in E_0 from left to right immediately provides a canonical rewrite system R such that no reduction ordering wrt E_0 containing R can be extended to a complete reduction ordering wrt E_0;
ii) When feeded with E_0 and a complete reduction ordering wrt E_0, ordered completion diverges. (Hence, no finite canonical rewrite system R for E_0 can be contained in a complete reduction ordering wrt E_0).

Proof: *i)* Let R be the rewrite system

$$R = \{fgix \rightarrow gifgjx, f'g'jx \rightarrow g'jf'g'ix, hgix \rightarrow gix, h'g'jx \rightarrow g'jx, fa \rightarrow a, f'a \rightarrow a, ga \rightarrow a, g'a \rightarrow a, ha \rightarrow a, h'a \rightarrow a, ia \rightarrow a, ja \rightarrow a\}$$

and $>$ be the ordering on $\mathcal{T}(\mathcal{F},\mathcal{X})$ defined by the transitive closure $\rightarrow_R^+$ of the R-rewrite relation. $>$ is a reduction ordering on $\mathcal{T}(\mathcal{F},\mathcal{X})$ if and only if R is terminating. Let c be the interpretation function from $\mathcal{T}(\mathcal{F},\mathcal{X})$ on the set of ordered pairs of natural numbers defined as follows : $\forall t \in \mathcal{T}(\mathcal{F},\mathcal{X}), c(t) = (c_1(t) + c_2(t), l(t))$, where :

- $c_1(t)$ is the total number of f in t appearing in a pattern $f^n gi$ of t, $n \geq 0$;
- $c_2(t)$ is the total number of f' in t appearing in a pattern $f'^m g'j$ of t, $n \geq 0$;
- $l(t)$ is the length of t.

We can easily see that, if s and t are two terms of $\mathcal{T}(\mathcal{F},\mathcal{X})$ such that $s \rightarrow_R t$, then $c(s) >_{lex} c(t)$, where $>_{lex}$ is the usual lexicographic ordering on ordered pairs of natural numbers : therefore, R is terminating. Moreover, since R is inter-reduced and left-hand sides do not overlap each other, R is actually a canonical rewrite system.
Since $>$ is equal to $\rightarrow_R^+$, every complete reduction ordering wrt E_0 containing R also contains $>$. Now, suppose that $>$ can be extended to a complete reduction ordering wrt E_0, called $\succ$. Since $ia =_{E_0} ja$, we should have either $ja \succ ia$ or $ia \succ ja$. $ja \succ ia$ would imply $fgja \succ fgia$ and, by the first rule, $fgja \succ gifgja$, which contradicts the property that $\succ$ is well-founded. $ia \succ ja$ would yield a similar contradiction. Consequently, $>$ cannot be extended to a complete reduction ordering wrt E_0.

ii) Let $>$ be a reduction ordering on $\mathcal{T}(\mathcal{F},\mathcal{X})$ that can be extended to a complete reduction ordering wrt E_0. We want to prove that no fair procedure of ordered completion starting with $(E_0, >)$ can yield a finite canonical rewrite system R. Thanks to Lemma 1, we can consider without loss of generality that $>$ is already a complete reduction ordering wrt E_0, and that all equations in E_0 or deduced during completion have to be oriented whenever possible. Consequently, all equations in E_0 except the two first ones are necessarily oriented by $>$ from left to right. Considering the two first equations, we know that we cannot orient both of them from left to right. Hence, two cases may occur : either we have $gifgjx \rightarrow fgix$ and $f'g'jx \rightarrow g'jf'g'ix$ (the case of $fgix \rightarrow gifgjx$ and $g'jf'g'ix \rightarrow f'g'jx$ is similar), or we have $gifgjx \rightarrow fgix$ and $g'jf'g'ix \rightarrow f'g'jx$. In the first case, the starting rules in Ω_0 are :

(1) $gifgjx \rightarrow fgix$	(5) $fa \rightarrow a$	(9) $ha \rightarrow a$
(2) $f'g'jx \rightarrow g'jf'g'ix$	(6) $f'a \rightarrow a$	(10) $h'a \rightarrow a$
(3) $hgix \rightarrow gix$	(7) $ga \rightarrow a$	(11) $ia \rightarrow a$
(4) $h'g'jx \rightarrow g'jx$	(8) $g'a \rightarrow a$	(12) $ja \rightarrow a$

Deduce applied with a ground rule always yields a confluent critical pair. Hence, the only interesting inference rule is obtained when overlapping $gifgjx$ on $hgix$; after simplification, it generates the equation $hfgix \approx fgix$ which can be oriented into the rule $hfgix \rightarrow fgix$. Hence $\Omega_1 = \Omega_0 \cup \{hfgix \rightarrow fgix\}$. Again, only one new non-confluent critical pair can be computed by overlapping $gifgjx$ on $hfgix$, which gives the new rule $hf^2gix \rightarrow f^2gix$. By induction, we get that $\Omega_\infty = \Omega_0 \cup \bigcup_{n>0}\{hf^ngix \rightarrow f^ngix\}$.
In the second case, the starting rules in Ω_0 are :

(1) $gifgjx \rightarrow fgix$	(5) $fa \rightarrow a$	(9) $ha \rightarrow a$
(2) $g'jf'g'ix \rightarrow f'g'jx$	(6) $f'a \rightarrow a$	(10) $h'a \rightarrow a$
(3) $hgix \rightarrow gix$	(7) $ga \rightarrow a$	(11) $ia \rightarrow a$
(4) $h'g'jx \rightarrow g'jx$	(8) $g'a \rightarrow a$	(12) $ja \rightarrow a$

The difference, now, is that, at each step, **Deduce** can be applied in two different ways, because (1) and (2) are respectively overlapping (3) and (4). But, we remark that, for $i \geq 0$, Ω_i is split into two independant sets of rules (the first one containing normal function symbols and the other one primed function symbols). Moreover, no inference rule uses a rule of one set to transform the other set. Consequently, by using a symmetry argument and the same reasoning as above, we get $\Omega_\infty = \Omega_0 \cup \bigcup_{n>0}\{hf^ngix \rightarrow f^ngix\} \cup \bigcup_{n>0}\{h'f'^ng'jx \rightarrow f'^ng'jx\}$.
Hence, whenever all equations derived by a fair procedure of ordered completion are rules, an infinite canonical rewrite system is computed. As a consequence of Lemma 1, any other reduction ordering wrt E_0 that can be extended to a complete one cannot yield a finite canonical rewrite system. □

This example gives an answer to a question raised by Nachum Dershowitz [3] : quasi-reduction orderings total on ground terms do not suffice to compute a given rewrite system by using ordered completion.

We therefore have supplied an example of an equational presentation E_0 which cannot be conveniently dealt with ordered completion. In our sense, ordered completion fails on this example, whatever complete reduction ordering is chosen. On the other hand, this example belongs to the very specific and nice class of linear equations. In the next section, we present a new completion procedure that solves the case of linear equations. For this reason, we call it *linear completion.*

5 Linear completion

Here is the system of inference rules, called **(LC)** (standing for linear completion) :

Deduce : $\Omega \cup \{l \approx r\} \cup \{g \approx d\} \;\vdash\; \Omega \cup \{l \approx r\} \cup \{g \approx d\} \cup \{r\sigma \approx l\sigma[d\sigma]_p\}$
if $\sigma = mgu(l|_p, g)$ and $[(l > r \text{ and } d \not\geq g) \text{ or } (g > d \text{ and } r \not\geq l)]$

Delete : $\Omega \cup \{l \approx l\} \;\vdash\; \Omega$

Simplify : $\Omega \cup \{l \approx r\} \cup \{g \approx d\} \;\vdash\; \Omega \cup \{u \approx r\} \cup \{g \approx d\}$
if $g > d$, $l \longrightarrow_{\{g \approx d\}} u$ and $(l \blacktriangleright g \text{ if } l > r)$

Remarks :
• We use standard rewriting, as in standard completion. We do not split Ω into two sets E

of equations and R of rules, as in standard completion. Such a distinction would increase the number of inference rules. As a consequence, **Simplify** sums up the inference rules called *Simplify*, *Compose* and *Collapse* by Dershowitz and Jouannaud [4].

• **Deduce** is applied when a starting equation is oriented (i.e., is a rule) and when the other one is either non-orientable, or oriented in such a way that we obtain a standard critical pair. In the sequel, $(r\sigma, l\sigma[d\sigma]_p)$ is called a *linear critical pair* (in order to distinguish with (standard) critical pairs in standard completion and ordered critical pairs in ordered completion, where overlaps between arbitrary equations are computed).

As in Bachmair [1], we develop a notion of proof and a rewrite relation on proofs. Let Ω_i be a set of equations obtained at step i of linear completion, and s and t be two terms in $\mathcal{T}(\mathcal{F},\mathcal{X})$. An *elementary* Ω_i*- proof* between s and t is a tuple $(s,t,l\approx r,p,\sigma)$, where $l \approx r$ is an equation in Ω_i, p is a position and σ is a substitution, such that $s|_p = l\sigma$ and $t|_p = r\sigma$. We write $s -_{p,\sigma,l\approx r} - t$. Three separate cases may occur and be distinguished by different notations :

• if $l > r$, we may write $s -_{p,\sigma,l\approx r} \rightarrow t$ and the corresponding pattern is $\rightarrow$;

• if $r > l$, we may write $s \leftarrow_{p,\sigma,l\approx r} - t$ and the corresponding pattern is $\leftarrow$;

• otherwise, we may write $s \leftarrow_{p,\sigma,l\approx r}\rightarrow t$, and the pattern is $\leftrightarrow$.

Note that we differ from Bachmair's notion of proof which is based on the ordered rewrite relation $\rightarrow_{\succ}$.

Concatenation of proofs is defined as follows : $s -_{p,\sigma,l\approx r} - t$ and $s' -_{p',\sigma',l'\approx r'} - t'$ can be concatenated if t and s' are syntactically equal. Concatenation is associative (whenever defined) and has an identity denoted by nil, standing for the elementary proof $s \leftarrow_{\emptyset,\emptyset,\emptyset}\rightarrow s$. An Ω_i-proof $\Pi(s,t)$ between s and t is a concatenation of elementary Ω_i-proofs. Then, we can define a notion of subproof just as we define the notion of factor in a word.

Notation : When a proof $\Pi(s,t)$ is a concatenation of elementary proofs using the same equation $l \approx r$ in the same direction, with the same substitution σ, at pairwise parallel positions $p_1,\cdots,p_n$, we write $s -_{\{p_1,\cdots,p_n\},\sigma,l\approx r} - t$.

In order to prove termination of rewrite rules on proofs, we define a proof ordering by associating a complexity measure to each proof Π, starting with the elementary proofs :

• $c(nil) = [\bot,\bot,\bot]$;

• $c(s -_{p,\sigma,l\approx r} \rightarrow t) = [\{s\}, l, t]$;

• $c(s \leftarrow_{p,\sigma,l\approx r} - t) = [\{t\}, r, s]$;

• $c(s \leftarrow_{p,\sigma,l\approx r}\rightarrow t) = [\{s,t\},\bot,\bot]$.

$\bot$ is added to $\mathcal{T}(\mathcal{F},\mathcal{X})$ and required to be minimal wrt both $>$ and $\blacktriangleright$ than any term. Triples are compared lexicographically, using $>_{mul}$ for the first component, $\blacktriangleright$ for the second and $>$ for the third. Arbitrary proofs, viewed as multisets of elementary proofs, are compared according to the multiset extension of this ordering. We denote by $\gg$ the corresponding ordering on proofs. This ordering is similar to the one used by Bachmair, Dershowitz and Hsiang [2], although proof patterns are different. $\gg$ is well-founded as a multiset and lexicographic combination of well-founded orderings. Moreover, $\gg$ is monotonic in the following sense : if $\Pi(s,t)$ is greater than $\Pi'(s,t)$, then every proof $\Psi(u,v)$ having $\Pi(s,t)$ as a subproof is greater than $\Psi'(u,v)$, where $\Psi'(u,v)$ is equal to $\Psi(u,v)$ in which $\Pi(s,t)$ has been replaced by $\Pi'(s,t)$.

We denote by $LC(\Omega_i)$ the union of all sets that can be obtained by applying at most one inference rule in **(LC)** to Ω_i, and by $\Rightarrow_i$ the rewrite relation on proofs restricted to $LC(\Omega_i)$-proofs. Indexing the rewrite relation by the current set of equations is needed because completion is not monotonic.

We can now describe the rewrite rules on proofs that we use : they work on proofs of length one and two, and transform them into smaller proofs wrt $\gg$. We may distinguish two kinds of rewrite rules on proofs : those between Ω_i-proofs and those between an Ω_i-proof and an Ω_{i+1}-proof. The first ones are due to an extension of Huet's lemma [8] :

LEMMA 2 *Let u,s and t be in $\mathcal{T}(\mathcal{F},\mathcal{X})$ and $l \approx r$ and $g \approx d$ be equations in Ω_i such that $u|_p = l\sigma$, $s = u[r\sigma]_p$ and $u|_q = g\tau$, $t = u[d\tau]_q$, where $q \notin p.\mathcal{GP}(l)$ and $p \notin q.\mathcal{GP}(g)$. Consider the proof $\Pi(s,t) = s -_{p,\sigma,r\approx l} - u \,.\, u -_{q,\tau,g\approx d} - t$. Since l and g do not overlap in u, three distinct cases may occur :*

- *if $p \parallel q$, then we can construct the proof $\Pi'(s,t) = s -_{q,\tau,g\approx d} - v \,.\, v -_{p,\sigma,r\approx l} - t$;*
- *if $q = poq'$, where $l|_o = x \in \mathcal{X}$, then we can construct the proof*

$\Pi'(s,t) = s -_{Q,\tau,g\approx d} - v \,.\, v -_{p,\sigma',r\approx l} - w \,.\, w -_{P,\tau,d\approx g} - t$

$$\begin{aligned} \textit{where}\ \ P &= \{po'q' \mid l|_{o'} = x, o' \neq o\},\\ Q &= \{po''q' \mid r|_{o''} = x\},\\ \sigma' &= (\sigma - \{x \to x\sigma\}) \cup \{x \to x\sigma[d\tau]_{q'}\};\end{aligned}$$

- *if $p = qop'$, where $g|_o = x \in \mathcal{X}$, then we can construct the proof*

$\Pi'(s,t) = s -_{P,\sigma,l\approx r} - w \,.\, w -_{q,\tau',g\approx d} - v \,.\, v -_{Q,\sigma,r\approx l} - t$

$$\begin{aligned} \textit{where}\ \ P &= \{qo'p' \mid g|_{o'} = x, o' \neq o\},\\ Q &= \{qo''p' \mid d|_{o''} = x\},\\ \tau' &= (\tau - \{x \to x\tau\}) \cup \{x \to x\tau[r\sigma]_{p'}\}.\end{aligned}$$

(In fact, every elementary proof appearing in these proofs uses one of the three patterns $\leftarrow$, $\rightarrow$, or $\leftrightarrow$, but we do not want to enumerate all possibilities). Note that, for each case, $\Pi'(s,t)$ is also an Ω_i-proof. The rewrite rule on Ω_i-proofs consists of substituting $\Pi'(s,t)$ for $\Pi(s,t)$ in a larger Ω_i-proof, as soon as $\Pi(s,t) \gg \Pi'(s,t)$.

The rewrite rules from an Ω_i-proof to an Ω_{i+1}-proof are of course related to the inference rules :

- If Ω_{i+1} is obtained from Ω_i by application of **Deduce**, there exists $l \approx r$ and $g \approx d$ in Ω_i such that g overlaps l at position p with *mgu* σ, and either $l > r$ and $d \not\gtrsim g$, or $g > d$ and $r \not\gtrsim l$. Hence, $\Omega_{i+1} = \Omega_i \cup \{r\sigma \approx l\sigma[d\sigma]_p\}$. Let us consider u, s and t in $\mathcal{T}(\mathcal{F},\mathcal{X})$ such that $u|_q = l\sigma\tau$, $s = u[r\sigma\tau]_q$ and $t = u[l\sigma\tau[d\sigma\tau]_p]_q$ for some substitution τ and position q. We have the Ω_i-proof $\Pi(s,t) = s -_{q,\sigma\tau,r\approx l} - u \,.\, u -_{qp,\sigma\tau,g\approx d} - t$. By using the new equation in Ω_{i+1}, we can construct an Ω_{i+1}-proof $\Pi'(s,t) = s -_{q,\tau,r\sigma\approx l\sigma[d\sigma]_p} - t$.

- If Ω_{i+1} is obtained from Ω_i by application of **Delete**, $\Pi(s,s) = s \leftarrow_{q,\tau,l\approx l}\rightarrow s$ and $\Pi'(s,t) = nil$.

- If Ω_{i+1} is obtained from Ω_i by application of **Simplify**, there exists $l \approx r$ and $g \approx d$ in Ω_i such that $l|_p = g\sigma$ and $g > d$. Hence $\Omega_{i+1} = (\Omega_i \backslash \{l \approx r\}) \cup \{l[d\sigma]_p \approx r\}$. Let s and t be in $\mathcal{T}(\mathcal{F},\mathcal{X})$ such that $s|_q = l\tau$ and $t|_q = r\tau$. We now have the Ω_i-proof $\Pi(s,t) = s -_{q,\tau,l\approx r} - t$ and the Ω_{i+1}-proof $\Pi'(s,t) = s -_{qp,\sigma\tau,g\approx d} \rightarrow v \,.\, v -_{q,\tau,l[d\sigma]_p\approx r} - t$.

In all three cases, we can easily prove that, for every orientation of the elementary proofs such that the corresponding inference rule can be applied, we have $\Pi(s,t) \gg \Pi'(s,t)$. The rewrite relation $\Rightarrow_i$ consists of substituting $\Pi'(s,t)$ for $\Pi(s,t)$ in a larger Ω_i-proof, as soon as the corresponding inference rule is applied. We say that $\Rightarrow_i$ *reflects* the set of inference rules. That $\Pi \gg \Pi'$ for these three rewrite rules is a proof of completeness of linear completion.

We call *proof in normal form* an Ω_∞-proof which cannot be rewritten by the previous rewrite rules on proofs, and *rewrite proof* a proof whose pattern is $\rightarrow^* \leftarrow^*$. If the ordered completion procedure computes a canonical system R from $(E_0, >)$, (i.e. $\Omega_\infty = R$), then proofs in normal form are rewrite proofs.

In the sequel, we will require the control to be *fair* :

Definition 2 *A derivation $\Omega_0 \vdash_{LC} \Omega_1 \vdash_{LC} \cdots$ is said to be* fair *if, whenever an Ω_i-proof Π_i is not in normal form wrt $\Rightarrow_i$, then there exists some j, $j \geq i$, and some Ω_j-proof Π_j, such that $\Pi_i \Rightarrow_i \Pi_j$. A control is said to be* fair *if it selects fair derivations.*

6 Linear completion of linear equations

6.1 Computation of a canonical set

We call *linear (equational) presentation* a set E_0 of axioms whose elements are linear equations. We first have a straightforward property of stability of linearity during the completion procedure :

Lemma 3 *Let E_i be a set of equations obtained by completion from E_0 and $>$. If E_0 is linear, then E_i is also linear.*

Definition 3 *A* valley-proof *is a proof of the form $\rightarrow^* \leftrightarrow^* \leftarrow^*$.*

Proposition 2 *Let E_0 be a linear presentation and $>$ be a reduction ordering. Then, every Ω_∞-proof in normal form is a valley-proof.*

Proof: There are nine proof patterns of length two : (1) $\rightarrow\leftarrow$, (2) $\rightarrow\rightarrow$, (3) $\leftarrow\leftarrow$, (4) $\rightarrow\leftrightarrow$, (5) $\leftrightarrow\leftarrow$, (6) $\leftrightarrow\leftrightarrow$, (7) $\leftarrow\rightarrow$, (8) $\leftarrow\leftrightarrow$ and (9) $\leftrightarrow\rightarrow$. One has to show that the last three patterns cannot occur in an Ω_∞-proof in normal form.
In each case, the pattern corresponds either to an overlap or no overlap. If both equations in a pattern number (7) (respectively number (8), number (9)) do not overlap each other, the corresponding proof can be transformed into the pattern number (1) (respectively number (5), number (4)), since the equations are linear, and this latter proof is smaller wrt $\gg$. If they overlap, there exists a proof using the associated linear critical pair and this proof is also smaller wrt $\gg$, as already noticed. If this linear critical pair belongs to Ω_∞, we are done. Otherwise, since Ω_∞ is the inductive limit of the Ω_i, the linear critical pair is rewritten into a smaller Ω_∞-proof and the same holds for the initial proof.
In all cases, an Ω_∞-proof containing such a pattern cannot be in normal form. Consequently, every Ω_∞-proof in normal form is a valley-proof. □

As a consequence, we obtain the following result :

Theorem 2 *Let R be a canonical system for a linear presentation E_0 and $>$ be a reduction ordering containing R. Every fair control on the system of inference rules* **(LC)** *will generate for input $(E_0, >)$ a derivation such that Ω_∞ and R are equal up to renaming.*

Proof: Let $l \to r$ be a rule in R. Thanks to the previous proposition, there exists an Ω_∞-proof $l \to^*_{\Omega_\infty} \leftrightarrow^*_{\Omega_\infty} \leftarrow^*_{\Omega_\infty} r$ between l and r. Since r is equal to $r\downarrow_R$, we easily get that the last step of this proof cannot use any of both patterns $\leftarrow$ and $\leftrightarrow$. Consequently, the proof has the form $l \to^*_{\Omega_\infty} r$. Since $l > r$, we have an Ω_∞ rewriting step from l : there exists $g \to d$ in Ω_∞ such that $l \trianglerighteq g$. However, l can only be rewritten by $l \to r$ in R. Consequently, so does g ; hence, g is equal to l up to renaming. The proof has the form $l -_{\Lambda,\sigma,g\approx d} \to d\sigma \;.\; d\sigma \to^*_{\Omega_\infty} r$, where σ is a renaming. Now, since Ω_∞ is inter-reduced, $d\sigma$ is equal to r. Therefore, the rule $l \to r$ belongs to Ω_∞ up to renaming. Using again that Ω_∞ is inter-reduced, we conclude that all equations in Ω_∞ belong to R up to renaming. □

6.2 Refutational completeness

We now prove that linear completion is refutationally complete for linear equational theories. The method and notations are borrowed from Bachmair [1].

Let $\mathcal{K}$ be a set of Skolem constants which is disjoint from $\mathcal{F}$ and $>$ be a reduction ordering on $\mathcal{T}(\mathcal{F} \cup \mathcal{K}, \mathcal{X})$. We add two new constants *true* and *false* and a new binary operator *eq* to $\mathcal{F}$, and consider the new set of terms $\mathcal{T}'$ made of *true*, *false* and all terms $eq\,(s,t)$, where s and t are in $\mathcal{T}(\mathcal{F} \cup \mathcal{K}, \mathcal{X})$. We extend $>$ to a reduction ordering $\succ$ on $\mathcal{T}' \cup \mathcal{T}(\mathcal{F} \cup \mathcal{K}, \mathcal{X})$ in the following way : for every s, t, s', t' in $\mathcal{T}(\mathcal{F} \cup \mathcal{K}, \mathcal{X})$, $s \succ t$ if and only if $s > t$ and $eq\,(s,t) \succ eq\,(s',t')$ if and only if $s \geq s'$ and $t \geq t'$, but $(s,t) \neq (s',t')$. Moreover, we use the precedence $eq \succ_p false \succ_p true$, so that *false* is smaller than any other term in $\mathcal{T}'$. We will consider both equations between terms in $\mathcal{T}(\mathcal{F} \cup \mathcal{K}, \mathcal{X})$ and equations between terms in $\mathcal{T}'$.

A derivation is said to be a *refutation* if it generates the equation $true \approx false$.

THEOREM 3 *Let E_0 be a linear presentation and s, t be two terms in $\mathcal{T}(\mathcal{F}, \mathcal{X})$ such that $s =_{E_0} t$. Let $\hat{s}$ and $\hat{t}$ be skolemized versions of respectively s and t, $\mathcal{K}$ be the set of Skolem constants appearing in $\hat{s}$ and $\hat{t}$, and $>$ be a reduction ordering on $\mathcal{T}(\mathcal{F} \cup \mathcal{K}, \mathcal{X})$. Let $\mathcal{T}'$ and $\succ$ be defined as above. Then, from input $(E_0 \cup \{\, eq\,(\hat{s},\hat{t}) \approx false \,\} \cup \{\, eq\,(x,x) \approx true \,\}, \succ)$, every fair control on* **(LC)** *will generate a refutation.*

Proof: The rule $eq\,(x,x) \to true$ cannot be simplified by any rule generated by the procedure. Moreover, it could only simplify a term of the form $eq\,(t,t)$ or generate a linear critical pair by overlapping a term of the same form. Hence, this rule will be useful only when an equation $g \approx d$ such that g matches $eq\,(x,x)$ will have been computed.

Such an equation $g \approx d$ can only come from computations with the rule $eq\,(\hat{s},\hat{t}) \to false$. As an immediate consequence, $g \approx d$ is actually a rule of the form $eq\,(\hat{w},\hat{w}) \to false$, where $\hat{w}$ is a ground term in $\mathcal{T}(\mathcal{F} \cup \mathcal{K})$. Once this rule is created, by simplifying it with the rule $eq\,(x,x) \to true$, we get the rule $false \to true$.

It remains to prove that a rule $eq\,(\hat{w},\hat{w}) \to false$ can be generated by any fair derivation for input E_0', where E_0' is $E_0 \cup \{\, eq\,(\hat{s},\hat{t}) \approx false \,\}$. Since E_0 is linear, we know that there is an Ω_∞-proof of $\hat{s} =_{E_0} \hat{t}$ of the form $\hat{s} \to^* \hat{u} \;.\; \hat{u} \leftrightarrow^* \hat{v} \;.\; \hat{v} \leftarrow^* \hat{t}$. Hence, there exists an E_0'-proof between *false* and $eq\,(\hat{u},\hat{u})$. Since E_0' is linear, this proof will be rewritten into a valley-proof $false \to^* \leftrightarrow^* \leftarrow^* eq\,(\hat{u},\hat{u})$. Now, *false* is smaller than any term whose head symbol is *eq*. Then, this valley proof is actually of the form $eq\,(\hat{u},\hat{u}) \to^+ false$. But, since û is in normal form, this proof must be a one-step proof $eq\ (\hat{u}, \hat{u}) \rightarrow$ *false*, hence $eq\ (\hat{u}, \hat{u}) \rightarrow$ *false* must be a rule. □

The proof is different from Bachmair's, since we do not require a total ordering on ground terms in $\mathcal{T}(\mathcal{F} \cup \mathcal{K}) \cup \mathcal{T}'$; the crucial point here is that ground terms are linear.

7 Limits of linear completion

Though Theorem 2 assumes linear axioms, finding a canonical rewrite system actually happens in a more general case.

For example, let E_0 be the set $\{1*(-x+x) \approx 0, 1*(x+-x) \approx x+-x, -x+x \approx y+-y\}$. Standard completion fails, whatever the reduction ordering is. Let however $>$ be the recursive path ordering with precedence $1 >_p 0$: linear completion computes the canonical system $\{1 * 0 \rightarrow 0, -x + x \rightarrow 0, x + -x \rightarrow 0\}$, in which two rules are not linear.

In this section, we want to extend the result of Theorem 2, by relaxing the hypothesis of linearity. We first prove an easy lemma whose generality goes beyond the peculiar context of linear completion :

LEMMA 4 *Let R be a canonical rewrite system, $l \rightarrow r$ be a rule in R, and $>$ be a reduction ordering containing R. Let E be a set of equations defining the same equational theory than R and v be a term such that $v =_E r$ and $v \neq r$. Then every E-proof of $v =_E r$ has a subproof of the form $u \rightarrow_{\{s \approx t\}} r$.*

Proof : If t is not in R-normal form, then neither is r. Hence, $s \rightarrow_R^+ t$, since R is confluent, and we have $s > t$. □

We can now prove the following result :

PROPOSITION 3 *Let R be a canonical system for a presentation E_0, $>$ be a reduction ordering containing R and Ω_∞ be the inductive limit for input $(E_0, >)$ with any fair control. A rule $l \rightarrow r$ in R is not in Ω_∞ (up to renaming) if and only if all Ω_∞-proofs in normal form of $l =_{E_0} r$ have a subproof of the form $v \leftarrow_{p,\sigma,s\approx t} \rightarrow u\ .\ u -_{q,\tau,g\approx d} \rightarrow w\ .\ w \rightarrow^* r$, such that $p = qop'$, where $g|_o = x \in \mathcal{X}$ and x occurs at least twice in g or in d (in other words, the non-orientable equation $s \approx t$ and the rule $g \rightarrow d$ must be such that t applies inside the substitution part of $g\tau$ under a non-linear variable of $g \rightarrow d$).*

Proof : The *only if* direction is straightforward by contraposition. We also prove the converse by contraposition : suppose that we have an Ω_∞-proof Π in normal form of $l =_{E_0} r$ which does not contain a subproof as described in the theorem. Using Lemma 4, we know that Π has a subproof of the form $u \rightarrow^+ r$. Three cases may then occur : either $l = u$ and then Π has the form $l \rightarrow^+ r$ (case (1)); or $l \neq u$ and then Π has a subproof of the form $v \leftarrow_{p,\sigma,s\approx t} - u\ .\ u -_{q,\tau,g\approx d} \rightarrow w\ .\ w \rightarrow^* r$ (case (2)) or of the form $v \leftarrow_{p,\sigma,s\approx t} \rightarrow u\ .\ u -_{q,\tau,g\approx d} \rightarrow w\ .\ w \rightarrow^* r$ (case (3)). We now show that a proof of the form (2) or (3) cannot be in normal form, because it can be reduced by using rewrite rules on proofs. Let us denote by Π' the subproof between v and w. If t and g overlap, then Π' can be rewritten. Otherwise, in case (2), by using Lemma 2, there is a proof of the form $v \rightarrow^* \leftarrow^* w$ which is smaller wrt $\gg$ than Π'. In case (3), when t and g do not overlap, three cases may occur, for which we can apply Lemma 2 :

• $p \parallel q$: there is a proof of the form $v \rightarrow\leftrightarrow w$ which is smaller than Π'.
• $q = poq'$, where $t|_o = x \in \mathcal{X}$: there is a proof of the form $v \rightarrow^*\leftrightarrow\leftarrow^* w$ which is smaller than Π'.
• $p = qop'$, where $g|_o = x \in \mathcal{X}$: according to our hypothesis, x is a linear variable in the rule $g \rightarrow d$; hence, there is a proof of the form $v \rightarrow w$ or $v \rightarrow\leftrightarrow w$: in both cases, this proof is smaller than Π'.
In summary, in cases (2) or (3), Π' can be rewritten by using rewrite rules on proofs. Hence, Π cannot be in normal form. Consequently, only case (1) may occur : Π is of the form $l \rightarrow^+_{\Omega_\infty} r$. Using the same arguments as in the proof of Theorem 2, we conclude that $l \rightarrow r$ belongs to Ω_∞ up to renaming. □

8 Conclusion

Ordered completion and linear completion cannot be easily compared. As a semi-decision procedure, ordered completion is more general than linear completion. But it may fail in computing a finite canonical system when such one exists.

Linear completion is adapted to linear presentations. In that case, the use of a complete reduction ordering is not required. Furthermore, there are less critical pairs to be computed, since critical pairs between non-orientable equations are ruled out. Note also that no useless unifier or match is computed, since, e.g., the condition to be satisfied when adding a new critical pair requires the unifier to be computed beforehand.

Acknowledgements : We thank Jean-Pierre Jouannaud and Hubert Comon for their help.

References

[1] L. Bachmair, "Proofs Methods for Equational Theories", *Thesis : University of Illinois at Urbana-Champaign,1987.*

[2] L. Bachmair, N. Dershowitz and J. Hsiang, "Orderings for Equational Proofs", *Symposium on Logic in Computer Science, Boston, June 1986.*

[3] N. Dershowitz, "Personal Communication", *Montréal, June 1990.*

[4] N. Dershowitz and J.-P. Jouannaud, "Term Rewriting Systems", *in Handbook for Theoretical Computer Science, North-Holland, to appear.*

[5] N. Dershowitz and J.-P. Jouannaud, "Notations for Rewriting", *to appear*

[6] N. Dershowitz, L. Marcus and A. Tarlecki, "Existence, Uniqueness, and Construction of Rewrite Systems", *Technical Report.*

[7] J. Hsiang and M. Rusinowitch, "On Word Problems in Equational Theories", *14th ICALP, 1987.*

[8] G. Huet, "Confluence Reductions : Abstract Properties and Applications to Term Rewritings, *JACM, 27, 1980, pp.797-821.*

[9] G. Huet, "A Complete Proof of Correctness of Knuth-Bendix Completion Algorithm", *JCSS, 23, 1981, pp.11-21.*

[10] G. Huet and D.C. Oppen, "Equations and Rewrite Rules : A Survey", *Formal Langages : Perspectives and Open Problems, R. Book, Academic Press, 1980.*

[11] D.E. Knuth and P.B. Bendix, "Simple Word Problems in Universal Algebra", *Computational Algebra, J. Leach, Pergamon Press, 1970, pp.263-297.*

Clausal Rewriting

Robert Nieuwenhuis
Fernando Orejas
Universidad Politécnica de Cataluña
Dept. Lenguajes y Sistemas Informáticos
Pau Gargallo 5, 08028 Barcelona, Spain
E-mail: eanieuw@ebrupc51 (bitnet)

Abstract: The techniques of clausal rewriting and completion are presented. The methods described provide several theorem proving techniques in first order logic with equality. The clausal completion procedure generates few new axioms and is compatible with the powerful clausal rewriting technique for simplification and deletion of redundant axioms.

1. Introduction

One way to reduce the search space in equational proofs is by allowing only steps in which equations are used in a *reductive* way, i.e. by replacing subterms that match with the "big" side of the equation by the corresponding instantiation of the "small" side of the equation, with respect to some ordering. A proof for $t = t'$ would consist of *rewriting* it in this way until an equality $t'' = t''$ is obtained. This method is not complete in general. However, the Knuth-Bendix completion procedure [KB 70] allows to transform sets of equations into *complete* sets of rewrite rules, such that rewriting using these rules becomes a complete deduction method.

In general, completion can be seen as a process that tranforms a set of axioms in such a way that, using the final *complete* set of axioms, certain kinds of *normal form* proofs with a smaller search space (e.g. *linear* proofs) can be obtained. For instance, in the equational case these normal form proofs are rewrite proofs. A nice additional property of completion procedures can be its refutational completeness, which many times becomes even more important than the aim to obtain a complete set of axioms.

Completion-based methods have been studied for conditional equations and Horn clauses by [BDJ 78, Gan 87, Kap 85, KR 87, RZ 85, Nav 87]. For general clauses with equality several completion-like refutational proof methods have been described in [HR 89, Rus 87, BG 90a]. However, until very recently, for general clauses no framework was defined in which powerful simplification methods were included. In [BG 90b], completion consists of the computation of a closure up to *redundancy* wrt. a set of inference rules. This notion of redundancy includes some known simplification and elimination methods, but, up to now, it is not known whether it is powerful enough to obtain complete systems for non-trivial examples.

In this paper we define completion-based deduction methods for *restricted equality clauses:* clauses of the form $l_1 \vee \ldots \vee l_n$ or $l_1 \vee \ldots \vee l_n \vee E = E'$, where $l_1, \ldots, l_n$ are positive or negative non-equality literals. By means of an equality predicate, defined by the

This work has been partially supported by the ESPRIT Project PROSPECTRA ref. no. 390, under subcontract with Alcatel SESA.

axioms $eq(x,x)$ and $\neg eq(x,y) \vee x = y$, this language can be made equivalent to full first order logic with equality. However, due to problems with the second axiom in most deduction methods, this is normally not done. On one hand, even without these axioms, this language still has more expressive power than Horn clauses, since it has no restrictions on the number of positive non-equality literals. On the other hand it is less powerful because no negative equality literals are allowed, although in practice normally only equations with boolean conditions are needed.

The *clausal rewriting* relation is defined on sets of clauses: if $(L \to R) \vee l_1 \vee \ldots \vee l_n$ is a rewrite rule and the clause C rewrites into C' using $L \to R$ with a substitution σ (in the usual sense) then C can be replaced in a set by the clauses: $\{\ C' \vee l_1\sigma \vee \ldots \vee l_n\sigma,\ \ C \vee \neg l_1\sigma\ , \ldots,\ C \vee \neg l_n\sigma\ \}$.
This type of rewriting is finitely terminating if the rules are *reductive* [Kap 85] and certain unnecessary steps are not made. The rewrite systems produced by our completion procedure are complete in the sense that a clause is provably true if and only if its set of normal forms contains only tautologies. We therefore obtain a decision procedure for the corresponding validity problem. Rewriting with the rules generated by non-terminating completion provides a semi-decision procedure. Moreover, clausal completion can be used for refutational theorem proving in several ways.

Powerful simplification techniques can be applied during completion, which is well-known to be a crucial aspect. In fact, another important advantage of our methods wrt. conditional rewriting, the one that makes them behave well in practice, is that also *conditional* axioms can be proved effectively and thus eliminated if they are shown to be redundant. The practical performance of our techniques has also been shown by means of an implementation using the TRIP system [Nie 88, NOR 90, Nie 90], and its integration within the CEC system [BGS 88a, BGS 88b] at the University of Dortmund.

Three additional applications of clausal rewriting for theorem proving in partial specifications, for efficient prototyping and for proving sufficient completeness, have been described and proved correct in [NO 90, Nie 90].

This paper is structured as follows. In section 3 a sound and complete "equational" deduction method, called *clausal deduction*, is defined. In section 4 it is shown that the clausal completion procedure transforms the axioms in such a way that any clausal deduction proof can be reduced to a clausal rewrite proof. We obtain our main results on refutational and consequence-finding theorem proving in section 5. Due to the lack of space most proofs have been omitted, but all of them can be found in detail in [Nie 90]. In the last section some examples of complete systems are shown.

2. Basic notions and notations

Clauses are (finite) multisets of literals. An equation is a multiset $\{E, E'\}$, denoted by $E \doteq E'$. Literals will be denoted by $l, c, d, \ldots$; clauses will be denoted by $A, C, D, \ldots$; sets of clauses will be denoted by symbols $\mathcal{A}, \mathcal{M}, \mathcal{C} \ldots$, etc. Deduction of a clause C from a set of clauses $\mathcal{A}$ will be denoted by $\mathcal{A} \vdash C$. A multiset containing n times an element e is denoted by $\{e\}^n$, or by $\{e^n\}$. A literal l has the same complexity in orderings as its negation $\neg l$. Substitutions are denoted by σ, σ', etc, and their application to a term t by $t\sigma$. Occurrences in terms are denoted by u, u', and their concatenation is denoted by $u \cdot u'$. The expression $t[s]$ means that s is a subterm of t. If u is an occurrence of a subterm s in a term t, then t/u denotes s, and $t[u \leftarrow t']$ denotes the result of the replacement in t of the subterm at occurrence u by t'. The left and right hand sides of rules are denoted by lhs and rhs respectively. The set of variables of a term T is denoted by $vars(T)$.

The recursive path ordering (RPO) is obtained by lifting a given well-founded ordering on a (possibly infinite) set of operator symbols to paths in terms [Der 87]. RPO is a simplification ordering, i.e. it is well-founded, compatible with operators, stable under substitutions and any term is greater than any of its proper subterms. We supose $>$ to be a simplification ordering on terms and literals that can be extended to a total simplification ordering $>_t$ on ground terms, which is the case for most general purpose orderings used in practice. We will denote the multiset extensions of $>$ and $>_t$ by $\gg$ and $\gg_t$ respectively.

3. Clausal Deduction

In this section we define a deduction method called *clausal deduction*, for theories expressed by restricted equality clauses. We will distinguish between *paramodulation* axioms, which have an equality literal, and axioms without equality literal, called *resolution* axioms.

Definition 3.1: Let $\mathcal{M}$ and $\mathcal{M}'$ be sets of clauses. Then $\mathcal{M}'$ can be obtained from $\mathcal{M}\cup\{C\}$ by one *clausal deduction step* with an axiom of $\mathcal{A}$, denoted by $\mathcal{M}\cup\{C\} \rightarrow_{\mathcal{A}} \mathcal{M}'$, in one of the following ways:

a) (*Paramodulation step*) There is a paramodulation axiom $(E \doteq E')\vee c_1\vee\ldots\vee c_m$ in $\mathcal{A}$, and $\mathcal{M}'$ is $\mathcal{M} \cup \{C'\vee c_1\sigma\vee\ldots\vee c_m\sigma,\ C\vee\neg c_1\sigma,\ldots,C\vee\neg c_m\sigma\}$, where $C/u = E\sigma$ and $C' = C[u \leftarrow E'\sigma]$.

b) (*Resolution step*) There is a resolution axiom $l\vee c_1\vee\ldots\vee c_m$ and C is $D\vee l\sigma$. In this case $\mathcal{M}'$ is $\mathcal{M} \cup \{D\vee true\vee c_1\sigma\vee\ldots\vee c_m\sigma,\ D\vee l\sigma\vee\neg c_1\sigma,\ \ldots,\ D\vee l\sigma\vee\neg c_m\sigma\}$.

c) (*Case Analysis step*) In this case $\mathcal{M}'$ is $\mathcal{M} \cup \{C\vee l_1\vee\ldots\vee l_m,\ C\vee\neg l_1,\ldots,C\vee\neg l_m\}$, for some set of non-equality literals $l_1,\ldots,l_m$.

Note that case analysis steps do not depend on $\mathcal{A}$. In all these steps, the clauses $C\vee\neg c_1\sigma,\ldots,C\vee\neg c_m\sigma$ (resp. $C\vee\neg l_1,\ldots,C\vee\neg l_m$) are called the *complementary* clauses, since in some sense they contain the complementary conditions with respect to $c_1\sigma\vee\ldots\vee c_m\sigma$ (resp. $l_1,\ldots,l_m$). The literals $\neg c_i\sigma$ (resp. $\neg l_i$) are called the *complementary literals* of the step. A clause C is proved by clausal deduction iff by applying clausal deduction steps to the set $\{C\}$, another set of clauses containing only tautologies can be obtained. The reflexive and transitive closure of $\rightarrow_{\mathcal{A}}$ is denoted by $\rightarrow^*_{\mathcal{A}}$. The following theorem is proved in [Nie 90]:

Theorem 3.2: (Soundness and completeness of clausal deduction) Let $\mathcal{A}$ be a set of restricted equality clauses, and let C be a clause without negative equality literals. Then $\mathcal{A} \vdash C$ iff $\{C\} \rightarrow^*_{\mathcal{A}} \mathcal{M}$, where $\mathcal{M}$ is a set of tautologies.

Note that by clausal deduction no negative equality literals can appear, and that a clause without negative equality literals is a tautology iff it contains a literal *true*, a literal ($t \doteq t$) or a pair of literals $\{l, \neg l\}$. The clauses proved by clausal deduction can be considered as ground clauses, since their variables are treated as constants.

4. Clausal Completion

In this section we define a framework that allows to find *normal form proofs* for the clausal deduction proofs given above. We will show that a clausal completion procedure is able to modify sets of axioms and *clausal rewrite rules* in such a way, that normal form proofs, i.e. proofs by clausal rewriting, using these sets are obtained. This method is similar to the transformation of equational proofs into rewrite proofs by completion [BDH 86].

4.1. Clausal Rewriting

Definition 4.1: Let C be a clause, and let l be a literal in C. We say that l is *maximal* in C wrt. $>$ iff there is no other literal l' in C with $l' > l$. The literal l is *strictly maximal* in C wrt. $>$ iff there is no other literal l' in C with $l' \geq l$. A literal l is maximal (strictly maximal) in C for a substitution σ wrt. $>$ iff there is no other literal l' in C with $l'\sigma > l\sigma$ (resp. $l'\sigma \geq l\sigma$).

Definition 4.2: Clausal rewrite rules are axioms of the form:

a) *paramodulation* rules: $(L \rightarrow R) \vee c_1 \vee \ldots \vee c_m$

where L and R, the left and right hand sides of the rule respectively, are terms, and the (non-equality) literals $c_1 \vee \ldots \vee c_m$ are called the *condition* of the rule. These rules are *reductive* as conditional rewrite rules in the sense of [Kap 85], i.e. $\{L\} \gg \{R, c_1, \ldots, c_m\}$.

b) *resolution* rules: $(L_1 \vee \ldots \vee L_k \rightarrow \mathit{true}) \vee c_1 \vee \ldots \vee c_m$

where there do not exist a simplification ordering $>_t$ extending $>$, a ground substitution σ, and a literal c_i in $\{c_1, \ldots, c_m\}$ such that c_i is maximal in the rule wrt. $>_t$ for σ.

Definition 4.3: Let $\mathcal{M}$ be a set of clauses and let r be a paramodulation rule $(L \rightarrow R) \vee c_1 \vee \ldots \vee c_m$. Then $\mathcal{M} \cup \{C\}$ can be rewritten by a step with r into

$$\mathcal{M} \cup \{C' \vee c_1\sigma \vee \ldots \vee c_m\sigma,\ C \vee \neg c_1\sigma, \ldots, C \vee \neg c_m\sigma\}$$

where, for some substitution σ, in C a subterm $L\sigma$ is replaced by $R\sigma$, obtaining C' in the case of a *reductive* step, or $R\sigma$ is replaced by $L\sigma$ in a *reverse* step.

Definition 4.4: Let $\mathcal{M}$ be a set of clauses, let $\succ$ be a simplification ordering, and let r be a resolution rule with literals $\{l, c_1, \ldots, c_m\}$. Then $\mathcal{M} \cup \{C \vee l\sigma\}$ can be rewritten wrt. $\succ$ by a step with r into

$$\mathcal{M} \cup \{C \vee \mathit{true} \vee c_1\sigma \vee \ldots \vee c_m\sigma,\ C \vee l\sigma \vee \neg c_1\sigma,\ \ldots,\ C \vee l\sigma \vee \neg c_m\sigma\}$$

where l is maximal in r for σ wrt. $\succ$. If l is strictly maximal in r for σ wrt. $\succ$, then the step is called *reductive*.

The notion of clausal rewriting with resolution rules has been defined with respect to a simplification ordering $\succ$. In general, when rewriting ground clauses, this ordering will be some total ordering $>_t$. Note that there are no reverse steps with resolution rules, since rewriting the literal *true*, i.e. rewriting a tautology, is useless.

Example 4.5: Below we show a very simple proof by clausal rewriting of $P(f(a))$ using the rules: 1) $(f(x) \rightarrow b) \vee Q(x)$, 2) $\neg Q(a) \rightarrow true$ and 3) $(P(b) \rightarrow true) \vee Q(a)$. The proof is expressed as a tree whose leafs are tautologies.

```
                                P(f(a))
                              /    | 1)
                             /     |
             P(b) v Q(a)        P(f(a)) v -Q(a)
              /     |                  |
             /   3) |                  | 2)
            /       |                  |
 true v Q(a)    P(b) v Q(a) v -Q(a)   P(f(a)) v true
```

Notation: If a set of clauses $\mathcal{M}$ is rewritten into $\mathcal{M}'$ by one reductive clausal rewrite step with a rule r of a set of clausal rewrite rules $\mathcal{R}$, then this is denoted by $\mathcal{M} \rightarrow_{\mathcal{R}} \mathcal{M}'$. The reflexive transitive closure of $\rightarrow_{\mathcal{R}}$ is denoted by $\rightarrow^*_{\mathcal{R}}$.

4.2. Inference rules for Clausal Completion

Definition 4.6: A paramodulation axiom $(E \doteq E') \vee c_1 \vee \ldots \vee c_m$ can be *oriented* into a rewrite rule $(L \rightarrow R) \vee c_1 \vee \ldots \vee c_m$ if E is L and E' is R or vice versa. A resolution axiom $L_1 \vee \ldots \vee L_k \vee c_1 \vee \ldots \vee c_m$ can be oriented into a rewrite rule $(L_1 \vee \ldots \vee L_k \rightarrow true) \vee c_1 \vee \ldots \vee c_m$.

Note that not every paramodulation rule can be oriented into a rule that fulfils the reductivity conditions of definition 4.2, and that resolution axioms can always be oriented in at least one way. In order to minimize in resolution rules the number of literals that could be used as left hand side, the technique of variable analysis [MN 90] can be useful:

Example 4.7: If we have $P > Q > R$, and the axiom a to be oriented is $P(x) \vee Q(y) \vee R(x,y)$, then there is no ground substitution σ and $>_t$ s.t. $R(x,y)\sigma$ is maximal in $a\sigma$. This is shown by analyzing the cases $x\sigma >_t y\sigma$, $y\sigma >_t x\sigma$, and $x\sigma = y\sigma$. Therefore a can be oriented into the rule $(P(x) \vee Q(y) \rightarrow true) \vee R(x,y)$.

The *clausal critical pairs* defined below are closely related to the ones used in conditional rewriting, but they also include refined versions of the inference rules of first order logic of resolution and factoring:

Definition 4.8: A *clausal critical pair* between two clausal rewrite rules is an axiom

(i) $((L[u \leftarrow R'] \doteq R) \vee c_1 \vee \ldots \vee c_m \vee d_1 \vee \ldots \vee d_n)\sigma$

if $(L \rightarrow R) \vee c_1 \vee \ldots \vee c_m$ and $(L' \rightarrow R') \vee d_1 \vee \ldots \vee d_n$ are paramodulation rules, and, for the mgu σ, $L\sigma/u = L'\sigma$ and L/u is not a variable.

(ii) $(L_1[u \leftarrow R] \vee L_2 \vee \ldots \vee L_k \vee c_1 \vee \ldots \vee c_m \vee d_1 \vee \ldots \vee d_n)\sigma$

if $(L \rightarrow R) \vee c_1 \vee \ldots \vee c_m$ is a paramodulation rule, $(L_1 \vee \ldots \vee L_k \rightarrow true) \vee d_1 \vee \ldots \vee d_n$ is a resolution rule r, and, for an mgu σ, $L_1\sigma/u = L\sigma$, L/u is not a variable, and there is a ground substitution σ' and a total extension $>_t$ of $>$ such that $L_1\sigma\sigma'$ is strictly maximal wrt. $>_t$ in $r\sigma'$.

(iii) $\quad (L_2 \vee \ldots \vee L_p \vee L'_2 \vee \ldots \vee L'_q \vee c_1 \vee \ldots \vee c_m \vee d_1 \vee \ldots \vee d_n)\sigma$

if $(L_1 \vee \ldots \vee L_p \rightarrow true) \vee c_1 \vee \ldots \vee c_m$ and $(L'_1 \vee \ldots \vee L'_q \rightarrow true) \vee d_1 \vee \ldots \vee d_n$ are resolution rules r and r', $L_1\sigma = \neg L'_1\sigma$, for an mgu σ, and there is a ground substitution σ' and a total extension $>_t$ of $>$ such that $L_1\sigma\sigma'$ and $L'_1\sigma\sigma'$ are strictly maximal wrt. $>_t$ in $r\sigma'$ and $r'\sigma'$ respectively.

(iv) $\quad (L_1 \vee L_3 \vee \ldots \vee L_k \vee c_1 \vee \ldots \vee c_m)\sigma$

if $(L_1 \vee \ldots \vee L_k \rightarrow true) \vee c_1 \vee \ldots \vee c_m$ is a resolution rule r, $L_1\sigma = L_2\sigma$, for an mgu σ, and there is a ground substitution σ' and a total extension $>_t$ of $>$ such that $L_1\sigma\sigma'$ is maximal wrt. $>_t$ in $r\sigma'$.

Example 4.9: The technique of case analysis on variables is also very useful for reducing the number of critical pairs. If we have $P > Q > R$, then it is not necessary to consider the axiom $P(x) \vee Q(y)$ as a critical pair of type (iii) between $P(x) \vee Q(y) \vee R(x,y,z) \rightarrow true$ and $\neg R(u,v,v) \rightarrow true$, since $R(x,y,y)\sigma'$ cannot be strictly maximal wrt. $>_t$ in $P(x)\sigma' \vee Q(y)\sigma' \vee R(x,y,y)\sigma' \rightarrow true$.

Definition 4.10: An axiom a can be *simplified* by $\mathcal{R}$, denoted by $\{a\} \rightarrow^s_{\mathcal{R}} \{a_0, \ldots, a_m\}$ if $\{a\} \rightarrow^*_{\mathcal{R}} \{a_0, \ldots, a_m\}$, where all the resolution steps are applied with respect to the ordering $>$, and for every a_i in $\{a_0, \ldots, a_m\}$ we have $a > a_i$ or else a_i is a tautology.

Definition 4.11: The *Clausal Completion (CC)* proof theory consists of the following inference rules modifying the axioms of $\mathcal{A}$ and $\mathcal{R}$:

(CC1) Orienting an axiom:

$$\frac{(\mathcal{A} \cup \{a\}, \mathcal{R})}{(\mathcal{A}, \mathcal{R} \cup \{r\})} \quad \text{if } r \text{ is obtained by orienting } a$$

(CC2) Adding clausal critical pair:

$$\frac{(\mathcal{A}, \mathcal{R})}{(\mathcal{A} \cup \{a\}, \mathcal{R})} \quad \text{if } a \text{ is a critical pair between two rules in } \mathcal{R}$$

(CC3) Simplifying an axiom:

$$\frac{(\mathcal{A} \cup \{a\}, \mathcal{R})}{(\mathcal{A} \cup \{a_0, \ldots, a_m\}, \mathcal{R})} \quad \text{if } \{a\} \rightarrow^s_{\mathcal{R}} \{a_0, \ldots, a_m\}$$

(CC4) Deleting a trivial axiom:

$$\frac{(\mathcal{A} \cup \{a\}, \mathcal{R})}{(\mathcal{A}, \mathcal{R})} \quad \text{if } a \text{ is a tautology}$$

Definition 4.12: A *CC-derivation* is a sequence $(\mathcal{A}_0, \mathcal{R}_0), (\mathcal{A}_1, \mathcal{R}_1), \ldots$, where a pair $(\mathcal{A}_{i+1}, \mathcal{R}_{i+1})$ is obtained from $(\mathcal{A}_i, \mathcal{R}_i)$ by applying one inference rule of the proof theory CC, denoted $(\mathcal{A}_i, \mathcal{R}_i) \vdash_{CC} (\mathcal{A}_{i+1}, \mathcal{R}_{i+1})$. An ordering $>_p$ on proofs is *compatible* with the proof theory CC if $(\mathcal{A}, \mathcal{R}) \vdash_{CC} (\mathcal{A}', \mathcal{R}')$ implies that for any proof P_1 of a clause C using $(\mathcal{A}, \mathcal{R})$, there is a proof P_2 of C using $(\mathcal{A}', \mathcal{R}')$ with $P_1 \geq_p P_2$.

Lemma 4.13: The proof theory CC is sound, i.e. if $(\mathcal{A}, \mathcal{R}) \vdash_{CC} (\mathcal{A}', \mathcal{R}')$, then, for any clause C, we have $\mathcal{A}' \cup \mathcal{R}' \vdash C \iff \mathcal{A} \cup \mathcal{R} \vdash C$.

Definition 4.14: Let CP_i be the set of critical pairs between rules in $\mathcal{R}_i$. A CC-derivation $(\mathcal{A}_0, \mathcal{R}_0), (\mathcal{A}_1, \mathcal{R}_1), \ldots$, is called *fair* iff

(i) $\forall j \bigcap_{i>j} \mathcal{A}_i = \emptyset$

(ii) whenever $C \in CP_i$, for some i, then $C \in \mathcal{A}_j$, for some j.

Definition 4.15: A *clausal completion procedure* is any procedure accepting as input a simplification ordering $>$, a set of axioms $\mathcal{A}$, and a set of rules $\mathcal{R}$, and, using inference rules of CC as the only basic computation steps, generates a fair CC-derivation $(\mathcal{A}_0, \mathcal{R}_0), (\mathcal{A}_1, \mathcal{R}_1), \ldots$ with $(\mathcal{A}_0, \mathcal{R}_0) = (\mathcal{A}, \mathcal{R})$. It *terminates* with output $\mathcal{R}_n$ if it generates a pair $(\mathcal{A}_n, \mathcal{R}_n)$ s.t. fairness does not require further inference steps, i.e. $\mathcal{A}_n = \emptyset$ and no more critical pairs have to be computed. If the empty clause $\Box$ is in some $\mathcal{A}_i$, then (by lemma 4.13) $\mathcal{A} \cup \mathcal{R}$ is inconsistent, and completion terminates with output $\Box$. Since a fair derivation is not always possible (due to the unorientability of some paramodulation rules), as in the purely equational case, there is a possibility of *failure*.

Definition 4.16: A proof is a *rewrite proof* if all its steps are reductive rewrite steps.

4.3. A proof ordering for clausal completion

A detailed proof of the following result, which is the basis of all the theorem proving techniques given in section 5, can be found in [Nie 90]. Because of the lack of space, in this section we can only define the ordering on proofs and outline how it has been used.

Theorem 4.17: Let $(\mathcal{A}_0, \mathcal{R}_0), (\mathcal{A}_1, \mathcal{R}_1), \ldots$ be a fair CC-derivation with $(\mathcal{A}_0, \mathcal{R}_0) = (\mathcal{A}, \mathcal{R})$. Then $\mathcal{A} \cup \mathcal{R} \vdash C$ iff the derivation generates a pair $(\mathcal{A}_n, \mathcal{R}_n)$ s.t. there is a rewrite proof for C using $\mathcal{R}_n$.

Let us consider a signature whose operators are all the possible proof steps with all axioms and rules, and whose constants are tautologies (they are leafs in proof trees). Below we define a well-founded ordering on these operators and we use the recursive path ordering to lift it, obtaining an ordering on proof terms, as done in [Gan 87].

Notation: The operators corresponding to proof steps will be denoted:

$\rightarrow_r$ for a reductive application of a rule r
$\leftarrow_r$ for a reverse application of a paramodulation rule r
$\rightarrow_{r,n}$ for a non-reductive application of a resolution rule r with n occurrences of the maximal literal
$\leftrightarrow_a$ for an application of an axiom a
$\leftrightarrow_{ca}$ for a case analysis step

Definition 4.18: Let M and M' be multisets. Then the *maximum union* of M and M', denoted by $M \cup_m M'$ is a multiset defined by $M \cup_m M'(l) = max\{M(l), M'(l)\}$.

Definition 4.19: Let C be a ground clause, and let l be maximal in C wrt. $>_t$. Then we define the multiset of maximal literals in C wrt. $>_t$, denoted by $Max(C)$, as $\{l\}^n$, where n is the multiplicity of l in C.

In a node of a proof tree, not always all the literals are really necessary for the subproof below it. For instance, if there is a proof for a clause C, then the same proof is also a proof of the clause $C \vee D$, by not using or ignoring the literals of the subclause D. This fact is taken into account by the following notion of *used* literals:

Definition 4.20: Let P be a proof of a clause C. The multiset of literals of C *used* in P, denoted by $Used(C,P)$ or simply C_u is the multiset containing the literals of the smallest subclause C' of C s.t. P is a proof for C'.

Below the proof steps are written by indicating the clause to which they are applied and the clauses obtained by applying the step. The complementary clauses obtained are not shown.

Definition 4.21: The complexity of the operators in proof terms is defined as follows:

Step:			Complexity:
$C \vee l$	$\rightarrow_r$	$C \vee l' \vee c_1\sigma \vee \ldots \vee c_m\sigma$	$M \cup_m \{l\}$
$C \vee l$	$\rightarrow_{r,n}$	$C \vee l' \vee c_1\sigma \vee \ldots \vee c_m\sigma$	$M \cup_m \{l\}^{2n}$
$C \vee l$	$\leftarrow_r$	$C \vee l' \vee c_1\sigma \vee \ldots \vee c_m\sigma$	$M \cup_m \{l'\}$
$C \vee l$	$\leftrightarrow_a$	$C \vee l' \vee c_1\sigma \vee \ldots \vee c_m\sigma$	$M \cup_m \{l, l', c_1\sigma, \ldots, c_m\sigma\}^2$
C	$\leftrightarrow_{ca}$	$C \vee c_1 \vee \ldots \vee c_m$	$Max(C_u \vee c_1 \vee \ldots \vee c_m) \cup_m \{c_1, \ldots, c_m\}^2$

where M denotes $Max(C_u \vee l \vee l' \vee c_1\sigma \vee \ldots \vee c_m\sigma)$. The complexities of constants (tautologies, which are leafs in proof trees) are defined as:

$C \vee true$	complexity:	$true$
$C \vee (t{=}t)$	complexity:	$\{t\}^2$
$C \vee l \vee \neg l$	complexity:	$\{l\}^2$

Notation: The ordering on operator symbols of proof terms that is obtained by comparing their complexities by the multiset extension of the ordering $>_t$ will be denoted by $>_o$. The ordering on proofs, obtained by using RPO to lift the ordering $>_o$ on operators to an ordering on proof terms will be denoted by $>_{CC}$.

Lemma 4.22: The ordering on proofs $>_{CC}$ fulfills the subterm property (every proof is greater wrt $>_t$ than any of its proper subproofs), it is well-founded, and it is monotonic provided only subproofs P for a clause C are replaced by subproofs P' for C with $Used(C,P) \supseteq Used(C,P')$.

The monotonicity property of $>_{CC}$ means that, when in a proof P some subproof is replaced by a smaller subproof, the result is a proof P' which is smaller than P. This property does not hold in general here, because of the notion of *used* literals, that determines the complexity of proof steps. The reason is that, by using literals in the new subproof that were not used in the old subproof, the complexity of the operators in other parts of P can change. Therefore, we will only apply those proof transformations for which we can guarantee that this is not the case.

Lemma 4.23: The ordering $>_{CC}$ is compatible with Clausal Completion.

Proof. We have to show that if $(\mathcal{A},\mathcal{R}) \vdash_{CC} (\mathcal{A}',\mathcal{R}')$, then for any proof P_1 of a clause C using $(\mathcal{A},\mathcal{R})$, there is a proof P_2 of C using $(\mathcal{A}',\mathcal{R}')$ with $P_1 \geq_{CC} P_2$. Compatibility of $>_{CC}$ with each inference rule has to be proved. Obviously, if the inference rule applied is the addition of a critical pair, then the result holds trivially, since P_2 can be P_1. Here we only show one transformation, corresponding to the deletion of a tautology. The proof trees are displayed horizontally, each level being represented by a column. The expression $C \quad \lhd P$ denotes the fact that P is a proof for C.

If an axiom a with $(E \doteq E)$ as a literal is deleted, then we have

$$(\mathcal{A} \cup \{a\},\mathcal{R}) \vdash_{CC} (\mathcal{A},\mathcal{R})$$

where a is an axiom of the form $(E \doteq E) \vee c_1 \vee \ldots \vee c_m$. Let P be the following subproof of P_1 containing a step with the axiom a applied to a clause C:

$$\begin{array}{llll} P: & C \leftrightarrow_a & C \vee c_1\sigma \vee \ldots \vee c_m\sigma & \lhd P_0 \\ & & C \vee \neg c_1\sigma & \lhd P_1 \\ & & \ldots & \\ & & C \vee \neg c_m\sigma & \lhd P_m \end{array}$$

The step using the tautology a can be simulated by a case analysis step, obtaining the following proof P':

$$\begin{array}{llll} P': & C \leftrightarrow_{ca} & C \vee c_1\sigma \vee \ldots \vee c_m\sigma & \lhd P_0 \\ & & C \vee \neg c_1\sigma & \lhd P_1 \\ & & \ldots & \\ & & C \vee \neg c_m\sigma & \lhd P_m \end{array}$$

By considering the complexities of the operators involved, and by applying RPO, we can show that $P >_{CC} P'$, and therefore, that P can be replaced in P_1 by P', obtaining a smaller proof P_2. ∎

The following *Clausal Critical Pair Lemma* states, roughly speaking, that for every clause obtained by applying a reductive rewrite step to a clause C, there will be a proof that is simpler or equal than the simplest proof for C. The proofs of this lemma and of theorem 4.17 (cf. [Nie 90]), consist of proof transformations applying the different types of critical pairs generated during completion.

Lemma 4.24: (*Clausal Critical Pair Lemma*)
Let $(\mathcal{A}_0,\mathcal{R}_0),(\mathcal{A}_1,\mathcal{R}_1),\ldots$ be a fair CC-derivation, and let P be a rewrite proof of a clause $C \vee l$ using $\mathcal{R}_i$. If we have $\{C\} \rightarrow_{\mathcal{R}_i} \{C_1,\ldots,C_m\}$, then for each clause C' in $\{C_1,\ldots,C_m\}$ there is a proof P' using $(\mathcal{A}_j,\mathcal{R}_j)$ with $j \geq i$ and $P \geq_{CC} P'$.

5. Main results

5.1. Refutational theorem proving by Clausal Completion

Theorem 5.1: (refutational completeness up to failure) Let CCP be a clausal completion procedure with input $>$ and $(\mathcal{A}, \mathcal{R})$ s.t. $\mathcal{A} \cup \mathcal{R} \vdash \square$. If CCP does not fail, then the empty clause $\square$ will be generated as a critical pair during the completion procedure.

Proof. If $\mathcal{A} \cup \mathcal{R} \vdash \square$, and the completion procedure does not fail, then, by theorem 4.17, it has to generate a rewrite proof for $\square$. This is impossible, since no reductive rewrite steps can be applied to $\square$. The only remaining possibility is that the empty clause belongs to some set $\mathcal{A}_i$. ∎

If $\mathcal{R}$ is a complete system of clausal rewrite rules, then fairness does not require any critical pair between these rules to be computed. Therefore, $\mathcal{R}$ can be used to do refutational theorem proving in a very efficient way.

5.2. Consequence-finding proofs by clausal rewriting

Clausal rewriting can be used as an efficient decision procedure if the resolution rules of the complete system used can be applied *deterministically*, i.e. the substitution is determined by the instantiation of the literal that is used as left hand side in each step.

Definition 5.2: A clausal rewrite rule is called *deterministic* if it is a paramodulation rule, or else, if it is a resolution rule r of the form $(L_1 \vee \ldots \vee L_k \rightarrow \mathit{true}) \vee c_1 \vee \ldots \vee c_m$, where $vars(L_i) = vars(r)$, for $1 \leq i \leq k$.

A *clausal rewrite tree* T of a clause C using $\mathcal{R}$ is any tree where C is its root, and where the children of each node C' are the clauses obtained by applying a reductive rewrite step to C' with a rule of $\mathcal{R}$.

A *redundant* step in T is any step applied to a node C' s.t. (i) C' is a tautology or (ii) the step creates a complementary clause $C' \vee l$, where l belongs to C' or to an ancestor of C' in T.

It is useless to add complementary literals that have already belonged to an ancestor, since, roughly speaking, the information is already present in the clause. Therefore, these steps can be forbidden without loss of completeness, as proved in [Nie 90]:

Theorem 5.3: Let $(\mathcal{A}_0, \mathcal{R}_0), (\mathcal{A}_1, \mathcal{R}_1), \ldots$ be a fair CC-derivation with $(\mathcal{A}_0, \mathcal{R}_0) = (\mathcal{A}, \mathcal{R})$. Then $\mathcal{A} \cup \mathcal{R} \vdash C$ iff the derivation generates a pair $(\mathcal{A}_n, \mathcal{R}_n)$ s.t. there is a rewrite proof without redundant steps for C using $\mathcal{R}_n$.

Theorem 5.4: There is no infinite clausal rewrite tree without redundant steps using a finite set of deterministic rules $\mathcal{R}$.

Proof. (Sketch) The proof is based on a well-founded ordering on nodes of clausal rewrite trees. The complexities of nodes are defined as a pairs, compared lexicographically, whose first component is the (finite!) set of literals that can still be added as complementary literals to the node in non-redundant steps, and whose second component is the multiset of literals of the node. It is easy to see that the children of each node D are strictly smaller under this ordering than D. ∎

Definition 5.5: A clausal rewrite tree T is closed with respect to a set of rules $\mathcal{R}$ if there are no redundant rewrite steps applied in T, and there is no leaf of T to which a non-redundant step with a rule of $\mathcal{R}$ can be applied.

Theorem 5.6: Let $\mathcal{R}$ be a finite complete set of deterministic clausal rewrite rules, let T be a clausal rewrite tree that is closed with respect to $\mathcal{R}$, and let C be the root of T. Then $\mathcal{R} \vdash C$ iff every leaf of T is a tautology.

Building such a closed rewrite tree consists simply in applying, in any order, non-redundant rewrite steps while this is possible. Since this process is finitely terminating, it provides a decision procedure. A semi-decision procedure can be obtained even if the completion procedure does not terminate, by clausal rewriting using the rules the diverging pocedure is generating:

Definition 5.7: Let Der be a fair infinite CC-derivation $(\mathcal{A}_0, \mathcal{R}_0), (\mathcal{A}_1, \mathcal{R}_1), \ldots$, let m and n be natural numbers with $n \geq m$, and let C be a clause. The n-th incremental closed rewrite tree of C in Der starting at m, denoted by $T_n^m(C)$ is defined as follows:

(i) $T_m^m(C)$ is any closed rewrite tree of C using R_m

(ii) $T_{i+1}^m(C)$, with $i \geq m$, is the rewrite tree obtained by replacing in $T_i^m(C)$ every non-tautology leaf D by a closed rewrite tree of D using R_{i+1}.

Theorem 5.8: Let Der be a fair infinite CC-derivation $(\mathcal{A}_0, \mathcal{R}_0), (\mathcal{A}_1, \mathcal{R}_1), \ldots$, and let C be a clause without negative equality literals. Then $(\mathcal{A}_0, \mathcal{R}_0) \vdash C$ iff $\forall m \exists n$ s.t. every $T_n^m(C)$ is a proof, i.e. it has only tautologies as leafs.

This method can be used in practice by generating successive incremental rewrite trees, starting from any desired point of the completion procedure. The only requirement is that the rewrite trees have to be kept closed after each new rule is generated.

5.3. Proving general first order clauses

A combination of the methods outlined above can be used for proving general first order clauses. This will be done by using the following result (cf. [Nie 90]):

Let C and D be general clauses, and let C be of the form $l_1 \vee \ldots \vee l_n$. Then we have $\mathcal{A} \cup \{\neg\bar{l}_1, \ldots, \neg\bar{l}_n\} \vdash \overline{D} \iff \mathcal{A} \vdash l_1 \vee \ldots \vee l_n \vee D$, where $\bar{l}$ and $\overline{D}$ denote the application of a Skolemization substitution for the variables of C to l and D repectively.

Now suppose we want to prove such a general clause $C \vee D$, where C is the subclause formed by the negative equality literals $l_1 \vee \ldots \vee l_n$, and that we want to deduce this theorem from a set of axioms $\mathcal{A}$ and a complete set of rewrite rules $\mathcal{R}$. Then we can start a completion procedure with input $(\mathcal{A} \cup \{\neg\bar{l}_1, \ldots, \neg\bar{l}_n\}, \mathcal{R})$. In this process we do not need to compute critical pairs between rules in $\mathcal{R}$.

If this completion procedure does not fail, and it generates the empty clause, then C, and therefore $C \vee D$ has been proved. Otherwise, it may terminate, and then there is a decision procedure for the clause $\overline{D}$: it can be proved or disproved by computing a closed rewrite tree for it. If it does not terminate, then the method outlined above can be used in order to obtain a semi-decision procedure for the clause $\overline{D}$.

6. Examples

These examples are complete systems of clausal rewrite rules, produced by the implementation within the TRIP system of our completion procedure. We show some theorems that can be proved by clausal rewriting, by refutation, or by a combination of both (cf. section 5.3). Note that the negation is expressed by the minus symbol. The first example describes the parity of integers with addition:

```
1)  s(p(x))         --> x
2)  p(s(x))         --> x
3)  x + 0           --> x
4)  0 + x           --> x
5)  x + s(y)        --> s(x + y)
6)  s(x) + y        --> s(x + y)
7)  x + p(y)        --> p(x + y)
8)  p(x) + y        --> p(x + y)
9)   even(0)        --> true
10) -even(s(x))     --> true  v  -even(x)
11)  even(s(x))     --> true  v   even(x)
12) -even(p(x))     --> true  v  -even(x)
13)  even(p(x))     --> true  v   even(x)
14) -odd(x)         --> true  v  -even(x)
15)  odd(x)         --> true  v   even(x)
16)  even(x + y)    --> true  v  -even(y)  v  -even(x)
17)  even(x + y)    --> true  v   even(y)  v   even(x)
18) -even(x + y)    --> true  v   even(y)  v  -even(x)
19) -even(x + y)    --> true  v   even(x)  v  -even(y)

Theorems:   odd(s(0))                          even(x+x)
            odd(s(x)+x)                        even(x+y) v -(x=y)
            even(s(s(x))) v  odd(x)            even(s(x)) v  even(x)
            even(y) v odd((x+y)+x)
```

The second example is a complete system for ordered lists (with Prolog notation) of natural numbers. It includes maximality, the maximal element of a list, and an operation for inserting elements in ordered lists.

```
1)  0 =< x                       --> true
2)  x =< x                       --> true
3)  -(s(x) =< 0)                 --> true
4)  -(s(x) =< s(y))              --> true      v   x =< y
5)    s(x) =< s(y)               --> true      v -(x =< y)
6)  -(s(x) =< y)                 --> true      v   x =< y
7)        x =< s(y)              --> true      v -(x =< y)
8)  x =< y    v    y =< x        --> true
9)  max(x,y)                     --> y         v -(x =< y)
10) max(x,y)                     --> x         v   x =< y
11)  ordered( [] )               --> true
12)  ordered( [x|[]] )           --> true
13)  ordered( [x|[y|z]] )        --> true      v -(y =< x) v -ordered([y|z])
14) -ordered( [x|[y|z]] )        --> true      v   y =< x
15) insert( x, [] )              --> [x|[]]
16) insert( x, [y|z] )           --> [x|[y|z]]            v  -(y =< x)
17) insert( x, [y|z] )           --> [y|insert(x,z)]      v    y =< x
18) maxl( [] )                   --> 0
19) maxl( [x|y] )                --> x         v -(maxl(y) =< x)
20) maxl( [x|y] )                --> maxl(y) v    maxl(y) =< x

Theorem: -ordered([X|L])  v  ordered([ maxl([X|L]) |[X|L] ])
```

7. References

[BDH 86] L. Bachmair, N. Dershowitz, J. Hsiang: Orderings for equational proofs. In Proc. Symp. Logic in Computer Science, 346-357, Boston (Massachusetts USA), 1986.

[BDJ 78] D. Brand, J.A. Darringer, W.H. Joyner: Completeness of conditional reductions. IBM T.J. Watson Res. Center Report RC-7404 (1978).

[BG 90a] L. Bachmair, H. Ganzinger: On restrictions of ordered paramodulation with simplification. In Proc. 10th Int. Conf. om Automated Deduction. LNCS, pp 427-441.

[BG 90b] L. Bachmair, H. Ganzinger: Completion of first order clauses with equality. 2nd Intl. Workshop on Conditional and Typed Term Rewriting, Montreal (1987). to appear in LNCS.

[BGS 88a] H. Bertling, H. Ganzinger, R. Schafers: A collection of specifications completed by the CEC-system Report, PROSPECTRA-Project, U. Dortmund, 1988.

[BGS 88b] H. Bertling, H. Ganzinger, R. Schafers: CEC: A system for conditional equational completion — User Manual (Version 1.0). Report, PROSPECTRA-Project, U. Dortmund, 1988.

[Der 87] N. Dershowitz: Termination of rewriting. Journal of Symbolic Computation, 69-116, 1987.

[Gan 87] H. Ganzinger: A completion procedure for conditional equations. 1st Intl. Workshop on Conditional Term Rewriting, Orsay (1987), also to appear in J. of Symb. Computation.

[HR 89] J. Hsiang, M. Rusinowitch: Proving refutational completeness of theorem proving strategies: The transfinite semantic tree method. Submitted for publication.

[KB 70] D.E. Knuth, P.B. Bendix: Simple word problems in universal algebras. J. Leech, editor, Computational Problems in Abstract Algebra, 263-297, Pergamon Press, Oxford, 1970.

[KR 87] E. Kounalis, M. Rusinowitch: On word problems in Horn logic. 1st Intl. Workshop on Conditional Term Rewriting, Orsay (1987).

[Kap 85] S. Kaplan: Fair conditional term rewriting systems: unification, termination and confluence. Recent Trends in Data Type Specification, Springer IFB 116 (1985).

[MN 90] U. Martin, T. Nipkow: Ordered rewriting and confluence. In Proc. 10th Int. Conf. om Automated Deduction. LNCS, pp 366-380.

[Nav 87] M.L. Navarro Gomez: Tecnicas de Reescritura para especificaciones condicionales. PhD Thesis, Barcelona, 1987.

[Nie 88] R. Nieuwenhuis: Algunas técnicas de reescritura y su implementacion en el sistema TRIP. Master thesis, UPC Barcelona, 1988.

[Nie 90] R. Nieuwenhuis: Theorem proving in first order logic with equality by clausal rewriting and completion. PhD thesis, UPC Barcelona, 1990.

[NOR 90] R. Nieuwenhuis, F. Orejas, A. Rubio: TRIP: an implementation of clausal rewriting. In Proc. 10th Int. Conf. om Automated Deduction. LNCS 449, pp 667-668.

[NO 90] R. Nieuwenhuis, F. Orejas: Clausal Rewriting: Applications and Implementation. In Proc. Int. Workshop on Abstract Data Types. 1990, to appear in LNCS.

[Rus 87] M. Rusinowitch: Theorem proving with resolution and superposition: An extension of the Knuth-Bendix completion procedure as a complete set of inference rules. Report 87-R-128, CRIN, Nancy, 1987.

[RZ 85] J.L. Remy, H. Zhang: Contextual rewriting. Proc. Conf. on Rewriting Techniques and Applications, Dijon (1985).

Chapter 4

Combined Systems, Combined Languages and Modularity

Adding Algebraic Rewriting to the Calculus of Constructions : Strong Normalization Preserved

Franco Barbanera
Dipartimento di Informatica
Corso Svizzera, 185 10149 Torino, Italy

Abstract. In this paper strong normalization is proved for terms of systems obtained by combining strong normalizing algebraic term rewriting systems with the Calculus of Constructions. The proof of the main result exploits the observation that the proof of strong normalization for the Calculus of Constructions given by Geuvers and Nederhof partly applies to this calculus extended with algebraic rewriting.

Introduction

Recently some attention has been paid to the interactions between the kind of computation formalized by means of rewrite rules for first order algebraic terms and that which proceeds by performing β-reductions on λ-terms. No doubt one of the most recent and interesting results has been obtained in [Breazu-Tannen,Gallier 90] where it is proved that the terms of the system obtained by combining a strongly normalizing many-sorted algebraic term rewriting system with the polymorphic second order λ-calculus are strongly normalizable according the β-reduction and the algebraic reductions.
As pointed out in [Breazu-Tannen,Gallier 90], it would be interesting and useful to see that this good interaction holds even for type disciplines more complicated then the second order polymorphic λ-calculus like the Calculus of Constructions ([Coquand, Huet 88]). What is proved in the present paper is that this is true if we restrict only to β-conversion the rule stating that if a term has a type A and this type is convertible to the type A', then the term has type A' as well.
This result of strong normalization is achieved by means of different modular steps. The first one consists in proving that the system obtained by mixing a strongly normalizable term rewriting system with the polymorphic λ-calculus of order ω is strongly normalizable according to the β-reduction and the algebraic reductions. This result is obtained by using a similar result, obtained in [Barba 89], for a type assignment system. In the second step it is proved that the result obtained in the first one implies the strong normalization for the system obtained by mixing a strongly normalizable term rewriting system with the Calculus of Constructions. This implication is proved by means of an adaptation to the present situation of the proof of strong normalization for the Calculus of Constructions given by Geuvers and Nederhof in [Geuvers et al.90], a proof which uses the strong normalization of the polymorphic λ-calculus of order ω and a clever reduction-preserving traslation from the former to the latter. Geuvers and Nederhof's method itself generalizes a similar method by Harper, Honsell and Plotkin [Harper et al.87] used to prove the strong normalization of their system LF. The translation and the proofs of its properties given in this paper (most of section 3) are modifications of those given in [Geuvers et al.90], mostly by adding the additional cases in proofs and definitions by structural induction.

1. Adding algebraic features to the Calculus of Constructions and to the typed λ-calculus of order ω.

Definition 1.1

Let S be a set of *sorts* and Σ an S-sorted algebraic signature. Each function symbol $f \in \Sigma$ has an *arity*, which is a string $s_1 ... s_n \in S^*$, $n \leq 0$, and a *sort* $s \in S$.
The intention is that each symbol in Σ names some heterogeneous operation which takes arguments of sorts (in order) $s_1, ..., s_n$ and return a result of sort s.

We define now the extensions with algebraic rewriting features of the typed λ-calculus of order ω and the Calculus of Constructions. The names we shall use to denote them will be $\lambda_{R\omega}$ and λ_{RC}, respectively. When it is not specified, the definitions are intended to be valid for both the systems, proviso the fact that the notions we refer to in them are to be tought of as belonging to $\lambda_{R\omega}/\lambda_{RC}$ if the definitions are seen as definitions for $\lambda_{R\omega}/\lambda_{RC}$.

Definition 1.2 (*Pseudo-expressions and Declarations*)
(i) The systems $\lambda R\omega$ and λ_{RC} are based on a set of *pseudo-expressions* T defined by the following abstract syntax :

$$T ::= x \mid * \mid \square \mid f \mid s \mid TT \mid \lambda x{:}T.T \mid \Pi x{:}T.T$$

where x is the category of variables (divided in two groups : Var* and $Var^{\square}$), * and □ are special constants, s ranges over a given set of sorts S and f ranges over the function symbol of a given S-sorted signature Σ.
(ii) A *declaration* for a variable x is an expression of the form x:A , where A∈T.
(iii) A declaration x:A is *algebraic* if A∈S.

On pseudo-expressions the notions of β-reduction and β-conversion are defined by the contraction rule :

$$(\lambda u{:}A.B)C \to_\beta B[C/u] .$$

Γ will denote in the following a finite ordered sequence of declarations. The empty sequence will be denoted by <> and if $\Gamma = \langle x_1{:}A_1,...,x_n{:}A_n\rangle$ then $\Gamma,x{:}B=\langle x_1{:}A_1,...,x_n{:}A_n,x{:}B\rangle$.

Definition 1.3 (*Term Formation Rules*)
The term-formation rules are given in two groups. Particular subsets of the second one will differentiate $\lambda R\omega$ and λ_{RC}

General axioms and rules

Axioms
$\langle\rangle \vdash * : \square$
$\langle\rangle \vdash s : *$ for each s∈S
$\langle\rangle \vdash f : s_1\to...\to s_n\to s$ for each f∈Σ, where $s_1...s_n$ is the arity and s is the sort of f.
($s_1\to...\to s_n\to s$ is short for $\Pi x_1{:}s_1.\Pi x_2{:}s_2...\Pi x_n{:}s_n.s$)

Start rule
$$\frac{\Gamma \vdash A : p}{\Gamma,u{:}A \vdash u{:}A} \quad p\in\{*,\square\}, u\in Var^p, u \text{ does not belong to } \Gamma$$

Weakening rule
$$\frac{\Gamma \vdash A : B \qquad \Gamma \vdash C : p}{\Gamma,u{:}C \vdash A : B} \quad p\in\{*,\square\}, u\in Var^p, u \text{ does not belong to } \Gamma$$

Application rule
$$\frac{\Gamma \vdash F : \Pi u{:}A.B \qquad \Gamma \vdash C : A}{\Gamma \vdash FC : B[C/u]}$$

Conversion rule
$$\frac{\Gamma \vdash A : B \qquad \Gamma \vdash B' : p \qquad B =_\beta B'}{\Gamma \vdash A : B'} \quad p\in\{*,\square\}$$

Specific rules
The specific rules are all introduction rules and are parametrised by * and □. Let $p_1,p_2\in\{*,\square\}$.

(p_1,p_2) rules

$$\frac{\Gamma \vdash A : p_1 \qquad \Gamma,u{:}A \vdash B : p_2}{\Gamma \vdash \Pi u{:}A.B : p_2}$$

$$\frac{\Gamma \vdash A : p_1 \quad \Gamma,u{:}A \vdash b : B \quad \Gamma,u{:}A \vdash B : p_2}{\Gamma \vdash \lambda u{:}A.b : \Pi u{:}A.B}$$

These last two parametrised rules will be called in the sequel also Π-intro and λ-intro respectively.

Notice that the conversion rule is the very same of the Calculus of Constructions, i.e. the notion of conversion used in the rule is the β-conversion and not the βR-conversion as one could expect. We have to give this restriction as in our proof of βR-strong normalization for λ_{RC} the calculus is assumed to be confluent. This means that the proof works if the conversion used in the rule is actually the β- one, for which confluence is known to hold. This is not a real drawback because it is easy to check that, as soon as confluence is proved for βR-conversion, our proof leads immediately to a proof of βR-strong normalization for λ_{RC} with βR-conversion used in the conversion rule.

It is not difficult to check that the specific term-formation rules have the following informal meanings. (*,*) allows forming terms depending on terms; (*,□) allows forming types depending on terms; (□,*) allows forming terms depending on types ; (□,□) allows forming types depending on types.

Definition 1.4 (*The systems $\lambda R\omega$ and λRC*)
The systems $\lambda R\omega$ and λRC are the systems which have as term formation rules the general rules plus a specific subset of the set of the specific rules, namely
$\lambda R\omega$: {(*,*), (□,*), (□,□) } λRC : { (*,*), (□,*), (*,□), (□,□) }.

Definition 1.5 (*Well-formed Terms*)
A *well-formed term* for $\lambda R\omega/\lambda RC$ of type B in Γ is a pseudo-expression A such that it is possible to derive $\Gamma\vdash A : B$ by a derivation which uses the term formation rules for $\lambda R\omega/\lambda RC$. We say that A is simply a well-formed term if there exist B and Γ such that A is a well-formed term of type B in Γ.

The notion of substitution of a terms for a variable is defined in the usual way. One, in substituting a term for a variable in a term A has of course to be careful in performing the sustitution also in the type as well as in the sequence of declarations, a thing correct by the following Substitution Lemma.
The simultaneous substitution $M[N_1/x_1,..,N_m/x_m]$ will be sometimes denoted by $M[\varphi]$, where φ is the (substitution-) function from variables to terms such that $\varphi(y)=N_i$ if $y\equiv x_i$ (i=1,..,m) and $\varphi(y)=y$ if $y\notin\{x_1,..,x_m\}$. The squared brakets in a substitution will be used both on the right and on the left of the term.

CONVENTIONS
(i) We shall use the notation $\Gamma\vdash_{Ri} M : A$ to stress the fact that we can derive $\Gamma\vdash M : A$ in the system λRi i=ω/C.
(ii) $\Pi u{:}A.B$ with $u\notin FV(B)$ will be denoted by $A \to B$.
(iii) $\Gamma\vdash A : B : C$ will stand for $\Gamma\vdash A : B$ and $\Gamma\vdash B : C$.
(iv) The set of terms of λRi (i = ω/C) will be denoted in the sequel by Term(λRi)
(v) We shall denote with Context(λRi) the set of all sequences of declarations for λRi.

Definition 1.6 (*Algebraic Terms*)
A (well-formed) term is *algebraic* if it is of the following forms :
- x of type $s\in S$
- $fM_1..M_n$ of type $s\in S$, $f\in\Sigma$ with arity $s_1..s_n$ and sort s and M_1 of type s_1,..., M_n of type s_n are algebraic terms

Then if $\Gamma\vdash M : A$ and M is an algebraic term $A \equiv s$ for some $s\in S$.

Definition 1.7 (*Algebraic Rewrite Rules, Reductions and Decompositions*)
(i) An *algebraic rewrite rule* is a pair <A,B> of algebraic terms which have the same type in the same context and the type is a sort. Moreover $FV(B) \subseteq FV(A)$ and A is not a variable.
(ii) Given an algebraic rewrite rule r ≡ <A,B>, we define the *algebraic reduction relation* $\to^r$ on well-formed terms by means of the following reduction rule:

$$A[\varphi] \to^r B[\varphi] \quad \text{where } \varphi \text{ is a substitution.}$$

Given a set of algebraic rewrite rules R we define $\to_R = \cup_{r\in R} \to^r$.
From now on we shall assume that the given set R is such that the algebraic terms are R-strongly normalizable.
(iii) An *algebraic decomposition* , if it exists, of $M\in$ Term(λRi) is a pair (A,φ), where A is an algebraic term and φ a substitution for A (i.e. $\varphi(x)\neq x$ iff $x\in FV(A)$) such that $M\equiv A[\varphi]$. In a decomposition (A,φ) the variables of A have to be all distinct. Besides, algebraic decompositions are considered modulo renamings of free variables.

Definition 1.8 (*Objects, Constructors, Kinds*)
In λRi :
(i) Let $\Gamma\vdash A : B : *$. Then A is called an *object* and B a *type*. The set of all the objects in λRi will be denoted in the sequel by Object(λRi) and the set of all types by Type(λRi).
(ii) Let $\Gamma\vdash A : B : \Box$. Then A is called a *constructor* and B a *kind*. The set of all the constructors in λRi will be denoted in the sequel by Constructor(λRi) and the set of all kinds by Kind(λRi).
Notice that * is a particular kind and that the types are particular constructor. It can be shown that a term is an object, a constructor or a kind.

From now on the word "context" will be an overloaded one, in fact it will be used not only to denote a sequence of declarations but to denote also the usual notion of "term with a hole" which could be defined in the following way.

Definition 1.9

A *context* for objects {for constructor} for λR_i is a term C of λR_i such that :
(i) there is a particular free variable $¢ \in Var^*$ {$\$ \in Var^\square$} used only in contexts and occurring only once in C (the traditional notation for a context C is C[ ]);
(ii) if we substitute an object {a constructor} A for ¢ {\$} in C we do not have to care about the renaming of free variables of A. The traditional notation for this particular substitution is C[A].

It will be always clear, if it is not precisely stated, if a particular context is for objects or for constructor.
It has been chosen to overload the word contex because on one hand it is commonly used in literature for both its meanings, and on the other hand because its "current" meaning will always be clear.

All the lemmas in the rest of this section follow easily from the corresponding proofs for the pure systems (see [Geuvers et al.90])

Lemma 1.10
Let $\Gamma \in Context(\lambda R\omega)$, $t \in Term(\lambda R\omega)$ and K be the set inductively defined in the following way :

- $* \in K$
- $k_1,k_2 \in K \Rightarrow \quad k_1 \to k_2 \in K$.

Then $t \in Kind(\lambda R\omega) \quad \Rightarrow \quad t \in K$ and $\vdash_{R\omega} t : \square$.

Lemma 1.11
Let $A, t \in Term(\lambda R\omega)$. Then
$A \in Constructor(\lambda R\omega)$ and t subexpression of A $\Rightarrow$ $t \in Kind(\lambda R\omega)$ or $t \in Constructor(\lambda R\omega)$.

Lemma 1.12 (SUBSTITUTION LEMMA)
Let $\Gamma_1,x{:}D,\Gamma_2 \in Context(\lambda R_i)$ and $B,C,D,d \in Term(\lambda R_i)$.
$\Gamma_1,x{:}D,\Gamma_2 \vdash_{Ri} B : C,\ \Gamma_1 \vdash_{Ri} d : D \quad \Rightarrow \quad \Gamma_1,\Gamma_2[d/x] \vdash_{Ri} B[d/x] : C[d/x]$.

Lemma 1.13 (STRIPPING LEMMA)
Let $\Gamma \in Context(\lambda R_i)$ and $M,N,R \in Term(\lambda R_i)$.

(i) $\Gamma \vdash_{Ri} p : R$ with $p \in \{*\} \cup S$ $\Rightarrow$ $\exists p' \in \{*,\square\}[\, R =_\beta p'$ and $p{:}p'$ is an axiom]
(ii) $\Gamma \vdash_{Ri} f : R$ with $f \in \Sigma$ $\Rightarrow$ $\exists P\ R =_\beta P$ and $f{:}P$ is an axiom
(iii) $\Gamma \vdash_{Ri} x : R$ with x variable $\Rightarrow$ Let $\Gamma = \langle x_1{:}A_1,\ldots,x_n{:}A_n \rangle$.
$\exists j \le n\ \exists p \in \{*,\square\}[x \equiv x_j \in Var^p$ and $R =_\beta A_j$ and $\Gamma' \vdash_{Ri} A_j{:}p]$ where $\Gamma' = \langle x_1{:}A_1,\ldots,x_{j-1}{:}A_{j-1} \rangle$
(iv) $\Gamma \vdash_{Ri} \Pi x{:}M.N : R$ $\Rightarrow$ $\Gamma \vdash_{Ri} M : p_1$, $\Gamma,x{:}M \vdash_{Ri} N{:}p_2$ and $R =_\beta p_2$ for some rule (p_1,p_2)
(v) $\Gamma \vdash_{Ri} \lambda x{:}M.N : R$ $\Rightarrow$ $\Gamma \vdash_{Ri} M : p_1$, $\Gamma,x{:}M \vdash_{Ri} B{:}p_2$, $\Gamma,x{:}M \vdash_{Ri} N{:}B$, $\Gamma \vdash_{Ri} \Pi x{:}M.B : p_2$ and $R =_\beta \Pi x{:}M.B$ for some rule (p_1,p_2)
(vi) $\Gamma \vdash_{Ri} MN : R$ $\Rightarrow$ $\Gamma \vdash_{Ri} M : \Pi x{:}A.B$, $\Gamma \vdash_{Ri} N : A$ and $R =_\beta B[N/x]$ for some terms A,B of R_i
(vii) $\Gamma \vdash_{Ri} P : R$ $\Rightarrow$ $\exists p \in \{*,\square\}[\, R \equiv p$ or $\Gamma \vdash_{Ri} R : s]$.

Lemma 1.14 (SUBJECT REDUCTION)
Let $\Gamma \in Context(\lambda R_i)$ and $B,B',C \in Term(\lambda R_i)$, then
$\Gamma \vdash_{Ri} B : C$ and $B \twoheadrightarrow_{\beta R} B' \quad \Rightarrow \quad \Gamma \vdash_{Ri} B' : C$.

REMARK 1.15
It is not difficult to check that $C \in Term(\lambda R_i)$, $C =_\beta p$ with $p \in \{*,\square\} \quad \Rightarrow \quad C \equiv p$.

Lemma 1.16 (CLASSIFICATION LEMMA)
(i) $Kind(\lambda R_i) \cap Type(\lambda R_i) = \varnothing$
(ii) $Constructor(\lambda R_i) \cap Object(\lambda R_i) = \varnothing$.

Definition 1.17
Let $\Gamma \in Context(\lambda R_i)$, $\Pi z:B_1.B_2, \lambda z:B_1.B_2, B_1B_2 \in Term(\Gamma)$.
(i) $\Pi z:B_1.B_2$ is *formed by* (p_1,p_2) in Γ iff
there are the specific rules (p_1,p_2) in λR_i,
$\Gamma \vdash_{Ri} B_1: p_1$ and $\Gamma, x:B_1 \vdash_{Ri} B_2: p_2$.
(ii) $\lambda z:B_1.B_2$ is *formed by* (p_1,p_2) in Γ iff
there are the specific rules (p_1,p_2) in λR_i,
$\Gamma \vdash_{Ri} \lambda z:B_1.B_2 : \Pi z:C_1.C_2$ and $\Pi z:C_1.C_2$ is formed by (p_1,p_2) in Γ.
(iii) B_1B_2 is *formed by* (p_1,p_2) in Γ iff
there are the specific rules (p_1,p_2) in λR_i,
$\Gamma \vdash_{Ri} B_1 : \Pi z:C_1.C_2$ and $\Pi z:C_1.C_2$ is formed by (p_1,p_2) in Γ.

Lemma 1.18 (UNICICY OF FORMATION LEMMA)
Let $\Gamma \in Context(\lambda R_i)$, $B \in Term(\lambda R_i)$,
B formed by (p_1,p_2) in Γ, B formed by (p_1',p_2') in Γ' $\Rightarrow$ $p_1 \equiv p_1'$, $p_2 \equiv p_2'$.

The Unicity of Formation Lemma allows us to use the terminology "formed by" without mentioning the context Γ.

2. *βR-strong normalization for terms of* $\lambda_{R\omega}$

In order to prove that each term of λR_ω is βR-strongly normalizable we introduce a type assignment system, namely the intersection type assignment system developed in [Coppo et al.81] to which we add algebraic features. The fact that terms typable in this type assignment system are strongly normalizable according to β-reduction and the algebraic reduction relation induced by the R-reduction on algebraic terms of λR_ω given in the previous section will imply the βR-strong normalization for λR_ω. This wil be achieved by means of a type erasing function on terms of λR_ω by proving that erased terms are typable in our type assignment system and that a βR-reduction on a term of λR_ω induces a reduction on the erased term.

The intersection type assignment system with algebraic rewriting ($\wedge TR$)

Type assignment systems in general are formal systems which allow us to assign types to untyped terms. These systems are defined by specifying a set of terms, a set of types we assign to terms and a set of type-inference rules.

Definition 2.1 (*Terms and Types of* $\wedge TR$)
(i) The set $T_\wedge$ of (*intersection*)*types* of $\wedge TR$ is inductively defined by :
1. $a,b,... \in T_\wedge$ (type variables,whose set well be denoted by V_T)
2. $s \in T_\wedge$ for $s \in S$ (S is the set of sorts used in the definition of λR_ω)
3. $\sigma,\tau \in T_\wedge \Rightarrow \sigma \to \tau \in T_\wedge$, $\sigma \wedge \tau \in T_\wedge$.

(ii) The *terms* of $\wedge TR$ we assign types to are the untyped λ-terms (Λ_Σ) built out of a denumerable set of (untyped) variables and a set of constant. The set of constant we consider has exactly the function symbol of Σ as elements, where Σ is the same S-sorted signature used to define λR_ω.
(iii) A *statement* is an expression of the form M : σ where $\sigma \in T_\wedge$ and $M \in \Lambda_\Sigma$. M is the *subject* of the statement.
(iv) A *basis* (the set of assumptions a statement depends on) is a set of statements with only variables as subjects. Without losing anything we assume that in a basis there are no two statements with the same subject.

If x does not occur in the basis B, then B,x: σ denotes $B \cup \{ x: \sigma \}$.

In this system we define inductively a preorder relation $\leq$ over types which formalizes the subset relation (on the interpretation of types). This relation will be used in a type assignment rule modeling the subsumption relation.

Definition 2.2
(i)The relation $\leq$ is the smallest relation satisfying

$$\tau \leq \tau \qquad \sigma \wedge \tau \leq \sigma \ , \ \sigma \wedge \tau \leq \tau$$
$$\tau \leq \tau \wedge \tau \qquad (\sigma \to \rho) \wedge (\sigma \to \tau) \leq \sigma \to (\rho \wedge \tau)$$
$$\sigma \leq \tau \leq \rho \Rightarrow \sigma \leq \rho \qquad \sigma \leq \sigma', \tau \leq \tau' \Rightarrow \sigma \wedge \tau \leq \sigma' \wedge \tau'$$
$$\sigma' \leq \sigma, \tau \leq \tau' \Rightarrow (\sigma \to \tau) \leq (\sigma' \to \tau').$$

(ii)The relation $\sim$ is defined as:

$$\sigma \sim \tau \Leftrightarrow \sigma \le \tau \le \sigma .$$

We give now the axioms and rules defining our type assignment system. Notice that, in spite of the similarity, these are type assignment rules, i.e. a thing conceptually very different from term-formation rules like the ones given for the systems λR_{ω} and λ_{RC}. In fact in this case the symbol $\vdash^{\wedge}$ can be thought of as a real deduction symbol while in term-formation rules it is only a separation mark between the part containing the informations about the types of the free variables of a term from the rest of the term.

Definition 2.3 (*The rules of* $\wedge$*TR*)
The axioms and type inference rules of the type assignment system $\wedge$TR are the following :

(Ax) $\quad B, x{:}\sigma \vdash^{\wedge} x{:}\sigma$

$\quad B \vdash f : s_1 \rightarrow \ldots \rightarrow s_n \rightarrow s$ $\quad$ for each $f \in \Sigma$ where $s_1 \ldots s_n$ is the arity and s is the sort of f.

$$(\rightarrow I)\ \frac{B, x{:}\sigma \vdash^{\wedge} M{:}\tau}{B \vdash^{\wedge} \lambda x.M : \sigma \rightarrow \tau} \qquad (\rightarrow E)\ \frac{B \vdash^{\wedge} M : \sigma \rightarrow \tau \quad B \vdash^{\wedge} N : \sigma}{B \vdash^{\wedge} (MN) : \tau}$$

$$(\wedge I)\ \frac{B \vdash^{\wedge} M : \sigma \quad B \vdash^{\wedge} M : \tau}{B \vdash^{\wedge} M : \sigma \wedge \tau} \qquad (\wedge E)\ \frac{B \vdash^{\wedge} M : \sigma \wedge \tau}{B \vdash^{\wedge} M : \sigma} \quad \frac{B \vdash^{\wedge} M{:}\sigma \wedge \tau}{B \vdash^{\wedge} M : \tau}$$

$$(\le)\ \frac{B \vdash^{\wedge} M : \sigma \quad \sigma \le \tau}{B \vdash^{\wedge} M : \tau}$$

Definition 2.4
A term M is *typable* <u>iff</u> there exist a basis B and a type σ such that $B \vdash M{:}\ \sigma$.
$\Lambda_{\wedge TR}$ will denote the set of terms typable in $\wedge$TR.

On terms of $\Lambda_{\wedge TR}$ the notion of β-reduction ($\rightarrow_{\beta(\wedge TR)}$) is defined in the usual way.

Theorem 2.5 ([Pottinger 80])
Each term of the pure λ-calculus is strongly normalizable (sn) <u>iff</u> it is typable in the type assignment system without the algebraic features.

Using $\wedge$***TR to prove βR-strong normalization for*** λR_{ω}

Let us define, by induction on the structure of the terms, a "type erasing" function from Object(λR_{ω}) to the elements of Λ_{Σ}.

Definition 2.6
The map $|\text{-}|$: Object(λR_{ω}) $\rightarrow \Lambda_{\Sigma}$ is inductively defined by

(i) $|x| = x$ $\quad$ if $x \in Var^*$
(ii) $|f| = f$ $\quad$ if $f \in \Sigma$
(iii) $|\lambda x{:}A.q| = \lambda x.|q|$ $\quad$ if $\lambda x{:}A.q$ is formed by (*,*)
(iv) $|\lambda\alpha{:}A.q| = |q|$ $\quad$ if $\lambda\alpha{:}A.q$ is formed by ($\Box$,*)
(v) $|pq| = |p||q|$ $\quad$ if pq is formed by (*,*)
(vi) $|pQ| = |p|$ $\quad$ if pQ is formed by ($\Box$,*).

Lemma 2.7
If $\Gamma \vdash M : s$ is an algebraic term in λR_{ω} then $|M| \in \Lambda_{\wedge TR}$.Besides $|M|$ is typable with type s from a basis B such that if $x{:}s' \in \Gamma$ then $x{:}s' \in B$.

Definition 2.8
Each algebraic rewrite rule for λR_{ω} : $r \equiv$ <A ,B> induces a reduction relation $\rightarrow^{r}_{\wedge TR}$ on terms of $\Lambda_{\wedge TR}$ in the following way :

$|A|[\varphi] \rightarrow^{r}_{\wedge TR} |B|[\varphi]$ where φ is a substitution for $\wedge$TR such that, if Γ is the context of A and B, and $x{:}s' \in \Gamma$ then $\varphi(x)$ is typable with type s'.

Notice that Lemma 2.7 assure us that this is a good definition.
Given a set of algebraic rewrite rules R for $\lambda R\omega$ we define $\rightarrow R(\wedge TR) = \cup_{r \in R} \rightarrow^{r}{}_{\wedge TR}$.
$\rightarrow\beta R(\wedge TR) \equiv \rightarrow\beta(\wedge TR) \cup \rightarrow R(\wedge TR)$.

In the sequel, since it will be always clear by the context, we shall denote a reduction on terms of $\lambda R\omega$ and the corresponding one on terms of $\Lambda_{\wedge TR}$ in the same way, i.e. we shall omit the subscript "$_{\wedge TR}$".

Theorem 2.9 ([Barba 89])
If the set R of algebraic rewrite rules of $\lambda R\omega$ is strongly normalizing for the set of algebraic terms of $\lambda R\omega$ then $\rightarrow\beta R$ is strongly normalizable for all terms in $\Lambda_{\wedge TR}$.

This theorem will be fundamental to prove that $\lambda R\omega$ terms are βR-strongly normalizable. In fact in the following we prove that $|M| \in \Lambda_{\wedge TR}$ in case $M \in$ Object($\lambda R\omega$) and that if a term M βR-reduces to M' so does |M| with respect to |M'|.

Definition 2.10
Let B, B' two bases.
$B \cup^{\wedge} B' =$
$\{x{:}\sigma\wedge\tau \mid x{:}\sigma \in B \text{ and } x{:}\tau \in B'\} \cup \{x{:}\sigma \mid x{:}\sigma \in B \text{ and } x \notin FV(B')\} \cup \{x{:}\tau \mid x \notin FV(B) \text{ and } x{:}\tau \in B'\}$

It is straightforward to check that if $B \vdash M{:}\ \beta$ then we have also $B \cup^{\wedge} B' \vdash M{:}\ \beta$ for all B'.

Lemma 2.11
Let $M \in$ Object($\lambda R\omega$) and let |M| be in β-normal form.
Then $\exists\, B,\sigma \quad B \vdash |M| : \sigma$.
Besides if in $\lambda R\omega$ the type of M is $\alpha \in S$ then $\exists\, B \quad B \vdash |M| : \alpha$.
Proof
By induction on the structure of |M|, by considering that, since |M| is in β-normal form it has to be $|M| \equiv \lambda x_1..x_n.gN_1..N_m$ where the N_i's have the same "shape" of |M|. (See [Hindley et al.86] cap.1 for a precise formulation and a proof of this statement). It is easy to check that this fact, valid if we do not consider algebraic features, still holds in our case.

Lemma 2.12
Let $M \in$ Object($\lambda R\omega$) $|M| \equiv [P/x]Q$. Let P be a term such that if $x \notin FV(Q)$ P is in β-normal form and $\exists\, P' \in$ Object($\lambda R\omega$) $P \equiv |P'|$. Then

$$(\exists B,\sigma \;\; B \vdash [P/x]Q : \sigma) \;\Rightarrow\; (\exists B' \;\; B' \vdash (\lambda x.Q)P : \sigma) .$$

Proof
We consider two cases according to whether x does occur in Q or not. We can assume that x does not occur in B.
1. x occurs in Q.
We will consider only the occurrences of P in [P/x]Q which replace occurrences of x in Q.
Let $B_i \vdash P : \delta_i$ ($i \in I$) the set of all the conclusions whose subject are such P's in the deduction of $B \vdash [P/x]Q : \sigma$.
Then it is not difficult to obtain a deduction of $B_1 \cup^{\wedge}..\cup^{\wedge} B_n \vdash P{:}\ \wedge_{i\in I}\, \delta_i$.
Moreover we can obtain a deduction of $B,x{:}\wedge_{i\in I}\,\delta_i \vdash Q : \sigma$ by extending the basis B and replacing in the proof of $B \vdash [P/x]Q : \sigma$ the deductions of $B,x{:}\wedge_{i\in I}\,\delta_i \vdash x{:}\delta_i$ (obtained using rules (Ax) and ($\wedge$E)) for the subdeductions of $B_j \vdash P : \delta_j$. So by rule ($\rightarrow$I) we have that $B \vdash \lambda x.Q : (\wedge_{i\in I}\delta_i)\rightarrow\sigma$ and since we have
$B_1 \cup^{\wedge}..\cup^{\wedge} B_n \vdash P : \wedge_{i\in I}\delta_i$ we can obtain
$B' \vdash (\lambda x.Q)P : \sigma$ where $B' = B_1 \cup^{\wedge}..\cup^{\wedge} B_n \cup^{\wedge} B$.
2. x does not occur in Q.
By the hypothesis' and by Lemma 2.11 we have that $\exists B_1,\delta \;\; B_1 \vdash P : \delta$.
We know that $\exists B,\sigma \;\; B \vdash [P/x]Q : \sigma$ i.e. $B \vdash Q : \sigma$ (x is not free in Q) then it is possible to prove $B,x{:}\delta \vdash Q : \sigma$ and from this $B \vdash \lambda x.Q : \delta\rightarrow\sigma$. Is is easy now to construct a proof of $B' \vdash (\lambda x.Q)P : \sigma$ where $B' = B \cup^{\wedge} B_1$.

Lemma 2.13
Let $M[M'/x] \in$ Object($\lambda R\omega$) where $x \in$ Var*. Then $|\, M[M'/x]\, | \equiv |M|[|M'|/x]$.

Lemma 2.14
Let $M \in$ Object($\lambda R\omega$). Then

$$|M| \rightarrow_\beta N \;\;\Rightarrow\;\; \exists\, M' \in \text{Object}(\lambda R\omega) \text{ such that } N \equiv |M'| \text{ and } M \rightarrow_\beta M'.$$

Lemma 2.15
Let $M \in Object(\lambda_{R\omega})$. Then there exist B, σ such that $B \vdash |M|:\sigma$.
Proof
|M| is β-strongly normalizable by Theorem 2.9 and by Lemma 2.14.Then all the reduction strategies allow us to get the β-normal form of |M|, in particular the reduction strategy according to which a contraction $(\lambda x.Q)P \rightarrow_\beta [P/x]Q$ is performed only if P is in β-normal form (i.e.the rightmost-innermost evaluation). Hence we can find B and σ using Lemma 2.11 on the β-normal form of |M| and iterating Lemma 2.12 backward along the chain of applicative order reduction steps which leads from |M| to its β-normal form.

CONVENTION
A β-reduction on a $\lambda_{R\omega}$ term such that the β-redex has been formed using rule (p_1,p_2) will be called in the sequel a $\beta(p_1,p_2)$-reduction.

Definition 2.16
Let φ be a substitution for $\lambda_{R\omega}$. We define the substitution $|\varphi|$ for terms of $\Lambda_{\wedge TR}$ in the following way: $Dom|\varphi| = Dom\varphi$ and for each $x \in Dom|\varphi|$ $|\varphi|(x) = |\varphi(x)|$.

Lemma 2.17
Let $M,N \in Object(\lambda_{R\omega})$, then
$M \rightarrow_{\beta R} N \Rightarrow |M| \rightarrow_{\beta R} |N|$ or $|M| \equiv |N|$ if the reduction is actually a $\beta(\square,\square)$- or a $\beta(\square,*)$-reduction.
Proof
We have that that $M \equiv C[R]$ where R is a β- or a R-redex and C[ ] is a suitable context. The proof can be simply given by checking the statement of the lemma for each notion of reduction, distinguishing the cases for the β-reduction according to which rule the redex has been formed.

Lemma 2.18 ([Girard 72])
If $M \in Term(\lambda_{R\omega})$ then M is β-strongly normalizing.

Actually Girard's theorem is about $\lambda_{R\omega}$ without the algebraic features, but in β-reductions the function symbols of Σ behave as free variables.

We can now prove the main theorem of this section.
Theorem 2.19
For each $M \in Term(\lambda_{R\omega})$ M is βR-strongly normalizing.
Proof
By Lemma 1.10 we have to consider only the cases $M \in Constructor(\lambda_{R\omega})$ or $M \in Object(\lambda_{R\omega})$.
If $M \in Constructor(\lambda_{R\omega})$ then, since in $\lambda_{R\omega}$ objects cannot occur into constructors by Lemma 1.11, it follows that each reduction in M is actually a β-reduction and hence the strong normalization of M follows from Theorem 2.18.
If $M \in Object(\lambda_{R\omega})$ we know from Lemma 2.15 that $|M| \in \Lambda_{\wedge TR}$ and then by Theorem 2.9 |M| is βR-strongly normalizing. To prove the thesis of the theorem it suffices to show that the βR-strong normalizability of |M| implies the βR-strong normalizability of M. To prove this implication by contradiction let us assume there exists an infinite sequence of reductions out of M. Let s be this sequence. We have two cases.
1. The number of $\beta(*,*)R$-reductions in s is infinite.
Then by using Lemma 2.17 it is possible to build out of s an infinite sequence of βR-reduction starting at |M|, contradicting the fact |M| is βR-strongly normalizing.
2. The number of $\beta(*,*)R$-reductions in s is finite.
Let M' be the term obtained by the last $\beta(*,*)R$-reduction of the sequence. We have therefore an infinite sequence of $\beta(\square,\square)$- and $\beta(\square,*)$-reductions out of M', a fact that contradicts Lemma 2.18 since, by the Subject Reduction Lemma even M' belongs to $Term(\lambda_{R\omega})$.

3. βR-strong normalization for $\lambda_{R\omega}$ implies βR-strong normalization for λ_{RC}

Definition 3.1
The map $\rho : \{\square\} \cup Kind(\lambda_{RC}) \rightarrow Kind(\lambda_{R\omega})$ is inductively defined by:
   (i) $\rho(*) = \rho(\square) = *$
   (ii) $\rho(\Pi\alpha:M.N) = \rho(M) \rightarrow \rho(N)$ if $\Pi\alpha:M.N$ is formed by $(\square,\square)$
        $\rho(\Pi x:M.N) = \rho(N)$ if $\Pi x:M.N$ is formed by $(*,\square)$.

The case distinction in the definition is correct by Lemma 1.18.

REMARK 3.2

The following properties of the map ρ will be used.

(i) $k_1, k_2 \in \mathrm{Kind}(\lambda_{RC}), k_1 =_{\beta R} k_2 \Rightarrow \rho(k_1) \equiv \rho(k_2)$

(ii) $k \in \mathrm{Kind}(\lambda_{RC}), u \in \mathrm{Var}^* \cup \mathrm{Var}^{\square}, A \in \mathrm{Term}(\lambda_{RC}) \Rightarrow \rho(k) \equiv \rho(k[u:=A])$.

Now we choose one of the variables of $\mathrm{Var}^{\square}$ to act as a fixed constant, i.e. it will not be used as a bound variable in an abstraction. This variable will be denoted by 0.

Definition 3.3

The map $\tau : \{\square\} \cup \mathrm{Kind}(\lambda_{RC}) \cup \mathrm{Constructor}(\lambda_{RC}) \to \mathrm{Constructor}(\lambda_{R\omega})$ is inductively defined by

(i) $\tau(*) = \tau(\square) = 0$

(ii) $\tau(\alpha) = \alpha$ if $\alpha \in \mathrm{Var}$

$\tau(s) = s$ if $s \in S$

(iii) $\tau(\Pi\alpha{:}M.N) = \Pi\alpha{:}\rho(M).\tau(M) \to \tau(N)$ if $\Pi\alpha{:}M.N$ is formed by $(\square,\square)$ or $(\square,*)$

$\tau(\Pi x{:}M.N) = \Pi x{:}\tau(M).\tau(N)$ if $\Pi x{:}M.N$ is formed by $(*,\square)$ or $(*,*)$

(iv) $\tau(\lambda\alpha{:}M.N) = \lambda\alpha{:}\rho(M).\tau(N)$ if $\lambda\alpha{:}M.N$ is formed by $(\square,\square)$

$\tau(\lambda x{:}M.N) = \tau(N)$ if $\lambda x{:}M.N$ is formed by $(*,\square)$

(v) $\tau(MN) = \tau(M)\tau(N)$ if MN is formed by $(\square,\square)$

$\tau(MN) = \tau(M)$ if MN is formed by $(*,\square)$.

The definition by cases is correct by Lemma 1.18. That the range of τ is really $\mathrm{Constructor}(\lambda_{R\omega})$ can be easily proved by checking the definition and by Lemma 3.7.

Lemma 3.4

If A is the traslation in λ_{RC} of a curryfied type of a functional symbol in the signature Σ then $\tau(A) = A$.

Lemma 3.5

Let $B, B' \in \mathrm{Kind}(\lambda_{RC}) \cup \mathrm{Constructor}(\lambda_{RC})$.

(i) $x \in \mathrm{Var}^*, A \in \mathrm{Object}(\lambda_{RC}) \Rightarrow \tau(B[A/x]) \equiv \tau(B) \equiv \tau(B)[A/x]$

(ii) $\alpha \in \mathrm{Var}^{\square}, A \in \mathrm{Constructor}(\lambda_{RC}) \Rightarrow \tau(B[A/\alpha]) \equiv \tau(B)[\tau(A)/\alpha]$

(iii) $B \to_{\beta R} B' \Rightarrow \tau(B) \to_{\beta R} \tau(B')$ or $\tau(B) \equiv \tau(B')$.

Proof

By induction on the structure of B.

The basis of the induction is easy.

(i) (ii) . After applying the Stripping Lemma and the induction hypothesis it follows immediately from remark 3.2 .

(iii) Let us consider separately the two reduction relations.

$B \to_\beta B'$. If $B \equiv \Pi u{:}M.N$, $B \equiv \lambda u{:}M.N$ or $B \equiv MN$ with $B' \equiv \Pi u{:}M'.N'$, $B' \equiv \lambda u{:}M'.N'$ or $B' \equiv M'N'$, then $\tau(B) \to_\beta \tau(B')$ or $\tau(B) \equiv \tau(B')$ follows immediately by induction hypothesis. The interesting situation occurs when $B \equiv (\lambda u{:}M.N)P$, $B' \equiv N[P/u]$. If $M \in \mathrm{Type}(\lambda_{RC})$ then $\tau((\lambda u{:}M.N)P) \equiv \tau(N) \equiv \tau(N[P/u])$ by definition of τ and by (i). If $M \in \mathrm{Kind}(\lambda_{RC})$ then $\tau((\lambda\alpha{:}M.N)P) \equiv (\lambda\alpha{:}\rho(M).\tau(N))\tau(P) \to_\beta \tau(N)[\tau(P)/\alpha] \equiv \tau(N[P/\alpha])$ by definition of τ and by (ii).

$B \to_R B'$. Because of the fact $B, B' \in \mathrm{Kind}(\lambda_{RC}) \cup \mathrm{Constructor}((\lambda_{RC})$ it has to be necessarily that $B \equiv C[A]$ and $B' \equiv C[A']$ for a suitable context for objects $C[\ ]$, where $A, A' \in \mathrm{Object}(\lambda_{RC})$ and $A \to_R A'$ (obviously A and A' have a sort as type). From (i) it follows $\tau(B) \equiv \tau(B')$. Notice that in (i) the usual substitution is considered, not the particular one for the variable ¢ of contexts. However it is quite staightforward to check that (i) is valid even for the substitutions for ¢ in contexts.

In order to map $\mathrm{Context}(\lambda_{RC})$ into $\mathrm{Context}(\lambda_{R\omega})$ we choose for each variable α of $\mathrm{Var}^{\square}$ a connected variable x^α in Var^*, such that no two variables of $\mathrm{Var}^{\square}$ are connected to the same variable of Var^*. We extend now the map τ in such away it acts also on $\mathrm{Context}(\lambda_{RC})$ yelding elements of $\mathrm{Context}(\lambda_{R\omega})$.

Definition 3.6

(i) Let $A \in \mathrm{Kind}(\lambda_{RC}) \cup \mathrm{Type}(\lambda_{RC})$.

$\tau(x{:}A) = x{:}\tau(A)$ if $x \in \mathrm{Var}^*$

$\tau(\alpha{:}A) = \alpha{:}\rho(A), x^\alpha{:}\tau(A)$ if $\alpha \in \mathrm{Var}$

(ii) Let $\Gamma = u_1{:}A_1, u_2{:}A_2, \ldots, u_n{:}A_n \in \mathrm{Context}(\lambda_{RC})$.

$\tau(\Gamma) = 0{:}*, d{:}\bot, \tau(u_1{:}A_1), \tau(u_2{:}A_2), \ldots, \tau(u_n{:}A_n)$.

The reason for putting $0{:}*$ and $d{:}\bot \equiv \Pi\alpha{:}*.\alpha$ in the context is that in the following definition of [ ] on terms of λ_{RC} it will be necessary to have a canonical inhabitant for every type and kind. If $\tau(\Gamma) \vdash_{R\omega}$

B:* or $\tau(\Gamma) \vdash_{R\omega}$ B: □ , we want $\tau(\Gamma) \vdash_{R\omega} c^B{:}B$ for a c^B which does not depend on the structure of Γ. Now if $\tau(\Gamma) \vdash_{R\omega}$ B:* we shall put $c^B \equiv dB$ and if $\tau(\Gamma) \vdash_{R\omega}$ B:□ a canonical inhabitant of B is defined inductively by

(i) $B \equiv * \Rightarrow c^* = 0$

(ii) $B \equiv k_1 \rightarrow k_2 \Rightarrow c^{k1\rightarrow k2} = \lambda\alpha{:}k_1.c^{k2}$

Notice that $c^B[N/u] \equiv c^{B[N/u]}$ for all kinds and constructors B, variables u and terms N in $\lambda_{R\omega}$.

Lemma 3.7
Let $\Gamma \in$ Context(λ_{RC}), M,N∈ Term(λ_{RC}).

$$\Gamma \vdash_{RC} M{:}N{:}\square \text{ or } \Gamma \vdash_{RC} M{:}N{\equiv}\square \quad \Rightarrow \quad \tau(\Gamma) \vdash_{R\omega} \tau(M){:}\rho(N).$$

Definition 3.8
The map [ ] : Kind(λ_{RC})∪Constructors(λ_{RC})∪Object(λ_{RC})→Objects($\lambda_{R\omega}$) is inductively defined by

(i) $[*] = c^0$ (ii) $[x] = x$ if $x\in$ Var*
(iii) $[\alpha] = x^\alpha$ if $\alpha\in \mathrm{Var}^\square$ (iv) $[s] = c^0$ if $s\in$ Sort
(v) $[f] = f$ if f belongs to the signature Σ
(vi) $[\Pi x{:}M.N] = c^{0\rightarrow 0\rightarrow 0}[M][N][c^{\tau(M)}/x]$
if Πx:M.N is formed by (*,*) or (*,□)
$[\Pi\alpha{:}M.N] = c^{0\rightarrow 0\rightarrow 0}[M][N][c^{\tau(M)}/x, c^{\rho(M)}/\alpha]$
if Πα:M.N is formed by (□,*) or (□,□)
(vii) $[\lambda x{:}M.N] = (\lambda z{:}0.\lambda x{:}\tau(M).[N])[M]$ where z is a fresh variable
if λx:M.N is formed by (*,*) or (*,□)
$[\lambda\alpha{:}M.N] = (\lambda z{:}0.\lambda\alpha{:}\rho(M).\lambda x^\alpha{:}\tau(M).[N])[M]$ where z is a fresh variable
if λα:M.N is formed by (□,*) or (□,□)
(viii) $[MN] = [M][N]$ if MN is formed by (*,*) or (*,□)
$[MN] = [M]\tau(M)[N]$ if MN is formed by (□,*) or (□,□).

This definition by cases is allright by the Stripping Lemma . The following theorem states that the range of [ ] is really Object($\lambda_{R\omega}$).

Lemma 3.9
Let $\Gamma\in$ Context(λ_{RC}), M,N∈ Term(λ_{RC}).

$$\Gamma \vdash_{RC} M{:}N \quad \Rightarrow \quad \tau(\Gamma) \vdash_{R\omega} [M]{:}\tau(N).$$

Proof
By induction on the structure of M. By Lemma 3.7 we know that τ(Γ) is a legal context in $\lambda_{R\omega}$.
- M ≡*, N ≡ □, then $[*] \equiv c^0{:}\ 0 \equiv \tau(\square)$ in τ(Γ).
- M ≡ s∈ Sort, N $=_\beta$ *. We have that $[s] \equiv c^0$ and that N ≡ * by Remark 1.15. Then $[s] \equiv c^0{:}\ 0 \equiv \tau(N)$ in τ(Γ) since τ(*) ≡ 0.
- M ≡ u∈ Var*∪$\mathrm{Var}^\square$, then u:A∈ Γ, with A $=_{\beta R}$ N.
If u ≡ α∈ $\mathrm{Var}^\square$ then $\tau(\alpha{:}A) \equiv \alpha{:}\rho(A), x^\alpha{:}\tau(A) \in \tau(\Gamma)$, if u ≡ x∈ Var* then $\tau(x{:}A) \equiv x{:}\tau(A) \in \tau(\Gamma)$. In both cases $\tau(\Gamma) \vdash_{R\omega} [u]{:}\tau(A)$. By $\tau(\Gamma) \vdash_{R\omega} \tau(N){:}*$ and $\tau(N) =_{\beta R} \tau(A)$, by Lemma 3.5, we find $\tau(\Gamma) \vdash_{R\omega} [u]{:}\tau(N)$.
- M ≡ f∈ Σ, with N $=_\beta$ A where A is the type of f in the signature Σ (more precisely the translation in λ_{RC} of the type of f). We have that [f] ≡ f : A in τ(Γ). By Lemma 3.4 it follows trivially [f] ≡ f : τ(A) in τ(Γ) and then by Lemma 3.5 and one application of the rule conversion we find [f] ≡ f : τ(N) in τ(Γ).
- M ≡ $\Pi u{:}B_1.B_2$, then $\Gamma \vdash_{R\omega} B_1{:}p_1$ and $\Gamma, u{:}B_1 \vdash_{R\omega} B_2{:}p_2$ for suitable $p_1,p_2 \in \{*,\square\}$. By induction hypothesis $\tau(\Gamma) \vdash_{R\omega} [B_1]{:}0$ and $\tau(\Gamma, u{:}B_1) \vdash_{R\omega} [B_2]{:}0$. By Lemma 3.7 $\tau(\Gamma) \vdash_{R\omega} \tau(B_1){:}*$ and so $\tau(\Gamma) \vdash_{R\omega} c^{\tau(B_1)}{:}\tau(B_1)$.
If $p_1 \equiv *$, u ≡ x∈ Var*, then $\tau(\Gamma, x{:}B_1) = \tau(\Gamma), x{:}\tau(B_1)$, so by the Substitution Lemma $\tau(\Gamma) \vdash_{R\omega} [B_2][c^{\tau(B_1)}/x]{:}0$. By using two times the application rule we get $\tau(\Gamma) \vdash_{R\omega} c^{0\rightarrow 0\rightarrow 0}[B_1][B_2][c^{\tau(B_1)}/x]{:}0$.
If $p_1 \equiv \square$, u ≡ α∈ $\mathrm{Var}^\square$, then $\tau(\Gamma, \alpha{:}B_1) = \tau(\Gamma), \alpha{:}\rho(B_1), x^\alpha{:}\tau(B_1)$, so with the Substitution Lemma $\tau(\Gamma) \vdash_{R\omega} [B_2][c^{\tau(B_1)}/x^\alpha, c^{\rho(B_1)}/\alpha]{:}0$. By using two times the application rule we get $\tau(\Gamma) \vdash_{R\omega} c^{0\rightarrow 0\rightarrow 0}[B_1][B_2][c^{\tau(B_1)}/x, c^{\rho(B_1)}/\alpha]{:}0$.
In both cases $\tau(\Gamma) \vdash_{R\omega} [\Pi u{:}B_1.B_2]{:}0$.
- M ≡ $\lambda u{:}B_1.B_2$, then $\Gamma \vdash_{R\omega} B_1{:}p_1$, $\Gamma, u{:}B_1 \vdash_{R\omega} B_2{:}C_2{:}p_2$ for suitable $C_2 \in$ Term(λ_{RC}), $p_1,p_2 \in \{*,\square\}$, with N $=_\beta \Pi u{:}B_1.C_2$.
By induction hypothesis $\tau(\Gamma) \vdash_{R\omega} [B_1]{:}0$ and $\tau(\Gamma, u{:}B_1) \vdash_{R\omega} [B_2]{:}\tau(C_2)$. By Lemma 3.7 $\tau(\Gamma) \vdash_{R\omega} \tau(B_1){:}*$ and $\tau(\Gamma, u{:}B_1) \vdash_{R\omega} \tau(C_2){:}*$.
If $p_1 \equiv *$, u ≡ x∈ Var*, then $\tau(\Gamma, x{:}B_1) = \tau(\Gamma), x{:}\tau(B_1)$. With two applications of λ-intro and one of the application rule we get
$\tau(\Gamma) \vdash_{R\omega} (\lambda z{:}0.\lambda x{:}\tau(B_1)[B_2])[B_1]{:}\Pi x{:}\tau(B_1).\tau(C_2)$.

If $p_1 \equiv \Box$, $u \equiv \alpha \in Var^{\Box}$, then $\tau(\Gamma,x:B_1) = \tau(\Gamma),\alpha:\rho(B_1),x^{\alpha}:\tau(B_1)$. With three applications of λ-introduction and one of the application rule we get
$\tau(\Gamma)\vdash_{R\omega}(\lambda z:0.\lambda\alpha:\rho(B_1).\lambda x^{\alpha}:\tau(B_1)[B_2])[B_1]:\Pi\alpha:\rho(B_1).\tau(B_1)\to\tau(C_2)$.
In both cases $\tau(\Gamma)\vdash_{R\omega}[\lambda u:B_1.B_2]:\tau(\Pi u:B_1.C_2)$ and so $\tau(\Gamma)\vdash_{R\omega}[\lambda u:B_1.B_2]:\tau(N)$.
- $M \equiv B_1B_2$ then $\Gamma\vdash_{R\omega}B_1:\Pi u:C_1.C_2$ and $\Gamma\vdash_{R\omega}B_2:C_1$ for some $C_1,C_2\in Term(\lambda_{RC})$ with $N =_{\beta R} C_2[B_2/u]$.
By induction hypothesis $\tau(\Gamma)\vdash_{R\omega}[B_1]:\tau(\Pi u:C_1.C_2)$ and $\tau(\Gamma)\vdash_{R\omega}[B_2]:\tau(C_1)$. By Lemma 3.7 $\tau(\Gamma)\vdash_{R\omega}\tau(C_1):*$.
If $C_1\in Type(\lambda_{RC})$, $u \equiv x\in Var^*$, then $\tau(\Pi x:C_1.C_2) \equiv \Pi x:\tau(C_1).\tau(C_2)$, so
$\tau(\Gamma)\vdash_{R\omega}[B_1][B_2]:\tau(C_2)[[B_2]/x] \equiv \tau(C_2[[B_2]/x]) \equiv \tau(C_2) \equiv \tau(C_2[B_2/x])$ (by Lemma 3.5(i)).
If $C_1\in Kind(\lambda_{RC})$, $u \equiv \alpha\in Var^{\Box}$, then $\tau(\Pi\alpha:C_1.C_2) \equiv \Pi\alpha:\rho(C_1).\tau(C_1)\to\tau(C_2)$, so
$\tau(\Gamma)\vdash_{R\omega}[B_1]\tau(B_2)[B_2]:\tau(C_2)[[B_2]/x^{\alpha},\tau(B_2)/\alpha] \equiv \tau(C_2[B_2/\alpha])$ (by Lemma 3.5).
In both cases $\tau(\Gamma)\vdash_{R\omega}[B_1B_2]:\tau(C_2[B_2/u])$ and then $\tau(\Gamma)\vdash_{R\omega}[B_1B_2]:\tau(N)$.

Lemma 3.10
Let $M\in Term(\lambda_{RC})$.
(i) $x\in Var^*$, $N\in Object(\lambda_{RC})$ $\Rightarrow$ $[M][[N]/x] \equiv [M[N/x]]$
(ii) $\alpha\in Var^{\Box}$, $N\in Constructor(\lambda_{RC})$ $\Rightarrow$ $[M][\tau(N)/\alpha,[N]/x^{\alpha}] \equiv [M[N/\alpha]]$
Proof
(i) By induction on the structure of M. If $M\in Var^*$ or $M\in Var^{\Box}$ or $M\in\Sigma$ or $M\equiv *$ we are immediately done. If M is a composed term we are done by induction hypothesis and the fact that $\tau(Q[N/x]) \equiv \tau(Q)[[N]/x]$ and $\rho(Q[N/x]) \equiv \rho(Q)[[N]/x]$ (by Remark 3.2 and Lemma 3.5) and so $c^{\tau(B)}[[N]/u] \equiv c^{\tau(B[N/u])}$.
(ii) By induction on the structure of M. If $M\in Var^*$ or $M\in Var^{\Box}$ or $M\in\Sigma$ we are immediately done. If M is a composed term we are done by induction hypothesis and the facts
$$\tau(Q)[\tau(N)/\alpha,[N]/x^{\alpha}] \equiv \tau(Q[N/\alpha]),\quad \rho(Q)[\tau(N)/\alpha,[N]/x^{\alpha}] \equiv \rho(Q[N/\alpha]),$$
$$c^{\tau(Q)}[\tau(N)/\alpha,[N]/x^{\alpha}] \equiv c^{\tau(Q[N/\alpha])},\quad c^{\rho(Q)}[\tau(N)/\alpha,[N]/x^{\alpha}] \equiv c^{\rho(Q[N/\alpha])},$$
which are true by Remark 3.2, Lemma 3.5 and the fact that $x^{\alpha}\notin FV(\rho(Q))\cup FV(\tau(Q))$.

Definition 3.11
Let φ be a substitution for λ_{RC}. We define the substitution [φ] for $\lambda_{R\omega}$ in the following way:
$Dom[\varphi] = Dom\varphi$ and for each $x\in Dom[\varphi]$ $[\varphi](x) = [\varphi(x)]$.

This definition is allright by Lemma 3.9.

Lemma 3.12
Let $M\in Object(\lambda_{RC})$ such that it has an algebraic decomposition (A,φ). Then
$[M] \equiv A[[\varphi]]\in Object(\lambda_{R\omega})$.

Lemma 3.13
Let $M,M'\in Term(\lambda_{RC})$.
$M \to_{\beta R} M' \Rightarrow [M] \to_{\beta R} [M']$
Recall that $\to_{\beta R}$ denotes a non-zero-step reduction.
Proof
By induction on the structure of M .
The basis of the induction is easy.
If $M \equiv \Pi u:B_1.B_2$, $M \equiv \lambda u:B_1.B_2$ or $M \equiv B_1B_2$ with $M'\equiv B_1B_2'$ or $B_1'B_2$, then $[M]\to_{\beta R}[M']$ follows immediately from the induction hypothesis. Let us distinguish now the cases M is a β- or a R-redex.
If $M \equiv (\lambda u:B_1.B_2)C$, $M' \equiv B_2[C/u]$ then we distinguish two cases, according to the rule by which M has been formed and apply Lemma 3.9.
• $(\lambda u:B_1.B_2)C$ formed by $(*,*)$. or $(*,\Box)$. Then
$[M] \equiv [(\lambda u:B_1.B_2)C] \equiv (\lambda z:0.\lambda u:\tau(B_1).[B_2])[B_1][C] \to_{\beta} [B_2][[C]/x]$ (recall that in the translation z is a fresh variable) $\equiv$
$\equiv [B_2[C/x]]$ by Lemma 3.10 (i)
$\equiv [M']$.
• $(\lambda u:B_1.B_2)C$ formed by $(\Box,*)$. or $(\Box,\Box)$. Then
$[M] \equiv [(\lambda u:B_1.B_2)C]$
$\equiv (\lambda z:0.\lambda u:\rho(B_1).\lambda x^{u}:\tau(B_1).[B_2])[B_1]\tau(C)[C] \to_{\beta} [B_2][\tau(C)/u,[C]/x^{u}]$
$\equiv [B_2[C/u]]$ by Lemma 3.10 (ii)
$\equiv [M']$.
If M is a R-redex then it has to be $M \equiv A[\varphi]$, $M' \equiv A'[\varphi']$ where (A,φ) and (A',φ') are algebraic decompositions of M and M' respectively, such that $\langle Q,T\rangle$ is an algebraic reduction and $Q\equiv A$, $T\equiv A'$ by a suitable renaming of the variables of A and A'.

Notice that the algebraic decomposition of M is (A',φ') and not (A',φ) by the fact that if (S,ζ) is an algebraic decomposition, then the variables of S are all different. In fact if A≡fxyz and fxyz $\rightarrow^r$ gxxx we have that A'≡guvz. This fact implies that, formally, φ' differ from φ.
By Lemma 3.10 [M] ≡ A[[φ]] and [M'] ≡ A'[[φ']]. It is straightforward to check now that A[[φ]] $\rightarrow^r$ A'[[φ']].

Using the previous lemmas we can now easily prove the main theorem of this section.

Theorem 3.14
βR-strong normalization for terms of $\lambda_{R\omega}$. implies βR-strong normalization for terms of λ_{RC}
Proof
Let us assume the βR-strong normalization property for $\lambda_{R\omega}$. By contradiction let M_1 be a term of λ_{RC} and $M_1 \rightarrow_{\beta R} M_2 \rightarrow_{\beta R} M_3 \rightarrow_{\beta R}$.. be an infinite βR-reduction sequence. By Lemma 3.13 $[M_1] \rightarrow_{\beta R} [M_2] \rightarrow_{\beta R} [M_3] \rightarrow_{\beta R}$..., so there is an infinite βR-reduction sequence in $\lambda_{R\omega}$. This is impossible by hypothesis, so there is no infinite βR-reduction sequence in λ_{RC}.

Finally we can prove:
Main Theorem
Each term of λ_{RC} is βR-strongly normalizable.
Proof
By Theorem 2.19 and Theorem 3.14.

Acknoledgments
"nothing that is worth knowing can be taught"[O.Wilde], but this is not always true for Prof. Mariangiola Dezani.
I wish to thank also Stefano Berardi for his useful suggestions.
Particular thanks for her support to Antonietta Cocci.

References
[Barba 89] F.Barbanera. Combining term rewriting and type assignment systems. *Proceedings Third Italian Conference on Theoretical Computer Science*. pp.71-84. Mantova.Bertoni, Boehm, Miglioli Eds. Nov. 1989.

[Breazu-Tannen, Gallier 90] Val Breazu-Tannen, Jan Gallier. Polymorphic rewriting conserves algebraic strong normalization. To appear in TCS.

[Coquand, Huet 88] T.Coquand, G.Huet. The Calculus of Constructions. *Information and Control*, pp.76-95, 120, 1988.

[Coppo et al.81] M.Coppo, M. Dezani-Ciancaglini, B.Venneri. Functional character of solvable terms. *Zeit.Math.Logik und Grund.Math.* 27, pp.45-58, 1981.

[Geuvers et al.90] H.Geuvers, M.J.Nederhof. A simple modular proof of the strong normalization for the Calculus of Constructions. To appear in "Journal of Functional Programming".

[Harper et al.87] R.Harper, F.Honsell, G.Plotkin. A framework for defining logics. *Proceedings of the symposium on Logic in Computer Science* . Ithaca, New York, IEEE, Washinton D.C. 1987.

[Pottinger 80] G.Pottinger. A type assignment for the strongly normalizable λ-terms. In *To H.B.Curry : Essays on Combinatory Logic, Lambda Calculus and Formalism*, Academic press, pp.561-578, 1980.

On Sufficient completeness of conditional specifications

Wadoud Bousdira, Jean-Luc Rémy
CRIN & INRIA Lorraine
BP 239 F54506 Vandœuvre Cedex
FRANCE
e-mail : bousdira, remy@loria.crin.fr
tel : (33) 83-59-30-21

Abstract

We present an algorithm which, given a conditional rewriting system and a set of distinguished operators, guarantees in case of success, that the normal forms of every ground term are primitive, i.e. built only from the distinguished operators. This property is crucial for completeness of definition and very helpful in inductive proofs. It was comparatively few investigated for conditional systems. The algorithm combines structural analysis, using patterns on distinguished operators, and case analysis, using sets of preconditions involved when reducing the instances of patterns. The algorithm works for reductive systems that constitute a large class of systems, as reductivity is assumed when proving most of the properties of conditional systems. The correctness proof of the algorithm and several examples are also presented.

1 Introduction

Specifications algebraically describe objects and operations of abstract data types. Traditionally, operations and rules of a specification $SPEC = (S, F, E)$ are partitioned in two parts. The first one $\langle BF, BE\rangle$ respectively contains the set C of constructors and the relations that can exist between them. The second one $\langle F - BF, E - BE\rangle$ contains function symbols called defined operations such that the specification satisfies a definition principle of the second part with respect to the first one.

An essential property has to be checked by the set F of specification symbols, namely the definition of each defined operation has to be complete with respect to the set of constructors C. In this case, the specification $SPEC$ is said to be complete with respect to C. Completeness amounts to check that each normal form of a ground term is a primitive term, i.e. a term only built from constructors. This property is essential for algebraic specifications because, together with the *consistency* property, it allows to prove the correctness of the instanciation of parameterized specifications. Also, it is useful in proof aspects in the sense that it allows to make proofs by induction (see for instance [15]).

Completeness of specifications is not decidable in general. However, some syntactic criteria allow to check it. Most of them are based on rewriting methods. Consistent works have already be made in order to study completeness of unconditional algebraic specifications and decidability conditions are stated in [8, 23, 9], [22, 4, 27], [16, 17] for some classes of specifications and more recently in [14, 21] and [20, 18].

In the context of conditional systems, the art is less developed. This is mostly due to the fact that the problem is much harder ; actually, the problem is undecidable even for one-rule conditional systems[1]. Our purpose in this paper is to study sufficient conditions to guarantee completeness.

Our sufficient conditions are effective and we propose an algorithm to check them. The main idea behind the algorithm is to mix structural analysis (in terms of patterns of constructors) and case analysis (in terms of set of conditions used in rewriting). Structural analysis builds a tree of patterns. Case analysis proves that the set of conditions allows to reduce all the instances of a pattern.

Another feature of our sufficient conditions is that they need not the systems to be checked satisfy many assumptions. For instance, we need not separately deal with operations used in the preconditions and

[1] For a proof, consider unconditional systems with undecidable inductive theory and conditional rule with the form $M = N \Rightarrow f(x_1, \ldots, x_n) \to 0$ where $x_1, \ldots, x_n$ are the variables in M and N ; f is completely defined if and only if the equation $M = N$ is valid for all the ground instances of its variables, that is belongs to the inductive theory [private communication with H. Comon].

operations defined by the conditional rules, as we did in previous works on hierarchical systems [3]. Instead we simply assume that the systems are reductive in the sense that the preconditions and the right hand sides are less complex than the corresponding left hand sides. This assumption has been made in many results on conditional systems, [13, 6] for instance. In addition, it is satisfied by most of the systems used in practice. It allows us to use an inductive argument when proving the correctness of our algorithm.

Let us now present the principle of our method on a simple, still typical, example. Let us consider the following simple four rule example on natural numbers:

$$pos(x) \rightarrow false,\ pos(succ(x)) \rightarrow true,\ f(0) \rightarrow 0,\ pos(x) = true \Rightarrow f(x) \rightarrow 0$$

It is impossible to prove sufficient completeness here, just looking at the rules. Indeed, in the conditional rule, the case $pos(x) = false$ seems to be missing. Therefore, we start to split the too general pattern $f(x)$ into two more specialized patterns, namely $f(0)$ and $f(succ(x))$. Now the former is dealt with by the unconditional rule. The latter is dealt with by the conditional rule as, now, the precondition evaluates to *true*. So every pattern is dealt with and the proof is achieved.

As another example, let us consider the following four rule system defining a predicate which compares a given element with all the elements of a list.

$$pp(x,\ empty\text{-}l) \rightarrow true,\ x \leq y = true\ \wedge\ pp(x,\ l) = true \Rightarrow pp(x,\ add(y,\ l)) \rightarrow true,$$
$$x \leq y = false \Rightarrow pp(x,\ add(y,\ l)) \rightarrow false,\ pp(x,\ l) = false \Rightarrow pp(x,\ add(y,\ l)) \rightarrow false.$$

Again we can split the general pattern $pp(x,\ l)$ into two patterns, namely $pp(x,\ empty\text{-}l)$ and $pp(x,\ add(y,\ l))$. The latter is not reducible, except if we assume something about preconditions. We do that by considering three contextual terms, one for each of the three conditional rules, and by proving that the disjunction of the preconditions is a tautology. At this stage of our research, we use a validity checker for propositional formulas, which is enough here. As a matter of fact, we can notice here that the operator to be checked occurs in two of the three preconditions. Because these preconditions are smaller than the pattern, we are able to use an inductive argument.

The paper is organized as follows: Section 2 formalizes the framework and presents the basic definitions. Section 3 introduces operational sufficient completeness (**o-s-c-** for short) and gives conditions under which this property is equivalent to algebraic sufficient completeness. Section 4 presents the sufficient condition that we state in order to check **o-s-c** in some classes of rewrite systems. This section includes three parts and each of them defines the verification tools namely, the *well-developedness*, the *contextual rewriting* and the *structural schemes*. Section 5 presents the **o-s-c** algorithm and in the last one, we conclude. We present in the appendix some interesting examples.

2 Basic definitions

Completeness being a term often used for different notions, we prefer for clarity to use the term sufficient completeness.

We consider the class of algebras A the boolean carrier of which is only constituted from ***true*** and ***false*** that are the interpretations of the boolean constants *true* and *false* in A. Moreover, we suppose that *true* and *false* are distinct irreducible constants in every rewriting system R.

In the following, $\oplus$ and $\ominus$ stand respectively for the disjoint union of sets and for the difference of sets.

Definition 1 Conjunction of boolean equations *CBE*

We call conjunction of boolean equations (CBE for short) a boolean expression $\vec{s}$ *of* $T(F,\ X)_{bool}$ *of the form* $s_1 = t_1\ \wedge \ldots \wedge\ s_n = t_n$ *such that for every* $i = 1, \ldots,\ n,\ t_i \in \{true,\ false\}$.

We consider conditional equations of the form $\vec{p} \Rightarrow g = d$ where $g,\ d$ are terms of $T(F,\ X)_s$ and $\vec{p}$ is a CBE.

For a given conditional specification, we define the equational congruence relation $=_E$ on ground terms t and t' of same sort as follows: let $\vec{p} \Rightarrow g = d$ be an equation in E, u be an occurrence of a subterm in t and σ be a substitution (a mapping from variables to terms, extended to a morphism from terms to terms). Then we say that $t =_E t'$ iff for every index $i,\ 1 \leq i \leq n,\ \sigma p_i =_E p'_i,\ t_{/u} = \sigma g$ then $t' = t[u \leftarrow \sigma d]$ or $t_{/u} = \sigma d$ then $t' = t[u \leftarrow \sigma g]$. For a given rewrite system R, we define the rewrite relation if $t_{/u} = \sigma g,\ \forall i,\ 1 \leq i \leq$

n, $\sigma p_i \rightarrow_R^* p'_i$, $t' = t[u \leftarrow \sigma d]$, where $\vec{p} \Rightarrow g \rightarrow d$ is a rule in R. $\rightarrow_R$ is said to be terminating iff there is no infinite derivation of form

$$t_1 \rightarrow_R t_2 \rightarrow_R \ldots \rightarrow_R t_n \rightarrow_R \ldots$$

$\rightarrow_R$ is said to be ground confluent on $T(F)$ iff for each ground terms t, t_1, $t_2 \in T(F)$ such that $t \rightarrow_R^* t_1$ and $t \rightarrow_R^* t_2$, there is a term $t' \in T(F)$ such that $t_1 \rightarrow_R^* t'$ and $t_2 \rightarrow_R^* t'$.
A linear variable of a term t is a variable that appears once in t. For any term t, $Var(t)$ is the set of variables of t.

Notation : Let $\vec{p}$ be the CBE of the form $\wedge_{i=1}^n p_i = p'_i$; for pratical reasons, we define $\vec{p} =_E TT$ iff for each index i, $1 \leq i \leq n$, $p_i =_E p'_i$ and $\vec{p} =_E FF$ iff there is at least an index i in $[1 \ldots n]$ such that if $p'_i = true$ (resp $false$) then $p_i =_E false$ (resp $true$).

Similarly, we define $\vec{p} \rightarrow_R^* TT$ if for every index i, $1 \leq i \leq n$, $p_i \rightarrow_R^* p'_i$ and $\vec{p} \rightarrow_R^* FF$ if there exists at least an index i in $[1 \ldots n]$ such that if $p'_i = true$ (resp $false$) then $p_i \rightarrow_R^* false$ (resp $true$).

In other words, TT is the abbreviation of the expressions $true = true$ and $false = false$ and FF is the one of $true = false$ and $false = true$.

The objects of algebraic abstract data types are defined from particular function symbols called constructors of the type. It is therefore natural to formalize the structure of specifications by splitting the set of operators as follows:

Let $BF \oplus DF$ be a partition of a set of function symbols such that BF contains the set C of constructors.
A BF-equation is an equation $\vec{p} \Rightarrow g = d$ which terms $\vec{p}$, g and d are in $T(BF, X)$ and such that the variables of $\vec{p}$ appear either in g or in d.
A BF-rule is an oriented BF-equation $\vec{p} \Rightarrow g \rightarrow d$ such that $Var(g)$ contains $Var(d)$ and $Var(\vec{p})$.
A DF-equation is an equation $\vec{p} \Rightarrow g = d$ one of its terms g or d is in $T(F, X) \ominus T(BF, X)$ and such that the variables of $\vec{p}$ appear either in g or in d.
A DF-rule is an oriented DF-equation $\vec{p} \Rightarrow g \rightarrow d$ which the left hand side is a term in $T(F, X) \ominus T(BF, X)$ and such that $Var(g)$ contains $Var(d)$ and $Var(\vec{p})$.

Definition 2 Structured specifications and systems

Let $F = BF \oplus DF$ be a set of function symbols such that BF contains a set C of constructors. A specification (S, F, E) is said to be structured based on (S, BF, BE) iff $E = BE \oplus DE$ where BE is a set of BF-equations and DE is a set of DF-equations. Moreover, for each BF-equation $\vec{p} \Rightarrow g = d$, if g and d are in $T(C, X)$ then $\vec{p}$ also is in $T(C, X)$.

A rewrite system (S, F, R) is structured based on (S, BF, BR) if $R = BR \oplus DR$ such that BR is a set of BF-rules and DR is a set of DF-rules. Moreover, for each BF-rule $\vec{p} \Rightarrow g \rightarrow d$ such that g is in $T(C, X)$, d and $\vec{p}$ also are in $T(C, X)$. In what follows, we more simply say that R is structured based on BR.

DF is the set of *defined operations*. Terms only built from constructors (resp from operators of BF) are called primitive terms (resp basic terms).
In what follows, all the specifications and systems are supposed *structured*.

Definition 3 Reductive conditional system

Let R be a conditional rewriting system and let $>$ be any reductive ordering; R is reductive if for each rule $\vec{p} \Rightarrow g \rightarrow d$ in R such that $\vec{p}$ is a CBE of form $\wedge_{i=1}^n p_i = p'_i$, for every substitution σ, $\sigma g > \sigma d$ and $\sigma g > \sigma p_i$, for all $i = 1, \ldots n$.

Definition of reductive rewrite systems allows to avoid infinite derivations. Works related to termination problems can be found in several papers, let us cite among them [5, 11, 12].

3 Operational sufficient completeness and related properties

Now, we formally define the sufficient completeness property of algebraic specifications. The first part recalls the results stated about sufficient completeness in a general frame. The reader can see them for instance in [7, 16, 14]. Sufficient completeness yields two concepts depending on whether we deal with the algebraic congruence or with the operational rewriting relation. This part principally states the equivalence between both concepts.

Definition 4 Algebraic and operational sufficient completeness

Let $C \subseteq BF$ be a set of constructors. $SPEC = (S,\ BF \oplus DF,\ E)$ is algebraically sufficiently complete (**a-s-c** *for short) with respect to C iff*

$$\forall t \in T(BF \oplus DF),\ \exists t_0 \in T(C) \text{ such that } t =_E t_0.$$

$SPEC$ is operationally sufficiently complete (**o-s-c** *for short) with respect to C iff*

$$\forall t \in T(BF \oplus DF),\ \exists t_0 \in T(C) \text{ such that } t \rightarrow^*_R t_0.$$

In this case, we say that R is **o-s-c** *with respect to C.*

a-s-c and **o-s-c** are consequently two distinct notions. While the former is defined from the equational congruence, the latter is operational because it is bound to the rewriting relation. The example that follows shows the difference between both concepts.

Example 1 *Let the specification $SPEC = (S,\ F,\ E)$ with:*
$S = \{elem\}$, $BF = \{e :\rightarrow elem\}$, $DF = \{a,\ b :\rightarrow elem\}$, $E = \{b = a,\ b = e\}$ *and* $C = BF$;

$SPEC$ is **a-s-c** *but not* **o-s-c**. ◇

This first proposition gives two conditions under which both concepts of sufficient completeness are equivalent, namely ground confluence of $\rightarrow_R$ and preservation of $T(C)$ by $\rightarrow_R$ which we define as follows:

Definition 5 *The closure of a set T of terms by $\rightarrow_R$ is equal to T if for each term t in T, if there is a term t' such that $t \rightarrow^*_R t'$, t' is also in T.*

Proposition 1 *If $(S,\ F,\ E)$ is structured based on $(S,\ BF,\ BE)$ then the closure of $T(C)$ by $\rightarrow_R$ is equal to $T(C)$.*

Proposition 2 *Let $(S,\ F,\ E)$ be a structured specification and let R be the associated system; if $\rightarrow_R$ is ground confluent on $T(F)$ then*

$(S,\ F,\ E)$ *is* **a-s-c** *with respect to C iff R is* **o-s-c** *with respect to C.*

Sufficient completeness of specifications is simple to formulate, but known decision methods are somewhat complex. Our research is partially related to Kounalis' works [17] stated for unconditional specifications. We consider rewriting systems that may include relations between constructors because it is often useful to define this kind of relations to reduce contexts to their normal forms. This important fact is illustrated in the following example:

Example 2 [29]
Let $SPEC = (S,\ F,\ E)$ be a structured specification based on $BSPEC = (S,\ BF,\ BE)$:
$S = \{bool,\ \mathcal{Z}\}$;
$BF = \{true,\ false :\rightarrow bool,\ 0 :\rightarrow \mathcal{Z},\ succ,\ pred : \mathcal{Z} \rightarrow \mathcal{Z}\}$, $C = BF$;
$BE = \{succ(pred(x)) = x,\ pred(succ(x)) = x\}$
$F \ominus BF = \{even : \mathcal{Z} \rightarrow bool\}$
$E \ominus BE = \{even(0) = true,\ even(succ(0)) = false,$
$even(succ(succ(x))) = even(x),\ even(pred(x)) = even(succ(x))\}$

We can prove that the system is **o-s-c** with respect to the constructors. However, we notice that if BE is empty, it is not possible to reduce the term $even(succ(pred(0)))$ to $even(0)$. In this case, the **o-s-c** of R is not checked. ◇

Another important property strongly related to **a-s-c** is inductive reducibility. We give the definition and the main result (see [10, 14] for details).

Definition 6 Inductive reducibility
A term t in $T(F,\ X)$ is inductively reducible by $\rightarrow_R$ if any ground instance of t is reducible by $\rightarrow_R$.

Theorem 1 *[16]*

Let $(S,\ F,\ R)$ be a structured rewriting system based on $(S,\ BF,\ BR)$ such that $\rightarrow_R$ is terminating. R is **o-s-c** *with respect to C iff for every f in $F \ominus BF$, $f(x_1,\dots,\ x_n)$ is inductively reducible by $\rightarrow_R$.*

And lastly, the definition completeness is also related to the operational sufficient completeness. Intuitively, if each function symbol is completely defined with respect to the constructors, the system is operationally sufficiently defined. $f \in F \ominus C$ is completely defined with respect to the constructors if for each terms $t_1,\dots,\ t_n \in T(C)$, there is $\bar{t} \in T(C)$ such that $f(t_1,\dots,\ t_n) \rightarrow^*_R \bar{t}$.

We consider now conditional rewrite systems that are structured. In the following, we study syntactic conditions in order to state a sufficient condition for checking sufficient completeness. Because of termination, we consider reductive systems. Moreover, we are interested in the operational approach to sufficient completeness since it is equivalent to the algebraic one if the system is structured and ground confluent. Therefore, we may wonder whether the system obtained after completion of a structured algebraic specification is still structured. It is probably necessary to define an ordering such that each symbol in C is smaller than any operator in BF which is itself smaller than any operator of $F \ominus BF$.

We characterize now a sufficient condition for the **o-s-c** property with respect to the constructors from the notion of *well-developed systems.*

4 A sufficient condition to check operational sufficient completeness for conditional specifications

This section is devoted to study the **o-s-c** property and to present a sufficient condition which allows to test this property in reductive structured rewrite systems.

4.1 Well-developed systems

Definition 7 Well-developed systems

Let C be a set of constructors. R is well-developed iff any ground term of the form $f(t_1,\dots,\ t_n)$, $t_i \in T(C)_{s_i}$, $\forall i,\ 1 \le i \le n$ is reducible by $\rightarrow_R$.

We do not forbid use of relations between defined operators. If we define the operators ***ordered*** and ***insert*** in the type ***list*** which constructors are [ ] and ***cons***, a relation between these two operators can be written:

$$ordered(insert(x,\ u)) = ordered(u)$$

Proposition 3 *Let $(S,\ F,\ R)$ be a structured conditional rewriting system based on $(S,\ BF,\ BR)$. Let us suppose that $\rightarrow_R$ is terminating;*
If any ground term in normal form is in $T(C)$ then R is **o-s-c** *with respect to C.*

proof: let t in $T(F) \ominus T(BF)$. Let $\bar{t}$ be the normal form of t. $\bar{t}$ exists since $\rightarrow_R$ is terminating. By hypothesis, $\bar{t} \in T(C)$. Therefore, $\bar{t}$ is the term t_0 used in the definition of **o-s-c** of R with respect to C. □

Theorem 2 *Let R be a structured conditional rewriting system based on BR; assume that $\rightarrow_R$ is terminating and that BR is* **o-s-c** *with respect to C. If R is well-developed, R is* **o-s-c** *with respect to C.*

Proof: Let t be a ground term of $T(F) \ominus T(BF)$ and $\bar{t}$ its normal form. $\bar{t}$ exists since $\rightarrow_R$ is terminating. By proposition 3, it is enough to show that $\bar{t}$ is in $T(C)$. Assume that $\bar{t} \notin T(C)$; since BR is **o-s-c** with respect to the constructors, there is a subterm t' of $\bar{t}$ such that t' is of form $f(t_1,\dots,\ t_n)$ with f in $F \ominus BF$ and for each i, $1 \le i \le n$, t_i is in $T(C)$. Because R is well-developed, there is a rule $\vec{p} \Rightarrow g \rightarrow d$ in R and a ground substitution σ such that $t' = \sigma g$ and $\sigma\vec{p} \rightarrow^*_R TT$. t' is therefore reducible and $\bar{t}$ likewise; this a contradiction with the fact that $\bar{t}$ is a normal form. □

4.2 Contextual rewriting and covering set of CBE

Of course, to check the reducibility of a term, the difficulty consists in checking if $\sigma\vec{p} \rightarrow^*_R TT$ which is undecidable. As precised in the following, this test is done by a strongest condition (in the sense that it is sufficient but not necessary): the covering property. In order to formalize these notions, we need to define the contextual rewriting relation.

Definition 8 Covering set of preconditions

Let $Cond = \{\vec{c}_1, \ldots, \vec{c}_n\}$ *be a set of* CBE. *Cond is said to be covering if the formula* $\overline{Cond}$ *is a tautology in the propositional calculus on the alphabet* $\overline{At}$, *where* $\overline{Cond}$ *and* $\overline{At}$ *are defined in the following way:*

let At *be the set of boolean atoms occurring in Cond and let* $\overline{At}$ *be an alphabet in bijection with* At. *Let* $Prop(\overline{At})$ *be the propositional calculus defined on the alphabet* $\overline{At}$ *and let* $\tilde{c}_1, \ldots, \tilde{c}_m$, $\overline{Cond}$ *be the following formulas in* $Prop(\overline{At})$

- $\overline{Cond} = \tilde{c}_1 \vee \ldots \vee \tilde{c}_m$
- *and for each* i *in* $[1 \ldots m]$, *with* $\vec{c}_i$ *of form* $\wedge_{j=1}^k t_j = t'_j$, $\tilde{c}_i$ *is the expression* $\wedge_{j=1}^k \epsilon_j \tilde{t}_j$ *where, for each* j *in* $[1 \ldots k]$, $\tilde{t}_j$ *is the symbol corresponding to the atom* t_j *and* ϵ_j *is the empty word if* t'_j *is true and the negation symbol* $\neg$ *if* t'_j *is false.*

Theorem 3 *Let Cond be a set of* CBE *and let* $At = \{a_1, \ldots, a_n\}$ *be the set of the boolean atoms occurring in Cond. Let* σ *be a ground substitution. Assume that the instanciated atoms satisfy:*

$$(**)\ \forall i \in [1 \ldots n],\ \exists t \in \{true,\ false\} \text{ such that } \sigma a_i \rightarrow^*_R t$$

Then, if Cond is covering, there exists an index k *in* $[1 \ldots m]$ *such that* $\sigma\vec{c}_k \rightarrow^*_R TT$.

Proof (by contradiction): Because of the hypothesis, we have for each k in $[1 \ldots m]$, $\sigma\vec{c}_k \rightarrow^*_R TT$ or $\sigma\vec{c}_k \rightarrow^*_R FF$. Let us assume that, for each k in $[1 \ldots m]$, $\sigma\vec{c}_k \rightarrow^*_R FF$. For each i in $[1 \ldots n]$, let b_i be a term in $\{true,\ false\}$ such that $\sigma a_i \rightarrow^*_R b_i$.
Let $B = \{\overline{true},\ \overline{false}\}$ be the boolean algebra and let μ the valuation of At in $\{true,\ false\}$ defined by

$$\mu(\tilde{a}_i) = \bar{b}_i\ \forall i = 1,\ \ldots,\ n$$

Then, saying that $\sigma\vec{c}_k$ reduces to FF exactly means that $\mu(\tilde{c}_k)$ is the value $\overline{false}$. Since that holds for each k, it turns that $\mu(\overline{Cond}) = \overline{false}$; in contradiction with the fact that $\overline{Cond}$ is a tautology. □

Theorem 3 gives a way to decompose the proof that some instanciated precondition in a set rewrites to *true*: on one side prove, once for all, that the overall set of preconditions is covering; on the other side, prove that the computation of the instanciated atoms of the precondition always terminate to *true* or *false*.

Our goal, in the result we prove here, is to get rid off the second part of the proof. For that, we use an induction argument, as property (**) is a particular consequence of **o-s-c** we are checking.

Now we give a way to generate a set of preconditions candidate for evaluation when we try to reduce the ground instance of a given term.

Definition 9 Contextual terms

A contextual term $(\vec{c} :: t)$ *or c-term is a pair of terms such that* t *is of any sort and* $\vec{c}$ *is a* CBE. *We call* $\vec{c}$ *a context.*

Definition 10 Contextual rewriting $\overset{c}{\rightarrow}_R$

We note the contextual rewriting relation by $\overset{c}{\rightarrow}_R$ *and we define it on contextual terms as follows: let* $\vec{c}_1$ *be a* CBE;

$(\vec{c}_1 :: t_1) \overset{c}{\rightarrow}_R (\vec{c}_2 :: t_2)$ *iff there is a rewrite rule* $\vec{p} \Rightarrow g \rightarrow d$ *in* R, *a substitution* σ *and an occurrence* u *of* t_1 *such that* $t_{1/u} = \sigma g$, *then*

$$t_2 = t_1[u \leftarrow \sigma d] \text{ and } \vec{c}_2 = \vec{c}_1 \wedge \sigma\vec{p}$$

Definition 11 Reducibility by case

A term t is reducible by case in R iff there is a set $\{(\vec{c_i} :: t_i)\}_{i\in I}$ *of contextual terms such that for each i in I,* $(TT :: t) \xrightarrow{c}_R (\vec{c_i} :: t_i)$ *and there is a covering set* $\{\vec{c}'_i\}_{i\in I}$ *such that* $\forall i \in I,\ \vec{c_i} \to^*_R \vec{c}'_i$.

Particularly, the unconditional reducibility is a reducibility by case.

From now on, we suppose that every conditional specification contains the boolean specification the only constructors of which are the constants *true* and *false*.

Proposition 4 *Let t be a term and let* $(\vec{c_1} :: t_1), \ldots, (\vec{c_m} :: t_m)$ *be a set of contextual reductions at one step of* $(TT :: t)$; *let* σ *be a ground substitution. If there is an index i in* $[1\ldots m]$ *such that* $\sigma\vec{c_i} \to^*_R TT$ *then* σt *is reducible.*

Proof: The proof is trivial since if $\sigma\vec{c_i} \to^*_R TT$, $\sigma t \to_R \sigma t_i$. Then, σt is reducible.

4.3 Structural scheme trees

First we present a decidability method for well-developedness inspired from Kounalis [16, 18] which computes a set of patterns. The main idea behind the algorithm of **o-s-c** is to compute for every f in $F \ominus BF$, a tree of patterns that contains terms of form $f(u_1, \ldots, u_n)$ with $u_i \in T(C, X)_{s_i}$, for all i, and in testing for every one of these terms contextual reducibility by case. The root of the pattern tree is labelled by the term $f(x_1, \ldots, x_n)$. As several trees can be generated from this term, we may wonder how to build a tree which captures the structure of the system R. For this, we use the following notions:

Definition 12 Structural scheme

$Sh_s = \{g(x_1, \ldots, x_n),\ g \in C_{\omega,s}\}$ *is a structural scheme of sort s where* $(x_1, \ldots, x_n)$ *is a list (eventually empty if g is a constant symbol) of appropriate sorts and where* ω *is the sequence of these sorts.*

Example 3 *If* $C = \{0 :\to int,\ pred,\ succ : int \to int\}$, *then* $Sh_{int} = \{0,\ pred(x),\ succ(x)\}$

If $C = \{true,\ false :\to bool\}$, *then* $Sh_{bool} = \{true,\ false\}$.

If $C = \{0 :\to nat,\ succ : nat \to nat,\ [\,] :\to list,\ cons : nat \times list \to list\}$;
$Sh_{nat} = \{0,\ succ(x)\}$, $Sh_{list} = \{[\,],\ cons(x,\ u)\}$. ◇

Definition 13 *Let t be a linear term and let u be an occurrence of t such that* $t_{/u} = x \in X_s$. t^c_u *is the term that we get by replacing in t the variable x by a structural scheme c (after renaming the variables) of* Sh_s. t^c_u *is a pattern of t in u. The set* $\{t^c_u,\ c \in C_s\}$ *of patterns of t in u is noted* $Word(t,\ u)$, *the transformation operation of t to* $Word(t,\ u)$ *is called graft to the occurrence u. Obviously,* $Word(t,\ u)$ *and* Sh_s *have the same number of elements.*

Example 4 *Let* $BF = \{0 :\to int,\ succ :\ int \to int\}$ *and* $C = BF$; $DF = \{+ : int \times int \to int\}$;
$Sh_{int} = \{0,\ succ(x)\}$ *and let* $t = succ(x_1) + succ(succ(x_2))$, *then*
$Word(t,\ 11) = \{succ(0) + succ(succ(x_2)),\ succ(succ(x_3)) + succ(succ(x_2))\}$.

Let $BF = \{0 :\to nat,\ succ : nat \to nat,\ empty{-}s :\to stack,\ push : nat \times stack \to stack\}$,
and $C = BF$, $F \ominus BF = \{top : stack \to int\}$,
$Sh_{nat} = \{0,\ succ(x)\}$, $Sh_{stack} = \{empty{-}s,\ push(x,\ u)\}$;
Let $t = top(push(x,\ u))$,
$Word(t,\ 11) = \{top(push(0,\ u)),\ top(push(succ(x_1),\ u))\}$
$Word(t,\ 12) = \{top(push(x,\ empty{-}s)),\ top(push(x,\ push(x_2,\ u_2)))\}$. ◇

Let $Def_R(f)$ be the set of rules of R the left member of which is of the form $f(\omega_1, \ldots, \omega_n)$ with $\omega_i \in T(C,\ X)_{s_i}$. This set intuitively defines the operator f.

Definition 14 *Let t be a term in* $T(F,\ X) \ominus T(BF,\ X)$; *we define* $Occ(f)$ *as follows:*

$Occ(f) = \{u$ such that there is $\vec{p} \Rightarrow g \to d \in Def_R(f)$ and $g_{/u} \in F$ or $g_{/u}$ is a non-linear variable in $g\}$.

Example 5 *If* $R = \{f(x, y, x) \to y, f(x, x, y) \to y, f(y, x, x) \to y\}$ *then,* $Def_R(f) = R$ *and* $Occ(f) = \{\epsilon, 1, 2, 3\}$.

From any node of the tree labelled by the term $t = f(u_1, \ldots, u_n)$, with $u_i \in T(C, X)_{s_i}$ for each $i = 1, \ldots, n$, we build the sons of this node by choosing an occurrence u of t in $Free(t) \cap Occ(f)$ and by making a graft at this occurrence. So, each son is labelled by an element of $Word(t, u)$. In this case, we say that t is splittable. The height of the pattern tree is bounded because we assume that the rules of R which have the function symbol f at the top is finite. This involves that the set $Occ(f)$ is finite. In the other hand, consecutive grafts in a same branch of the tree are made at occurrences deepest and deepest. We can then state that the set $Free(t) \cap Occ(f)$ decreases during the construction of the tree. Consequently, the height of the tree is bounded.

Again, it is useless to continue the graft process when we meet a node labelled by a term t which is reducible by case. Therefore, we can describe the construction of the tree in the following way: from the tree initially constituted from the root $f(x_1, \ldots, x_n)$, we check the **o-s-c** by testing reducibility by case of $f(x_1, \ldots, x_n)$. In the other way, we build at every step the sons of the tree by making a graft operation on the node at an appropriate occurrence. The tree is totally built if each of its sons either is reducible by case or we cannot no more split it. Computation of $Occ(f)$ allows to take into account the structure of R.

5 The algorithm and the correctness proof

5.1 The algorithm for operational sufficient completeness

The algorithm: (S, F, R) is a reductive structured rewriting system based on (S, BF, BR) such that BR is **o-s-c** with respect to C. Suppose moreover that the only constructors of the boolean sort are the constants *true* and *false*.
R is a global variable of all these procedures. However, it is not present in the algorithms because it do not change in them.

```
o-s-c
var Candidates: set of terms, Reg: set of contextual terms, well: bool;
begin:
for each f in F ⊖ BF do
    Candidates := {f(x1,..., xn)};
    well := well-reduce(Candidates);
    if well = true then write(R is o-s-c)
      else write(we cannot conclude) end-if;
end-for;
end.
```

```
Well-reduce input(Candidates: set of terms) output(bool)
begin:
  if Candidates is empty then return(true)
  else t := choose-term(Candidates);
      % test if t is reducible by case %
      if case-reducible(t)
        then return(well-reduce(delete-term(t, Candidates))
          else % test if t can be split %
            if there is an occurrence u in Var(t) ∩ Occ(f)
              then return(well-reduce(delete-term(t, Candidates) ∪ Word(t, u)))
              else return(false);
            end-if;
      end-if;
  end-if;
end.
```

The *choose–term*(*Candidates*) procedure chooses any term in the set *Candidates* without deleting it and the *delete–term*(t, *Candidats*) procedure deletes the term t from the set *Candidates*.

case–reducible(t) : **input**(t : term) **output**(bool)
begin :
 if t reducible by $\xrightarrow{c}_R$ **then**
  let $\{(\vec{c_i} :: t_i),\ i = 1, \ldots, m\}$ be the set of contextual terms such that for each i in $[1 \ldots m]$,
  $(TT :: t) \xrightarrow{c}_R (\vec{c_i} :: t_i)$; let $\{\vec{c}'_i\}_{i=1}^m$ be the set such that $\forall i \in [1 \ldots m]$, $\vec{c_i} \rightarrow^*_R \vec{c}'_i$;
  cover := *wsp*($\{\vec{c}'_1, \ldots, \vec{c}'_m\}$)
  if *cover* = *true* **then** *return*(*true*) **else** *return*(*false*) **end-if**
 end-if
end.

The *wsp* procedure checks whether the set $\{\vec{c}'_1, \ldots, \vec{c}'_m\}$ is covering or not. The algorithm can be any of the classical algorithms for checking validity in propositional calculus.

We formulate now the principal theorem on **o-s-c**. To consider reductive conditional rewriting systems allows stating a simple proof of this theorem by structural induction on ground terms.

In abstract terms, the algorithm builds for $t = f(x_1, \ldots, x_n)$, a pattern tree $\mathcal{T}$ such that, for every leaf t_n two cases are possible

- either $(TT :: t_n)$ is reducible by case.
- or t_n is no longer splittable.

The algorithm returns *success* if and only if each leaf is reducible by case. The following result, proven in [16] says that the tree $\mathcal{T}$ is complete with respect to C.

Proposition 5 [16]

Let t be the term $f(x_1, \ldots, x_n)$ with f in $F \ominus BF$, let $\mathcal{T}$ be the tree built by the algorithm and let σ be a ground substitution. There exists a leaf t' of $\mathcal{T}$ and a substitution σ' such that $\sigma t = \sigma' t'$.

5.2 Main result of correctness

The following theorem uses an ordering denoted by $\prec_{st}$ which is the transitive closure of any reduction ordering $<$ and the strict subterm ordering $<_{st}$, that is $\prec_{st} = (< \cup <_{st})^+$. This ordering is well-founded [11].

Theorem 4 *Let R be a reductive conditional rewrite system based on BR; let us suppose that BR is* **o-s-c** *with respect to C; if the algorithm* **o-s-c** *stops with success, then R is* **o-s-c** *with respect to C.*

Proof: we have to prove the property $\mathcal{P}$ stated as follows:

$$(\mathcal{P})\quad \forall \omega \in T(F),\ \exists \bar{\omega} \in T(C) \text{ such that } \omega \rightarrow^*_R \bar{\omega}.$$

Let us suppose that the property $\mathcal{P}$ is valid for each terms ω' such that $\omega' \prec_{st} \omega$. If ω is in $T(BF)$, the result follows from the **o-s-c** property of BR with respect to the constructors. Otherwise, we may consider a subterm ω" of ω of the form $f(\omega_1, \ldots, \omega_n)$ with f in DF and $\omega_1, \ldots, \omega_n$ in $T(BF)$. We first prove that ω" is reducible. If some term ω_i belongs to $T(BF) \ominus T(C)$, that follows of the **o-s-c** of BR. Now we assume that $\omega_1, \ldots, \omega_n$ belong to $T(C)$.

Let $t = f(x_1, \ldots, x_n)$ and σ_0 be a ground substitution such that ω" $= \sigma_0 t$. By proposition 5, there is a leaf t' in the pattern tree built by the algorithm for t, and a substitution σ' such that $\sigma_0 t = \sigma' t'$.

If the algorithm stops with success, there exists a sequence of contexts (or CBE) $\vec{q_1}, \ldots, \vec{q_m}$ and a set of terms $\alpha_1, \ldots, \alpha_m$ such that

for every i in $[1 \ldots m]$, $(TT :: t') \xrightarrow{c}_R (\vec{q_i} :: \alpha_i)$ and the set of CBE, $\{\vec{q'_1}, \ldots, \vec{q'_m}\}$ is covering where $\forall i$, $\vec{q}'_i$ is the $\rightarrow_R$-normal form of $\vec{q_i}$.

By definition, there is for every i, a rule $\vec{p_i} \Rightarrow g_i \rightarrow d_i$, an occurrence v of t' and a substitution τ such that $t'_{/v} = \tau g$ and $\vec{q_i} = \tau \vec{p_i}$. Then, it holds that $\vec{q_i} = \tau \vec{p_i} <^2 \tau g <_{st} t'$. Also, $\vec{q'_i} \leq \vec{q_i}$ since $\vec{q_i} \rightarrow^*_R \vec{q}'_i$.

[2]We consider here the multiset extension of $<$ in order to compare $\vec{p_i}$ with g or $\vec{q'_i}$ with $\vec{q_i}$

Now, let $a_1, \ldots, a_n$ be the atoms occurring in $\vec{q'_1}, \ldots, \vec{q'_m}$. We still have $a_k \leq \vec{q'}_i <_{st} t'$ if a_k occurs in $\vec{q'_i}$. Therefore, $\sigma' a_k \prec_{st} \sigma' t' = \sigma_0 t$. By induction hypothesis applied to $\sigma' a_k$, there is a term a in $T(C)_{bool}$ such that $\sigma' a_k \rightarrow^*_R a$. By hypothesis, a is *true* or *false*.

Since $\{\vec{q'_1}, \ldots, \vec{q'_m}\}$ is covering, by theorem 3, there is an index i in $[1 \ldots m]$ such that $\sigma' \vec{q'_i} \rightarrow^*_R TT$. Also, $\sigma' \vec{q}_i \rightarrow^*_R \sigma' \vec{q'_i} \rightarrow^*_R TT$. Therefore, by proposition 4, $\sigma' t'$ is reducible, which means $\omega'' = \sigma_0 t = \sigma' t'$ is reducible into some term t''. Now, let ω''' be the term ω where ω'' is replaced by t''. Since $\omega'' \rightarrow^+_R t''$, we have $t'' < \omega''$ and $\omega''' < \omega$ as $<$ is a reduction ordering. By induction hypothesis applied to ω''', there is a term $\overline{\omega'''}$ in $T(C)$ such that $\omega''' \rightarrow^*_R \overline{\omega'''}$. Therefore, $\omega \rightarrow^*_R \omega''' \rightarrow^*_R \overline{\omega'''}$. □

6 Conclusion

We have proven in this paper a sufficient condition to check the sufficient completeness property of conditional specifications. In the conditional framework, few results have been stated and we feel that our test algorithm is relevant to progress in this area. Indeed, our method is effective for reductive structured conditional systems and these two properties are not restricting and they are often assumed by the authors in the conditional framework. The algorithm allows to show that a function symbol is completely defined with respect to the constructors of an algebraic specification simply by building a tree of patterns and by checking the inductive reducibility of the leafs by the contextual relation which is powerful because it includes case reasoning. This tree constitues as a matter of fact a *test set* of terms that we have to reduce as defined in [10]. The algorithm uses an inductive argument for dealing with the completeness of the precondition calculus as well as with the completeness of other definitions and this is a progress with respect to hierarchical processings. Also, by using a propositional calculus prover, no syntactic form is imposed to the preconditions. Finally, the relations between the constructors are not forbidden.

In the unconditional framework, various approaches have been considered in order to state sufficient conditions for checking the completeness property. We may mention Padawitz's works where the author introduces the notion of *w-genericity* of a predicate defined by induction on the constructors, and that imposes the system be left-linear [24]. This restriction has been deleted by Dershowitz which has implemented an algorithm based on the calculus of a test set for checking the inductive reducibility of a term. This test set also is computed by Thiel in [27], which proposes an algorithm that uses an unification process. However, this algorithm does not consider specifications that contain relations between constructors and Lazrek in his thesis [20], and also together with Thiel and Lescanne in [21] suggests a method more flexible that allows to deal with them. By another way, Kounalis states an algorithm which computes a tree of patterns and which constitutes an improvement of Dershowitz' method. This is the method that we have choosen to apply to the conditional specifications.

For conditional specifications, some ideas are just suggested by Rémy in [25] based on the construction of structurally basic sets of preconditions. Zhang [29] has also stated completeness results principally founded on contextual rewriting and normal form calculus. His algorithm, implemented in REVEUR4 [26], supposes more properties than ours, for example, the *well-covering* property and the *hierarchy* of the rewrite system, and moreover, it does not take into account the type of the operators. Later on, Rémy took up again his method together with Uhrig in [28]. The authors state a test of sufficient completeness of reductive systems, corresponding to specific classes of conditional specifications. The idea of particularly considering sufficient completeness of the specification with respect to its boolean part is inspired from their works. The authors analyze the rules of the system statically and this choice makes their algorithm ineffective for some classes of specifications. We present in this paper, a method which strictely includes the one of Rémy and Uhrig and which is based on contextual rewriting and case reasoning. Also, the sufficient completeness has been studied by Koulanis and Rusinowitch in [19] indirectly by inductive proof methods. The authors compute a *test-set*: set of terms which describes the initial model of the corresponding theory. With this test-set, they state a method which checks if a term is *pseudo-reducible* and can conclude that all the ground instances of this term are reducible.

An experimental prototyping has been developed at the laboratory of Nancy in CAML on SUN III systems. It checks o-s-c of ground confluent conditional rewrite systems. It is also a helpful tool for building complete specifications. The implementation is flexible because it allows to display the decomposition tree [1].

Some extensions can be made in order to apprehend a larger class of conditional specifications. We intend to improve our algorithm by combining it with an inductive prover. For instance, if we consider again the one rule example $M = N \Rightarrow f(x_1, \ldots, x_n) \to 0$, we know that the operator f is sufficiently defined if an only if the equation $M = N$ is an inductive theorem. Also we know that sufficient completeness is an helpful property for inductive proofs. We also want to use the informations the algorithm can give, if slightly transformed, in case of failure. In such cases, we can get a set of patterns and preconditions for which there may not exist defining rules. The user should then be responsible for checking whether the incompleteness is real (in which case he will propose a definition) or not. Such situations occur for instance when there exist superpositions between patterns or when some precondition is unsatisfiable. A third extension will concern the way of reducing the preconditions before checking that they form a covering set. We will use contextual rewriting as we did when reducing leaves of the decomposition trees. We are working on these techniques in a parallel research on completion algorithm for conditional systems [2].

References

[1] A. Bouhoula. Implantation d'un algorithme de test de complétude suffisante d'une spécification conditionnelle et d'aide à la construction de définitions structurelles complètes. Rapport de Stage de Fin d'Etudes, Centre de Recherche en Informatique de Nancy, 1990.

[2] W. Bousdira. Etude des propriétés des systèmes de réécriture conditionnelle. Mise en œuvre de deux algorithmes de test de confluence sur les termes clos., 1990. Thèse de l'Institut National Polytechnique de Lorraine.

[3] W. Bousdira and J-L. Rémy. Hierarchical Contextual Rewriting with Several Levels. In R. Cori and M. Wirsing, editors, *Lecture Notes in Computer Science*, volume 294, pages 193–206, Bordeaux, France, 1988. 5th Synposium on Theoretical Aspects of Computer Science, Springer-Verlag.

[4] N. Dershowitz. Computing with Term Rewriting Systems. In *Proceedings of An NSF Workshop On The Rewrite Rule Laboratory*, 1984.

[5] N. Dershowitz. Termination. In J-P. Jouannaud, editor, *Lecture Notes in Computer Science*, volume 202, Dijon, France, 1985. 1st Conference on Rewriting Techniques and Applications, Springer-Verlag. Also in Journal os jsc 3, special issue on Rewriting Techniques and Applications, 69-115, 1987.

[6] N. Dershowitz and M. Okada. Conditional Equational Programming and the Theory of Conditional Term Rewriting. In *Proceedings of the International Conference on Fifth Generation Computer Systems*, pages 337–346, Tokyo, Japan, November 1988. ICOT.

[7] J. V. Guttag and J. J. Horning. The Algebraic Specification of Abstract Data Types. *Acta Informatica*, 10:27–52, 1978.

[8] J.V. Guttag. Abstract Data Types and Software Validation. *Communications of the Association for Computing Machinery*, 21:1048–1064, 1978.

[9] G. Huet and J-M. Hullot. Proofs by Induction in Equational Theories with Constructors. *Journal of Computer and System Sciences*, 25(2):239–266, October 1982. Preliminary version in Proceedings 21st Symposium on Foundations of Computer Science, IEEE, 1980.

[10] J-P. Jouannaud and E. Kounalis. Automatic Proofs by Induction in Theories without Constructors. Technical Report 295, Université de Paris-Sud, Centre d'Orsay, September 1986.

[11] J-P. Jouannaud and B. Waldmann. Reductive Conditional Term Rewriting Systems. In M. Wirsing, editor, *Elsevier Science Publishers*, pages 223–244, Ebberup, Denmark, 1986. 3rd IFIP Working Conference on Formal Description of Programming Concepts.

[12] S. Kaplan. Simplifying Conditional Term Rewriting Systems : Unification, Termination and Confluence. *Journal of Symbolic Computation*, 1987.

[13] S. Kaplan and J-L. Rémy. Completion Algorithms for Conditional Rewriting Systems. In H. Ait-Kaci and M. Nivat, editors, *Resolution of Equations in Algebraic Structures*, volume 2, Austin, Texas, 1987. MCC-INRIA Colloquium, Academic Press, INC.

[14] D. Kapur, P. Narendran, and H. Zhang. On Sufficient Completeness and Related Properties of Term Rewriting Systems. *Acta Informatica*, 24:395–415, 1987.

[15] H. Kirchner. A General Inductive Algorithm and Application to Abstract Data Types. In R. Shostak, editor, *Proceedings 7th International Conference on Automated Deduction*, pages 282–302. Springer-Verlag, Lecture Notes in Computer Science, 1984.

[16] E. Kounalis. Completeness in Data Type Specifications. In B. Buchberger, editor, *Proceedings EUROCAL Conference*, volume 204 of *Lecture Notes in Computer Science*, Linz (Austria), 1985. Springer-Verlag.

[17] E. Kounalis. Validation des Spécifications Algébriques par Complétion Inductive, 1985. Thèse de l'Université de Nancy 1.

[18] E. Kounalis. Testing for inductive (co)-reducibility. In A. Arnold, editor, *Proceedings 15th CAAP*, volume 431 of *Lecture Notes in Computer Science*, pages 221–238. Springer-Verlag, May 1990.

[19] E. Kounalis and M. Rusinowitch. Mechanizing Inductive Reasoning. *EATCS Bulletin*, 41:216–226, 1990.

[20] A. Lazrek. Etude et Réalisation de Méthodes de Preuve par Récurrence en Logique Equationnelle, décembre 1988. Thèse de l'Institut National Polytechnique de Lorraine.

[21] A. Lazrek, P. Lescanne, and J-J. Thiel. Tools for Proving Inductive Equalities, Relative Completeness and ω-Completeness. Technical Report 88-R-131, Centre de Recherche en Informatique de Nancy, 1988. to be published in Information and Computation.

[22] T. Nipkow and G. Weikum. A Decidability Result about Sufficient Completeness of Axiomatically Specified Abstract Data Types. In *6th GI Conference*, volume 145, pages 257–268. Springer-Verlag, Lecture Notes in Computer Science, 1983.

[23] P. Padawitz. New Results on Completeness and Consistency of Abstract Data Types. In *Lecture Notes in Computer Science*, volume 88, pages 460–473. Proceedings 9th Symposium on Mathematical Foundations of Computer Science, Springer-Verlag, 1980.

[24] P. Padawitz. Correctness, Completeness and Consistency of Equational Data Type Specifications. Technical Report Bericht Nr 83-15, Technische Universitat, Berlin, West Germany, 1983.

[25] J-L. Rémy. Etude des Systèmes de Réécriture Conditionnelle et Applications aux Types Abstraits Algébriques, 1982. Thèse d'Etat, Institut National Polytechnique de Lorraine, Nancy.

[26] J-L. Rémy and H. Zhang. REVEUR4 : a System for Validating Conditional Algebraic Specifications of Abstract Data Types. In *ECAI*, pages 563–572, Pisa, Italy, 1984. 6th ECAI Conference.

[27] J-J. Thiel. Stop loosing sleep over Incomplete Data Type Specifications. In *Proceeding 11th ACM Symp. on Principles of Programming Languages*, pages 76–82. Association for Computing Machinery, 1984.

[28] S. Uhrig and J-L. Rémy. An Algorithm for Testing Suffisent Completeness of a Simple Class of Conditional Specifications. Technical Report 88-R-155, Centre de Recherche en Informatique de Nancy, 1988.

[29] H. Zhang. REVEUR4 : Etude et mise en œuvre de la Réécriture Conditionnelle, 1984. Thèse de 3ème cycle, Université de Nancy 1.

A Examples

The interest of this section is to present our contribution with respect to the one already existing stated by Rémy and Uhrig [28] by two examples.

Example 6 *Let the following specification:*

$S = \{nat\}$,

$BF = \{0 :\rightarrow nat,\ succ : nat \rightarrow nat,\ > : nat \times nat \rightarrow bool\}$, $F \ominus BF = \{f : nat \rightarrow nat\}$

$BR = \{0 > x \rightarrow false,\ succ(x) > 0 \rightarrow true,\ succ(x) > succ(y) \rightarrow x > y\}$

$R \ominus BR = \{f(0) \rightarrow 0,\ x > 0 = true \Rightarrow f(x) \rightarrow 0\}$.

and let the constructors be the set $C = \{true,\ false,\ 0,\ succ\}$.

We can remark that BR *is o-s-c with respect to* C. *Let* $Occ(f) = \{\epsilon,\ 1\}$ *and* $Sh_{nat} = \{0,\ succ(x)\}$ *and let the term* $t = f(x_1)$. *By applying our method, we build the following tree:*

Decomposition tree of the operator f:

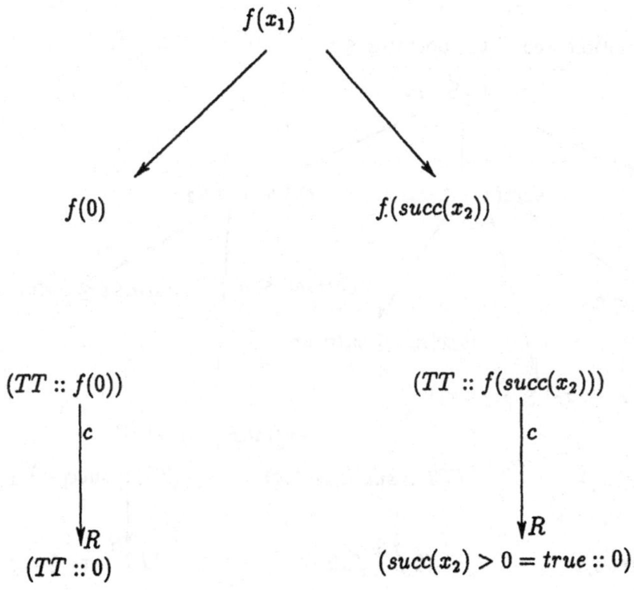

$succ(x_2) > 0 = true \rightarrow_R true = true \rightarrow_R TT$

Example 7 *This example is a specification of ordered lists of integers bounded by a constant entmax. The expression pp(x, l) means that x is smaller than every element of the list l. The specification is as follows:*

$SP = (S,\ F,\ E); S = \{bool,\ nat,\ list\};$

$BF = \{true,\ false : \rightarrow bool,\ 0,\ entmax : \rightarrow nat,\ succ : nat \rightarrow nat,$
$\quad l\text{–}empty : \rightarrow list,\ add : nat \times list \rightarrow list\}$

and $C = BF;$

$F \ominus BF = \{\leq : nat \times nat \rightarrow bool,\ pp : nat \times list \rightarrow bool\}.$

$BR = \{succ(entmax) \rightarrow entmax\},$

$R \ominus BR = \{0 \leq x \rightarrow true,\ succ(x) \leq 0 \rightarrow false,$
$\quad succ(x) \leq succ(y) \rightarrow x \leq y,\ x \leq entmax \rightarrow true,$
$\quad entmax \leq 0 \rightarrow false,\ entmax \leq succ(y) \rightarrow entmax \leq y,$

$\quad pp(x,\ l\text{–}empty) \rightarrow true,$
$\quad x \leq y = true \ \wedge\ pp(x,\ l) = true \Rightarrow pp(x,\ add(y,\ l)) \rightarrow true,$
$\quad x \leq y = false \Rightarrow pp(x,\ add(y,\ l)) \rightarrow false,$
$\quad pp(x,\ l) = false \Rightarrow pp(x,\ add(y,\ l)) \rightarrow false\}.$

Let

$$Sh_{bool} = \{true,\ false\},\ Sh_{nat} = \{0,\ succ(x),\ entmax\},\ Sh_{list} = \{l\text{–}empty,\ add(x,\ l)\}.$$

Let us show that R is o-s-c *with respect to C.*

Decomposition tree of the operator $\leq$:

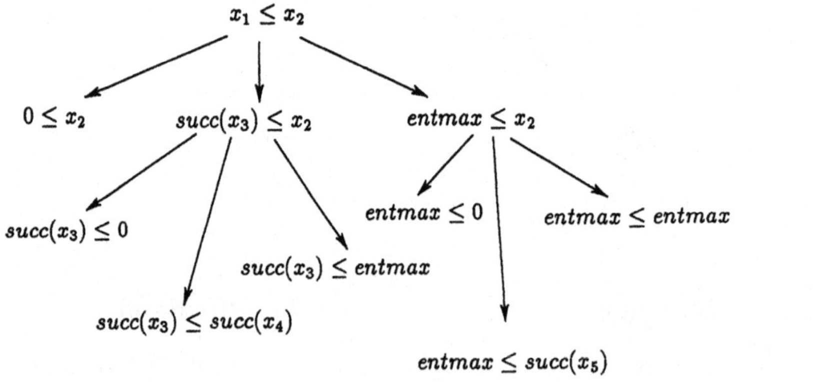

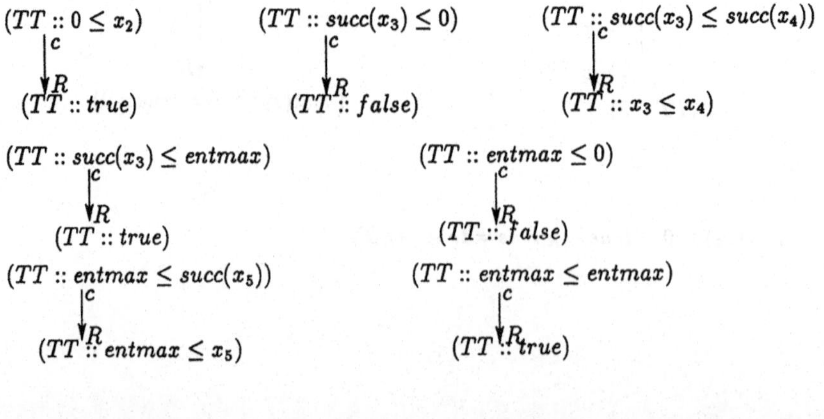

Decomposition tree of the operator *pp*:

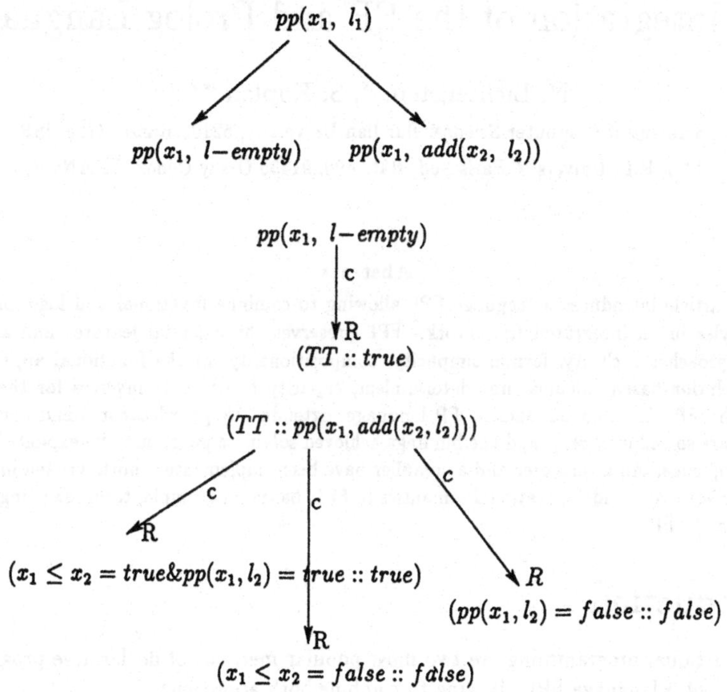

Now, it is easy to check that the formula

$$(x_1 \leq x_2 = true \wedge pp(x_1,\ l_2) = true) \vee (x_1 \leq x_2 = false) \vee (pp(x_1,\ l_2) = false) = true$$

is a tautology in the propositional calculus.

The second example of the appendix shows that any syntactic form is imposed to the preconditions of the system unlike to Rémy and Uhrig's works [28]. Effectively, if a term t which the top symbol is f in $F \ominus BF$ is not reducible by case and if it has non-linear variable occurrences included in the set $Occ(f)$, we build a set of patterns from t by choosing an appropriate occurrence to check the reducibility by case of the new terms. This method allows to have more flexible form of preconditions.

FPL : Functional Plus Logic Programming
An Integration of the FP and Prolog Languages

N. Lichtenstein *, S. Kaplan ** *

* Department of Computer Science, Bar-Ilan University, 52100 Ramat-Gan, ISRAEL

** L.R.I., Université Paris-Sud, Bât. 490, 91405 Orsay Cédex, FRANCE

Abstract

This article introduces a language FPL allowing to combine functional and logic programming styles in an integrated framework. FPL preserves the essential features and assets of both approaches : clarity, formal simplicity, compositionality for the functional approach — and resolution-based calculus, non-determinism, capacity to compute inverses for the logical approach. FPL is based on Backus' FP language, extended by guarded command constructs. Guards are solved in Prolog, and the bindings achieved solving a guard may be exploited within its consequence. An interpreter and a compiler have been implemented, both written in Prolog. We show how to provide operational semantics to FPL, based on a simple, term rewriting system semantics for FP.

1 Introduction

Logic and Functional programming are two most popular methods of declarative programming. In paper, we present a language FPL allowing to combine both approaches.

Recent proposals for integrating logic and functional styles can be roughly partitioned into two main classes : the *logic plus functional* approach, and the *functional plus logic* approach.

- In the former, emphasis is on logic, with extensions to deal with functions. The theoretical framework is generally first-order logic with equality. For instance, LOGLISP [RS 82] combines logic with LISP. Other languages are EQLOG [GM 84] were functions are handled by narrowing–, and FRESH [Smolka 86] which integrates logic programming features into a higher-order functional programming environment.

- In the second approach, integration means addition of a resolution mechanism (pure resolution, narrowing) to a functional language. Examples are LEAF [BBLM 86], FUNLOG [SY 84], HASL [Abramson 84] QUTE [SS 84] and APPLOG [Cohen 86]. All integrate functional programming style into a logic programming environment.

This article introduces a new language FPL, that belongs to the *functional plus logic* approach. The functional notation and structure are the leading ones in the mixed language : the user focuses on *values* rather than on *goals*. The logic part is used to constrain values to satisfy particular properties, or even to "discover" particular values and later use them. We implemented an *interpreter* for FPL, written in Prolog–, and a *compiler*, written in Prolog and generating Prolog code.

It is assumed that the reader is acquainted with Backus' FP [Backus 78], and to its operational semantics based on term rewriting systems (e.g. [HWWW 84]).

This article is organized as follows: section 2 is a description of FPL, section 3 gives selected FPL examples, section 4 presents a compiler for FPL, while section 4 addresses semantics issues.

2 Description of FPL

2.1 Overview

FPL is an extension of Backus' FP language [Backus 78]. Logic capabilities are added to FP, to obtain FPL. In our setting, a programmer may use ***extended conditional forms***, with boolean guards in Prolog. Solving the guards (by standard SLD-resolution) binds logic variables, that may then be used in the conclusion of the condition.

The most basic, new functional form is `P->F`, where `P` is an FPL goal clause, similar to a Prolog goal clause (cf. under)–, and `F` is a FPL functional form. Applying this form to an object means :

> *If* `P` *succeeds as a goal, then apply* `F` *with the bindings obtained during the resolution of P — else, failure.*

Actually, P may be a conjunction of clauses, that may themselves contain FPL expressions.

Example : Consider the following FPL function definition :

```
h := p(X,^g) -> add1 @ $X.
```

The meaning of the request to apply `h` to an object `obj`,

```
h : obj ?
```

is then :

> *Compute* `y = g:obj`. *If there is a substitution* `{X\x}` *that satisfies the goal* `p(X,y)`, *the result is* `x+1`. *Otherwise, the result is* $\perp$ *(bottom).*

Here, "`@`" stands for regular FP (FPL) composition. The symbol "`^`" may be seen as a call to FPL evaluation from within a guard; "`$X`" is the FP(L) constant function returning `X` (for the actual value `X = x`)

An FPL environment comprises the following :

- Objects : the same as Prolog data objects (atoms, lists, application of a functor) that are built on FP atoms
- Primitive functions – as in FP
- Functional forms – cf. below
- Predicates
- Logical variables
- Goal clauses – cf. below
- Function definitions – as in FP
- Application – as in FP
- Queries

2.2 FPL functional forms

FPL *functional forms* are expressions denoting a function; they are built by application of FPL *functionals* to more basic functional forms. FPL has all FP functionals, and a few more. The *construction, composition, insert, map* and *constant* functionals are left untouched.

FPL's new functionals are the *extended conditionals.* These are :

- The *If-Then* construct : `P->Q`. Its intuitive meaning is as follows (cf. section 4 for further details): `P->Q : obj` = *if there exists a substitution* σ *that satisfies the guard* `P`*, then* `Q`σ`:obj`. *More precisely, since the resolution process is non-deterministic, we restrict to* Prolog*-like, SLD-resolution – we actually look for a* σ *such that* $\vdash^{SLD}$ `P`σ. *Otherwise, if* Prolog fails *(finitely) to satisfy* `P`, $\perp$. *Otherwise (that is if* $\vdash^{SLD}$ `P` *does not terminate), then undefined.*

- The *If-Then-Else* construct : `P->Q;R`. Then, `P->Q;R : obj` = *if there exists a substitution* σ *that such that* $\vdash^{SLD}$ `P`σ*, then* `Q`σ`:obj`. *Otherwise, if* Prolog fails *to satisfy* `P`*, then* `R:obj`. *Otherwise, undefined.*

- The *guarded command* : `P1->Q1#...#Pn->Qn`. An *Else*-part "`;R`" may be appended. Its meaning is, as usual, a non-deterministic choice among the `Qi`σ_i`:obj` such that $\vdash^{SLD}$ `Pi`σ_i

- The *While* functional form : `Q while P`

A syntactic constraint is that if a logic variable `X` appears in a *then* part (`Q`), then this must be under a "`$`" sign, and `X` should appear in the corresponding condition (`P`). Formal semantics is given in section 4.

Example : Consider the expression :

```
f := p(^id,X) -> plus @ [$X,$1]; id
```

where `p` is a predicate defined by :

```
p(X,Y):- X < 10, Y is 2*X.
```

We recall that `id` is FP identity function, and that `$1` is the FP(L) constant function returning 1. Then the query :

```
f : 5?
```

yields 11, since `^id` is replaced by 5, `X` is bound to 10 in order to satisfy `p(5,X)`, and `plus @ $[X,1]` evaluates to 11. On the other hand, the query :

```
f : 12?
```

yields 12 : the goal `p(12,X)` is unsatisfiable, and the *Else*-part `id:12` is reduced into 12.

2.3 FPL goal clauses

Goal clauses are Prolog-like goals, used in the guards of the conditional forms. They are constructed by composition of Prolog predicates, constants and variables, and of FPL expressions (that must appear under a "`^`" sign). There are of three kinds :

- A goal unit clause `p(t1,...,tn)`, where it is assumed that `p` is a predicate that has received a Prolog definition. The `ti` are usual Prolog objects, except that they may contain sub-expression of the form `^(...)`. As mentioned, this is actually a call to an FPL evaluation.

- A boolean unit clause `^f`. It is assumed that `f` is an FPL function returning a boolean.

- A conjunction of the two previous cases `c1,...,cn`.

An example of a goal clause is `p(X,^g),^f@[$X,id,Y]`. When applied to an object `obj`, then, successively :

- `g:obj` is evaluated, yielding `o'`
- Prolog tries to satisfy `p(X,o')`, which binds `X` to a value `x`
- Prolog tries to satisfy `^f@[$x,id,Y]`, i.e. to find a substitution `{Y\y}` such that `f:[x,obj,Y]` evaluates to `true` (`f` is a boolean-valued function).

3 FPL selected examples

We present several FPL examples, in order to demonstrate FPL expressive power and conceptual simplicity (for a user acquainted to Prolog and FP !), and to show several other features of FPL.

3.1 Quick-Sort

The following definition implements a quicksort algorithm.

```
qsort := ^null -> [];
         apnd @ [    qsort @ (X : member(X,^tl), X<  ^1)),
                 [1 | qsort @ (X : member(X,^tl), X>= ^1))]] .
```

`tl` is FP 'tail' function. We used the special *all* form `(X : P[X])`, to denote the constant function returning the list of all `X`'s such that `P[X]`, applied to the argument of `qsort`, succeeds. We believe that, in this example, FPL compares very favorably with pure Prolog or pure FP.

3.2 Rational Plus

We use an infix functor "/" to construct rational numbers : if `P`, `Q` are integers ($Q \neq 0$), the expression `P/Q` denotes their rational ratio. The following `rat-plus` function computes the sum of two rationals.

```
rat-plus  :=  ^id = [P1/Q1,P2/Q2],
              Z is P1*Q2+P2*Q1,
              R is Q1*Q2,
              GCD = ^ gcd @ $[Z,R],
              P is Z/GCD,
              Q is R/GCD                -> $ (P/Q)
```

with

```
gcd       :=  ^lt        -> gcd@[2,1] ;
                         MOD=^mod  -> (MOD=0 -> 2 ;
                                              gcd@[2,$MOD])
```

`lt` is the FP 'less-than' predicate. `1` and `2` are accessor functions ; we recall that, in general, `i:[o1,...,on] = oi` if $1 \leq i \leq n$ (and $\perp$ otherwise). The query `rat-plus:[4/6,15/20] ?` yields the result 17/12. Both examples show that, if needed, we may simulate *local variables* in FPL in order to avoid useless re-computations (`GCD`, `MOD`).

3.3 Compilation of term rewriting systems

It is easy to compile *term rewriting systems* (conditional, with priorities, etc.) into FPL, using unification to perform pattern-matching. The TRS `{add(X,0) -> X, add(X,s(Y)) -> s(add(X,Y))}` is automatically compiled into :

```
add :=  ^2 = 0    -> 1;
        ^2 = s(Y) -> s @ add @ [1,$Y].
```

where "s" (ambiguously) denotes the Prolog predicate defined by :

```
s(X,s(X)).
```

Thus, constructors (i.e. symbols s.a. `s` that do not appear at the root of a left handside) are implemented as predicates, whereas non-constructors are compiled into FPL definitions. The request `add:[s(s(0)),s(s(s(0)))] ?` then yields `s(s(s(s(0))))`.

3.4 Extension to higher-order guards

Several Prolog environments have higher-order facilities. Higher-order terms are variables bound to some goal at run time, and that may be executed as goals. We can take profit of this in FPL, as shown by the following example.

```
combine := (P =.. [^1,Z,^2] , Q =.. [^1,Z,^3] ,
            P               , Q               ) -> $Z
```

`combine` uses as an argument a function or a predicate name, later invoking it with some input. An example of the use of `combine` is in the following `non-disjoint` function, which tests whether two list are disjoint :

```
non-disjoint := combine @ [$member | id].
```

Then, the evaluation of `non-disjoint:[[1,2,3],[2,5,6]] ?` proceeds as follows :

- the request is transformed into `combine:[member | [[1,2,3],[2,5,6]]]` = `combine:[member,[[1,2,3],[2,5,6]]]`
- `^1` and `^2` are applied to `[member,[[1,2,3],[2,5,6]]`, yielding respectively `member` and `[1,2,3]`. Then Prolog solves the goal `P =.. [member,Z,[1,2,3]]`, i.e. `P` is bound to `member(Z,[1,2,3])`
- Similarly, `Q` is bound to `member(Z,[2,5,6])`
- The evaluation proceeds with solving `P,Q`, which binds `Z` to 2
- the final result is `$2:[...]` = 2.

4 An interpreter and a compiler for FPL

4.1 An interpreter for FPL

In order to evaluate queries, we developed an interpreter for FPL written in Prolog. The interpreter is a Prolog procedure that applies functions (or functional forms) to objects, using FPL semantic rules (cf. section 4). The evaluation may be considered as a TRS normalization process. The interpreter is further described in [Licht 90].

4.2 A compiler for FPL

An FPL compiler was also developed, that statically translates FPL definitions into Prolog predicates. All FP(L) primitive function are mapped into Prolog binary predicates. Then an FPL functional form `f` is translated into equivalent Prolog predicates `ff`, that uses sub-expressions as subgoals. Now, the correctness of the compilation process means the following. It should be the case that `f:obj` $\stackrel{FPL^*}{\rightarrow}$ `o'` if $\vdash^{SLD}$`ff(obj,o')`.[1] When `o'` = $\perp$, then $\vdash^{SLD}$`ff(obj,X)` should either succeed with `{X\`$\perp$`}`, or fail.[2] Lastly, if the normalization of `f:obj` does not terminate (unfolding of a non-terminating definition), then $\vdash^{SLD}$`ff(obj,X)` should not terminate.

[1] $\stackrel{FPL}{\rightarrow}$ is defined in section 4.

[2] All over this article, *fail* means *finite* failure.

Example : The following function definition :

```
f := p(^g,X) -> q @ $X
```

is compiled into the following Prolog program :

```
ff(_5, _6) :- f3(_5, _6, []).              % main predicate
f3(_14 , _15 , [] ) :-                     % for p(^g,X) -> q @ $X:
            g(_14, _78),                   %     invoke g
            p(_78, _0),                    %     execute Prolog goal
            f2(_14, _15, [_0]).            %     consequence q @ $X
f2(_161, _162, [_0]) :-                    % for q @ $X
             f1(_161, _169, [_0]),         %     call to $X
             q(_169, _162).                %     call to q
f1(_194, _0  , [_0]) :- true.              % for $X
```

5 Semantics of FPL

It is classical (cf. e.g. [HWWW 84]) to assign operational semantics to Backus' *FP* by means of conditional, positive/negative [CTRS 88] rewrite rules, such as :

	$f:\perp$	$\stackrel{FP}{\rightarrow}$	$\perp$	(strictness)
	$(f@g):obj$	$\stackrel{FP}{\rightarrow}$	$f:(g:obj)$	(composition)
	$[f_1,\ldots,f_n]:obj$	$\stackrel{FP}{\rightarrow}$	$[f_1:obj,\ldots,f_n:obj]$	(construction)
$obj \neq \perp \Rightarrow$	$\$c:obj$	$\stackrel{FP}{\rightarrow}$	c	(constant application)

Operational semantics for *FPL* may be defined similarly :

o Every FPL-expression that reduces via the rules of FP (quantified on FPL) reduces also in FPL :

$$\frac{P \stackrel{FP^*}{\rightarrow} P'}{P \stackrel{FPL}{\rightarrow} P'}$$

o In order to assign semantics to the extended conditional forms, we define $\Gamma_{\mathtt{obj}}(\mathtt{P})$ (for a goal P and an object obj) as being P where all occurrences of an expression of the form ^f are replaced by the result of evaluating f:obj). For instance, if P = p(X,^g), then $\Gamma_{\mathtt{obj}}(\mathtt{P}) = \mathtt{p}(\mathtt{X},\mathtt{o}')$, where o' stands for the result of evaluating g:obj.

If-Then(-Else) forms are then defined by :

$$\frac{\vdash^{SLD}\Gamma_{\mathtt{obj}}(P)\sigma}{(P\rightarrow Q):obj \stackrel{FPL}{\rightarrow} Q\sigma:obj} \qquad \frac{\vdash^{SLD}\Gamma_{\mathtt{obj}}(P)\sigma}{(P\rightarrow Q;R):obj \stackrel{FPL}{\rightarrow} Q\sigma:obj}$$

$$\frac{\text{Prolog fails on } \Gamma_{\mathtt{obj}}(P)}{(P\rightarrow Q):obj \stackrel{FPL}{\rightarrow} \perp} \qquad \frac{\text{Prolog fails on } \Gamma_{\mathtt{obj}}(P)}{(P\rightarrow Q;R):obj \stackrel{FPL}{\rightarrow} R:obj}$$

$$\frac{\text{Prolog loops on } \Gamma_{\mathtt{obj}}(P)}{(P\rightarrow Q):obj \stackrel{FPL}{\rightarrow} (P\rightarrow Q):obj} \qquad \frac{\text{Prolog loops on } \Gamma_{\mathtt{obj}}(P)}{(P\rightarrow Q;R):obj \stackrel{FPL}{\rightarrow} (P\rightarrow Q;R):obj}$$

Again, "fails" actually means finite failure.

o *Guarded commands* are defined by :

$$\frac{\vdash^{SLD}\Gamma_{\mathbf{obj}}(P_i)\sigma}{(P_1 \rightarrow Q_1 \sharp \cdots \sharp P_n \rightarrow Q_n) : obj \overset{FPL}{\rightarrow} Q_i\sigma : obj} \qquad \frac{\vdash^{SLD}\Gamma_{\mathbf{obj}}(P_i)\sigma}{(P_1 \rightarrow Q_1 \sharp \cdots \sharp P_n \rightarrow Q_n); R : obj \overset{FPL}{\rightarrow} Q_i\sigma : obj}$$

$$\frac{\forall i, \text{ Prolog fails on } \Gamma_{\mathbf{obj}}(P_i)}{(P_1 \rightarrow Q_1 \sharp \cdots \sharp P_n \rightarrow Q_n) : obj \overset{FPL}{\rightarrow} \bot} \qquad \frac{\forall i, \text{ Prolog fails on } \Gamma_{\mathbf{obj}}(P_i)}{(P_1 \rightarrow Q_1 \sharp \cdots \sharp P_n \rightarrow Q_n); R : obj \overset{FPL}{\rightarrow} R : obj}$$

$$\frac{\begin{array}{c}\exists i, \text{ Prolog loops on } \Gamma_{\mathbf{obj}}(P_i) \\ \text{and, } \forall j \neq i, \\ \text{either Prolog loops on } \Gamma_{\mathbf{obj}}(P_j) \\ \text{or Prolog fails on } \Gamma_{\mathbf{obj}}(P_j)\end{array}}{\begin{array}{c}(P_1 \rightarrow Q_1 \sharp \cdots \sharp P_n \rightarrow Q_n) : obj \overset{FPL}{\rightarrow} \\ (P_1 \rightarrow Q_1 \sharp \cdots \sharp P_n \rightarrow Q_n) : obj\end{array}} \qquad \frac{\begin{array}{c}\exists i, \text{ Prolog loops on } \Gamma_{\mathbf{obj}}(P_i) \\ \text{and, } \forall j \neq i, \\ \text{either Prolog loops on } \Gamma_{\mathbf{obj}}(P_j) \\ \text{or Prolog fails on } \Gamma_{\mathbf{obj}}(P_j)\end{array}}{\begin{array}{c}(P_1 \rightarrow Q_1 \sharp \cdots \sharp P_n \rightarrow Q_n); R : obj \overset{FPL}{\rightarrow} \\ (P_1 \rightarrow Q_1 \sharp \cdots \sharp P_n \rightarrow Q_n); R : obj\end{array}}$$

o *Function definition* is defined by :

$$(f{:=}E[f]) : obj \overset{FPL}{\rightarrow} E[f \backslash f{:=}E[f])] : obj \qquad (*)$$

o *While* forms are defined by :

$$(Q \ while \ P) : obj \overset{FPL}{\rightarrow} \left[P \rightarrow \left((Q \ while \ P)@Q\right); id\right] : obj$$

Note : the reader may convince himself that the above rules actually describe the corresponding constructs. However, we feel that the following points deserve further attention :

- Establish that $\overset{FPL}{\rightarrow}$ is *strongly uniquely terminating*. By this, we mean that if an expression t admits a (finite) normal form, then this normal form is unique –, and that every rewrite sequence starting from t is finite.

- Investigate the case of infinite reductions. For instance, establish that a reduction sequence is infinite if and only if it contains infinitely many applications of the above rule (*) for definition unfolding. Also, which conditions should be introduced in order to ensure that, when an expansion is infinite, an infinite, ω-normal form (in the sense of [DKP 90]) is actually computed ?

- Is there a need for lazy evaluation strategies, in order to avoid unnecessary infinite expansions –, and what are they ?

We believe that it is still an open challenge to prove these points, in a *non ad hoc* fashion. That is, we would like to (easily ?) solve these questions using general setting ; analysis of λ-calculus under the angle of TRS [Klop 87b], or ω-rewriting [DKP 90] may be relevant to this respect.

References

[Abramson 84] H. Abramson, *a Prological Definition of HASL, a Purely Functional Language with Unification Based Conditional Binding Expressions*, New Generation Computing **2**,1, pp. 3-35 (1984)

[Backus 78] J. Backus, *Can Programming be liberated from Von Neumann Style?*, CACM **21**,8, pp. 613-641 (1978)

[BBLM 86] R. Barbuti, M. Belllia, G. Levi, M. Montanari, *LEAF: a Language that Integrates Logic Equations and Functions*, in Logic Programming: Functions, Relations and Equations, Degroot, Lindstrom Eds., Prentice Hall, pp. 201-238 (1986)

[Cohen 86] S. Cohen, *The APPLOG Language*, in Logic Programming: Functions, Relations and Equations, Degroot, Lindstrom Eds., Prentice Hall, pp. 239-276 (1986)

[CTRS 88] Proc. of the Int. Conf. on Conditional Term Rewriting Systems, Kaplan, Jouannaud Eds., LNCS 308, Springer Verlag (1988)

[DKP 90] N. Dershowitz, S. Kaplan, D. Plaisted, *Infinite Normal Forms*, to appear in Theoretical Computer Science (1990)

[GM 84] J.A. Goguen, J. Meseguer, *Equality, Types, Modules and (why not?) Generics for Logic Programming*, Journal of Logic Programming **6**,2, pp. 179-210 (1984)

[HWWW 84] J. Halpern, J.H. Williams, E.L. Wimmers, T.C. Winkler, *Denotational Semantics and Rewrite Rules for FP*, Proc. of the POPL'84 Conf., pp. 108-120 (1984)

[Klop 87] J.W. Klop, *Term rewriting systems : a tutorial*, Bulletin of the EATCS, **32**, pp. 143-183 (1987)

[Klop 87b] J.W. Klop, *Rewrite Rule Systems*, unpublished manuscript, CWI, Amsterdam (1987)

[Licht 90] N. Lichtenstein, *FPL: Functional Plus Logic Programming. An integration of the FP and Prolog Languages*, M. Sc. Thesis, Bar-Ilan University, Ramat-Gan, Israel (1990)

[RS 82] J.A. Robinson, E.E. Sibert, *LOGLISP: Motivation, Design and Implementation*, in Logic Programming, Clark, Tarnlund Eds., Academic Prss, pp. 299-314 (1982)

[Smolka 86] G. Smolka, *FRESH: a Hiher-Order Language with Unification and Multiple Results*, in Logic Programming: Functions, Relations and Equations, Degroot, Lindstrom Eds., Prentice Hall, pp. 469-524 (1986)

[SS 84] M. Sato, T. Takurai *QUTE: a Functional Language based on Unification*, Int. Conf. Fifth Generation Computer Systems, ICOT, pp. 157-165 (1984)

[SY 84] P.A. Subrahmanyam, J.H. You, *FUNLOG: a Computational Model Integrating Logic Programming and Functional Programming*, Proc. of the IEEE International Conference on Logic Programming, pp. 144-153 (1984)

Confluence of the Disjoint Union of Conditional Term Rewriting Systems

Aart Middeldorp

Centre for Mathematics and Computer Science,
Kruislaan 413, 1098 SJ Amsterdam.
email: ami@cwi.nl

ABSTRACT

Toyama proved that confluence is a modular property of term rewriting systems. This means that the disjoint union of two confluent term rewriting systems is again confluent. In this paper we extend his result to the class of conditional term rewriting systems. In view of the important role of conditional rewriting in equational logic programming, this result may be of relevance in integrating functional programming and logic programming.

Introduction

Two directions can be distinguished in the use of conditional term rewriting systems. Bergstra and Klop [1], Kaplan [10] and Zhang and Rémy [25] studied conditional term rewriting as a means of enhancing the expressiveness in the algebraic specification of abstract data types. Recently, serious efforts have been initiated for integrating functional and logic programming. It has been recognized that conditional term rewriting systems provide a natural computational mechanism for this integration, see Dershowitz and Plaisted [5, 6], Fribourg [7] and Goguen and Meseguer [8].

For ordinary term rewriting systems a sizeable amount of theory has been developed. Only a small part has been extended to conditional term rewriting systems, notably sufficient conditions for confluence and termination ([1], [3], [4], [9], [11]). In this paper we extend a result of Toyama [22], which states that if a term rewriting system can be partitioned into two confluent systems with disjoint alphabets then the original system is confluent, to conditional term rewriting systems.

Conditional term rewriting is introduced in the next section. In Section 2 we consider disjoint unions of term rewriting systems. In Section 3 we prove that confluence is a modular property of join systems, a particular form of conditional term rewriting introduced in the next section. In Section 4 we observe that confluence is also a modular property of so-called semi-equational and normal systems and we conclude with suggestions for further research.

Note: Research partially supported by ESPRIT BRA project nr. 3020, Integration.

1. Conditional Term Rewriting Systems: Preliminaries

Before introducing conditional term rewriting, we review the basic notions of unconditional term rewriting. Term rewriting is surveyed in Klop [12] and Dershowitz and Jouannaud [2].

Let $\mathcal{V}$ be a countably infinite set of *variables*. An *unconditional term rewriting system* (TRS for short) is a pair $(\mathcal{F}, \mathcal{R})$. The set $\mathcal{F}$ consists of *function symbols*; associated with every $f \in \mathcal{F}$ is its arity $n \geq 0$. Function symbols of arity 0 are called *constants*. The set of terms built from $\mathcal{F}$ and $\mathcal{V}$, notation $\mathcal{T}(\mathcal{F}, \mathcal{V})$, is the smallest set such that $\mathcal{V} \subset \mathcal{T}(\mathcal{F}, \mathcal{V})$ and if $f \in \mathcal{F}$ has arity n and $t_1, \ldots, t_n \in \mathcal{T}(\mathcal{F}, \mathcal{V})$ then $f(t_1, \ldots, t_n) \in \mathcal{T}(\mathcal{F}, \mathcal{V})$. Terms not containing variables are *ground* terms. The set $\mathcal{R}$ consists of pairs (l, r) with $l, r \in \mathcal{T}(\mathcal{F}, \mathcal{V})$ subject to two constraints:

(1) the left-hand side l is not a variable,

(2) the variables which occur in the right-hand side r also occur in l.

Pairs (l, r) are called *rewrite rules* or *reduction rules* and will henceforth be written as $l \rightarrow r$. We usually present a TRS as a set of rewrite rules, without making explicit the set of function symbols.

A *substitution* σ is a mapping from $\mathcal{V}$ to $\mathcal{T}(\mathcal{F}, \mathcal{V})$ such that the set $\{x \in \mathcal{V} \mid \sigma(x) \not\equiv x\}$ is finite (the symbol $\equiv$ stands for syntactic equality). This set is called the *domain* of σ and will be denoted by $\mathcal{D}(\sigma)$. Substitutions are extended to morphisms from $\mathcal{T}(\mathcal{F}, \mathcal{V})$ to $\mathcal{T}(\mathcal{F}, \mathcal{V})$, i.e. $\sigma(f(t_1, \ldots, t_n)) \equiv f(\sigma(t_1), \ldots, \sigma(t_n))$ for every n-ary function symbol f and terms $t_1, \ldots, t_n$. We call $\sigma(t)$ an *instance* of t. An instance of a left-hand side of a rewrite rule is a *redex* (reducible expression). Occasionally we present a substitution σ as $\sigma = \{x \rightarrow \sigma(x) \mid x \in \mathcal{D}(\sigma)\}$. The *empty* substitution will be denoted by ε (here $\mathcal{D}(\varepsilon) = \varnothing$).

A *context* $C[\,,\ldots,\,]$ is a 'term' which contains at least one occurrence of a special symbol $\Box$. If $C[\,,\ldots,\,]$ is a context with n occurrences of $\Box$ and $t_1, \ldots, t_n$ are terms then $C[t_1, \ldots, t_n]$ is the result of replacing from left to right the occurrences of $\Box$ by $t_1, \ldots, t_n$. A context containing precisely one occurrence of $\Box$ is denoted by $C[\,]$. A term s is a *subterm* of a term t if there exists a context $C[\,]$ such that $t \equiv C[s]$.

The *rewrite relation* $\rightarrow_{\mathcal{R}} \subset \mathcal{T}(\mathcal{F}, \mathcal{V}) \times \mathcal{T}(\mathcal{F}, \mathcal{V})$ is defined as follows: $s \rightarrow_{\mathcal{R}} t$ if there exists a rewrite rule $l \rightarrow r$ in $\mathcal{R}$, a substitution σ and a context $C[\,]$ such that $s \equiv C[\sigma(l)]$ and $t \equiv C[\sigma(r)]$. The transitive-reflexive closure of $\rightarrow_{\mathcal{R}}$ is denoted by $\twoheadrightarrow_{\mathcal{R}}$; if $s \twoheadrightarrow_{\mathcal{R}} t$ we say that s *reduces* to t. We write $s \leftarrow_{\mathcal{R}} t$ if $t \rightarrow_{\mathcal{R}} s$; likewise for $s \twoheadleftarrow_{\mathcal{R}} t$. The symmetric closure of $\rightarrow_{\mathcal{R}}$ is denoted by $\leftrightarrow_{\mathcal{R}}$. The transitive-reflexive closure of $\leftrightarrow_{\mathcal{R}}$ is called *conversion* and denoted by $=_{\mathcal{R}}$. If $s =_{\mathcal{R}} t$ then s and t are *convertible*. Two terms t_1, t_2 are *joinable*, notation $t_1 \downarrow_{\mathcal{R}} t_2$, if there exists a term t_3 such that $t_1 \twoheadrightarrow_{\mathcal{R}} t_3 \twoheadleftarrow_{\mathcal{R}} t_2$. Such a term t_3 is called a *common reduct* of t_1 and t_2. The relation $\downarrow_{\mathcal{R}}$ is called *joinability*. We often omit the subscript $\mathcal{R}$.

A term s is a *normal form* if there are no terms t with $s \rightarrow t$. A TRS is *terminating* or *strongly normalizing* if there are no infinite reduction sequences $t_1 \rightarrow t_2 \rightarrow t_3 \rightarrow \ldots$. In other words, every reduction sequence eventually ends in a normal form. A TRS is *confluent* or has the *Church-Rosser* property if for all terms s, t_1, t_2 with $t_1 \twoheadleftarrow s \twoheadrightarrow t_2$ we have $t_1 \downarrow t_2$. A well-known equivalent formulation of confluence is that every pair of convertible terms is joinable $(t_1 = t_2 \Rightarrow t_1 \downarrow t_2)$.

The rewrite rules of a *conditional term rewriting system* (CTRS) have the form

$$l \to r \Leftarrow s_1 = t_1, \ldots, s_n = t_n$$

with $s_1, \ldots, s_n, t_1, \ldots, t_n, l, r \in \mathcal{T}(\mathcal{F}, \mathcal{V})$. The equations $s_1 = t_1, \ldots, s_n = t_n$ are the *conditions* of the rewrite rule. Depending on the interpretation of the =-sign in the conditions, different rewrite relations can be associated with a given CTRS. In this paper we restrict ourselves to the three most common interpretations.

(1) In a *join* CTRS the =-sign in the conditions is interpreted as *joinability*. Formally: $s \to t$ if there exists a conditional rewrite rule $l \to r \Leftarrow s_1 = t_1, \ldots, s_n = t_n$, a substitution σ and a context $C[\,]$ such that $s \equiv C[\sigma(l)]$, $t \equiv C[\sigma(r)]$ and $\sigma(s_i) \downarrow \sigma(t_i)$ for all $i \in \{1, \ldots, n\}$. Rewrite rules of a join CTRS will henceforth be written as

$$l \to r \Leftarrow s_1 \downarrow t_1, \ldots, s_n \downarrow t_n.$$

(2) *Semi-equational* CTRS's are obtained by interpreting the =-sign in the conditions as *conversion*.

(3) In a *normal* CTRS $\mathcal{R}$ the rewrite rules are subject to the additional constraint that every t_i is a ground normal form with respect to the unconditional TRS obtained from $\mathcal{R}$ by omitting the conditions. The rewrite relation associated with a normal CTRS is obtained by interpreting the equality sign in the conditions as *reduction* ($\twoheadrightarrow$).

This classification originates essentially from Bergstra and Klop [1]. The nomenclature stems from Dershowitz, Okada and Sivakumar [4].

The restrictions we impose on CTRS's $\mathcal{R}$ in any of the three formulations are the same as for unconditional TRS's: if $l \to r \Leftarrow s_1 = t_1, \ldots, s_n = t_n$ is a rewrite rule of $\mathcal{R}$ then l is not a single variable and variables occurring in r also occur in l. In particular, extra variables in the conditions are perfectly acceptable. In Section 4 we will discuss the technical problems associated with a possible relaxation of this requirement.

Sufficient conditions for the termination of CTRS's were given by Kaplan [11], Jouannaud and Waldmann [9] and Dershowitz, Okada and Sivakumar [4]. Sufficient conditions for confluence can be found in Bergstra and Klop [1] and Dershowitz, Okada and Sivakumar [3].

EXAMPLE 1.1. The semi-equational CTRS

$$\mathcal{R}_1 = \begin{cases} a \to b \\ a \to c \\ b \to c \Leftarrow b = c \end{cases}$$

is easily shown to be confluent. So conversion in that system coincides with joinability. However, the corresponding join CTRS

$$\mathcal{R}_2 = \begin{cases} a \to b \\ a \to c \\ b \to c \Leftarrow b \downarrow c \end{cases}$$

is not confluent: the reduction step from b to c is no longer allowed.

The following inductive definition of $\rightarrow_{\mathcal{R}}$ is fundamental ([1, 3, 4]) for analyzing the behaviour of CTRS's.

DEFINITION 1.2. Let $\mathcal{R}$ be a join, semi-equational or normal CTRS. We inductively define TRS's $\mathcal{R}_i$ for $i \geq 0$ as follows ($\square$ denotes $\downarrow$, $=$ or $\twoheadrightarrow$):

$$\mathcal{R}_0 = \{l \rightarrow r \mid l \rightarrow r \in \mathcal{R}\}$$

$$\mathcal{R}_{i+1} = \{\sigma(l) \rightarrow \sigma(r) \mid l \rightarrow r \Leftarrow s_1 \square t_1, \ldots, s_n \square t_n \in \mathcal{R} \text{ and } \sigma(s_j) \square_{\mathcal{R}_i} \sigma(t_j) \text{ for } j = 1, \ldots, n\}.$$

Observe that $\mathcal{R}_i \subseteq \mathcal{R}_{i+1}$ for all $i \geq 0$. We have $s \rightarrow_{\mathcal{R}} t$ if and only if $s \rightarrow_{\mathcal{R}_i} t$ for some $i \geq 0$. The *depth* of a rewrite step $s \rightarrow_{\mathcal{R}} t$ is defined as the minimum i such that $s \rightarrow_{\mathcal{R}_i} t$. Depths of conversions $s =_{\mathcal{R}} t$ and valleys $s \downarrow_{\mathcal{R}} t$ are similarly defined.

2. Modular Properties

In this section we review some of the results that have been obtained concerning the disjoint union of TRS's. We will also give the necessary technical definitions and notations for dealing with disjoint unions. These are consistent with [22, 24, 15].

DEFINITION 2.1. Let $(\mathcal{F}_1, \mathcal{R}_1)$ and $(\mathcal{F}_2, \mathcal{R}_2)$ be CTRS's with disjoint alphabets (i.e. $\mathcal{F}_1 \cap \mathcal{F}_2 = \varnothing$). The *disjoint union* $\mathcal{R}_1 \oplus \mathcal{R}_2$ of $(\mathcal{F}_1, \mathcal{R}_1)$ and $(\mathcal{F}_2, \mathcal{R}_2)$ is the CTRS $(\mathcal{F}_1 \cup \mathcal{F}_2, \mathcal{R}_1 \cup \mathcal{R}_2)$.

DEFINITION 2.2. A property $\mathcal{P}$ of CTRS's is called *modular* if for all CTRS's $\mathcal{R}_1$, $\mathcal{R}_2$ the following equivalence holds:

$$\mathcal{R}_1 \oplus \mathcal{R}_2 \text{ has the property } \mathcal{P} \quad \Leftrightarrow \quad \text{both } \mathcal{R}_1 \text{ and } \mathcal{R}_2 \text{ have the property } \mathcal{P}.$$

All previous research on modularity has been carried out in the world of unconditional TRS's. This research can be characterized by the phrase "simple statements, complicated proofs". Confluence was the first property for which the modularity has been established.

THEOREM 2.3 (Toyama [22]). *Confluence is a modular property of* TRS's. $\square$

Toyama also gave the following simple example showing that termination is not modular.

EXAMPLE 2.4 (Toyama [23]). Let $\mathcal{R}_1 = \{F(0, 1, x) \rightarrow F(x, x, x)\}$ and

$$\mathcal{R}_2 = \begin{cases} or(x, y) & \rightarrow & x, \\ or(x, y) & \rightarrow & y. \end{cases}$$

Both systems are terminating, but in $\mathcal{R}_1 \oplus \mathcal{R}_2$ the term $F(or(0, 1), or(0, 1), or(0, 1))$ reduces to itself.

Other modularity results are presented in [13, 14, 15, 16, 21, 23, 24]. Middeldorp [20] contains a comprehensive survey.

Let $(\mathcal{F}_1, \mathcal{R}_1)$ and $(\mathcal{F}_2, \mathcal{R}_2)$ be disjoint CTRS's. Every term $t \in \mathcal{T}(\mathcal{F}_1 \cup \mathcal{F}_2, \mathcal{V})$ can be viewed as an alternation of $\mathcal{F}_1$-parts and $\mathcal{F}_2$-parts. This structure is formalized in Definition 2.5 and illustrated in Figure 1.

NOTATION. We abbreviate $\mathcal{T}(\mathcal{F}_1 \cup \mathcal{F}_2, \mathcal{V})$ to $\mathcal{T}$ and we will use $\mathcal{T}_i$ as a shorthand for $\mathcal{T}(\mathcal{F}_i, \mathcal{V})$ $(i = 1, 2)$.

DEFINITION 2.5.

(1) The *root symbol* of a term t, notation $root(t)$, is defined by

$$root(t) = \begin{cases} F & \text{if } t \equiv F(t_1, \ldots, t_n), \\ t & \text{otherwise.} \end{cases}$$

(2) Let $t \equiv C[t_1, \ldots, t_n] \in \mathcal{T}$ with $C[\,, \ldots, \,] \not\equiv \square$. We write $t \equiv C[\![t_1, \ldots, t_n]\!]$ if $C[\,, \ldots, \,]$ is a $\mathcal{F}_a$-context and $root(t_1), \ldots, root(t_n) \in \mathcal{F}_b$ for some $a, b \in \{1, 2\}$ with $a \neq b$. The t_i's are the *principal* subterms of t.

(3) If $t \in \mathcal{T}$ then

$$rank(t) = \begin{cases} 1 & \text{if } t \in \mathcal{T}_1 \cup \mathcal{T}_2, \\ 1 + \max\{rank(t_i) \mid 1 \le i \le n\} & \text{if } t \equiv C[\![t_1, \ldots, t_n]\!]. \end{cases}$$

(4) A subterm s of t is *special* if $s \equiv t$ or s is a special subterm of a principal subterm of t.

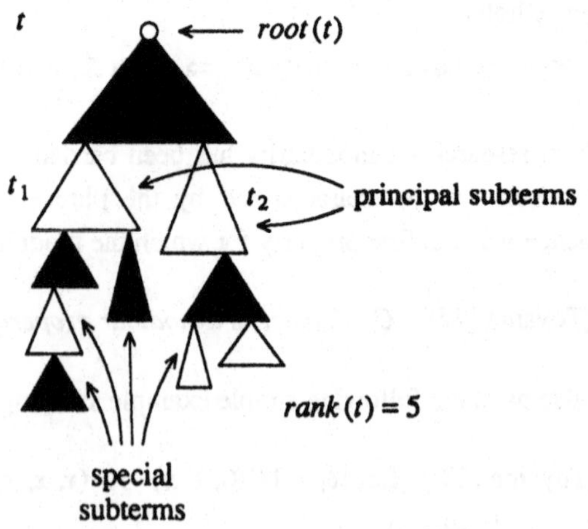

FIGURE 1.

To achieve better readability we will call the function symbols of $\mathcal{F}_1$ *black* and those of $\mathcal{F}_2$ *white*. Variables have no colour. A black (white) term does not contain white (black) function symbols, but may contain variables. In examples, black symbols will be printed as capitals and white symbols in lower case.

PROPOSITION 2.6. *If* $s \twoheadrightarrow_{\mathcal{R}_1 \oplus \mathcal{R}_2} t$ *then* $rank(s) \geq rank(t)$.

PROOF. Straightforward. □

DEFINITION 2.7. Let $s \to t$ by application of a rewrite rule $l \to r$. We write $s \to^i t$ if $l \to r$ is being applied in one of the principal subterms of s and we write $s \to^o t$ otherwise. The relation $\to^i$ is called *inner* reduction and $\to^o$ is called *outer* reduction.

DEFINITION 2.8. Suppose σ and τ are substitutions. We write $\sigma \propto \tau$ if $\sigma(x) \equiv \sigma(y)$ implies $\tau(x) \equiv \tau(y)$ for all $x, y \in \mathcal{V}$. Notice that $\sigma \propto \varepsilon$ if and only if σ is injective. We write $\sigma \twoheadrightarrow \tau$ if $\sigma(x) \twoheadrightarrow \tau(x)$ for all $x \in \mathcal{V}$. Clearly $\sigma(t) \twoheadrightarrow \tau(t)$ whenever $\sigma \twoheadrightarrow \tau$.

DEFINITION 2.9. A substitution σ is called *black* (*white*) if $\sigma(x)$ is a black (white) term for every $x \in \mathcal{D}(\sigma)$. We call σ *top black* (*top white*) if the root symbol of $\sigma(x)$ is black (white) for every $x \in \mathcal{D}(\sigma)$.

Notice the subtle difference in handling variables: the substitution $\sigma = \{x \to F(y), y \to x\}$ is black but not top black. The following proposition is frequently used in the next section.

PROPOSITION 2.10. *Every substitution* σ *can be decomposed into* $\sigma_2 \circ \sigma_1$ *such that* σ_1 *is black* (*white*), σ_2 *is top white* (*top black*) *and* $\sigma_2 \propto \varepsilon$.

PROOF. Let $\{t_1, \ldots, t_n\}$ be the set of all maximal subterms of $\sigma(x)$ for $x \in \mathcal{D}(\sigma)$ with white root. Choose distinct fresh variables $z_1, \ldots, z_n$ and define the substitution σ_2 by $\sigma_2 = \{z_i \to t_i \mid 1 \leq i \leq n\}$. Let $x \in \mathcal{D}(\sigma)$. We define $\sigma_1(x)$ by case analysis.

(1) If the root symbol of $\sigma(x)$ is white then $\sigma(x) \equiv t_i$ for some $i \in \{1, \ldots, n\}$. In this case we define $\sigma_1(x) \equiv z_i$.
(2) If $\sigma(x)$ is a black term then we take $\sigma_1(x) \equiv \sigma(x)$.
(3) In the remaining case we can write $\sigma(x) \equiv C[\![t_{i_1}, \ldots, t_{i_k}]\!]$ for some $1 \leq i_1, \ldots, i_k \leq n$ and we define $\sigma_1(x) \equiv C[z_{i_1}, \ldots, z_{i_k}]$.

By construction we have $\sigma_2 \propto \varepsilon$, σ_1 is black and σ_2 is top white. □

3. Modularity of Confluence for Join Systems

In this section we show that confluence is a modular property of join CTRS's. To this end, we assume that $\mathcal{R}_1$ and $\mathcal{R}_2$ are disjoint confluent join CTRS's. We assume furthermore that all rewrite relations introduced in this section are defined on $\mathcal{T}$, unless stated otherwise. The same assumption is made for terms.

The fundamental property of the disjoint union of two unconditional TRS's $\mathcal{R}_1$ and $\mathcal{R}_2$, that is to say $s \to_{\mathcal{R}_1 \oplus \mathcal{R}_2} t$ implies either $s \to_{\mathcal{R}_1} t$ or $s \to_{\mathcal{R}_2} t$, does no longer hold for CTRS's, as can be seen from the next example.

EXAMPLE 3.1. Let $\mathcal{R}_1 = \{F(x, y) \to G(x) \Leftarrow x \downarrow y\}$ and $\mathcal{R}_2 = \{a \to b\}$. We have $F(a, b)$

$\rightarrow_{\mathcal{R}_1 \oplus \mathcal{R}_2} G(a)$ because $a \downarrow_{\mathcal{R}_1 \oplus \mathcal{R}_2} b$, but neither $F(a, b) \rightarrow_{\mathcal{R}_1} G(a)$ nor $F(a, b) \rightarrow_{\mathcal{R}_2} G(a)$.

The problem is that when a rule of one of the CTRS's is being applied, rules of the other CTRS may be needed in order to satisfy the conditions. So the question arises how the rewrite relation $\rightarrow_{\mathcal{R}_1 \oplus \mathcal{R}_2}$ is related to $\rightarrow_{\mathcal{R}_1}$ and $\rightarrow_{\mathcal{R}_2}$. In the example above we have $F(a, b) \rightarrow_{\mathcal{R}_2} F(b, b) \rightarrow_{\mathcal{R}_1} G(b) \leftarrow_{\mathcal{R}_2} G(a)$. This suggests that $\rightarrow_{\mathcal{R}_1 \oplus \mathcal{R}_2}$ corresponds to joinability with respect to the union of $\rightarrow_{\mathcal{R}_1}$ and $\rightarrow_{\mathcal{R}_2}$. However, it turned out that $\rightarrow_{\mathcal{R}_1} \cup \rightarrow_{\mathcal{R}_2}$ is not an entirely satisfactory relation from a technical viewpoint. For instance, confluence of $\rightarrow_{\mathcal{R}_1} \cup \rightarrow_{\mathcal{R}_2}$ is not easily proved (cf. Lemma 3.6). We will define two more manageable rewrite relations $\rightarrow_1$ and $\rightarrow_2$ such that:

(1) their union is confluent (Lemma 3.6),

(2) reduction in $\mathcal{R}_1 \oplus \mathcal{R}_2$ corresponds to joinability with respect to $\rightarrow_1 \cup \rightarrow_2$ (Lemma 3.8).

From these two properties the modularity of confluence for join CTRS's is easily inferred (Theorem 3.9). The proof of the first property is a more or less straightforward reduction to Toyama's confluence result for the disjoint union of TRS's. The proof of the second property is rather technical but we believe that the underlying ideas are simple.

DEFINITION 3.2. The rewrite relation $\rightarrow_1$ is defined as follows: $s \rightarrow_1 t$ if there exists a rewrite rule $l \rightarrow r \Leftarrow s_1 \downarrow t_1, \ldots, s_n \downarrow t_n$ in $\mathcal{R}_1$, a context $C[\ ]$ and a substitution σ such that $s \equiv C[\sigma(l)]$, $t \equiv C[\sigma(r)]$ and $\sigma(s_i) \downarrow_1^o \sigma(t_i)$ for $i = 1, \ldots, n$, where the superscript o in $\sigma(s_i) \downarrow_1^o \sigma(t_i)$ means that $\sigma(s_i)$ and $\sigma(t_i)$ are joinable using only *outer* $\rightarrow_1$-reduction steps. Notice that the restrictions of $\rightarrow_1$ and $\rightarrow_{\mathcal{R}_1}$ to $\mathcal{T}_1 \times \mathcal{T}_1$ coincide. The relation $\rightarrow_2$ is defined similarly.

EXAMPLE 3.3. Let

$$\mathcal{R}_1 = \begin{cases} F(x, y) & \rightarrow & G(x) & \Leftarrow & x \downarrow y \\ A & \rightarrow & B \end{cases}$$

and suppose $\mathcal{R}_2$ contains an unary function symbol g. We have $F(g(A), g(B)) \rightarrow_{\mathcal{R}_1} G(g(A))$ but not $F(g(A), g(B)) \rightarrow_1 G(g(A))$ because $g(A)$ and $g(B)$ are different normal forms with respect to $\rightarrow_1^o$. The terms $F(g(A), g(B))$ and $G(g(A))$ are joinable with respect to $\rightarrow_1$: $F(g(A), g(B)) \rightarrow_1 F(g(B), g(B)) \rightarrow_1 G(g(B)) \leftarrow_1 G(g(A))$.

NOTATION. The union of $\rightarrow_1$ and $\rightarrow_2$ is denoted by $\rightarrow_{1,2}$ and we abbreviate $\rightarrow_{\mathcal{R}_1 \oplus \mathcal{R}_2}$ to $\rightarrow$.

PROPOSITION 3.4. *If $s \rightarrow_{1,2} t$ then $s \rightarrow t$.*

PROOF. Trivial. □

The next proposition states a desirable property of $\rightarrow_1^o$-reduction. The proof however is more complicated than the analogical statement for TRS's (cf. Lemma 3.2 in [22]).

PROPOSITION 3.5. *Let s, t be black terms and suppose σ is a top white substitution such that $\sigma(s) \rightarrow_1^o \sigma(t)$. If τ is a substitution with $\sigma \propto \tau$ then $\tau(s) \rightarrow_1^o \tau(t)$.*

PROOF. We prove the statement by induction on the depth of $\sigma(s) \to_1^o \sigma(t)$. The case of zero depth is straightforward. If the depth of $\sigma(s) \to_1^o \sigma(t)$ equals $n+1$ ($n \geq 0$) then there exist a context $C[\,]$, a substitution ρ and a rewrite rule $l \to r \Leftarrow s_1 \downarrow t_1, \ldots, s_m \downarrow t_m$ in $\mathcal{R}_1$ such that $\sigma(s) \equiv C[\rho(l)]$, $\sigma(t) \equiv C[\rho(r)]$ and $\rho(s_i) \downarrow_1^o \rho(t_i)$ for $i = 1, \ldots, m$ with depth less than or equal to n. Proposition 2.10 yields a decomposition $\rho_2 \circ \rho_1$ of ρ such that ρ_1 is black, ρ_2 is top white and $\rho_2 \propto \varepsilon$. The situation is illustrated in Figure 2. We define the substitution ρ^* by

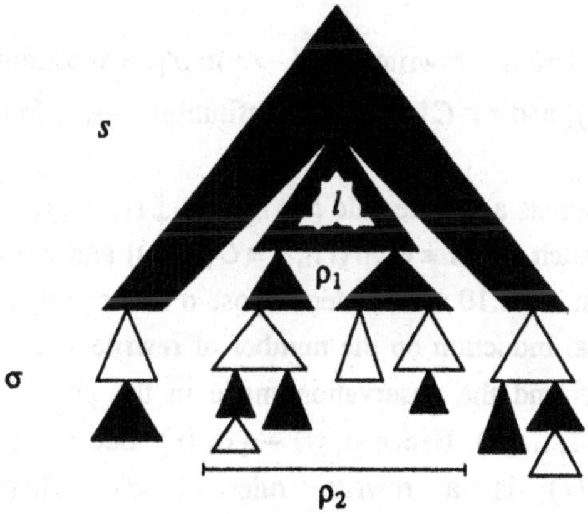

FIGURE 2.

$\rho^*(x) \equiv \tau(y)$ for every $x \in \mathcal{D}(\rho_2)$ and $y \in \mathcal{D}(\sigma)$ satisfying $\rho_2(x) \equiv \sigma(y)$. Notice that ρ^* is well-defined by the assumption $\sigma \propto \tau$. We have $\rho_2 \propto \rho^*$ since $\rho_2 \propto \varepsilon$ and $\varepsilon \propto \rho^*$. Combined with $\rho_2(\rho_1(s_i)) \downarrow_1^o \rho_2(\rho_1(t_i))$, the induction hypothesis and the observation that if $\rho_2(u_1) \to_1^o u_2$ and u_1 is a black term then $u_2 \equiv \rho_2(u_3)$ for some black term u_3, we obtain $\rho^*(\rho_1(s_i)) \downarrow_1^o \rho^*(\rho_1(t_i))$ by a straightforward induction on the length of the conversion $\rho_2(\rho_1(s_i)) \downarrow_1^o \rho_2(\rho_1(t_i))$ for $i = 1, \ldots, m$ (see Figure 3). Hence $\rho^*(\rho_1(l)) \to_1^o \rho^*(\rho_1(r))$. Let $C^*[\,]$ be the context obtained

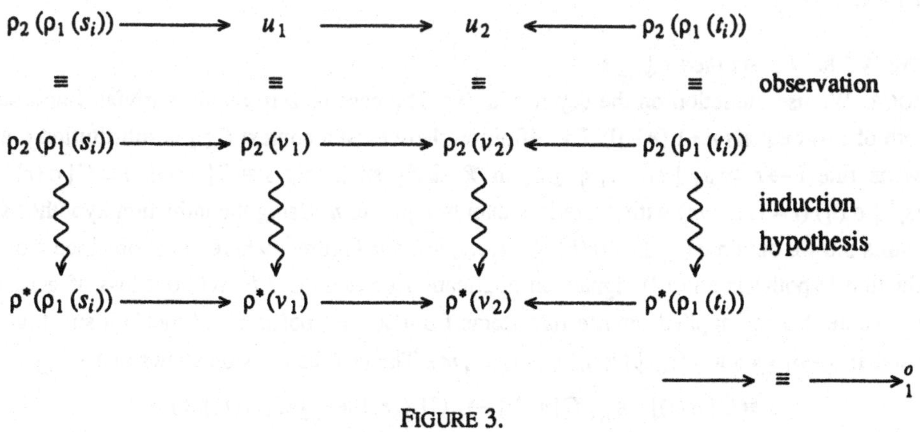

FIGURE 3.

from $C[\,]$ by replacing every principal subterm, which has the form $\sigma(x)$ for some variable $x \in \mathcal{D}(\sigma)$, by the corresponding $\tau(x)$. We leave it to the motivated reader to show that $\tau(s) \equiv C^*[\rho^*(\rho_1(l))]$ and $\tau(t) \equiv C^*[\rho^*(\rho_1(r))]$. We conclude that $\tau(s) \to_1^o \tau(t)$. □

LEMMA 3.6. *The rewrite relation* $\rightarrow_{1,2}$ *is confluent.*

PROOF. Define the unconditional TRS's $\mathcal{S}_1$ and $\mathcal{S}_2$ by ($i = 1, 2$)

$$\mathcal{S}_i = \{s \rightarrow t \mid s, t \in \mathcal{T}_i \text{ and } s \rightarrow^i t\}.$$

With some effort we can show that the restrictions of $\rightarrow_{\mathcal{S}_i}$, $\rightarrow_i$ and $\rightarrow_{\mathcal{R}_i}$ to $\mathcal{T}_i \times \mathcal{T}_i$ are the same†. Therefore $\mathcal{S}_1$ and $\mathcal{S}_2$ are confluent TRS's. Theorem 2.3 yields the confluence of $\mathcal{S}_1 \oplus \mathcal{S}_2$. We will show that $\rightarrow_{\mathcal{S}_i}$ and $\rightarrow^i$ coincide (on $\mathcal{T} \times \mathcal{T}$). Without loss of generality, we only consider the case $i = 1$.

$\subseteq$ If $s \rightarrow_{\mathcal{S}_1} t$ then there exists a rewrite rule $l \rightarrow r$ in $\mathcal{S}_1$, a substitution σ and a context $C[\,]$ such that $s \equiv C[\sigma(l)]$ and $t \equiv C[\sigma(r)]$. By definition $l \rightarrow_1 r$, from which we immediately obtain $s \rightarrow_1 t$.

$\supseteq$ If $s \rightarrow_1 t$ then there exists a rewrite rule $l \rightarrow r \Leftarrow s_1 \downarrow t_1, \ldots, s_n \downarrow t_n$ in $\mathcal{R}_1$, a substitution σ and a context $C[\,]$ such that $s \equiv C[\sigma(l)]$, $t \equiv C[\sigma(r)]$ and $\sigma(s_i) \downarrow_1^o \sigma(t_i)$ for $i = 1, \ldots, n$. According to Proposition 2.10 we can decompose σ into $\sigma_2 \circ \sigma_1$ such that σ_1 is black, σ_2 is top white and $\sigma_2 \propto \varepsilon$. Induction on the number of rewrite steps in $\sigma(s_i) \downarrow_1^o \sigma(t_i)$ together with Proposition 3.5 and the observation made in the proof of Proposition 3.5 yields $\sigma_1(s_i) \downarrow_1^o \sigma_1(t_i)$ ($i = 1, \ldots, n$). Hence $\sigma_1(l) \rightarrow_1 \sigma_1(r)$. Because $\sigma_1(l)$ and $\sigma_1(r)$ are black terms, $\sigma_1(l) \rightarrow \sigma_1(r)$ is a rewrite rule of $\mathcal{S}_1$. Therefore $s \equiv C[\sigma_2(\sigma_1(l))] \rightarrow_{S_1} C[\sigma_2(\sigma_1(r))] \equiv t$.

Now we have $\rightarrow_{\mathcal{S}_1 \oplus \mathcal{S}_2} = \rightarrow_{\mathcal{S}_1} \cup \rightarrow_{\mathcal{S}_2} = \rightarrow_1 \cup \rightarrow_2 = \rightarrow_{1,2}$. Therefore $\rightarrow_{1,2}$ is confluent. □

Due to space limitations, the reader is referred to the full version [17] of this paper for the complicated proof of the next proposition. The proof can also be found in [20].

PROPOSITION 3.7. *Let* $s_1, \ldots, s_n, t_1, \ldots, t_n$ *be black terms. For every substitution* σ *with* $\sigma(s_i) \downarrow_{1,2} \sigma(t_i)$ *for* $i = 1, \ldots, n$ *there exists a substitution* τ *such that* $\sigma \twoheadrightarrow_{1,2} \tau$ *and* $\tau(s_i) \downarrow_1^o \tau(t_i)$ *for* $i = 1, \ldots, n$. □

LEMMA 3.8. *If* $s \rightarrow t$ *then* $s \downarrow_{1,2} t$.

PROOF. We use induction on the depth of $s \rightarrow t$. The case of zero depth is trivial. Suppose the depth of $s \rightarrow t$ equals $n+1$ ($n \geq 0$). By definition there exist a context $C[\,]$, a substitution σ and a rewrite rule $l \rightarrow r \Leftarrow s_1 \downarrow t_1, \ldots, s_m \downarrow t_m$ in $\mathcal{R}_1 \oplus \mathcal{R}_2$ such that $s \equiv C[\sigma(l)]$, $t \equiv C[\sigma(r)]$ and $\sigma(s_i) \downarrow \sigma(t_i)$ ($i = 1, \ldots, m$) with depth less than or equal to n. Using the induction hypothesis and Lemma 3.6 we obtain $\sigma(s_i) \downarrow_{1,2} \sigma(t_i)$ ($i = 1, \ldots, m$), see Figure 4 where (1) is obtained from the induction hypothesis and (2) signals an application of Lemma 3.6. Without loss of generality we assume that the applied rewrite rule stems from $\mathcal{R}_1$. Proposition 3.7 yields a substitution τ such that $\sigma \twoheadrightarrow_{1,2} \tau$ and $\tau(s_i) \downarrow_1^o \tau(t_i)$ ($i = 1, \ldots, m$). The next conversion shows that $s \downarrow_{1,2} t$:

$$s \equiv C[\sigma(l)] \twoheadrightarrow_{1,2} C[\tau(l)] \rightarrow_1 C[\tau(r)] \twoheadleftarrow_{1,2} C[\sigma(r)] \equiv t.$$

□

† A minor technical complication is caused by rewrite rules containing extra variables in the conditions.

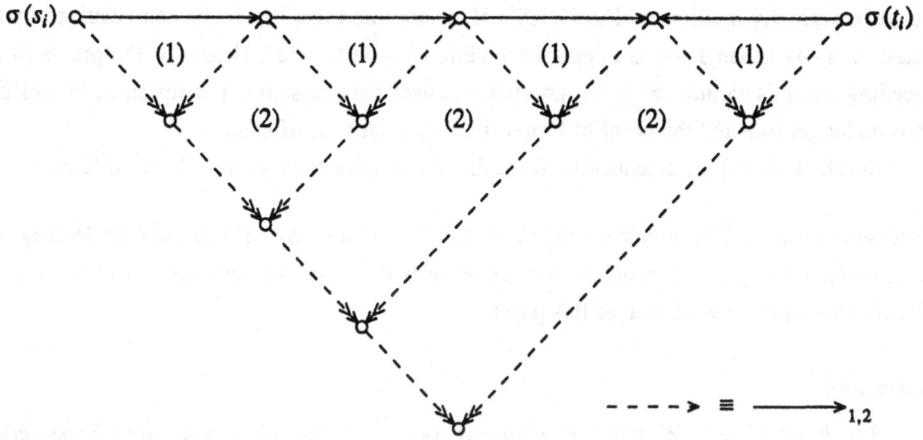

FIGURE 4.

THEOREM 3.9. *Confluence is a modular property of join* CTRS's.

PROOF. Suppose $\mathcal{R}_1$ and $\mathcal{R}_2$ are disjoint join CTRS's. We must prove the following equivalence: $\mathcal{R}_1 \oplus \mathcal{R}_2$ is confluent $\Leftrightarrow$ both $\mathcal{R}_1$ and $\mathcal{R}_2$ are confluent.

$\Rightarrow$ Trivial.

$\Leftarrow$ Easy consequence of Proposition 3.4, Lemma 3.6 and Lemma 3.8.

□

4. Concluding Remarks

In the previous section we have shown that confluence is modular property of join CTRS's. Since every normal CTRS can be viewed as a join CTRS, this result also holds for normal CTRS's. Confluence is also a modular property of semi-equational CTRS's. The proof has exactly the same structure, apart from the proof of Proposition 3.5, which is more complicated because the observation made in order to make the second induction hypothesis applicable is no longer sufficient. Details can be found in [17] or [20]. It is conceivable that we might prove a more general theorem from which we not only immediately obtain the above results, but also

(1) the modularity of confluence for other kinds of CTRS's like *normal-join* systems or *meta-conditional* systems (see [4]), and

(2) confluence results for the disjoint union of two different kinds of CTRS's.

This matter clearly has to be further pursued.

Another point which needs investigation is the syntactic restrictions imposed on the rewrite rules. From a programming point of view the assumption of a rewrite rule $l \to r \Leftarrow s_1 = t_1, \ldots, s_n = t_n$ satisfying the requirement that r only contains variables occurring in l, is too restrictive. A semi-equational CTRS like ([3])

$$\mathcal{R} = \begin{cases} Fib(0) & \to \ <0, S(0)> \\ Fib(S(x)) & \to \ <z, A(y, z)> \ \Leftarrow \ Fib(x) = <y, z> \end{cases}$$

should be perfectly legitimate. The CTRS's $\mathcal{R}$ we are interested in, can be characterized by the phrase "if $s \to_{\mathcal{R}} t$ then $s \to t$ is a legal unconditional rewrite rule". However, the proofs in the preceding sections cannot easily be modified to cover these systems. For instance, Proposition 2.6 is no longer true and the proof of Proposition 3.5 seems insufficient.

In [18] and [19] we extended several other modularity results for TRS's to CTRS's.

Acknowledgements. The author would like to thank Roel de Vrijer for discussions leading to a better understanding of the problem and Jan Willem Klop and Vincent van Oostrom for carefully reading a previous version of this paper.

References

1. J.A. Bergstra and J.W. Klop, *Conditional Rewrite Rules: Confluence and Termination*, Journal of Computer and System Sciences **32**(3), pp. 323-362, 1986.
2. N. Dershowitz and J.-P. Jouannaud, *Rewrite Systems*, to appear in: Handbook of Theoretical Computer Science (ed. J. van Leeuwen), North-Holland, 1989.
3. N. Dershowitz, M. Okada and G. Sivakumar, *Confluence of Conditional Rewrite Systems*, Proceedings of the 1st International Workshop on Conditional Term Rewriting Systems, Orsay, Lecture Notes in Computer Science **308**, pp. 31-44, 1987.
4. N. Dershowitz, M. Okada and G. Sivakumar, *Canonical Conditional Rewrite Systems*, Proceedings of the 9th Conference on Automated Deduction, Argonne, Lecture Notes in Computer Science **310**, pp. 538-549, 1988.
5. N. Dershowitz and D.A. Plaisted, *Logic Programming cum Applicative Programming*, Proceedings of the IEEE Symposium on Logic Programming, Boston, pp. 54-66, 1985.
6. N. Dershowitz and D.A. Plaisted, *Equational Programming*, in: Machine Intelligence 11 (eds. J.E. Hayes, D. Michie and J. Richards), Oxford University Press, pp. 21-56, 1987.
7. L. Fribourg, *SLOG: A Logic Programming Language Interpreter Based on Clausal Superposition and Rewriting*, Proceedings of the IEEE Symposium on Logic Programming, Boston, pp. 172-184, 1985.
8. J.A. Goguen and J. Meseguer, *EQLOG: Equality, Types and Generic Modules for Logic Programming*, in: Logic Programming: Functions, Relations and Equations (eds. D. DeGroot and G. Lindstrom), Prentice-Hall, pp. 295-363, 1986.
9. J.-P. Jouannaud and B. Waldmann, *Reductive Conditional Term Rewriting Systems*, Proceedings of the 3rd IFIP Working Conference on Formal Description of Programming Concepts, Ebberup, pp. 223-244, 1986.
10. S. Kaplan, *Conditional Rewrite Rules*, Theoretical Computer Science **33**(2), pp. 175-193, 1984.
11. S. Kaplan, *Simplifying Conditional Term Rewriting Systems: Unification, Termination and Confluence*, Journal of Symbolic Computation **4**(3), pp. 295-334, 1987.
12. J.W. Klop, *Term Rewriting Systems*, to appear in: Handbook of Logic in Computer

Science, Vol. I (eds. S. Abramsky, D. Gabbay and T. Maibaum), Oxford University Press, 1989.

13. M. Kurihara and I. Kaji, *Modular Term Rewriting Systems: Termination, Confluence and Strategies*, Report, Hokkaido University, Sapporo, 1988. (Abridged version: *Modular Term Rewriting Systems and the Termination*, Information Processing Letters **34**, pp. 1-4, 1990.)

14. M. Kurihara and A. Ohuchi, *Modularity of Simple Termination of Term Rewriting Systems*, Journal of IPS Japan **31**(5), pp. 633-642. 1990.

15. A. Middeldorp, *Modular Aspects of Properties of Term Rewriting Systems Related to Normal Forms*, Proceedings of the 3rd International Conference on Rewriting Techniques and Applications, Chapel Hill, Lecture Notes in Computer Science **355**, pp. 263-277, 1989.

16. A. Middeldorp, *A Sufficient Condition for the Termination of the Direct Sum of Term Rewriting Systems*, Proceedings of the 4th IEEE Symposium on Logic in Computer Science, Pacific Grove, pp. 396-401, 1989.

17. A. Middeldorp, *Confluence of the Disjoint Union of Conditional Term Rewriting Systems*, Report CS-R8944, Centre for Mathematics and Computer Science, Amsterdam, 1989.

18. A. Middeldorp, *Termination of Disjoint Unions of Conditional Term Rewriting Systems*, Report CS-R8959, Centre for Mathematics and Computer Science, Amsterdam, 1989.

19. A. Middeldorp, *Unique Normal Forms for Disjoint Unions of Conditional Term Rewriting Systems*, Report CS-R9003, Centre for Mathematics and Computer Science, Amsterdam, 1990.

20. A. Middeldorp, *Modular Properties of Term Rewriting Systems*, Ph.D. thesis, Vrije Universiteit, Amsterdam, 1990.

21. M. Rusinowitch, *On Termination of the Direct Sum of Term Rewriting Systems*, Information Processing Letters **26**, pp. 65-70, 1987.

22. Y. Toyama, *On the Church-Rosser Property for the Direct Sum of Term Rewriting Systems*, Journal of the ACM **34**(1), pp. 128-143, 1987.

23. Y. Toyama, *Counterexamples to Termination for the Direct Sum of Term Rewriting Systems*, Information Processing Letters **25**, pp. 141-143, 1987.

24. Y. Toyama, J.W. Klop and H.P. Barendregt, *Termination for the Direct Sum of Left-Linear Term Rewriting Systems* (preliminary draft), Proceedings of the 3rd International Conference on Rewriting Techniques and Applications, Chapel Hill, Lecture Notes in Computer Science **355**, pp. 477-491, 1989.

25. H. Zhang and J.L. Rémy, *Contextual Rewriting*, Proceedings of the 1st International Conference on Rewriting Techniques and Applications, Dijon, Lecture Notes in Computer Science **202**, pp. 46-62, 1985.

Implementing Term Rewriting by Graph Reduction: Termination of Combined Systems *

Detlef Plump
Fachbereich Mathematik und Informatik
Universität Bremen, Postfach 33 04 40
D-2800 Bremen 33

Abstract

It is well known that the disjoint union of terminating term rewriting systems does not yield a terminating system in general. We show that this undesirable phenomenon vanishes if one implements term rewriting by graph reduction: given two terminating term rewrite systems $\mathcal{R}_0$ and $\mathcal{R}_1$, the graph reduction system implementing $\mathcal{R}_0 + \mathcal{R}_1$ is terminating. In fact, we prove the stronger result that the graph reduction system for the union $\mathcal{R}_0 \cup \mathcal{R}_1$ is terminating provided that the left-hand sides of $\mathcal{R}_i$ have no common function symbols with the right-hand sides of $\mathcal{R}_{1-i}$ $(i = 0, 1)$.

The implementation is complete in the sense that it computes a normal form for each term over the signature of $\mathcal{R}_0 \cup \mathcal{R}_1$.

1 Introduction

The operational semantics of algebraic specification languages are usually based on term rewriting (see, e.g., [BCV 85], [BHK 89], [EM 85], [FGJM 85], [GH 86]). In this context, confluence and termination are particularly relevant properties of term rewriting systems. Hence, when specifications are structured as combinations of smaller subspecifications, the question arises whether these properties are preserved by a given combination mechanism.

Recently, research in this direction has been started by considering the disjoint union $\mathcal{R}_0 + \mathcal{R}_1$ of two term rewriting systems $\mathcal{R}_0$ and $\mathcal{R}_1$: the rule set of $\mathcal{R}_0 + \mathcal{R}_1$ is the union of the rules of $\mathcal{R}_0$ and $\mathcal{R}_1$ where the function symbols occurring in $\mathcal{R}_0$ and $\mathcal{R}_1$ are disjoint (or are made disjoint by renaming). Toyama [Toy 87a] proves that $\mathcal{R}_0 + \mathcal{R}_1$ is confluent

*Work supported by ESPRIT project #390, *PROSPECTRA*, and by ESPRIT Basic Research Working Group #3264, *COMPASS*.

if both components are confluent. However, the corresponding result for termination does not hold, as the following counterexample [Toy 87b] shows:

$$\begin{aligned} \mathcal{R}_0: &\ \mathtt{f(0,1,x)} \rightarrow \mathtt{f(x,x,x)} \\ \mathcal{R}_1: &\ \mathtt{or(x,y)} \rightarrow \mathtt{x}, \\ &\ \mathtt{or(x,y)} \rightarrow \mathtt{y} \end{aligned}$$

$\mathcal{R}_0$ and $\mathcal{R}_1$ are terminating, while $\mathcal{R}_0 + \mathcal{R}_1$ is not: `f(or(0,1),or(0,1),or(0,1))` reduces in three steps to itself. Examples given by Barendregt and Klop (see [Toy 87b]) and Drosten [Dro 89] show that not even the disjoint union of convergent (i.e. confluent and terminating) systems is guaranteed to terminate.

Sufficient conditions for the termination of the disjoint union of terminating term rewrite systems are given by Ganzinger and Giegerich [GG 87], Rusinowitch [Rus 87], Drosten [Dro 89], Middeldorp [Mid 89a], and Toyama, Klop and Barendregt [TKB 89]. These authors impose various restrictions upon the involved rewrite systems. (See also [Mid 89b] for results on conditional term rewriting systems.)

In this paper we show that these restrictions can be dropped if term rewriting is implemented by graph reduction. More precisely, we use the framework of *jungle evaluation* [HP 88] where terms are represented by (hyper-)graphs and term rewrite rules are translated into graph reduction rules such that graph reduction steps correspond to (parallel) term rewriting steps. Let $\mathcal{G}(\mathcal{R})$ denote the graph reduction system corresponding to a term rewriting system $\mathcal{R}$. Then $\mathcal{G}(\mathcal{R})$ implements $\mathcal{R}$ in the following sense: if a term t is represented by a graph G and $G \overset{*}{\underset{\mathcal{G}(\mathcal{R})}{\Longrightarrow}} H$ is a reduction with H in normal form (i.e., no rule in $\mathcal{G}(\mathcal{R})$ is applicable), then H represents a normal form of t.

In [HP 88] it is shown that $\mathcal{G}(\mathcal{R})$ is terminating if $\mathcal{R}$ is terminating. However, the converse does not hold and we will see that $\mathcal{G}(\mathcal{R}_0 + \mathcal{R}_1)$ is terminating whenever $\mathcal{R}_0$ and $\mathcal{R}_1$ are terminating, no matter whether $\mathcal{R}_0 + \mathcal{R}_1$ is terminating or not. As an example, consider the term rewriting systems $\mathcal{R}_0$ and $\mathcal{R}_1$ shown above. Starting with the tree representation of `f(or(0,1),or(0,1),or(0,1))` we get the following graph reduction steps[1]:

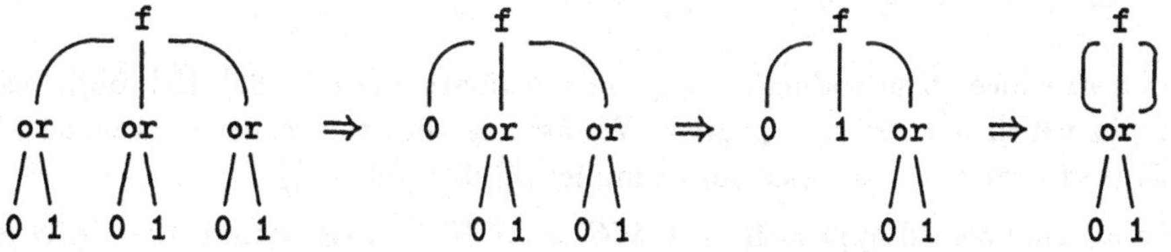

At this stage only one further step is possible: either 

or 

is computed, both representing normal forms of the input term.

The crucial point for the termination of the above reduction is the application of the graph reduction rule corresponding to $\mathtt{f(0,1,x)} \rightarrow \mathtt{f(x,x,x)}$. While two copies of the

[1]Garbage arising from reduction steps is not depicted.

subterm matched by `x` are produced in the term case, graph reduction only introduces two additional pointers to this argument. As a consequence, the three occurrences of `or(0,1)` in `f(or(0,1),or(0,1),or(0,1))` are represented only once and cannot be rewritten independently. Actually, Rusinowitch [Rus 87] shows that the disjoint union of term rewriting systems preserves termination if no such duplicating rules are present.

The disjoint union is a rather restrictive combination mechanism for term rewriting systems, in view of the disjointness requirement. This requirement may be relaxed by considering the union $\mathcal{R}_0 \cup \mathcal{R}_1$ of two term rewriting systems. Sufficient conditions for the termination of $\mathcal{R}_0 \cup \mathcal{R}_1$ are given by Bachmair and Dershowitz [BD 86]. They show that the termination of $\mathcal{R}_0$ and $\mathcal{R}_1$ implies the termination of $\mathcal{R}_0 \cup \mathcal{R}_1$ provided that $\mathcal{R}_0$ is left-linear, $\mathcal{R}_1$ is right-linear, and there is no overlap between the left-hand sides of $\mathcal{R}_0$ and the right-hand sides of $\mathcal{R}_1$. (The overlap condition implies that no right-hand side in $\mathcal{R}_1$ is a single variable.)

In section 3 we consider the union $\mathcal{R}_0 \cup \mathcal{R}_1$ where the left-hand sides of $\mathcal{R}_i$ have no common function symbols with the right-hand sides of $\mathcal{R}_{1-i}$ $(i = 0, 1)$. Our main result states that in this case the termination of $\mathcal{R}_0$ and $\mathcal{R}_1$ guarantees the termination of $\mathcal{G}(\mathcal{R}_0 \cup \mathcal{R}_1)$ without further restrictions. In fact, the termination of $\mathcal{R}_0$ and $\mathcal{R}_1$ is not necessary for the termination of $\mathcal{G}(\mathcal{R}_0 \cup \mathcal{R}_1)$: it suffices that $\mathcal{G}(\mathcal{R}_0)$ and $\mathcal{G}(\mathcal{R}_1)$ are terminating.

This result shows that terminating graph reduction implementations can, with respect to termination, be safely combined as long as the left- and right-hand sides of the corresponding term rewrite systems are disjoint in the described manner. The resulting system is guaranteed to terminate although the underlying term rewrite system may fail to terminate. In particular, this result together with the termination theorem from [HP 88] (see section 2 below) shows that the union $\bigcup_{i=1}^{n} \mathcal{R}_i$ of a family of terminating term rewrite systems $(\mathcal{R}_i)_{i=1,\dots,n}$ has a terminating graph reduction implementation, provided these systems satisfy the above disjointness condition pairwise.

2 Jungle Evaluation

This section provides an introduction to jungle evaluation ([HKP 88], [HP 88]), being the setting in which our result is proved. We assume that the reader is familiar with basic notions of term rewriting (see for example [DJ 90], [Klo 90]).

We pursue a many-sorted approach. Let $SIG = (S, F)$ be a signature, that is, S is a set of sorts and F is a set of function symbols such that each $f \in F$ has a string of argument sorts $s_1 \dots s_n \in S^*$ and a result sort $s \in S$. A function symbol f is written $f : s_1 \dots s_n \to s$ when the argument and result sorts matter.

A ***hypergraph*** $G = (V_G, E_G, s_G, t_G, l_G, m_G)$ over SIG consists of a finite set V_G of nodes, a finite set E_G of hyperedges (or edges for short), two mappings $s_G : E_G \to V_G^*$ and $t_G : E_G \to V_G^*$, assigning a string of source nodes and a string of target nodes to each hyperedge, and two mappings $l_G : V_G \to S$ and $m_G : E_G \to F$, labeling nodes with sorts and hyperedges with function symbols.

A hypergraph G over SIG is a *jungle* if

- the labeling of G is compatible with SIG, that is, for each $e \in E_G$, $m_G(e) = f : s_1 \dots s_n \to s$ implies $l_G^*(s_G(e)) = s$ and $l_G^*(t_G(e)) = s_1 \dots s_n$, [2]
- $outdegree_G(v) \leq 1$ for each $v \in V_G$,
- G is acyclic.

Example. Assume that SIG contains a sort nat and function symbols

$$\begin{aligned} &\mathtt{0} : \to \mathbf{nat}, \\ &\mathtt{succ}, \mathtt{fib} : \mathbf{nat} \to \mathbf{nat}, \\ &+ : \mathbf{nat}\ \mathbf{nat} \to \mathbf{nat}. \end{aligned}$$

Then the following hypergraph is a jungle over SIG.

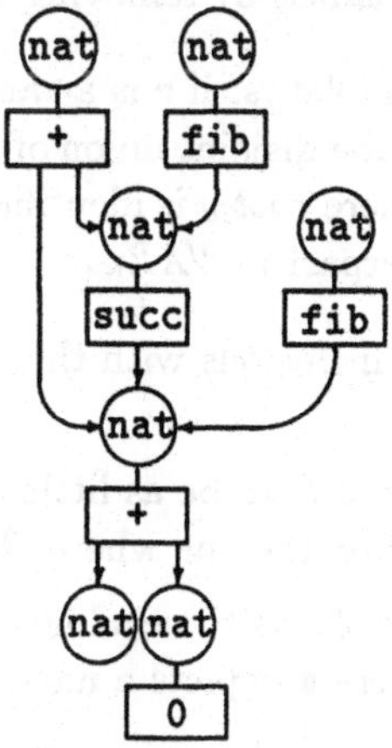

Nodes are drawn as circles and hyperedges as boxes, both with their labels written inside. A line without arrow-head connects a hyperedge with its unique source node, while arrows point to the target nodes (if there are any). The arrows are arranged from left to right in the order given by the target mapping. □

When SIG-terms shall be represented by jungles, nodes without outgoing hyperedge serve as variables. The set of all such nodes in a jungle G is denoted by VAR_G. (Note that as a consequence of this definition each variable occurs only once in a jungle.) The term represented by a node v is then defined by

$$term_G(v) = \begin{cases} v & \text{if } v \in VAR_G, \\ m_G(e)term_G^*(t_G(e)) & \text{otherwise, where } e \text{ is the unique} \\ & \text{hyperedge with } s_G(e) = v. \end{cases}$$

For instance, if v is the left one of the two topmost nodes in the above example, then $term_G(v) = (\mathtt{x}+\mathtt{0})+\mathtt{succ}(\mathtt{x}+\mathtt{0})$ (in infix notation) where $\mathtt{x}$ is the unique variable node in G.

[2] Given a mapping $g : A \to B$, the extension $g^* : A^* \to B^*$ is defined by $g^*(\lambda) = \lambda$ and $g^*(aw) = g(a)g^*(w)$ for all $a \in A$, $w \in A^*$. Here λ denotes the empty string.

For defining reduction rules and reduction steps we need structure preserving mappings between jungles. A *jungle morphism* $g : G \to H$ consists of two mappings $g_V : V_G \to V_H$ and $g_E : E_G \to E_H$ which preserve sources, targets, and labels, that is, $s_H \circ g_E = g_V \circ s_G$, $t_H \circ g_E = g_V^* \circ t_G$, $l_H \circ g_V = l_G$, and $m_H \circ g_E = m_G$.

Now we are ready to translate term rewrite rules into *evaluation rules*. For motivations and further details, however, we refer to [HP 88].

Let $\mathcal{R}$ be a term rewriting system over SIG, i.e., each rule $l \to r$ in $\mathcal{R}$ consists of two SIG-terms l and r of equal sort such that l is not a variable and all variables in r occur already in l. Given a rule $l \to r$ from $\mathcal{R}$, the corresponding evaluation rule $(L \hookleftarrow K \xrightarrow{b} R)$ consists of three jungles L,K,R and jungle morphisms $K \hookrightarrow L$ and $b : K \to R$ such that:

- L is a variable-collapsed tree[3] with $term_L(root_L) = l$.
- K is the subjungle of L obtained by removing the edge outgoing from $root_L$.
- R is constructed from K as follows: if r is a variable, then $root_L$ is identified with the node r; otherwise, R is the disjoint union of K and a variable-collapsed tree R' with $term_{R'}(root_{R'}) = r$ where $root_{R'}$ is identified with $root_L$ and each $x \in VAR_{R'}$ is identified with its counterpart in VAR_K.
- $K \hookrightarrow L$ and $b : K \to R$ are inclusions with the possible exception that b identifies $root_L$ with some variable.

Remark. For simplicity we choose R to be as little collapsed as possible. Other forms of R may be used as well, including the one where R is "fully collapsed" (cf. [HP 88]).

Example. The following picture shows the evaluation rule for the rewrite rule `succ(n) + m` $\to$ `succ(n + m)` (where • depicts a nat-labeled node):

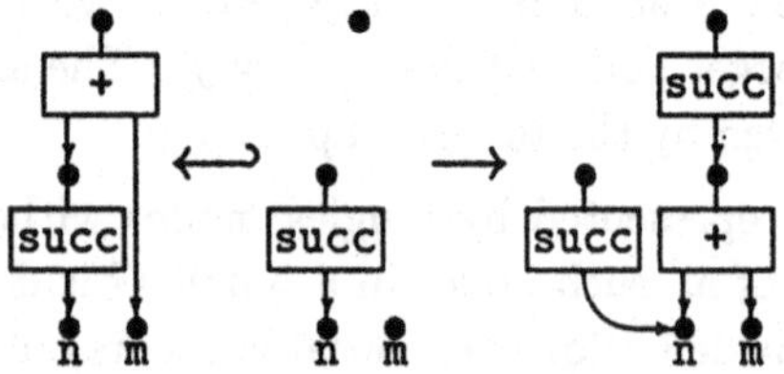

□

The application of an evaluation rule $(L \hookleftarrow K \xrightarrow{b} R)$ to a jungle G requires to find an occurrence of L in G and works then analogously to the construction of K and R described above:

- Find a jungle morphism $g : L \to G$.
- Remove the edge outgoing from $g_V(root_L)$, yielding a jungle D.

[3]That is, there is a unique node $root_L$ with $indegree_L(root_L) = 0$ and only variables may have an indegree greater than 1.

- The resulting jungle H is obtained from D like R is constructed from K: if r is a variable, then $g_V(root_L)$ is identified with $g_V(r)$; otherwise, H is the disjoint union of D and the variable-collapsed tree R' where $root_{R'}$ is identified with $g_V(root_L)$ and each $x \in VAR_{R'}$ is identified with $g_V(x)$.

The set of all evaluation rules for rewrite rules in $\mathcal{R}$ is denoted by $\mathcal{E}$. We write $G \underset{\mathcal{E}}{\Longrightarrow} H$ if H is the result of applying a rule from $\mathcal{E}$ to G and call this an *evaluation step*. For the resulting jungle H only its structure and labeling matters. So each jungle H' isomorphic[4] to H is allowed as a result of this evaluation step as well. We require, however, that variable nodes are not renamed.

Evaluation steps rewrite certain terms represented by a jungle. But there is a problem with evaluation rules for non-left-linear rewrite rules: since the left-hand side L of such a rule is not a tree, there may be no jungle morphism from L to a jungle G although the underlying term rewrite rule is applicable to a term represented by G. This happens if a shared variable in L corresponds to different occurrences of a subterm in G. To overcome this problem we introduce *folding rules* which allow to compress a jungle such that equal subterms are represented by the same node.

For each function symbol in F there is a folding rule ($L \hookleftarrow K \xrightarrow{b} R$):

For each constant symbol $c :\rightarrow s$:

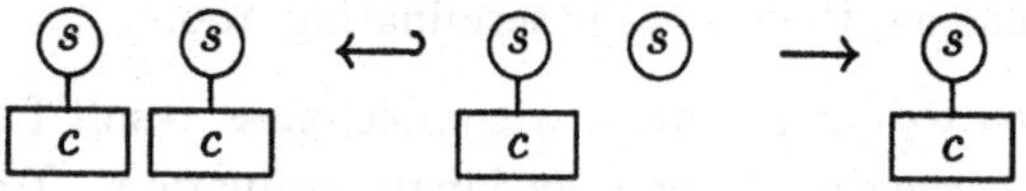

For each function symbol $f : s_1 \ldots s_n \rightarrow s$:

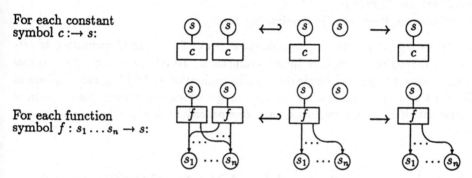

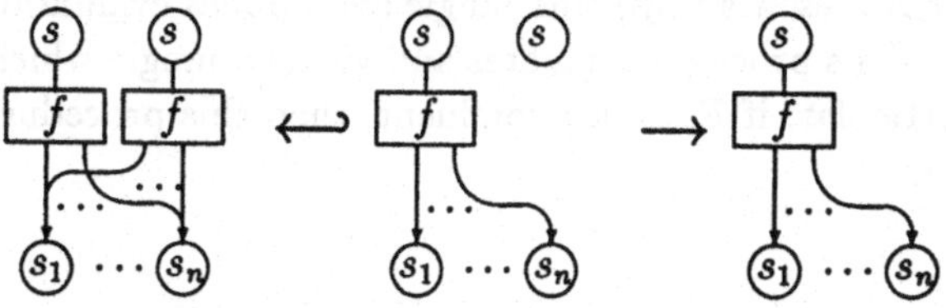

$\mathcal{F}$ denotes the set of all folding rules for F. The application of a folding rule to a jungle G is called a *folding step* and works as follows:

- Find a jungle morphism $g : L \rightarrow G$ with g_E injective.
- Remove $g_E(e)$, where e is the unique edge in $L - K$.
- Identify $g_V(v)$ and $g_V(v')$, where v and v' are the two source nodes in L.

If H is the resulting jungle, then this folding step is denoted by $G \underset{\mathcal{F}}{\Longrightarrow} H$. Again we allow each jungle isomorphic to H as a result of this folding step (but require that variables are not renamed).

The jungles involved in the construction of an evaluation or folding step are related by jungle morphisms in the following way:

[4]Two jungles G and H are said to be *isomorphic*, denoted by $G \cong H$, if there is a jungle morphism $g : G \rightarrow H$ with g_V and g_E bijective.

$$\begin{array}{ccccc} L & \hookleftarrow & K & \xrightarrow{b} & R \\ g\downarrow & & \downarrow & & \downarrow \\ G & \hookleftarrow & D & \xrightarrow{c} & H \end{array}$$

In fact, this diagram represents two pushouts in a category of hypergraphs and hypergraph morphisms (see [Ehr 79] for a category theoretic approach to graph rewriting).

Evaluation and folding steps are called *reduction steps* in the following and for each such reduction step we have a mapping $track : V_G \rightarrow V_H$ which coincides with c_V in the above diagram (note that $V_G = V_D$).

We conclude this section with two main theorems from [HP 88]. Let $\mathcal{G}(\mathcal{R}) = \mathcal{E} \cup \mathcal{F}$ be the set of all evaluation and folding rules for $\mathcal{R}$ and let $\underset{\mathcal{G}(\mathcal{R})}{\Longrightarrow}$ and $\overset{*}{\underset{\mathcal{G}(\mathcal{R})}{\Longrightarrow}}$ denote the reduction step relation and its transitive reflexive closure, respectively.

2.1 Normal Form Theorem
Let $G \overset{*}{\underset{\mathcal{G}(\mathcal{R})}{\Longrightarrow}} H$ be a reduction such that H is in normal form (i.e. no rule in $\mathcal{G}(\mathcal{R})$ is applicable). Then for each node v in G, $term_H(track(v))$[5] is a normal form of $term_G(v)$.

2.2 Termination Theorem
If $\mathcal{R}$ is terminating, then $\mathcal{G}(\mathcal{R})$ is terminating, too.

Both theorems together provide a completeness result for the implementation of terminating term rewrite systems by jungle evaluation. To rewrite a term t to normal form one represents t by some jungle and applies evaluation and folding rules as long as possible. This process terminates and yields a jungle which represents a normal form of t. In particular, if $\mathcal{R}$ is also confluent, then this procedure delivers the unique normal form of t.

3 Termination of Combined Reduction Systems

In this section we prove that combined jungle reduction systems inherit the termination property from their components. For $i = 0, 1$ let $\mathcal{R}_i$ be a term rewriting system with signature $SIG_i = (S_i, F_i)$ and let $\mathcal{G}_i = \mathcal{E}_i \cup \mathcal{F}_i$ be the set of all evaluation and folding rules for $\mathcal{R}_i$. All jungles considered in the following are jungles over the combined signature $(S_0 \cup S_1, F_0 \cup F_1)$.[6]

3.1 Main Theorem
Let the left-hand sides of $\mathcal{R}_i$ have no common function symbols with the right-hand sides of $\mathcal{R}_{1-i}$ $(i = 0, 1)$. Then $\mathcal{G}_0 \cup \mathcal{G}_1$ is terminating if and only if $\mathcal{G}_0$ and $\mathcal{G}_1$ are terminating. □

[5]Here *track* denotes the obvious extension of the *track*-function to sequences of reduction steps.

[6]We assume that each $f \in F_0 \cup F_1$ has unique argument and result sorts in $S_0 \cup S_1$.

Since the "only if"-direction is obvious we assume from now on that $\mathcal{G}_0$ and $\mathcal{G}_1$ are terminating (and that $\mathcal{R}_0$ and $\mathcal{R}_1$ satisfy the disjointness condition of the theorem).

For $i = 0,1$ let $SIG_i^\ominus$ be the subsignature of SIG_i consisting of all function symbols occurring in the rules of $\mathcal{R}_i$ and all associated argument and result sorts; $LEFT(\mathcal{R}_i)$ denotes the set of all function symbols occurring in the left-hand sides of $\mathcal{R}_i$. For each jungle G and $i = 0,1$ let

- G^i be the subjungle[7] consisting of all nodes and edges with label in $SIG_i^\ominus$,
- $\mathcal{L}(G^i)$ be the subjungle consisting of all edges with label in $LEFT(\mathcal{R}_i)$ and all source and target nodes of these edges (hence $\mathcal{L}(G^i)$ is a subjungle of G^i), and
- $G^\#$ be the number of edges with label in $LEFT(\mathcal{R}_0) \cap LEFT(\mathcal{R}_1)$.

For proving theorem 3.1 we exploit the fact that in each reduction over $\mathcal{G}_0 \cup \mathcal{G}_1$ the number $G^\#$ remains constant after a finite number of steps (lemma 3.2). From this point on the application of $\mathcal{G}_i$-rules affects G^{1-i} in a way described by an extension of $\underset{\mathcal{G}_{1-i}}{\Longrightarrow}$ which is still terminating. As a consequence, every infinite reduction over $\mathcal{G}_0 \cup \mathcal{G}_1$ can be transformed into an infinite sequence of extended $\underset{\mathcal{G}_i}{\Longrightarrow}$-steps for some $i \in \{0,1\}$.

3.2 Lemma

For all jungles G, H:

$$G \underset{\mathcal{G}_0 \cup \mathcal{G}_1}{\Longrightarrow} H \text{ implies } G^\# \geq H^\#.$$

Proof

Let $G \underset{\mathcal{G}_0 \cup \mathcal{G}_1}{\Longrightarrow} H$ through a rule $r = (L \hookleftarrow K \xrightarrow{b} R)$. Then $H^\# = (G^\# - L^\#) + R^\#$. So it suffices to show $L^\# \geq R^\#$. Folding rules have this property by construction. If r is an evaluation rule, then $L^\# \geq K^\# = R^\#$ since, by assumption, no symbol in $LEFT(\mathcal{R}_0) \cap LEFT(\mathcal{R}_1)$ occurs in any right-hand side of $\mathcal{R}_0 \cup \mathcal{R}_1$. □

The key to the proof of theorem 3.1 is to consider the "fusion" of two jungle nodes, where at least one of the nodes is a variable. We define a relation $\rightsquigarrow$ on jungles as follows:

$G \rightsquigarrow H$ if H is (isomorphic to a jungle) obtained from G by identification of a variable with some other node.

3.3 Lemma

For $i = 0,1$, the relation $\underset{\mathcal{G}_i}{\Longrightarrow} \cup \rightsquigarrow$ is terminating on SIG_i-jungles.

Proof

Each $\rightsquigarrow$-step decreases the number of variables in a jungle by one while $\underset{\mathcal{G}_i}{\Longrightarrow}$-steps do not change this number. Hence each sequence of $\underset{\mathcal{G}_i}{\Longrightarrow}$- and $\rightsquigarrow$-steps contains only a finite number of $\rightsquigarrow$-steps and the proposition follows from the termination of $\underset{\mathcal{G}_i}{\Longrightarrow}$. □

[7] A jungle U is called a ***subjungle*** of a jungle G, denoted by $U \subseteq G$, if $V_U \subseteq V_G$, $E_U \subseteq E_G$, and if s_U, t_U, l_U, m_U are restrictions of the corresponding mappings of G.

In the following $\mathcal{F}$ stands for the set $\mathcal{F}_0 \cup \mathcal{F}_1$ of all folding rules. It is clear that $\underset{\mathcal{F}}{\Longrightarrow}$ is terminating since each folding step decreases the number of nodes by one.

The next lemma reveals the basic idea for the proof of theorem 3.1. It describes the effect of an $\mathcal{G}_i$-step on the SIG_{1-i}-part of a jungle.

3.4 Lemma

Let $G \underset{\mathcal{E}_i \cup \mathcal{F}}{\Longrightarrow} H$ and $U \subseteq G^{1-i}$ for some $i \in \{0,1\}$. If $G^\# = H^\#$, then there is $\tilde{U} \subseteq H^{1-i}$ such that

$$U \cong \tilde{U} \text{ or } U \rightsquigarrow \tilde{U} \text{ or } U \underset{\mathcal{F}_{1-i}}{\Longrightarrow} \tilde{U}.$$

Proof

Case 1: $G \underset{\mathcal{E}_i}{\Longrightarrow} H$. Let e be the edge in $\mathcal{L}(G^i)$ which is deleted. By assumption, the evaluation step does not create edges with label in $LEFT(\mathcal{R}_{1-i})$. With $G^\# = H^\#$ follows $e \notin \mathcal{L}(G^{1-i})$, as otherwise $G^\# > H^\#$.

We conclude $e \notin G^{1-i}$ since the label of e does not occur in any right-hand side of $\mathcal{R}_{1-i}$. So the evaluation step can change the structure of U only through the identification of two nodes. (That is, if no such identification takes place, then H^{1-i} contains an isomorphic copy $\tilde{U}$ of U.) Assume that there are nodes v_1,v_2 in U with $v_1 \neq v_2$ but $track(v_1) = track(v_2)$. Then either v_1 or v_2 is the source node of e. But this node is a variable in U because $e \notin G^{1-i}$. Hence $U \rightsquigarrow \tilde{U}$ with $\tilde{U} \subseteq H^{1-i}$.

Case 2: $G \underset{\mathcal{F}}{\Longrightarrow} H$. There is a unique surjective jungle morphism $fold : G \rightarrow H$ with $fold_V = track$. Let e_1,e_2 be the two edges in G with $e_1 \neq e_2$ but $fold(e_1) = fold(e_2)$. If $e_1, e_2 \in U$, then the applied folding rule is in $\mathcal{F}_{1-i}$ and we have $U \underset{\mathcal{F}_{1-i}}{\Longrightarrow} fold(U)$ where $fold(U) \subseteq H^{1-i}$. Otherwise, a simple case analysis shows that either $U \cong fold(U)$ or $U \rightsquigarrow fold(U)$. □

To assemble the proof of the main theorem we need one more lemma which allows to restrict evaluation steps to subjungles.

3.5 Lemma

Let $G \underset{\mathcal{E}_i}{\Longrightarrow} H$ and $U \subseteq G^i$ for some $i \in \{0,1\}$. If $\mathcal{L}(G^i) \subseteq U$, then $U \underset{\mathcal{E}_i}{\Longrightarrow} X$ where $\mathcal{L}(H^i) \subseteq X \subseteq H^i$.

Proof

If $\mathcal{L}(G^i) \subseteq U$, then the occurrence of the left-hand side of the applied rule lies in U. Hence there is a restricted step $U \underset{\mathcal{E}_i}{\Longrightarrow} X \subseteq H$ (for a proof see the CLIP-theorem in [Kre 77]). Clearly we have $X \subseteq H^i$. Moreover, $\mathcal{L}(H^i) \subseteq X$ holds since all edges produced by the evaluation step belong to X. □

3.6 Proof of the Main Theorem

Suppose that there is an infinite sequence

$$G_1 \underset{\mathcal{G}_0 \cup \mathcal{G}_1}{\Longrightarrow} G_2 \underset{\mathcal{G}_0 \cup \mathcal{G}_1}{\Longrightarrow} G_3 \underset{\mathcal{G}_0 \cup \mathcal{G}_1}{\Longrightarrow} \cdots$$

The sequence contains infinitely many $\underset{\mathcal{E}_i}{\Longrightarrow}$-steps for some $i \in \{0,1\}$, since $\underset{\mathcal{F}}{\Longrightarrow}$ is terminating. Without loss of generality we may assume that there are infinitely many

$\underset{\mathcal{E}_0}{\Longrightarrow}$-steps. Lemma 3.2 implies that $G_n^{\#}$ remains constant after a finite number of steps. Hence there is an infinite sequence

$$H_1 \underset{\mathcal{E}_0}{\Longrightarrow} H_2 \overset{*}{\underset{\mathcal{E}_1 \cup \mathcal{F}}{\Longrightarrow}} H_3 \underset{\mathcal{E}_0}{\Longrightarrow} H_4 \overset{*}{\underset{\mathcal{E}_1 \cup \mathcal{F}}{\Longrightarrow}} \dots$$

such that $H_n^{\#} = H_{n+1}^{\#}$ for $n \geq 1$.

Consider some step $H \underset{\mathcal{E}_1 \cup \mathcal{F}}{\Longrightarrow} \overline{H}$ in the above sequence and some subjungle U of H satisfying $\mathcal{L}(H^0) \subseteq U \subseteq H^0$. By lemma 3.4 there is $\tilde{U} \subseteq \overline{H}^0$ such that $U \cong \tilde{U}$ or $U \rightsquigarrow \tilde{U}$ or $U \underset{\mathcal{F}_0}{\Longrightarrow} \tilde{U}$. We also have $\mathcal{L}(\overline{H}^0) \subseteq \tilde{U}$ because the reduction step does not create edges with label in $LEFT(\mathcal{R}_0)$. Combined with lemma 3.5, this shows that there is an infinite sequence

$$U_1 \underset{\mathcal{E}_0 \cup \mathcal{F}_0}{\Longrightarrow} U_2 \overset{*}{\rightsquigarrow} U_3 \underset{\mathcal{E}_0 \cup \mathcal{F}_0}{\Longrightarrow} U_4 \overset{*}{\rightsquigarrow} \dots$$

where $U_1 = H_1^0$ and where $\overset{*}{\rightsquigarrow}$ denotes the transitive isomorphic closure of $\rightsquigarrow$. But this contradicts the termination of $\underset{\mathcal{E}_0 \cup \mathcal{F}_0}{\Longrightarrow} \cup \rightsquigarrow$ (lemma 3.3) since all the U_n are SIG_0-jungles. □

References

[BCV 85] M. Bidoit, C. Choppy, F. Voisin: *The ASSPEGIQUE Specification Environment: Motivations and Design.* In H.-J. Kreowski (ed.): *Recent Trends in Data Type Specification.* Informatik-Fachberichte 116, Springer-Verlag (1985)

[BHK 89] J. Bergstra, J. Heering, P. Klint: *Algebraic Specification.* ACM Press (1989)

[BD 86] L. Bachmair, N. Dershowitz: *Commutation, Transformation, and Termination.* Proc. 8th CADE, Lecture Notes in Comp. Sci. 230, 5-20 (1986)

[DJ 90] N. Dershowitz, J.-P. Jouannaud: *Rewrite Systems.* In J. van Leeuwen (ed.): *Handbook of Theoretical Computer Science.* Vol. B, North-Holland (1990)

[Dro 89] K. Drosten: *Termersetzungssysteme.* Informatik-Fachberichte 210, Springer-Verlag (1989) (In German)

[Ehr 79] H. Ehrig: *Introduction to the Algebraic Theory of Graph Grammars.* Proc. 1st Graph Grammar Workshop, Lecture Notes in Comp. Sci. 73, 1-69 (1979)

[EM 85] H. Ehrig, B. Mahr: *Fundamentals of Algebraic Specification 1 – Equations and Initial Semantics.* Springer-Verlag (1985)

[FGJM 85] K. Futatsugi, J. Goguen, J.P. Jouannoud, J. Meseguer: *Principles of OBJ2.* Proc. 1985 Symposium on Principles of Programming Languages, 52-66 (1985)

[GG 87] H. Ganzinger, R. Giegerich: *A Note on Termination in Combinations of Heterogeneous Term Rewriting Systems.* EATCS Bulletin 31, 22-28 (1987)

[GH 86] A. Geser, H. Hußmann: *Experiences with the RAP System: A Specification Interpreter Combining Term Rewriting and Resolution.* Proc. ESOP '86, Lecture Notes in Comp. Sci. 213, 339-350 (1986)

[HKP 88] A. Habel, H.-J. Kreowski, D. Plump: *Jungle Evaluation.* Proc. Fifth Workshop on Specification of Abstract Data Types, Lecture Notes in Comp. Sci. 332, 92-112 (1988)

[HP 88] B. Hoffmann, D. Plump: *Jungle Evaluation for Efficient Term Rewriting.* Proc. Algebraic and Logic Programming, Akademie-Verlag, Berlin (GDR), 191-203 (1988). Also in Lecture Notes in Comp. Sci. 343, (1989). Long version available as technical report no. 4/88, Universität Bremen

[Klo 90] J.W. Klop: *Term Rewriting Systems - From Church-Rosser to Knuth-Bendix and Beyond.* Proc. ICALP '90, Lecture Notes in Comp. Sci. 443, 350-369 (1990)

[Kre 77] H.-J. Kreowski: *Manipulationen von Graphmanipulationen.* Dissertation, TU Berlin (1977)

[Mid 89a] A. Middeldorp: *A Sufficient Condition for the Termination of the Direct Sum of Term Rewriting Systems.* Proc. 4th IEEE Symposium on Logic in Computer Science, 396-401 (1989)

[Mid 89b] A. Middeldorp: *Termination of Disjoint Unions of Conditional Term Rewriting Systems.* Report CS-R8959, Centre for Mathematics and Computer Science, Amsterdam (1989)

[Rus 87] M. Rusinowitch: *On Termination of the Direct Sum of Term Rewriting Systems.* Information Process. Lett. 26, 65-70 (1987)

[TKB 89] Y. Toyama, J.W. Klop, H.P. Barendregt: *Termination for the Direct Sum of Left-Linear Term Rewriting Systems.* Proc. Rewriting Techniques and Applications '89, Lecture Notes in Comp. Sci. 355, 477-491 (1989)

[Toy 87a] Y. Toyama: *On the Church-Rosser Property for the Direct Sum of Term Rewriting Systems.* Journal of the ACM 34, 128-143 (1987)

[Toy 87b] Y. Toyama: *Counterexamples to Termination for the Direct Sum of Term Rewriting Systems.* Information Process. Lett. 25, 141-143 (1987)

Chapter 5

Architecture, Compilers and Parallel Computation

Compiling Concurrent Rewriting onto the Rewrite Rule Machine*

Hitoshi Aida,[†] Joseph Goguen[‡] and José Meseguer
SRI International, Menlo Park CA 94025

Abstract

Following a brief review of the of the Rewrite Rule Machine's concurrent rewriting model of computation, this paper describes a technique for transforming rule sets which eliminates non-left linear and conditional rules. By Theorem 1, this transformation preserves termination and result; it is used as a first phase of compilation. After a brief review of RRM architecture, the second phase of compilation is described, which transforms unconditional left linear rule sets into ensemble micro code. By Theorem 2, this transformation preserves the time complexity of a set of rewrite rules executed in parallel. The paper concludes with a discussion of the ensemble controller's role.

1 Introduction

The Rewrite Rule Machine (RRM) Project resembles many other projects in its goal of building a high performance parallel graph reduction machine for declarative languages. However, it differs significantly from other projects in its choice of architecture and languages.

The RRM architecture is a multi-level hierarchy that encompasses both fine-grain SIMD and coarse-grain MIMD computation, and eliminates the von Neumann bottleneck between processor and memory by processing data where it is stored. The lowest level of the RRM is the **cell**, which combines processing, storage, and communication capabilities. Many cells and a controller are laid out on a chip called an **ensemble**, which operates in SIMD mode. The total RRM is a network of ensembles operating in MIMD mode.

The RRM model of computation, called **concurrent rewriting**, is designed to mediate between this architecture and various programming languages. The software side of the RRM project has been exploring declarative **multi-paradigm** languages which support various combinations of functional, object oriented, and logic programming.

This paper presents some techniques, embodied in a prototype compiler, for compiling equations into SIMD ensemble code, which is then distributed across many ensembles to perform a computation in a globally MIMD mode. Since the techniques are developed for the abstract model of computation, they should be applicable to compiling other languages onto other parallel machines.

The three major stages of the compilation process are as follows:

1. Translate source code into a set of (conditional) rewrite rules. The relatively straightforward translation techniques for functional languages extend to object-oriented and (relational) logic programming and various combinations [18].

2. Eliminate conditions and left-non-linearities from the rules. This is very important for SIMD parallel execution, since finding a match for an (unconditional) left-linear rule only requires local structural checks that can be easily achieved by broadcasting a sequence of SIMD instructions, but this is not at all the case for more general rules; see Section 3.

*Supported by Office of Naval Research Contracts N00014-86-C-0450 and N00014-90-C-0210, and NSF Grants CCR-8707155 and CCR-9007010.

[†]Department of Electrical Engineering, University of Tokyo, 7-3-1 Hongo, Bunkyo-ku, Tokyo 113 Japan; work done at SRI International.

[‡]Oxford University, Programming Research Group, Oxford OX1 3QD, United Kingdom.

3. Translate the resulting rule set into SIMD ensemble code for matching and replacement (as described in Section 5), and for detecting termination (see Section 6).

The main focus of [18] was on step (1) of the process described above; this paper focuses on steps (2) and (3) and provides a backend compiler which can accept as input the translation into rewrite rules of not only functional programs but also of programs in multi-paradigm declarative languages such as those described in Section 1.1 below, and even of programs in conventional languages.

Section 2 discusses concurrent rewriting. Section 3 gives a general technique for eliminating conditional and non-left-linear rules and states precise conditions for its correctness. Section 4 describes the RRM architecture and the SIMD instruction set into which rewrite rules are compiled. Section 5 explains the actual compilation process for both matching and replacement. The parallel complexity of the original program is preserved by the compilation process, as explained in Section 5.3. Section 6 describes the strategy used by the controller to broadcast instructions corresponding to different rules and to detect termination.

1.1 Some Literature

The main difference between the RRM [25, 18, 1, 12] and other proposed parallel architectures for functional languages, including dataflow architectures (see the survey [3]), Alice [6, 20], and Grip [21], is that the RRM has no physical separation between processors and memory, and thus avoids the latency and scalability problems often caused by that separation. Indeed, although there are some other architectures that resemble the RRM, they are not designed for functional languages. These include the ASP machine of Lea [24], and the IUA [31] from Hughes Research Labs, which has a processor called CAAP that plays a role similar to that of the ensemble in the RRM. Both of these machines, like the RRM, have a hierarchical MIMD/SIMD structure, using custom VLSI associative processors. However, these machines were designed for homogeneous applications, particularly signal processing, and cannot be expected to achieve optimum performance for non-homogeneous applications. Also, both machines are programmed with fairly conventional low level languages, and hence can be expected to have difficulties with software reliability, portability and maintenance.

Graph reduction models of computation have been widely used for implementing functional languages; see [23] for a collection of fairly recent papers, and [22] for a general survey. Although much of this work concerns the sequential case, recent work is focusing on parallel models of computation, in order to exploit the full potential of declarative programming. Work in this area includes Lean [5], Dactl [11], and Id [4]; see also [10]. The concurrent rewriting model of computation [14] generalizes the term rewriting model to computations that perform many reductions concurrently, in a way that also supports object-oriented programming and systems programming applications. [18] gives the foundations of some compilation techniques for functional languages like OBJ [8, 13], as well as for multi-paradigm languages, like Eqlog [15, 16] (which combines functional and logic programming), FOOPS (which combines functional with object oriented programming), and FOOPlog [17] (which combines all three paradigms).

More recently, in work by Meseguer [28, 26], concurrent rewriting has been identified with deduction in a logic called *rewriting logic*; a new language called Maude based directly on rewriting logic and containing OBJ as a functional sublanguage has also been proposed in [28] and [27] as a way of unifying functional programming with concurrent object-oriented programming.

The declarative style of these languages is reflected in the fact that the fundamental statements of OBJ and FOOPS are equations, while those of FOOPlog are Horn clauses with equality. In Maude, statements are rewrite rules, which are the basic sentences in rewriting logic.

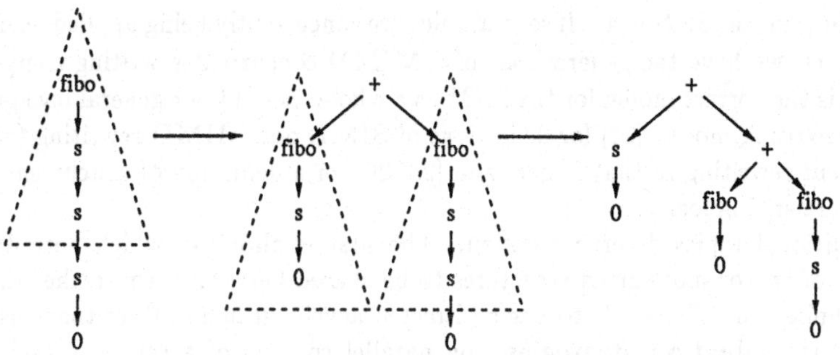

Figure 1: Concurrent rewriting of Fibonacci expressions

2 Concurrent Rewriting Model of Computation

In the term rewriting model of computation, data are **terms**, constructed from a given set of constant and function symbols, and a program is a set of equations that are interpreted as left to right **rewrite rules**. The lefthand side (abbreviated, LHS) and righthand side (RHS) of a rewrite rule are constructed from variables as well as function symbols. A variable can be instantiated with any term of the appropriate sort[1], and a set of instantiations for variables is called a **substitution**.

A term rewriting computation starts with a given term as **data** and a given set of rewrite rules as **program**. Applying a rewrite rule has two phases: **matching** and **replacement**. The matching phase attempts to find a substitution which yields a subterm of the input term when applied to the rewrite rule's lefthand side. Then, in the replacement phase, the matched subterm, called a **redex**, is replaced by the righthand side of the rule, instantiated with the same substitution. Rules are applied to the term until no more matches can be found; the resulting term is called **reduced** and considered to be the final result.

In the concurrent rewriting model of computation, more than one rule can be applied at once, and each rule can be applied to many subterms of the given term at once. Let us explain this by example. Here is a simple program to compute the Fibonacci numbers using basic Peano notation[2].

```
fibo(0) = 0                                      (1)
fibo(s(0)) = s(0)                                (2)
fibo(s(s(N))) = fibo(N) + fibo(s(N))             (3)
```

If you give `fibo(s(s(s(0))))` (i.e. $fibo(3)$) as data, the top node will match rule (3), thus the whole term will be replaced by

`fibo(s(0)) + fibo(s(s(0)))` i.e. $fibo(1) + fibo(2)$.

In the next step, the first `fibo` node will match rule (2), and the second `fibo` will match rule (3) again, thus simultaneous application of these rules yields

`s(0) + (fibo(0) + fibo(s(0)))` i.e. $1 + (fibo(0) + fibo(1))$

in just one step of concurrent rewriting. Figure 1 illustrates these two concurrent rewriting steps for the tree representation of the expressions.

We say that a concurrent rewriting computation is SIMD, when only one rewrite rule is being applied concurrently at each moment; in the RRM, this style of concurrent rewriting is realized by

[1]Although this aspect is not emphasized in this paper, in general, rewrite rules may have type restrictions.

[2]Peano notation is of course inefficient and is used here only for expository purposes. Each RRM cell has an ALU to perform basic arithmetic operations; efficient parallel computation of arbitrary precision arithmetic can also be supported on the RRM [33].

an **ensemble** chip (see Section 4). If several rules are concurrently being applied, each to possibly many instances, we have the general case of a MIMD concurrent rewriting computation; this general case is the correct model for the RRM as a whole. See [14] for general background on the concurrent rewriting model, [12] for definitions of SIMD and MIMD rewriting (called *parallel* and *concurrent* rewriting in that paper) and [28, 26] for a definition of concurrent rewriting as deduction in rewriting logic.

Two additional topics deserve mention. The first is **sharing**, which permits a common substructure of two or more given structures to be shared between them, rather than requiring that it be duplicated. This leads to dag's (directed acyclic graphs) rather than just trees. The second topic is **evaluation strategies**. on parallel consists of a set $\{i_1, ..., i_k\} \subseteq \{1, ..., n\}$ indicating the argument places that should be already reduced before a rewrite rule match for f is attempted. For example, `if_then_else_fi` is typically computed with strategy $\{1\}$, and integer addition with "bottom-up" strategy $\{1, 2\}$. Under relatively mild assumptions, such strategies do not change the semantics of functional computations. Evaluation strategies can also be extended to concurrency control for support systems programming applications, and lazy evaluation [14].

3 Transforming Conditional Rewrite Rules

Matching of conditional and non-left-linear rules involves complicated checks that make such rules difficult to compile directly into SIMD instructions. This section gives a general technique for transforming a program containing rules of this kind into an equivalent one in which such problematic rules have been eliminated.

A **left-linear** rule is one with no repeated variables in its LHS. A non-left-linear rule is

```
N + (- N) = 0
```

which requires checking that the two arguments matched by the variable `N` are identical before performing the replacement. To avoid non-left-linear rules, we can transform them into conditional rewrite rules with an equality operator in their conditions. For example,

```
N + (- M) = 0 if N == M
```

However, conditional rewrite rules also have a similar problem, since the computation of their condition can take an indefinite amount of time. Of course, directly testable conditions, such as arithmetic comparison of simple numbers, can be straightforwardly implemented by RRM instructions. The general solution to this problem is to transform conditional rewrite rules into ordinary rewrite rules by the method described below. We illustrate the process of transforming a program P into an equivalent program $\tilde{P}$ not containing any conditional or non-left-linear rules by an example. A detailed description of the transformation $P \mapsto \tilde{P}$ will be given in [2]; this transformation has some similarities with other transformation techniques for eliminating if-then-else's or conditions proposed previously [7, 9, 32].

Suppose we have the following program:

```
f(0) = 0
f(g(X)) = h(X) if p(X)
f(f(X)) = f(X) if q(X)
```

To transform conditional rewrite rules into ordinary ones, we first introduce a rewrite rule to replace the function symbol `f` by a new function symbol `f'`, with one additional argument corresponding to each condition of one of its conditional rewrite rules. These two extra arguments are initialized to `nil` and are ignored when checking equality.

```
f(X) = f'(X,nil,nil)
```

Next, all occurrences of the function f in the LHS's in the original program are replaced by f' with new free variables in their extra arguments:

```
f'(0,Y,Z) = 0
f'(g(X),Y,Z) = h(X) if p(X)
f'(f'(X,Y',Z'),Y,Z) = f(X) if q(X)
```

Then, we replace each conditional rule by two ordinary rewrite rules: one replaces the nil value in the extra argument position by its corresponding condition; the other does the actual rewrite if the corresponding extra argument is true.

```
f'(g(X),nil,Z) = f'(g(X),p(X),Z)
f'(g(X),true,Z) = h(X)
f'(f'(X,Y',Z'),Y,nil) = f'(f'(X,Y',Z'),Y,q(X))
f'(f'(X,Y',Z'),Y,true) = f(X)
```

The general conditions required of our original program P for the correctness of the transformation $P \mapsto \tilde{P}$ are:

(1) For any conditional rule $t(\vec{x}) = t'(\vec{x})$ if $p(\vec{x})$ in P, and for any ground substitution θ, the term $\theta p(\vec{x})$ always reduces to a unique normal form; the rule is applied only when such a normal form is the constant true[3].

(2) If an n-ary function symbol f is the top symbol of the LHS of a conditional rule in P, then f has a bottom-up strategy $\{1, ..., n\}$ (so that the variable substitution does not change during the evaluation of the condition.)

The correctness of this transformation technique is expressed by the following theorem. Its proof uses noetherian induction and case analysis and will appear in [2].

> **Theorem 1.** Let P be a program satisfying conditions (1) and (2). If we can prove that P terminates, then the transformed program $\tilde{P}$ also terminates, and gives the same results as the original program P if we map each occurrence of the newly introduced function symbols (f' in the example) back to the original symbols (f in the example) by disregarding their extra arguments. □

Condition (2) can be dropped. In this case, the transformation $P \mapsto \tilde{P}$ is somewhat more complicated and yields rules that are potentially nonterminating and must be executed under appropriate fairness assumptions. In this more general context, the treatment of non-left-linear rules presents some difficulties, because testing syntactic equality of two subterms matched by the same variable becomes difficult when they are not reduced and are perhaps being concurrently rewritten at the same time.

4 Rewrite Rule Machine Architecture

The following describes the hierarchical structure of the RRM, and briefly explains the SIMD instructions of an ensemble chip; such SIMD instructions are the elementary operations of an abstract SIMD machine which provides the target for our rewrite rule compiler.

[3] Note that, since the condition itself has to be reduced by rewriting, the rewriting relation for conditional rules is defined recursively.

4.1 Hierarchy of the RRM

The most basic computational element in the RRM is the **cell** [25, 1], which stores one data item with pointers to other cells, and also provides basic computational and communication power; thus cells mix storage, computation and communication. A cell consists of:

- Several **registers** including:
  - `token`, which encodes the function or constant symbol of a data node,
  - `left` and `right`, which point to the descendant argument nodes[4],
  - `marks`, which holds volatile control information (similar to condition codes),
  - `flags`, which holds less volatile information, such as type and reduction status, and
  - four general purpose temporary registers, `ntoken`, `nleft`, `nright` and `nflags`.
- An **ALU** to operate on and test the contents of registers.
- **Interfaces** to communication channels and the controller.

Several (currently 4 is assumed) cells and message buffers make up a **tile**, in which communication channels to other tiles are shared. If two cells in a tile simultaneously attempt to use a communication channel, then there is a conflict, and only one will succeed.

An **ensemble** is a two-dimensional toroidal array of tiles, with a common **controller** to broadcast instructions to all its cells. Each tile is connected to its four horizontal and vertical neighbors via communication channels. Each ensemble is implemented as a VLSI chip. A complete RRM is a network of many ensembles.

A single ensemble yields very fast fine-grain SIMD rewriting, but RRM execution is coarse-grain MIMD at the RRM level, since each ensemble independently executes its own rewrites on its own data, communicating with other ensembles when necessary. This realization of multi-grain concurrency yields high performance without sacrificing programmability.

Recent simulation results show that a single RRM ensemble chip is about 50 times faster than a SUN-3/60 on average, which gives a performance estimate of about 150 MIPS. See [19] for more detailed discussion about the performance of the RRM.

4.2 Cell Operations and the Instruction Set

Each cell in an ensemble executes the SIMD instructions broadcast from the controller. However, only active cells will perform the operation. Some instructions may suffer from communication conflicts or allocation failures during execution. In such cases, the cell is deactivated and a specific bit in the `marks` register (the `failure` bit) is set to indicate the failure.

Each cell also performs some tasks as **autonomous processes**, including automatic data relocation and garbage collection. Generally, SIMD operations have precedence over these autonomous operations. However, reference count management requires that the contents of temporary registers should be kept unchanged until all necessary acknowledgements are received after a `commit` operation (see below). To maintain consistency, SIMD operations that attempt to modify the contents of temporary registers will fail in such cases.

To get information about the status of the computation, the controller can request global feedback from all the cells in the ensemble. Those cells that are active at the moment set a flip-flop, and the controller will get the OR of all the outputs of such flip-flops. In our current design, global feedback will take (less than) three clock cycles to complete.

Here is a short description of SIMD instructions for an ensemble chip, summarized by category:

[4] Unary functions only use `left`, and n-ary functions for $n > 2$ are decomposed into binary ones.

- **Arithmetic/logic instructions** compares the content of a register against another register, perform addition, subtraction, bitwise boolean operations, or simply move data between registers. For example, `(is-equal token '`*value*`)` compares the contents of `token` against *value*; the cell remains active only when they are equal[5].

- **Mark instructions** test or modify a bit in the `marks` register. For example, `(test-mark 'a)` tests a bit called 'a' in the `marks` register; the cell remains active only when it is set.

- **Inter-cell communication instructions** get or put the contents of a register, or test or modify a bit in the `marks` register, of a cell pointed at by one of the registers. For example, `(move token@left ntoken)` copies the content of the `token` register in the cell pointed at by the `left` register into the `ntoken` register (this is a fetch operation). Also, `(test-mark 'a @left)` tests a bit called 'a' in the `marks` register of the cell pointed at by `left` register; the original cell remains active only when it is set. If the `failure` bit in the `marks` register of the cell pointed at by `left` register is set, then the `failure` bit of the original cell is also set.

  In this category of instructions, if the pointed at cell is not in an adjacent tile, the original cell is deactivated with the `failure` bit set and an autonomous data relocation operation is initiated to copy the contents of the pointed at cell to an adjacent tile.

- **Allocate instructions** allocate new cells. For example, `(allocdir nleft ANY)` allocates a new cell in any of the adjacent tiles and stores a pointer to it in the `nleft` register.

- **The `commit` instruction** replaces the contents of `token`, `left`, `right`, and `flags` registers by temporary registers `ntoken`, `nleft`, `nright`, and `nflags`, respectively. This instruction causes autonomous garbage collection to take place. Depending on the choice of the garbage collection mechanism, `commit` may require some operands to maintain correctness.

- **Other instructions** reset the contents of cells, activate cells, etc. For example, `(init)` re-activates all allocated cells.

5 Compiling Rewrite Rules

We describe the compilation of rewrite rules into SIMD ensemble instructions. Note that, using the techniques of Section 3, we may assume that all conditional and non-left-linear rules have already been transformed into unconditional left-linear rules. The two main phases of rewriting are matching of LHS's and replacement of a match by the corresponding RHS instance.

5.1 Matching Phase

For example, to find the LHS pattern of the rewrite rule (3) in Section 2, the following sequence of steps is used:

1. Find all nodes whose function symbol is `fibo`, mark their descendant with 'a' and mark themselves with 'b'.
2. Find all nodes marked 'a' whose function symbol is s, mark their descendant with 'c' and mark themselves with 'd'.
3. Find all nodes marked 'c' whose function symbol is s and mark them with 'e'.
4. Find all nodes marked 'd' with its descendant marked 'e', and mark those nodes with 'f'.

[5] Actually, this is a macro instruction; it decomposes into a sequence of two machine instructions.

5. Find all nodes marked 'b' with its descendant marked 'f', then, those nodes are the root of an instance of the LHS pattern.

or in RRM code,

```
(init) (is-equal token 'fibo) (set-mark 'a @left) (set-mark 'b)
(init) (test-mark 'a) (is-equal token 's) (set-mark 'c @left) (set-mark 'd)
(init) (test-mark 'c) (is-equal token 's) (set-mark 'e)
(init) (test-mark 'd) (test-mark 'e @left) (set-mark 'f)
(init) (test-mark 'b) (test-mark 'f @left) (set-mark 'g)
```

'g' marks roots of the LHS pattern. If an evaluation strategy is provided for the top-level function symbol of the LHS, checking whether corresponding descendant nodes are reduced or not would be included in the code sequence.

In this sequence of code, the possibility of communication conflict is greatly reduced especially when we add code to test whether the top node is reduced or not and perform matching only if the answer is negative. Some further optimizations can be done to decrease the number of marks required in this matching, etc. This top-down matching technique improves upon a previous bottom-up matching technique for an RRM-like abstract machine proposed by Montanari and Goguen [29], that can create many more communication conflicts when implemented on the RRM.

A function symbol's evaluation strategy affects the compiled code by delaying the application of the rewrite rules associated with it until the corresponding argument positions become reduced. This can be easily implemented by checking the 'reduced flag' (see Section 6) of the appropriate descendant nodes in the matching phase.

5.2 Replacement Phase

After matching a LHS, the corresponding RHS structure will be created using allocate instructions, and the old structure will be replaced by the new structure. To be sound, the old structure should be replaced only when the RHS is correctly constructed, and should be replaced atomically using a `commit` instruction.

The currently proposed garbage collection scheme for the RRM requires that all pointers between newly allocated cells should be adjacent. (Pointers from new cells to old cells need not be adjacent.) To satisfy this restriction, the construction of the RHS is done in a top-down manner:

1. Allocate cells for the first level structure at the root of the redex into `nleft` and/or `nright` and mark them.
2. Go to marked cells and allocate one more level into `left` and/or `right` and mark them again.
3. Repeat 2 until all needed cells are allocated.
4. Propagate success or failure of allocation up to the root.

In the example of rewrite rule (3), this yields the code:

```
(init) (test-mark 'g) (allocdir nleft ANY) (move 'fibo token@nleft)
       (allocdir nright ANY) (move 'fibo token@nright) (set-mark 'h @nright)
       (set-mark 'i)
(init) (test-mark 'h) (allocdir left ANY) (move 's token@left) (set-mark 'j)
(init) (test-mark 'i) (test-mark 'j @nright) (set-mark 'k)
```

Alternatively, if a substructure of the LHS also appears in the RHS, the old structure can be reused instead of allocating and building that substructure again.

After allocation of the RHS cells, the pointers to the structures matched by the variables are transferred from LHS to RHS. Again, to preserve correctness, such transfer of variables should be checked for success. For the above example, the code is as follows:

```
(init) (test-mark 'f) (move left@left ntoken) (set-mark 'l)
(init) (test-mark 'k) (test-mark 'l @left) (move ntoken@left ntoken)
       (move ntoken left@nleft) (move ntoken ntoken@nright) (set-mark 'm)
(init) (test-mark 'j) (move ntoken left@left) (set-mark 'n)
(init) (test-mark 'm) (test-mark 'n @nright) (set-mark 'o)
```

The last step of the replacement phase is actual replacement using a commit instruction:

```
(init) (test-mark 'o) (move '+ ntoken) (move flags nflags) (commit 2)
```

5.3 Correctness and Computational Complexity of Compiled Programs

The formal relationship between SIMD rewriting in the model of computation and its realization in an ensemble using the corresponding compiled code can be studied by means of a series of intermediate models of increasing complexity. A useful model capturing many of the essential features of an ensemble is an idealized *cell machine* of the kind described in [12] which has arbitrarily large size and direct connections among all its cells, i.e., each cell can communicate with any other cell without any communication conflicts. Such cells can contain function symbols, pointers to the corresponding arguments and state information to react appropriately to SIMD instructions. Call S the set of all possible states of such a cell machine and I the set of SIMD instructions. The broadcasting of a SIMD instruction causes a change of state according to a state transition function $\delta : I \times S \longrightarrow S$, which extends in the usual way to a transition function on instruction sequences $\delta^b : I^* \times S \longrightarrow S$. In essence, such a transition function is just like the one exhibited by an ensemble, except that in an actual ensemble there will be *communication failures* so that some cells may fail in their attempt at executing an instruction, or size limitations may be exceeded, whereas no such communication failures or space limitations arise in an idealized cell machine.

For each term t there is a set $rep(t)$ of states of the cell machine representing that term in the obvious way, i.e., the tree connections are represented by pointers; of course, for $t \neq t'$ we have $rep(t) \cap rep(t') = \emptyset$. The lack of communication conflicts means that SIMD rewrites will succeed everywhere; this is connected with the notion of a *standard* SIMD rewrite. For a rewrite rule whose LHS does not *overlap* with itself (i.e., does not unify with any of its nonvariable subterms) we say that one step of concurrent rewriting using such a rule is **standard** if all the redexes of the rule are rewritten; this notion can also be defined for rules whose LHS overlaps with itself, but the details are more involved. The correctness of the compilation process relative to the cell machine model can be formulated as follows. Let $t_0 = t_1$ be a rewrite rule, let $w \in I^*$ be the sequence of SIMD instructions for matching and replacement into which such a rule has been compiled, let t be a term, and let $s \in rep(t)$. We say that the compiler is *correct* if for any such rewrite rule, term t, and state $s \in rep(t)$, if t rewrites to t' in one step of standard SIMD rewriting using the rule $t_0 = t_1$, then $\delta^b(w, s) \in rep(t')$. A detailed proof of the correctness of the compilation process in the case of bottom-up matching has been given by P.M. Sewell in [30]; the paper [12] gives formal specifications for the cell machine and discusses semantic issues.

We now turn to computational complexity issues. Although a discussion of MIMD computation would also be very interesting, we restrict ourselves to SIMD computation at the level of the concurrent rewriting model and also at that of an idealized cell machine; using the program transformation techniques proposed in this paper we may assume that all our rewrite rules are unconditional and left linear. We assume that a step of SIMD rewriting takes constant time and say that f is a bound for the time complexity of the SIMD execution of a set of unconditional left-linear rules P if and only if for an input of size[6] n there is a sequence of standard SIMD rewrite steps reducing the input in time $O(f(n))$. We assume that each SIMD machine instruction takes constant time. The compiled version $P^\diamond$ of our original program P consists of a finite set of code segments, i.e., strings $w \in I^*$, one for each rewrite rule in P. We say that f is a bound for the time complexity of $P^\diamond$ running on an idealized cell machine if for any term t of size n there is a sequence of code segments in $P^\diamond$ whose total number of instructions is bounded by $O(f(n))$ and whose execution beginning in a state $s \in rep(t)$ ends in a state s' representing a reduced form of t. Let k be the maximum length code segments in $P^\diamond$ corresponding to rewrite rules in P. The correctness of the code then guarantees that the idealized cell machine will simulate any sequence of standard SIMD rewrites of P with a slowdown factor smaller or equal to k. Therefore, we have

> **Theorem 2.** If f bounds the time complexity of the SIMD execution of a set of rewrite rules P, then it also bounds the time complexity of the execution of $P^\diamond$ on an idealized cell machine. □

A natural question to ask is how realistic such a theoretical result is for approximating the behavior of an actual ensemble. The first obvious observation is that the result is only meaningful for *inputs of limited size* such that they and their intermediate rewrites can fit within the limits of an actual ensemble; for bigger inputs the discussion must be carried out in the context of MIMD computations happening across a network of ensembles. A second factor to take into account is the slowdown caused by communication conflicts and the need for relocating data given that in the actual VLSI implementation communication resources must be shared and no direct physical connection exists among an arbitrary pair of cells. Such factors complicate the picture considerably and are therefore hard to study analytically. Our approach has been to explore them by means of detailed simulations of the ensemble's behavior. Our experimental results in this area have been reported in [19]; they suggest that—always assuming a limited size for the input—communication conflicts and the need of relocating data do not essentially affect the basic SIMD complexity.

6 Controller Strategy and Checking Termination

Because a program usually consists of many rewrite rules, the selection of which rule to apply affects the efficiency and completeness of the execution. To avoid broadcasting unnecessary rules, the controller keeps a list of possibly existing function symbols in the ensemble. Let's call this list L. L will be maintained in such a way that no symbol not appearing in L can appear in the ensemble. L is initially created by analyzing the input data. The controller uses the following strategy:

1. Select a function symbol f from L[7].
2. Broadcast `(init)` `(test-token` f`)` and request a global feedback.
3. If the result of feedback is negative, delete f from L and go to 1.
4. Select a rule for f in which all the function symbols appearing in the LHS are in L.

[6] The size of a term is the number of its nodes.

[7] We don't discuss here the choice of strategy for selecting a function symbol; however, the *fairness* of the selection should always be guaranteed for such a strategy.

5. Broadcast the matching phase.
6. At the end of matching phase, request a global feedback for successful match. If there is no successful match, go to 4. (or 1.).
7. Broadcast the replacement phase.
8. At the end of replacement phase, request a global feedback. If there are successful replacement, add the function symbols appearing in the RHS to L.
9. Go to 4. (or 1.).

Both for efficient execution and to implement evaluation strategy, nodes which cannot be rewritten any more are flagged as *reduced*. Such a flag is first placed on the leaves of the data and is propagated upwards. The function symbols can be classified in two categories:

1. **constructors** which have no corresponding rewrite rules, and
2. real **functions** for which there are one or more rewrite rules in which they appear at the root of the LHS.

Constructor nodes become reduced simply when all of their descendants are reduced. Function nodes should be flagged reduced only when all their descendants become reduced and no rewrite rules are applicable to them. This check can be performed easily and simply, because we can know whether a rewrite rule is possibly applicable or not by testing the `failure` bit when a match is not found; if the `failure` bit is not set, the rule is not applicable.

The computation ends when the root of the data is flagged reduced. The controller can also terminate the execution if L contains only constructor symbols.

7 Future Work

This paper has described the techniques for parallel SIMD execution of rewrite rules, embodied in our prototype compiler. Many interesting topics require further work. One is the extension of these techniques to support object-oriented programming and logic programming. The ideas described in [18] and in [27] provide a good basis, but much more needs to be done.

Another important task ahead is investigating the computational mechanisms at the inter-ensemble level of the RRM. At that level, the major problem is how and when to distribute the data across ensembles. We have recently begun simulation studies of inter-ensemble communication and are currently exploring the necessary communication protocols to support such inter-ensemble computations.

Acknowledgements

We thank Prof. Ugo Montanari for his early contributions to the treatment of parallel matching in the RRM; we also thank our fellow members of the Rewrite Rule Machine project, Mr. Timothy Winkler, Dr. Sany Leinwand and Mr. Patrick Lincoln, for their help in many ways in the process of developing the ideas presented in this paper.

References

[1] Hitoshi Aida, Sany Leinwand, and José Meseguer. Architectural design of the rewrite rule machine ensemble. To appear in Proc. Workshop on VLSI for Artificial Intelligence and Neural Networks, Oxford, September 1990.

[2] Hitoshi Aida and José Meseguer. Transforming conditional rewrite rules into unconditional ones. In preparation.

[3] Arvind and D.E. Culler. Dataflow architectures, 1986. Annual Reviews in Computer Science.

[4] Arvind, Rishiyur Nikhil, and Keshav Pingali. I-structures: Data structures for parallel computing. In Joseph Fasel and Robert Keller, editors, *Graph Reduction*, pages 337–369. Springer-Verlag, 1987. Lecture Notes in Computer Science, Volume 279.

[5] H.P. Barendregt, M.C.J.D. van Eekelen, J.R.W. Glauert, J.R. Kennaway, M.J. Plasmeijer, and M.R. Sleep. Towards an intermediate language based on graph rewriting. In *Proceedings, PARLE Conference*. Springer-Verlag, 1987. Lecture Notes in Computer Science, Volume 259.

[6] J. Darlington and M.J. Reeve. Alice: A multiprocessor reduction machine for the parallel evaluation of applicative languages. In *ACM Conference on Functional Programming Languages and Computer Architecture*. ACM, 1981.

[7] Nachum Dershowitz and David A. Plaisted. Equational programming. In J. Richards, editor, *Machine Intelligence 11: The logic and acquisition of knowledge*, pages 21–56. Oxford University Press, 1988.

[8] Kokichi Futatsugi, Joseph Goguen, Jean-Pierre Jouannaud, and José Meseguer. Principles of OBJ2. In Brian Reid, editor, *Proceedings of 12th ACM Symposium on Principles of Programming Languages*, pages 52–66. ACM, 1985.

[9] E. Giovannetti and C. Moiso. Notes on the elimination of conditions. In Jean-Pierre Jouannaud and Stephane Kaplan, editors, *Proceedings, Conference on Conditional Term Rewriting, Orsay, France, July 8-10, 1987*, pages 91–97. Springer-Verlag, Lecture Notes in Computer Science No. 308, 1988.

[10] J.R.W. Glauert and J.R. Kennaway. A categorical construction for generalised graph rewriting. Submitted to PARLE 89 conference.

[11] J.R.W. Glauert, J.R. Kennaway, and M.R. Sleep. Specification of Dactl. Technical report, School of Information Systems, University of East Anglia, 1987.

[12] J.A. Goguen. Semantic specifications for the rewrite rule machine. In A. Yonezawa, W. McColl, and T. Ito, editors, *Concurrency: Theory, Language and Architecture.* Springer LNCS, 1990.

[13] Joseph Goguen, Claude Kirchner, Hélène Kirchner, Aristide Mégrelis, José Meseguer, and Timothy Winkler. An introduction to OBJ3. In Jean-Pierre Jouannaud and Stephane Kaplan, editors, *Proceedings, Conference on Conditional Term Rewriting, Orsay, France, July 8-10, 1987*, pages 258–263. Springer-Verlag, Lecture Notes in Computer Science No. 308, 1988.

[14] Joseph Goguen, Claude Kirchner, and José Meseguer. Concurrent term rewriting as a model of computation. In R. Keller and J. Fasel, editors, *Proc. Workshop on Graph Reduction, Santa Fe, New Mexico*, pages 53–93. Springer LNCS 279, 1987.

[15] Joseph Goguen and José Meseguer. Eqlog: Equality, types, and generic modules for logic programming. In Douglas DeGroot and Gary Lindstrom, editors, *Logic Programming: Functions, Relations and Equations*, pages 295–363. Prentice-Hall, 1986. An earlier version appears in *Journal of Logic Programming*, Volume 1, Number 2, pages 179-210, September 1984.

[16] Joseph Goguen and José Meseguer. Models and equality for logical programming. In Hartmut Ehrig, Giorgio Levi, Robert Kowalski, and Ugo Montanari, editors, *Proceedings, 1987 TAPSOFT*, pages 1–22. Springer-Verlag, 1987. Lecture Notes in Computer Science, Volume 250; extended version to appear in *J. Logic Programming*.

[17] Joseph Goguen and José Meseguer. Unifying functional, object-oriented and relational programming with logical semantics. In Bruce Shriver and Peter Wegner, editors, *Research Directions in Object-Oriented Programming*, pages 417–477. MIT Press, 1987. Preliminary version in *SIGPLAN Notices*, Volume 21, Number 10, pages 153-162, October 1986; also, Technical Report CSLI-87-93, Center for the Study of Language and Information, Stanford University, March 1987.

[18] Joseph Goguen and José Meseguer. Software for the rewrite rule machine. In *Proceedings, International Conference on Fifth Generation Computer Systems 1988*, pages 628–637. Institute for New Generation Computer Technology (ICOT), 1988.

[19] Joseph Goguen, José Meseguer, Sany Leinwand, Timothy Winkler, and Hitoshi Aida. The rewrite rule machine. Technical Report SRI-CSL-89-6, SRI International, Computer Science Lab, March 1989.

[20] P.G. Harrison and M.J. Reeve. The parallel graph reduction machine, alice. In R. Keller and J. Fasel, editors, *Proc. Workshop on graph reduction, Santa Fe, New Mexico*, pages 181–202. Springer LNCS 279, 1987.

[21] Simon L Peyton Jones, Chris Clack, Jon Salkild, and Mark Hardie. GRIP - a high-performance architecture for parallel graph reduction. In Kahn, editor, *Proceedings, IFIP Conference on Functional Programming Languages and Computer Architecture*, pages 98–112. Springer-Verlag, Portland, September 1987. Lecture Notes in Computer Science, Volume 274.

[22] Simon Peyton Jones. *The Implementation of Functional Programming Languages.* Prentice-Hall, 1987.

[23] Robert Keller and Joseph Fasel, editors. *Proceedings, Graph Reduction Workshop.* Springer-Verlag, 1987. Lecture Notes in Computer Science, Volume 279.

[24] R.M. Lea. ASP modules: Cost-effective building blocks for real-time dsp systems. *J. of VLSI Signal Processing*, 1:69–84, 1989.

[25] S. Leinwand, J.A. Goguen, and T. Winkler. Cell and ensemble architecture for the rewrite rule machine. In *Proceedings of the International Conference on Fifth Generation Computer Systems, Tokyo, Japan*, pages 869–878. ICOT, 1988.

[26] José Meseguer. Conditional rewriting logic: deduction, models and concurrency. This volume.

[27] José Meseguer. A logical theory of concurrent objects. In *ECOOP-OOPSLA'90 Conference on Object-Oriented Programming, Ottawa, Canada, October 1990*, pages 101–115. ACM, 1990.

[28] José Meseguer. Rewriting as a unified model of concurrency. In *Proceedings of the Concur'90 Conference, Amsterdam, August 1990*, pages 384–400. Springer LNCS Vol. 458, 1990.

[29] Ugo Montanari and Joseph Goguen. An abstract machine for fast parallel matching of linear patterns. Technical Report SRI-CSL-87-3, Computer Science Lab, SRI International, May 1987.

[30] P.M. Sewell. Cell machine correctness via parallel jungle rewriting. MSc Thesis, Programming Research Group, University of Oxford, 1990.

[31] D.B. Shu, J.G. Nash, and C.C. Weems. A multiple-level heterogeneous architecture for image understanding. In *Parallel Architectures and Algorithms for Image Understanding.* Academic Press, 1990. To appear.

[32] G. Sivakumar. *Proofs and Computations in Conditional Equational Theories.* PhD thesis, CS Dept., U. Illinois at Urbana, 1989.

[33] Timothy Winkler. Numerical computation on the RRM. Technical report, SRI International, Computer Science Lab, November 1988. Technical Note SRI-CSL-TN88-3.

Design Strategies for Rewrite Rules

Sergio Antoy

Virginia Polytechnic Institute and State University
Northern Virginia Graduate Center
Falls Church, VA 22042

Abstract: Term rewriting systems employed for the specification of abstract data types and for very high level programming languages based on rewrite rules are often required to have properties, such as confluence, termination, and sufficient-completeness, which are undecidable. Rather than attempting, as it is usually done, to check these properties in a system *a posteriori*, i.e. after the rules have been designed, we propose two strategies for addressing this problem *a priori*, i.e. during the design phase of rules. We propose the concepts of under- and over-specification, determine sufficient and/or necessary conditions to avoid them, show how to obtain these conditions in a constructive way, and relate them to a number of desirable properties which have appeared in the literature. Our approach is based on the completeness and parsimony properties of sets of tuples of terms and on a recursive mechanism which extends primitive recursion from natural numbers to abstract data types. We prove a number of results, illustrate their application to the design of rewriting systems by means of examples, and discuss the power and the limitations of our approach.

Keywords: Rewriting systems, Algebraic specifications, Under- and Over-specification, Completeness, Parsimony, Termination, Confluence, Linearity, Sufficient-completeness, Binary choice procedure, Recursive reduction, Rewrite rule design.

1. Introduction

Equational axioms and rewrite rules are used for the specification of algebraic structures, such as monoids or groups [KNUT70, EHRI85], abstract data types [GOGU78, GUTT78b], and very high level programming languages, such as OBJ [FUTA84, GOGU88]. For applications in computer science, specifications are often required to have some desirable properties, such as confluence [HUET80b], termination [DERS85], sufficient-completeness [GUTT78c, KAPU87], etc. Unfortunately, many of these properties are undecidable. We present and analyze two methods for designing rewrite rules. The problem of determining whether a system $\mathcal{R}$ satisfies some desirable properties can be dealt with more easily when $\mathcal{R}$'s rules have been designed with the methods we propose.

This paper is organized as follows. In Section 2 we introduce terminology and notation. In particular we define a term rewriting system as a 6-tuple and introduce a minimal language for the presentation in a more readable form of the system's components. In Section 3 we introduce the concepts of under- and over-specification, we argue why these conditions should be avoided in a system, we relate these concepts to two properties of sets of tuples of term, i.e. completeness and parsimony, and we propose a strategy for the design of complete and parsimonious sets. In Section 4 we introduce a strategy for the definition of recursive operations and we prove that operations designed through the combined use of this and the previous strategy are neither under- nor over-specified. In Section 5 we discuss additional results concerning the design strategies and relate them to other properties of term rewriting systems. These results focus on what is possible and what is impossible to do through the design strategies and are an attempt to evaluate their power and their limitations. In Section 6 we compare our results to similar or related ideas which have appeared in the literature. The example presented in the discussion are selected by merit of simplicity. In the appendix we show some more significant examples.

This material is based upon work supported by the National Science Foundation Grant No. CCR-8908565.

Current address: Dept. of Computer Science, Portland State University, P.O.Box 751, Portland, OR 97207.

2. Preliminary Concepts

A term rewriting system $\mathcal{R}$ is a 6-tuple $\langle \mathcal{S}, \mathcal{C}, \mathcal{D}, \tau, \mathcal{Q}, \mathcal{A} \rangle$, where the various components are defined as follows. $\mathcal{S}$ is a set of symbols called *sorts*. $\mathcal{C}$ is a set of symbols called *constructors*. $\mathcal{D}$ is a set of symbols called *(defined) operations*. $\mathcal{C}$ and $\mathcal{D}$ are disjoint. τ is a mapping from $\mathcal{C} \uplus \mathcal{D}$ into $\mathcal{S}^+$ called *arity*, where $\mathcal{S}^+$ denotes the set of non-null strings over $\mathcal{S}$. With τ in mind, $\mathcal{C} \uplus \mathcal{D}$ is denoted by Σ and referred to as a *signature*. For any f in Σ, $\tau(f)$ is denoted as $s_1 \times \ldots \times s_n \to s$ or simply as s when $n = 0$. s and $s_1 \times \ldots \times s_n$ are respectively called *range* and *domain* of f. $\mathcal{Q}$ is a set of rewrite rules, called *quotient rules*, in which there are no occurrences of defined operations. $\mathcal{A}$ is a set of rewrite rules, called *defining rules*, characterized by the property discussed below. $\mathcal{Q}$ and $\mathcal{A}$ are disjoint and their union is denoted by $\mathcal{E}$.

We recall that all the variables occurring in the right side of a rewrite rule must occur in the left side too. $T(\bullet)$ denotes the set of *terms* built over a signature denoted by the symbol $\bullet$. $T(\bullet)_s$ is the subset of $T(\bullet)$ of all the terms of sort s. $\mathcal{X}$ denotes a set of sorted *variables*. An element in $T(\mathcal{C} \cup \mathcal{X})$ is a *constructor term*. A term t of the form $f(x_1, \ldots, x_n)$, where f is an operation and the arguments are constructor terms is an *f-rooted constructor term* and f is the *root* of t. A rule *defining* f is characterized by having an f-rooted constructor term as left side. We assume that all terms respect τ (are well formed).

The presentation of a rewrite system in the form of a 6-tuple is difficult to understand, because conceptually close pieces of information are textually separated. We use a small language for presenting a system in a more accessible way. A sort is declared together with its constructors, the domain of each constructor (the range is the sort in which it is declared), and quotient rules, if any. A defined operation is declared together with its arity and all its defining rules. Natural and conventional notations for the signature symbols are preserved by allowing the use of infix and "mixfix" [FUTA84, GUTT85a] operators. The declaration of the domain of any such symbol supplies a template for these characteristics. We ignore problems of precedence, associativity, and parsing, since they are irrelevant to our discussion. We do not declare the sort of variables for reasons which will become clear later.

sort *nat*	(1)
constructors	(2)
0;	(3)
succ(*nat*).	(4)
operation *max*(*nat*, *nat*) → *nat*	(5)
axioms	(6)
max(*0*, *0*) → *0*;	(7)
max(*succ*(*X*), *0*) → *succ*(*X*);	(8)
max(*succ*(*X*), *succ*(*Y*)) → *succ*(*max*(*X*, *Y*)).	(9)
operation *min*(*nat*, *nat*) → *nat*	(10)
axioms	(11)
min(*0*, *Y*) → *0*;	(12)
min(*X*, *0*) → *0*;	(13)
min(*succ*(*X*), *succ*(*Y*)) → *succ*(*min*(*X*, *Y*)).	(14)

Figure 1. Definition of the sort *nat*, modeling the natural numbers, and and the operations *max* and *min* intended for specifying or computing the maximum and minimun of two numbers. The defining rules of both operations are somewhat flawed.

Any variable which occurs only once in a rule is called *anonymous* and is optionally denoted by an underscore symbol. Generic sorts are defined through the use of some parameter sorts. A sort s is a *parameter* of a set of rules when there are no occurrences of the constructors of s in $\mathcal{E}$. This form of modularity is rather rudimentary when compared with those suggested, e.g., in OBJ [FUTA84, GOGU88], the Larch shared language [GUTT85a], ACT ONE [EHRI85], or ASF [BERG89], and it is proposed here only to show how our ideas accommodate the needs of parameterization. For example, the sort *list* (Figure 2) is parameterized by the sort *element* — the sort of the elements of a list.

3. Completeness and Parsimony — Binary Choices

In Figure 1 the operation *max* is not specified enough in the sense

sort *integer*	(1)
constructors	(2)
0;	(3)
succ(*integer*);	(4)
pred(*integer*);	(5)
quotient	(6)
succ(*pred*(*X*)) → *X*;	(7)
pred(*succ*(*X*)) → *X*.	(8)
sort *element* **parameter**.	(9)
sort *list*	(10)
constructors	(11)
[];	(12)
[*element*|*list*].	(13)

Figure 2. Additional features of the small language: quotient rules, lines 6-8, and parameter sorts, line 9. The constructors of *list* are denoted in standard Prolog notation. In our discussion, the symbols *0* and *succ* are unrelated to the symbols with the same name of Figure 1.

that, e.g., from $max(0, succ(0))$ we cannot derive a constructor term of sort *nat*.

Definition 1. An operation f is *under-specified* if there exists some ground f-rooted constructor term which cannot be rewritten to some ground constructor term. An *under-specified* system is one containing an under-specified operation.

Lemma 2. If a system $\mathcal{R}$ is not under-specified, then any ground term in $\mathcal{R}$ is reducible to a constructor term.

Proof. Let t be a term in $T(\Sigma)$. The proof is by strong arithmetical induction on the height of t. Base case: either $t = c$ for some constant contructor c or $t = f$ for some constant operation f. If $t = c$, then the claim is already satisfied, If $t = f$, then since f is not under-specified, t is reducible to a constructor term. Inductive case: Let t be a term of positive height and assume the claim for all its proper subterms. Without loss of generality we assume that $t = f(t_1, \ldots, t_k)$, for some defined operation f and subterms $t_1, \ldots, t_k$. By the inductive hypothesis, each t_i, $1 \le i \le k$ is reducible to a constructor term t'_i. Thus t is reducible to $f(t'_1, \ldots, t'_k)$ which is a ground f-rooted constructor term and since f is not under-specified is reducible to a constructor term. □

Over-specification is somewhat the opposite of under-specification. It arises when we say too much, rather than too little, about an operation and it is sometimes the consequence of a naive approach to avoid under-specification. The operation *min* is over-specified in the sense that the term $min(0, 0)$ can be reduced by two distinct rules defining *min*. If the right side of the rule at line (12) were, e.g., $succ(succ(0))$ the confluence of the system would be lost and the completion [KNUT70] of the system would require the additional rule $succ(succ(0)) \to 0$.

Definition 3. An operation f is *over-specified* if there exists some ground f-rooted constructor term matched by two distinct defining rules. An *over-specified* system is one containing an over-specified operation.

We now introduce two key concepts of our discussion, they are useful to address under- and over-specification in a constructive way.

Definition 4. Given a sequence of sorts $s = s_1 \ldots s_k$, $k > 0$ a set $R = \{t_1, \ldots, t_n\}$, $n \ge 0$, of k-tuples $t_i = \langle t_{i1}, \ldots, t_{ik} \rangle$ of constructor terms such that the sort of t_{ij} is s_j, $1 \le i \le n$, $1 \le j \le k$, is *complete for s* (we drop the qualification "for s" when s is understood) if every k-tuple $u = (u_1, \ldots u_k)$ of ground constructor terms such that the sort of u_j is s_j, $1 \le j \le k$, is an instance of at least one element of R. R is *parsimonious* if u is an instance of at most one element of R. An operation f is *complete* or *parsimonious* if the set of tuples of terms consisting of the arguments of f in the left sides of f's defining rules is respectively complete or parsimonious for the domain of f. A system is *complete* or *parsimonious* if all its defined operations are respectively complete or parsimonious.

The relationships between completeness and under-specification and between parsimony and over-specification are captured by the following lemmas whose proofs are immediate.

Lemma 5. The completeness of a system is a necessary condition to avoid under-specification.

Lemma 6. The parsimony of a system is a sufficient condition to avoid over-specification.

Completeness alone is not strong enough to avoid under-specification. For an example, consider a binary operation $f : s \times s \to s$, for some sort s, whose only defining rule is:

$$f(X,Y) \to f(Y,X) \qquad (3.3)$$

f is clearly complete, but grossly under-specified if $T(C)_s$ is not empty. Parsimony takes care of over-specification, but is not strong enough to provide confluence. Sufficient conditions to avoid under-specification and/or to provide confluence need further work and will be discussed in the next sections.

The definitions of completeness and parsimony are non-constructive. We now describe a procedure for generating complete and parsimonious sets of tuples. The procedure is non-deterministic, since it contains some indefinite selections and choices. Also, the termination of the procedure for arbitrary choices is not guaranteed. In practical applications, these selections and choices correspond to design decisions and the termination of the procedure is an expected consequence of the designer's strategy. □ is a distinguished symbol, called *place*, which is neither in Σ nor in $\mathcal{X}$. The symbol □ is overloaded, since its *occurrences* are sorted. We resolve the overloading by explicitly stating the sort of a place when it cannot be inferred from the context.

Binary Choice Procedure.

Input:	S:	a set of sorts,
	C:	a signature of constructors,
	$s_1, \ldots, s_k$:	a sequence of sorts.
Output:	R:	a set of k-tuples of constructor terms.

1. [Initialize] Initialize R to a set consisting of a single k-tuple whose components are all places. Let s_i be the sort of the i-th place, for $1 \le i \le k$.
2. [Test] If there are no occurrences of places anywhere in R, then halt with output R.
3. [Select] Select a place in some tuple t of R. Let u be the occurrence of the selected place and s its sort.
4. [Choose] Choose either "variable" or "inductive" for the place at u.
5. [Update] If the binary choice is "variable", then replace in t the place at u with a fresh variable, else remove t from R and for each constructor c, of arity $s_1 \times \ldots \times s_n \to s$, for some $s_1, \ldots, s_n$ in S, insert in R a tuple obtained by replacing in t the place at u with the term $c(□, \ldots, □)$ in which there are exactly n places and the sort of the i-th place, for $1 \le i \le n$, is s_i.
6. [Iterate] Go to step 2).

The initial template, 1, results from the domain of *length*.

length(■) (1)

Templates 2 and 3 come from 1 with choice "inductive".

length([]) (2)
length([■|□]) (3)

Template 5 comes from 3 with choice "variable", 4 is 2.

length([]) (4)
length([X|■]) (5)

Template 7 comes from 5 with choice "variable", 6 is 4.

length([]) (6)
length([X|Y]) (7)

Figure 3. Coding of the left sides of the rules defining the length of a list using an implementation of the *binary choice* procedure. At every iteration of the interactive process an occurrence of □, called *cursor* and denoted with ■, is selected and the user prompted for a choice. The implementation selects places in lexicographic order, but the user may override this strategy. The coding is completed in just three keystrokes and both the completeness and parsimony of the operation are guaranteed.

We use an implementation of the binary choice procedure to show its application to the design of the left sides of defining rules. The procedure has been implemented, within the *GNU Emacs* editor [STAL87] in the style of [HALM88], as a template-driven interactive process. A simplified session of the design of the operation computing the length of a list is presented in Figure 3. For clarity, the definition of the right sides of the rules is omitted from the description. The right sides of the rules will be completed later. The initial template is automatically generated from the domain of *length*. The constructors of *list* are derived from its declaration, see Figure 2. A heuristic for guiding the choices is as follows. Ideally, we would like to fill all the places of a template with variables. If doing so prevents us from defining the right side of a rule, we select a place, choose "inductive" for it and try again.

Our interest in the binary choice procedure is motivated by the next theorem. First, we discuss the termination of the procedure.

Lemma 7. If the number of "inductive" choices is bounded, the binary choice procedure terminates.

Proof. At every iteration of the procedure the number of occurrences of places in R is either decremented by 1 if the choice is "variable" or incremented by a finite amount if the choice is "inductive". If the number of "inductive" choice is bounded, eventually the number of occurrences of places in R will decrease to 0 and the procedure terminate. □

Theorem 8. The output of the binary choice procedure is a complete and parsimonious set.

Proof. Let R denote the output of the binary choice procedure.
– Completeness: [HUET82] proposes an inductive, *operative* definition of completeness for linear tuples of constructor terms. R

length([]) → *0*	(1)
length([X|Y]) → *succ*(!)	(2)

Figure 4. Rules for the operation *length* whose left sides were defined in Figure 3. The symbol "!" in line 2 denotes the *recursive reduction* of the left side and stands for *length*(Y). The names of the variables of rule 2 do not convey any information, since they occur in only one occurrence. Our notation makes X and Y anonymous, hence they should be replaced by an underscore symbol.

satisfies this definition, hence by [HUET82, Lemma 3] R also satisfies our definition of completeness.

– Parsimony: Let t_1 and t_2 be tuples of R with a common instance, and let σ_1 and σ_2 be substitutions such that $\sigma_1(t_1) = \sigma_2(t_2)$. If $t_1 \neq t_2$, then for some occurrence u, selected in step 3 of the binary choice procedure, the roots of t_1 and t_2 at u are different constructors, hence $\sigma_1(t_1) \neq \sigma_2(t_2)$. Thus, $t_1 = t_2$ and R is parsimonious. □

4. Termination — Recursive Reductions

In this section, we relate under-specification to recursion and termination.

Definition 9. A sort s is *recursive* if there exists a constructor c such that $\tau(c) = s_1 \times \ldots \times s_n \rightarrow s$ and $s_i = s$ for some i such that $1 \leq i \leq n$. c and its i-th argument are called *recursive* too.

Example 10. The sorts *nat* and *list* previously presented are recursive — *succ* and the mixfix symbol "[|]" denote their respective recursive constructors. Only the second argument of "[|]" is recursive.

In order to employ recursion in the design of a rewrite rule there must be an occurrence of a recursive constructor in the rule's left side. Initially, for simplifying discussion and notation, we impose the additional condition that each recursive constructor has only one recursive argument. This restriction will be relaxed later.

Definition 11. We call *recursive reduction* the function $\rho : T(\mathcal{C} \cup \mathcal{X}) \rightarrow T(\mathcal{C} \cup \mathcal{X})$ defined as follows:

$$\rho(t) = \begin{cases} t, & \text{if } t \text{ is a variable} \\ & \quad \text{or a constant constructor;} \\ \rho(t'), & \text{if } t = c(\ldots, t', \ldots), \text{ where} \\ & \quad c \text{ is a recursive constructor and} \\ & \quad t' \text{ is its recursive argument;} \\ c(\rho(t_1), \ldots, \rho(t_k)), & \text{if } t = c(t_1, \ldots, t_k) \text{ and} \\ & \quad c \text{ is not recursive.} \end{cases} \tag{4.1}$$

The recursive reduction of a term t "strips away" all occurences of recursive constructors in t. More precisely, each subterm of t whose root is a recursive constructor is (recursively) replaced by its recursive argument. The definition (4.1) is extended by (4.2) to include in its domain operation-rooted constructor terms, in particular, the left sides of defining rules designed with the binary choice procedure.

$$\rho(f(x_1, \ldots, x_k)) = f(\rho(x_1), \ldots, \rho(x_k)) \tag{4.2}$$

We also extend with this concept the language we use for presenting rewriting systems. The symbol "!" on the right side of a rule $\alpha \rightarrow \beta$, such that α has some occurrences of recursive constructors, denotes the recursive reduction of the left side, see Figure 4. If there are no occurrences of recursive constructors in α the use of "!" in β is illegal.

We now discuss sufficient conditions to avoid under-specification and show how the combined use of the binary choice and the recursive reduction strategies provides these conditions.

Theorem 12. If f is a defined operation such that: 1) f is complete, 2) any operation g different from f occuring in the right side of a rule defining f is not under-specified, and 3) any f-rooted (sub)term occurring in the right side of a rule defining f is the recursive reduction of the corresponding left side, then f is not under-specified.

Proof. The proof is by strong arithmetical induction on the number of occurrences of recursive constructors in a ground f-rooted constructor term. If t is a ground f-rooted constructor term, then r_t denotes the number of occurrences of recursive constructors in t. Notice that the first condition on f implies that any f-rooted constructor term is matched by some rule, say $\alpha \rightarrow \beta$, defining f. Base case: If $r_t = 0$, then, by the third condition on f t is reduced to some term t' in which there are no occurrence of f and, by the second condition on f and Lemma 2, t' is eventually reduced to some ground constructor term. Inductive case: Assume the claim true for any term in which there are up to r occurrences of recursive contructors, $r \geq 0$. Let t be a term such that $r_t = r+1$. If there are no occurrences of f in β, then, by the second condition on f and Lemma 2, t is eventually reduced to some ground constructor term. Otherwise, t is reduced to a term t' in which there are some occurrences of recursive reductions. If t'' is the subterm at any such occurrence, by definition of recursive reduction, $r_{t''} \leq r$. By the inductive hypothesis, t'' can be reduced to a ground constructor term, thus also t' and t can be reduced to a ground constructor term. Hence, f is not under-specified. □

Corollary 13. A system $\mathcal{R}$ designed with the binary choice procedure and with recursive reductions as the only recursion mechanism is neither under- nor over-specified.

Proof. By Theorem 8, $\mathcal{R}$ is complete and parsimonious. Thus, by Lemma 6, $\mathcal{R}$ is not over-specified. Any operation in $\mathcal{R}$ satisfies the three conditions of Theorem 12, thus $\mathcal{R}$ is not under-specified. □

The notion of recursive reduction extends that of primitive recursion from natural numbers to abstract data types. When recursive reductions are applied to complete operations the resulting mechanism is at least as powerful as primitive recursion. The template proposed in [MINS67, p. 175] to define primitive recursion can be trivially reformulated using our strategies. Although primitive recursion bans mutual recursion, this is not a substantial limitation. We share [GUTT78c]'s viewpoint that "In programming, the occasions when one has need of a non-primitive recursive function seem to be very rare indeed."

When constructors have multiple recursive arguments, Definition 11 is not appropriate since the second case of equation 4.2 would imply a non-deterministic choice of one recursive argument. For constructors with multiple recursive arguments it seems more appropriate to define the recursive reduction of a term as a set of terms. Consequently, the use of recursive reductions in a rule requires a convenient notation to indicate the selection of one element in this set. Figure 5 proposes a plausible notation for this situation. The results we have stated in this section continue to hold when some constructors have multiple recursive arguments.

5. Power and Limitations of the Strategies

The design strategies we have presented in the previous sections are suited for an incremental methodology for the design

sort *node* **parameter**; (1)

sort *bin_tree* (2)

constructors (3)

empty; (4)

make(*node*, *bin_tree*, *bin_tree*). (5)

operation *leaf_count*(*bin_tree*) → *nat* (6)

axioms (7)

leaf_count(*empty*) → *succ*(*0*); (8)

leaf_count(*make*(_, *L*, *R*)) → !*L*+!*R*. (9)

Figure 5. Specification of the sort modeling binary trees and the operation "count the number of leaves." The recursive constructor *make* has two arguments of sort *bin_tree*, a condition which violates an assumption made in the definition of *recursive reduction*. In this situation the left side of rule 9 has two distinct recursive reductions. They are denoted with !*L* and !*R* which stand respectively for *leaf_count*(*L*) and *leaf_count*(*R*).

of software abstractions and their specifications by a rewriting system. The methodology is called "Stepwise Specification by Extensions" in [EHRI85]. With the strategies we proposed the initial step consists in enumerating the sorts of a data abstraction, and for each sort its constructors and any mutual relationship existing between them. The initial step results in a system without operations and consequently defining rules. Subsequent steps consist in extending or enriching [GOGU78], with new defined operations, a system obtained in a previous step. This extension should be conservative, i.e. complete and consistent in the following sense. Intuitively, in a complete extension no new data elements are added in the initial semantics. If the extension is consistent, then in the initial semantics no old data elements are identified [EHRI85]. Incompleteness, and particularly inconsistency, are almost invariably unacceptable, e.g. see the clauses *protecting* and *extending* in OBJ [FUTA84, GOGU88]. The relationships between the conservativeness of an extension and the strategies we propose will be discussed shortly.

For non-trivial data abstractions, the effort required in the initial step is generally small relative to the total effort required for completing the whole abstraction. Many types involved in a data abstraction are based on well-understood models, such as numbers, lists, trees, sets, etc. whose formalization is sometimes already available, e.g. consult [MANN85]. Our methodology is not concerned with the design of sorts, in particular with the quotient rules. Notice that these rules are a small fraction of the rules of a data abstraction and for them conditions of under- and over-specification are meaningless.

As a preliminary observation we notice that if $T(C)_s$ is empty, for some sort s, then any defined operation whose domain contains s is neither under- nor over-specified, regardless of its defining rules. In fact, in this circumstance, both conditions become vacuously true since there are no ground terms of sort s. Confluence and termination of a rewriting system extended with the design strategies we propose partially depend on some characteristics of the quotient rules. Both confluence and termination are undecidable properties, hence any attempt to determine if they hold for a given system is non-trivial. In this section, unless otherwise specified, we assume that any system $\mathcal{R}$ is obtained by extending an initial system $\mathcal{R}_0 = \langle S, C, \Phi, \tau, \mathcal{Q}, \Phi\rangle$ with defined operations and we assume that the system's defining rules are designed using exclusively non-recursive functional composition and the binary choice and the recursive reduction strategies.

The next corollary, reduces the problem of the termination of a possibly large set of rules to that of the termination of a smaller, possibly empty, subset.

Theorem 14. If $\mathcal{Q}$ is terminating, then $\mathcal{E}$ is terminating.

Proof. Let $\mathcal{R}_i$ be the system obtained after the i-th extension, $i \geq 0$. We prove the claim by induction on i. Base case: If $i = 0$, then $\mathcal{E} = \mathcal{Q}$ and the claim is satisfied. Inductive case: Let $\mathcal{R}_{i+1}$ be the extension of $\mathcal{R}$ obtained with step $i+1$, $i \geq 0$, and assume $\mathcal{R}_i$ terminating. By [DERS85 Lemma 1] it suffices to prove the termination of any term t such that t is an f-rooted constructor term for some f not in $\mathcal{R}_i$. The proof, by induction on r_t, the number of occurrences of recursive constructors in t, is identical to Theorem 12 with "non under-specified" replaced by "terminating". □

We now turn our attention to confluence. The next lemma is the basis for a condition sufficient to provide confluence and suggests a more efficient approach to detect the critical pairs [KNUT70] and eventually determine the confluence of a system when the binary choice strategy is employed.

Lemma 15. No critical pair originates from any two distinct rules in $\mathcal{A}$.

Proof. Let $\alpha_1 \to \beta_1$ and $\alpha_2 \to \beta_2$ be two distinct rules of $\mathcal{R}$. If they define different operations, then the root of α_1 does not occur in α_2 and vice versa. If they define the same operation, α_1 could superimpose α_2 (and vice versa) only at the root, but the parsimony property prevents this from happening. Thus in both cases no critical pair originates from the rules. □

Corollary 16. If $\mathcal{Q} = \Phi$, then $\mathcal{E}$ is confluent.

Proof. Lemma 15 guarantees that no critical pair may originate from the left sides of two defining rules. The absence of quotient rules implies that there exist no critical pairs at all in the system. Thus, by Lemma 3.1 of [HUET80b] $\mathcal{E}$ is locally confluent. By Theorem 14 $\mathcal{E}$ is terminating. Thus, by Lemma 2.4 of [HUET80b] $\mathcal{E}$ is confluent. □

The assumption that in a system there are no quotient rules, hence no quotient sorts, is sometimes too restrictive. In these situations it is still necessary to check the system for critical pairs, but our strategies make this check more efficient. Let d denote the number of defining rules of a system and q the number of quotient rules. If the defining rules are designed without any restriction on their left sides, the superposition algorithm must check any pair of rules, thus $(d+q)^2$ terms must be considered. However, if the defining rules have been designed with the binary choice strategy, then Lemma 15 suggests that it is useless to check defining rules against each other, thus one needs to consider only $(d+q)q$ terms. Since d is usually much larger than q, the gain may be substantial.

We now discuss the completeness and consistency of extensions. The equivalence of terms is $\overset{*}{\leftrightarrow}$, the reflexive, simmetric, transitive closure of $\to$. The next corollaries imply that any extension designed with our strategies is always complete and if the extended system is confluent, then the extension is consistent too. The confluence of the system is addressed by Corollary 16.

Corollary 17. $\mathcal{R}$ is a complete extension of $\mathcal{R}_0$.

Proof. By Lemma 2, for any t in $T(\Sigma)$ there exists a t' in $T(C)$ such that $t \overset{*}{\to}_{\mathcal{R}} t'$. Then, by [EHRI85, 6.13.1], $\mathcal{R}$ is a complete extension of $\mathcal{R}_0$. □

Corollary 18. If $\mathcal{E}$ is confluent, then $\mathcal{R}$ is a consistent extension of $\mathcal{R}_0$.

Proof. Assume x and y are in $T(C)$ and $x \overset{*}{\leftrightarrow}_{\mathcal{E}} y$. Since $\mathcal{E}$ is confluent and constructor terms are reduced to constructor terms only, there exists a term z in $T(C)$ such that $x \overset{*}{\rightarrow}_{Q} z$ and $y \overset{*}{\rightarrow}_{Q} z$, thus $x \overset{*}{\leftrightarrow}_{Q} y$. Then, by [EHRI85, 6.13.2], $\mathcal{R}$ is a consistent extension of $\mathcal{R}_0$. □

To assess the limitations of the binary choice procedure we explore the existence of complete and parsimonious sets of tuples which cannot be generated by the binary choice procedure. To understand the characteristics of these sets a few definitions and preliminary results are needed.

Lemma 19. If $R = \{\langle x_{11},\ldots x_{1k}\rangle,\ldots\langle x_{n1},\ldots x_{nk}\rangle\}$, for some $k > 0$ and $n > 0$, is a set of k-tuples generated by the binary choice strategy, then either for some i, such that $1 \le i \le k$, and for all j, such that $1 \le j \le n$, the term x_{ji} is a variable, or for some i, such that $1 \le i \le k$, and for all j, such that $1 \le j \le n$, the term x_{ji} is not a variable.

Proof. The index i is the occurrence of the place selected in step 3 during the first iteration of the binary choice procedure on the initial template. □

We call i-th *column* of R the tuple of elements $\langle x_{1i},\ldots x_{ni}\rangle$. We call *pure* any column in which either all or none of the elements are variables. The previous lemma then suggests to look into sets of k-tuples without pure columns. One such set is provided by the arguments in the left sides of the rules of the operation $\wedge$ specified in Figure 6.

sort *Boolean*	(1)
constructors	(2)
false; *true*.	(3)
operation *Boolean* $\wedge$ *Boolean* $\rightarrow$ *Boolean*	(4)
axioms	(5)
true $\wedge$ *false* $\rightarrow$ *false*;	(6)
false $\wedge$ *true* $\rightarrow$ *false*;	(7)
$X \wedge X \rightarrow X$.	(8)

Figure 6. Specification of the operation $\wedge$ (logic conjunction). The operation $\wedge$ has not been designed with the binary choice strategy.

Notice that the rule at line (8) of Figure 6 is non-linear. The next result relates linearity to both completeness and parsimony and is important for understanding sets of tuples without pure columns. Through the remainder of this section we assume that sorts are always strict.

Theorem 20. If R is a complete set of k-tuples of constructor terms, t is a non-linear element of R, and the sort of a non-linear variable of t is infinite, then R is not parsimonious.

Proof. To simplify the proof we prove the theorem for $k = 2$ and $S = \{nat\}$ (see Figure 1) and address the problem of its generalization later. For any term x, the expression $succ^i(x)$, where i is a non-negative integer, stands for x when $i = 0$ and for $succ(succ^{i-1}(x))$ otherwise. Any instance of *nat* is representable by $succ^i(x)$ where either $x = 0$ or $x = X$ for some non-negative integer i and variable X. If R is complete, then, since *nat* is infinite and R is finite, by simple case analysis, R must contain some element t' of the form $\langle succ^{v_1}(X), succ^{v_2}(Y)\rangle$ for some non-negative integers v_1 and v_2 and distinct variables X and Y. If t is non-linear, then $t = \langle succ^{u_1}(Z), succ^{u_2}(Z)\rangle$ for some u_1, u_2, and Z. The pair $\langle succ^{max(u_1,v_1)}(0), succ^{max(u_2,v_2)}(0)\rangle$ is an instance of both t and t', hence R is not parsimonious.

When the proof is generalized, the notation becomes rather cumbersome since the sort of a non-linear variable may have more than one recursive constructor and these may have in turn more than one argument, thus the simple notation $succ^i$ cannot be used. Apart from this, no other substantial difficulty arises. The hypothesis that the sort of one non-linear variable is infinite is a necessary one as Figure 6 proves. □

The next display shows a set R of 3-tuples in which the sort of each component is *nat*. R is complete and parsimonious. Since there are no pure columns, R cannot be generated by the binary choice procedure.

$$R = \{\ \langle succ(X),\ 0,\ succ(Z)\rangle\ \langle 0,\ succ(Y),\ 0\rangle\ \langle 0,\ 0,\ Z\rangle\ \langle X,\ succ(Y),\ succ(Z)\rangle\ \langle succ(X),\ Y,\ 0\rangle\ \} \tag{5.1}$$

The fact that R has 3 *columns* is necessary. We elaborate on this in the next corollary. Notice that the hypothesis of linearity in the statement of the corollary is automatically implied by those of completeness and parsimony (Theorem 20) for infinite sorts.

Corollary 21. Any complete and parsimonious set R of linear k-tuples of terms in $T(C \cup \mathcal{X})$, for $k = 1$ or $k = 2$ and any set of constructors C, has a pure column.

Proof. The proof is trivial for $k = 1$ and when the cardinality of R is less than 2, thus let $R = \{\langle x_{11}, x_{12}\rangle,\ldots\langle x_{n1}, x_{n2}\rangle\}$ and $n \ge 2$. Let $\bar{x}_m$, for $1 \le m \le n$, denote $\langle x_{m1}, x_{m2}\rangle$. Suppose that neither column of R is pure. Then, there exist two indices, i and j, such that x_{i1} and x_{j2} are variables. If $i \ne j$, then $\langle x_{j1}, x_{i2}\rangle$ is an instance of both $\bar{x}_i$ and $\bar{x}_j$. If $i = j$, then, since $x_{i1} \ne x_{j2}$, any 2-tuple of R is an instance of $\bar{x}_i$. In any case R is not parsimonious, thus at least one column of R must be pure. □

Theorem 20 and Corollary 21 imply that when $k = 1$ or $k = 2$ any complete and parsimonious set of k-tuples of constructor terms of sort *nat* is generated by the binary choice procedure. A consequence of this fact is discussed in Example 23. A simple generalization of the above statement to any sort is not possible. In fact, Corollary 21 considers only the "outer" structure of the set. Even in a set of 1-tuples we can recreate conditions similar to those of (5.1), if in the system there is some constructor which takes 3 arguments of sort *nat*.

The above results do not support a definite conclusion on the limitations of the binary choice strategy. While in some simple situations this strategy is substantially the only approach to completeness and parsimony, in slightly more complex situations other alternatives seem available. We have not found any circumstance in which the binary choice procedure did not appear to be a natural and effective approach for designing the left sides of an operation's rules. Our experience seems to indicate that the circumstances in which the binary choice procedure is inappropriate to specify abstract data types should be quite scarce.

6. Related work and concluding remarks

The notion of completeness is proposed in [HUET82] through an inductive definition for testing whether a set of linear tuples is complete for a set of constructors. Completeness is a condition supporting a principle of definition assumed by an "inductionless" method of proof by induction. [THIE84] and [KOUN85] extend this definition to sets of non-linear tuples. They address the detection of completeness and/or parsimony; non-parsimonious sets, called ambiguous in [THIE84] and redundant in [KOUN85]. However, both papers seem unaware of the relationship between completeness, parsimony, and linearity that we stated in Theorem 20.

The question of deciding whether a set of rules indeed defines an operation has been repeatedly addressed in the literature. [HUET82] proposes a *Principle of Definition* based on equations. [KAPU87] reformulates for rewrite rules a notion of *sufficient-completeness* originally proposed by [GUTT78c]. In the following, we compare both these conditions to under-specification. Our comparison is performed in the framework of rewriting. Accordingly, we interpret the congruence relation of the principle of definition to be the derivation relation established by $\mathcal{E}$, see [HUET82, Section 2]. The simplest characterization of under-specification, we believe, is offered by the following theorem.

Theorem 22. If a system $\mathcal{R}$ satisfies Huet's Principle of Definition, then $\mathcal{R}$ is not under-specified, and if $\mathcal{R}$ is not under-specified, then $\mathcal{R}$ is sufficiently-complete.

Proof. The above conditions are all centered on how the terms of a system are related to constructor terms. Let t be a term of the system $\mathcal{R}$. If $\mathcal{R}$ satisfies the principle of definition, then the normal form of t is a constructor term, hence $\mathcal{R}$ is not under-specified. If $\mathcal{R}$ is not under-specified, then for some constructor term c we have $t \xrightarrow{*} c$, hence $t \xleftrightarrow{*} c$ and $\mathcal{R}$ is sufficiently-complete. □

[KAPU87] shows that the sufficient-completeness property of a term rewriting system is undecidable even under relatively strong assumptions. However, for constructor-preserving systems sufficient-completeness is equivalent to the quasi-reducibility [KAPU87] of any term of the form $f(X_1,\ldots,X_k)$ where f is a defined operations and X_i, $1 \leq i \leq k$, is a variable. If a system is not under-specified, then, as an immediate consequence of Lemma 2, any term of the above form is quasi-reducible

The sufficient-completeness of a system seems too weak a property for specifying data abstractions or for programming in very high level programming languages. Figure 7 rephrases an example presented in [KAPU87]. This system is sufficiently-complete, but clearly under-specified. An intuitively equivalent, simpler, and not under-specified system consists of the following single rule.

$$g(X) \to c \tag{6.1}$$

With this system, the defining rules at lines (4) and (5) of Figure 7 become theorems which are easily provable from (6.1).

sort s **constructors** c.	(1)
operation $g : s \to s$	(2)
axioms	(3)
$g(g(X)) \to g(X)$;	(4)
$g(g(X)) \to c$.	(5)

Figure 7. This system [KAPU87] is sufficiently-complete, but under-specified.

Strong assumptions on the characteristics of algebraic specifications and rewriting systems are found frequently in the literature. This almost always occurs when specifications are involved in some process which requires the interpretation, compilation, or implementation of a specification. For example, consult recent efforts [EMDE87, TOGA87] directed toward the integration of functional and/or equational programs into logic programs. The conditions often required in a specification to be implemented originate from the desire of a precise separation between the concepts of value and computation, for a discussion see [ANTO90], so that a computation can always return a value for any legal input. In particular, for the application of a translation scheme from rules to logic programs [EMDE87] requires a term rewriting system to be canonical (or *strongly-canonical*) and [TOGA87] requires any computation to be *successfully terminating*. In both cases the question of whether a system is indeed adequate for the application for which it is considered is left unanswered. Although quite disturbing, this is only mildly surprising since these questions are undecidable.

A more pragmatic approach is taken by the authors of OBJ, an evolving family of very high level programming languages based on rewriting, which we quote below: "we have found that experienced programmers *usually* write rules that satisfy these properties *[Church-Rosser and termination]*" [FUTA84], and "We conjecture that users *almost always* write equations for abstract data types that are canonical, because they tend to think of equations as programs, and then write primitive recursive definitions for operations" [GOGU88].

Our work is an attempt to address this unsatisfying state of things on two fronts: 1) operations designed with our strategies satisfy each of the requirements mentioned above, thus our effort is somewhat complementary to that of other researchers, and 2) our approach improves the design of certain systems by simultaneously increasing the reliability of the result and decreasing both the skill of the designer and the time for the design, three conditions which are usually pair-wise conflicting.

Acknowledgment.

During the preparation of this work the author was on leave of absence from the Istituto per la Matematica Applicata of the Consiglio Nazionale delle Ricerche.

References

[ANTO90] Antoy S., P. Forcheri, and M.T. Molfino, "Specification-based Code Generation", 23rd Hawaii Int'l Conf. on System Sciences, Kona, Hawaii, Jan. 3-5, 1990, 165-173.

[BERG89] Bergstra J.A., J. Heering, and P. Klint, (eds.) Algebraic Specifications, Addison-Wesley, Wokingham, England, 1989.

[BURS69] Burstall L., "Proving Properties of Programs by Structural Induction", *Computer Journal*, 21-1, 1969, 41-48.

[DERS85] Dershowitz N., "Termination", Proc. Rewriting Techniques and Applications, Dijon, France, Springer-Verlag, May 1985, 180-223.

[EHRI85] Ehrig M. and B. Mahr, Fundamentals of Algebraic Specification 1, Springer-Verlag, Berlin, 1985.

[EMDE87] van Emden M. and K. Yukawa, "Logic Programming with Equations", *The Journal of Logic Prog.*, 4, 265-288, 1987.

[FUTA84] Futatsugi K., J.A. Goguen, J. P. Jouannaud, and J. Meseguer, "Principles of OBJ2", in 12th Annual Symp. on Principles of Prog. Languages, 52-66, SIGPLAN and SIGACT, 1984.

[GARL88] Garland S.J. and J.V. Guttag, "Inductive Methods for Reasoning about Abstract Data Types" Proc. of the 15th ACM Conf. on Principle of Prog. Lang., 219-228, 1988.

[GOGU78] Goguen J.A., J.W. Thatcher, and E.G. Wagner, "An Initial Algebra Approach to the Specification, Correctness, and Implementation of Abstract Data Types", in Current Trends in Programming Methodology 4, 80-149 (Ed. R.T. Yeh), Prentice-Hall, Englewood Cliff, NJ, 1978.

[GOGU88] Goguen J.A. and T. Winkler, "Introducing OBJ3", *SRI-CSL-88-9*, SRI International, Menlo Park, CA, 1988

[GUTT78a] Guttag J.V., E. Horowitz, and D. Musser, "The Design of Data Type Specifications", in Current Trends in Programming Methodology 4, 60-79 (Ed. R.T. Yeh), Prentice-Hall, Englewood Cliff, NJ, 1978.

[GUTT78b] Guttag J.V., E. Horowitz, and D. Musser, "Abstract Data Types and Software Validation", *CACM*, 21, 1978, 1048-1064.

[GUTT78c] Guttag J.V. and J.J. Horning, "The Algebraic Specifications of Abstract Data Types", *Acta Informatica*, 10, 1978, 27-52.

[GUTT85a] Guttag J.V., J.J. Horning, and J.M. Wing, "The Larch Family of Specification Languages", *IEEE Software*, Sept 1985, 24-36.

[GUTT85b] Guttag J.V. and Horning J.J., "A Larch Shared Language Handbook" *Science of Computer Programming* 6-2, 135-157, 1985.

[HALM88] Halme H. and J. Heilanen, "GNU Emacs as a Dynamically Extensible Programming Environment", *Software—Practice and Experience*, 18-10, 1988, 999-1009.

[HOPC79] Hopcroft J.E. and J.D. Ullman, Introduction to Automata Theory, Languages, and Computations, Addison-Wesley, Reading, MA, 1979.

[HUET80a] Huet G. and D. Oppen, "Equations and rewrite rules: A survey", in Formal Language Theory (R. Book, ed.), Academic Press, 1980, 349-405.

[HUET80b] Huet G., "Confluent Reductions: Abstract Properties and Applications to Term-Rewriting Systems", *JACM*, 27, 1980, 797-821.

[HUET82] Huet G. and J.-M. Hullot, "Proofs by Induction in Equational Theories with Constructors", *JCSS* 25, 1982, 239-266.

[KAPU87] Kapur D., P. Narendran, and H. Zhang, "On Sufficient-Completeness and Related Properties of Term Rewriting Systems", *Acta Informatica*, 24, 1987, 395-415.

[KFOU82] Kfoury A., R. Moll, and M. Arbib, A Programming Approach to Computability, Springer-Verlag, New York, NY, 1982.

[KNUT70] Knuth D.E. and P.B. Bendix, "Simple Word Problems in Universal Algebras", in Computational Problems in Abstract Algebras, (J. Leech, ed.), Pergamon Press, New York, 1970, 263-297.

[KOUN85] Kounalis E., "Completeness in Data Type Specifications", EUROCAL '85, LNCS 204, 1985, 348-362.

[MANN85] Manna Z. and R. Waldinger, The Logical Basis for Computer Programming, Addison-Wesley, Reading, MA, 1985.

[MINS67] Minsky M. L., Computation: Finite and Infinite Machines, Prentice-Hall, Englewood Cliffs, NJ, 1967.

[STAL81] Stallman R., "Emacs: the Extensible, Customizable, Self-documenting Display Editor", Proc. ACM SIGPLA/SIGOA Symp. on Text Manipulation, Portland, OR, 1981.

[STAL87] Stallman R., *GNU Emacs Manual*, Sixth ed., Version 18, Free Software Foundation, Cambridge, MA, 1987.

[THIE84] Thiel J.J., "Stop Losing Sleep over Incomplete Data Type Specifications", in 11th Annual Symp. on Principles of Prog. Languages, 76-82, ACM, 1984.

[TOGA87] Togashi A. and S Noguchi, "A Program Transformation from Equational Programs into Logic Programs", *The Journal of Logic Prog.*, 4, 85-103, 1987.

Appendix

Example 23.

This example concerns the specification of the Ackermann's function. We show how to design the left sides of the specification rules using the binary choice procedure. Since this function is not primitive recursive, recursive reductions are not appropriate for the rules' right sides. The Ackermann's function has exactly two arguments of sort *nat*, thus any complete and parsimonious specification of this function has to be designed with the binary choice or an equivalent procedure. In the following rules we abbreviate $succ^i(0)$ with i

The arity of the operation *acker* is $nat \times nat \rightarrow nat$. The initial template is

$acker(■,□)$ (1)

With choice "inductive" we obtain

$acker(0,■)$ (2)

$acker(succ(□),□)$ (3)

Now we choose "variable" for the cursor, we define the left side, and we select the next place for the cursor. We obtain

$acker(0,Y) \rightarrow 1$ (4)

$acker(succ(□),■)$ (5)

With choice "inductive" we obtain

$acker(0,Y) \rightarrow 1$ (6)

$acker(succ(■),0)$ (7)

$acker(succ(□),succ(□))$ (8)

Now we choose "inductive" and we define the left side. We obtain

$acker(0,Y) \rightarrow 1$ (9)

$acker(1,0) \rightarrow 2$ (10)

$acker(succ(succ(■)),0)$ (11)

$acker(succ(□),succ(□))$ (12)

All the remaining selections are in lexicographic order and the choice is "variable" in each case. As soon as there are no places in the left side of a rule we define the right side. We obtain

$acker(0,Y) \rightarrow 1$ (13)

$acker(1,0) \rightarrow 2$ (14)

$acker(succ(succ(X)),0) \rightarrow X+4$ (15)

$acker(succ(X),succ(Y)) \rightarrow acker(acker(X,succ(Y)),Y)$

This specification is substantially identical to a set of rules proposed in [HOPC79, p. 175] for the definition of the Ackermann's function. Notice that the completeness and parsimony of the original definition as well as of this specification are not immediately apparent.

Example 24.

This example concerns the car pool of a car rental agency and the determination of the cars available in the pool. A car is specified by a tuple of attributes. To keep the example manageable we consider two attributes only: a generic *id*, which is left unspecified, and a *state*, which is either *avail*, *rented*, or *broken*. Cartesian product sorts, i.e. sorts whose elements are tuples, have only one constructor. By convention, we use the same identifier for the sort and its constructor, since they never occur in the same context.

sort *id* **parameter**. (1)
sort *state* **constructors** *avail*; *rented*; *broken*. (2)
sort *car* **constructors** *car*(*id*, *state*). (3)

The set of cars in the pool is described by the sort *pool*. The first quotient rule specifies that the order in which cars are added to the pool is irrelevant. The second quotient rule specifies that there are no duplicate cars in the pool. Notice that the quotient rules of *pool* are non-terminating.

sort *pool* (4)
constructors (5)
nocars; (6)
add(*pool*, *car*); (7)
quotient (8)
$add(add(X,Y),Z) \rightarrow add(add(X,Z),Y)$; (9)
$add(add(X,Y),Y) \rightarrow add(X,Y)$. (10)

We imagine that, during time, cars are added to or removed from the pool and that their states are occasionally updated. Now we design an operation for counting how many available cars there are in the pool. The arity of *count* is *pool* → *nat*. The initial template is

count(■) (11)

We choose "inductive" and define the first right side. We obtain

count(*nocars*) → *0* (12)
count(*add*(■,□) (13)

With choice "variable" we obtain

count(*nocars*) → *0* (14)
count(*add*(*X*,■) (15)

With choice "inductive" we obtain

count(*nocars*) → *0* (16)
count(*add*(*X*, *car*(■,□)) (17)

Since the sort *id* is a parameter, the choice is automatically "variable".

count(*nocars*) → *0* (18)
count(*add*(*X*, *car*(*Y*,■)) (19)

With choice "inductive" we obtain

count(*nocars*) → *0* (20)
count(*add*(*X*, *car*(*Y*, *avail*)) (21)
count(*add*(*X*, *car*(*Y*, *rented*)) (22)
count(*add*(*X*, *car*(*Y*, *broken*)) (23)

We now complete the right sides using recursive reductions

count(*nocars*) → *0* (24)
count(*add*(*X*, *car*(*Y*, *avail*)) → *succ*(!) (25)
count(*add*(*X*, *car*(*Y*, *rented*)) → ! (26)
count(*add*(*X*, *car*(*Y*, *broken*)) → ! (27)

Finally, notice that the variable names occur in only one occurrence, thus they serve no purpose. All the variables could be as well anonymous.

A Simplifier for Untyped Lambda Expressions

Louis Galbiati
Quickturn Systems, Inc.
Mountain View, CA

Carolyn Talcott
Stanford University
Stanford, CA
CLT@SAIL.STANFORD.EDU

Abstract

Many applicative programming languages are based on the call-by-value lambda calculus. For these languages tools such as compilers, partial evaluators, and other transformation systems often make use of rewriting systems that incorporate some form of beta reduction. For purposes of automatic rewriting it is important to develop extensions of beta-value reduction and to develop methods for guaranteeing termination. This paper describes an extension of beta-value reduction and a method based on abstract interpretation for controlling rewriting to guarantee termination. The main innovations are (1) the use of rearrangement rules in combination with beta-value reduction to increase the power of the rewriting system and (2) the definition of a non-standard interpretation of expressions, the *generates* relation, as a basis for designing terminating strategies for rewriting.

1. Introduction

The original motivation for this work came from a project to compile programs by transformation to continuation-passing style [Steele 1976]. This program transformation in its simplest form tends to introduce extraneous lambda-applications. Instead of complicating the transformation to avoid introducing these lambda-applications it seemed preferable to use it in conjunction with a general purpose simplifier. The idea being that such a simplifier could be shared by many automatic program manipulation tools as well as being useful in interactive program manipulation systems. For example, such a simplifier can be used for optimizing programs built by combining many components, since inlining procedure calls (call unfolding) and many peep-hole optimizations are instances of beta-reduction. It could also serve as a tool for building semantics directed compilers and partial evaluators.

Our simplifier is composed of a reduction system and a method for limiting application of reductions to insure termination. The basic reduction system can be used in combination with other control strategies and the analysis underlying our method for limiting reduction should work for variants of the reduction system.

The target language for our simplifier is that of the lambda calculus [Barendregt 1981]. The reduction system consists of the beta-value (beta-v) reduction rule together with two rearrangement rules designed to create additional sites for beta reduction. The beta-v rule is the restriction of the standard beta conversion rule to applications in which the operand is a value expression, e.g. a variable, constant, or lambda abstraction. Thus $(\lambda x.f\ x)z$ is a beta-v reduction site (reducing to $f\ z$), while $(\lambda x.f\ x)(g\ z)$ is a beta reduction site but not a beta-v reduction site. The beta-v rule corresponds to call-by-value semantics for a programming language and [Plotkin 1975] shows that this rule is adequate to evaluate closed expressions. However there are many programs that are equivalent under a wide class of observations that cannot be proved equivalent in the lambda-v calculus. One example is the evaluated position context theorem: $C[e]$ is equivalent to $\texttt{let}\{x := e\}C[x]$ where C is any expression with a unique hole occurring in a position that will be evaluated before any other serious computation takes place [Talcott 1989]. The rearrangement rules of our reduction system are corollaries of this theorem expressing the fact that a let-binding (application of a lambda abstraction) occurring in the function position of an application or in the argument position of a application in which the function position contains a value can be moved outside of the application. Thus $(\texttt{let}\{f := g\,z\}\lambda x.f\,x)y$ rearranges to $\texttt{let}\{f := g\,z\}(\lambda x.f\,x)y$ and $(\lambda x.f\,x)\texttt{let}\{g := h\,z\}\lambda y.g\,y$ rearranges to $\texttt{let}\{g := h\,z\}(\lambda x.f\,x)(\lambda y.g\,y)$. Note that in both cases the expression before rearrangement has no beta-v reduction site, while the expression after rearrangement does have a beta-v reduction site. The rearrangement rules have the effect of moving expressions that intervene between a function and its argument to the outside. They define a canonical form in which functions are more likely to appear directly applied to their arguments. The rearrangement rules by themselves form a confluent, terminating system. They are not derivable in the beta-v calculus and hence our reduction system is more powerful than one based purely on beta-v reduction.

[Moggi 1989] introduces the notion of computational monad as a framework for axiomatizing features of programming languages. Computational monads accomodate a wide variety of language features including assignment, exceptions, and control abstractions. An extension of the lambda-v calculus called the lambda-c calculus is presented and shown to be valid in all computational monads. Our rearrangement rules are derivable in the lambda-c calculus and thus are valid for any language whose semantics can be modeled as a computational monad.

Writing a simplifier based on rules that include beta reduction is made difficult by the fact that unrestricted application of these rules can lead to infinite reduction sequences. Thus a strategy is needed for limiting beta reduction. One possible strategy is to fix a maximum number of reduction steps and perform reductions at random until this limit is reached. This strategy has the disadvantage that it treats all reduction steps the same way, rather than favoring those which simplify the expression over those which wander aimlessly. A second strategy is to beta reduce a lambda-application $(\lambda x.e)v$ only if the bound variable x occurs free at most once in the body e or if the operand v is atomic. Call this the *reduces-size* strategy. It guarantees that each beta reduction step decreases the size of the overall expression. This strategy can be overly conservative, since some expressions can be simplified only by first performing steps which increase the size of the expression, e.g. unfold and simplify. Note that neither of these strategies are confluent. This is obvious in the case of limiting the number of steps. To see this for the reduces size strategy we observe that for any lambda abstraction v $(\lambda x.\lambda z.(\lambda y.y(y\,z))x)v$ reduces to $(\lambda z.(\lambda y.y(y\,z))v)$ and to $(\lambda x.\lambda z.x(x\,z))v$.

In this paper we describe a new strategy, *statically-limited* rewriting, in which we compute a subset B of lambda-nodes in the initial expression such that any rewriting of that expression is guaranteed to terminate if beta reduction is restricted to descendants of nodes in **B**. (The descendant relation is the natural relation between nodes in an expression and nodes in the result of rewriting that expression.)

We use a form of abstract interpretation (cf. [Abramsky and Hankin 1987]) to compute a suitable set **B**. First we define a non-standard interpretation of expressions, the *generates* relation **xgen** and the notion of a set of lambda nodes being an **xgen**-cycle. We then show that limiting reduction to descendants of a subset of lambda-nodes containing no **xgen**-cycle guarantees termination. Given an initial expression $e_{\mathtt{init}}$, **xgen** is a relation on reduction paths and pairs of lambda-nodes of $e_{\mathtt{init}}$ defined as follows. Let a and b be lambda nodes in the initial expression and let q be a reduction sequence beginning with $e_{\mathtt{init}}$. We say that a generates b in the final step of q (and write **xgen**(q, a, b)), if the final step of q is a beta-v reduction at a site whose operator is a descendant of a, and this reduction step entails (in the case $a \neq b$) an increase in the number of descendants of b, or (in the case $a = b$) no decrease. We say a generates b along p if **xgen**(q, a, b) for some prefix q of p. A set of lambda nodes $a_0, \ldots, a_n$ in the initial expression is an **xgen**-cycle if, roughly, there is a reduction sequence along which a_i generates a_{i+1} for $i < n$ and a_n generates a_0.

For example consider the expression $(\lambda^1 x.x\,x)\,(\lambda^2 x.x\,x)$ where the superscripts are used to associate names with lambda-nodes. Here there is a single reduction path along which 1 generates 2 and 2 generates 2. Limiting beta-reduction to descendants of node 1 guarantees termination (after one step!). As another example consider the expression $(\lambda^1 p.p\,p\,z)(\lambda^2 x.\lambda^3 y.\lambda^4 s.s\,x\,y)$ For this expression there are reduction paths along which 1 generates 2, 3, 4 and there are no other generates instances. Since there are no cycles all reduction sequences must terminate. Note that the reduces-size strategy mentioned earlier does not permit any reduction.

In general **xgen** can be an infinite relation. Thus we want to find a finite, computable approximation that serves the same purpose. Using the methodology of abstract interpretation we say that a relation together with a corresponding notion of cycle is a safe approximation to **xgen** if it preserves the "no-cycles implies termination" property. As a first step we define a binary relation **gen** on lambda nodes that is a safe finite approximation of **xgen** using the usual notion of cycle induced by a binary relation. **gen** is the set of pairs a, b such that for some reduction sequence q beginning with $e_{\mathtt{init}}$, a generates b in the final step of q.

We are still not done, as we have no general (uniformly terminating) algorithm for computing **gen**. Instead we define a safe computable approximation **gen**$'$ of **gen**. The computation of **gen**$'$ is based on computing upper bounds to the sets of nodes in the initial expression whose descendants can occupy certain kinds of positions (cf. control flow analysis [Shivers 1988] and closure analysis [Bondorf 1990]) and on computing an upper bound to the set of lambda nodes in the initial expression that are "doublers", i.e. have a descendant with more than one free occurrence of the bound variable in the body. Then **gen**$'$ is roughly the set of all pairs (a, b) of `lam` nodes such that a is a doubler and there is some c such that a descendant of a is applied to a descendant c, and a descendant of b can become a subexpression of a descendant of c.

To summarize, given an expression to simplify, we proceed as follows: (i) compute gen′; (ii) choose a set **B** with no gen′-cycles; (iii) perform **B**-limited reduction until termination. Limited rewriting is in fact locally confluent. Thus we are free to apply the rules in whatever order we like; the final outcome will be the same.

Although usually less conservative than the reduces-size strategy, the new strategy is sometimes still overly conservative. A less conservative alternative strategy, *dynamically-limited* rewriting, is the following. Instead of computing **gen′**, we merely apply rules, accumulating a relation consisting of the pairs (a, b) such that a has generated b in some step of the rewriting so far, and disallowing any step which would cause this relation to contain a cycle. The alternative strategy guarantees termination but fails to preserve the confluence property. Nevertheless it may be the more appropriate strategy for a practical simplifier. Our static and dynamic strategies have an analogue in two approaches to partial evaluation. The static strategy corresponds to the use of binding time analyis and other static analyses performed to determine which applications should be left to run time and which are to be carried out at partial-evaluation time (cf. [Jones, Sestoft, and Søndergaard 1989], and [Bondorf 1990]). The dynamic strategy is more in the spirit of [Weise and Ruf 1990] where a call stack is maintained during partial evaluation and used for potential loop detection.

In this paper we present the reduction system, introduce statically- and dynamically-limited rewriting, and state the key results. Formal definitions and proofs can be found in [Galbiati and Talcott 1990].

2. Syntax

We use standard lambda calculus syntax [Barendregt 1981]. To define and analyze reduction rules it is convenient to represent expressions as labeled trees where each node of the tree corresponds to an occurrence of a subexpression. In this section we define the set of expressions and their representation as labeled trees.

We assume given a countably infinite set **Var** of variables. Then the set **Exp** of expressions is the least set containing the variables and closed under lambda abstraction and application. That is, **Exp** is the least set satisfying the following equation.

$$\mathbf{Exp} = \mathbf{Var} \cup \lambda\mathbf{Var}.\mathbf{Exp} \cup \mathbf{Exp}\,\mathbf{Exp}$$

We let $x, x_0, \ldots$ range over **Var** and $e, e_0, \ldots$ range over **Exp**. Expressions of the form x, $\lambda x.\, e$, and $e_1\, e_2$ are called atomic expressions, abstractions, and applications, respectively. In an abstraction $\lambda x.\, e$, we call x the bound variable and e the body. In an application $e_1\, e_2$, we call e_1 the operator and e_2 the operand. We let **Vxp** be the set $\mathbf{Var} \cup \lambda\mathbf{Var}.\,\mathbf{Exp}$ of atomic expressions and abstractions; expressions in **Vxp** are called value expressions. We let $v, v_0, \ldots$ range over **Vxp**.

Free and bound variables are defined as usual and expressions identical up to alpha conversion we regard as indistinguishable. We write $e_1\{x := e_2\}$ for the result of substituting e_2 for all free occurrences of x in e_1. Here we assume that alpha variants are chosen "hygienically" so that no trapping of free variables occurs. $\mathtt{let}\{x := e_0\}e_1$ abbreviates $(\lambda x.e_1)e_0$. We adopt the usual conventions for disambiguating written expressions, namely that (1) application associates left, so that $e_1\, e_2\, e_3$ is $(e_1\, e_2)\, e_3$, and

(2) the body of an abstraction or let extends as far right as possible, so that $\lambda x.\, e_1\, e_2$ is $\lambda x.(e_1\, e_2)$. Parentheses may be used to override the default grouping as in $e_0(e_1\, e_2)$ or $(\lambda x.e_0)\, e_1$.

The tree structure of an expression is the abstract syntax tree modified to replace each bound variables by a pointer to the node in the tree corresponding to its binding lambda (cf. [deBruijn 1972]). Each node in the tree structure of an expression corresponds to a (unique) subexpression occurrence. Nodes are labeled by the constructor of the corresponding subexpression and edges are labeled by component selectors. A pointer is represented by a path (sequence of edges) relative to a top-level expression. To make this precise we define selectors, locations, and tags as follows. A selector is an element of the set $\{\mathtt{L}, \mathtt{R}, \mathtt{B}\}$. Selectors name immediate subexpressions of an expression and label the edges of a tree. B names the body of an abstraction and L and R name the operator (left) and operand (right) components of an application. The set **Loc** of locations is the set of finite sequences with elements taken from the set of selectors.

$$\mathbf{Loc} = \{\mathtt{L}, \mathtt{R}, \mathtt{B}\}^*$$

Locations represent paths or nodes of a tree and are used to name occurrences of subexpressions. The set **Tag** of tags is defined by

$$\mathbf{Tag} = \{\mathtt{app}, \mathtt{lam}\} \cup \mathtt{atx}(\mathbf{Loc})$$

Tags label nodes of a tree. A nodes tag identifies the constructor of the corresponding subexpression and in the case of a bound variable the location of its binding abstraction.

We let $c, c_0, \ldots$ range over $\{\mathtt{L}, \mathtt{R}, \mathtt{B}\}$, $l, l_0, \ldots$ range over **Loc**, and $t, t_0, \ldots$ range over **Tag**. $\square$ is the empty sequence and selectors are considered to be singleton sequences. We write $l.l'$ for the concatenation of the sequences l and l' and $l.c$ for the extension of l by c. If $l = l_0.l_1$ then l_0 is called a prefix of l.

For simplicity we will assume outermost expressions are closed (by adding lambdas if necessary). This is not a serious restriction, it just eliminates the need for a special case for free variables. For an outermost expression e, the locations, $\mathtt{locs}(e)$, the subexpression $(e)_l$ at location l and its tag $\mathtt{tag}(e, l)$ are defined by induction on the construction of e in the obvious way. As an example let $e = \lambda f.\lambda x.f\, x$. The tree written as a term would be

$$\mathtt{lam}(\mathtt{B} : \mathtt{lam}(\mathtt{B} : \mathtt{app}(\mathtt{L} : \mathtt{atx}(\square), \mathtt{R} : \mathtt{atx}(\mathtt{B}))))$$

where component selectors are made explicit using key-word argument syntax. Further we have

$$\mathtt{locs}(e) = \{\square, \mathtt{B}, \mathtt{B.B}, \mathtt{B.B.L}, \mathtt{B.B.R}\}$$

$$\mathtt{tag}(e, \square) = \mathtt{lam} \qquad \mathtt{tag}(e, \mathtt{B}) = \mathtt{lam} \qquad \mathtt{tag}(e, \mathtt{B.B}) = \mathtt{app}$$

$$\mathtt{tag}(e, \mathtt{B.B.L}) = \mathtt{atx}(\square) \qquad \mathtt{tag}(e, \mathtt{B.B.R}) = \mathtt{atx}(\mathtt{B})$$

Let l be a location in e. If l has tag lam (i.e. $\mathtt{tag}(e, l) = \mathtt{lam}$), we say l is a lam-node of e. If l has tag app we say l is a app-node of e. If l has tag $\mathtt{atx}(l')$ we say that l is an atx-node bound at l' in e.

3. Reduction

An expression is simplified stepwise by applying one of three reduction rules.

(1) $(\lambda x.\, e_0)\, e_1\, e_2 \mapsto_1 (\lambda x.\, e_0\, e_2)\, e_1$ provided x is not free in e_2.

(2) $v((\lambda x.\, e_0)\, e_1) \mapsto_2 (\lambda x.\, v\, e_0)\, e_1$ provided x is not free in v.

(3) $(\lambda x.\, e_0)\, v \mapsto_3 e_0\{x := v\}$

The stepwise reduction relation $e \longrightarrow e'$ is the congruence closure of the union of the three reduction rules viewed as binary relations. That is, $e \longrightarrow e'$ just if for some $(r, l) \in \{1, 2, 3\} \times \mathbf{Loc}$, and some e_0, e_1 we have that $(e)_l = e_0$, $e_0 \mapsto_r e_1$, and e' is obtained from e by replacing the occurrence of e_0 at l by e_1. (Note that this is replacement, not substitution, and free variables of e_1 may be trapped by abstractions above l.) Pairs (r, l) for $r \in \{1, 2, 3\}$ and $l \in \mathbf{Loc}$ are called rule applications. We write $e \xrightarrow{(r,l)} e'$ to make the rule application explicit and we call l a site (in e) for application of rule r.

A reduction sequence is a sequence of stepwise reductions. We let p, p_0, ..., q, q_0, ...range over sequences of rule applications (r, l) and write $e \xrightarrow{p} e'$ if $p = (r_1, l_1), \ldots, (r_n, l_n)$ $e = e_0$, $e' = e_n$, and $e_{i-1} \xrightarrow{(r_i,l_i)} e_i$ for $1 \leq i \leq n$.

Rule 3 is the beta-v reduction rule [Plotkin 1975]. Rules 1 and 2, called left-rearrangement and right-rearrangement respectively, would be superfluous in a system with unlimited beta-reduction and beta-expansion. However with only call-by-value beta-reduction, these rules can create sites for application of rule 3 which would not otherwise be created. Rearrangement merely rearranges the nodes in a tree, while beta-reduction may duplicate some subtrees and destroy others. The reduction rules preserve operational equivalence (cf. [Plotkin 75]). with respect to a call-by-value evaluator They are also valid in a wide range of extensions of the basic language including control abstractions [Talcott 1989] and memory operations [Mason and Talcott 1989a,b] and are valid for the λ_c theory of [Moggi 1989].

Theorem (Rearrangement is canonical): The reduction system generated by the rearrangement rules (the reflexive transitive congruence closure of $\mapsto_1 \cup \mapsto_2$) is terminating and confluent. Thus every expression has a unique normal form with respect to rearrangement.

In order to analyze properties of reduction sequences, we need to be able to trace the ancestry of nodes in an expression resulting from applying a sequence of reductions. For $e \xrightarrow{(r,l)} e'$ and $l' \in \mathtt{locs}(e')$ there is a natural predecessor of l' in e, $\mathtt{pred}(e, (r, l), l')$. Consider an application of rule 1. Making the relevant tree structure explicit we have

$$e = \mathtt{app}^1(\mathtt{app}^2(\mathtt{lam}^3(x, e_0), e_1), e_2) \mapsto_1 \mathtt{app}^2(\mathtt{lam}^3(x, \mathtt{app}^1(e_0, e_2), e_1) = e'.$$

The predecessor of a node in the subexpression e_0, e_1, or e_2 of e' is the corresponding node the subexpression e_0, e_1, or e_2 of e. The predecessors of the remaining nodes of e' are given by the superscripts. The predecessor function for applications of rule 2 or rule 3 is analogous. For beta reduction our notion of predecessor coincides with that of [Wadsworth 1978].

The following lemma is a direct consequence of the definitions. It expresses the key structural properties of reductions and points out the crucial distinction between rearrangements and beta reduction.

Lemma (pred): The predecessor function is 1-1 and onto except in the case of a rule 3 reduction where the application and abstraction nodes of the reduction site have no successors and nodes of the value may have zero or more successors.

The ancestor function anc generalizes the predecessor function to sequences of reduction steps mapping locations in the final expression of a reduction sequence to locations in the initial expression from which they derive. If $\mathtt{anc}_e(p, l) = a$ then we say that l is a p-descendant of a. The following lemma shows that, via the ancestor relation, tag types and binding relations are preserved by reduction.

Lemma (tag preservation): Let $e \xrightarrow{p} e'$, $l' \in \mathtt{locs}(e')$, and $\mathtt{anc}_e(p, l') = l$. If $\mathtt{tag}(e, l) \in \{\mathtt{app}, \mathtt{lam}\}$ then $\mathtt{tag}_e(p, l') = \mathtt{tag}(e, l)$. If $\mathtt{tag}(e, l) = \mathtt{atx}(l_0)$ then $\mathtt{tag}_e(p, l') = \mathtt{atx}(l'_0)$ where l'_0 is the (unique) location in e' such that l'_0 is a prefix of l' and $\mathtt{anc}_e(p, l'_0) = l_0$.

4. Limited rewriting

In this section the relation gen is introduced, two forms of limited rewriting are defined and shown to terminate, and it is shown that one form of limited rewriting is confluent while the other is not. Finally we discuss limited rewriting as a basis for a practical rewrite-control strategy.

To simplify the definitions, for the remainder of the paper we fix an initial expression $e_{\mathtt{init}}$. **A** will denote the set of locations in $e_{\mathtt{init}}$ ($\mathbf{A} = \mathtt{locs}(e_{\mathtt{init}})$) and $a, b, a_0, \ldots$ will range over **A**. $\mathbf{A}_{\mathtt{lam}}$ will denote the set of lam locations in $e_{\mathtt{init}}$ ($\mathbf{A}_{\mathtt{lam}} = \{l \in \mathbf{A} \mid \mathtt{tag}(e_{\mathtt{init}}) = \mathtt{lam}\}$). Having fixed $e_{\mathtt{init}}$ we specialize the ancestor functions to $e_{\mathtt{init}}$ and omit the subscript. We let **Rseq** be the set of rule application sequences starting from $e_{\mathtt{init}}$, that is, sequences p such that $e_{\mathtt{init}} \xrightarrow{p} e$ for some e. For brevity, in situations where an expression is required a sequence p in **Rseq** may be used to denote the (unique) e such that $e_{\mathtt{init}} \xrightarrow{p} e$. In particular we will write $\mathtt{tag}(p, l)$ for $\mathtt{tag}(e, l)$.

4.1. The gen relation and limited rewriting

We begin by defining the *generates* relations xgen on $\mathbf{Rseq} \times \mathbf{A}_{\mathtt{lam}} \times \mathbf{A}_{\mathtt{lam}}$ and gen on $\mathbf{A}_{\mathtt{lam}} \times \mathbf{A}_{\mathtt{lam}}$.

Let a and b be lam nodes in $\mathbf{A}_{\mathtt{lam}}$ and let q be a rule-application sequence in **Rseq**. We say that a generates b in the final step of q (and write $\mathtt{xgen}(q, a, b)$), if the final step of q is a rule-3 reduction at a site whose operator is a descendant of a, and this final step entails (in the case $a \neq b$) an increase in the number of descendants of b, or (in the case $a = b$) no decrease.

Definition (xgen): $\mathtt{xgen}(q, a, b)$ just if $a, b \in \mathbf{A}_{\mathtt{lam}}$, $q \in \mathbf{Rseq}$, and there are p, e, e', l such that $q = p.(3, l)$ and (i-iii) hold.

(i) $e_{\mathtt{init}} \xrightarrow{p} e \xrightarrow{(3,l)} e'$

(ii) $\mathrm{anc}(p, l.\mathrm{L}) = a$

(iii) $n_b < n'_b$ if $a \neq b$ and $n_b \leq n'_b$ if $a = b$; where n_b is the number of locations l' in e such that $\mathrm{anc}(p, l') = b$ and n'_b is the number of locations l' in e' such that $\mathrm{anc}(p.(3,l), l') = b$.

We say that a generates b (and write $\mathrm{gen}(a, b)$) if a generates b in some step of some reduction sequence beginning with $e_{\mathtt{init}}$.

Definition (gen): $\mathrm{gen}(a, b)$ just if there is some $q \in \mathbf{Rseq}$ for which $\mathtt{xgen}(q, a, b)$.

We now define two forms of limited rewriting.

Definition (R-limited rewriting): Given a relation $\mathbf{R}$ on $\mathbf{A}_{\mathtt{lam}} \times \mathbf{A}_{\mathtt{lam}}$, we define an $\mathbf{R}$-limited rewriting to be any reduction sequence $e_{\mathtt{init}} \longrightarrow e_1 \longrightarrow \ldots$ starting with $e_{\mathtt{init}}$, and satisfying the restriction that a step in which some node a generates some node b is allowed only if $(a, b) \in \mathbf{R}$. That is, if $e_{\mathtt{init}} \xrightarrow{p} e \xrightarrow{(3,l)} e'$ is an initial segment of such a sequence and $\mathtt{xgen}(p.(3,l), a, b)$, then $(a, b) \in \mathbf{R}$.

Definition (B-limited rewriting): Given a subset $\mathbf{B}$ of $\mathbf{A}_{\mathtt{lam}}$ we define a B-limited rewriting to be any reduction sequence $e_{\mathtt{init}} \longrightarrow e_1 \longrightarrow \ldots$ starting with $e_{\mathtt{init}}$, and satisfying the restriction that a beta reduction step is allowed only if the operator is a descendant of a location in B. Thus if $e_{\mathtt{init}} \xrightarrow{p} e \xrightarrow{(3,l)} e'$ is an initial segment of such a sequence then $\mathrm{anc}(p, l.\mathrm{L}) \in \mathbf{B}$.

4.2. Termination of limited rewriting

In this subsection we show that under suitable conditions each of the two forms of limited rewriting is guaranteed to terminate. We say that a binary relation R on a set X has no cycles if there is no sequence $x_0, \ldots, x_n$ of elements of X such that $x_0 = x_n$ and $R(x_i, x_{i+1})$ for $0 \leq i < n$.

Theorem (R-limited rewriting terminates): Let $\mathbf{R}$ be a relation on $\mathbf{A}_{\mathtt{lam}} \times \mathbf{A}_{\mathtt{lam}}$ with no cycles. Then any R-limited rewriting must be finite.

In fact (**R-limited rewriting terminates**) holds for any extension of the beta-v rule by the addition of a terminating collection of rules with the property that application of one of these rules never increases the number of descendants of a node.

Corollary (B-limited rewriting terminates): Let B be a subset of $\mathbf{A}_{\mathtt{lam}}$ such that the restriction $\mathrm{gen}_{\mathbf{B}}$ of gen to B has no cycles. Then any B-limited rewriting must be finite.

4.3. Confluence of limited rewriting

In the previous section we showed that for certain subsets B of $\mathbf{A}_{\mathtt{lam}}$, B-limited rewriting terminates. In this section we show that for *any* subset B of $\mathbf{A}_{\mathtt{lam}}$, B-limited rewriting is locally confluent. R-limited rewriting, however, is not confluent.

Theorem (B-limited rewriting is locally confluent): If B is any subset of $\mathbf{A}_{\mathtt{lam}}$ then B-limited rewriting is locally confluent. That is, if $e_{\mathtt{init}} \xrightarrow{p} e \xrightarrow{(r_k,l_k)} e_k$ is a B-limited rewriting for $k \in \{\alpha, \beta\}$ then we can find p_k and e' such that $e_{\mathtt{init}} \xrightarrow{p} e \xrightarrow{(r_k, l_k)} e_k \xrightarrow{p_k} e'$ is a **B**-limited rewriting for $k \in \{\alpha, \beta\}$.

Corollary (B-limited rewriting is canonical): Each expression $e_{\mathtt{init}}$ has a unique simplified form with respect to **B**-limited rewriting for any $\mathbf{B} \subseteq \mathbf{A_{lam}}$ such that $\mathbf{gen_B}$ contains no cycles.

One might suppose that **R**-limited rewriting is canonical for any $\mathbf{R} \subseteq \mathbf{A_{lam}} \times \mathbf{A_{lam}}$ with no cycles. This conjecture is false. For example, take

$$e_{\mathtt{init}} = (\lambda^1 z.(\lambda^2 x.x\,x)\,(\lambda^3 p.z))\,(\lambda^4 w.w)$$

and $\mathbf{R} = \{(2,3)\}$. There are two choices for the first step of **R**-limited reduction, and the resulting expressions have no common reachable expression.

4.4. Strategies for controlling rewriting

The results of this section suggest the following strategies for controlling rewriting.

(1: Statically-limited) Compute **gen**, choose a maximal subset **B** of $\mathbf{A_{lam}}$ with no gen-cycles, and perform **B**-limited rewriting until termination.

(2: Dynamically-limited) Instead of computing **gen**, merely apply rules, accumulating information about the **xgen** relation as the set of pairs (a, b) such that a has generated b in some step of the rewriting so far, and disallowing any step which would cause this relation to contain a cycle. Since any reduction sequence generated by this method is **R**-limited for some **R** with no cycles, no infinite reduction sequence can be generated.

The first strategy has some obvious advantages. First, it is fully specified in the sense that it terminates with the same final result regardless of the order in which rules are applied. This means that it is simpler to analyze. Another advantage of strategy (1) is that it does not require computing generation pairs (a, b) at each beta reduction step. In practice, since we have no algorithm for computing **gen**, strategy (1) will be implemented using some safe approximation **gen**$'$ of **gen**. One particular approximation is based on a decomposition of **gen** suggested by the following characterization: $\mathbf{gen}(a, b)$ holds just if there is some rewriting e of $e_{\mathtt{init}}$ with a site for application of the beta-v rule such that the ancestor of the abstraction·component is a, the bound variable of that abstraction occurs at least twice in the body, and there is a location within the value component with ancestor b. (For details see [Galbiati and Talcott 1990].)

Let us say that one rewrite-control strategy is *always as powerful* as another if every reduction sequence allowed by the first is allowed by the second. Otherwise we say that the first is *sometimes less powerful* than the second (and the second *sometimes more powerful* than the first). It is interesting to compare the power of strategies (1) and (2) with that of the reduces-size strategy mentioned in the introduction.

Both of the strategies (1) and (2) are sometimes more powerful than the reduces-size strategy (for example consider the second example given in the introduction). Reduces-size rewriting is identical with **R**-limited rewriting with **R** the empty relation. So strategy (2) is always as powerful as the reduces-size strategy.

Strategy (1) is sometimes less powerful than the reduces-size strategy. For example if the initial expression is

$$\mathtt{let}^1\{f := \lambda^3 y.\mathtt{let}^4\{x := y\}\mathtt{let}^5\{w := x\}w\,w\}f\,\lambda^2 z.f$$

then gen includes the cycle $4 \longrightarrow 4$. (To see this, reduce the application of 1; then reduce leftmost applications of 3, 5, and 4. Node 4 generates itself in the last step.) This means that our choice of **B** for statically-limited rewriting cannot include node 4. Thus, statically-limited rewriting will not allow reduction of 4-applications. However the reduces-size strategy allows reduction of a 4-application as the first step.

5. Towards a general purpose simplifier

The ultimate goal of this work is to develop simplifiers which are of practical use as as automatic program manipulation tools. The work presented here provides a foundation for developing general-purpose expression simplifiers. We have extended the beta-v reduction rule by adding rearrangememt rules that substantially increase the simplification power. These rules remain valid for a wide range of extensions of the lambda calculus by primitive operations to permit embedding of traditional programming languages. We have seen that there are trade-offs between maintaining confluent systems and increasing simplification power. What remains to be done is to work out a variety of substantial examples to test the practical applicability of the various strategies and to determine what are the limiting factors in practical situations. In this section we discuss potential deficiencies and possible improvements of our analysis.

5.1. Preserving context information

We separated simplification from the continuation-passing transformation in order to simplify the basic transformation and to develop a generic simplifier that could be shared among a variety of program manipulation tools. Of course this means loss of information. For example a continuation-passing transformer can carry out beta reductions based on knowledge about whether the application came from the original program or was introduced by the transformation. This approach has been successfully used in developing a continuation-passing transformation program [Danvy, private communication].

We gained simplicity by considering only the language of the pure lambda calculus. Following [Landin 1966] we can represent (by adding primitive constants and syntactic sugar) a wide range of language features (block structure, loops, recursive definition, branching, assignment, goto, escape, labels, ...) without invalidating our reduction rules. In fact any set of rules that are valid in the lambda-c calculus will have this property. Again we lose information in translating from a richer language to the lambda calculus and we may want to consider more refined simplification mechanisms based on richer languages. For example [Moggi 1989] treats `let` as a construct distinct from lambda-application and gives a normalizing system of let-reductions. The system includes the analog of beta-value reduction and many instances (but not all) of our rearrangement rules. It also includes rules such as $\mathtt{let}\{x := e\}x \mapsto e$ which are not derivable in our system. It will require further investigation to determine the relative merits of the two sets of rules (and other alternatives) as the basis of simplification systems.

To improve the usefulness of a generic simplifier a language is needed for expressing information such as that discussed above. One such language is the two-level lambda calculus [Nielson 1988]. Here there are two copies of each syntactic construct. The distinction can be interpreted as compile-time vs run-time or as expressing binding time information [Jones *et al.* 1989]. To account for the wide range of information we need to express will require a more general annotation language.

5.2. Adding new rules

In addition to extending the capabilities of a simplifier by increasing the information and lambda rules available one may also wish to add constants to the language and add corresponding delta-rules. These might include rewriting rules for an abstract data type, rules for conditional expressions, rules for updating operations [Mason and Talcott 1989a], or rules for control operations [Talcott 1989, 1990]. In general the combination of two or more terminating rewriting systems does not produce a terminating system. However, [Breazu-Tannen and Gallier 1989] studies combinations of algebraic term rewriting systems and polymorphic lambda term rewriting and shows that properties such as strong normalization and confluence are preserved for a number of combinations.

Acknowledgements

The authors would like to thank Ian Mason for carefully reading earlier versions of this paper, and pointing out numerous obscurities and mistakes. Conversations with Neil Jones and Olivier Danvy have substantially improved our understanding of issues relating to partial evaluation and abstract interpretation.

This research was partially supported by DARPA contract N00039-84-C-0211.

6. References

Abramsky, S. and Hankin, C. (eds.) [1987] *Abstract interpretation of applicative languages* (Michael Horwood, London).

Barendregt, H. [1981] *The lambda calculus: its syntax and semantics* (North-Holland, Amsterdam).

Bondorf, A. [1990] *Automatic Autoprojection of Higher Order Recursive Equations*, ESOP'90.

Breazu-Tannen, V. and Gallier, J. [1989] Polymorphic rewriting conserves algebraic strong normalization and confluence, in: *16th International colloquium on Automata, Languages, and Programming*, Lecture Notes in Computer Science **372**, pp. 137–150.

Galbiati, L. and Talcott, C. [1990] A Simplifier for Untyped Lambda Expressions, Computer Science Department, Stanford University Technical Report.

Jones, N. D., Gomard, C., Bondorf, A., Danvy, O., and Mogensen, T. [1989] A Self-Applicable Partial Evaluator for the Lambda-Calculus, *IEEE Computer Society 1990 International Conference on Computer Languages.*

Jones, N. D., Sestoft, P., and Søndergaard, H. [1989] Mix: A self–applicable partial evaluator for experiments in compiler generation, *Lisp and Symbolic Computation*, **2**, pp. 9–50.

Landin, P. J. [1966] The next 700 programming languages, *Comm. ACM*, **9**, pp. 157–166.

Mason, I. A. and Talcott, C. L. [1989a] A sound and complete axiomatization of operational equivalence between programs with memory, *Fourth annual symposium on logic in computer science*, (IEEE).

Mason, I. A. and Talcott, C. L. [1989b] Programming, transforming, and proving with function abstractions and memories, in: *16th International colloquium on automata, languages, and programming*, Lecture Notes in Computer Science **372**, pp. 574–588.

Moggi, E. [1989] Computational lambda-calculus and monads, *Fourth annual symposium on logic in computer science*, (IEEE).

Nielson, F. [1988] A formal type system for comparing partial evaluators, in Bjørner, D., Ershov, A.P., and Jones, N. D. (editors) *Partial Evaluation and Mixed Computation* (North-Holland) pp. 349–384.

Plotkin, G. [1975] Call-by-name, call-by-value and the lambda calculus, *Theoretical Computer Science*, **1**, pp. 125–159.

Shivers O. [1988] Control Flow Analysis in Scheme, *Proceedings of SIGPLAN '88 Conference on Programming Language Design and Implementation.*

Steele, G. L. [1976] Lambda: the ultimate declarative, Artificial Intelligence Laboratory, Massachusetts Institute of Technology, Technical Report 379.

Talcott, C. [1989] *Programming and proving with function and control abstractions*, Stanford University Computer Science Department Report No. STAN-CS-89-1288.

Talcott, C. [1990] A theory for program and data type specification, to appear in: *International Symposium on Design and Implementation of Symbolic Computation Systems (DISCO 90), Capri, Italy* (Springer-Verlag).

Wadsworth, C. [1978] Approximate reduction and lambda calculus models, *Siam J. Comput.* 7 pp. 337–356.

Weise, D. and Ruf, E. [1990] Computing Types During Program Specialization, CSL-TR-90-441, Computer Systems Laboratory, Stanford University.

PARALLEL GRAPH REWRITING ON LOOSELY COUPLED MACHINE ARCHITECTURES

M.C.J.D. van Eekelen, M.J. Plasmeijer, J.E.W. Smetsers.[1]
Faculty of Mathematics and Computer Science,
University of Nijmegen, Toernooiveld 1, 6525 ED Nijmegen, The Netherlands,
October 1990.

Abstract

Graph rewriting models are very suited to serve as the basic computational model for functional languages and their implementation. Graphs are used to share computations which is needed to make efficient implementations of functional languages on sequential hardware possible. When graphs are rewritten (reduced) on parallel loosely coupled machine architectures, subgraphs have to be copied from one processor to another such that sharing is lost. In this paper we introduce the notion of lazy copying. With lazy copying it is possible to duplicate a graph without duplicating work. Lazy copying can be combined with simple annotations which control the order of reduction. In principle, only interleaved execution of the individual reduction steps is possible. However, a condition is deduced under which parallel execution is allowed. When only certain combinations of lazy copying and annotations are used it is guarantied that this so-called non-interference condition is fulfilled. Abbreviations for these combinations are introduced. Now complex process behaviours, such as process communication on a loosely coupled parallel machine architecture, can be modelled. This also includes a special case: modelling multi-processing on a single processor. Arbitrary process topologies can be created. Synchronous and asynchronous process communication can be modelled. The implementation of the language Concurrent Clean, which is based on the proposed graph rewriting model, has shown that complicated parallel algorithms which can go far beyond divide-and-conquer like applications can be expressed.

1 Introduction

Ideally, a computational model of a language is a formal model as close as possible to both its semantics and its implementation, still it models only the essential aspects of them. In the following paragraphs it is explained why Graph Rewriting Systems (GRS's) are suited to serve as a computational model of functional languages and their implementations. After that, GRS's are extended in order to deal with parallel evaluation.

Graph rewriting systems and functional languages

Our prime interests are functional languages and their implementation on sequential and parallel hardware. Traditionally, the pure lambda calculus (Church (1932/3), Barendregt (1984)) is considered to be a suitable model for these languages. However, in our opinion, some important aspects of functional languages and the way they are usually implemented, cannot be modelled with this calculus. In particular, the calculus itself lacks pattern matching and the notion of sharing of computations. Patterns contain important information for strictness analyzers (Nöcker (1988)). Sharing of computations is essential to obtain efficient implementations on traditional hardware (Fasel & Keller (1986)).

Graph Rewriting systems are based on pattern matching and sharing. We believe that compared to the λ-calculus graph rewriting systems (Barendregt *et al.* (1987a,b)) are better suited to serve as computational model for functional languages. In the past we have defined and implemented the intermediate language Clean (Brus *et al.* (1987)) based on graph rewriting systems with a functional evaluation strategy and we have shown that efficient state-of-the-art implementations on sequential hardware can be obtained by compiling functional languages to Clean (Koopman & Nöcker (1988)).

Parallel evaluation

At any stage during its evaluation a functional program may contain more than one function application that can be rewritten (reducible expression or shorter redex). If in this context redexes are rewritten in any order, the normal form (if it exists) will always be the same. This uniqueness of normal forms offers the theoretical possibility to reduce redexes in parallel. So, functional languages are often considered to be well suited for parallel computation.

Two kinds of parallelism are distinguished: fine grain and coarse grain. In principle, with fine grain parallelism any redex (grain) is a candidate for parallel evaluation. Fine grain machine architectures try to exploit this parallelism fully. Unfortunately, these architectures, such as data flow machines (Gurd *et al.* (1985), Arvind *et al.* (1987)), are very complex and not yet commercially available.

For coarse grain parallelism loosely coupled machine architectures, such as Transputer systems, are available on a wide scale. But now one of the major problems is that most reductions of function applications will not contain a sufficient amount of

[1] This research is partially supported by the ESPRIT basic research action 3074: Semagraph (Semantics and Pragmatics of Generalised Graph Rewriting).

computation compared with the overhead costs caused by the inter-processor communication (grain size problem). Therefore, for these architectures only redexes that yield a large amount of computation are suited to be evaluated in parallel.

In analogy with the concurrent imperative languages, a parallel functional language should provide a way to create concurrent entities (processes) in a program, preferably without violating the functional semantics. Arbitrary communications between processes have to be definable in a general way. Special language constructs have been proposed to make process creation and communication possible (see section 6). Mostly, these constructs are either rather ad hoc or have limited expressive power. We are looking for powerful but elegant basic components needed to realize dynamic process creation with arbitrary communication.

Parallel graph rewriting

In order to deal with parallelism, the graph rewriting model is extended with two issues: a way to control the evaluation order and a way to regulate the distribution of data.

By denoting subgraphs on which reduction processes have to be created, the reduction order in graph rewriting can influenced. Reduction processes which evaluate an indicated subgraph are created dynamically.

A subclass of GRS's in which reducers can be created explicitly will be prefixed with P. So the abbreviation for general graph rewriting systems with explicit parallelism is *P-GRS*.

The distribution of data, which in the context of graph rewriting involves copying of graphs, can be regulated by means of a so called *lazy copying* mechanism. Intentionally, sharing is used to prevent that the same computation is performed more than once. With lazy copying it is possible to make a copy without loosing this advantage. Although in implementations generally some kind of copying/sharing scheme is used, up to now it has never been incorporated in graph rewriting models. For all these reasons we have given a more firm basis to lazy copying by explicitly incorporating it in graph rewriting systems. As we shall see, with lazy copying one can model all the major aspects of data distribution on parallel machine architectures. It is possible to specify whether synchronous or asynchronous inter-processor links are used and also the kind of and the moment at which data is communicated. To handle all these aspects it is unavoidable that the lazy copying mechanism has become very complex. This in contrast with the rather obvious way of creating parallel reducers.

A subclass of GRS's which is extended with lazy copying will be prefixed with C. So the abbreviation for general graph rewriting systems with lazy copying is *C-GRS*.

In this paper it will be shown that arbitrary process structures with various forms of inter-process communication can be modelled in *PC-GRS's* (GRS's with explicit parallelism and lazy copying). In particular, loosely coupled parallel evaluation is defined wherein any process structure can be expressed. In order to illustrate the expressive power examples will be given of some non-trivial parallel algorithms.

Structure of this paper

This paper has not the intention to give a full formal description of a parallel graph rewriting model. As can be seen in Smetsers (1991) such a description becomes very complex. We believe that going into too many technical details at this stage will not help the reader to understand the fundamental issues of what we want to address.

The next section introduces graph rewriting briefly. After that in section 3 process creation is incorporated in graph rewriting. In section 4 lazy copying is introduced. The power of the combination of lazy copying and process creation is shown in section 5. In particular, the use of the system to model parallel graph reduction on loosely coupled parallel architectures is demonstrated. In section 6 comparisons with related work, implementation aspects and directions for future research are given.

2 Graph Rewriting

In graph rewriting systems (Barendregt *et al.* (1987b)) a program is represented by a set of rewrite rules. Each *rewrite rule* consists of a left-hand-side graph (the *pattern*), an optional right-hand-side graph (the *contractum*) and one or more redirections. A *graph* is a set of nodes. Each node has a defining node-identifier (the *nodeid*). A *node* consists of a symbol and a (possibly empty) sequence of applied nodeid's (the *arguments* of the symbol). Applied nodeid's can be seen as references (arcs) to nodes in the graph, as such they have a *direction*: from the node in which the nodeid is applied to the node of which the nodeid is the defining identifier. Starting with an initial graph the graph is rewritten according to the rules. When the pattern matches a subgraph, a *rewrite* can take place which consists of building the contractum and doing the redirections. A *redirection* of one nodeid to another nodeid means that all applied occurrences of one nodeid are replaced by occurrences of the other. The part of the graph that matches the pattern is sometimes called a *redex*.

A *reduction strategy* is a function that indicates one or more of the available redexes. A *reducer* is a process that reduces redexes which are indicated by the strategy. The result of a reducer is reached as soon as the reduction strategy does not indicate redexes anymore. A reducer chooses either deterministically or non-deterministically one of the redexes that are indicated by the strategy. In this paper only deterministic reducers (i.e. reducers which make their choices deterministically) are used. A graph is in *normal form* if none of the patterns in the rules match any part of the graph. A graph is said to be in *root normal form* when the root of a graph is not the root of a redex and can never become the root of a redex. The root normal form property is in general undecidable (Plasmeijer & Eekelen (1991)). Even if a graph has only one unique normal form, this graph may be reduced to several root normal forms depending on how far the subgraphs are reduced.

An important subclass of graph rewriting systems is the class which is defined by the following restrictions:

- all graphs are connected;
- every rule has exactly one redirection which is a redirection from the root of the pattern to the root of the contractum (or when there is no contractum, to the root of a subgraph indicated in the pattern);
- no rule is comparing (rewriting systems where multiple occurrences of variables on left-hand-sides are allowed are called *comparing* or *non left-linear*). No multiple occurrence of variables implies that it is impossible to pattern match on equivalency of nodeid's (sharing). In fact, a left-hand-side is always a graph without sharing (like a term).
- a special reduction strategy is used: the *functional* reduction strategy. Reducing graphs according to this strategy resembles very much the way execution proceeds in lazy functional languages (a full formal definition of this strategy can be found in Smetsers (1991)).

This class will be called: *Functional Graph Rewriting Systems* (*FGRS's*). In an FGRS every rewrite implies that the root of the redex is redirected to another graph. Every node that after the rewrite is not connected to the root of the graph, is considered to be non-existent (*garbage*).

FGRS's serve as basis for Clean. Clean is an experimental functional language based on graph rewriting (Brus *et al.* (1987)). The language is designed to provide a firm base for functional programming. In particular, Clean is suitable and used as an intermediate language between functional languages and sequential machine architectures. Every Clean program is an FGRS.

Although the proposed extensions are also meaningful in more general graph rewriting systems, throughout the rest of this paper it will be assumed that FGRS's are used. In all examples the Clean syntax will be used. The extensions to graph rewriting proposed in this paper are incorporated in a new intermediate language: Concurrent Clean (Eekelen *et al.* (1989)).

For an intuitive understanding of what follows it is not necessary to know all details of FGRS's. Some general knowledge about graphs and functional languages will be sufficient. A few examples of FGRS's:

```
Hd (Cons a b)                     ->  a                                ;

Fib 0                             ->  1                                |
Fib 1                             ->  1                                |
Fib n                             ->  +I (Fib (--I n)) (Fib (-I n 2))  ;

Second (Pair x y:(Cons a b))      ->  y                                ;

Ones                              ->  x: Cons 1 x                      ;
```

Every expression is actually a graph consisting of nodes. Each node contains a symbol and a possibly empty sequence of argument nodeid's. If these nodeid's are implicit, an ordinary tree structure is assumed. Using them explicitly before a :, one can define any graph structure. The last rule in the example is a typical graph rewrite rule containing a cycle in the right-hand-side.

In many cases, the functional graph rewrite rules can intuitively be seen as ordinary function definitions. Each function has one or more alternatives which are distinguished by patterns on the left-hand-side of the definition. Symbols other than function symbols are called *constructors* because they are usually used as data structures (i.e. constructs for defining new data types). For practical reasons some types are assumed to be predefined, such as `INT` or `BOOL`. Furthermore, some functions for arithmetic are assumed to be defined on these types, such as `++I` (i.e. integer increment) or `*I` (i.e. integer multiplication).

3 Extending FGRS's with Dynamic Process Creation: P-FGRS's

The creation of parallel reduction processes, also called *reducers*, can be seen as a special case of influencing the order of evaluation. In FGRS's the reduction order is changed by means of annotations. These annotations, which have the form of a string placed between curly braces, can be assigned to both nodes and nodeid's. When the reduction strategy encounters an annotation it changes its default reduction order which will influence the way in which a result is achieved. Changing the reduction order is important if one wants to optimise the time and space behaviour of the reduction process.

In sequential FGRS's only one annotation is defined indicating that the reduction of the annotated argument of a symbol (function or constructor) is demanded. From an operational point of view, this annotation, denoted by {!}, will force the evaluation of the corresponding argument before it is tried to rewrite the graph according to a rule definition of the symbol. Note that these annotations may make the reduction strategy partially eager instead of lazy.

In formal reasoning about programs with {!} annotations on the left-hand-side it will always be true that the annotated argument will be in root normal form when the corresponding rule is applied. The semantics of annotations on the right-hand-side can be explained via transformations to sets of rules with left-hand-side annotations only. Intuitively, the transformation involves introducing an extra internal reduction with an annotated left-hand-side which forces evaluation before the rule is applied. The precise transformation for {!} can be found in Smetsers (1991).

Example of a rule with a {!} on the right-hand-side:

```
Gen n -> Cons n (Gen {!}(++I n)) ;
```

which is transformed into:

```
Gen n          -> Gen' n (++I n)    ;
Gen' n {!} m -> Cons n (Gen m)      ;
```

Process creation

A single sequential reducer repeatedly chooses one of the redexes which are indicated by the reduction strategy and rewrites it. Interleaved reduction can be obtained by incarnating several sequential reducers which reduce different parts of the same graph.

By using {!} annotations it is possible to influence the order in which the redexes are reduced by a single reducer. Now a new annotation is introduced: {!!}, to indicate that a new sequential reducer has to be created with the following properties:

- the new reducer reduces the corresponding graph to root normal form after which the reducer dies;
- the new reducer can proceed interleaved with the original reduction process;
- all rewrites are assumed to be indivisible actions;
- if for pattern matching or reduction a reducer needs access to a graph which is being rewritten by another reducer, the first reducer will wait until the second one has reduced the graph to root normal form.
- for determining its redexes it a uses the functional reduction strategy parameterized with {!} annotations.

Considered operationally, if a {!!} annotation is encountered in the right-hand-side by a reducer, a new reducer is created after the redirection has been done (and if there is copying, also after the copying). If a {!!} annotation is specified on the left-hand-side, a new reducer is created just before the original reducer would reduce the corresponding function application.

In reasoning about programs with {!!} annotations on the left-hand-side it will always be true that the annotated argument will have been reduced (by another reducer) to root normal form when the corresponding rule is applied. The meaning of {!!} on a left-hand-side can be explained via transformations to sets of rules with right-hand-side annotations only (Smetsers (1991)).

Example of {!!} on the right-hand-side:

```
Fib 0          ->   1                                          |
Fib 1          ->   1                                          |
Fib n          ->   +I ({!!} Fib (-I n 1)) ({!!} Fib (-I n 2))  ;
```

Another way of looking at {!!} annotations is that they influence the overall reduction order. In this view, {!!} annotations are parameters of the overall reduction strategy. This overall reduction strategy will then indicate possibly more than one redex (every process may have a redex). The global reducer will make a non-deterministic choice out of the redexes indicated by the global strategy. So, in this way parallel reduction is modelled via (non-deterministic) interleaved execution of the individual reducers. In section 5 we will see how real parallel evaluation can be made possible.

4 Extending FGRS's with Lazy Copying: C-FGRS's

A notion of graph copying is necessary if one wants to express explicitly the distribution of data in parallel environments. One would expect however that it is already possible to express graph copying in graph rewriting systems. Although this is indeed the case, it is rather complex.

A function has to be defined which duplicates its argument. Evidently, the following definition only produces two pointers to the argument but it does not duplicate the argument itself!

A graph sharing example:

```
Duplicate x -> Pair x x                ;
```

the left graph

```
@1: Duplicate @2,
@2: Pair @3 @2,
@3: 1;
```

@x is a denotation for a nodeid

reduces to the right graph

```
@4:  Pair @2 @2,
@2:  Pair @3 @2,
@3:  1;
```

which is illustrated in the following picture

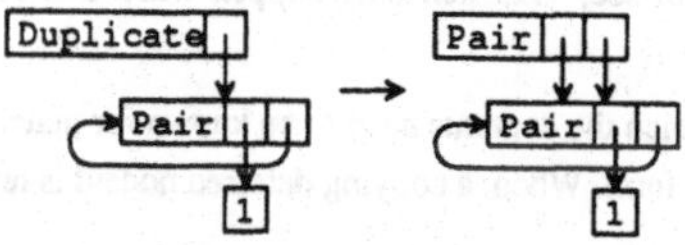

The only way to access the structure of the argument is to use pattern matching. The only way to duplicate a constructor is to match on it on the left-hand-side and to create a new node with the same constructor on the right-hand-side. Such a rewrite rule is needed for every constructor that may appear. Furthermore, on the right-hand-side the graph structure of the argument has to be duplicated. To detect sharing multiple occurrences of the same nodeid on the left-hand-side should be introduced in FGRS's. Then, with many of such left-comparing rules (and a special strategy that handles left-comparing rules) a structure can be copied. The rules that define copying, are themselves part of the system which makes it difficult to reason about them because the copying gets intertwined with the rest of the evaluation. As a consequence, if such a structure contains redexes that have to be copied too the reduction strategy has to be changed again in order to prevent that the strategy indicates these redexes.

So, extending the semantics of FGRS's with a special mechanism to explicitly copy graphs (possibly containing redexes) would considerably increase the expressive power of these graph rewriting systems.

Eager copying

To denote a graph g that should be copied, the node identifier that refers to the root of g is attributed with a subscript c. The c subscript can be placed on nodeid's of the right-hand-side only. The copying takes place <u>after</u> the contractum is built but <u>before</u> the root of the redex is redirected to the root of the contractum. All copies of one right-hand-side are simultaneously.

Copying a graph g implies that an equivalent graph g' is made which has no nodes in common with the original graph g. During the copying no rewriting takes place. So, for every node of g (also for redexes) there is an equivalent node in g'.

A graph copying example:

$\texttt{Duplicate x -> Pair x } x_c \quad ;$

the left graph

```
@1: Duplicate @2,
@2: Pair @3 @2,
@3: 1;
```

reduces to the right graph

```
@4: Pair @2 @12,
@3: 1,
@2: Pair @3 @2,
@12: Pair @13 @12,
@13: 1;
```

which is also illustrated in the following picture

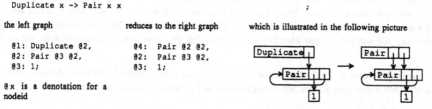

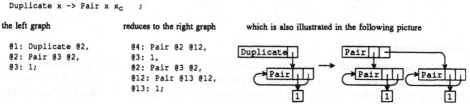

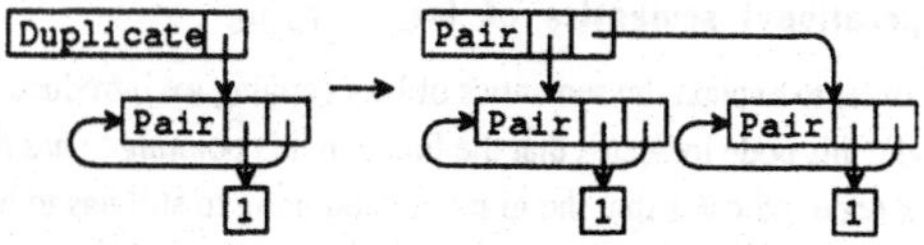

The new nodeid's are chosen in such a way that the structure is easily seen.

This way of copying is also called *eager copying* in contrast to lazy copying which is defined in the following sections.

Lazy copying

Take a graph containing redexes. The extension of explicit copying to graph rewriting introduces the possibility to copy this graph including all its redexes. We had already the possibility of sharing the graph. Unfortunately, there is nothing in between. However, duplication of work can be avoided by maintaining the sharing with the original graph as long as the corresponding function applications have not been evaluated. If after the evaluation to root normal form the copying is continued, the graph is duplicated after the work is done. But, also it can be useful to break up the sharing. Take for example a function application that delivers a large structure after relatively few reduction steps. If a graph containing such a function application is submitted to another processor then it is preferable not to reduce this application before the submission.

Copying with the choice of maintaining or breaking up the sharing is called *lazy copying*.

A node on which copying will be stopped temporarily, is called a *deferred node*. To denote a deferred node it is attributed with a subscript d. Because every node has an explicit symbol, it is syntactically convenient to attach the attribute of the node to the symbol of the node.

Lazy copying implies that when a copy action hits a deferred node, the copying itself is deferred. The applied occurrence of the nodeid of the deferred node of which the (now deferred) copy was being made, will be administered as being a *copying deferred nodeid*. When a deferred node is in root normal form, the node will not longer be deferred. The actual copying may continue, but, as we shall see, this will only happen when this copy is demanded. The actual copy of a deferred node will not be deferred.

Nodeid's of which the contents need to be known for matching, are according to the functional reduction strategy first reduced to root normal form. When a copying deferred nodeid is reduced this will trigger the continuation of the copying.

A lazy copying example:

```
Start           -> Duplicate (Fac_d 6)        ;

Duplicate x     -> Pair x x_c                 ;

Fac 0           -> 1                          |
Fac n           -> *I n (Fac (--I n))         ;
```

the following rewrites occur:

```
@1:  Start;         ->     @2:  Duplicate @3,   ->     @4:  Pair @3 @3_c,   ->>
                           @3:  Fac_d 6;               @3:  Fac_d 6;

@4:  Pair @5 @5_c,  ->>    @4:  Pair @5 @5_c,   ->     @4:  Pair @5 @15,
@5:  720_d;                @5:  720;                   @5:  720,
                                                       @15: 720;
```

The nodeid attribute c in the graph is used to denote that that the nodeid is a copying deferred nodeid. Note that the c attribute was inherited when the node @3 was redirected to @5 which corresponded with the reduction of the node. Do not confuse the c attribute in the graph with the c attribute in the rules which denotes that a copy action has to be started. The deferred attribute of the node @5 is taken away when it is recognized that the node is in root normal form.

The rewrites are also shown in the following picture:

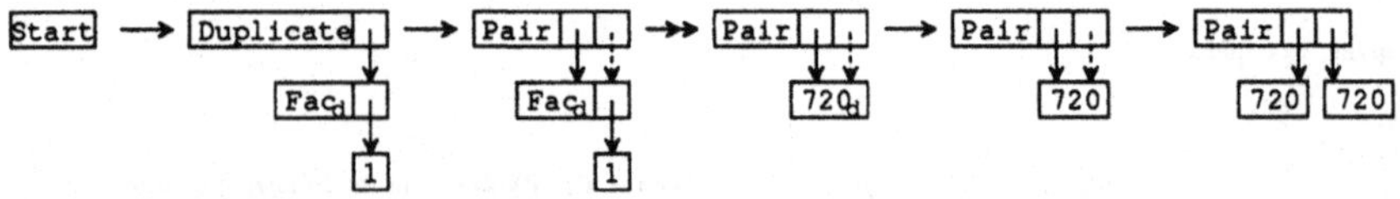

Operational semantics of lazy copying

In order to explain the semantics of lazy copying we introduce two special kind of indirection nodes are introduced: a D(eferred) node: this node indicates that the function it is pointing to has the deferred attribute. And a C(opy of such a deferred node) node: this node indicates that the graph it is pointing to still has to be copied: the copying is deferred. If on a right-hand-side nodeid n is attributed with the c subscript, all nodes accessible from n have to be copied such that the new graph structure is copy-equivalent with the old one. However, if the copy action hits on a D-node, a C-node which refers to the D-node is created and the subgraph to which the D-node refers is not copied. If the copy action hits on a C node, a new C node is created which has the same argument as the original C node. After the copying has been performed this way, this rewrite is finished and reduction continues as usual. D and C-nodes can be rewritten via the following *internal* reduction rules:

```
D (!) x          -> x          ;
C (!) x          -> x_c        ;
```

The strict annotations provide the property that a function application is "deferred" or "not yet copied" property is inherited by all intermediate function results until finally a root normal form is reached. Hereafter the reducer is able to apply the special rewrite rules for D and C which will make these nodes disappear.

If the previous example is considered again, it should be more clear what the semantics are:

```
@1:  Start;            ->       @2:  Duplicate @i,   ->       @4:  Pair @i @j,      ->>
                                @i:  D @3,                    @i:  D @3,
                                @3:  Fac 6;                   @j:  C @i,
                                                              @3:  Fac 6;

@4:  Pair @i @j,       ->       @4:  Pair @5 @j,     ->       @4:  Pair @5 @15,
@i:  D @5,                      @j:  C @5,                    @5:  720,
@j:  C @i,                      @5:  720;                     @15: 720;
@5:  720;
```

which is also illustrated in the following picture:

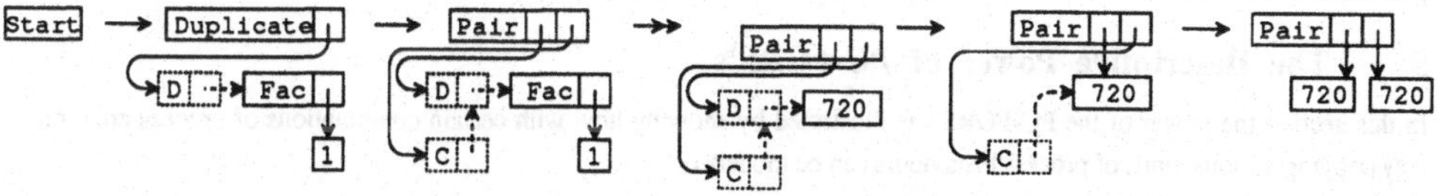

Note that when the deferred copy turns out to be not needed by the reduction strategy, the C rule will never be executed, so the copying will not be continued.

Properties of lazy copying

An interesting aspect of lazy copying is that normal forms do not contain defer or copying deferred attributes. In a normal form every subgraph is trivially in root normal form. Evaluation of nodes to root normal form eliminates the defer attributes. Evaluation to root normal form and/or the attempt to access a node will cause the deferred copying to continue.

Normal forms:

With the following rule: / this will be the normal form of `Start`:

```
Start -> x: Pair  1 x ;       @1: Pair 1 @1;          || a cycle.

Start -> x: Pair_d 1 x ;      @1: Pair 1 @1;          || a cycle.

Start -> x: Pair  1 x_c;      @1: Pair 1 @11,         || a cycle with a copy of it:
                              @11:Pair 1 @11;         || a once unravelled cycle.

Start -> x: Pair_d 1 x_c;     @1: Pair 1 @11,         || every copy contains a copying deferred
                              @11:Pair 1 @1_c;        || nodeid which leads to more unravelling
                                                      || yielding an infinite normal form!
```

Lazy copying does influence the normal forms in the graph world. Sharing may be broken up when a cycle is copied which contains deferred nodes. The result will be partly unravelled with respect to a full copy. A typical example is given by the latter rewrite rule of the previous example.

In C-FGRS's the normal form is also influenced by the order of evaluation (and hence by annotations). If the deferred nodes are not reduced before an attempt to copy them is made, the result will be partly unravelled. A typical example is given below.

With the following rules: / the normal form is: / without the `{!}` the normal form is:

```
Start    ->    r: A (F x),         @1 : A @13,        @1   : A @13,
               x: B y,             @13: B @15,        @13  : B @15,
               y: {!}I_d z,        @15: C @13;        @15  : E @113,
               z: E x;                                @113: B @15;
```

```
F x        ->    x_c;
I x        ->    x;
```

Note that an extra rule had to be introduced in order to delay the copying.

The unravelling of the normal forms of a rule system with lazy copying will always be the same as the unravelling of the normal forms of the same rule system without lazy copying. In other words unravelling is invariant with respect to lazy copying. This is an interesting property for the implementation of functional languages and for term graph rewriting as in Barendregt *et al.* (1987a). It seems that it enables the proof of the soundness and completeness of implementations which use sharing and copying via term graph rewriting. Lazy copying and term graph rewriting is a very promising topic for further research.

With the `copy` indication and the `defer` indication lazy copying is introduced in graph rewriting systems. By introducing the possibility to use subtle combinations of sharing and copying this greatly improves the expressive power of graph rewriting systems. Furthermore, in the next section it will be shown that lazy copying can also be the basis for communication in a parallel environment.

5 The Descriptive Power of PC-FGRS's

In this section the power of the PC-FGRS's is illustrated by showing how with certain combinations of process creation and lazy copying various kinds of process behaviours can be modelled.

There are several kinds of behaviours one may be interested in, such as fine and coarse grain parallelism, all kinds of process topologies (hierarchical and non-hierarchical process topologies), synchronous and asynchronous communication between processes, etcetera.

At first glance it may seem easy to specify these behaviours in PC-GRS's, since there is the possibility to create reducers dynamically. However, note that a rewriting step is considered to be indivisible and without this assumption reasoning about rewriting systems is in general not possible. Still, of course, one would like to be able to create reducers of which the rewriting steps can be performed in *parallel* instead of *interleaved.* However, it should be clear that, without any restrictions, parallel rewriting causes problems. Imagine that a copy of a subgraph is made while another reducer is working on that subgraph. Problems may also arise when redirections are performed in parallel. Probably there will not be a problem when two reducers are running on subgraphs which have no node in common and no reference to each other.

To call a reducer a *parallel* reducer with respect to another reducer it has to be proven that the constraint that a rewrite step is an indivisible action can be weakened. More precisely, it has to be proven that the corresponding rewrite steps cannot interfere with each other and therefore may be *considered* as being indivisible so that they actually can be performed in parallel. This condition that has to be proven is also called the *non-interference* condition. Hence, the claim that parallel computations can be expressed in our model can only be justified by proving that, under specific conditions, certain reducers are parallel reducers with respect to certain other reducers.

As in the introduction, we consider loosely coupled parallel machine architectures (each processor has its private memory) as the most interesting class of architectures. It should be clear that the kind of architecture the reduction is performed on influences the rewriting model. For instance, in a shared memory machine graph copying may be superfluous.

5.1 Modelling parallel rewriting on loosely coupled parallel architectures

A loosely coupled parallel computer is defined as a multiprocessor system that consists of a number of self-contained computers, i.e. processors with their private memory attached to each other by a sparsely connected network. An important property of such system is that for each processor it is more efficient to access objects located in its own local memory than to use the communication medium to access remote objects. In order to achieve an efficient implementation it is necessary to map the computation graph to the physical processing elements in such a way that the communication overhead due to the exchanging of data is relatively small. Therefore, the computation graph is divided into a number of subgraphs (*grains*) which have the property that the intermediate links are sparsely used.

Unfortunately, it is undecidable how much work the reduction of a subgraph involves. Furthermore, there are no well-established heuristics for dividing a graph into grains. So, this partition of the graph cannot automatically be performed. Therefore, in the program it has to be explicitly indicated what is expected to represent a large amount of reductions relative to the expected communication overhead. In this way the program can be tuned to a particular parallel machine architecture.

The annotations and indications in the PC-FGRS have to be used in such a way that non-interference can be proven for reducers which might be executed on different processors. In order to avoid the need for a proof for every PC-FGRS methods of annotating and indicating will be developed. Using these methods will guarantee that parallel execution of groups of reducers is allowed.

Divide-and-Conquer evaluation

An obvious method to get safe parallelism is to create a reducer on a copy of an indicated subgraph. Such a copied subgraph has the property that it is *self-contained*, i.e. the root of the subgraph is the only connection between the subgraph and the rest of the graph. This will make it possible that the copied subgraph is reduced in parallel on another processor. When it is reduced to root normal form the result will be copied back to the father processor. So, copying is performed twice: one copy is made of the task for the off-loading of the task and one copy is made of the result to communicate it to the father.

A self-contained subgraph will be regarded as a *virtual processor* because it has the property that it may be reduced on another processor.

It is easy to prove that on a self-contained subgraph it is allowed to weaken the interleaving restriction to parallelism: the self-contained subgraph can only be accessed by other reducers via the root and the semantics of P-FGRS's does not allow reducers to access a node on which another reducer is running.

Example of a divide-and-conquer algorithm:

```
Fib 0     ->   1                                  |
Fib 1     ->   1                                  |
Fib n     ->   +I left_c right_c,
               left:  {!!}Fib_d (-I n_c 1),
               right: {!!}Fib_d (-I n_c 2)       ;
```

The {!!} annotations combined with the `copy` and `defer` indications specify that both calls of `Fib` can be evaluated in parallel. The graph on which each process runs is self-contained because the root of the graph on which a process is started, is built with copies of subgraphs as arguments. The father reducer is already started with copying the result but this is immediately deferred. The copying of the result can continue each time when an argument of `+I` is in root normal form.

The following picture illustrates the virtual processor structure after one reduction of `Fib 5`:

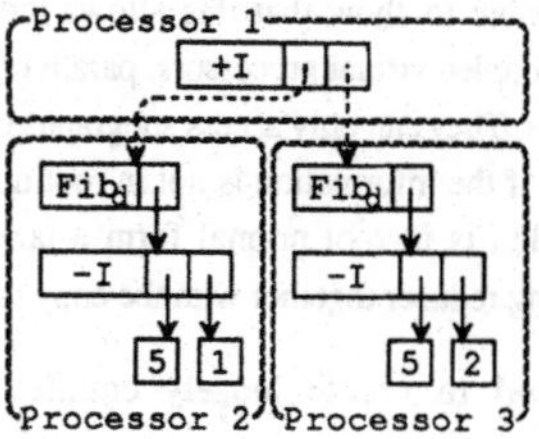

This way of modelling divide-and-conquer algorithms relies on the fact that the subgraph to be reduced is self-contained and that after the reduction to root normal form, the result is also self-contained. Unfortunately, self-containedness is an undecidable property, for, if a lazy copy of a certain graph is made this graph <u>may</u> contain deferred nodes. But, as will be shown in the next section, it is possible to use a graph property that is on the one hand weaker than the property of self-containedness and, on the other hand, strong enough not to violate the non-interference requirement.

Modelling loosely coupled evaluation

A method which makes it possible to model process behaviours that are more general than divide-and-conquer, must provide a way to define arbitrary connections between processes and processors. The lazy copy scheme introduced in section 4 provides a way to make a self-contained copy on a lazy manner. Such a lazy copy is a self-contained subgraph with the exception of copying deferred nodeid's, which are references to deferred nodes in the graph. These deferred nodes will be copied *later* if they are in root normal form and needed for the evaluation. So, copying deferred nodeid's are natural candidates for serving as inter-connections between parallel executing processes because they induce further copying when they are accessed. Therefore, communication between parallel processes can be realized via copying deferred nodeid's. In this context copying deferred nodeid's are also called *communication channels* or just *channels.*

A subgraph is *loosely connected* if channels (copying deferred nodeid's) are the only connections between the subgraph and the rest of the graph. Note that this implies that a self-contained subgraph is loosely connected if its root is a channel. Also a loosely connected subgraph is regarded as a virtual processor because it has the property that it may be reduced on another processor. From now on the notion virtual processor stands for loosely connected subgraph. Several processes (reducers) can run on such a virtual processor. Processes running on the same virtual processor are running interleaved. So, there is interleaved multiprocessing on each virtual processor. Processes running on different virtual processors run in parallel.

The semantics of copying deferred nodeid's implies that channels have the following properties. The flow of data through a channel is the reverse of the direction of the copying deferred nodeid in the graph. Since channels are nodeid's, they can be passed as parameters or copied. Now, suppose that a parallel process is reducing a loosely connected subgraph. This process may need the reduction of a channel connected to another processor. Of course, this channel cannot be reduced by the demanding process. It has to be reduced by another process running on the virtual processor which contains the graph whereto the channel refers. Now, the demanding process will be suspended until the result has been calculated by the process running on the other processor. A channel can be used to retrieve an (intermediate) result in a demand-driven way, i.e. as soon as the result of a sub-reduction is needed a request for the result is made. This request will be answered if the corresponding result is in root normal form. Note that the channel vanishes after the result has been returned. Because the copying is lazy new channels may have come into existence.

The question is now when the non-interference condition is fulfilled for reducers running on different virtual processors such that they can run in parallel instead of interleaved. The non-interference condition is satisfied if it can be guarantied throughout the execution of the program that when a parallel reducer is demanding information from a channel which refers to another virtual processor,

- this subgraph is either in root normal form (such that it can be lazy-copied to the demanding process) or,
- there is a process running on the other virtual processor which is reducing the subgraph if it is not yet in root normal form (such that the demanding process will wait until the information has been reduced to root normal form).

Virtual processors which satisfy these conditions are called *loosely coupled virtual processors.*

It is possible to show that this allows the weakening of the restriction of interleaving to parallelism with respect to the loosely coupled virtual processors: parallel reducers running on different virtual processors work on different loosely connected subgraphs. They can only access subgraphs on other processors via copying deferred nodeid's (channels). The demanding reducer will wait if the information is not in root normal form because in that case another process is reducing the information. If the information is in root normal form a lazy copy is made. In that case the resulting graph, i.e. the original graph of the demanding reducer together with the copy that has been made, is also loosely connected.

A method to create loosely coupled virtual processors

The obvious way to guarantee that virtual processors are loosely coupled, is to create a reducer on every deferred node. Hence, when a deferred node is created, at the same time also a process is started which reduces the deferred node. So, when via a copy a channel will be created to the node, the node will already be in root normal form or a reducer is still reducing it to root normal form.

First, we introduce two abbreviations `{e}` and `{i}` that can be put on a node `n`.

Example:

```
Fib n      ->     +I left right,
                  left:  {i} Fib (-I n 1),
                  right: {e} Fib (-I n 2)
```

The `{e}` abbreviation (e for external) will create a new loosely coupled virtual processor together with an external reducer which reduces the corresponding loosely connected subgraph in parallel. To realize this, a channel to a lazy copy of the subgraph is made and a process is created to reduce this copy. The channel provides that a (lazy) copy of the result is returned if its value is demanded on other processors. In particular a lazy copy of the result is returned to the father process if it demands its value.

The `{i}` abbreviation (i for internal) will create a new internal reducer on the same virtual processor which reduces the corresponding subgraph interleaved with the other processes on the same virtual processor. A deferred node to this subgraph is created which provides that a (lazy) copy of the result is returned if its value is demanded on other virtual processors (since all virtual processor are created via lazy copies, this demand will come via a channel). To realize this, a deferred node to the indicated subgraph is made and a process is created to reduce it.

The `{e}` and `{i}` abbreviations may be used on the same positions as annotations. When an `{e}` or `{i}` abbreviation is put on a nodeid, this is equivalent with putting it on the node the nodeid belongs to. For each occurrence on a node a simple program transformation is made as follows:

Each occurrence of: will be substituted by:

```
n : {e} Sym a1 .. an        n : I xc,
                            x : {!!} Id yc,
                            y : Sym a1 .. an
```

A reducer is created by the {!!} annotation, it will reduce a node which contains the identity function of a lazy copy of the annotated node `Sym a1 .. an`. The node on which the reducer is started, is itself deferred and a channel is immediately created to it via the copy in the new definition of the node n.

```
n : {i} Sym a1 .. an        n : {!!} Id x,
                            x : Sym a1 .. an
```

A reducer is created on a deferred node. All sharing is maintained.

The nodeid's x and y in the substitution rules stand for nodeid's not used elsewhere in the rewrite rule.

`I` is just the identity function: `I x -> x;`

The indirection nodes are only created to see to it that the copies are made correctly. In the following they are considered to be internal nodes.

It can be proven that, when using the {e} and {i} abbreviations only (i.e. neither other defer or copy attributes nor other process annotations), it is guarantied that each subgraph supplied with an {e} denotes a loosely coupled virtual processor. So, the proposed abbreviations provide a method to create loosely coupled virtual processors which allows real parallel execution.

5.2 Examples of the use of the proposed method

In this section some small examples are given illustrating the expressive power of the method for loosely coupled evaluation.

Non-hierarchical process topology

With the {e} abbreviation parallel (sub)reduction can be created and distributed over a number of virtual processors. With the creation of internal processes by using {i}, multiprocessing can be realized on each virtual processor. The only way to refer to such an internal process is via its channel. If such a channel node is passed (via a lazy copy) to another virtual processor, a communication channel between this processor and the reducer on the original processor is established. In this way any number and any topology of communication channels between processes and processors can be set up. For instance, it is possible to model a cycle of virtual processors. An example of this is given at the end of this section. In the following example a simple non-hierarchical process topology is demonstrated. It serves the purpose of explaining how such process topologies can be expressed (it does not realistically implement the Fibonacci function).

The `Fib` example using a non-hierarchical process structure (which is very unconventional for `Fib`): the second call of `Fib` will be executed on another virtual processor but the argument of that call is reduced internally on the virtual processor that also does the first call of `Fib`.

```
Fib 0     -> 1                        |
Fib 1     -> 1                        |
Fib n     -> +I (Fib (-I n 1)) m,
             m: {e} Fib o,
             o: {i} -I n 2            ;
```

which is equivalent to:

```
Fib 0     -> 1                        |
Fib 1     -> 1                        |
Fib n     -> +I (Fib (-I n 1)) m,
             m : I xc,
             x : {!!} Id yc,
             y : Fib o,
             o : {!!} Id z,
             z : -I n 2               ;
```

So the following process topology is obtained (a snapshot of the program execution of `Fib 5` is given):

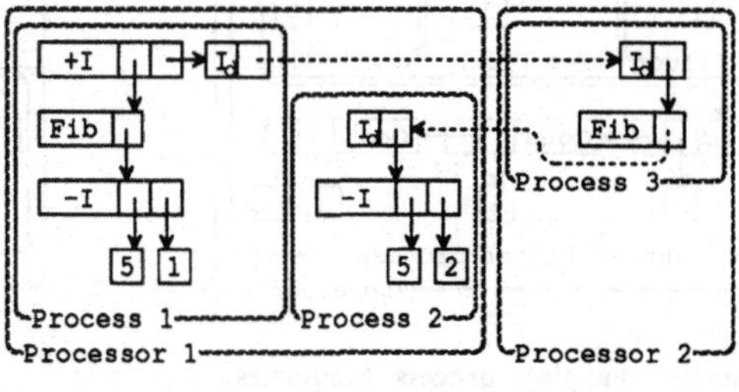

In the picture it is shown how the graph is distributed over two virtual processors. Channels are dashed. Note that the direction of the flow of data through a channel is the reverse of the direction of the corresponding reference in the graph. In the following, internal indirection nodes are not shown in pictures and their defer indications are added to the nodes they refer to.

Asynchronous virtual processor communication with streams

It is possible to model asynchronous communication between virtual processors, i.e. a virtual processor is already computing the next data before the previous data is communicated. To achieve this a family of internal processes has to be created connected to the communication channel between the processors. Each process computes a partial result which can be sent across the channel. Just before a process delivers the partial result (and dies) it creates a new process chained via a new channel to the delivered result. This new member of the family will compute the next partial result on the same way. For convenience sake, such a cascaded family of processes is often regarded as being one (asynchronously) sending process with some family name. The chain of channels is then regarded to be one channel. The total result which is copied, is sometimes called a *stream*. Note that this kind of stream is capable of sending over more than one node (a *burst*) at the same time. Furthermore, these streams can contain cyclic graphs such that cycles can be sent to another processor.

A virtual processor may contain several such families each producing a stream via a chain of channels. In the case of the following filter example the virtual processor contains exactly one such process: `Filter`. It sends a stream via the channel to the process `Print`.

The following example describes an asynchronous communication behaviour with streams:

```
Start list                  ->   Print s,
                                 s: {e} Filter list 2        ;

Filter Nil         pr       ->   Nil                         |
Filter (Cons f r) pr        ->   IF (=I (MOD f pr) 0)
                                    (Filter r pr)
                                    (NewFilter f r pr)       ;

NewFilter f r pr            ->   Cons f rest,
                                 rest: {i} Filter r pr       ;
```

The main virtual processor creates a new virtual processor on which the `Filter` process is started. The channel `s` is the communication channel between the two processors. The function `Filter` removes from its first argument, which is a list, all the elements which are divisible by the number `n`. A part of the stream becomes available as soon as `Filter` has computed an element of the result list and a new interleaved `Filter` process has been created. It may start already computing the next element of the stream before the first is asked to be communicated. The partial stream result is a list containing the first element and a new channel reference to the new filtering process.

Assume that the list to be filtered is the list containing the natural numbers from 1 to 7. Then the following situations can arise:

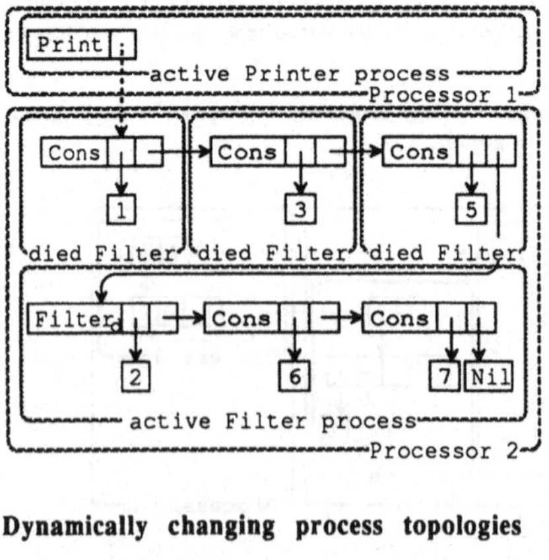

Now the three list elements, root normal forms yielded by successive filter processes, can be shipped with one lazy copy action.

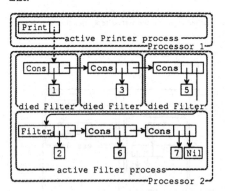

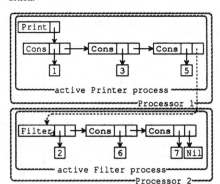

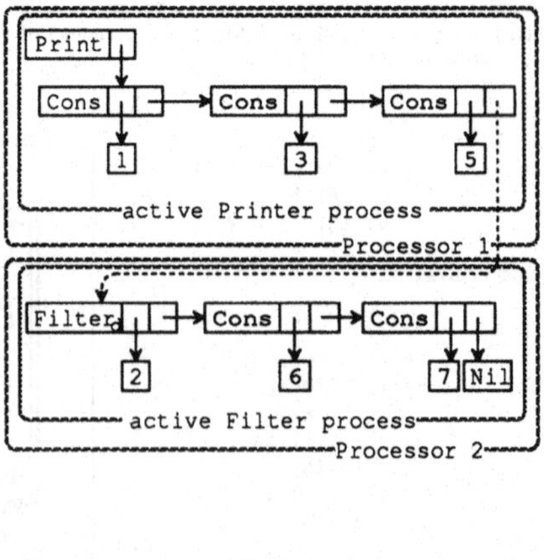

Dynamically changing process topologies

The sieve of Eratosthenes is a classical example which generates all prime numbers. A pipeline of virtual processors is created. On each processor a `Sieve` process (a family of processes actually) is running. Those Sieves hold the prime numbers in ascending order, one in each `Sieve`. Each `Sieve` accepts a stream of integers as its input. Those integers are not divisible by any of the foregoing primes in the pipeline. If an incoming integer is not divisible by the local prime as well, it is send to the next Sieve. A newly created Sieve accepts the first incoming integer as its own prime and outputs this prime and the channel of the next Sieve to a printing processor. After that it starts sieving. A virtual processor called Gen sends a stream of integers greater than one to the first Sieve.

The Gen process and every Sieve process proceed in more or less the same way as the Filter process of the previous example. They all are actually families of processes servicing chains of channels. They are regarded as single processes. Every chain of channels is regarded as one channel.

So Sieve1 holds 2 as its own prime, Sieve2 holds 3, Sieve3 holds 5, and so on. The printing process one by one receives the channel identifications from these sieves and collects the corresponding primes. Seen through the time this can be illustrated as follows (all arrows indicate flow of data on channels):

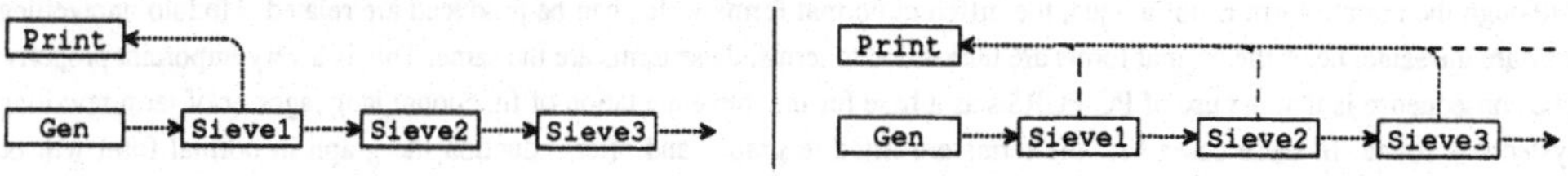

The Sieve program:

```
Start                       ->    Print s,
                                  s: {e} Sieve g,
                                  g: {e} Gen 2                  ;

Sieve (Cons pr stream)      ->    Cons pr s,
                                  s: {e} Sieve f,
                                  f: {i} Filter stream pr       ;

Gen n                       ->    Cons n rest,
                                  rest: {i} Gen {!} (++I n)     ;

Filter (Cons f r) pr        ->    IF (=I (MOD f pr) 0)
                                     (Filter r pr)
                                     (NewFilter f r pr)         ;

NewFilter f r pr            ->    Cons f rest,
                                  rest: {i} Filter r pr         ;
```

Note that when the {!} annotation in Gen would be left out, the increments of the integers would not be evaluated by Gen but by the first Sieve. Even worse: because the result of Gen is copied, the Sieve would have to recalculate every new integer by increments starting from 2.

Cyclic process structures

The next example shows how a cyclic process structure, i.e. a number of parallel reducer that are mutual dependent, can be created. This example has been extracted from quite a large program that implements Warshall's solution for the shortest path problem. The full algorithm can be found in Eekelen (1988).

First the intended reducer topology is given in a picture:

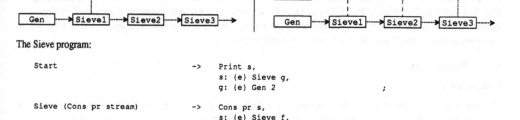

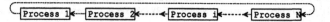

This reducer structure can be specified directly in the following way:

```
Start                   -> last:CreateProcs NrOfProcs last        ;

CreateProcs 1 left      -> Process 1 left                         |
CreateProcs pid left    -> CreateProcs (--I pid) new,
                           new: {e} Process pid left              ;
```

CreateProcs is responsible for the generation of all the parallel reducers. This process, which will finally become the first reducer, has initially a reference to itself in order to make it possible to expand it to a cycle of reducers. Each reducer is connected to the next one, i.e. the one with the next pid number, by means of a channel. During the creation of the processes this channel is passed as a parameter called left

5.3 Properties of the proposed method

With the proposed abbreviations arbitrary process structure can be expressed clearly. Still one has to be careful with their use. Normally the abbreviations will be used to obtain a parallel version of an ordinary sequential program. In general the sequential program has to be transformed to create the wanted processes and process topologies. If the abbreviations of any parallel program are regarded as comments, again a sequential version of the program is obtained. In the given examples such a sequential version would yield the same result as the parallel version. Unfortunately, in general the normal form is not unique. In section 4 it was that the normal form in a C-FGRS depends on the order of evaluation. In section 3 it was explained that the overall reduction strategy of a P-FGRS is non-deterministic. Hence the normal form PC-FGRS will in general depend on the choices made by the reducer.

Although the normal form is not unique, the different normal forms which can be produced are related. Modulo unravelling they are the same, i.e. if the normal forms are unravelled to terms, these terms are the same. This is a very important property. The consequence is that the use of PC-FGRS's as a base for the implementation of functional languages or of term rewriting systems is sound. In these cases first the terms are lifted to graphs and after reduction the graph in normal form will be unravelled to a term again. Then, always the same term will be yielded.

6 Discussion

Related work

The idea to use annotations (Burton (1987), Glauert *et al.* (1987), Goguen *et al.* (1986), Hudak & Smith (1986)) or special functions (Kluge (1983), Vree & Hartel (1988)) which control the reduction order is certainly not new. Some of them are introduced on the level of the programming language (Burton (1987), Hudak & Smith (1986), Vree & Hartel (1988)) while others are introduced on the level of the computational model (Glauert *et al.* (1987), Goguen *et al.* (1986), Kluge (1983)). They all express that an indicated expression has to be shipped to another (or to some concrete) processor. Most annotations (Hudak & Smith (1986), Goguen *et al.* (1986), Kluge (1983), Vree & Hartel (1988)) are only capable of generating strict hierarchical "divide-and-conquer parallelism". Non-hierarchical process structures are possible in Burton's proposal. He proposes a call-by-name parameter passing mechanism (which must involve copying of some nodes) between mutual recursive functions. In DACTL (Glauert *et al.* (1987)), also based on Graph Rewriting Systems (Barendregt *et al.* (1987b)) there is no overall reduction strategy. This means that the reduction order is completely controlled by the annotations in the rewrite rules. This makes DACTL very suited for fine grain parallelism, but makes it very hard to reason about the overall behaviour of the program. In all proposals copying graphs from one processors to another and back is implicit and cycles cannot be copied.

Some annotations (Burton (1987), Hudak & Smith (1986)) are not only used to control parallelism but also to control the actual load distribution. Annotations for load distribution are not yet incorporated in the model, primarily because virtual processors can be freely created on the level of the computational model. However, the specification of load distribution will be investigated in the future.

Implementation aspects

Efficient implementation of FGRS's is possible on sequential hardware (Brus *et al.* (1987), Smetsers (1989)). The ideas introduced in this paper are incorporated in the language Concurrent Clean (van Eekelen *et al.* (1989)). Type information (Plasmeijer & van Eekelen (1991)) and strictness information ((Nöcker (1988), (Nöcker & Smetsers (1990)) play an important role.

To investigate parallel programming a simulator for Concurrent Clean has been developed simulating multi-processing running. This simulator runs in any sequential C environment. Experiments with this implementation indicate that PC-FGRS's are in principle very suited for implementation on loosely coupled parallel architectures. Most problems that have to be solved are of a general nature: "How can a graph (with cycles) be shipped fast from one processor to another?", "What is the best suited algorithm for distributed garbage collection?", "What happens if one of the processors is out of memory or is completely out of order?". The efficiency of a parallel implementation will strongly depend on the solutions found for these general type of problems. These problems have to be solved for other kinds of concurrent languages too. Perhaps it is possible to adopt existing solutions. But also alternative solutions which take the special behaviour of GRS's into account are thinkable.

Future work

Besides the concepts introduced in this paper (lazy copying, annotations for dynamic process creation, abbreviations) we will add annotations for load distribution and add predefined rules such that frequently used process topologies (pipelines, array of processes) can easily be defined. Efficient implementation of Concurrent Clean are planned on loosely coupled multiprocessor systems (e.g. a Transputer rack). Developing an efficient implementation will also involve research to load balancing and garbage collection (without stopping all processors).

The theoretical properties of PC-FGRS's will be further investigated. Especially in the context of term graph rewriting new results are envisaged. Using sharing and lazy copying, different ways of lifting term rewriting systems to graph rewriting systems can be investigated. Other strategies than the functional strategy may be interesting (van Eekelen & Plasmeijer (1986)). For instance, adding reducers following a non-deterministic strategy may be useful for the specification of process control, including scheduling and interrupts.

7 Conclusions

In this paper two extensions of Functional Graph Rewriting Systems are presented: lazy copying and annotations to control the order of evaluation. The extensions are simple and elegant. The expressive power of a FGRS extended with both notions is very high. Multi-processing can be modelled as well as graph reduction on loosely coupled systems. Arbitrary process and processor topologies can be modelled, as well as synchronous and asynchronous process communication. The introduced abbreviations guarantee that the indicated subgraphs can be evaluated in parallel instead of interleaved. The abbreviations directly correspond with the notion of processes and processors and they are therefore relatively simple to use. The user-friendliness can be increased by creating libraries with functions which can create often used processor topologies like pipelines and arrays of processors. Efficient implementation of the proposed model on loosely coupled parallel architectures should be possible. Actual implementations are started.

PC-FGRS's are very suited to serve as a base for the implementation of functional languages. Sequential functional languages can efficiently be implemented by translating them to FGRS's. The expressive power of the proposed abbreviations in PC-FGRS's and the properties of these systems will now make it also possible to exploit the potential parallelism in the programs successfully.

References

Arvind, Nikhil, Rishiyur S. (1987), 'Executing a Program on the MIT Tagged Token Dataflow Architecture', Proceedings of Parallel Architectures and Languages Europe (PARLE), part I, Eindhoven, The Netherlands. *Springer Lec. Notes Comp. Sci.* 258, 1-29.

Barendregt, H. P. (1984), 'The Lambda Calculus, its Syntax and Semantics', Studies in Logic and the Foundations of Mathematics 103, North-Holland.

Barendregt, H.P., Eekelen, M.C.J.D. van, Glauert, J.R.W., Kennaway, J.R., Plasmeijer, M.J., Sleep, M.R. (1987a), 'Term Graph Reduction', Proceedings of Parallel Architectures and Languages Europe (PARLE), part II, Eindhoven, The Netherlands. *Springer Lec. Notes Comp. Sci.* 259, 141-158.

Barendregt, H.P., Eekelen, M.C.J.D. van, Glauert, J.R.W., Kennaway, J.R., Plasmeijer, M.J., Sleep, M.R. (1987b), 'Towards an Intermediate Language based on Graph Rewriting', Proceedings of Parallel Architectures and Languages Europe (PARLE), part II, Eindhoven, The Netherlands. *Springer Lec. Notes Comp. Sci.* 259, 159-175.

Brus, T., Eekelen M.C.J.D. van, Leer M. van, Plasmeijer M.J. (1987), 'Clean - A Language for Functional Graph Rewriting', Proceedings of the Third International Conference on Functional Programming Languages and Computer Architecture (FPCA '87), Portland, Oregon, USA. *Springer Lec. Notes Comp. Sci.* 274, 364-384.

Burton, F.W. (1987), 'Functional Programming for Concurrent and Distributed Computing'. *The Computer Journal* 30-5, 437-450.

Church, A. (1932/3), 'A Set of Postulates for the Foundation of Logic', *Annals of Math. (2)* 33, 346 - 366 and 34, 839-864.

Eekelen, M.C.J.D. van (1988), 'Parallel Graph Rewriting, Some Contributions to its Theory, its Implementation and its Application', Ph.D. Thesis, University of Nijmegen.

Eekelen, M.C.J.D. van, Nöcker, E.G.J.M.H., Plasmeijer, M.J., Smetsers, J.E.W. (1989), 'Concurrent Clean, Version 0.5', Technical Report 89-18, University of Nijmegen.

Eekelen, M.C.J.D. van, Plasmeijer, M.J. (1986), 'Specification of rewriting strategies in Term Rewriting Systems', Proceedings of the Workshop on Graph Reduction, Santa Fe, New Mexico. *Springer Lec. Notes Comp. Sci.* 279, 215-239.

Fasel, J.H., Keller, R.M. (1986). Proceedings of the Workshop on Graph Reduction, Santa Fe, New Mexico. *Springer Lec. Notes Comp. Sci.* 279.

Glauert, J.R.W., Kennaway, J.R., Sleep, M.R. (1987), 'DACTL: A Computational Model and Compiler Target Language Based on Graph Reduction', *ICL Technical Journal* 5, 509-537.

Goguen, J., Kirchner, C., Meseguer, J. (1986), 'Concurrent tem rewriting as a model of computation', Proceedings of the Workshop on Graph Reduction, Santa Fe, New Mexico. *Springer Lec. Notes Comp. Sci.* 279, 53-94.

Gurd, J.R., Kirkham, C.C., Watson, I. (1985), 'The Manchester Prototype Dataflow Computer'. *Communications of the ACM*. 28-1, 34-52.

Hudak, P., Smith, L. (1986), 'Para-functional Programming: A Paradigm for Programming Multiprocessor Systems', *12th ACM Symp. on Principles of Programming Languages*, 243-254.

Johnson Th. (1984), 'Efficient compilation of lazy evaluation', Proceedings of the ACM SIGPLAN '84, Symposium on Compiler Construction. *SIGPLAN Notices* 19/6.

Klop, J.W. (1987), 'Term rewriting systems: a tutorial', Center for Mathematics and Computer Science, CWI Amsterdam. Note CS-N8701.

Kluge, W.E. (1983), 'Cooperating reduction machines', *IEEE Transactions on computers* C-32/11, 1002-1012.

Koopman, P., Nöcker, E. (1988). 'Compiling Functional Languages to Term Graph Rewrite Systems', University of Nijmegen. Technical Report 88-1.

Nöcker, E.G.J.M.H., (1988), 'Strictness Analysis Based on Abstract Reduction of Term Rewriting Systems', Proceedings of the Workshop on Implementation of Lazy Functional Languages, Göteborg, Sweden. *Programming Methodology Group*. 53, 451-463.

Nöcker, E.G.J.M.H., (1989), 'The PABC Simulator v0.5. Implementation Manual', University of Nijmegen, Technical Report 89-19.

Nöcker, E.G.J.M.H., Smetsers, J.E.W., (1990), 'Partially Strict Data Types', Proceedings of the Second International Workshop on Implementations of Functional Languages on Distributed Architectures, University of Nijmegen, November 1990.

Plasmeijer, M.J., Eekelen, M.C.J.D. van (1991), 'Functional Programming and Parallel Graph Rewriting', Lecture notes, University of Nijmegen, to appear at Addison Wesley 1991.

Smetsers, J.E.W., (1989), 'Compiling Clean to Abstract ABC-Machine Code',University of Nijmegen, Technical Report 89-20.

Smetsers, J.E.W., (1991), 'Operational semantics of Concurrent Clean', University of Nijmegen. Technical report: in preparation.

Vree, W.G., Hartel, P.H. (1988), 'Parallel graph reduction for divide-and-conquer applications; Part I - programme transformations', University of Amsterdam. Internal Report D-15.

Chapter 6

Basic Frameworks for Typed and Order-Sorted Systems

Typed Equivalence, Type Assignment, and Type Containment[1]

Roberto M. Amadio[2]

Abstract

The study of models based on *partial equivalence relations* (per) and the analysis of the interpretation of *inheritance* that arises in these structures (Bruce&Longo[88]) leads us to reconsider the classical problem of type-assignment in this framework. Moreover we introduce:

(1) A natural generalization of type-assignment systems to *typed-equivalence systems* that is suggested by the per-models.

(2) A specialization of the type-assignment system to a *type-containment system* that is motivated by the search for a "complete" theory of inheritance.

In the last section we show that a fragment of such theory of inheritance can be *fully automated.*

0. Introduction

The study of models based on partial equivalence relations and the analysis of the interpretation of *inheritance* that arises in these structures (Bruce&Longo[88]) leads us to reconsider the classical problem of type-assignment in this framework (Curry&al.[58/72]).

Various theories of inheritance have been proposed in the literature on software engineering (see, e.g., Cardelli[88], Liskov[88]). Their principal aim is to support a certain cycle of software development where programs evolve during the time, they are maintained, restructured and new functionalities are added. Such theories support an incremental design of software systems and establish under which conditions the programmer is allowed to *reuse* previously created modules.

Such reuse may require the introduction of explicit or implicit *coercions* whose effect on the semantics of the program has to be clearly understood by the programmer. A formalization of this concept in the context of *typed languages* can be given in two steps:

(a) Introduce a relation of subtype denoted by $<$. If α and β are types, the intuitive interpretation of $\alpha < \beta$ (read as α is a subtype of β) is: every α-value can be coerced to a β-value.

(b) Specify nature and use of such coercions.

In this paper, following Bruce&Longo[88], we will say that α is a subtype of β if for every term M of type α (denoted with M: α) and for every possible choice of a run-time code d for M (henceforth we will say that d is a realizer for M), there is a unique term N: β (up to semantic equivalence) that has d among its realizers. From a pragmatic point of view this decision will have rather dramatic consequences:

(1) The coercions are uniquely determined.

(2) Coercions do not produce run-time code, hence there is no need for recompilation.

(3) The specific "implementation" of a data type becomes a relevant issue.

For instance records and products are isomorphic structures in our interpretations i.e. $\alpha \times \beta \cong (l_1: \alpha, l_2: \beta)$ and also from a computational point of view it is easy to simulate one data structure in the other, i.e. the isomorphisms are definable in the language, nevertheless their theories of inheritance are quite different, the one for records being much more interesting (this motivates their inclusion in the language described in §2).

At this point we should make clear that we are considering *a theory of inheritance for typed functional languages* which should not be confused with other notions or intuitions of inheritance that appear in the literature on object-oriented programming.

Realizability structures give a mathematical framework to make precise what we just said, indeed the present approach to inheritance arose from an attempt of giving meaning to programming languages combining certain rules of inheritance used in "object-oriented" languages with higher and higher order types (see also Martini[88] and Breazu-Tannen&al.[89] for other models and/or approaches).

Familiarity with type theory and realizability models is assumed (see, e.g., Longo& Moggi[88]).

An *overview* of the paper goes as follows:

§1 is a brief introduction to various realizability structures that provide models for higher order λ-calculi (with the possibility of considering particular structures that interpret recursive operators) and determine a neat notion of subtype.

In §2 we formalize a second order typed λ-calculus extended with a functional version of records and define its interpretation in a "type structure".

In §3 we observe that such type structures can be seen as models for both the typed language and for a "typed equivalence system". Typed equivalence systems are generalizations of type assignment system a'la Curry where the atomic judgment establishes the equivalence of two λ-terms in a type. In particular the notion of term equivalence and subtyping can be expressed in a typed equivalence system. The typed equivalence system is shown to be sound and complete for the class of type structures without empty types. In §4 we prove similar results for a type assignment system.

---

1 Parts of this chapter appeared as TR 28/89 "Formal Theories of Inheritance for Typed Functional Languages", Dipartimento di Informatica, Universita' di Pisa and was presented at the EEC Jumelage meeting on typed λ-calculi (Nijmegen, November 1988) and at the Workshop on Conditional and Typed Rewriting Systems '90, Montreal.

2 Work done while on leave from Dipartimento di Informatica, Universita' di Pisa. Author's current address: LIENS, 45 rue d'Ulm, 75230 Paris, Cedex 05 (France). e-mail: amadio@dmi.ens.fr .

In §5 we develop a sound system for reasoning on Horn clauses of subtyping assertions. We observe the definability in the typed calculus of the associated coercions and the completeness of the system for a particular class of clauses.

In §6 we undertake a study of the *proof theoretic properties* of the theory of inheritance introduced. The central observation is a simple *cut elimination* theorem for a fragment of the theory of subtyping. An immediate application is the *decidability* of the theory of subtyping considered. A more interesting fall-out is an elegant approach to the problem of an *automatic inference of coercions.*

1 A model theoretic approach

All our constructions will take place over a *partial combinatory algebra* (see, e.g., Barendregt[84], Bethke[88]) denoted by $\mathbf{D} \equiv (D, \cdot, k, s)$. A selected countable set $L \subseteq D$ is going to interpret the collection of records' labels.

This algebraic structure will provide a universe of *realizers* to be thought of as the run-time representations of the typed programs. Note however that such representatives are still largely "extensional" in nature, e.g. they are not distinguished by their efficiency.

1.1 Data types

Data types are going to be interpreted as objects in full subcategories of the category of *partial equivalence relations* over D (henceforth per_D ; see, e.g., Hyland[88]).

1.1.1 Definition *(per_D)*

Given a pca D the category of partial equivalence relations over D is defined as follows:

$$Ob_{perD} =_{\Delta} \{A \,/\, A \subseteq D \times D \text{ and } A \text{ is symmetric and transitive}\}$$
$$per_D[A, B] =_{\Delta} \{f{:}D/A \to D/B \,/\, \exists \phi \in D\ \forall d \in D\ dAd \Rightarrow \phi d \in f([d]_A)\}$$

where conventionally: $dAe \equiv (d,e) \in A$, $[d]_A =_{\Delta} \{e \in D \,/\, dAe\}$, $D/A =_{\Delta} \{[d]_A \,/\, dAd\}$, $Dom(A) =_{\Delta} \{d \in D \,/\, dAd\}$.

As remarked in the introduction our theories of inheritance will depend on a specific choice of the data type constructors. In this section we will consider some possible choices that can be seen as *specific implementations* in per_D of general categorical specifications by means of universal properties.

1.1.2 Functional types

The functional space from the per A to the per B is represented by the per B^A defined as follows:

$$fB^Ag \Leftrightarrow_{\Delta} \forall d, e.\ dAe \Rightarrow fdBge.$$

This is the *simple* (quotient) intepretation of the functional space. If $D \equiv (D, k, s, \varepsilon, \cdot)$ is a λ-model then we may consider a different one, let $F =_{\Delta} \{d \in D \,/\, \varepsilon d = d\}$, where ε is the choice element, and define $(B^A)_F =_{\Delta} B^A \cap F^2$. This is also known as *F-semantics.*

Note that: $B^A \cong (B^A) \cap F^2$ in per_D

(Hint: the isomorphisms are realized by skk and ε).

With these definitions of the exponent per_D can be made into a ccc.

1.1.3 Polymorphic types

Given $T \subseteq per_D$ and a set theoretic operator $F: T \to T$, define $\Pi(F) =_{\Delta} \cap_{A \in T} F(A)$. It can be easily shown that this leads to a sound interpretation of second order type quantification. The definition can be also shown to fit categorical definitions of models for the second order λ-calculus (see, e.g., Longo& Moggi[88]).

1.1.4 Record types

The implementation of a record can be expressed via exponent and intersection.

The F-semantics of records is given using the F-exponent.

Let $l \in L$, $L \subseteq D$ $\quad rec_l(A) =_{\Delta} A^{\{(l,\, l)\}}$.

A record with more then one field is expressed as the intersection of the component records.

If a predicate of equality on labels, say eq_L, is definable i.e.

$$eq_L(l_i, l_j) = \text{true if } i=j \text{ and false if } i \neq j$$

then one can show that records and products are isomorphic:

$$rec_{l1}(A_1) \cap \ldots \cap rec_{ln}(A_n) \cong A_1 \times \ldots \times A_n \quad \text{for } l_i \neq l_j \text{ if } i \neq j.$$

Hint: the isomorphisms are realized by the following combinators (in metanotation)

$\lambda r. <..< r\, l_1, r\, l_2>, \ldots, r\, l_n>$, *(from record to product type)*

$\lambda p.\lambda l.$ if $eq_L(l_1, l)$ then $\pi_1(\ldots(\pi_1\ p)..)..$

if $eq_L(l_n, l)$ then $\pi_2 p$ else $\perp$

(from product to record type)

where: $\times$ associates to the left and $< , >$, π_1, π_2 are pairing and projections.

1.2 Type structures

A *type structure* **T** is made up of two components:

- A *partial combinatory algebra* $(D, \cdot, k, s)$ enriched by a countable set $L \subseteq D$ that interprets the collection of records' labels.

- A *type frame* $(T, [T \Rightarrow T])$ where $T \subseteq per_D$ and $[T \Rightarrow T]$ is a collection of set-theoretic functionals over T satisfying the following *closure properties:*

(1) $A, B \in T \Rightarrow B^A \in T$.

(2) $F \in [T \Rightarrow T] \Rightarrow \cap_{A \in T} F(A) \in T$.

(3) $A_i \in T$ for $i \in In \Rightarrow \cap_{i \in In} A_i^{\{(l_i,\, l_i)\}} \in T$.

As long as T is closed under arbitrary intersections we assume $[T \Rightarrow T]$ to be the collection of all set-theoretic functions. Then per_D is a type structure. Here is another related example.

1.2.1 Complete pers (cper)

Let D s.t. $D^D < D$ (and consequently $D \times D < D$) in the category of cpos. The full subcategory of complete pers is determined by the collection of $A \in per_D$ s.t.

(1) $\perp_D A \perp_D$ and (2) $X \subseteq A$ directed $\Rightarrow \cup X \in A$.

The resulting category has fixed-points (see Amadio[88]). The cpers admitting exactly one equivalence class determine the familiar type structure of *ideals* (Mac Queen&al.[86]). In Amadio[89] we define a full subcategory of cper that also interprets recursive domain equations. The issue of the containment of recursive types is discussed in Amadio[89(a)] and further developed in Amadio&Cardelli[90] (see Cardone&Coppo[89] for a system of type-assignment).

1.2.2 Reducibility candidates (Tait&Girard)

The collection of reducibility candidates is not exactly a type-structure but enjoys the closure properties of a type-frame.

Denote with Λ the collection of λ-terms and set:

$$SN =_{\Delta} \{M \,/\, M \text{ is strongly normalizing}\}.$$

$X \subseteq SN$ is a *reducibility candidate* iff

(1) $xN_1 \ldots N_n \in X$ if $N_i \in SN$

(2) $(\lambda x.M)NN_1 \ldots N_n \in X$ if $[N/x]MN_1 \ldots N_n \in X \wedge N \in SN$.

Take: $T =_{\Delta} \{X^2 \,/\, X \text{ is a reducibility candidate}\}$ (i.e. trivial

pers), $L =_\Delta \{\lambda xy.x^n y \,/\, n \in \omega\}$ (Church's numerals).
This leads to a relevant interpretation entailing the strong normalization of the calculus introduced in §2).

1.3 Models for type assignment and typed equivalence systems

The type structures considered so far were introduced to model the typed language defined in §2, nevertheless they are also well-suited for the interpretation of type assignment systems (tas) a'la Curry, indeed they are extensions of what in the literature on the semantics of type assignment are known as *quotient set interpretations* (Scott[76]).

In §3 we will assume D is a λ-model, this is a simplifying assumption as systems for the assignment of types to type-free λ-terms have nice properties and are easier to handle and to relate to the explicitly typed ones.

Let T be a type structure, α a closed type and P a closed type free λ-term then as usual the type assignment $P{:}\alpha$ is sound in the model T iff $[P]^D \in \mathrm{Dom}([\alpha]^T)$.

Besides we will also consider atomic formulas of the shape $P\alpha Q$. Formal systems with this kind of assertions are said *typed equivalence systems (tes)*.
A (closed) formula $P\alpha Q$ is valid in the type structure T iff $[P]\,[\alpha]\,[Q]$, where: $[P], [Q] \in D$, and $[\alpha] \in T \subseteq per_D$.

1.4 The notion of subtype

Let T be a type structure, $A, B \in T$. Say that:
A is a *subtype* of B $\Leftrightarrow_\Delta \exists f{:}A \to B$ (id ⊩ f)
(or equivalently $dAe \Rightarrow dBe$, i.e. $A \subseteq B$).

Note that the morphism realized by the identity, if it exists, is *uniquely determined.*
Given α, β closed formal types: $T \models \alpha < \beta \Leftrightarrow_\Delta [\alpha] \subseteq [\beta]$.

2. A Typed λ-calculus

We introduce a language including the second order lambda calculus and a functional version of records. The language is chosen mainly for didactic purposes as it allows to point out the main methodological and mathematical issues without obscuring them with the details of an overwhelming syntactic notation.

2.1 Types

Denote with t, s... type variables, with l_1, l_2... labels and with α, β... types.
Types are (informally) given by the following production rules:

$$\alpha ::= t \mid (\alpha \to \beta) \mid \{(l_i : \alpha_i)\}_{i \in In} \mid (\forall t.\alpha)$$

Conventions

Assume a bijective function $l : \omega \to L$, denote with In a subset of ω of cardinality n.$(l_i : \alpha_i)_{In} \equiv_{abr.} \{(l(i) : \alpha_i)\}_{i \in In}$.
An analogous convention holds for records' values.

$\Rightarrow$ separates premisses and conclusions of an inference rule, premisses are separated by a comma, $\vdash$ denotes formal derivability of certain judgments that can be expressed in the system, namely:
(1) a term M has a type α: $M{:}\alpha$
(2) two typed terms M, N are convertible: $M{=}N$
(3) α is subtype of β under the subtyping assumptions Σ: $\Sigma \supset \alpha < \beta$.

2.2 Typed terms

M, N denote terms and x^α, y^β term variables (equipped with their type label).

(var) $x^\alpha : \alpha$

($\to$I) $M : \beta \Rightarrow (\lambda x^\alpha.M) : (\alpha \to \beta)$

($\to$E) $M : (\alpha \to \beta)$, $N : \alpha \Rightarrow (MN) : \beta$

(Rec I) $\forall i \in In\ M_i : \alpha_i \Rightarrow (l_i = M_i)_{In} : (l_i : \alpha_i)_{In}$

(Rec E) $M : (l_i : \alpha_i)_{In}$, $j \in In \Rightarrow (M.l_j) : \alpha_j$

($\forall$ I) $M : \alpha$, t not free in M $\Rightarrow (\lambda t.M) : (\forall t.\alpha)$

($\forall$ E) $M : (\forall t.\alpha) \Rightarrow (M\,\beta) : [\beta/t]\alpha$

Conversion and Reduction

Types: variable redenomination for $\forall$.
Terms: variable redenomination for λ and

(β) $(\lambda x^\alpha.M)\,N = [N/x]M$

(β_{Rec}) $(l_i : M_i)_{In}.l_j = M_j$ for $j \in In$

(β_t) $(\lambda t.M)\,\alpha = [\alpha/t]M$

Plus the rules that turn = into a congruence. All terms are assumed well-typed.
Reduction, denoted with -->, is given by orienting from left to right the axioms and by making their compatible closure.
Let -->> $=_\Delta$ (-->)*.

2.4 Facts *(General properties of the system , basically Girard[72]).*

(1) *Subject reduction.*
$M : \alpha$, $M \dashrightarrow M' \Rightarrow M' : \alpha$
(2) *Confluence for typed terms.*
If $M : \alpha$, M -->> M', M -->> M" then
$\exists$ N (M' -->> N <<-- M").
(3) *Decidability of type-checking.*
Given a term M it is decidable if there exists α such that $M{:}\alpha$.
(4) *Strong normalization* .
The calculus is strongly normalizing.

2.5 Interpretations

The (in-)equational theories for types and terms generated by the following interpretation are studied in Amadio[89], we just point-out that the conversion relation specified in 2.2 is adequately modelled.

Types:
Given a type-environment: η: Tvar$\to$ T, the *type interpretation* is defined as follows:

$$[\,t\,]_\eta = \eta(t)$$
$$[\,\alpha \to \beta\,]_\eta = [\beta]_\eta{}^{[\alpha]_\eta} \quad (\cap F^2)$$
$$[\,(l_i{:}\alpha_i)_{In}\,]_\eta = \cap_{i \in In}\,[\alpha_i]_\eta{}^{\{(l_i, l_i)\}} \quad (\cap F^2)$$
$$[\,\forall t.\alpha\,]_\eta = \cap_{A \in T}\,[\alpha]_{\eta[A/t]}$$

$(\cap F^2)$ gives the F-semantics. The clause for polymorphic types is well-defined iff
$\forall\alpha\forall\eta\ \lambda A{\in}T.[\alpha]_{\eta[A/t]} \in [T \Rightarrow T]$. In the following when referring to a type structure we will always assume that this condition is verified.

Terms:
Given a type structure T define term and type application as follows:

$\forall A, B \in per_D \quad \bullet_{A,B} : D/B^A \times D/A \to D/B$

$\qquad [f]_{B^A} \bullet_{A,B} [d]_A =_\Delta [fd]_B$

$\forall F \in [T \Rightarrow T] \quad \bullet_F : (D/\cap F) \times T \to \cup_{A \in T} D/F(A)$

$\qquad [f]_{\cap F} \bullet_F A =_\Delta [f]_{F(A)}.$

Let ρ: TypedVar $\to \cup_{A\in T} D/A$, η : TypeVar $\to T$ and $\rho\downarrow\eta$, where: $\rho\downarrow\eta \Leftrightarrow_\Delta \forall x^\alpha\ \rho(x^\alpha) \in D/[\alpha]_\eta$.
For each syntactic label l_i assume a corresponding element $l_i \in L \subseteq D$. The *term interpretation* is inductively defined on well-typed terms as follows:

$[x^\alpha]_{\rho\eta} = \rho(x^\alpha)$

$[\lambda x^\alpha.M]_{\rho\eta} = \{\phi \in D / (d\,[\alpha]_\eta\, d \Rightarrow \phi d \in [M]_{\rho[([d][\alpha]\eta)/x^\alpha]\eta})\}$ $(\cap F^2)$

$[MN]_{\rho\eta} = [M]_{\rho\eta} \bullet_{A,B} [N]_{\rho\eta}$
where: $M:\alpha\to\beta$, $N:\alpha$, $A =_\Delta [\alpha]_\eta$ and $B =_\Delta [\beta]_\eta$

$[(l_i = M_i)_{In}]_{\rho\eta} = \{\phi \in D / \forall i \in In\ \phi l_i \in [M_i]_{\rho\eta}\}$ $(\cap F^2)$

$[M.l_j]_{\rho\eta} = [M]_{\rho\eta} \bullet_{A,B} \{l_j\}$
where: $M:(l_i:\alpha_i)_{In}$, $j \in In$, $A =_\Delta [(l_i:\alpha_i)_{In}]_\eta$, $B =_\Delta \{(l_j, l_j)\}$

$[\lambda t.M]_{\rho\eta} = \{\phi \in D / \forall A \in T\ \forall d \in D\ \ \phi d \in [M]_{\rho\eta[A/t]}\}$

$[M\beta]_{\rho\eta} = [M]_{\rho\eta} \bullet_F [\beta]_\eta$
where: $M:\forall t.\alpha$, $F =_\Delta \lambda A \in T.[\alpha]_{\eta[A/t]}$

2.6 Semantic entailment

Fix a class C of type structures and let J denote a generic judgment.

$C \models J \Leftrightarrow_\Delta \forall T \in C\ (T \models J)$

$T \models M:\alpha \Leftrightarrow_\Delta \forall \rho\eta\ (\rho\downarrow\eta \Rightarrow \rho\eta \models M:\alpha)$,
$\quad \rho\eta \models M:\alpha \Leftrightarrow_\Delta [M]_{\rho\eta} \in D/[\alpha]_\eta$

$T \models M = N \Leftrightarrow_\Delta \forall \rho\eta\ (\rho\downarrow\eta \Rightarrow \rho\eta \models M = N)$,
$\quad \rho\eta \models M = N \Leftrightarrow_\Delta [M]_{\rho\eta} = [N]_{\rho\eta}$

$T \models \Sigma \supset \alpha < \beta \Leftrightarrow_\Delta$
$\qquad \forall \eta\ ((\forall \alpha_i < \beta_i \in \Sigma\ \ \eta \models \alpha_i < \beta_i) \Rightarrow \eta \models \alpha < \beta)$
$\quad \eta \models \alpha < \beta \Leftrightarrow_\Delta [\alpha]_\eta \subseteq [\beta]_\eta$

3. Typed Equivalence

We have observed in §1 that a type structure can be seen both as a model for the typed calculus and as a quotient set model for type inference systems.
We develop a system christened as *typed equivalence system* (tes) whose atomic assertions state that two type-free λ-terms are equivalent w.r.t. a certain type, written as $P\alpha Q$.

The system is sound and complete (via a Henkin-like term model construction) for deriving *Horn clauses* of typed equivalences w.r.t. a class of type structures that does not allow empty types.
A standard type assignment P: α is still expressible as $P\alpha P$.
We make a point of using no ad hoc proof-theoretic techniques in the proofs of soundness and completeness.

The choices of *Horn clauses* as the logic relating the atomic formulas, and of *type-structures without empty types* as models, are justified by the desire of giving a *greater flexibility* to the formalism while *preserving simplicity and standard properties* e.g. existence of an initial model and classical reasoning (cnfr. Mahr&Makowsky[84] and Meyer&al.[87]).
In particular empty types force the introduction of disjunctive reasoning and therefore the distinction between classical and intuitionistic models (see also Mitchell&Moggi[87]).

3.1 Type-free calculus

Denote with P, Q... terms of the type free lambda calculus over a countable set of constants L extended with record's constructor and selector, henceforth $\lambda\beta l$.
Informally: $P ::= x \mid l \mid (PQ) \mid \lambda x.P \mid (l_i = P_i)_{In} \mid P.l$ with the obvious *β-conversion rules:*

$(\lambda x.P)Q = [Q/x]P$ and $(l_i = P_i)_{In}.l_j = P_j$ if $j \in In$.

3.2 Typed equivalence system

A basis of assumptions B is a set $\{P_i\alpha_iQ_i / i \in I\}$.
A basic judgment is of the shape: $B \supset P\,\alpha\,Q$.
$\Rightarrow$ separates premisses and conclusions of an inference rule.
Types are considered up-to-redenomination of bound variables.
E is a $\lambda\beta l$-theory.

(Asmp) $P\alpha Q \in B \Rightarrow B \supset P\alpha Q$

(Eq) $B \supset P'\alpha Q'$, $\vdash_E P = P'$, $\vdash_E Q = Q' \Rightarrow B \supset P\alpha Q$

(weak) $B \supset P\,\beta\,Q \Rightarrow B \cup B' \supset P\beta Q$

(rmv) $B \cup \{x\alpha y\} \supset P\,\beta\,Q \Rightarrow B \supset P\beta Q$
for x,y not free in B, P, Q

(sym) $B \supset P\,\alpha\,Q \Rightarrow B \supset Q\,\alpha\,P$

(trans) $B \supset P\,\alpha\,P'$, $B \supset P'\alpha\,P'' \Rightarrow B \supset P\,\alpha\,P''$

($\to$I) $B \cup \{x\alpha y\} \supset P\beta Q \Rightarrow B \supset \lambda x.P\ \alpha\to\beta\ \lambda y.Q$
for $x \neq y$, $x,y \notin FV(B)$, $x \notin FV(Q)$, $y \notin FV(P)$

($\to$Iη) $B \supset \lambda x.Px\ \alpha\to\beta\ \lambda x.Qx \Rightarrow B \supset P\alpha\to\beta Q$
for , $x \notin FV(PQ)$

($\to$E) $B \supset P\,\alpha\to\beta\,Q$, $B \supset P'\alpha\,Q' \Rightarrow B \supset PP'\,\beta\,QQ'$

(RecI) $B \supset P_i\alpha_iQ_i\ \forall i \in In \Rightarrow$
$\qquad B \supset (l_i = P_i)_{In}\,(l_i : \alpha_i)_{In}\,(l_i = Q_i)_{In}$

(RecIη) $B \supset Pl_i\,\alpha_i\,Ql_i\ \forall i \in In \Rightarrow B \supset P\,(l_i : \alpha_i)_{In}\,Q$

(RecE) $B \supset P\,(l_i : \alpha_i)_{In}\,Q$, $j \in In \Rightarrow B \supset P.l_j\,\alpha_j\,Q.l_j$

($\forall$I) $B \supset P\,\alpha\,Q$, t not free in B $\Rightarrow B \supset P\forall t.\alpha\,Q$

($\forall$E) $B \supset P\,\forall t.\alpha\,Q \Rightarrow B \supset P\,[\beta/t]\alpha\,Q$

If we wish to consider a type assignment system for the F-semantics then we can add the following rules, where it is intended that the type-operators subscripted with F are interpreted according to the F-semantics:

($\to_F$) $B \supset P\alpha\to\beta Q$, $\vdash_E \varepsilon P = P$, $\vdash_E \varepsilon Q = Q \Rightarrow$
$\qquad B \supset P\,\alpha\to_F\beta\,Q$

(Rec$_F$) $B \supset P\,(l_i : \alpha_i)_{In}\,Q$, $\vdash_E \varepsilon P = P$, $\vdash_E \varepsilon Q = Q \Rightarrow$
$\qquad B \supset P\,(l_i :_F \alpha_i)_{In}\,Q$

3.3 Convention

Let E be a λ–theory and B a set of typed equivalence assumptions then
$\vdash E, B \supset P\alpha Q$ if there is a finite derivation of $B \supset P\alpha Q$ in the formal system just specified such that all the applications of the rule (Eq) refer to the λ-theory E.
We simply write $\vdash B \supset P\alpha Q$ when E is clear from the context.
Notice that a standard compactness result holds, namely:
$\vdash E, B \supset P\alpha Q \Leftrightarrow \exists B_0 \subseteq B$ (B_0 finite and $\vdash E, B_0 \supset P\alpha Q$).

3.4 Term interpretation

Given a type structure T we have just to define the intepretations of the type-free terms in an enriched λ-model. In the following we *refer to type structures whose domains D are λ-models.*
Besides we *assume the existence of a family of combinators*:

$\{\varepsilon_{In}\}_{In\subset\omega} \subseteq D$ satisfying $\varepsilon_{In} d_1 \ldots d_n l_j = d_j$
for $j \in In$, $In = \{i_1, \ldots, i_n\}$, $i_1 < \ldots < i_n$.

We just define the clauses for record constructor and selector (see, e.g., Barendregt[84] for the definition of the other clauses). Let θ: Var $\rightarrow$ D then:

$$[\, (l_i = P_i)_{In} \,]_\theta = \varepsilon_{In} [P_1]_\theta \ldots [P_n]_\theta$$
$$[\, P.l_j \,]_\theta = [P]_\theta l_j$$

3.5 Semantic entailment

Let C be a class of type structures.

$C \models E, B \supset P\alpha Q \Leftrightarrow_\Delta \forall T \in C \ (T \models E, B \supset P\alpha Q)$
where assuming θ: Var $\rightarrow$ D :

$T \models E, B \supset P\alpha Q \Leftrightarrow_\Delta T \models E \Rightarrow$
$\forall \theta\, \eta \ (\theta\, \eta \models B \Rightarrow \theta\, \eta \models P\alpha Q)$.
$\theta\, \eta \models B \Leftrightarrow_\Delta \forall P\alpha Q \in B \ (\theta\eta \models P\alpha Q)$,
$\theta\, \eta \models P\alpha Q \Leftrightarrow_\Delta [P]_\theta [\alpha]_\eta [Q]_\theta$.

3.6 Typed vs. type-free: semantic correspondences

Let be given a type structure: T, typed terms: M, N, types: α, β and a collection of subtyping assumptions: Σ. Let er(M) denote the type free λ-term obtained by removing all the type informations. We have the following correspondences:

(1) $[M^\alpha]_{\rho\eta}{}^T = [\, [er(M)]_\theta{}^D \,]_{[\alpha]\eta}{}^T$,
where: $\rho\downarrow\eta$ and $\theta(er(x^\alpha)) \in \rho(x^\alpha)$
The proof consists of a natural induction on the structure of M (cnfr. Mitchell[86]).

(2) If M: α and N: α then
$T \models M{=}N \Leftrightarrow T \models \{x_1 \alpha_1 x_1, \ldots, x_n \alpha_n x_n\} \supset er(M)\alpha\, er(N)$
where: $\{x_1{}^{\alpha_1}, \ldots, x_n{}^{\alpha_n}\} = FV(M) \cup FV(N)$.

(3) $T \models \Sigma \supset \alpha{<}\beta \Leftrightarrow$
$T \models \{\lambda x.x\ (\alpha' \rightarrow \beta')\ \lambda x.x \,/\, \alpha'{<}\beta' \in \Sigma\} \supset \lambda x.x\ (\alpha \rightarrow \beta)\ \lambda x.x$.

3.7 Theorem *(Soundness for tes)*

Let $C_{\neg\emptyset}$ be the class of type structures without empty types. Then:

$$\vdash E, B \supset P\alpha Q \Rightarrow C_{\neg\emptyset} \models E, B \supset P\alpha Q$$

Proof

Let $T \in C_{\neg\emptyset}$, and suppose $T \models E$. By induction on the lenght of the derivation we show that
$\vdash E, B \supset P\alpha Q \Rightarrow T \models E, B \supset P\alpha Q$.
The soundness of the rules (Asmp), (Eq), (weak), (sym) and (trans) is immediate.

(rmv) Suppose $\theta\eta \models B$. Then define $\theta' =_\Delta [d/x][d/y]\theta$, where $d[\alpha]_\eta d$, such d exists by the assumption $T \in C_{\neg\emptyset}$. Observe $\theta'\eta \models B \cup \{x\alpha y\}$, by ind. hyp. $\theta'\eta \models P\beta Q$ and by $x, y \notin FV(PQ)$ we conclude $\theta\eta \models P\beta Q$.

($\rightarrow$I) Suppose $\theta\eta \models B$. Recall that:
$\theta\eta \models \lambda x.P\ \alpha{\rightarrow}\beta\ \lambda y.Q \Leftrightarrow_{def.}$
$\forall de\, (d[\alpha]_\eta e \Rightarrow [P]_{[d/x]\theta} [\beta]_\eta [Q]_{[e/y]\theta})$.
Observe: $d[\alpha]_\eta e \Rightarrow [d/x][e/y]\theta\, \eta \models B \cup \{x\alpha y\} \Rightarrow$ $[d/x][e/y]\theta\, \eta \models P\beta Q \Rightarrow [P]_{[d/x]\theta} [\beta]_\eta [Q]_{[e/y]\theta}$.
Notice that the side conditions on variables are essential.

($\rightarrow$Iη) Immediate by the definition of the simple semantics.

($\rightarrow$E) Immediate by the definition of the interpretations of terms and types.

(RecI), (RecIη), (RecE) The proof of soundness of these rules follows the pattern of the corresponding rules for $\rightarrow$. Notice that the rule (RecI) can be derived from the rule (RecIη) via (Eq) and (RecE).

($\forall$I) Suppose $\theta\eta \models B$. Recall: $\theta\eta \models P\forall t.\alpha\ Q \Leftrightarrow_{def.}$ $\forall A \in T\ \theta[A/t]\eta \models P\ \alpha\ Q$.
Since $\forall A \in T\ \theta[A/t]\eta \models B$ we have the thesis by ind. hyp. .

($\forall$E) By a simple substitution lemma $[\, [\beta/t]\alpha \,]_\eta = [\alpha]_{[B/t]\eta}$, where $B =_\Delta [\beta]_\eta$. Suppose $\theta\eta \models B$ then $\forall A \in T$ $\theta[A/t]\eta \models P\ \alpha\ Q$, in particular $\theta[B/t]\eta \models P\ \alpha\ Q$ that is $\theta\eta \models P\ [\beta/t]\alpha\ Q$.

($\rightarrow$F), (RecF). By the definition of F-semantics. □

3.8 Theorem. *(Completeness for tes)*

Given a λ-theory E and a collection of assumptions B there is a type structure $T \in C_{\neg\emptyset}$ such that:

$$T \models E \quad \text{and} \quad T \models E, B \supset P\alpha Q \Leftrightarrow \vdash E, B \supset P\alpha Q.$$

Proof

The main point of the term model construction is to show that we can conservatively adjoin to the set of assumptions B an infinite number of typing assumptions on independent variables (e.g. see Hindley[83] and Mitchell[88] for a similar proof tecnique).

Given a judgment $B \supset P\alpha Q$ let σ be an injective substitution from (type-)variables to (type-)variables (recall that the usual definition of substitution takes care of name-clashes of bound and free variables). We call such substitution a free-variable redenomination.

Note that: $\vdash E, B \supset P\alpha Q \Leftrightarrow \vdash E, \sigma B \supset \sigma P\alpha Q$, where σB and $\sigma P\alpha Q$ have the obvious componentwise definition. The proof is immediate by observing that the inference rules do not depend on the name of the free variables but only on their respective relationships.

Given B define a free-variable redenomination σ such that $Var \backslash \sigma(Var)$ and $Tvar \backslash \sigma(Tvar)$ are infinite.
Let $B' =_\Delta \{\sigma(P\alpha Q) \,/\, P\alpha Q \in B\} \cup \{x_{i\alpha}\alpha y_{i\alpha} \,/\, i \in \omega, \alpha \text{ type}\}$ where $x_{i\alpha}, y_{i\alpha} \in Var \backslash \sigma(Var)$ and are all distinct.
We claim:

(1) $C_{\neg\emptyset} \models E, B \supset P\alpha Q \Leftrightarrow C_{\neg\emptyset} \models E, B' \supset \sigma(P\alpha Q)$
(2) $E, B' \supset \sigma(P\alpha Q) \Rightarrow \vdash E, B \supset P\alpha Q$.

(1) Given η, θ let η', θ' s.t. $\eta'(t) = \eta(\sigma^{-1}(t))$ if $t \in range(\sigma)$, $\theta'(x) = \theta(\sigma^{-1}(x))$ if $x \in range(\sigma)$ and $\theta'(x) = c_\alpha$ if $x \equiv x_{i\alpha}$ (or $y_{i\alpha}$) where $c_\alpha \in Dom([\alpha]_{\eta'})$. Observe:

(a) $[\alpha]_\eta = [\sigma\alpha]_{\eta'}$, (b) $[P]_\theta = [\sigma P]_{\theta'}$,
(c) $\theta\eta \models P\alpha Q \Leftrightarrow \theta'\eta' \models \sigma(P\alpha Q)$,
(d) $\theta\eta \models B \Rightarrow \theta'\eta' \models B'$.

On the other hand given η, θ let $\eta''(t) =_\Delta \eta(\sigma t)$, $\theta''(x) =_\Delta \theta(\sigma x)$ and observe:

(a) $[\sigma\alpha]_\eta = [\alpha]_{\eta''}$, (b) $[\sigma P]_\theta = [P]_{\theta''}$,
(c) $\theta\eta \models \sigma(P\alpha Q) \Leftrightarrow \theta''\eta'' \models P\alpha Q$,
(d) $\theta\eta \models B' \Rightarrow \theta''\eta'' \models B$.

We can now prove (1).

($\Rightarrow$): $\theta\eta \models B' \Rightarrow \theta''\eta'' \models B \Rightarrow \theta''\eta'' \models P\alpha Q \Rightarrow$ $\theta\eta \models \sigma(P\alpha Q)$.

($\Leftarrow$): $\theta\eta \models B \Rightarrow \theta'\eta' \models B' \Rightarrow \theta'\eta' \models \sigma(P\alpha Q) \Rightarrow$ $\theta\eta \models P\alpha Q$.

(2) Suppose: $\llcorner$ E, B' $\supset \sigma(P\alpha Q)$.

Then by compactness:

$\exists B'_0 \subseteq_{finite} B' (\vdash E, B'_0 \supset \sigma(P\alpha Q))$.

By iterating the rule (rmv) on the adjoint variables $x_{i\alpha}$ we conclude:

$\exists B_0 \subseteq_{finite} B \ \vdash E, \{\sigma(P'\beta Q') / P'\beta Q' \in B_0\} \supset \sigma(P\alpha Q)$.

Apply the property of free-variable redenomination to have:

$\vdash E, B_0 \supset P\alpha Q$ and by (weak): $\vdash E, B \supset P\alpha Q$.

We are now ready to build a term model over the collection of assumptions B'.

As domain D take the term λ-model $\lambda\beta I / =_E$. We just point-out that the record combinator ε_{In} is given by $[\lambda x_1 ... \lambda x_n. \{l_{i1} = x_1, ..., l_{in} = x_n\}]$ where [ ] denotes the equivalence class w.r.t. E and In $=_\Delta \{i_1, ..., i_n\}$.

Types are taken up to α-redenomination. T $\subseteq per_D$ is composed exactly of the relations $[\alpha]$ associated to each type α defined as follows:

$$[P] [\alpha] [Q] \Leftrightarrow_\Delta \ \vdash E, B' \supset P\alpha Q$$

Using the rule (Eq), (sym) and (trans) it is easy to show that $[\alpha] \in per_D$.

$[T \Rightarrow T] =_\Delta \{F: T \rightarrow T / \exists \beta \ F([\alpha]) = [\ [\alpha/t]\beta\ ]\}$

i.e. just the definable transformations.

In order to show that this is a type structure (denote it with T and observe T $\models$ E and T$\in C_{\neg\emptyset}$ by the assumptions on B') it remains to check the *closure properties*.

- $[P] [\alpha \rightarrow \beta] [Q] \Leftrightarrow$

$\forall [P'], [Q'] ([P'] [\alpha] [Q'] \Rightarrow [PP'][\beta][QQ'])$

($\Rightarrow$) If $\llcorner B_0 \supset P\alpha \rightarrow \beta Q$ and $\llcorner B_1 \supset P'\alpha Q'$ then $\llcorner B_0 \cup B_1 \supset PP'\beta QQ'$ by (weak) and ($\rightarrow$E).

($\Leftarrow$) For some $x_{i\alpha} \equiv_{abr.} x$, $y_{i\alpha} \equiv_{abr.} y \notin FV(PQ)$

$\llcorner B_0 \supset P x \beta Q y$.

By (weak) and ($\rightarrow$I) $\vdash B_0 \supset \lambda x.Px \ \alpha \rightarrow \beta \ \lambda y.Qy$ and by ($\rightarrow$Iη) $\llcorner B_0 \supset P \ \alpha \rightarrow \beta \ Q$.

- $[P] (l_i : \alpha_i)_{In} [Q] \Leftrightarrow \forall i \in In \ [Pl_i] [\alpha_i] [Ql_i]$.

($\Rightarrow$) Simply by (RecE).

($\Leftarrow$) If $\forall i \in In \ \vdash B_i \supset Pl_i \alpha_i Ql_i$ then by (weak) and (RecIη) $\vdash \cup_{i \in In} B_i \supset P(l_i : \alpha_i)_{In} Q$.

- $[P] [\forall t.\alpha] [Q] \Leftrightarrow \forall \beta \ [P] [\ [\beta/t]\alpha\ ] [Q]$.

($\Rightarrow$) If $\vdash B_0 \supset P\forall t.\alpha Q$ then $\vdash B_0 \supset P[\beta/t]\alpha Q$ by ($\forall$E).

($\Leftarrow$) Fixed P, Q and α pick up a type variable s such that s$\in$ Tvar\σ(Tvar) and s does not occur free in some type γ associated to a variable $x_{i\delta}$ occurring free in PQ.

By hyp. $\exists B_0 \subseteq_{finite} B' \ \vdash E, B_0 \supset P[s/t]\alpha Q$, by (rmv) we can eliminate all the adjoined variables except those occurrying free in PQ and then we can apply ($\forall$I) on s and (weak).

In general $\vdash E, B \supset P\alpha Q \ \Rightarrow T \models E, B \supset P\alpha Q$ follows by soundness.

Viceversa observe: $T \models B' \supset P\alpha Q \Rightarrow \vdash E, B' \supset P\alpha Q$.

As usual let $\theta_o(x) =_\Delta [x]$ and $\eta_o(t) =_\Delta [t]$. Then $\theta_o \eta_o \models B'$ by well-known properties of substitution and (Asmp). Hence $\theta_o \eta_o \models P\alpha Q$, i.e. $\vdash E, B' \supset P\alpha Q$.

The thesis is obtained by the following chain of implications:

$T \models E, B \supset P\alpha Q \Leftrightarrow T \models E, B' \supset \sigma(P\alpha Q) \Rightarrow$

$\vdash E, B' \supset \sigma(P\alpha Q) \Rightarrow \vdash E, B \supset P\alpha Q$.

Eventually notice that the system is complete also for the F-semantics as the rules ($\rightarrow_F$) and (Rec_F) read semantically express a necessary and sufficient condition. It is also interesting to observe that the collection of valid subtyping judgments: $\Sigma \supset \alpha < \beta$ is invariant w.r.t. the choice simple/F semantics as $\varepsilon(\lambda x.x) = \lambda x.x$ (cnfr. semantic correspondences).

□

3.9 Derived rules

From the completeness theorem we can infer the following derived rules whose soundness can be easily checked.

(Sbst) $\vdash E, B \supset P\alpha Q \ \Rightarrow \ \vdash E, \sigma B \supset \sigma P\alpha Q$
where σ is a substitution.

(MP) $\vdash E, B \cup \{P'\beta Q'\} \supset P\alpha Q \quad \vdash E, B \supset \{P'\beta Q'\} \Rightarrow \vdash E, B \supset P\alpha Q$.

4. Type Assignment

We develop a *type-assignment system* (tas) that is sound and complete w.r.t. to the class of type structures without empty types.

We should point-out that in general the *soundness of tas* for quotient set semantics is not evident. Observe that:

$\models x: \alpha \supset x:\beta \Leftrightarrow \models Dom(\alpha) < Dom(\beta)$ whereas

$\models \lambda x.x: \alpha \rightarrow \beta \Leftrightarrow \models \alpha < \beta$.

Since by the standard rule ($\rightarrow$I) for tas we have: $x: \alpha \supset x:\beta \Rightarrow \lambda x.x: \alpha \rightarrow \beta$ we can conclude that to have a sound and complete tas we need to prove:

$\models Dom(\alpha) < Dom(\beta) \Rightarrow \models \alpha < \beta$.

In logical terms the problems come from the asimmetry between the interpretations of $\supset$ and $\rightarrow$, the latter being stronger.

4.1 Type assignment system

A well formed basis of assumptions B is a set $\{P_i:\alpha_i / i \in I\} \cup \{x_j:\alpha_j / j \in J\}$ where the λ-terms P_i are closed and the variables x_j are all distinct.

Apart for this restriction the same conventions as for the typed equivalence systems hold. Besides the rules for F-semantics and the derived rules (3.9) have a straightforward adaptation.

(Asmp) $P: \alpha \in B \ \Rightarrow \ B \supset P: \alpha$

(Eq) $B \supset P': \alpha \quad \vdash_E P = P' \ \Rightarrow \ B \supset P: \alpha$

(weak) $B \supset P: \alpha \quad B \cup B'$ well formed $\Rightarrow B \cup B' \supset P: \alpha$

(rmv) $B \cup \{x:\alpha\} \supset P: \beta \quad x \notin FV(P) \ \Rightarrow \ B \supset P: \beta$

($\rightarrow$I) $B \cup \{x:\alpha\} \supset P: \beta \ \Rightarrow \ B \supset \lambda x.P: \alpha \rightarrow \beta$

($\rightarrow$Iη) $B \supset \lambda x.Px: \alpha \rightarrow \beta \quad x \notin FV(P) \ \Rightarrow \ B \supset P: \alpha \rightarrow \beta$

($\rightarrow$E) $B \supset P: \alpha \rightarrow \beta \quad B \supset Q: \alpha \ \Rightarrow \ B \supset PQ: \beta$

(RecI) $B \supset P_i: \alpha_i \quad i \in In \ \Rightarrow \ B \supset (l_i = P_i)_{In}: (l_i : \alpha_i)_{In}$

(RecIη) $B \supset P.l_i: \alpha_i \quad i \in In \ \Rightarrow \ B \supset P: (l_i : \alpha_i)_{In}$

(RecE) $B \supset P: (l_i : \alpha_i)_{In} \quad j \in In \Rightarrow B \supset P.l_j:\alpha_j$

($\forall$I) $B \supset P: \alpha \quad t \notin FV(B) \ \Rightarrow \ B \supset P: \forall t.\alpha$

($\forall$E) $B \supset P: \forall t.\alpha \ \Rightarrow \ B \supset P: [\beta/t]\alpha$

4.2 Theorem *(Soundness for tas)*

Let $C_{\neg\emptyset}$ be the class of type structures without empty types. Then: $\vdash E, B \supset P: \alpha \ \Rightarrow \ C_{\neg\emptyset} \models E, B \supset P: \alpha$

Proof

Let T be a type structure such that T $\models$E and T$\in C_{\neg\emptyset}$. Conventionally $\vdash E, B \supset P:\alpha \equiv_{abr.} \llcorner B \supset P: \alpha$ and $T \models E, B \supset P: \alpha \equiv_{abr.} \models B \supset P: \alpha$. Given a judgment $B \supset P: \alpha$ let $b =_\Delta \{x_1:\alpha_1, ..., x_n:\alpha_n\}$ and $f =_\Delta \{y_1:\beta_1, ..., y_m:\beta_m\}$ such that: $n, m \geq 0$, $b \subseteq B$, $y_j \notin FV(B)$

for j=1,..., m. Let $\lambda x_b.\lambda y_f.P : \alpha_b \to \beta_f \to \alpha \equiv_{abr.}$ $\lambda x_1...\lambda x_n.\lambda y_1...\lambda y_m.P$: $\alpha_1 \to (...(\alpha_n \to (\beta_1 \to (... \to (\beta_m \to \alpha)..)$

Generalize the thesis as follows:
if $\vdash B \supset P: \alpha$ then for all b, f satisfying the previous conditions we have:
$\models B \supset \lambda x_b.\lambda y_f.P : \alpha_b \to \beta_f \to \alpha$ (*). The thesis is obtained for n=m=0.

Preliminary remarks:
(a) The validity of (*) is invariant under consistent permutations of λ-abstractions and types.
(b) If $\models B \supset P: \alpha$ and $\{x_1,...,x_n\} \cap FV(P) = \emptyset$ then for any choice of types $\alpha_1,...,\alpha_n$ we have:
$\models B \supset \lambda x_1...\lambda x_n.P: \alpha_1 \to (...(\alpha_n \to \alpha)..)$.
(c) The right generalization was found as follows:
say (1) the naive statement $\models B \supset P: \alpha$, say (2) the (half-)idea of introducing b (implicit in Hindley[83]) and say (3) the (half-)idea of introducing f.
This is what we needed to have the induction go through as the following dependencies arise at the subscribed rules:
(1) $\to_{(\to I)}$ (2) $\to_{(weak)} \to$ (3).

We now proceed by induction on the lenght of the deduction of $B \supset P: \alpha$ denoting with b' and f' the sets b, f on which we apply the inductive hypothesis. Referring to the rules in 3.2.1 we have:

(Asmp) If $P \equiv x_i$ then $\lambda x_b.\lambda y_f.x_i$ is a projection of the right type o.w. $\lambda x_b.\lambda y_f.P$ is constant in x_b, y_f (either a closed term or a variable not in $b \cup f$) and (b) applies.

(Eq) Apply ind. hyp. on P' with b'=b and f'=f.

(weak) Apply ind. hyp. with $b' = b \setminus B'$ and $f' = f \cup (b \setminus b')$ where $FV(B) \cap FV(B') = \emptyset$.

(rmv) Apply ind. hyp. In particular if $x:\gamma \in f$ then consider b'=b, $f' = f \setminus \{(x:\gamma)\}$ and apply (b) to x.

(→I) Let $b' = b \cup \{(x:\alpha)\}$ and $f' = f \setminus \{(x:\alpha)\}$, apply (a), (b).

(→Iη) By ind. hyp. and the condition on the interpretation.

(→E) By ind. hyp. on P and Q with the same b and f.

(RecI), (RecIη), (RecE), (∀I), (∀E) follow by an easy application of the ind. hyp. and the conditions on the interpretation. □

4.3 Theorem *(Completeness for tas)*

Given a λ-theory E and a collection of assumptions B there is a type structure $T \in C_{\to\emptyset}$ such that: $T \models E$ and $T \models E, B \supset P:\alpha \Leftrightarrow \vdash E, B \supset P:\alpha$.

Proof

Repeat the steps of the proof 3.1.6 with the following notable differences:

$$B' =_\Delta \{\sigma(P:\alpha) \ / \ P:\alpha \in B\} \cup \{x_{i\alpha}:\alpha \ / \ i \in \omega, \ \alpha \text{ type}\}$$

and

$$[P][\alpha][Q] \Leftrightarrow_\Delta \vdash E, B' \supset P:\alpha \text{ and } \vdash E, B' \supset Q:\alpha$$

i.e. a term model with trivial pers. □

5. Type Containment

In this section we introduce a system for reasoning on Horn clauses of subtyping assertions that is sound for every type structure. We observe that to every provable subtyping judgment is possible to associate a typed λ-term that denotes a coercion in the model. Moreover we observe that a result of J.C. Mitchell implies the completeness of the system for a a particular class of clauses.

5.1 Subtyping system

Σ is a finite set of subtyping assumptions. Some of the following rules belong to the folklore of the subject. The rules of substitution and distribution are inspired by Mitchell[88], they represent one of the challenges in the study of the proof theoretic properties of the system along the lines described in §6.

(T.<.Asmp) $\alpha < \beta \in \Sigma \ \Rightarrow \ \Sigma \supset \alpha < \beta$

(T.<.MP$_{id}$) $\Sigma \supset (\alpha \to \beta) < (\alpha' \to \beta')$, $\Sigma \supset \alpha < \beta \ \Rightarrow$ $\Sigma \supset \alpha' < \beta'$

(T.<.Refl) $\vdash \alpha = \beta \ \Rightarrow \Sigma \supset \alpha < \beta$

(T.<.Trans) $\Sigma \supset \alpha < \beta$, $\Sigma \supset \beta < \gamma \ \Rightarrow \ \Sigma \supset \alpha < \gamma$

(T.<.→) $\Sigma \supset \alpha' < \alpha$, $\Sigma \supset \beta < \beta' \ \Rightarrow$ $\Sigma \supset (\alpha \to \beta) < (\alpha' \to \beta')$

(T.<.Rec) $\forall i \in Im \ \Sigma \supset \alpha_i < \beta_i$, $Im \subseteq In \ \Rightarrow$ $\Sigma \supset (l_i: \alpha_i)_{In} < (l_i: \beta_i)_{Im}$

(T.<.∀) $\Sigma \supset \alpha < \beta$, t not free in $\Sigma \Rightarrow$ $\Sigma \supset (\forall t.\alpha) < (\forall t.\beta)$

Conventions

$\beta_n \equiv_{abr.} \beta_1 \cdots \beta_n$. $t_n \equiv_{abr.} t_1 \cdots t_n$.
$\forall t_n \equiv_{abr.} \forall t_1 ... \forall t_n$, an analogous convention holds for the operator λ. $[\beta/t]_n \equiv_{abr.} [\beta_1/t_1,..., \beta_n/t_n]$.
$\Sigma \supset \alpha = \beta \equiv_{abr.} \Sigma \supset \alpha < \beta$ and $\Sigma \supset \beta < \alpha$.

(T.<.Sub) $\Sigma \supset \forall t_n.\alpha < \forall s_m.[\beta/t]_n\alpha$ for s not free in $\forall t.\alpha$; n,m≥0.

(T.<.Dist.Arrow) $\Sigma \supset \forall t.(\alpha \to \beta) < (\forall t.\alpha \to \forall t.\beta)$

(T.<.Dist.Rec) $\Sigma \supset \forall t.(l_i : \alpha_i)_{In} = (l_i : \forall t.\alpha_i)_{In}$

5.2 Theorem. *(Soundness of the subtyping system).*

$\vdash \Sigma \supset \alpha < \beta \ \Rightarrow T \models_{(F)} \Sigma \supset \alpha < \beta$.

where: T is a generic type structure; the subscript (F) means that (if T is built over a λ-model) the rules for subtyping are sound also for an F-interpretation of functional and record types.

Proof

By a rather standard induction on the lenght of the proof.□

5.3 Combining type theory and coercions: preliminary discussion

The results presented so far support an interactive use of coercions in the process of type-checking, namely given a term M: α that we want to use as having type β we may ask if α<β under the assumptions Σ and if the answer is positive then we may directly use the code associated to M as code of a (unique) corresponding term of type β.

Towards a complete automatization two problems arise:

(1) Given Σ, α, β is it *decidable* if $\vdash \Sigma \supset \alpha < \beta$?

(2) We may look for an *automatic insertion of coercions*, namely given a term M ask if modulo the insertion of appropriate coercions it can be well-typed. Of course problems of ambiguity can arise. Partial answers to these problems will be given in §6.

The next theorem shows that indeed the coercions are directly definable in the calculus. This fact can be useful in an *axiomatization* of the behaviour of explicit coercions and in proving results of *conservativity* of extended calculi over the pure one.

5.4 Theorem. *(Coercions are definable)*

If $\vdash \Sigma \supset \alpha < \beta$ then there exists a term M such that:

(a) $\vdash M: \alpha \to \beta$

(b) $FV(M) \subseteq \{x_i^{\alpha_i \to \beta_i} \,/\, \alpha_i < \beta_i \in \Sigma\}$

(c) $\forall$ T type structure

$(\eta \models \Sigma \Rightarrow \exists \rho\, (\rho \downarrow \eta\, ,\, id \in [M]_{\rho\eta})$.

Proof

Given a proof Π of $\Sigma \supset \alpha < \beta$ we define a term $M(\Pi)$ satisfying the given specifications.

Of course if $\eta \models \Sigma$, $x^{\alpha \to \beta} \in FV(M)$, and $\alpha < \beta \in \Sigma$ we define $\rho(x^{\alpha \to \beta}) =_\Delta [id]_{[\alpha \to \beta]\eta}$. Observe that if Π and Π' are both proofs of $\Sigma \supset \alpha < \beta$ and $\eta \models \Sigma$ we have a result of *coherence* at semantic level, namely: $\rho\eta \models M(\Pi) = M(\Pi')$.

With reference to the rules for subtyping define inductively on the structure of the proof Π (type labels of bounded variables are omitted, the λ-abstractions always refer to fresh variables):

(Asmp) $M \equiv x^{\alpha \to \beta}$

(MP$_{id}$) $M \equiv M'_\gamma M''_{\alpha \to \beta}$, $\gamma \equiv (\alpha \to \beta) \to (\alpha' \to \beta')$

(Refl) $M \equiv \lambda x^\alpha . x$

(Trans) $M \equiv \lambda x^\alpha .\, M_{\beta,\gamma}(M_{\alpha,\beta}\, x)$

(Rec) $M \equiv \lambda x^{(l_i:\, \alpha_i)In}.\, (l_i = M_{\alpha i, \beta i}\, (x.l_i))_{Im}$

($\to$) $M \equiv \lambda f^{\alpha \to \beta}.\, \lambda x^{\alpha'}. M_{\beta,\beta'}(f\, (M_{\alpha',\alpha} x))$

($\forall$) $M \equiv \lambda f^{\forall t.\beta}.\, \lambda t. M_{\beta,\beta'}(f\, t))$

(Sub) $M \equiv \lambda x^{\forall t_n . \alpha}.\, \lambda s_m .\, x\beta_n$

(Dist.$\to$) $M \equiv \lambda f^{\forall t.\alpha \to \beta}.\, \lambda y^{\forall t.\alpha}.\, \lambda t.\, (f\, t\, (y\, t))$

(Dist.Rec) $M \equiv \lambda f^{\forall t.(l_i\, :\, \alpha_i)In}\, .\, (l_i = (\lambda t.((ft).l_i))\,)_{In}$
$M \equiv \lambda f^{(l_i\, :\, \forall t.\alpha_i)In}\, .\, \lambda t.\, (l_i = (f.l_i)t)_{In}$

Using the definitions of types' and terms' interpretation the reader can check (boring but easy) by induction on the structure of the proof Π that $id \in [M]_{\rho\eta}$.

Note. This coherence result admits a nice *proof-theoretic presentation*. Given a term M inductively defined on the structure of a proof Π of a judgment $\Sigma \supset \alpha < \beta$ with free variables $x_1, \ldots, x_n$ we have:

$[\lambda x.x/x_1, \ldots, \lambda x.x/x_n]er(M) \to\!\!\!\to_{\beta\eta} \lambda x.x.$

It is intended that *er* is the usual "erasing types" function returning a term in the type-free, record-enriched $\lambda\beta l$-calculus defined in 3.1 coming with extensional rules:

(η) $\lambda x.Px = P$ (if $x \notin FV(P)$) and (η_{Rec}) $(l_i = P.l_i)_{In} = P$.

The proof is a simple induction on the construction of M. □

5.5 Theorem *(cnfr. Mitchell[88], pp 243, cor. 19)*

The system composed of the inference rules: (T.<.Trans), (T.<.$\to$) and (T.<.$\forall$) and axioms: (T.<.Sub) and (T.<.Dist. Arrow) (see 5.1) is (sound and) complete for deriving subtyping judgments of the shape $\emptyset \supset \alpha < \beta$ with α, β pure types of the second order λ-calculus (i.e. functional and polymorphic types) that are valid in all type structures.

Proof *(Schema)*

Starting from the fragment of the tas referring to pure second order types one considers the minimal λ-theory $\lambda\beta$ for which a well-known result of "subject reduction" is available. In this way we get rid of the rule (Eq) for terms in normal form; this is relevant as we are interested in characterizing:

$T \equiv \{(\alpha,\beta) \,/\, \vdash \emptyset \supset \lambda x.x: \alpha \to \beta\}$ and $\lambda x.x$ is in n.f. .

Next Mitchell introduces an equivalent, "syntax-driven" tas that produces as a fall-out the characterization of the set T by means of the mentioned theory.

Eventually the general completeness result follows by having shown completeness of the formal system for a particular term model. It is interesting to notice that for this fragment of the subtyping system the restriction to type-structures without empty types and based on λ-models does not entail any new valid subtyping. □

5.6 Remark

Mitchell's result heavily relies on the peculiar proof-theoretic properties of type-assignment systems without the rule (Eq) and with a *normal subject basis* (in a β-normal subject basis every term is in normal form and does not begin with a λ, see Hindley&Seldin[86], pp 210). In particular one exploits *strong normalization* of the typable terms and *subject reduction*.

Unfortunately both of these properties fail for a basis of the kind: $\{id: \alpha_i \to \beta_i \,/ i \in I\}$. It is not difficult to recover strong normalization by imposing suitable restrictions on the structure of $\alpha_i \to \beta_i$. On the other hand for getting subject reduction is necessary to strenghten the system with some rule that, at least, takes into account the equivalence of $(\lambda x.x)$ P with P, e.g.:

(id) $B \supset \lambda x.x: \alpha \to \beta\ ,\ B \supset P: \alpha \ \Rightarrow\ B \supset P: \beta$

6 Coercions' Inference

The decidability of the theory of subtyping, even starting with an empty set of assumptions, is an open problem (as far as we know). If we drop the axioms for substitution and distribution and assume that the collection of assumptions is empty then the resulting theory admits a cut-elimination theorem.

Convention

We need a linear notation for representing proof trees. Π, Π_1... denote proofs. π, π_1,... denote ambigously proofs, side conditions or multi-sets of proofs and conditions.

If we want to explicit the root of a proof we write: $\pi: \alpha < \beta$.

If we are also interested in the assumptions $\pi_1, \ldots, \pi_n$ and in the last rule being applied, say R, we write:

$(\pi_1, \ldots, \pi_n) :_R \alpha < \beta$.

Proviso

Let the system S include the rules (Refl), (Trans), ($\to$), (Rec) and ($\forall$). Types are considered up-to-redenomination of bound variables.

Proofs' reduction rules

We define rewrite rules on proof trees that transform a legal proof $\pi: \alpha < \beta$ into another legal proof $\pi': \alpha < \beta$. Denote with $\dashrightarrow_\Pi$ ($\dashrightarrow\!\!\to_\Pi$) the derived one-step (multi-step) reduction relation.

(Eq) $\pi:_R \alpha < \beta$ *reduces to:* $\alpha = \beta : \alpha < \beta$ if $\vdash_S \alpha = \beta$. (1)

(Eql) $(\alpha = \beta :_{Refl} \alpha < \beta,\, \pi: \beta < \gamma): \alpha < \gamma$ *reduces to:* $\pi': \alpha < \gamma$ (2)

(Eqr) $(\pi: \alpha < \beta,\, \beta = \gamma :_{Refl} \beta < \gamma): \alpha < \gamma$ *reduces to:* $\pi': \alpha < \gamma$ (2)

($\to$,trans) $((\pi_{11}: \beta_1 < \alpha_1,\, \pi_{12}: \alpha_2 < \beta_2) :_\to \alpha_1 \to \alpha_2 < \beta_1 \to \beta_2,$
$(\pi_{21}: \gamma_1 < \beta_1,\, \pi_{22}: \beta_2 < \gamma_2) :_\to \beta_1 \to \beta_2 < \gamma_1 \to \gamma_2)\ )$
$:_{trans}\ \alpha_1 \to \alpha_2 < \gamma_1 \to \gamma_2$ *reduces to:*
$((\pi_{21}: \gamma_1 < \beta_1, \pi_{11}: \beta_1 < \alpha_1) :_{trans} \gamma_1 < \alpha_1,$
$(\pi_{12}: \alpha_2 < \beta_2, \pi_{22}: \beta_2 < \gamma_2) :_{trans} \alpha_2 < \gamma_2) :_\to \alpha_1 \to \alpha_2 < \gamma_1 \to \gamma_2$

(∀,trans) $((\pi_1:\alpha<\beta) :_{\forall} \forall t.\alpha< \forall t.\beta ,$
$(\pi_2:\beta<\gamma) :_{\forall} \forall t.\beta< \forall t.\gamma) :_{trans} \forall t.\alpha<\forall t.\gamma$
reduces to:
$(((\pi_1:\alpha<\beta) , (\pi_2:\beta<\gamma)) :_{trans} \alpha<\gamma) :_{\forall} \forall t.\alpha<\forall t.\gamma$

(Rec,trans)
$((\pi_i:\alpha_i<\beta_i$ for $i\in Im, Im\subseteq In) :_{Rec} (l_i : \alpha_i)_{In}<(l_i : \beta_i)_{Im},$
$(\pi'_i:\beta_i<\gamma_i$ for $i\in Ip, Ip\subseteq Im) :_{Rec} (l_i : \beta_i)_{Im}<(l_i : \gamma_i)_{Ip})$
$:_{trans} (l_i : \alpha_i)_{In}<(l_i : \gamma_i)_{Ip}$ *reduces to:*
$((\pi_i:\alpha_i<\beta_i , \pi'_i:\beta_i<\gamma_i) :_{trans} \alpha_i<\gamma_i$ for $i\in Ip , Ip\subseteq In)$
$:_{Rec} (l_i : \alpha_i)_{In}<(l_i :\gamma_i)_{Ip}$

(1) The rule R is not (Eq).
(2) It is intended that the other rule being applied is not Refl.
π' is obtained by π via a suitable redenomination of bounded variables.

6.1 Proposition *(Trans-elimination and normal proofs)*

Given a proof $\pi:\alpha<\beta$ there is a reduction strategy such that $\pi:\alpha<\beta$ -->>_{Π} $\pi':\alpha<\beta$ and the latter is in normal form. Besides the structure of a normal proof of a judgment $\alpha<\beta$ is completely determined by the structure of α and β and it is decidable if $\vdash_S \alpha<\beta$.

Proof

Determine a subproof Π such that:
$\Pi \equiv (\pi_1:_{R1}\alpha<\beta, \pi_2:_{R2}\beta<\gamma) :_{trans} \alpha<\gamma$ and π_1, π_2 are trans-free. Now analyse the sixteen possible combinations of the rules R1 and R2 being applied. Show that we can either eliminate (trans) directly using rules (Eq(l)(r)) or apply one of the remaining rules.

If we are in the second case observe that we obtain a new proof of $\alpha<\gamma$ where (trans) appears at a lower level. Therefore iterating this process we get a trans free proof of $\alpha<\gamma$. Proceeding from the leaves towards the root we eventually obtain a trans-free proof of the initial subtyping judgment. To get a normal proof apply the reduction rule (Eq).

The proof of the second assertion is immediate. □

6.2 Remarks

(1) *(Structural subtyping)* Note that: $\vdash_S \alpha<\beta$ iff α and β have the same outmost type constructors and certain obvious conditions hold on the respective components i.e.
$\vdash_S \alpha_1\to\alpha_2<\beta_1\to\beta_2 \Rightarrow \vdash_S \beta_1<\alpha_1$ and $\vdash_S \alpha_2<\beta_2$.
$\vdash_S (l_i : \alpha_i)_{In}<(l_i : \beta_i)_{Im} \Rightarrow \vdash_S \alpha_i<\beta_i$ for $i\in Im$ and $Im\subseteq In$.
$\vdash_S \forall t.\alpha<\forall t.\beta \Rightarrow \vdash_S \alpha<\beta$.
Observe that the rule for records makes this theory not completely trivial. Besides the rule MP_{id} is derived as follows:

$\alpha_1\to\alpha_2<\beta_1\to\beta_2$ and $\alpha_1<\alpha_2 \Rightarrow$
$\beta_1<\alpha_1, \alpha_1<\alpha_2$ and $\alpha_2<\beta_2 \Rightarrow \beta_1<\beta_2$.

(2) *(A stronger coherence property)* For the theory S it is possible to prove a stronger coherence result than that given in §5 (failing for rules (Sub) and (Dist)).
Namely, given $\pi:\alpha<\beta$ and $\pi':\alpha<\beta$ denote with M and M' the terms associated to the proofs in §5 then $\vdash M =_{ext} M'$ where $=_{ext}$ denotes the relation of convertibility of the calculus enriched with three (η-)axioms expressing extensionality of (parametric)functions and records:

(η) $\lambda x^{\alpha}.Mx = M$ (if $x\notin FV(M)$),
(η_{Rec}) $(l_i = M.l_i)_{In} =M$,
(η_t) $\lambda t.Mt = M$ (if $t\notin FV(M)$)

There is a rather elegant and easy proof of this fact. Using the previous proposition it is enough to show that proof transformations preserve the convertibility of the associated terms. If $\pi:\alpha<\beta \text{-->}_{\Pi} \pi':\alpha<\beta$ via (→/∀/Rec, trans) then $\exists N\, (M \text{-->>} N \text{<<--} M')$ i.e. the terms have a common reduct (without using η-rules). For the rule (Eq) it is enough to check that, assuming the hereditary coercions convertible to the identity, the resulting coercion is (η-)reducible to the identity.

6.3 Coercions' inference

Denote with $\lambda^{\forall}Rec$ the calculus introduced in §2 and let $\alpha<\beta \equiv_{abr.} \vdash_S \alpha<\beta$. Given a term M in $\lambda^{\forall}Rec$ we are interested in the problem of determining if it can be well-typed modulo the insertion of appropriate coercions. Besides if this is the case we want to investigate the existence of a "most general solution".

There are *two equivalent ways* of formalizing this idea:

(1) *Subsumption* . Add to the calculus $\lambda^{\forall}Rec$ the following typing rule:
(Sub) $\quad M: \alpha , \quad \alpha<\beta \Rightarrow M: \beta$
Denote with $\vdash_{Sub}$ formal derivability in this new system.

(2) *Explicit coercions.* Assume a collection of constants $\{c_{\alpha,\beta} / \alpha, \beta$ types$\}$ and add to the calculus $\lambda^{\forall}Rec$ the following typing rule:
(ExpCoer) $\quad M: \alpha , \quad \alpha<\beta \Rightarrow (c_{\alpha,\beta} M) : \beta$
Denote with $\vdash_c$ formal derivability in this new system.

6.4 Definition

Define er_c (mnemonic for *erase coercions*) as the obvious function that given a well-typed term erases all its explicit coercions returning a possibly ill-typed term in $\lambda^{\forall}Rec$.

6.5 Definition

The function CI (for *coercions' inference*) takes a term M in $\lambda^{\forall}Rec$ and either fails or returns a well typed term N in $(\lambda^{\forall}Rec)_c$ (i.e. with explicit coercions) s.t. $er_c(N) \equiv M$.

(var) $CI(x^{\alpha}) =_{\Delta} x^{\alpha}$

(→I) $CI(\lambda x^{\alpha}.M) =_{\Delta}$ if $CI(M): \beta$ then $\lambda x^{\alpha}.CI(M)$ else FAIL

(→E) $CI(MN) =_{\Delta}$ if $(CI(M):\alpha\to\beta, CI(N):\gamma, \gamma<\alpha)$ then $CI(M)(c_{\gamma,\alpha}CI(N))$ else FAIL

(∀I) $CI(\lambda t.M) =_{\Delta}$ if $CI(M): \alpha$ and $(t\notin FV(M))$ then $\lambda t.CI(M)$ else FAIL

(∀E) $CI(M\beta) =_{\Delta}$ if $CI(M): \forall t.\alpha$ then $CI(M)\beta$ else FAIL

(RecI) $CI((l_i=M_i)_{In}) =_{\Delta}$ if $(CI(M_i) : \alpha_i$, for $i\in In)$ then $(l_i=CI(M_i))_{In}$ else FAIL

(RecE) $CI(M.l) =_{\Delta}$ if $CI(M): (l : \alpha, l_i: \alpha_i)_{In}$ then $CI(M).l$ else FAIL

6.6 Proposition

Let M be a term in $\lambda^{\forall}Rec$. Then

(1) $\exists N\, (\vdash_c N: \alpha ,\ er_c(N) \equiv M)$ iff $\vdash_{Sub} M : \alpha$.

(2) If $CI(M) : \beta$ then $er_c(CI(M)) \equiv M$.

Proof

(1) The derivations can be directly mapped one into the other. To every introduction of an explicit coercion corresponds an application of the (Sub) rule and viceversa.

(2) Inductively on the definition of CI. □

6.7 Theorem

Let M be a term in $\lambda^{\forall}$Rec. If $\exists$N: α ($er_c(N) \equiv M$) (or equivalently $\vdash_{Sub} M:\alpha$) then for some β:
(1) $CI(M) : \beta$ *(completeness)* and (2) $\beta<\alpha$ *(least type)*.

Proof

By induction on the structure of M.
CN is a meta-notation for $c_{\alpha n-1,\alpha n}(...(c_{\alpha 1,\alpha 2}\ N))..)$ where: $n\geq 1$, N: α_1, $\alpha_i < \alpha_{i+1}$.

$M \equiv x^{\beta}$ Observe: $er_c(N) \equiv x^{\beta} \Rightarrow N \equiv C\ x^{\beta} : \alpha$ and $\beta<\alpha$. On the other hand: $CI(x^{\beta}) \equiv x^{\beta} : \beta$.

$M \equiv \lambda x^{\alpha}.M$ Observe: $er_c(N) \equiv \lambda x^{\alpha}.M \Rightarrow$
$N \equiv C(\lambda x^{\alpha}.N'): \gamma$, $er_c(N') \equiv M$ and $N':\beta'$.
By ind. hyp. $CI(M) : \beta$ and $\beta < \beta'$, hence by def.
$CI(\lambda x^{\alpha}.M) : \alpha\rightarrow\beta$. Note that:
(a) $\alpha\rightarrow\beta' < \gamma$ by def. of N and this implies (by 6.2.1)
$\gamma \equiv \gamma_1\rightarrow\gamma_2$, $\gamma_1 < \alpha$ and $\beta' < \gamma_2$.
(b) Besides $\beta < \beta' < \gamma_2$ implies $\alpha\rightarrow\beta < \gamma$.

$M \equiv M_1M_2$ $er_c(N) \equiv M_1M_2 \Rightarrow N \equiv C(N_1N_2) : \gamma$,
$er_c(N_i) \equiv M_i$ i=1,2, $N_1: \gamma_1\rightarrow\gamma_2$, $N_2:\gamma_1$.
By ind. hyp. $CI(M_i) : \beta_i$ i=1, 2, $\beta_1 < \gamma_1\rightarrow\gamma_2$ and $\beta_2 < \gamma_1$.
From (6.2.1) follows: $\beta_1 \equiv \beta_1' \rightarrow \beta_1''$, $\gamma_1 < \beta_1'$, $\beta_1'' < \gamma_2$.
By def. $CI(M_1M_2) : \beta_1''$ as $\beta_2 < \gamma_1 < \beta_1'$.
Eventually observe: $\beta_1'' < \gamma_2 < \gamma$.

$M \equiv \lambda t.M$. $er_c(N) \equiv \lambda t.M \Rightarrow N \equiv C(\lambda t.N') : \gamma$,
$er_c(N') \equiv M$, $N': \alpha'$. By ind. hyp. $CI(M) : \alpha$, $\alpha < \alpha'$.
By def. $CI(\lambda t.M) : \forall t.\alpha$ (o.w N cannot be well-typed).
Besides: $\forall t.\alpha < \forall t.\alpha' < \gamma$.

$M \equiv M\beta$. $er_c(N) \equiv M\gamma \Rightarrow N \equiv C(N'\gamma)$, $er_c(N') \equiv M$,
$N': \forall t.\alpha'$.
By ind. hyp. $CI(M) : \alpha$, $\alpha < \forall t.\alpha'$.
By the analysis in (6.2.1) $\alpha < \forall t.\alpha' \Rightarrow \alpha \equiv \forall s.\beta$,
$[t/s]\beta < \alpha'$.
Hence: $CI(M\gamma) : [\gamma/s]\beta$ and (by (6.2.1) again) $[t/s]\beta < \alpha'$
$\Rightarrow [\gamma/s]\beta < [\gamma/t]\alpha'$.

$M \equiv (l_i{=}M_i)_{In}$. $er_c(N) \equiv (l_i{=}M_i)_{In} \Rightarrow$
$N \equiv C((l_i{=}N_i)_{In})$, $er_c(N_i) \equiv M_i$, $N_i : \beta_i$.
By ind. hyp. $CI(M_i) : \alpha_i$, $\alpha_i < \beta_i$.
By def. $CI(M) : (l_i : \alpha_i)_{In}$ and $(l_i : \alpha_i)_{In} < (l_i : \beta_i)_{In}$.

$M \equiv M.l$ $er_c(N) \equiv M.l \Rightarrow N \equiv C(N'.l)$, $er_c(N') \equiv M$,
$N': (l : \alpha, l_i : \alpha_i)_{In}$.
By ind. hyp. $CI(M) : \gamma$, $\gamma < (l : \alpha, l_i : \alpha_i)_{In}$.
By (6.2.1) $\gamma \equiv (l : \beta, l_i : \beta_i)_{Im}$, $Im \supset In$, $\beta < \alpha$,
$\beta_i < \alpha_i$ $i \in In$. By def. $CI(M.l) : \beta$, $\beta < \alpha$. □

Theorem 6.7 shows that given a term M if the set
$$er_c^{-1}(M) =_{\Delta} \{N / \vdash_c N:\alpha,\ er_c(N) \equiv M\}$$
is not empty then the function CI applied to M returns a canonical element of least type in $er_c^{-1}(M)$. We may wonder if the elements in $er_c^{-1}(M)$ are in some sense equivalent.

If $\vdash_c N_1:\alpha$, $\vdash_c N_2:\alpha$, $er_c(N_1) \equiv er_c(N_2) \equiv M$ then the property 3.6.(1) of the per-models tells us that N_1 and N_2 receive the same interpretations. However a more refined result suggested to me by G. Longo is available.

Towards this aim given a term N such that $\vdash_c N:\alpha$ denote with N' a term resulting from substituting every explicit coercion $c_{\alpha,\beta}$ with a corresponding definable typed term $M(\Pi)$ as specified in thm. 5.4. Of course the term $M(\Pi)$ depends from the proof Π of $\alpha<\beta$, however we have remarked in 6.2.(2) that all these terms are equivalent in a theory of typed $\beta\eta$-conversion.

The result 6.7 suggests that the terms N_1' and N_2' mentioned above can be related by means of the term $(c_{\beta,\alpha}CI(M))'$. This leads to the following

6.8 Corollary

Suppose that $\vdash_c N:\alpha$ and consequently $CI(er_c(N)):\beta$, $\beta<\alpha$. Then $\vdash_{\beta\eta} N' = (c_{\beta,\alpha}\ CI(\ er_c(N)))'$
where $\vdash_{\beta\eta}$ is the theory of conversion in 6.2.(2).

Proof

The proof goes by induction on the structure of $M \equiv er_c(N)$. It is enough to follow the pattern of proof 6.7 and check the convertibility of the terms keeping in mind the structure of the definable coercions.
As an example let us consider the case $M \equiv M_1M_2$. By ind. hyp. we can conclude: $(c_{\beta 1'\rightarrow\ \beta 1'',\ \gamma 1\rightarrow\gamma 2}\ CI(M_1))' = N_1'$ and $(c_{\beta 2,\ \gamma 1}\ CI(M_2))' = N_2'$. It is enough to observe $(N_1N_2)' = (c_{\beta 1'',\ \gamma 2}\ CI(M))'$ exploiting the structure of $c_{\beta 1'\rightarrow\ \beta 1'',\ \gamma 1\rightarrow\gamma 2}$. □

A final remark. *On the interpretation of (Sub).*

The rule (sub) is soundly interpreted in those models where subtyping means set-theoretic containment. E.g. ideals and reducibility candidates do model it, in particular the second interpretation shows that the erasures of terms typable using rule (Sub) are still (strongly) normalizing. It seems fair to say that the rule is against the nature of type theory (and category theory) as for example the types of a term are not unique up-to-conversion. Besides general per-models do not give sound interpretations. On the other hand the previous result shows that each term M in $\lambda^{\forall}$Rec s.t. $\vdash_{Sub}$ M: α admits a *canonical interpretation*, namely let $[M] =_{\Delta} [CI(M)]$. This observation is sistematically developed in Bruce&Longo[88].

7. Conclusion.

The *inital motivation* of this paper was to deal with completeness and proof-theoretic issues for certain theories of inheritance based on per models hence complementing Amadio[89] and Bruce&Longo[88] that provide the model-theoretic foundation and Cardelli&Longo[90] that apply some of this research to the design of the Quest programming language and suggest some non-trivial extensions of the intepretation. A framework for dealing with recursive types is discussed in Amadio[89] and fully developed in Amadio&Cardelli[90].

The *main contributions* could be summarized as follows:
- We develop sound and complete typed equivalence and type assignement systems for the class of type structures without empty types (§3, §4).
- We present a sound system for reasoning on Horn clauses of subtyping assertions. We observe the definability of the associated coercions and the completeness of the system for a particular class of clauses (§5).
- We start an investigation of the proof theoretic properties of such theories of inheritance and give some hints for the derivation of type-checking algorithms that can deal in an automatic way with the relation of subtyping (§6). Curien&Ghelli[90] have recently extended this result to a version of system F including a notion of bounded quantification and Amadio&Cardelli[90] considers a suitable extension to recursive types.

The definability result in 5.4 suggests the possibility of giving an abstract axiomatization of the notion of coercion. If

D is the category of data types we can think of coercions as a subcategory C of D that is essentially a poset satisfying certain closure conditions as specified by the categorical translation of the λ-terms in proof 5.4. Perhaps this point of view could help in finding other interesting interpretations of the notion of coercion, different from the realizability ones .

Acknowledgments

I wish to thank my advisor Giuseppe Longo for many discussions on the main ideas of this research.

References.

Amadio R. [1988] "A fixed point extension of the second order lambda calculus: observable equivalences and models", 3rd IEEE LICS, Edinburgh.

Amadio R. [1988(a)] "Proof theoretic properties of a theory of inheritance", internal report, October '88, Universita' di Pisa, (presented at the EEC Jumelage meeting on Typed λ-calculi, Nijmegen, November 1988).

Amadio R. [1989] "Recursion over realizability structures", TR 1/89, Universita' di Pisa, to appear on Info.&Comp.

Amadio R. [1989(a)] "Formal theories of inheritance for typed functional languages", TR 28/89 Universita' di Pisa.

Amadio R. [1990] "Recursion and Subtyping in Lambda Calculi", PhD Thesis, Università di Pisa.

Amadio R., Cardelli L. [1990] "Subtyping Recursive Types", DEC-SRC TR #62, ext. abs in ACM-POPL91.

Barendregt H. [1984] "The lambda calculus; its syntax and semantics", Revised and expanded edition, North Holland.

Bethke I. [1988] "Notes on partial combinatory algebras", PhD thesis, University of Amsterdam.

Breazu-Tannen V., Coquand T., Gunter C., Scedrov A. [1989] "Inheritance and explicit coercion", 4th IEEE-LICS '89.

Bruce K., Longo G. [1988] "A modest model of records, inheritance and bounded quantification" 3rd IEEE LICS, Edinburgh (Expanded and improved version appeared in Info&Comp).

Bruce K., Meyer A.[1984] "The semantics of second order polymorphic lambda-calculus" , in Semantics of data types, Kahn et al. (eds.), SLNCS 173.

Cardelli L. [1988] "A semantics of multiple inheritance", Info.&Comp., 76, (138-164).

Cardelli L., Longo G. [1990] "A semantic basis for Quest", LISP&FP90, Nice.

Cardone F., Coppo M. [1989] "Type inference with recursive types: syntax and semantics", preprint, Dipartimento di Informatica, Università di Torino.

Curien P.L., Ghelli G. [1990] "Coherence of Subsumption", CAAP90, Copenhagen.

Curry H., Feys R. [1958] "Combinatory Logic", vol. 1, North Holland.

Curry H., Hindley R., Seldin J. [1972] "Combinatory Logic", vol. 2, North Holland.

Girard J.Y. [1971] " Une extension de l'interpretation de Gödel a l'analyse, et son application a l'elimination des coupures dans l'analyse et la theorie des types". In 2nd Scandinavian Logic Simposium, J.E. Festand ed., North-Holland, (63-92).

Henkin L. [1950] "Completeness in the theory of types", JSL, 15, 2, (81-91).

Hindley R. [1983] "The completeness theorem for typing λ-terms", TCS, 22, (1-17).

Hindley R., Seldin J. [1986] "Introduction to Combinators and λ-calculus", Cambridge University Press.

Hyland M. [1988] "A small complete category", APAL, 40, 2, (135-165).

Liskov B. [1988] "Data abstraction and hierarchy", Addendum Proc. OOPSLA '87, Sigplan notices, 23, 5, (17-34).

Longo G., Moggi E. [1988] "Constructive natural deduction and its modest interpretation", CMU TR CS-88-131.

Mac Queen D., Plotkin G., Sethi R. [1986] "An ideal model for recursive polymorphic types", Info.&Contr., 71, 1-2.

Mahr B., Makowski J. [1984] "Characterizing specification languages which admit initial semantics", TCS, 31, (49-59).

Martini S. [1988] "Bounded quantifiers have interval models", ACM Lisp and Funct. Progr. Conf., Snowbird.

Meyer A., Mitchell J., Moggi E., Statman R. [1987] "Empty types in polymorphic lambda calculus", 14th POPL '87, (253-262).

Mitchell J. [1986] "A type-inference approach to reduction properties and semantics of polymorphic expressions", Lisp and Functional Programming Conference.

Mitchell J. [1988] "Polymorphic type inference and containment", Info.&Comp., 76, (211-249).

Mitchell J., Moggi E. [1987] "Kripke-style models for typed lambda calculus", 2nd IEEE-LICS '87, (303-314).

Reynolds J. [1974] "Towards a theory of type structure", SLNCS19, (408-425).

Scott D. [1976] "Data types as lattices", SIAM J. Comp. 5, (522-587).

A Fixed-Point Semantics for Feature Type Systems

Martin Emele Rémi Zajac

Project Polygloss *
University of Stuttgart
IMS-CL/IfI-AIS, Keplerstraße 17,
D 7000 Stuttgart 1, West Germany
polygloss@is.informatik.uni-stuttgart.de

Abstract
The solution of a set of recursive feature type equations is defined in terms of a fixed-point semantics. Using the corresponding operational semantics, the result of the evaluation of a feature term is computed by successive continuous approximations. This semantics allows the usage of any order for evaluating a term, and also allows for cyclic feature terms, alleviating the need for an "occur-check".

1 Features, types, sets, and inheritance

Assume the existence of an (abstract) informational domain, for example the set of persons. *Feature terms* describe objects of this universe by specifying values for attributes of objects. More precisely, as feature terms can provide only partial information about the objects they describe, a feature term denotes a *set* of objects in this universe. Feature terms are partially ordered by a subsumption relation: a feature term t_1 subsumes another feature term t_2 iff t_1 provides *less information* than t_2: $t_1 \sqsupseteq t_2$. In our universe, this means that the set described by t_1 is *larger* than the set described by t_2. For example, a feature term with the feature-value [age: NATURAL] subsumes a feature term with the feature-value [age: 15]: the set of persons of age 15 is included in the set of persons.

As different sets of attribute-value pairs are appropriate for different kinds of objects, we divide our feature terms into different types, called *feature types*. These types are ordered by a subtype relation: a type T_1 is a subtype of another type T_2 iff T_1 provides more information than T_2. For example, if one assumes that a student is a person, then the set of students is included in the set of persons. If the symbol STUDENT denotes the set of students, the symbol PERSON denotes the set of persons, then STUDENT is a subtype of PERSON.

This description implies, of course, that if we know that somebody is a student, we can deduce that he is a person. This deduction mechanism is transported in our type system as *type inheritance*: if a feature term t is of type T and there exist supertypes of T, then t inherits all the attribute-value pairs of the feature terms associated with the supertypes.

In the rest of this paper, we first present a syntax and a fixed point semantics for feature types. We then give an operational semantics for a feature type system which reflects the denotational characterization of a solution of a type system. This operational semantics has two interesting properties: first, the order of evaluation is irrelevant for solving a type system, and several strategies can be explored; second, this semantics is also valid for cyclic feature terms, thus alleviating the need for an "occur-check".

*Research reported in this paper is partly supported by the German Ministry of Research and Technology (BMFT, Bundesminister für Forschung und Technologie), under grant No. 08 B3116 3. The views and conclusions contained herein are those of the authors and should not be interpreted as representing official policies.

paths	*types symbols*	*variables*
ϵ	STUDENT	#x
advisor	FACULTY	#z
advisor.assistant	STUDENT	#x
advisor.secretary	STAFF	#y
roommate	EMPLOYEE	#t
roommate.representative	STAFF	#y

#x=STUDENT[advisor: FACULTY[assistant: #x,
secretary:#y=STAFF],
roommate: EMPLOYEE[representative: #y]]

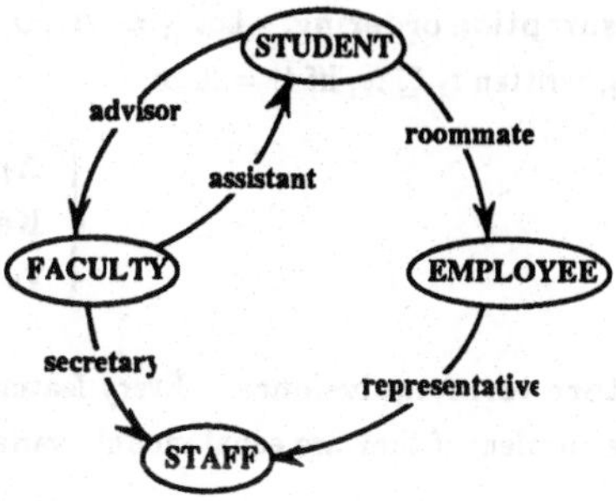

Figure 1: A feature term, its textual and its graphical representation.

2 Feature type systems

In this section, we give a characterization of the signature of type symbols, a syntax for feature terms, define a feature type system, and introduce the (syntactic) normal form of a feature type system.

2.1 Signature of type symbols

The user defines a preorder on a finite set of type symbols which reflects his categorization of an abstract informational domain. This partial order, called a *signature* Σ of type symbols, always implicitly contains the most general type $\top$ denoting the whole universe, and the most specific type $\bot$ denoting the empty set in this universe.

Let us assume that the signature has a lattice structure: two type symbols τ_1 and τ_2 always have a unique greatest common lower element $\tau_1 \sqcap \tau_2$ (greatest lower bound, GLB), and a unique least common upper element $\tau_1 \sqcup \tau_2$ (least upper bound, LUB).

In the following, we shall assume without loss of generality that the signature Σ is a complete distributive lattice: it is always possible to embed a partial order on a set of type symbols into a complete restricted power set which preserves existing GLBs (but not LUBs)[1]. The complete restricted power set of a partially ordered set P is $2^{[P]}$, the set of all non-empty subsets of incomparable elements of P. These subsets (called the *crowns* of P) are partially ordered by the Hoare subsumption relation:

$$X \sqsubseteq_H Y \Leftrightarrow \forall x \in X, \exists y \in Y, x \leq y.$$

2.2 Syntax of feature terms

We now introduce a syntax for feature terms which is taken from [Aït-Kaci 86].

Feature term. A feature term $t \in \mathcal{F}$, a set of feature terms, is a triple $\langle \Delta, \psi, \tau \rangle$ where Δ is a set of finite strings of labels from $\mathcal{L}$ (the set of label symbols) such that Δ is prefix-closed, finitely branching and regular: Δ is the set of all paths of a feature term. ψ is a function from Δ to Σ, a partially ordered signature of type symbols, which associates a type symbol with each path of the feature term. τ is a function from Δ to $\mathcal{V}$, a set of variables, such that $\mathrm{Ker}(\tau)$ is a right invariant equivalence relation: two paths having the same variable corefer (Fig. 1).

[1]See [Plotkin 76, Smyth 78], and also [Aït-Kaci 86] for a Hoare powerdomain construction.

Subsumption ordering. Let $t_1 = \langle \Delta_1, \psi_1, \tau_1 \rangle$ and $t_2 = \langle \Delta_2, \psi_2, \tau_2 \rangle$ be two feature terms. t_1 is subsumed by t_2, written $t_1 \sqsubseteq t_2$, iff $t_1 = \bot$ or

$$\begin{cases} \Delta_2 \subseteq \Delta_1 \\ \mathbf{Ker}(\tau_2) \subseteq \mathbf{Ker}(\tau_1) \\ \forall w \in \mathcal{L}^*, \psi_1(w) \sqsubseteq \psi_2(w) \end{cases}$$

Feature term equivalence. Every feature term that contains $\bot$ is equivalent to $\bot$. Two feature terms are equivalent if they are equal modulo variable renaming.

Unification. The unifier of two feature terms t_1, t_2 is defined as $t_1 \sqcap t_2$, the GLB of these terms. This is always well defined, because if Σ is a lattice, so is $\mathcal{F}$.

2.3 Feature type system

In the following, we define the notion of a feature type system as a formal characterization of a set of recursive feature type equations.

A feature type system. A feature type system $\mathcal{S}$ is a triple $\langle \Sigma, \mathcal{F}, f \rangle$ where Σ is a signature of type symbols, $\mathcal{F}$ a set of feature terms, and f a function from Σ to $\mathcal{F}$.

Example: Lists and append on lists.

The signature of type symbols		
{ NIL, CONS } < { LIST }.		
{ APPEND0, APPEND1 } < { APPEND }.		
The feature terms associated to type symbols		
LIST	→	LIST[].
NIL	→	NIL[].
CONS	→	CONS[first: ⊤, rest: LIST].
APPEND	→	APPEND[front: LIST, back: LIST, whole: LIST].
APPEND0	→	APPEND0[front: NIL, back: #l=LIST, whole: #l].
APPEND1	→	APPEND1[front: CONS[first: #f, rest: #l1], back: #l2=LIST, whole:CONS[first: #f, rest: #l3], cond:APPEND[front: #l1, back: #l2, whole: #l3]].

Consistency. A feature type system $\mathcal{S} = \langle \Sigma, \mathcal{F}, f \rangle$ is consistent iff it preserves meets: $f(\mathrm{T}_1 \sqcap \mathrm{T}_2) = f(\mathrm{T}_1) \sqcap f(\mathrm{T}_2)$ and $f(\mathrm{T}_1 \sqcup \mathrm{T}_2) = f(\mathrm{T}_1) \sqcup f(\mathrm{T}_2)$, and iff the feature term $f(\mathrm{T})$ is of the same type as T: $\forall \mathrm{T} \in \Sigma, \psi_{f(\mathrm{T})}(\varepsilon) = \mathrm{T}$.

Inheritance. Feature type inheritance is introduced through ϕ, defined as a monotonic extension of f:

$$\forall \mathrm{T} \in \Sigma, \phi(\mathrm{T}) = \sqcap_{\mathrm{T}_i \geq \mathrm{T}} f(\mathrm{T}_i)$$

If $\forall \mathrm{T} \in \Sigma, \phi(\mathrm{T}) \neq \bot$, we get an improved feature type system $\mathcal{S}' = \langle \Sigma, \mathcal{F}, \phi \rangle$ where all feature term information has been inherited. One can easily verify that in this case $\mathcal{S}'$ is consistent. In the following, we shall assume without loss of generality that a feature type system is consistent.

3 A fixed-point semantics

The fixed-point semantics formally defines the solution of a feature type system. Let

- $\mathcal{S} = \langle \Sigma, \mathcal{F}, \phi \rangle$ be a feature type system;
- ξ, a function from Σ to 2^{Σ} which maps a type T to its immediate subtypes:
  $\forall \mathrm{T} \in \Sigma, \xi(\mathrm{T}) = \{\mathrm{X} \in \Sigma \mid \mathrm{X} < \mathrm{T}$ and there is no $\mathrm{Y} \in \Sigma$ such that $\mathrm{X} < \mathrm{Y} < \mathrm{T}\}$;
- δ, a function from Σ to $2^{\mathcal{F}}$: $\forall \mathrm{T} \in \Sigma, \delta(\mathrm{T}) = \xi(\mathrm{T}) \sqcap \phi(\mathrm{T})$. $\delta(\mathrm{T})$ defines a compact form for the definition of feature type T as the meet between the associated feature term of T and its immediate subtypes.

We write a set of feature type definitions as a vector with $|\Sigma|$ components, each component being a pair of a type symbol T_i (left-hand side of a definition) and a join of feature terms $\delta_i = \delta(\mathrm{T}_i)$ (right-hand side of a definition):

$$\langle \langle \mathrm{T}_1, \delta_1 \rangle, \ldots, \langle \mathrm{T}_{|\Sigma|}, \delta_{|\Sigma|} \rangle \rangle$$

Term substitution. Now, we define a term substitution $subs(t, u, t')$, where a term t' is unified with a subterm starting at the path u in a term t:

$$subs(t, u, t') = t \sqcap u \cdot t'.$$

This is easily extended to joins of terms $\{t_1, \ldots, t_n\} = T$ and $\{t'_1, \ldots, t'_m\} = T'$:

$$subs(T, u, T') = T \sqcap u \cdot T' = \bigsqcup_{i=1}^{n} t_i \sqcap u \cdot (\bigsqcup_{j=1}^{m} t'_j) = \bigsqcup_{i=1}^{n} \bigsqcup_{j=1}^{m} t_i \sqcap u \cdot t_j.$$

Fixed point. The solution of a feature type system $\mathcal{S} = \langle \Sigma, \mathcal{F}, \phi \rangle$ is defined as the least fixed point of the function $\mathcal{M}$ from $(2^{\mathcal{F}})^{|\Sigma|}$ to $(2^{\mathcal{F}})^{|\Sigma|}$ which has its ith component defined as follows:

$$\mathcal{M}_i(T) = \bigsqcup_{d \in \delta_i} \sqcap_{u \in \Delta_d} subs(d, u, T).$$

In order to prove that this function has a least fixed point defined by

$$\mathcal{M}^*(\bot) = \bigsqcup_{k=0}^{\infty} \mathcal{M}^k(\bot)$$

we need to prove that $\mathcal{M}$ is monotonic and continuous (Tarski's Least Fixed Point Theorem).

- $\mathcal{M}$ is monotonic for the Hoare ordering: $\forall X \in 2^{\mathcal{F}}, X \sqsubseteq_H \mathcal{M}(X)$.
  According to its definition, $\mathcal{M}$ is a join of meets, and it is easy to verify that for all $i = 1, \ldots, |\Sigma|$, we have
  $$X \sqsubseteq_H \bigsqcup_{d \in \delta_i} \sqcap_{u \in \Delta_d} subs(d, u, X).$$
- $\mathcal{M}$ is continuous:
  $$\mathcal{M}(\bigsqcup_{k=0}^{\infty} T^k) = \bigsqcup_{k=0}^{\infty} \mathcal{M}(T^k).$$
  This is proven for each component of $\mathcal{M}$ using distributivity of joins on meets.

4 An operational semantics

In this section, we give an operational semantics of a feature type system: to evaluate a (set of) feature terms is to find the set of most specific feature terms which has the same denotation as the input. The result of an evaluation of a feature term t is $\mathcal{N}(t)$, called the normal form of the feature term t.

Rewriting. Let $\mathcal{S} = \langle \Sigma, \mathcal{F}, \phi \rangle$ be a feature type system. For all feature terms $t = \langle \Delta_t, \psi_t, \tau_t \rangle$, we first define a simple rewriting function $r(t, u)$, where u is a path in t ($u \in \Delta_t$), as a unifying substitution at the path u in t with the feature type definition defined for the type symbol occurring at path u:

$$\begin{aligned} r(t,u) &= subs(t, u, \delta(\psi_t(u))) \\ &= t \sqcap u \cdot \delta(\psi_t(u)) \\ &= t \sqcap u \cdot (\xi(\psi_t(u)) \sqcap \phi(\psi_t(u))). \end{aligned}$$

where $\psi_t(u)$ is the type symbol occurring at address u in the term t, and where the function ϕ computes inheritance and the function ξ narrows down one level in the type hierarchy.

The rewriting function $\mathcal{R}$ itself is now defined for all paths in t as:

$$\mathcal{R}(t) = \sqcap_{u \in \Delta_t} r(t, u).$$

$\mathcal{R}$ is extended to joins of terms $T = \{t_1, \ldots, t_n\}$ as follows:

$$\mathcal{R}(T) = \bigsqcup_{i=1}^{n} \mathcal{R}(t_i)$$

Monotonic approximation of the solution. The normal form of a feature term is computed as a sequence of successive steps approximating the final result. Each of these steps is defined as one application of the rewriting function $\mathcal{R}$ from $\mathcal{F}$ to $2^{\mathcal{F}}$ which makes each term resulting from an application of $\mathcal{R}$ more specific than the term computed at the previous step (the function $\mathcal{R}$ is continuously decreasing, see below). The normal form of a feature term t is then defined as the least upper bound of all terms that can be rewritten from t and contains only type symbols which are minimal elements (leaves) of the type hierarchy:

$$\mathcal{N}(t) = \bigsqcup \{t' \in \mathcal{R}^*(t) \mid \forall u \in \Delta_{t'}, \psi_{t'}(u) \in \text{minimal elements of } (\Sigma \setminus \bot)\}.$$

Correctness. To prove the correctness of this specification, we need to prove that the normal form of a feature type system $\mathcal{S} = \langle \Sigma, \mathcal{F}, \phi \rangle$ is equal to the least fixed point of the system:

$$\mathcal{N}(\delta) = \mathcal{M}^*(\bot).$$

This can be proven if we recall that rewriting was defined by substitution

$$r(t, u) = subs(t, u, \delta(\psi_t(u))).$$

Order of evaluation and cyclic feature terms. The rewriting function $\mathcal{R}$ is defined solely in terms of meets and joins, and by the associativity and the distributivity of these operations, we have:

$$\begin{aligned} \forall x \in X, \exists y \in \mathcal{R}(X) \mid x \sqsupseteq y &\Leftrightarrow X \sqsupseteq_{Hoare} \mathcal{R}(X) \\ \forall y \in \mathcal{R}(X), \exists x \in X \mid x \sqsupseteq y &\Leftrightarrow X \sqsupseteq_{Smyth} \mathcal{R}(X) \end{aligned}$$

which means that $\mathcal{R}$ is continuously decreasing for both Hoare and Smyth orderings.

Since the result of the evaluation of a term t is a fixed-point of $\mathcal{R}$, any order of substitution is well-defined (i.e. leads to the same result). This allows us to experiment with different strategies of evaluation.

5 Related work

The pioneering work in this context was [Aït-Kaci 86], who established the connection between "feature structures" [Kay 84] as used in computational linguistics, records, semantic networks and abstract data types. He proposed a method for solving systems of type equations based on term rewriting using a particular order of computation which did not handle cyclic feature terms.

[Pollard and Moshier 89] give linguistic motivation for introducing set values for feature terms and present a fixed-point logic for feature types with set values using a Smyth powerdomain construction.

Conclusion

Based on the semantics described in [Aït-Kaci 86] which we have extended to handle multiple inheritance, an implementation of an interpreter for feature type systems of the kind described in this paper has been carried out by the authors [Emele and Zajac 1989]. However, this semantics led to a complex implementation and was not entirely satisfactory, mainly because the original operational semantics as defined in [Aït-Kaci 86] did not clearly separate inheritance information from disjunctive information. The solutions of an evaluation were not subsumed by the input term (during computation of inheritance, only the most general type symbols were kept). One consequence was that cyclic feature terms were not allowed.

We presented a semantics for feature type systems. A type system defines an isomorphism between a lattice of type symbols and a lattice of feature terms. The evaluation proceeds as successive rewritings narrowing down the input term to reach the final solution: at each step, the elements of the approximation subsume the elements of the next step. Because any order of substitution is well-defined, cyclic feature terms are allowed.

The interpreter for feature type systems which is under development in our project makes use of this semantics.

References

[Aït-Kaci 86] Hassan Aït-Kaci. "An Algebraic Semantics Approach to the Effective Resolution of Type Equations". *Theoretical Computer Science 45*, 293-351.

[Emele and Zajac 1989] Martin Emele and Rémi Zajac. *Multiple Inheritance in RETIF*. ATR Technical Report TR-I-0114, ATR, Kyoto.

[Kay 84] Martin Kay. "Functional Unification Grammar: a formalism for machine translation". *Proc. of the 10th Intl. Conf. on Computational Linguistics, COLING-84*, Stanford.

[Plotkin 76] G.D. Plotkin. "A Powerdomain Construction". *SIAM J. of Computing* 5/3.

[Pollard and Moshier 89] Carl Pollard and Drew Moshier. "Unifiying partial descriptions of sets". In P. Hanson (ed.), *Information, Language and Cognition*, Vancouver Studies in Cognitive Science 1, University of British Columbia Press, Vancouver.

[Scott 82] Dana Scott. "Domains for Denotational Semantics". In: *ICALP-82*, LNCS 140, Springer-Verlag, Heidelberg.

[Smolka 88] Gert Smolka. *A Feature Logic with Subsorts*. LILOG Report 33, IBM Deutschland, Stuttgart.

[Smyth 78] M.B. Smyth. "Power Domains". *J. of Computer and System Sciences* 16, 23–36.

Unique-Sort Order-Sorted Theories : A Description as Monad Morphisms

John G. Stell
Departments of Mathematics and Computer Science
University of Keele, Keele, Staffs, ST5 5BG, England
email: john@uk.ac.kl.cs

Abstract

We present an apparently novel definition of order-sorted theory with the intention of clarifying various approaches to order-sorted algebra used in existing work on term rewriting and unification. If S is a poset we take an S-sorted theory to be a morphism between two monads, one in a category of S-sorted sets and one in **Set**. We describe various categories of S-sorted sets, but we concentrate on one appropriate for describing theories where the terms have unique sorts. One motivation for this work is to determine an appropriate category in which the existing category-theoretic description of unsorted non-equational unification may be extended to the order-sorted context.

We use our notion of theory to relate two superficially different ways of describing a collection of order-sorted terms and substitutions, which have appeared in the literature on order-sorted unification. This both clarifies the relationship between these two approaches and shows how far our definition accommodates existing approaches.

1 Introduction

If S is a partially ordered set, there are many applications of terms having sorts, or types, which are elements of S. While such collections of order-sorted terms are widely used, there does not seem to be a single agreed notion of order-sorted theory. By 'theory', we mean here a presentation-independent structure which can be described by giving a permissible collection of terms and substitutions, but which is more abstract than any particular description. The development of such a notion is important if we are to understand the relationships between superficially different approaches to order-sorted algebra.

For unsorted algebra, one abstract notion of theory is provided by the concept of a monad in **Set**, the category of sets. More generally, monads in other categories have been studied, and have various applications. It has been suggested, explicitly in [7] and implicitly in [6], that order-sorted theories can described in this framework by considering monads in a category of S-sorted sets. We claim that there are two respects in which this approach is in need of further development.

Firstly, there is not one but several categories of S-sorted sets. It seems that it is this diversity which accounts for much of the variety of approaches to order-sortedness which have appeared. We suspect that a full description of the relationships between these approaches will depend on a description of all possible categories of S-sorted sets and of their inter-relationships. As far as we are aware, such a comprehensive description has not yet been given.

The second respect in which the consideration of order-sorted theories as monads is in need of attention concerns the relation of sorted to unsorted theories. Presentations of order-sorted theories generally provide an underlying unsorted theory. We shall see that, to take a theory to be a monad in a category of S-sorted sets, may be inadequate as it is sometimes impossible to obtain the underlying unsorted theory in its entirety from this monad. This problem can be solved by taking order-sorted theories to be morphisms between monads. The domain of the morphism is a monad in a category of S-sorted sets, and the codomain is a monad in **Set**.

1.1 Relating two approaches to order-sorted algebra

Two approaches to order-sorted algebra are those of Meseguer, Goguen and Smolka [4, 5] and that of Schmidt-Schauß [11, 12]. In [4] the authors observed, in connection with [11], that "the technical definitions and general approach ... are so different from ours that we find it difficult to make a precise comparison". In this paper we show that our concept of order-sorted theory can be used as a means of establishing such a comparison. However, some further work is needed to provide a complete picture.

In section 2 we define several categories of S-sorted sets, all of which appear to have some rôle to play in S-sorted algebra. In this paper, we concentrate on one choice of category of S-sorted sets, the $\mathbf{Set}_S$ defined in section 2. Monad morphisms using $\mathbf{Set}_S$ are suited to describing theories where the terms have unique sorts. Monad morphisms using other categories defined in section 2 should be important for describing other types of order-sorted theory, but we do not treat these in this paper.

Both the approach of [4, 5] and that of [11, 12] have the notion of an underlying unsorted theory for any order-sorted theory in their sense. In the case of the work of Scmidt-Schauß, this unsorted theory cannot be entirely recaptured from knowledge of only a monad in a category of S-sorted sets. We are only able to obtain a subtheory of it by this route.

1.2 Order-sorted unification categorically

One motivation for developing a monadic approach to S-sorted theories, besides the clarification of order-sorted theories generally, is for the light thrown on order-sorted unification.

The original stimulus for the work presented here was to extend the existing category theoretic description of unsorted unification and term rewriting [9, 10] to the order-sorted case. This description represents substitutions by morphisms in a 'substitution category'. Unification problems are represented as parallel pairs of morphisms. This has certain advantages in the treatment of variables, and has been used to express unsorted non-equational unification as the construction of coequalizers. This has allowed an implementation of a unification algorithm using constructions from category theory.

To extend this work to the order-sorted context, it is necessary to obtain the appropriate substitution category. From a monad in $\mathbf{Set}_S$ we can obtain, via the Kleisli category construction, a category of sorted substitutions, which allows us to treat the unification algorithm in [16] in the spirit of [9]. This work will be described in detail in [15].

1.3 General points and acknowledgements

This paper represents the first stage in the programme of studying the various approaches to order-sorted algebra as outlined at the beginning of this introduction. We restrict ourselves to giving some indication of the diversity of categories of S-sorted sets and to using one type of S-sorted monad to relate two ways of presenting a collection of order-sorted terms. Further work is necessary in several areas.

Besides the literature on order-sorted theories, we assume a familiarity with some basic category theory and the elements of the monadic approach to universal algebra. Useful references for this material include [13, 3, 8, 1]. Within this context, most of our notation is fairly standard. Composition of morphisms in general and of functions in particular is usually written in the diagrammatic order. Thus fg means $\xrightarrow{f}\xrightarrow{g}$.

I would like to acknowledge the assistance of David Rydeheard in first drawing my attention to order-sorted unification and in discussing some portions of the work reported here. The departments of mathematics and of computer science at the University of Keele provided financial assistance which enabled me to attend the CTRS'90 workshop at Montreal.

2 Categories of S-sorted Sets

Most approaches to order-sorted algebra use the concept of S-sorted set as defined below. There is more than one way in which these entities form the objects of a category, since several notions of morphism can be defined. Other categories arise by using S-sorted sets which satisfy certain conditions. In this section we indicate the diversity of possibilities and introduce the particular category we use later in the paper.

2.1 Various categories

Definition 2.1 *An* S-sorted set J *consists of, for each* $s \in S$, *a set* J_s *such that if* $s_1 \leq s_2$ *then* $J_{s_1} \subseteq J_{s_2}$

Definition 2.2 *A* local *morphism between* S*-sorted sets* $\varphi : J \to K$ *has, for each* $s \in S$ *a* **Set** *morphism* $\varphi_s : J_s \to K_s$, *such that if* $s_1 \leq s_2$ *then the following diagram commutes*

$$\begin{array}{ccc} J_{s_1} & \xrightarrow{\varphi_{s_1}} & K_{s_1} \\ \big\downarrow & & \big\downarrow \\ J_{s_2} & \xrightarrow{\varphi_{s_2}} & K_{s_2} \end{array}$$

Definition 2.3 *A local morphism is* extensible *if for any* $s_1, s_2 \in S$ *whenever* $a \in J_{s_1} \cap J_{s_2}$ *we have* $a\varphi_{s_1} = a\varphi_{s_2}$.

Definition 2.4 *A* global *morphism between* S*-sorted sets* J *and* K *is a* **Set** *morphism* $\varphi : \cup_{s \in S} J_s \to \cup_{s \in S} K_s$.

We thus have three categories:

S_loc S-sorted sets and local morphisms.

S_ext S-sorted sets and extensible morphisms. In [6] this category is denoted $\mathbf{Set}^S$.

S_glob S-sorted sets and global morphisms.

Further possible categories having the same objects are obtained by taking the connected components of $(S, \leq)$ into consideration. We can have morphisms which are extensible or global only in a componentwise sense.

S-sorted sets which satisfy the least sorts condition provide another category.

Definition 2.5 *An* S*-sorted set* J has least sorts *if for any* $a \in \cup_{s \in S} J_s$ *the set* $\{s \in S | a \in J_s\}$ *has a least element.*

This gives the category **S_least** of least sort S-sorted sets and local morphisms. Observe that these morphisms are necessarily extensible.

A different approach to S-sorted sets is to construe the poset S as a category $\mathbf{S}$ by putting $s_1 \to s_2$ whenever $s_1 \leq s_2$, and to take S-sorted sets as objects of the functor category $\mathbf{Set}^{\mathbf{S}}$. In [7] some use is made of this functor category in connection with order-sorted theories as monads.

2.2 The category Set$_S$

We now introduce a category of S-sorted sets where elements have unique sorts. It may be helpful to observe that this is an example of a lax slice category.

Definition 2.6 *The category* Set$_S$ *has, for objects, Set morphisms* $\alpha : X \to S$, *often denoted* X_α. *The morphisms in* Set$_S$, *typically* $f : X_\alpha \to Y_\beta$, *are* **Set** *morphisms* $f : X \to Y$ *satisfying* $xf\beta \leq x\alpha$ *for all* $x \in X$. *Think of* α *as a sorting of* X. *We denote by* $U :$ Set$_S \to$ Set *the forget the sorting functor, i.e.* $X_\alpha \mapsto X$ *on objects.*

Given an object X_α of Set$_S$ it is often convenient to identify it with the set $\{\langle x, x\alpha \rangle | x \in X\}$, and to denote $\langle x, x\alpha \rangle$ by $x_{x\alpha}$. Thus if $S = \{A, B\}$, $X = \{x, y, z\}$ and α is $x \mapsto A\, y \mapsto B\, z \mapsto A$, then we may write X_α as $\{x_A, y_B, z_A\}$.

In [4, 5] the sets of sorted variables used to construct terms have least sorts. In [11] every variable is assigned a unique sort. These are equivalent since we have the following.

Proposition 2.7 *The categories* Set$_S$ *and S_least are isomorphic for any S.* $\quad\Box$

For being adequate to provide carriers for order-sorted algebraic theories and for relating-sorted to unsorted theories it is significant that

Proposition 2.8 *The category* Set$_S$ *has an initial object and binary coproducts. These are preserved by the functor U.* $\quad\Box$

For many S, Set$_S$ lacks coequalizers. For instance, take $S = \{A, B, C\}$ with $A < B$ and $C < B$ with A and C incomparable. The conditions under which Set$_S$ has coequalizers, or more generally, multicoequalizers are related to the conditions in [16] on the structure of S under which a non-equational S-sorted theory has most general unifiers or, more generally, most general sets of unifiers.

3 Monad Sortings

As mentioned in the introduction, we make use of morphisms between monads in different categories. First we recall the standard definition.

Definition 3.1 *If* $F : \mathbf{A}_1 \to \mathbf{A}_2$ *is a functor and* $\mathbf{T}_i = \langle T_i, \eta_i, \mu_i \rangle$ *is a monad in* $\mathbf{A}_i$ *for* $i = 1, 2$ *then a monad morphism along* F *is a natural transformation* $\chi : T_1 F \to F T_2$ *for which the following diagrams commute.*

$$
\begin{array}{ccc}
 & F & \\
 \eta_1 F \nearrow & & \nwarrow F\eta_2 \\
 T_1 F \xrightarrow{\quad \chi \quad} & & FT_2
\end{array}
\qquad
\begin{array}{ccc}
 T_1 T_1 F & \xrightarrow{(T_1\chi)(\chi T_2)} & FT_2 T_2 \\
 \mu_1 F \downarrow & & \downarrow F\mu_2 \\
 T_1 F & \xrightarrow{\quad \chi \quad} & FT_2
\end{array}
$$

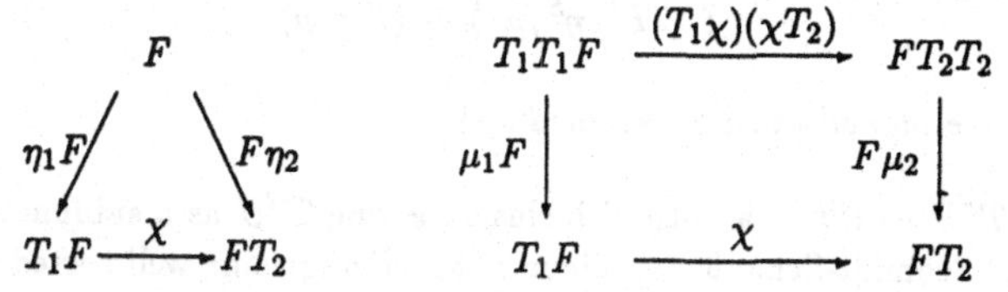

We apply this to the context where F is $U :$ Set$_S \to$ Set

Definition 3.2 *An S*-sorted monad *consists of a monad in* Set$_S$ *and a monad in* Set *together with a monad morphism along U between them.*

4 MGS Presentations

At present, lacking any better nomenclature, we refer to the two types of presentation in [4, 5] and in [11] by the initials of the respective authors.

4.1 Signatures

We recall the notion of presentation of an S-sorted theory due to Meseguer, Goguen and Smolka in [4, 5].

Definition 4.1 *An* MGS presentation Σ *is an $S^* \times S$-sorted set satisfying the regularity condition: for any $f \in \Sigma_{w,s}$ and $w_0 \leq w$ the set*

$$\{\langle w', s'\rangle \in S^* \times S \mid f \in \Sigma_{w',s'} \text{ and } w_0 \leq w'\}$$

has a least element.

If $f \in \Sigma_{s_1 \cdots s_n, s}$ we may write $f : s_1 \times s_2 \times \cdots \times s_n \rightarrow s$. For elements of $\Sigma_{\lambda,s}$ where λ denotes the empty string, this becomes $f : s$.

Although the authors of [4, 5] do define S-sorted signatures in general, without the regularity condition, the restriction is imposed for much of their work. Also the authors of this approach work with least sorts rather than unique sorts, but as we have observed in proposition 2.7 this is an equivalent viewpoint.

Given Σ as above we define the underlying unsorted signature $|\Sigma|$ by

$$|\Sigma|_n = \bigcup_{\langle w,s\rangle \in S^n \times S} \Sigma_{w,s}$$

for each natural number n. We usually use Ω for an unsorted signature.

Definition 4.2 *If Ω is an unsorted signature, an Ω* based MGS presentation *is an MGS presentation with underlying unsorted signature Ω.*

MGS presentations give rise to S-sorted monads. The definitions of the various parts of the S-sorted monad are straightforward, and rely partly on a rephrasing of the description of order-sorted terms in [4, 5] to take account of our use of unique sorts rather than least sorts. After checking several technical details we arrive at the following result.

Theorem 4.3 *An Ω based MGS presentation induces an S-sorted monad*

$$\zeta : \langle T^S, \eta^S, \mu^S\rangle \rightarrow \langle T, \eta, \mu\rangle$$

where $\langle T, \eta, \mu\rangle$ is the monad in **Set** *presented by Ω.* $\square$

Observe that $\zeta : T^S U \rightarrow UT$ is a natural inclusion giving $T^S U$ as a subfunctor of UT. This is the counterpart of Schmidt-Schauß' technique of specifying some well sorted terms as a subset of all Ω terms.

4.2 The sorted substitution category

The notion of sorted substitution used in [4, 5] can be recovered in a natural way from the above construction. If $\mathbf{K}$ is the Kleisli category of the monad in $\mathbf{Set}_S$ due to Σ, we define $\mathbf{Sub}^S_\Sigma$, the sorted substitution category, to be the full subcategory of $\mathbf{K}$ determined by the objects X_α where X is finite.

The objects of $\mathbf{Sub}_\Sigma^S$ may be thought of as finite sets of variables, each variable having a unique sort. The morphisms $X_\alpha \to Y_\beta$ in $\mathbf{Sub}_\Sigma^S$ are $\mathbf{Set}_S$ morphisms $X_\alpha \to Y_\beta T^S$. Alternatively these may be described as functions $\varphi : X \to Y_\beta T^S U$, for which $x\alpha \leq x\varphi\beta T^S$ for all $x \in X$.

Unlike the unsorted substitution category, arbitrary pairs of terms cannot be represented as parallel pairs in $\mathbf{Sub}_\Sigma^S$. A parallel pair of the form $\{x_A\} \rightrightarrows Y_\beta$ corresponds to a pair of terms, with variables in Y_β, and each term having a sort less than or equal to A. The terms need not have the same sort.

4.3 Equations

Recall that a poset is *directed* if every pair of elements has an upper bound. In [4, 5] equations are defined for *coherent* signatures. These are signatures of the form described above with the additional restriction that every connected component of the sort set be directed. For coherent signatures the following notion of equation is precisely equivalent to the definition of equation used in [4, 5].

If Σ is an MGS presentation, a Σ *equation* is a triple $\langle X_\alpha, t_1, t_2\rangle$ where $t_1, t_2 \in X_\alpha T^S U$ and where $t_1\, \alpha T^S$ and $t_2\, \alpha T^S$ have an upper bound in S.

The assumption that the sorts of t_1 and t_2 have an upper bound in S means that any equation can be represented, in at least one way, as a parallel pair in $\mathbf{Sub}_\Sigma^S$,

$$\{x_A\} \rightrightarrows X_\alpha$$

where A is an upper bound of the sorts of t_1 and t_2.

While it is interesting that equations of this form correspond to parallel pairs, it is more convenient here to characterize equations by reference to the congruence they generate than by their individual form. The ability to represent equations and unification problems as parallel pairs in the sorted substitution category does have particular relevance to categorical unification algorithms for sorted terms.

The most general notion of Σ equation would be that of a triple of the form $\langle X_\alpha, t_1, t_2\rangle$ where $t_1, t_2 \in X_\alpha T^S U$. Given a set E of equations in this sense, we obtain an equivalence relation $\sim_E$ on each $X_\alpha T^S U$ as the least equivalence for which

1. If $\langle X_\alpha, t_1, t_2\rangle \in E$ then $t_1 \sim_E t_2$
2. If $t_i \sim_E t'_i$ for $i = 1, \ldots, n$ and $f \in \Sigma_{s_1\cdots s_n, s}$ then $f(t_1, \ldots, t_n) \sim_E f(t'_1, \ldots, t'_n)$.
3. If $t \sim_E t'$ in $Y_\beta T^S U$ and $\varphi : Y_\beta \to X_\alpha$ in $\mathbf{Sub}_\Sigma^S$ then $t\varphi \sim_E t'\varphi$ in $X_\alpha T^S U$.

An arbitrary set of equations may lead to equivalence classes of terms to which there is no natural assignment of unique sorts which respects the original sorting.

Definition 4.4 *A set E of Σ equations is* regular *if for every X_α and every $t \in X_\alpha T^S U$ the set*

$$\{t'\, \alpha T^S \in S \mid t' \sim_E t\}$$

has a least element.

For an example of a set of equations which is not regular, consider the equation $b = c$ in the context where S is $\{A, B, C\}$ with $B < A$, $C < A$, and $b : B$ and $c : C$.

Given an MGS presentation Σ and arbitrary set E of Σ equations, we have an underlying unsorted equational presentation $(|\Sigma|, |E|)$. We can extend theorem 4.3 to equational presentations as follows

Theorem 4.5 *An MGS presentation Σ together with a regular set E of Σ equations, gives rise to an S-sorted monad*

$$\langle T^S, \eta^S, \mu^S\rangle \to \langle T, \eta, \mu\rangle$$

where $\langle T, \eta, \mu\rangle$ is the monad in **Set** *presented by $(|\Sigma|, |E|)$.* □

5 SS Presentations

The following is equivalent to Schmidt-Schauß' definition in [11]. We assume Ω is an unsorted signature.

Definition 5.1 *An Ω based* SS presentation *is a 4-tuple $(\Omega, V, \mathcal{W}, \mathsf{sort})$, where V is a countably infinite set of variable symbols, and where*

1. *$\mathcal{W} \subseteq VT$. Where T is the endofunctor of the monad presented by Ω, so VT is the set of Ω terms with variables in V. $\mathcal{W}$ is the set of* well sorted terms *and is required to satisfy*
   (a) $V \subseteq \mathcal{W}$.
   (b) $\mathcal{W}$ is closed under the formation of subterms.
2. *$\mathsf{sort} : \mathcal{W} \to S$ is a function satisfying the following conditions.*
   (a) For every $s \in S$ there are infinitely many $v \in V$ such that $v\,\mathsf{sort} = s$
   (b) The set of all substitutions $\sigma : V \to VT$ such that, for all $v \in V$, $v\sigma \in \mathcal{W}$ and $v\sigma\,\mathsf{sort} \leq v\,\mathsf{sort}$ is closed under composition.

Analogously to theorem 4.3 we can establish the following.

Theorem 5.2 *An Ω based SS presentation induces an S-sorted monad*

$$\zeta : \langle T^S, \eta^S, \mu^S \rangle \to \langle T, \eta, \mu \rangle$$

where $\langle T, \eta, \mu \rangle$ is the monad in **Set** *presented by Ω.* □

6 The Comparison of MGS and SS Presentations

We have taken order-sorted theories to be S-sorted monads, we now use this description of the nature of such theories to establish a relationship between the two modes of presentation defined above. SS presentations and MGS presentations provide two ways of giving an order-sorted theory, starting from an unsorted signature Ω. Intuitively we view these methods as follows. Firstly, as with SS presentations, we may use Ω to present an unsorted theory and then specify a sorting of this theory. Alternatively, in the technique of MGS presentations, we can first sort the signature and use this to present the sorted theory. We could say that an SS presentation is a sorting of a theory whereas an MGS presentation is a sorting of a theory presentation.

The precise relationship between the two approaches is established by developing the observation that an Ω based MGS presentation provides a sorting of the function symbols in Ω whereas an Ω based SS presentation provides a sorting of the derived function symbols.

6.1 SS presentation from MGS

For presentations without equations we have the following relationship. Further work is needed to extend this to the equational case.

Theorem 6.1 *Every non-equational MGS presentation has an associated SS presentation which presents the same S-sorted monad.*

Proof Suppose the MGS presentation Σ leads to the monad $\langle T^S, \eta^S, \mu^S \rangle$ in $\mathbf{Set}_S$ and that the underlying signature of Σ is Ω. Let V_α be any object of $\mathbf{Set}_S$ such that for every $s \in S$ there are infinitely many $v \in V$ such that $v\alpha = s$. Define $\mathcal{W}$ to be $V_\alpha T^S U$ and define sort to be $\alpha T^S : V_\alpha T^S U \to S$. The SS presentation $(\Omega, V, \mathcal{W}, \mathsf{sort})$ leads to the same S-sorted monad as the MGS presentation Σ. □

6.2 Derived function symbols

To describe how we may obtain an MGS presentation from an Ω based SS presentation, we need to review some material on the derived function symbols of Ω. A convenient notation for derived function symbols is obtained by taking them to be Ω terms of a certain form.

Definition 6.2 *An n-ary* derived Ω function symbol *is an Ω term with variables $\{1,\ldots,n\}$, in which each variable occurs at least once, and in which the order of first occurrence of the variables is $1,2,\ldots,n$.*

If Ω contains a binary function symbol f, then examples of derived function symbols are 1, $f(1,1)$, and $f(1,f(2,1))$ but not $f(2,1)$.

By considering all the derived function symbols we obtain a new signature $der\,\Omega$. This derived signature presents a different theory from Ω, but we can find a set E of $der\,\Omega$ equations such that $(der\,\Omega, E)$ presents the same theory as Ω. If Ω contains f as above then one of the equations which will need to follow from E is $f(1,1)(x) = f(1,2)(x,x)$.

To describe the required equivalence formally, note that we have a natural transformation $\psi : T_{der\,\Omega} \to T_\Omega$. The component of ψ at X, $\psi_X : XT_{der\,\Omega} \to XT_\Omega$, is defined by induction on term structure:

$$\begin{array}{rcl} [x] & \mapsto & [x] \text{ for } x \in X \\ d & \mapsto & d \text{ for any 0-ary derived function symbol } d \\ f(t_1,\ldots,t_n) & \mapsto & f[t_1\psi_X\backslash 1,\ldots,t_n\psi_X\backslash n] \end{array}$$

We also have, for each set X, $\varphi_X : XT_\Omega \to XT_{der\,\Omega}$ defined by

$$\begin{array}{rcl} [x] & \mapsto & [x] \text{ for } x \in X \\ d & \mapsto & d \text{ for any 0-ary derived function symbol } d \\ t & \mapsto & t[1\backslash x_1,\ldots,m\backslash x_m](x_1,\ldots,x_m) \text{ where the variables in } t \\ & & \text{in order of occurrence are exactly} (x_1,\ldots,x_m) \end{array}$$

Note that the φ_X are not the components of a natural transformation, and that the composite $\varphi_X\psi_X$ is the identity on XT_Ω.

To describe the required equations, take any countably infinite set V. If $d \in VT_{der\,\Omega}$, denote by $\overline{d}$ the equivalence class of d under the relation of bijective variable renaming. For each equivalence class $\overline{d}$ for which d is a term, distinct from $d\psi_V\varphi_V$, and of depth at most 2, we need an equation

$$d = d\psi_V\varphi_V$$

6.3 MGS presentation from SS

Suppose we have an SS presentation $P = (\Omega, V, \mathcal{W}, \mathsf{sort})$, which gives rise to the S-sorted monad

$$\langle T^S, \eta^S, \mu^S\rangle \to \langle T, \eta, \mu\rangle$$

The presentation provides a sorting of the derived Ω function symbols as follows.

Definition 6.3 *An n-ary derived Ω function symbol d is* well sorted at $s_1\cdots s_n \in S^n$ with sort s *if there is an injective function $\theta : \{1,\ldots,n\} \to V$ for which*

$$\begin{array}{rcl} i\theta\,\mathsf{sort} & = & s_i \\ d\,\theta T & \in & \mathcal{W} \\ d\,\theta T\,\mathsf{sort} & = & s \end{array}$$

It can be seen that sorts of terms are invariant under a sort preserving bijective renaming of the variables. Thus, if d is well sorted at $s_1 \cdots s_n$, then it has a unique sort there.

Construct an MGS presentation Σ by taking $\Sigma_{s_1 \cdots s_n, s}$ to consist of all the derived Ω function symbols which are well sorted at $s_1 \cdots s_n$ and have sort s there. It can be shown that Σ is regular. Thus we can construct a monad in $\mathbf{Set}_S$, including the functor $T_\Sigma : \mathbf{Set}_S \to \mathbf{Set}_S$. In the same way that the signature $der\,\Omega$ yields a different theory from Ω, T_Σ is not the same endofunctor as T^S. However by defining appropriate equations, we can obtain a presentation (Σ, E), which induces the same monad in $\mathbf{Set}_S$.

For each object X_α of $\mathbf{Set}_S$, we have morphisms

$$\psi_{X_\alpha} : X_\alpha T_\Sigma \to X_\alpha T^S$$

$$\varphi_{X_\alpha} : X_\alpha T^S \to X_\alpha T_\Sigma$$

defined analogously to $\psi_X : XT_{der\,\Omega} \to XT_\Omega$ and $\varphi_X : XT_\Omega \to XT_{der\,\Omega}$ in the unsorted case.

These have the following properties with respect to the sortings αT_Σ and αT^S

$$t\,\varphi_{X_\alpha}\,\alpha T_\Sigma = t\,\alpha T^S \quad \forall t \in X_\alpha T^S U$$

$$d\,\psi_{X_\alpha}\,\alpha T^S \leq d\,\alpha T_\Sigma \quad \forall d \in X_\alpha T_\Sigma U$$

Now we describe the set E of equations. In this description we shall write $\mathcal{V}$ for the S-sorted set V_{sort}. If $d \in \mathcal{V} T_\Sigma U$, denote by $\overline{d}$ the equivalence class of d under the relation of bijective sort preserving variable renaming. For each equivalence class $\overline{d}$, for which d is a term distinct from $d\psi_\mathcal{V}\varphi_\mathcal{V}$, and of depth at most 2, we need an equation

$$\langle \mathcal{V}, d, d\psi_\mathcal{V}\varphi_\mathcal{V} \rangle.$$

The equivalence classes into which $X_\alpha T_\Sigma U$ is decomposed by these equations take the form $\{d' \in X_\alpha T_\Sigma U | d\psi_{X_\alpha} = d'\psi_{X_\alpha}\}$. The equations are regular since the least sort assigned by αT_Σ to the equivalence class of d is $d\,\psi_{X_\alpha}\,\alpha T^S$. This is a lower bound since $d\,\psi_{X_\alpha}\,\alpha T^S \leq d\,\alpha T_\Sigma$. To find an element of the equivalence class having this sort, we may take $d\,\psi_{X_\alpha}\varphi_{X_\alpha}$.

There is a bijective sort preserving correspondence between the equivalence classes of $X_\alpha T_\Sigma U$ and the elements of $X_\alpha T^S U$. Working from this correspondence, we can establish the following

Theorem 6.4 *Given any SS presentation there is an associated equational MGS presentation which induces the same monad in* $\mathbf{Set}_S$. □

It is not, however, correct to say that the same S-sorted monad is obtained. The monad in **Set** due to the associated MGS presentation, may not be that due to the original SS presentation. This is due to the fact that the signature $|\Sigma|$ which underlies Σ, will, in general, be not $der\,\Omega$ but some subsignature of it.

6.3.1 Possibility of minimal MGS presentation from SS presentation

Although the above construction is sufficient to establish theorem 6.4, it can produce an signature Σ with infinitely many function symbols, and an infinite set of equations, in cases where it is possible to obtain an MGS presentation, with the appropriate properties, for which these are both finite. It appears from the second example below that there is a more sophisticated construction which produces as small an MGS presentation as possible. However, we do not at present, have the full details of the general case of such a construction.

6.3.2 Examples

The following simple examples should clarify the above construction.

Example 1 Consider the following SS presentation. Ω has one unary function symbol g and one constant symbol a. The sort set is $S = \{A\}$. The only non-variable terms in $\mathcal{W}$ are a and $g(a)$.

The associated MGS presentation Σ has two unrelated constant symbols $g(a) : A$ and $a : A$. There are no equations.

In this example we cannot recapture the monad in **Set** due to Ω from that in $\mathbf{Set}_S$ due to Σ.

Example 2 Ω has a binary function symbol f. The sort set is $S = \{A, B\}$ with $B < A$. All the terms are well sorted. sort satisfies

$$\begin{aligned} f(x_A, y_A)\ \mathsf{sort} &= A \\ f(x_A, x_A)\ \mathsf{sort} &= B \end{aligned}$$

In this case Σ contains

$$\begin{aligned} f(1,2) &: A \times A \to A \\ f(1,1) &: A \to B \end{aligned}$$

The set of equations will include $f(1,2)(x_A, x_A) = f(1,1)(x_A)$. The two Σ terms in this equation have sorts A and B respectively, they both represent an element of $\mathcal{W}$ with sort B.

In this example $|\Sigma|$ is $der\,\Omega$ and the SS and MGS presentations give not only the same monad in $\mathbf{Set}_S$, but also the same S-sorted monad. However, to take $|\Sigma|$ to be all of $der\,\Omega$ is unnecessary in this case. It is sufficient to have $|\Sigma|$ containing just two function symbols p and q, say. Set $\Sigma_{A,B} = \{p\}$ and $\Sigma_{AA,A} = \{q\}$, with $\Sigma_{w,s}$ empty otherwise. We need only one equation $q(x_A, x_A) = p(x_A)$. In this simple case it is easy to see that for these new values of Σ and E the theories presented by Ω and by $(|\Sigma|, |E|)$ are isomorphic.

7 Other Concepts of S-sorted Theory

There have been other proposals for a description of order-sorted theories as monads. In [7] S-sorted theories are treated as monads in the functor category $\mathbf{Set}^{\mathbf{S}}$. In [6] various notions of order-sorted algebra for an order-sorted theory are examined. The failure of the categories of algebras to be monadic in some cases is noted. In Poigné's work, it is only monadicity over the category **S_ext** which is considered.

While taking the base category to be $\mathbf{Set}^{\mathbf{S}}$ or **S_ext** is appropriate in some circumstances, there are reasons why our attention to $\mathbf{Set}_S$ is justified. Besides the utility of considering unique-sort theories on their own, the categorical properties of $\mathbf{Set}_S$ are important for understanding one approach to order-sorted unification. A treatment of order-sorted unification as in [16] in the spirit of [9] depends on $\mathbf{Set}_S$. To see this note that both $\mathbf{Set}^{\mathbf{S}}$ and **S_ext** are cocomplete, so, in particular, have coequalizers of arbitrary parallel pairs. Now, in the order-sorted unification described in [16] even the S-sorted theory with no function symbols may lack most general unifiers. This situation corresponds to the fact that, depending on the structure of S, $\mathbf{Set}_S$ may lack coequalizers.

Monads in **S_ext** appear to be appropriate for a treatment of some kinds of order-sorted unification other than the approach in [16].

An extensive treatment of order-sorted algebra and rewriting is provided by Smolka, Nutt, Goguen and Meseguer in [14]. However, their perspective is not that of theories as monads.

8 Categories of Order Sorted Algebras

The fact that there is not one generally agreed notion of the category of algebras of an order-sorted theory, is explained by the observation that, even if we agree that S-sorted sets are as defined as in definition 2.1, there are several possible definitions of morphism between such objects. There are the categories **S_loc**, **S_ext** and **S_glob** as defined in section 2 above; there are also those having morphisms which are extensible or global only on connected components. Besides the five categories of S-sorted sets just mentioned, which all have the same objects, there is $\mathbf{Set}_S$, on which this paper has concentrated.

In defining a category of algebras the morphisms are used in two places. Firstly in the interpretation of the function symbols, and secondly in the definition of homomorphism between algebras. Some definitions of algebra use morphisms from one category of S-sorted sets to interpret function symbols and define homomorphisms using morphisms from a different category. For example the type B algebras defined in [6], which apparently appeared in [2], use local morphisms to interpret function symbols and global morphisms to define homomorphisms. It is not surprising that such categories of algebras can fail to be monadic.

8.1 Algebras for MGS presentations

The authors of [4, 5] define a category $\mathbf{OSAlg}_\Sigma$ for their notion of presentation. The carriers of such algebras are arbitrary S-sorted sets, the elements of which need not have least sorts. This does not seem quite compatible with use of *regular* signatures, where terms, built from variables having least sorts, do have least sorts.

There is a general notion of algebras for any monad. Using this we can show that

Proposition 8.1 *Let Σ be an MGS presentation which induces the monad $\langle T^S, \eta^S, \mu^S\rangle$ in $\mathbf{Set}_S$. The category of algebras for this monad is the full subcategory of $\mathbf{OSAlg}_\Sigma$ determined by the objects where the carriers have least sorts.* □

9 Conclusions and Further Work

We have considered one possibility for the definition of order-sorted theory as monad morphism. The approach we have chosen has been used here as a framework to establish a comparison between two superficially different notions of presentation of order-sorted theory. The definition of S-sorted monad we have used also seems to be capable of capturing some of the term rewriting and unification aspects of such theories. Our approach, using monad morphisms and the category $\mathbf{Set}_S$ does not seem to have been used in previous accounts of the area.

Further work remains to be done in a number of areas. It seems likely that an MGS presentation with regular equations presents an S-sorted monad which can be presented by some form of SS presentation with equations. Monad morphisms using the category **S_ext** need to be studied. These appear to be suited to a treatment of the more general notion of presentation developed by Schmidt-Schauß in [12]. The treatment of algebras for monads needs to be extended to monads in **S_ext** and other categories of S-sorted sets.

In view of the discrepancy between the category $\mathbf{OSAlg}_\Sigma$ and the category of algebras for the monad $\langle T^S, \eta^S, \mu^S\rangle$ mentioned in section 8.1, it may be that not all aspects of order-sorted algebra can be treated entirely straightforwardly using the idea of theory as monad morphism. However, we believe that theories as monad morphisms can probably be used as a central concept, with reference to which many aspects of order-sorted algebra can usefully be described and inter-related.

References

[1] M. Barr and C. Wells. *Toposes, Triples and Theories*, volume 278 of *Grundlehren der Mathematischen Wissenschaften*. Springer Verlag, 1985.

[2] J. Goguen and J. Meseguer. Order sorted algebra I, partial and overloaded operations, errors and inheritance. Technical report, SRI, 1985.

[3] E. G. Manes. *Algebraic Theories*, volume 26 of *Graduate Texts in Mathematics*. Springer Verlag, 1976.

[4] J. Meseguer, J. Goguen, and G. Smolka. Order sorted unification. Technical Report CLSI-87-86, CLSI, 1987.

[5] J. Meseguer, J. Goguen, and G. Smolka. Order sorted unification. *Journal of Symbolic Computation*, 8:383–413, 1990.

[6] A. Poigné. Parameterization for order-sorted algebraic specification. *Journal of Computer and System Sciences*, 40(2):229–268, 1990.

[7] D. E. Rydeheard. *Applications of Category Theory to Programming and Program Specification*. PhD thesis, University of Edinburgh, 1981.

[8] D. E. Rydeheard and R. M. Burstall. Monads and theories: A survey for computation. In M. Nivat and J. C. Reynolds, editors, *Algebraic Methods in Semantics*, pages 576–605. Cambridge University Press, 1985.

[9] D. E. Rydeheard and R. M. Burstall. A categorical unification algorithm. In Pitt et al, editor, *Category Theory and Computer Programming*, volume 240 of *Lecture Notes in Computer Science*, pages 493–505. Springer Verlag, 1986.

[10] D. E. Rydeheard and J. G. Stell. Foundations of equational deduction: A categorical treatment of equational proofs and unification algorithms. In Pitt et al, editor, *Category Theory and Computer Science*, volume 283 of *Lecture Notes in Computer Science*, pages 114–139. Springer Verlag, 1987.

[11] M. Schmidt-Schauß. Unification in many sorted equational theories. In *8th Conference on Automated Deduction*, volume 230 of *Lecture Notes in Computer Science*, pages 538–552. Springer-Verlag, 1986.

[12] M. Schmidt-Schauß. *Computational Aspects of an Order-Sorted Logic with Term Declarations*, volume 395 of *Lecture notes in Computer Science*. Springer Verlag, 1989.

[13] H. Schubert. *Categories*. Springer Verlag, 1972.

[14] G. Smolka, W. Nutt, J. A. Goguen, and J. Meseguer. Order-sorted equational computation. In H. Aït-Kaci and M. Nivat, editors, *Resolution of Equations in Algebraic Structures, Volume 2, Rewriting Techniques*, pages 297–367. Academic Press, 1989.

[15] J. G. Stell. Some categorical aspects of unification and rewriting. Forthcoming, to be submitted as Ph.D. Thesis, University of Manchester.

[16] C. Walther. Many sorted unification. *Journal of the ACM*, 35:1–17, 1988.

Equational Logics*(Birkhoff's Method Revisited)

Yong SUN
Department of Computer Science
University of York
Heslington York YO1 5DD, U.K.
E-mail : yong%york.minster@nsfnet-relay.ac.uk

1 Introdution

We are interested in many-sorted theories which can be expressed by equations or axioms of first-order languages. This kind of equational theories, including their corresponding software system environments (say term rewriting systems), has developed dramatically in recent years, and is still rapidly growing. In these theories, *equational logics* play fundamental roles. However, there are confusions in the literature about the logics. Some of them are terminological or notational, and the others of them are theoretical. This paper is an attempt at a brief survey and subsequent clarification in the area.

Terminologically, there is no confusion in equational logics for pure equations. But the situations are different for the other two kinds of equational logics. One is for *equational implications* and the other is for *conditional equations*. For equational implications, some people refer them as "conditional equations", and even confuse them as conditional equations. For conditional equations, they are refered to either as *quasi-equations* or as *universal Horn Clauses* by different schools. Actually, the three kinds of equational logics have a close relationship between them. In order to avoid a future confusion let us use *dependent equations* and *quasi-dependent equations* for equational implications and conditional equations respectively.

We let Σ be a signature and S be $Sort(\Sigma)$; and we fix $\tilde{V}$ as the collection of (countably) infinite variables indexed by elements in S. Suppose $\mathbf{T}_\Sigma(\vec{X})$ is the term Σ-algebra with variables in $\vec{X} \subseteq \vec{V}$, we assume that

(a) t, u are terms in $\mathbf{T}_\Sigma(\vec{X})$,

(b) $\mathbf{A}$ is a Σ-algebra and

(c) α is a Σ-homomorphism from $\mathbf{T}_\Sigma(\vec{X})$ to $\mathbf{A}$, written as $\alpha : \mathbf{T}_\Sigma(\vec{X}) \to \mathbf{A}$.

Sometimes, we will omit the signature Σ in our terminology for simplicity. An *equation* $t \simeq_{\vec{X}} u$ is presented with a variable index. Standard notation uses universally-quantified equations instead, like $\forall X.t \simeq u$ (see [GM81, EM85, GM85]). However, we prefer variable-indexed equations to universally-quantified ones, simply because variable indices make more semantic sense than quantifiers on the unification of the three equational logics. We will come back to this point later. Without variable indices (or universal quantifiers), **true** $\simeq$ **false** is derivable, see Example 1 below.

Example 1 (necessity of variable indices [GM81, GM85]) : *Let signature* Σ *have sorts* $\{a, b\}$ *and* $\Sigma_{\epsilon,b} = \{\mathbf{True}, \mathbf{False}\}$, $\Sigma_{b,b} = \{\neg\}$, $\Sigma_{b\times b,b} = \{\mathbf{and}, \mathbf{or}\}$, $\Sigma_{a,b} = \{\mathbf{foo}\}$, *and other* Σ *be* $\emptyset$. *The axiom are the collection of*

$$\begin{array}{l} \neg\,\mathbf{True} \simeq \mathbf{False} \\ \neg\,\mathbf{False} \simeq \mathbf{True} \\ x\,\mathbf{or}\,\neg x \simeq \mathbf{True} \\ x\,\mathbf{and}\,\neg x \simeq \mathbf{False} \\ x\,\mathbf{and}\,x \simeq x \\ x\,\mathbf{or}\,x \simeq x \\ \mathbf{foo}(y) \simeq \neg\,\mathbf{foo}(y) \end{array}$$

where x *is a variable sorted by* b *and* y *is a variable sorted by* a. *We can derive* **True** $\simeq$ **False**

*A preliminary version of this paper has been presented in CTRS '90, Concordia University, Montréal Canada.

as follows.

True $\simeq$
foo(x) **or** $\neg$ **foo**$(x) \simeq$
foo(x) **or foo**$(x) \simeq$
foo$(x) \simeq$
foo(x) **and foo**$(x) \simeq$
foo(x) **and** $\neg$ **foo**$(x) \simeq$
False.

For equation $t \simeq_{\vec{X}} u$, we have the following.

1. $\mathbf{A}, \alpha \models t \simeq_{\vec{X}} u$ iff $\alpha(t) = \alpha(u)$; and
2. $\mathbf{A} \models t \simeq_{\vec{X}} u$ iff $\mathbf{A}, \alpha \models t \simeq_{\vec{X}} u$, for every $\alpha : \mathrm{T}_\Sigma(\vec{X}) \to \mathbf{A}$.

You should know that $\mathbf{A} \models t \simeq_{\vec{X}} u$ is a *semantic* property (or meta-property) of the algebra **A**. Intuitively, it says that t and u are *indistinguishable* by **A**. Obviously, it is more interesting if this property is universal. More precisely, t and u are *indistinguishable* iff for every algebra **A**, $\mathbf{A} \models t \simeq_{\vec{X}} u$. It is very important and convenient to have an inference-rule system, $\tilde{D}$, in which we can deduce all indistinguishable terms.

For example, given a collection Γ of equations, we would like to capture the property that for every **A**, if $\mathbf{A} \models \Gamma$ (which is the natural extension of $\mathbf{A} \models t \simeq_{\vec{X}} u$ to a collection of equations) then $\mathbf{A} \models t \simeq_{\vec{X}} u$, written as $\Gamma \models t \simeq_{\vec{X}} u$. In other words, we hope that $t \simeq_{\vec{X}} u$ *is derivable from* Γ, *written as* $\Gamma \vdash t \simeq_{\vec{X}} u$, reflects the property $\Gamma \models t \simeq_{\vec{X}} u$ above precisely. In the present paper, $\Gamma \vdash t \simeq_{\vec{X}} u$ is written as $t \simeq_{\vec{X}} u \in \sqcup\tilde{D}(\Gamma)$, which is the least fixpoint of $\tilde{D}$ containing Γ with the usual inclusion order when considering $\tilde{D}$ as a function for deductions from a collection of equations to another collection. When the latter $\Gamma \vdash t \simeq_{\vec{X}} u$ implies the former $\Gamma \models t \simeq_{\vec{X}} u$, we commonly say that $\tilde{D}$ is *sound*; when the former implies the latter, we commonly say that $\tilde{D}$ is *complete*.

Similarly, we can weaken the indistinguishability and obtain dependent indistinguishability and quasi-dependent indistinguishability as follows.

1. Dependent equation (or equational implication) $\gamma_{\vec{X}} \mapsto t \simeq_{\vec{X}} u$:
   (a) $\mathbf{A} \models \gamma_{\vec{X}} \mapsto t \simeq_{\vec{X}} u$ iff if $\mathbf{A} \models \gamma_{\vec{X}}$ then $\mathbf{A} \models t \simeq_{\vec{X}} u$;
   where $\gamma_{\vec{X}}$ ranges over collections of equations, and $\mathbf{A} \models \gamma_{\vec{X}}$ means $\mathbf{A} \models t' \simeq_{\vec{X}} u'$ for each $t' \simeq_{\vec{X}} u' \in \gamma_{\vec{X}}$.
   (b) $\Gamma^d \models \gamma_{\vec{X}} \mapsto t \simeq_{\vec{X}} u$ iff $\mathbf{A} \models \Gamma^d$ implies $\mathbf{A} \models \gamma_{\vec{X}} \mapsto t \simeq_{\vec{X}} u$ for every algebra **A**;
   where Γ^d range over collections of dependent equations and $\mathbf{A} \models \Gamma^d$ means for each $\gamma'_{\vec{X}} \mapsto t' \simeq_{\vec{X}} u' \in \Gamma^d$, $\mathbf{A} \models \gamma'_{\vec{X}} \mapsto t' \simeq_{\vec{X}} u'$.
2. Quasi-dependent equation (or conditional equations [GM82] or quasi-equations [M73]) $\gamma_{\vec{X}} \hookrightarrow t \simeq_{\vec{X}} u$:
   (a) $\mathbf{A} \models \gamma_{\vec{X}} \hookrightarrow t \simeq_{\vec{X}} u$ iff $\mathbf{A}, \alpha \models \gamma_{\vec{X}}$ implies $\mathbf{A}, \alpha \models t \simeq_{\vec{X}} u$ for every $\alpha : \mathrm{T}_\sigma(\vec{X}) \to \mathrm{T}_\sigma(\vec{X})$;
   where $\mathbf{A}, \alpha \models \gamma_{\vec{X}}$ means $\mathbf{A}, \alpha \models t' \simeq_{\vec{X}} u'$ for each $t' \simeq_{\vec{X}} u' \in \gamma_{\vec{X}}$.
   (b) $\Gamma^q \models \gamma_{\vec{X}} \hookrightarrow t \simeq_{\vec{X}} u$ iff $\mathbf{A} \models \Gamma^d$ implies $\mathbf{A} \models \gamma_{\vec{X}} \hookrightarrow t \simeq_{\vec{X}} u$ for every algebra **A**;
   where Γ^q range over collections of quasi-dependent equations (as well as dependent equations) and $\mathbf{A} \models \Gamma^q$ means for each $\gamma'_{\vec{X}} \hookrightarrow t' \simeq_{\vec{X}} u' \in \Gamma^q$, $\mathbf{A} \models \gamma'_{\vec{X}} \hookrightarrow t' \simeq_{\vec{X}} u'$.

Note that when $\gamma_{\vec{X}} = \emptyset$, both the dependent equation $\gamma_{\vec{X}} \mapsto t \simeq_{\vec{X}} u$ and the quasi-dependent equation $\gamma_{\vec{X}} \hookrightarrow t \simeq_{\vec{X}} u$ are the equation $t \simeq_{\vec{X}} u$.

Remark 2: *Henkin has recognized the difference between dependent equations and quasi-dependent equations in [H77]. He obtains two deduction systems for natural numbers corresponding to dependent and quasi-dependent implications.*

2 Equational Logic $\tilde{D}$

The soundness and completeness of many-sorted $\tilde{D}$, was believed to be a trivial extension of the single-sorted one, and was first claimed by

Goguen and Meseguer in [GM81] which demonstrated that the naive belief did not hold. The simple example provided by them is Example 1. Two special rules, *Abstraction* and *Concretion*

$$(ab)\ \frac{t \simeq_{\vec{X}} u}{t \simeq_{\vec{X}\cup\{y\}} u} \quad (co)\ \frac{T_\Sigma(\emptyset)_i \neq \emptyset, t \simeq_{\vec{X}} u, x \in \vec{X}_i}{t \simeq_{\vec{X}-\{x\}} u}$$

where y is not in $\vec{X}$ and x is not in t nor in u, are given in the deduction system (see [GM81, GM85]) besides the ordinary rules to emphasize the potentiality of empty sorts (or carriers). The full version of the soundness and completeness of their system appears in [GM85]. This proof involves building function spaces from term algebras and the verification that these spaces form clones (see [C80] for basic properties of clones). Since they only allow arbitrary finite quantification over equations, they use the co-limit result from category theory [M71] in eliminating quantification over equations, i.e. in obtaining a quantifier-free (or variable index-free) calculus. Ehrig and Mahr seem to follow the outline of the proof in [GM81] and provide another proof in [EM85] but they exclude the case of empty sorts. There has been some confusion between these two proofs, see [ELM86] and [GM86].

We take a different approach from that used by the authors above in our proof of soundness and completeness of $\tilde{D}$. Following Birkhoff (see [BS81, C80, G79]), we extend the proof of the single-sorted case by introducing the concept of *cross-fully invariant congruences*, which differs from fully-invariant congruences as in [BS81]. This approach removes the need to build function spaces from term algebras to form clones, and use clone properties to obtain the completeness result as in [GM85]; nor do we need to exclude algebras with empty carriers as in [EM85]. We should point out that there is no place for the new concept in the single-sorted case. The necessary and sufficient condition for variable index-free $\tilde{D}$ is derived within this approach.

Birkhoff's approach is conceptually simpler, and it is more coherent when we consider the extensions to include dependent equations and quasi-dependent equations. Also, information about models, e.g. what are equationally definable, dependent equationally definable and/or quasi-dependent equationally definable classes of algebras (see [CK73, M73] for examples about models of quasi-dependent equations), can easily be carried out. This kind of benefit is not easily available in the other approaches such as [GM85]. To understand this point more clearly, let us consider universally-quantified equations, like $\forall X.t \simeq u$. It is easy to extend universally-quantified equations to dependent equations, i.e.

$$(\forall X.\gamma) \mapsto (\forall X.t \simeq u)$$

where $\forall X.\gamma$ is $\{\forall X.t' \simeq u' | t' \simeq u' \in \gamma\}$. However, since every equations are universally-quantified, it is hard to represent a quasi-dependent equation, like

$$\forall X.(\gamma \hookrightarrow t \simeq u),$$

in this universally-quantified framework for equations. There is no such a problem in generalizing variable-indexed equations to other kinds of equational forms, see the following

$$t \simeq_{\vec{X}} u, \quad \gamma_{\vec{X}} \mapsto t \simeq_{\vec{X}} u, \quad \gamma_{\vec{X}} \hookrightarrow t \simeq_{\vec{X}} u.$$

3 Calculus $\tilde{D}^d$ for Dependent Equations

Birkhoff's approach provides further clarification on the soundness and completeness of $\tilde{D}^d$ and of $\tilde{D}^q$ which are deductive functions for dependent equations and quasi-dependent equations respectively. Essentially, a sound and complete calculus for dependent equations can be obtained by the following fact with completeness of $\tilde{D}$.

Fact 3: *$\Gamma \models \gamma \mapsto t \simeq u$ iff $\Gamma \models t \simeq u$ when $\Gamma \models \gamma$.*

Therefore, let $\tilde{D}^d$ have all the obvious rules[1] with an extra rule below

$$\frac{\emptyset \mapsto t \simeq u \in \tilde{D}^d(\{\emptyset \mapsto t' \simeq u' | t' \simeq u' \in \gamma\} \cup \Gamma)}{\gamma \mapsto t \simeq u \in \tilde{D}^d(\Gamma)}.$$

[1] There is only one requirement on these rules, i.e. they must form a sound and complete calculus with respect to pure equations. For this, we only need to point out that the following substitution rule is not sound

$$\frac{\gamma \mapsto t \simeq u}{\alpha(\gamma) \mapsto \alpha(t) \simeq \alpha(u)}$$

where α is a substitution map. The correct substitution

Thus, it is sound and complete.

4 Calculus for Quasi-dependent Equations

For quasi-dependent equations, there is one obvious temptation as follows.

(i) $Congr_{S,\Sigma}$ denotes the complete lattice of congruences on term algebra $T_\Sigma(\vec{X})$.

(ii) θ_{Γ^q} is the functional : $Congr_{S,\Sigma} \to Congr_{S,\Sigma}$ such that given a congruence $\equiv_1$, $\theta_{\Gamma^q}(\equiv_1)$ is the smallest congruence family $\equiv_2$ satisfying: for each $\gamma_{\vec{X}} \hookrightarrow t \simeq_{\vec{X}} u \in \Gamma^q$ and every $\alpha : \vec{X} \to T_\Sigma(\vec{X})$,

(ii.a) if $\alpha(t') \equiv_1 \alpha(u')$ for each $t' \simeq_{\vec{X}} u' \in \gamma_{\vec{X}}$ then $\alpha(t) \equiv_2 \alpha(u)$,

(ii.b) $\equiv_1 \subseteq \equiv_2$.

(iii) $\equiv_{\Gamma^q}$ is the least fixpoint of θ_{Γ^q}, i.e. $\equiv_{\Gamma^q} = \bigcup_{i \in Nat} \theta_{\Gamma^q}{}^{[i]}(\equiv)$ where $\equiv$ is the plain equality on $T_\Sigma(\vec{X})$. Note that $\equiv_\emptyset = \equiv$.

(iv) Replacing the $\equiv_i$ by $\equiv_{\Gamma^q}$ in the two conditions of (ii) above, we let the closure of Γ^q, written as $[\Gamma^q]$, be the least collection of quasi-dependent equations, like $\gamma_{\vec{X}} \hookrightarrow t \simeq_{\vec{X}} u$, satisfying: if $\alpha(t') \equiv_{\Gamma^q} \alpha(u')$ for each $t' \simeq_{\vec{X}} u' \in \gamma_{\vec{X}}$ then $\alpha(t) \equiv_{\Gamma^q} \alpha(u)$.

(v) Obviously, $\equiv_{\Gamma^q} = \equiv_{[\Gamma^q]}$.

Therefore, the new context can be achieved by adding a new rule

$$(d\text{-}q) \quad \frac{\{\alpha(\gamma_{\vec{X}}) \mapsto \alpha(t) \simeq_{\vec{Y}} \alpha(u) | \alpha \in Subst[\vec{X}, \vec{Y}]\}}{\gamma_{\vec{X}} \hookrightarrow t \simeq_{\vec{X}} u}$$

Actually, rule (*d-q*) is (v) above. However, this rule is not sound in general, but the calculus containing this rule is sound and complete with respect to equations and dependent equations. So, we name this calculus as $\tilde{D}^q_+$. A counter-example to soundness of $\tilde{D}^q_+$ (or more properly to the rule (*d-q*)) is as follows.

Example 4 (counter-example to soundness of (*d-q*)): *Let* **A** *be the same algebra as in Example 2, and* $\tilde{\Gamma}^q$ *contain the following axiom:*

$$\mathbf{true}(x) \simeq_{\{x,y\}} \mathbf{true}(y)$$

Thus,

(a) $\tilde{\Gamma}^q$ *is sound in* **A** *as well as* $\mathbf{true}(x) \simeq_{\{x,y\}} \mathbf{true}(y) \hookrightarrow \mathbf{true}(\mathbf{true}(x)) \simeq_{\{x,y\}} \mathbf{true}(y)$.

(b) $x \simeq_{\{x,y\}} y \hookrightarrow \mathbf{true}(x) \simeq_{\{x,y\}} y$ *is not sound in* **A** *(consider the assignment of both x and y to F), i.e.*

$$\mathbf{A} \not\models x \simeq_{\{x,y\}} y \hookrightarrow \mathbf{true}(x) \simeq_{\{x,y\}} y.$$

(c) However, for every substitution $\check{\alpha} : T_\Sigma(\vec{X}) \to T_\Sigma(\vec{X})$,

$$\check{\alpha}(x \simeq_{\{x,y\}} y) \mapsto \check{\alpha}(\mathbf{true}(x) \simeq_{\{x,y\}} y)$$

is derivable. So, $x \simeq_{\{x,y\}} y \hookrightarrow \mathbf{true}(x) \simeq_{\{x,y\}} y$ *is derivable by (d-q).*

For (c), we only need to show four simple cases. The general case is not hard to prove from the observation of these four, and is left out.

(i) when $\check{\alpha}$ *is a permutation of variables, say the identity substitution,*

$$x \simeq_{\{x,y\}} y \mapsto \mathbf{true}(x) \simeq_{\{x,y\}} y$$

is derivable because of the rule ($\mapsto$ *-introd);*

(ii) when $\check{\alpha}$ *assigns x to* **true**(*x*) *and y to y,*

$$\mathbf{true}(x) \simeq_{\{x,y\}} y \mapsto \mathbf{true}(\mathbf{true}(x)) \simeq_{\{x,y\}} y$$

is derivable for the same reason as (i);

(iii) when $\check{\alpha}$ *assigns x to x and y to* **true**(*y*),

$$x \simeq_{\{x,y\}} \mathbf{true}(y) \mapsto \mathbf{true}(x) \simeq_{\{x,y\}} \mathbf{true}(y)$$

is derivable by the weakening rule;

(iv) when $\check{\alpha}$ *assigns x to* **true**(*x*) *and y to* **true**(*y*), $\mathbf{true}(x) \simeq_{\{x,y\}} \mathbf{true}(y) \mapsto \mathbf{true}(\mathbf{true}(x)) \simeq_{\{x,y\}} \mathbf{true}(y)$ *is derivable by the obvious rule (q-d)*

$$(q\text{-}d) \quad \frac{\gamma_{\vec{X}} \hookrightarrow t \simeq_{\vec{X}} u}{\gamma_{\vec{X}} \mapsto t \simeq_{\vec{X}} u}.$$

Actually, rule (*d-q*) is not necessary. The key point to have a sound and complete calculus for quasi-dependent equations is the following two facts.

---

rule can be

$$\frac{\emptyset \mapsto t \simeq u}{\emptyset \mapsto \alpha(t) \simeq \alpha(u)}.$$

Fact 5: $\Gamma \models \gamma \mapsto t \simeq u$ *iff* $\Gamma \models \gamma \hookrightarrow t \simeq u$, *provided that there is no free variable in* γ *and* $t \simeq u$.

Fact 6: $\Gamma \models \gamma \hookrightarrow t \simeq u$ *iff* $\Gamma \models \gamma' \mapsto t' \simeq u'$, *where* γ' *and* $t' \simeq u'$ *are the results of the substitution of fresh constants for free variables*[2] *in* γ *and* $t \simeq u$.

From the above two facts, we can borrow "skomelization" technique from logic and apply it to here. This is examplified by the following rule:

$$\frac{\gamma' \mapsto t' \simeq u' \in \tilde{D}^q(\Gamma)}{\gamma \hookrightarrow t \simeq u \in \tilde{D}^q(\Gamma)}$$

where γ' and $t' \simeq u'$are the results of the substitution of fresh constants for free variables in them. This rule coupled with other obvious rules[3] forms a sound and complete calculus for quasi-dependent equations.

To be more explicit, we apply the above rule to Example 4 and demonstrate that $x \simeq y \hookrightarrow \mathbf{true}(x) \simeq y$ is not derivable.

Example 7: *After introducing two extra constants* **True** *and* **False** *into Example 4 to instantiate* x *and* y *respectively, we have that*

$$x \simeq_{\{x,y\}} y \hookrightarrow \mathbf{true}(x) \simeq_{\{x,y\}} y$$

is derivable iff

$$\frac{\mathbf{True} \simeq \mathbf{False}}{\mathbf{true}(\mathbf{True}) \simeq \mathbf{False}}.$$

However, $\mathbf{true}(\mathbf{True}) \simeq \mathbf{False}$ *is not derivable by adding an extra axiom* $\mathbf{True} \simeq \mathbf{False}$, *since every derivable equation has the property that either constructor* **true** *does not occur in both sides of* $\simeq$ *or it appears in both sides of* $\simeq$ *at the same time. Therefore,* $x \simeq_{\{x,y\}} y \hookrightarrow \mathbf{true}(x) \simeq_{\{x,y\}} y$ *is not derivable.*

On the other hand, $\mathbf{true}(x) \simeq_{\{x,y\}} y \hookrightarrow \mathbf{true}(\mathbf{true}(x)) \simeq_{\{x,y\}} y$ in Example 4 is derivable by the extra rule, see Example 8.

[2]This technique is commonly called "skomelization" in logic.

[3]As long as they form a sound and complete calculus with respect to dependent equations.

Example 8: *Similar to Example 7, by "skomelization" we have that*

$$\mathbf{true}(x) \simeq_{\{x,y\}} y \hookrightarrow \mathbf{true}(\mathbf{true}(x)) \simeq_{\{x,y\}} y$$

is derivable iff

$$\frac{\mathbf{true}(\mathbf{True}) \simeq \mathbf{False}}{\mathbf{true}(\mathbf{true}(\mathbf{True})) \simeq \mathbf{False}}.$$

In contrast to Example 7, $\mathbf{true}(\mathbf{true}(\mathbf{True})) \simeq \mathbf{False}$ *is derivable by adding an extra axiom* $\mathbf{true}(\mathbf{True}) \simeq \mathbf{False}$. *This is demonstrated as follows. Let* $(*)$ *be* $\mathbf{true}(\mathbf{True}) \simeq_{\{x,y\}} \mathbf{False}$, *then,*

(i) $\mathbf{true}(x) \simeq_{\{x,y\}} \mathbf{true}(y)$	
(ii) $\mathbf{true}(x) \simeq_{\{x,y\}} \mathbf{true}(\mathbf{True})$	*by* (i)
(iii) $\mathbf{true}(x) \simeq_{\{x,y\}} \mathbf{False}$	*by* (ii) *and* $(*)$
(iv) $\mathbf{true}(\mathbf{true}(\mathbf{True})) \simeq \mathbf{False}$	*by* (iii)

Thus, $\mathbf{true}(x) \simeq_{\{x,y\}} y \hookrightarrow \mathbf{true}(\mathbf{true}(x)) \simeq_{\{x,y\}} y$ *is derivable by the extra rule.*

Because of the application of "skomelization", the non-emptyness seems to be the crucial condition for our solution to work. However, this condition can be removed if we know the fact that the existence of empty sorts implies non-existence of variable assignments (or homomorphisms).

5 Comments

In first order logic, we know that the validation (or satisfaction) relation is actually the *quasi-dependent* implication. So, as a by-product of this paper, we know that first order logic without quantifiers[4] is sound and complete. Actually, this by-product is not easily foreseen.

For a complete treatment of first order logic in the approach taken by this paper, we need introduce *binding* into algebraic framework. This direction of research has been taken in [S89].

So far, we have briefly clarified the soundness and completeness of the three equational logics. We have deliberately omit details of the proofs for our results in favour of readability.

[4]I would like to suggest *predicate logic* for this kind of first order logic without quantifiers, although "predicate logic" has been used equivalently as first order logic in literature.

6 References

[BS81] S. Burris and H. P. Sankappanavar, "A Course in Universal Algebra", Graduate Texts in Mathematics (GTM), vol. , Springer-Verlag 1981.

[CK73] C. C. Chang and M. J. Keisler, "Model Theory", Studies in Logic Series vol.73, 1973.

[C80] P.M. Cohn, "Universal Algebra", 2nd ed., Harper and Row, New York 1980.

[EM85] H. Ehrig and B. Mahr, "Fundamentals of Algebraic Specification 1, Equations and initial semantics", EATCS Monographs on Theoretical Computer Science vol. 6, 1985.

[ELM86] H. Ehrig, J. Loeckx and B. Mahr, "A Remark on the Equational Calculus for Many-sorted Equational Logic", EATCS Bullettin, No.30 Nov. 1986.

[GM81] J. A. Goguen and J. Meseguer, "Completeness of Many-sorted Equational Logic", ACM SigPlan, 1981

[GM82] J. A. Goguen and J. Meseguer, "Universal Realization, Persistent Interconnection and Implementation of Abstract Modules", LNCS vol. 140, 1982.

[GM85] J. A. Goguen and J. Meseguer, "Completeness of Many-sorted Equational Logic", Houston Journal of Mathematics, 1985.

[GM86] J. A. Goguen and J. Meseguer, "Remarks on Remarks of Completeness of Many-sorted Equational Logic", EATCS Bullettin, No.30 Nov. 1986.

[G79] G. Gratzer, "Universal Algebra", 2nd. ed., Springer-Verlag, 1979.

[H77] L. Henkin, "Logic of Identities", Mathematical Monthly, American Math. Soc., 1977.

[M71] S. Maclane, "Categories for the Working Mathematician", GTM vol.5 , 1971.

[M73] A. I. Mal'cev, "Algebraic Systems", Die Grundlehren der mathematischen Wissenschaften in Einzeldarstellungen Band 192, 1973.

[S89] Y. Sun, "Equational characterization of binding", Dept. of Computer Science, University of Edinburgh, LFCS, report No.94, 1989.

Compatibility of Order-Sorted Rewrite Rules

Uwe Waldmann

Lehrstuhl Informatik V, Universität Dortmund
Postfach 50 05 00
W4600 Dortmund 50, Germany
E-Mail: uwe@ls5.informatik.uni-dortmund.de

Unlike in the unsorted case the application of an order-sorted rewrite rule to a term t may be prohibited although the left hand side of the rule matches with a subterm of t, since the resulting term would be ill-formed. A rewrite rule is called compatible, if it may be applied to a term, whenever its left hand side matches with a subterm. We show that compatibility in finite signatures is decidable.

1 Signatures and Algebras

Order-sorted specifications (i.e., many-sorted specifications with subsort relations) have been proved to be a useful tool for the description of partially defined functions and error handling in abstract data types [GJM85, GM89, Gog86, SNGM87]. Order-sorted rewriting differs substantially from unsorted or many-sorted rewriting: In general it may happen that the left hand side of an order-sorted rewrite rule matches with a subterm of a term t, but that nevertheless the rule must not be applied to t, since the resulting term would be ill-formed. A rewrite rule is called compatible, if it may be applied to a term, whenever its left hand side matches with a subterm. In this paper we prove that the compatibility of rewrite rules is decidable, provided that the signature is finite.

We shall first summarize some basic results about order-sorted signatures and algebras, following [SNGM87] and [Wal89].

Definition 1.1 *An order-sorted signature is a triple $(S, \leq, \Sigma)$, where S is a set of sorts, $\leq$ a partial ordering over S, and Σ a family $\{\, \Sigma_{w,s} \mid w \in S^*, s \in S \,\}$ of (not necessarily disjoint) sets of operator symbols.*

In order to make the notation simpler and more intuitive we shall often write $f : w \rightarrow s$ instead of $f \in \Sigma_{w,s}$ and $f : \ \rightarrow s$ instead of $f \in \Sigma_{\varepsilon,s}$. We shall also use Σ as an abbreviation for both $(S, \leq, \Sigma)$ and $\bigcup_{w,s} \Sigma_{w,s}$.

Definition 1.2 *An S-sorted variable set is a family $V = \{\, V_s \mid s \in S \,\}$ of disjoint sets.*

A variable $x \in V_s$ is written $x : s$. We shall use V as an abbreviation for $\bigcup_{s \in S} V_s$ and shall always assume that all variable sets are disjoint from Σ.

Definition 1.3 *The set $T_\Sigma(V)_s$ of terms over Σ and V with sort s is the least set with the following properties:*

(i) $x \in T_\Sigma(V)_s$ *if* $x : s_0 \in V$ *and* $s_0 \leq s$.

(ii) $f \in T_\Sigma(V)_s$ *if* $f : \rightarrow s_0$ *and* $s_0 \leq s$.

(iii) $f(t_1, \ldots, t_n) \in T_\Sigma(V)_s$ *if* $f : s_1 \ldots s_n \rightarrow s_0$ *such that* $s_0 \leq s$ *and* $t_i \in T_\Sigma(V)_{s_i}$ *for every* $i \in \{1, \ldots, n\}$.

$T_\Sigma(V) := \bigcup_{s \in S} T_\Sigma(V)_s$ *denotes the set of all terms over* Σ *and* V. *The set of all ground terms over* Σ *is* $T_\Sigma := T_\Sigma(\emptyset)$.

The set of all variables in a term $t \in T_\Sigma(V)$ is abbreviated by $\mathrm{Var}(t)$.

Sometimes we need a more general notion of terms, which does not have the sort constraints of the previous definition:

Definition 1.4 *The set $ET_\Sigma(V)$ of extended terms over Σ and V is the least set with the following properties:*

(i) $x \in ET_\Sigma(V)$ *if* $x : s_0 \in V$.

(ii) $f \in ET_\Sigma(V)$ *if* $f : \rightarrow s_0$.

(iii) $f(t_1, \ldots, t_n) \in ET_\Sigma(V)$ *if* $f : s_1 \ldots s_n \rightarrow s_0$ *and* $t_i \in ET_\Sigma(V)$ *for* $1 \leq i \leq n$.

The set $T_\Sigma(V)$ is a subset of $ET_\Sigma(V)$. If an extended term $t \in ET_\Sigma(V)$ is an element of $T_\Sigma(V)$, we say that t is a well-formed term, otherwise t is called ill-formed.

As usual occurrences of a term are denoted by strings of natural numbers [HO80]. The set of all occurrences of an extended term $t \in ET_\Sigma(V)$ is $\mathcal{O}(t)$; the subterm of t at the occurrence $o \in \mathcal{O}(t)$ is written t/o; the result of the replacement of the subterm at o in t by t' is denoted by $t[o \leftarrow t']$. Note that $t[o \leftarrow t']$ may be ill-formed, even if t and t' are well-formed terms.

Definition 1.5 *Let $(S, \leq, \Sigma)$ be an order-sorted signature. A $(S, \leq, \Sigma)$-algebra A consists of a family $\{ A_s \mid s \in S \}$ of sets and a function $A_f : D_f^A \rightarrow C_A$ for every $f \in \Sigma$ such that the following conditions are fulfilled:*

(i) $A_s \subseteq A_{s'}$, *if* $s \leq s'$.

(ii) D_f^A *is a subset of* $(C_A)^*$ *where* $C_A := \bigcup_{s \in S} A_s$.

(iii) *If* $f \in \Sigma_{w,s}$, *then* $A_w \subseteq D_f^A$ *and* $A_f(A_w) \subseteq A_s$.

We use $A_{s_1 \ldots s_n}$ as an abbreviation for $A_{s_1} \times \cdots \times A_{s_n}$, A_ε is some one-point set. (A function with domain A_ε may be considered as a constant.)

Obviously we can make T_Σ (which we shall abbreviate by T) a $(S, \leq, \Sigma)$-algebra by $T_s := T_{\Sigma,s}$; for $f \in \Sigma$ we define $D_f^T := \bigcup_{f:w \rightarrow s} T_w$ and $T_f(t_1, \ldots, t_n) := f(t_1, \ldots, t_n)$ for $(t_1, \ldots, t_n) \in D_f^T$.

Definition 1.6 *Let A and B be two $(S, \leq, \Sigma)$-algebras. An $(S, \leq, \Sigma)$-homomorphism $h : A \to B$ is a function $h : \mathrm{C}_A \to \mathrm{C}_B$ such that:*

(i) *$h(A_s) \subseteq B_s$ for each $s \in S$.*

(ii) *$h(\mathrm{D}_f^A) \subseteq \mathrm{D}_f^B$ and $h(A_f(a_1, \ldots, a_n)) = B_f(h(a_1), \ldots, h(a_n))$ for all $f \in \Sigma$ and $(a_1, \ldots, a_n) \in \mathrm{D}_f^A$.*

A homomorphism $h : A \to B$ is called an isomorphism, if a homomorphism $h' : B \to A$ exists satisfying $h' \circ h = \mathrm{id}_A$ and $h \circ h' = \mathrm{id}_B$.

For every fixed signature $(S, \leq, \Sigma)$ the Σ-algebras and Σ-homomorphisms make up the category OSA_Σ.[1]

As in the unsorted case, we have:

Theorem 1.7 *The term algebra T_Σ is the initial Σ-algebra, it is determined uniquely (up to isomorphism).*

Definition 1.8 *An assignment ν from a variable set V into a Σ-algebra A is a set of functions $\{ \nu_s : V_s \to A_s \mid s \in S \}$.*

Theorem 1.9 *The algebra $\mathrm{T}_\Sigma(V)$ is the free Σ-algebra generated by V, i.e., for every Σ-algebra A and every assignment ν from V to A there is exactly one homomorphism $\nu^* : \mathrm{T}_\Sigma(V) \to A$ that extends ν.*

Definition 1.10 *A substitution σ is an assignment from a variable set Y into the term algebra $\mathrm{T}_\Sigma(X)$. In general the uniquely determined extension $\sigma^* : \mathrm{T}_\Sigma(Y) \to \mathrm{T}_\Sigma(X)$ of σ will also be denoted by σ.*

A substitution $\sigma : \{x_1, \ldots, x_n\} \to \mathrm{T}_\Sigma(X)$ that maps the variables $x_1, \ldots, x_n$ to the terms $t_1, \ldots, t_n$, respectively, is written $\sigma = (x_1 \leftarrow t_1, \ldots, x_n \leftarrow t_n)$.

2 Rewriting

Definition 2.1 *A rewrite rule is a pair (l, r) where $l, r \in \mathrm{T}_\Sigma(Y)$.*

We shall usually write $l \to r$ instead of (l, r).

Definition 2.2 *A term $t \in \mathrm{T}_\Sigma(X)$ rewrites to $t' \in \mathrm{T}_\Sigma(X)$ with a rewrite rule $l \to r$ at the occurrence o if the following conditions are satisfied.*

(i) *There exists a substitution $\sigma : \mathrm{Var}(l) \cup \mathrm{Var}(r) \to \mathrm{T}_\Sigma(X)$ such that $\sigma(l) = t/o$.*

(ii) *$t' = t[o \leftarrow \sigma(r)]$.*

(iii) *t' is a well-formed term.*

We abbreviate this by $t \longrightarrow^{[o, l \to r]} t'$, possibly omitting the indices.

[1] We have defined order-sorted algebras and homomorphisms as in [SNGM87]. Note that for "overloaded" algebras as presented in [GJM85, GM89] more restrictive definitions of rewriting and compatibility are necessary [Wal89].

Order-sorted rewriting differs from many-sorted rewriting in that the well-formedness of the resulting term must in general be explicitly checked:

Example 2.3 *Let the signature Σ and the set of rewrite rules R be defined by*

$$(S, \leq) = \{ s_1 \leq s_0 \}$$
$$\Sigma = \{ a : \rightarrow s_1, \quad b : \rightarrow s_0, \quad f : s_1 \rightarrow s_0 \}$$
$$R = \{ a \rightarrow b \}$$

Although we have $a \longrightarrow^{[a \rightarrow b]} b$, the rule $a \rightarrow b$ must not be applied to the term $f(a)$, since rewriting would yield the ill-formed term $f(b)$.

Even worse, the usual Church-Rosser theorem does not hold for order-sorted rewriting, as shown by the following example due to G. Smolka et al. [SNGM87]:

Example 2.4 *Let the signature Σ and the set of rewrite rules R be defined by*

$$(S, \leq) = \{ s_1 \leq s_0 \}$$
$$\Sigma = \{ a : \rightarrow s_1, \quad a' : \rightarrow s_1, \quad b : \rightarrow s_0, \quad f : s_1 \rightarrow s_1 \}$$
$$R = \{ a \rightarrow b, \quad a' \rightarrow b \}$$

R is a confluent and terminating rewrite system, and $f(a)$ and $f(a')$ are R-equal. Nevertheless $f(a) \overset{*}{\longrightarrow}_R \overset{*}{\longleftarrow}_R f(a')$ or even $f(a) \overset{*}{\longleftrightarrow}_R f(a')$ does not hold, since $f(b)$ is not a well-formed Σ-term.

3 Compatibility

Definition 3.1 *A rewrite rule $l \rightarrow r$ is called compatible, if for every $t \in \mathrm{T}_\Sigma(X)$ and every $o \in \mathcal{O}(t)$ we have: If $\sigma l = t/o$ for some substitution $\sigma : \mathrm{Var}(l) \cup \mathrm{Var}(r) \rightarrow \mathrm{T}_\Sigma(X)$, then $t[o \leftarrow \sigma r]$ is a well-formed term.*

Definition 3.2 *A rewrite rule $l \rightarrow r$ is called sort decreasing, if we have for every substitution $\sigma : \mathrm{Var}(l) \cup \mathrm{Var}(r) \rightarrow \mathrm{T}_\Sigma(X)$ and every sort $s \in S$: If $\sigma l \in \mathrm{T}_\Sigma(X)_s$ then $\sigma r \in \mathrm{T}_\Sigma(X)_s$.*

If we regard a term rewrite system as a kind of nondeterministic computer program, compatibility corresponds to well-typedness of programs. G. Smolka et al. have shown that for compatible rewrite rules Church-Rosser property and confluence are equivalent, and they have demonstrated that every rewrite system can be transformed into a semantically equivalent system that is compatible [SNGM87]. Sort decreasingness implies compatibility, besides for sort decreasing rewrite rules the usual critical pair lemma holds. (Compatibility is not sufficient here.)

We shall now demonstrate that compatibility is actually decidable for finite signatures (though the existence of a polynomial time decision procedure is extremely improbable). To simplify notation we shall first define an extended rewriting relation that may produce ill-formed terms.

Definition 3.3 *A term $t \in \mathrm{T}_\Sigma(X)$ x-rewrites to $t' \in \mathrm{ET}_\Sigma(X)$ with a rewrite rule $l \to r$ at the occurrence o if the following conditions are satisfied.*

(i) *There exists a substitution $\sigma : \mathrm{Var}(l) \cup \mathrm{Var}(r) \to \mathrm{T}_\Sigma(X)$ such that $\sigma(l) = t/o$.*

(ii) $t' = t[o \leftarrow \sigma(r)]$.

We abbreviate this by $t \twoheadrightarrow^{[o,l\to r]} t'$, possibly omitting some of the indices.

The relation $\twoheadrightarrow$ differs from $\longrightarrow$ only with respect to the missing well-formedness test; thus we have $t \longrightarrow^{[o,l\to r]} t'$ if and only if $t \twoheadrightarrow^{[o,l\to r]} t'$ and $t' \in \mathrm{T}_\Sigma(X)$.

Lemma 3.4 *The rewrite rule $l \to r$ is compatible, if and only if $t \twoheadrightarrow^{[l\to r]} t'$ implies that t' is well-formed for all terms $t \in \mathrm{T}_\Sigma(X)$, $t' \in \mathrm{ET}_\Sigma(X)$.*

Proof Follows immediately from Definitions 3.1 and 3.3. □

Definition 3.5 *The spectrum $\mathrm{spctr}(t)$ of an extended term $t \in \mathrm{ET}_\Sigma(V)$ is the set of all sorts $s \in S$ such that $t \in \mathrm{T}_\Sigma(V)_s$. The set of all spectra of (well-formed) terms in $\mathrm{T}_\Sigma(V)$ is denoted by $\mathrm{spctr}_{\Sigma\cup V}$.*

The computation of $\mathrm{spctr}(t)$ is possible using the following recursion formulae:

$$\begin{aligned}
\mathrm{spctr}(x:\tilde{s}) &= \{\, s \in S \mid \tilde{s} \le s \,\} \\
\mathrm{spctr}(f) &= \{\, s \in S \mid f : \;\to \tilde{s},\ \tilde{s} \le s \,\} \\
\mathrm{spctr}\big(f(t_1,\ldots,t_n)\big) &= \big\{\, s \in S \;\big|\; f : s_1 \ldots s_n \to \tilde{s},\ \tilde{s} \le s, \\
&\qquad\qquad s_i \in \mathrm{spctr}(t_i) \text{ for } i \in \{1,\ldots,n\} \,\big\}
\end{aligned}$$

The following lemma is due to M. Schmidt-Schauß [SS88, Wal89].

Lemma 3.6 *If the signature $(S, \le, \Sigma)$ is finite, then $\mathrm{spctr}_{\Sigma\cup V}$ is finite and effectively computable. Besides there exists a finite and computable set $Q \subseteq \mathrm{T}_\Sigma(V)$, such that for every spectrum $M \in \mathrm{spctr}_{\Sigma\cup V}$ there is a term $t \in Q$ satisfying $M = \mathrm{spctr}(t)$.*

It is easy to see that the spectrum of an extended term $t \in \mathrm{ET}_\Sigma$ is non-empty if and only if t is well-formed. Note that $\emptyset$ (i.e., the spectrum of every ill-formed term) is not contained in $\mathrm{spctr}_{\Sigma\cup V}$.

Definition 3.7 *Given a rewrite rule $l \to r$ the rule spectrum $\mathrm{rs}(l \to r)$ of $l \to r$ is defined as*

$$\begin{aligned}
\mathrm{rs}(l\to r) := \big\{\, (M, M') \in 2^S \times 2^S \mid \exists t \in \mathrm{T}_\Sigma(X),\ t' \in \mathrm{ET}_\Sigma(X): \\
t \twoheadrightarrow^{[l\to r]} t',\ M = \mathrm{spctr}(t),\ M' = \mathrm{spctr}(t') \,\big\}
\end{aligned}$$

Lemma 3.8 *The rewrite rule $l \to r$ is compatible, if and only if $\mathrm{rs}(l \to r)$ does not contain a pair $(M, \emptyset)$.*

Proof Follows from Lemma 3.4. □

We shall now demonstrate that computing a significantly smaller set than $\mathrm{rs}(l \to r)$ is already sufficient to check for compatibility.

Definition 3.9 *Let $t \in \mathrm{T}_\Sigma(X)$ and $o \in \mathcal{O}(t)$. We say that t is maximally o-stripped, if either $o = \varepsilon$ or $o = j\pi$, $t = f(x_1, \ldots, x_{j-1}, t_j, x_{j+1}, \ldots, x_n)$, where $f : s_1 \ldots s_n \to s_0$, $x_i : s_i$ for $i \in \{1, \ldots, n\} \setminus \{j\}$, and $t_j \in \mathrm{T}_\Sigma(X)_{s_j}$ is maximally π-stripped.*

Lemma 3.10 *For every sort $s \in S$, every term $t \in \mathrm{T}_\Sigma(X)_s$, and for every occurrence $o \in \mathcal{O}(t)$ there exists a term $\bar{t} \in \mathrm{T}_\Sigma(V)_s$ with the following properties:*

(i) *$t = \rho\bar{t}$ for some substitution $\rho : V \to \mathrm{T}_\Sigma(X)$,*

(ii) *$\bar{t}/o = t/o$,*

(iii) *$\bar{t}$ is maximally o-stripped.*

Proof If $o = \varepsilon$, we define $\bar{t} := t$. If $o = j\pi$, let $f = f(t_1, \ldots, t_n)$ where $f : s_1 \ldots s_n \to s_0$, $s_0 \leq s$, and $t_i \in \mathrm{T}_\Sigma(X)_{s_i}$ for $i \in \{1, \ldots, n\}$. By the induction hypothesis there exists a maximally π-stripped term $\bar{t}_j \in \mathrm{T}_\Sigma(\hat{V})_{s_j}$ such that $\bar{t}_j/\pi = t_j/\pi$, and $t_j = \hat{\rho}\bar{t}_j$ for some $\hat{\rho} : \hat{V} \to \mathrm{T}_\Sigma(X)$. For $i \in \{1, \ldots, n\} \setminus \{j\}$ let $x_i : s_i$ be variables not occurring in $\hat{V} \cup X$, then $\bar{t} := f(x_1, \ldots, x_{j-1}, \bar{t}_j, x_{j+1}, \ldots, x_n)$ and $\rho := \hat{\rho} \cup (\, x_i \leftarrow t_i \mid 1 \leq i \leq n,\, i \neq j\,)$. □

Lemma 3.11 *The following two properties are equivalent:*

(i) *The rule $l \to r$ is compatible.*

(ii) *Whenever $\bar{t} \in \mathrm{T}_\Sigma(X)$ is a maximally o-stripped term, then $\bar{t} \twoheadrightarrow^{[o,l\to r]} \bar{t}'$ implies that $\bar{t}'$ is well-formed.*

Proof We shall prove at first that property (ii) implies (i). Let $t \in \mathrm{T}_\Sigma(X)_s$ be any term, $o \in \mathcal{O}(t)$, $\sigma : \mathrm{Var}(l) \cup \mathrm{Var}(r) \to \mathrm{T}_\Sigma(X)$, such that $\sigma l = t/o$. Let $\bar{t} \in \mathrm{T}_\Sigma(V)_s$ be a maximally o-stripped term so that $t = \rho\bar{t}$ and $\bar{t}/o = t/o$. Since $t/o = \sigma l$, by (ii) the term $\bar{t}' := \bar{t}[o \leftarrow \sigma r]$ is well-formed. Now we have $t[o \leftarrow \sigma r] = (\rho\bar{t})[o \leftarrow \sigma r] = \rho(\bar{t}[o \leftarrow \sigma r])$, thus $t[o \leftarrow \sigma r]$ is well-formed, too.

The proof of the (i)⇒(ii) part is trivial. □

Definition 3.12 *Given a rewrite rule $l \to r$ the restricted rule spectrum $\mathrm{rrs}(l \to r)$ of $l \to r$ is defined as*

$$\mathrm{rrs}(l \to r) := \{\, (M, M') \in 2^S \times 2^S \mid \exists t \in \mathrm{T}_\Sigma(X),\, t' \in \mathrm{ET}_\Sigma(X) : \\ t \text{ is maximally } o\text{-stripped},\ t \twoheadrightarrow^{[o,l\to r]} t', \\ M = \mathrm{spctr}(t),\ M' = \mathrm{spctr}(t'),\ M \not\subseteq M' \,\}$$

Lemma 3.13 *The rewrite rule $l \to r$ is compatible, if and only if $\mathrm{rrs}(l \to r)$ does not contain a pair $(M, \emptyset)$.*

Proof Since $\mathrm{rrs}(l \to r)$ is a subset of $\mathrm{rs}(l \to r)$, the proof of the "only if" part is rather trivial. To prove the "if" part suppose that $l \to r$ is not compatible. Hence there exist a maximally o-stripped term $t \in \mathrm{T}_\Sigma(X)$ and an ill-formed term $t' \in \mathrm{ET}_\Sigma(X)$ such that $t \twoheadrightarrow^{[o,l\to r]} t'$. Define $M := \mathrm{spctr}(t)$, $M' := \mathrm{spctr}(t')$, then obviously $M \neq \emptyset$, $M' = \emptyset$, thus $M \not\subseteq M'$ and $(M, \emptyset) \in \mathrm{rrs}(l \to r)$. □

Definition 3.14 *Given a rewrite rule $l \to r$, we define sets $\mathrm{rrs}_k(l \to r)$, $k \in \mathbf{N}$, as follows:*

$$\begin{aligned}
\mathrm{rrs}_0(l \to r) \;&:= \{ (M, M') \in 2^S \times 2^S \mid M \not\subseteq M',\ \exists \sigma : \mathrm{Var}(l) \cup \mathrm{Var}(r) \to \mathrm{T}_\Sigma(X): \\
&\qquad M = \mathrm{spctr}(\sigma l),\ M' = \mathrm{spctr}(\sigma r) \} \\
\mathrm{rrs}_{k+1}(l \to r) &:= \mathrm{rrs}_k(l \to r) \cup \{ (M, M') \in 2^S \times 2^S \mid \\
&\quad M \not\subseteq M', \\
&\quad \exists f : s_1 \ldots s_n \to s_0,\ 1 \le j \le n,\ (N, N') \in \mathrm{rrs}_k(l \to r): \\
&\qquad s_j \in N, \\
&\qquad (s \in M \iff f : \tilde{s}_1 \ldots \tilde{s}_n \to \tilde{s}_0,\ \tilde{s}_0 \le s, \\
&\qquad\qquad \tilde{s}_i \ge s_i \text{ for } i \in \{1, \ldots, n\} \setminus \{j\},\ \tilde{s}_j \in N), \\
&\qquad (s' \in M' \iff f : \hat{s}_1 \ldots \hat{s}_n \to \hat{s}_0,\ \hat{s}_0 \le s', \\
&\qquad\qquad \hat{s}_i \ge s_i \text{ for } i \in \{1, \ldots, n\} \setminus \{j\},\ \hat{s}_j \in N') \}
\end{aligned}$$

Lemma 3.15 *For every pair $(M, M') \in 2^S \times 2^S$ the following two properties are equivalent:*

(i) $(M, M') \in \mathrm{rrs}_k(l \to r)$.

(ii) *There are terms $t \in \mathrm{T}_\Sigma(X)$, $t' \in \mathrm{ET}_\Sigma(X)$ such that $M = \mathrm{spctr}(t)$, $M' = \mathrm{spctr}(t')$, $t \twoheadrightarrow^{[o,l\to r]} t'$, where t is maximally o-stripped and $\mathrm{length}(o) \le k$ and $M \not\subseteq M'$.*

Proof First we prove by induction on k that (i) implies (ii). Suppose that $k = 0$. Since $(M, M') \in \mathrm{rrs}_0(l \to r)$ implies that $M = \mathrm{spctr}(\sigma l)$ and $M' = \mathrm{spctr}(\sigma r)$ for some substitution $\sigma : \mathrm{Var}(l) \cup \mathrm{Var}(r) \to \mathrm{T}_\Sigma(X)$, and $M \not\subseteq M'$, we may simply set $t := \sigma l$ and $t' := \sigma r$.

Now let $(M, M') \in \mathrm{rrs}_{k+1}(l \to r)$. If $(M, M') \in \mathrm{rrs}_k(l \to r)$ then the proof is trivial. Otherwise let $f : s_1 \ldots s_n \to s_0$, $1 \le j \le n$, $(N, N') \in \mathrm{rrs}_k(l \to r)$, and $s_j \in N$, such that

$$s \in M \iff f : \tilde{s}_1 \ldots \tilde{s}_n \to \tilde{s}_0,\ \tilde{s}_0 \le s,\ \tilde{s}_i \ge s_i \text{ for } i \in \{1, \ldots, n\} \setminus \{j\},\ \tilde{s}_j \in N$$

and

$$s' \in M' \iff f : \hat{s}_1 \ldots \hat{s}_n \to \hat{s}_0,\ \hat{s}_0 \le s',\ \hat{s}_i \ge s_i \text{ for } i \in \{1, \ldots, n\} \setminus \{j\},\ \hat{s}_j \in N'.$$

By the induction hypothesis there exist terms $\bar{t} \in \mathrm{T}_\Sigma(X)$, $\bar{t}' \in \mathrm{ET}_\Sigma(X)$ such that $N = \mathrm{spctr}(\bar{t})$, $N' = \mathrm{spctr}(\bar{t}')$, $\bar{t} \twoheadrightarrow^{[\pi,l\to r]} \bar{t}'$, where $\bar{t}$ is maximally π-stripped and $\mathrm{length}(\pi) \le k$. Define $t := f(x_1, \ldots, x_{j-1}, \bar{t}, x_{j+1}, \ldots, x_n)$ and $t' := f(x_1, \ldots, x_{j-1}, \bar{t}', x_{j+1}, \ldots, x_n)$ where $x_i : s_i$ is a new variable for every $i \in \{1, \ldots, n\} \setminus \{j\}$. It is easy to check that $M = \mathrm{spctr}(t)$, $M' = \mathrm{spctr}(t')$, and since $s_j \in \mathrm{spctr}(\bar{t})$ we know that t is well-formed. Besides t is maximally $j\pi$-stripped and $t \twoheadrightarrow^{[j\pi,l\to r]} t'$.

To prove the (ii)$\Rightarrow$(i) part let first $k = 0$. Suppose that there are terms $t \in \mathrm{T}_\Sigma(X)$, $t' \in \mathrm{ET}_\Sigma(X)$ such that $M = \mathrm{spctr}(t)$, $M' = \mathrm{spctr}(t')$, $t \twoheadrightarrow^{[\epsilon,l\to r]} t'$, and $M \not\subseteq M'$. This implies $t = \sigma l$, $t' = \sigma r$ for some substitution $\sigma : \mathrm{Var}(l) \cup \mathrm{Var}(r) \to \mathrm{T}_\Sigma(X)$, hence (M, M') is contained in $\mathrm{rrs}_0(l \to r)$.

Now let $t \in \mathrm{T}_\Sigma(X)$, $t' \in \mathrm{ET}_\Sigma(X)$ such that $t \twoheadrightarrow^{[o,l\to r]} t'$, $M = \mathrm{spctr}(t)$, $M' = \mathrm{spctr}(t')$, $M \not\subseteq M'$, where $\mathrm{length}(o) \le k+1$ and t is maximally o-stripped. If $\mathrm{length}(o) \le k$ there is nothing to prove, so suppose that $o = j\pi$ and $\mathrm{length}(\pi) = k$. We know that $t = f(x_1, \ldots, x_{j-1}, t_j, x_{j+1}, \ldots, x_n)$ where $x_i : s_i$ for $i \in \{1, \ldots, n\} \setminus \{j\}$, $t_j \in \mathrm{T}_\Sigma(X)_{s_j}$,

$f : s_1 \ldots s_n \to s_0$, and t_j is maximally π-stripped. The term t' has the form $t' = f(x_1, \ldots, x_{j-1}, t'_j, x_{j+1}, \ldots, x_n)$. Note that $\mathrm{spctr}(t_j) \subseteq \mathrm{spctr}(t'_j)$ would imply $M \subseteq M'$, contradicting our assumption; so since $t_j \twoheadrightarrow^{[\pi, l \to r]} t'_j$ and $\mathrm{length}(\pi) = k$, the induction hypothesis yields that the pair $(\mathrm{spctr}(t_j), \mathrm{spctr}(t'_j))$ is contained in $\mathrm{rrs}_k(l \to r)$. As

$$s \in M \iff f : \tilde{s}_1 \ldots \tilde{s}_n \to \tilde{s}_0,\ \tilde{s}_0 \leq s,\\ \tilde{s}_i \geq s_i \text{ for } i \in \{1, \ldots, n\} \setminus \{j\},\ \tilde{s}_j \in \mathrm{spctr}(t_j)$$

and

$$s' \in M' \iff f : \hat{s}_1 \ldots \hat{s}_n \to \hat{s}_0,\ \hat{s}_0 \leq s',\\ \hat{s}_i \geq s_i \text{ for } i \in \{1, \ldots, n\} \setminus \{j\},\ \hat{s}_j \in \mathrm{spctr}(t'_j),$$

we can conclude that $(M, M') \in \mathrm{rrs}_{k+1}(l \to r)$. □

Corollary 3.16 *For every rewrite rule $l \to r$ we have*

$$\mathrm{rrs}(l \to r) = \bigcup_{k \geq 0} \mathrm{rrs}_k(l \to r).$$

Lemma 3.17 *If the signature $(S, \leq, \Sigma)$ is finite, then $\mathrm{rrs}(l \to r)$ is finite and effectively computable.*

Proof To compute $\mathrm{rrs}_{k+1}(l \to r)$ from $\mathrm{rrs}_k(l \to r)$ is obviously straightforward. The computation of $\mathrm{rrs}_0(l \to r)$ is more problematic, since there are in general infinitely many substitutions $\sigma : \mathrm{Var}(l) \cup \mathrm{Var}(r) \to T_\Sigma(X)$.

The following trick helps to overcome this problem: According to Lemma 3.6 we can compute a finite set $Q \subseteq T_\Sigma(V)$ such that for every spectrum $M \in \mathrm{spctr}_{\Sigma \cup X}$ there is a term $t \in Q$ that satisfies $M = \mathrm{spctr}(t)$. Now we shall prove that $\mathrm{rrs}_0(l \to r) = \{ (M, M') \mid M \not\subseteq M', \exists \bar{\sigma} : Y \to Q \colon M = \mathrm{spctr}(\bar{\sigma} l),\ M' = \mathrm{spctr}(\bar{\sigma} r) \}$, where Y denotes $\mathrm{Var}(l) \cup \mathrm{Var}(r)$: Suppose that the pair (M, M') is contained in $\mathrm{rrs}_0(l \to r)$, i.e., $M = \mathrm{spctr}(\sigma l)$ and $M' = \mathrm{spctr}(\sigma r)$ for some substitution $\sigma : Y \to T_\Sigma(X)$. Choose a substitution $\bar{\sigma} : Y \to Q$ such that $\mathrm{spctr}(\sigma y) = \mathrm{spctr}(\bar{\sigma} y)$ for every $y \in Y$, then by induction on the structure of terms we have $M = \mathrm{spctr}(\sigma l) = \mathrm{spctr}(\bar{\sigma} l)$ and $M' = \mathrm{spctr}(\sigma r) = \mathrm{spctr}(\bar{\sigma} r)$. The proof of the inverse inclusion is trivial. As the number of substitutions $\bar{\sigma} : Y \to Q$ is finite, $\mathrm{rrs}_0(l \to r)$ is computable.

The sets $\mathrm{rrs}_k(l \to r)$ form a monotonously increasing sequence of subsets of $2^S \times 2^S$. As $2^S \times 2^S$ is finite, there must be some k_0 satisfying $\mathrm{rrs}_{k_0}(l \to r) = \mathrm{rrs}_{k_0+1}(l \to r)$. We can conclude that $\mathrm{rrs}_{k_0}(l \to r) = \mathrm{rrs}_k(l \to r)$ holds for every $k \geq k_0$, hence we get

$$\mathrm{rrs}(l \to r) = \bigcup_{k \geq 0} \mathrm{rrs}_k(l \to r) = \mathrm{rrs}_{k_0}(l \to r).$$

Since all the sets $\mathrm{rrs}_k(l \to r)$ are effectively computable, $\mathrm{rrs}(l \to r)$ is computable as well. □

Corollary 3.18 *If the signature $(S, \leq, \Sigma)$ is finite, then the compatibility of a rewrite rule $l \to r$ is decidable.*

In fact, the algorithm to decide compatibility is an extension of the sort decreasingness check that was given in [Wal89]. We have:

$$l \to r \text{ is sort decreasing} \iff \mathrm{rrs}_0(l \to r) = \emptyset \iff \mathrm{rrs}(l \to r) = \emptyset.$$

4 Complexity

The algorithm to decide the compatibility of a rewrite rule that was described in the previous section may be extremely inefficient, if $\mathrm{spctr}_{\Sigma \cup V}$ is large. Since $\mathrm{spctr}_{\Sigma \cup V}$ is a subset of 2^S, there is obviously an exponential bound for $\mathrm{spctr}_{\Sigma \cup V}$. The following example demonstrates that the cardinality of $\mathrm{spctr}_{\Sigma \cup V}$ may in fact grow exponentially with respect to the size of the signature Σ.

Example 4.1 *For $n \in \mathbf{N}$, $n > 0$ consider the following signature:*

$$\begin{aligned}(S_n, \leq) &= \{ s_i \leq s_0 \mid 1 \leq i \leq n \} \\ \Sigma_n &= \{ a_i : \rightarrow s_i \mid 0 \leq i \leq n \} \\ &\quad \cup \{ f : s_i\, s_0 \rightarrow s_i, \\ &\qquad f : s_0\, s_i \rightarrow s_i \mid 0 \leq i \leq n \}\end{aligned}$$

The signature $(S_n, \leq, \Sigma_n)$ consists of $n+1$ sorts and $3n+2$ operator declarations. It is easy to check that $\mathrm{spctr}(a_i) = \mathrm{spctr}(x : s_i) = \{s_i, s_0\}$ and $\mathrm{spctr}(f(t,t')) = \mathrm{spctr}(t) \cup \mathrm{spctr}(t')$. Now we can prove that $\mathrm{spctr}_{\Sigma_n \cup V} = \{ M \cup \{s_0\} \mid M \subseteq \{s_1, \ldots, s_n\} \}$: Without loss of generality M is nonempty, so suppose that $M = \{s_{i_1}, \ldots, s_{i_k}\}$, where $k > 0$. Defining $t := f(a_{i_1}, f(a_{i_2}, \ldots f(a_{i_k}, a_0) \ldots))$ we have $M \cup \{s_0\} = \mathrm{spctr}(t)$. The proof of the inverse inclusion is trivial, since every term $t \in \mathrm{T}_{\Sigma_n}(V)$ has the sort s_0; thus we know that $|\mathrm{spctr}_{\Sigma_n \cup V}| = 2^n$.

Indeed both the Non-Compatibility problem and the Non-Sort-Decreasingness problem are PSPACE hard if the signature is considered part of the input [Wal90].

Most problems connected with order-sorted computation can be solved more efficiently if we restrict ourselves to regular signatures. For regular signatures there exists a bijection between spectra and sorts, hence $\mathrm{spctr}_{\Sigma \cup V}$ has the same size as the sort set S.

Definition 4.2 *A signature Σ is called regular, if every term $t \in \mathrm{T}_\Sigma(V)$ has a least sort, i.e., if every $M \in \mathrm{spctr}_{\Sigma \cup V}$ has a smallest element. (We assume that V contains at least one variable of every sort.)*

If the signature $(S, \leq, \Sigma)$ is regular, then $\mathrm{spctr}_{\Sigma \cup V}$ consists exactly of the spectra of the variables of V, i.e., of those subsets M of S that contain a sort s' such that $s \in M \iff s \geq s'$. As a consequence, the set Q in the proof of Lemma 3.17 may be chosen equal to V: In regular signatures, for every term t there exists a variable with exactly the same sorts as t. Nevertheless there are fixed, regular signatures for which the Non-Compatibility problem and the Non-Sort-Decreasingness problem are NP hard; this can be proved in nearly the same way as the NP completeness of order-sorted unification [SS88, SNGM87].

Acknowledgments: I am grateful to Harald Ganzinger for his comments on a previous draft of this paper.

References

[GJM85] J. A. Goguen, J.-P. Jouannaud, and J. Meseguer. Operational semantics for order-sorted algebra. In W. Brauer, editor, *12th International Colloquium on Automata, Languages and Programming. Proceedings. Nafplion, Greece*, LNCS 194, pages 221–231. Springer-Verlag, 1985.

[GM89] J. A. Goguen and J. Meseguer. Order-sorted algebra I: Equational deduction for multiple inheritance, overloading, exceptions and partial operations. Draft, Programming Research Group, University of Oxford and SRI International, 1989.

[Gog86] M. Gogolla. *Über partiell geordnete Sortenmengen und deren Anwendung zur Fehlerbehandlung in abstrakten Datentypen*. Dissertation, Universität Braunschweig, Naturwissenschaftliche Fakultät, 1986.

[HO80] G. Huet and D. C. Oppen. Equations and rewrite rules: A survey. In R. Book, editor, *Formal Languages: Perspectives and Open Problems*, pages 349–405. Academic Press, 1980.

[SNGM87] G. Smolka, W. Nutt, J. A. Goguen, and J. Meseguer. Order-sorted equational computation. SEKI-Report SR-87-14, Universität Kaiserslautern, Fachbereich Informatik, 1987. Revised version in H. Aït-Kaci and M. Nivat, editors, *Resolution of Equations in Algebraic Structures*, volume 2, pages 297–367. Academic Press, 1989.

[SS88] M. Schmidt-Schauß. *Computational Aspects of an Order-Sorted Logic with Term Declarations*. Dissertation, Universität Kaiserslautern, Fachbereich Informatik, 1988.

[Wal89] U. Waldmann. Semantics of order-sorted specifications. Forschungsbericht 297, Universität Dortmund, Fachbereich Informatik, 1989.

[Wal90] U. Waldmann. Complexities of order-sorted computations. In preparation, Universität Dortmund, Fachbereich Informatik, 1990.

Chapter 7

Extension of Unification and Narrowing Techniques

An universal termination condition for solving goals in equational languages

G. Aguzzi, U. Modigliani, M. C. Verri

Dipartimento di Sistemi e Informatica

Universita' di Firenze - Via S. Marta, 3 - 50139 Firenze Italy

e-mail EST@IFIIDG.BITNET

ABSTRACT

A decomposition procedure, called DP, operating on a "sorted" set of equations is here used as the operational semantics of CTRS. Then, a class of CTRS called *conic-flat*, is defined for which DP is shown to be universally terminating when solving the equation $t_1=_R t_2$, with t_1 or t_2 ground.

INTRODUCTION

The termination problem for a Decomposition Procedure, called DP, is here studied following the lines yet given in our previous papers [1,2]. In these works the termination of a slight different version of DP for solving equations in a theory expressed by a canonical TRS, has been studied. Such a procedure is similar to the ones defined in [7,8,10,15] for E-unification [9,11,12,18] in theories expressed by TRS's and is used as the operational semantics of a CTRS [14,17].

The main result of this paper is the characterization of a large class of CTRS, called Conic-Flat CTRS for which the Decomposition Procedure is terminating when solving the equation $t_1=_R t_2$ with respect to a system R, with either t_1 or t_2 ground. Namely, DP is terminating when simulating a reduction by solving $t_1=_R t_2$ with t_1 ground and t_2 in normal form, and when simulating a resolution to solve the equation $t_1=_R t_2$ with t_2 ground. It happens that the termination condition we give can also be used to study the termination property of Horn clause programs and goals corresponding to previous equations when the resolution procedure is applied with the computation rule that chooses the left-most atom [3,4,19,20] in the current goal.

The class of CTRS here described is called *flat* since any such system is a possible flattened version of an unconditional TRS.

For example, the system R called *Flat-Sum*:

1) $+(x,0) \to x$

2) $+(x,y) =_R z \supset +(x,s(y)) \to s(z)$

is the flattened version of the unconditional TRS

1') $+(x,0) \to x$

2') $+(x,s(y)) \to s(+(x,y))$

where the function symbols are the constructors s and 0 and the defined function symbol +.

Similarly, the conditional rule $c(x,y) =_R z \supset f(z) \to g(x,y)$ is the flattening of $f(c(x,y)) \to g(x,y)$.

In order to characterize a CTRS system R terminating with respect to the DP procedure, we need to introduce the *conicity* property of a term in R: a term is strictly conic with respect to its i-th parameter in a system R if, roughly speaking, every its ground instance can be computed, according to R, yielding a value greater than that of its i-th parameter. For example, the term s(+(x,y)) is strictly conic with respect to the parameter +(x,y) in the previous system *Flat-Sum*.

The CTRS *Flat-Sum* is *conic* since the right-hand side of the rule 2') is a strictly conic term with respect to the parameter constituting the recursive call, and the left-hand side of both rules 1') and 2') is greater than the right-hand side, according to a simplification ordering [5,6]. Then the system *Flat-Sum* is both flat and conic.

In order to define the conicity property, the ordering relation >, has been defined on the set of ground terms not containing any defined function symbol. The validity of $s_1 > s_2$ requires that s_2 be a proper subterm of the recursive parameter of s_1.

Let cons, nil, 1, 2 be constructor symbols, then cons(1,cons(2,nil)) > cons(2,nil) > nil, but cons(1,cons(2,nil)) $\not>$ cons(1,nil), and also cons(2,nil) $\not>$ 2 even if 2 is a subterm of cons(2,nil).

Testing the conicity property is crucial for deciding the termination property. This problem has been partially solved by giving sufficient conditions to be satisfied by the syntactic form of the R system rules. Such conditions involve knowing whether a function defined in the CTRS system R is increasing in its i-th parameter. This problem has been solved, too, by giving sufficient conditions for deciding such property.

DECOMPOSITION PROCEDURE

When using DP as the operational semantics of a CTRS R, as suggested in [13], R is considered as a logic programming system that is able to give the solution to a set of equations S in the form of substitutions making the pairs of terms of all equations simultaneously equal in the theory R. A set of equations is said to be in solved form if and only if all equations are of the form $x_i =_R t_i$, and every variable which makes up the left-hand side of some equation only occurs there. A set S of equations in solved form has the solution $sol(S) = \{(x_1/t_1), \cdots, (x_n/t_n)\}$.

This procedure applies equation transformations to an ordered set of equations until the set is constituted only by equations in solved form. The transformations that can be applied are called *variable substitution, term decomposition,* and *term rewriting.* We recall that using only variable substitution and term decomposition the m.g.u. of a set of equations, if it exists, is computed, as it is shown in [16]. The DP procedure non-deterministically applies any of the above transformations, if it is possible, to the first equation of the considered set. We show that DP computes a complete set of correct R-unifier, as defined in [14], for a given set of equations according to a CTRS R. We remark that the operational semantics for a CTRS given in [14] is instead based on the notion of narrowing [9,18].

The main difference between our Decomposition Procedure and the ones found in the literature [7,8,10,13,15] lies in the fact that the set of equations to be solved is sorted, and a computation rule is adopted which always selects the first equation of the set as the one to be refined.

An ordered set of equations is such that:

a) any syntactic equation precedes any semantic one (where the terms of a syntactic equation are constituted only by constructors and/or variables, while those of a semantic one also contain defined function symbols);
b) among the semantic equations any equation with a ground member precedes the ones with both non-ground members.

We remark that adopting our computation rule on a sorted set of equations, is drastically different from such a computation rule applied to an unsorted set: for example in the usual Prolog strategy of applying SLD-resolution, the computation rule selects the leftmost atom in the current goal without a previous sorting of its atoms. This implies the well known incompleteness of the strategy.

INCREASE AND CONICITY

Two theorems allow the decision of the function increase property and of the term conicity property with respect to a CTRS R, if R matches some constraints.
For example, the first of these theorems assures that the function append defined by the following system *Append* is increasing in its first parameter:

1) append(nil,y)→y

2) append(cons(x,y),z)→cons(x,append(y,z))

where the function symbols are the constructors cons and nil, and the defined function symbol append. In fact the normal form of append(cons(a,cons(b,nil)),cons(c,nil)) is cons(a,cons(b,cons(c,nil))) and the normal form of append(cons(b,nil),cons(c,nil)) is cons(b,cons(c,nil)), where cons(a,cons(b,nil)) > cons(b,nil) and cons(a,cons(b,cons(c,nil))) > cons(b,cons(c,nil)). We remark that a, b, c are any constant symbol.

Analogously, the second theorem assures, for example, that the term append(cons(x,y),z) is strictly conic with respect to z in the system *Append*, i.e. the normal form of any ground instance of append(cons(x,y),z) is greater than the normal form of the corresponding instance of z.

TERMINATION PROPERTY OF CONIC-FLAT CTRS

The main result is the finite termination property of the Decomposition Procedure applied to the goal $t_1 =_R t_2$ with either t_1 or t_2 ground, when R is a Conic-Flat CTRS.
For example, let R be the following system *Flat-append*:

1') append(nil,y) → y

2') append(y,z)$=_R$w ⊃ append(cons(x,y),z) → cons(x,w)

This system is called *flat* since it is equivalent to the above unconditional system *Append*. Moreover *Flat-append* is also *conic* since it satisfies the following constraints:

– the variables set occurring in the premise of rule 2') is a subset of the set of variables of the conclusion;

- no rule deletes variables since both rules 1) and 2) of *Append* do not delete variables;
- both rules 1) and 2) are left linear;
- for every substitution σ, rules 1) and 2) of *Append* satisfy the relation $\sigma(\lambda) >_s \sigma(\rho)$, where $>_s$ is any simplification ordering;
- the right member of the recursive rule 2) of *Append* is strictly conic with respect to its second parameter, i.e. the recursive one.

Then, DP applied to the goal append(x,y)$=_R$cons(a,cons(b,nil)) terminates yielding the three solutions {x/nil, y/cons(a,cons(b,nil))}, {x/cons(a,nil), y/cons(b,nil)}, and {x/cons(a,cons(b,nil)), y/nil}. The proof of the termination theorem is based on the observation that in each chain $S_0 \Rightarrow S_1 \Rightarrow \cdots$, where $\Rightarrow$ denotes a DP transformation step, a subchain $S_{i_0} \Rightarrow S_{i_1} \Rightarrow \cdots$ does exist such that the first equation of any S_{i_j} is greater than the first equation of $S_{i_{j+1}}$ according to a well founded ordering $\triangleright$.

REFERENCES

[1] G. Aguzzi and M. C. Verri, "On the Termination of a Unification Procedure", in Proceedings of 3rd Italian Conference on Theoretical Computer Science, World Scientific, 1989.

[2] G. Aguzzi and M. C. Verri, "On the termination of an algorithm for unification in equational theories", in Proceedings of IEEE TENCON '89, pp. 1034-1039

[3] G. Aguzzi and M.C. Verri, " A class of terminating logic programs", R.T. 2/90, D.S.I., Univ. of Florence, 1990.

[4] M. Baudinet, "Proving Termination Properties of Prolog Programs: a Semantic Approach", 3rd Annual Symp. on Logic in Computer Science, Edimburg, 1988, pp. 336-347.

[5] N. Dershowitz, "Orderings for Term Rewriting Systems", Theor. Comp. Sci. vol. 17, 1982, pp. 279-301.

[6] N. Dershowitz, "Termination", Lecture Notes in Computer Science, vol.202, 1985, pp. 180-224.

[7] N. Dershowitz and G. Sivakumar, "Solving Goals in Equational Languages", Lecture Notes in Computer Science vol. 308, Springer Verlag, 1988, pp 45-55.

[8] N. Dershowitz and J. P. Jouannaud, Handbook of Theoretical Computer Science chapter Rewrite Systems. Volume B, North Holland, 1990 (to appear).

[9] M. Dincbas and P. Van Hentenryck, "Extended Unification Algorithms for the Integration of Functional Programming into Logic Programming", J. Logic Programming, vol. 4, 1987, pp 199-227.

[10] J. H. Gallier and W. Snyder, "A General Complete E-Unification Procedure", Lecture Notes in Computer Science vol. 256, Springer Verlag, 1987, pp. 216-227.

[11] G. Huet and D.C. Oppen, "Equations and rewrite rules: A survey", in Formal Language Theory: Perspectives and open problems, R. Book Ed., Academic Press, New York, 1980, pp. 349-405.

[12] J-M. Hullot, "Canonical Forms and Unification", in Proceedings of the 5th Conference on Automated Deduction, 1980, pp. 318-334.

[13] B. Jayaraman, "Semantics of EqL", IEEE Transactions on Software Engineering, vol. 14, n. 4, april 1988, pp 472-480.

[14] S. Kaplan, "Simplifying Conditional Term Rewriting Systems: Unification, Termination and Confluence", J. Symbolic Computation, 4, 1987, pp. 295-334.

[15] A. Martelli, C. Moiso and G. F. Rossi, "An Algorithm for Unification in Equational Theories", 3rd IEEE Symp. on Logic Programming, Salt Lake City, Utah, 1986, pp. 180-186.

[16] A. Martelli and U. Montanari, "An Efficient Unification Algorithm", ACM Transaction on Programming Languages and Systems, 4, 2, 1982, pp 258-282.

[17] M. Okada, "A logical analysis on theory of conditional rewriting", Lecture Notes in Computer Science vol. 308, Springer Verlag, 1988, pp 179-196.

[18] U. S. Reddy, "Narrowing as the Operational Semantics of Functional Languages", in Proceedings on Logic Progr., IEEE, 1985, pp. 138-151.

[19] J. A. Robinson, "A Machine-Oriented Logic Based on the Resolution Principle", J. ACM vol. 12, 1965, pp. 23-41.

[20] T. Vasak and J. Potter, "Characterization of terminating logic programs", 3rd IEEE Symp. on Logic Programming, Salt Lake City, Utah, 1986, pp. 140-147.

Constrained Equational Deduction

John Darlington   Yike Guo
Department of Computing
Imperial College
180 Queen's Gate London SW7 2BZ U.K.
e-mail: {jd,yg}@doc.ic.ac.uk

Abstract

In the areas of automated deduction, algebraic specification and declarative programming, symbolic computation is always required to cooperate with computation in concrete mathematical systems. Therefore, to design a deduction mechanism within equational programming logic camp, symbolic equational deduction should be extended to exploit the semantic information behind abstract symbols. In this paper, we propose constrained equational deduction as a framework for such an extension within a general constraint equational logic programming setting. Constrained equational deduction takes advantage of the hierarchical constraint information within the equations and establishs a smooth link between symbolic equational deduction and various constraint solving mechanisms. We present a constructive approach to combine a constraint system in the domain of discourse with the symbolic equational constraints in the term space to establish the constraint equational logic programming paradigm. Constrained equational deduction models are then presented to be the computational model of the paradigm.

1 Introduction

Exploiting semantic information of abstract symbols within a deduction calculus has a long history. In [25], G.Plotkin suggested a modification of resolution by replacing the symbolic unification with an E-unification procedure to deal with the semantic properties, such as commutativity, associativity, of certain function symbols. More recently, Stickel's *theory resolution* [28] established a more general resolution scheme to use tailored inference rules performing semantic deduction for objects with their intended interpretations. The computational models in Jaffer and Lassez's [14] constraint logic programming and Smolka's work on polymorphically order-sorted typed logic programming can be regarded as further developments of general theory resolution by adopting the notion of constraints to abstract the intended semantic feature of objects. Therefore, deduction calculi over constraints can be uniformly regarded as constraint solving procedures in their intended semantic model.

In this paper, we present constraint equational logic programming (CELP) as a general scheme to integrate constraint solving procedures with the Horn equational logic (equational logic for short) programming paradigm. The scheme entails performing equational logic programming over a well understood mathematical system equipped with natural algebraic operations together with associated privileged predicates. The logic formulas in the system are abstractly regarded as constraints. Constraint solving in such a mathematical system is a decision procedure to determine the satisfiability of constraints. By incorporating constraint systems into the equational logic programming paradigm, constraint equational logic programming provides a general and powerful programming paradigm to exploit efficiency and expressiveness of constraint-based computation. A constrained equational deduction calculus is then constructed by combining the symbolic equational deduction with the underlying constraint solver to provide a computational model for CELP. The research aims to provide not only an expressive equational programming paradigm but also a design space for extending equational deduction calculi to exploit semantic knowledge.

In section 2, we define the notion of constraint systems. We abstract from various concrete constraint systems and give a general formalization to capture their common features. We then, in section 3, propose the notion of constraint logic programming. Its declarative semantics and the general computation model are presented there. In section 4, we present constrained equational deduction as the operational model of CELP. The conclusion and some related works are presented in the last section.

2 Equational Deduction and Constraint Systems

We assume familiarity with equational logic [12] and various equational deduction models, such as, term rewriting system [6], conditional equations, unification [18], E-unification[7] and paramodulation [31]. We start by stating informally the notion of equational logic programming systems using the framework of general logics [21].

2.1 Equational Logic Programming

Meseguer's framework of general logics provides a very broad view on the notion of a logic programming system [21]. Given a logic $\mathcal{L}$ in general logics, a logic programming system ("weak logic programming" in the sense of [21]) can be designed based on proof systems of the logic to justify the entailment relations between the sentences of the logic.

Let $\mathcal{E}$ be the Horn equational logic in the framework of general logics. Given a signature Σ, an Σ-conditional equation (Horn equation) is of the form: $s = t :- s_1 = t_1, \ldots, s_n = t_n$ where s, t, s_i, t_i are Σ-terms. An equational theory is defined by a finite set of Σ- conditional equations E consisting of all logic consequences of E under the entailment relation of $\mathcal{E}$. E forms the axioms of the theory. Therefore, a theory can be represented as a pair $T = (\Sigma, E)$.

An equational program is a set of (conditional) equations E being the set of axioms of an equational theory. A query $\varphi = \exists \vec{x} \wedge_{i=1}^{n} s_i = t_i$ is an existentially quantified conjunction of equations used for asking if some properties hold in the theory $T = (\Sigma, E)$. Therefore, computation in equational programming justifies the entailment relation between the theory $T = (\Sigma, E)$ and the query φ. Thus, when the equational program E is loaded, the user can pose φ as a query to start the equational deduction computation. If φ is a provable consequence of the program Γ, then the deduction will return a set of answers (substitutions) to justify the truth of φ. Each answer corresponds to a particular proof of φ. If the query φ is not provable from E then the computation will either stop to report "failure" or run forever.

This traditional viewpoint on equational deduction as a pure symbolic proving procedure is no more adequate when equational deduction is used for practical equational programming systems [4]. An obvious weakness of this "programming as theory and computation as proving" approach is that "there is no reference to the models that the theory is a linguistic device for"[21]. In a logic system, a theory may have a lot of models. However, when we are programming for a particular problem, we always have a particular model in mind. Such a model is the *intended* model of the theory. The program is just a linguistic description of this intended model. The model itself is primary. By this "semantic oriented " approach, computation becomes the mechanism to deduce properties of the intended model by taking the program equations as the specification (or axiomatization) of the intended model. This viewpoint on equational programming results in a *constraint equational logic programming (CELP)* paradigm which fits well with the *Assertional Programming* framework proposed in [5] as the general model of declarative programming languages. Before the discussion on CELP systems, we first define the notion of a constraint system.

2.2 Constraint System

Considering some well studied computational domains such as sets, boolean expressions, integers, rationals, etc. These domains are equipped with natural algebraic operations such as set intersection and disjunction, rational addition and multiplication, etc. They also have associated privileged predicates such as equality and various forms of inequalities. Such a computational domain together with their operators can be regarded as an algebra. The logic formulas of the algebra are abstractly regarded as *constraints*. Constraints provide a way to implicitly define objects by stating their logic relations. Thus, a constraint denotes a set of objects which can satisfy the relation. This notion is captured by the following formalism:

Definition 2.1 (Constraint System) *Given a computational domain A, equipped with a set of variables V. We can formally define a constraint system as a tuple:* $<\mathcal{A}, V, \Phi, \mathcal{I}>$ *where* Φ *is a decidable set of constraints.* $\mathcal{I}$ *consists of a solution mapping* $[]^{\mathcal{I}}$ *which maps every constraint* $\phi(\bar{x})$ *to* $[\phi(\bar{x})]^{\mathcal{I}}$ *of A–valuations, called the solutions of* $\phi(\bar{x})$*, such that:* $\forall \alpha \in [\phi(\bar{x})]^{\mathcal{I}}, A \models \forall\emptyset.\phi(\alpha\bar{x})$

We define Val_A as the set of all $\mathcal{A}$-valuations. A constraint is satisfiable in the constraint system iff its solution set is non-empty. A constraint ϕ is valid in I iff $[\phi]^I = Val_A$. For a set of constraints Φ, I is a model of Φ if all the constraints in Φ are valid in I. Assuming a set W of variables, the W-solutions of a constraint ϕ in an interpretation $\mathcal{I}$ is the set of solutions : $[\phi]^I_{|W} := \{\alpha_{|W} \mid \alpha \in [\phi]^I\}$. A constraint ϕ is equivalent to a constraint ϕ' iff $[\phi]^I = [\phi']^I$.

Definition 2.2 (Closed under Logic Connectives) *Given a constraint system :* $\mathcal{S}_{\mathcal{A}} :< \mathcal{A}, V, \Phi, \mathcal{I} >$*, the systems is closed under logic conjunction (resp. disjunction, implication, negation, existential qualification) if for any constraints* $\phi_1, \phi_2 \in \Phi$*, there exists a constraint* $\delta \in \Phi$ *such that* $\delta = \phi_1 \wedge \phi_2 \in \Phi$ *(resp.*$\phi_1 \vee \phi_2, \phi_1 :- \phi_2, \neg\phi_1, \exists x.\phi_1$*) and :*

$$[\phi_1 \wedge \phi_2]^I = [\phi_1]^I \cap [\phi_2]^I$$

$$[\phi_1 \vee \phi_2]^I = [\phi_1]^I \cup [\phi_2]^I$$

$$[\phi_1 :- \phi_2]^I = [\phi_1]^I \cup \{Val_A - [\phi_2]^I\}$$

$$[\neg\phi_1]^I = Val_A - [\phi_1]^I$$

$$[\exists x.\phi]^I = \{\alpha \in Val_A \mid \alpha_{|W-x} = \beta_{|W-x}, \beta \in [\phi]^I\}$$

where W is the set of variables in ϕ*.*

We assume that all concerned constraint systems in the paper are closed under conjunction and variable instantiation and contain equality constraints.

To solve constraints, we require the constraint system comes with a set of *solved forms* such that every constraint in solved form is satisfiable. For every satisfiable constraint G, there exists a complete set of solved forms Sol_G such that the disjunction of all constraints in the set is equivalent to G. That is, for all $G'_i \in Sol_G$ and $G' = \bigvee_{i=1}^{n} G'_i$, $[G]^I = [G]^I = \bigcup_{i=1}^{m} [G'_i]^I$. The justification for defining the solved forms comes from the easily verified procedure for checking their satisfiability and for deriving the explicit solutions. The procedure computing the solved forms of a constraint is called *constraint solver*. We use $\longrightarrow_c$ to denote the derivation relation for constraint solving in terms of a solver C.

Definition 2.3 (Soundness and Completeness) *Let* $<\mathcal{A}, V, \Phi, \mathcal{I}>$ *be a constraint system and C be a constraint solver of the system.*

Soundness: *C is sound iff for any constraint G:* $G \longrightarrow_c G' \Longrightarrow [G']^I \subseteq [G]^I$

Completeness: *C is complete iff for any constraint G,* $\forall\alpha \in [G]^I$*, there exists a constraint G′ which is in solved form such that:* $G \longrightarrow_c G'$ *and* $\alpha \in [G']^I$.

Constraint System $\mathcal{S}_{Real}$: Linear Equations over Real Numbers : Let $\mathcal{R}$ be the real number domain equipped with the arithmetic operations $< +, -, * >$ and Φ_R be the set of linear equations. $\mathcal{S}_{Real} = < \mathcal{R}, V, \Phi_R, \mathcal{I}_R >$ is a constraint system over real numbers where the interpretation $\mathcal{I}_R$ maps each linear equation to the (possibly infinite) set of its solutions. A set (or conjunction) of linear equations:

$$y_1 = a_{1,1}x_1 + \ldots + a_{1,n}x_n + c_1$$
$$\ldots$$
$$y_m = a_{m,1}x_1 + \ldots + a_{m,n}x_n + c_m$$

is said to be in solved form if the variables $y_1, \ldots, y_m$ and $x_1, \ldots, x_n$ are all distinct. The variables $y_1, \ldots, y_m$ are *eliminatable variables* and $x_1, \ldots, x_n$ are *parameters* . Algorithms for solving linear equations, such as the Gaussian elimination method, compute the solved forms of linear equations and constitute sound and complete constraint solvers in the system. Linear arithmetic constraints has been intensively studied by Lassez and Jaffer el.al as the primitive component in a constraint logic programming system[19] [14].

Constraint System $\mathcal{S}_\Sigma$: Unification Problem over First Order Terms : Let T_Σ be the set of all ground $\Sigma-$terms for a given signature Σ, V be a set of variables and Φ_Σ be the set of equations over $T_\Sigma(V)$. $\mathcal{S}_\Sigma = < T_\Sigma, V, \Phi_\Sigma, \mathcal{I}_\Sigma >$ is a constraint system over the first order Σ-terms where the interpretation $\mathcal{I}$ maps each term equation to the (possibly infinite) set of its ground unifiers. A conjunction of term equations ϕ is usually called a system of equations in the literature. As shown in [18], for any satisfiable system of equations, its unique solved form is a new system of equations in the form $\{x_1 = t_1, \ldots, x_n = t_n\}$ and x_i's are distinct variables which do not occur in any t_j's. Therefore, similarly to the constraint system $\mathcal{S}_{Real}$, variables $x_1, \ldots, x_n$ are *eliminatable variables* and variables in t_j's are *parameters*. Traditionally, the notion of the most general unifier (mgu) is used as the means of representation of solutions. The relation between the solved form and the mgu of a system of equations can be established by following theorem [18][7].

Let $\alpha = \{x_1 \rightarrow t_1, \ldots, x_n \rightarrow t_n\}$ be an idempotent substitution, α is the mgu of a system of equations ϕ iff the equation system:$\{x_1 = t_1, \ldots, x_n = t_n\}$ is the solved form of ϕ. Therefore, we can call the equational system as *idempotent substitution equations.*

Transformational unification algorithms, such as the following Martelli and Montanari algorithm [24], constitute complete and sound constraint solvers transforming a system of equational constraints to its solved form.

Trivial:

$$\frac{G_0 : \{t = t\} \wedge S}{G_1 : S}$$

Term Decomposition:

$$\frac{G_0 : \{f(t_1, \ldots, t_n) = f(s_1, \ldots, s_n)\} \wedge S}{G_1 : \{t_1 = s_1, \ldots, t_n = s_n\} \wedge S}$$

Variable Elimination:

$$\frac{G_0 : \{x = t\} \wedge S}{G_1 : \{x = t\} \wedge \sigma S}$$

where $x \notin var(t)$ and $\sigma = t/x$.

3 Constraint Equational Logic Programming

In a traditional equational specification system, its term initial model is taken as its standard (or intended) model. However, it is always the case that we need to consider some "built-in" data types within a program.

The subtheories corresponding to these types should be interpreted by their fixed interpretations to give the intended meaning. Therefore, the initiality of the semantic model is relativized to the underlying intended models of the fixed types. Originally, the problem arises from the semantics of parameterized modules, where a model of a parameter theory is extended to a model of the body theory. In general, this so called *free extension* problem [22] arises when we try to extend the fixed model for a built-in theory to initial models of an equational program where the built-in theory becomes "a data constraint" [3] in the sense it imposes restrictions on the interpretations of the program. "Built-in" theories always enhance the expressiveness as well as efficiency of equational programming systems. This motivate the notion of **constraint equational logic programming** (CELP) we propose here as a paradigm for practical equational logic programming.

3.1 Syntax

We first introduce the notion of *mixed terms* which are terms containing elements from a particular algebra giving the intended semantic meaning of some symbols.

Definition 3.1 (Mixed Term) *Let A be a Σ-algebra, $T_{\Sigma(A)}$ stands for the set of (ground) A-mixed terms and itself can be regarded as a Σ-algebra generated by A. Each element in $T_{\Sigma(A)}$ may be seen as a partially evaluated expression with abstract entities (Σ-operators) and concrete terms (element of A).*

When a particular algebra for a logic theory is given and we also have a constraint system over this algebra, then we can integrate the constraint system with an equational logic programming system where the constraint system becomes a " built-in" deduction engine for concrete objects. That is, given a constraint system $S_A :< \mathcal{A}, V, \Phi_A, \mathcal{I} >$ specifying the semantic properties of a particular theory, a CELP program $< E, \Phi_A >$ is a set of universally quantified S_A- constrained conditional equations of the form :

$$\forall \vec{x}(s = t :- \wedge_{i=1}^{n} s_i = t_i, \phi)$$

where $\phi \in \Phi_A$ is a A-constraint and $s_i = t_i$ are equations on the A-mixed Σ-terms. A query constraint φ is an S_A-constrained equation conjunction :

$$\{\phi_A : \wedge_{k=1}^{m} \phi_k\} \bigwedge \{\phi_E : \wedge_{i=1}^{n} s_i = t_i\}$$

3.2 Semantics

We use the notion of free extension to give the semantics of CELP programs. Consider a containment of equational programs $< \Sigma, E > \subseteq < \Sigma', E' >$ where $\Sigma \subseteq \Sigma'$ and $E \subseteq E'$. Then any Σ', E'-model $M_{E'}$ can be regarded as a Σ, E-model M_E by forgetting the extended functions in $\Sigma' - \Sigma$. The containment can be modelled as a specification morphism $J :< \Sigma, E > \rightarrow < \Sigma', E' >$ and the forgetting process can be modelled as a functor J^* between two model categories $Mod_{\Sigma',E'} \rightarrow Mod_{\Sigma,E}$. Conversely, by Lawvere's result [20], for every Σ, E-model A, there is a Σ', E'-model A^* and a homomorphism (called the universal morphism) $\eta : A \rightarrow J^*(A^*)$ and characterized by the following universal property: $\forall B \in Mod_{\Sigma',E'}$ and any morphism $f : A \rightarrow J^*(B) \in Mod_{\Sigma,E}$, there is a unique morphism $f^* : J^*(A^*) \rightarrow B$ such that $f = J^* \circ f^* \circ \eta$ in $Mod_{\Sigma,E}$. Such an object $A^* \in Mod_{\Sigma',E'}$ is called a *free extension* of A along the specification morphism J (Lawvere modeled this free construction property of a logic system as an algebraic character called liberality of the institution). The universal property determines A^* uniquely up to isomorphism in $Mod_{\Sigma',E'}$. The existence of free extensions for models in the Horn clause with equality was proved in[9]. As shown in [22], for any Σ, E-model M, we can explicitly construct its free extension by axiomatizating M in terms of its diagram determined by the graph of its operators. For a Σ, E-model M and $\bar{M}$ denotes its carrier, the signature $\Sigma(M)$ is given by adding M-operators and the element $\bar{M}$ disjointly as new functions and new constants respectively. The diagram $D(M)$ of M is the set of $\Sigma(\bar{M})$-equations of the form $f(a_1, \ldots, a_n) = f^M(a_1, \ldots, a_n)$ or $p(a_1, \ldots, a_n) = true$ for function symbols and predicates in Σ respectively (for simplicity, we regard predicates as boolean valued functions by attaching to every algebra the boolean value $\{true, false\}$ as constants). The following lemma gives an explicit construction of a free extension of a Σ, E-model M.

Lemma 3.1.1 *Given a specification morphism $J :< \Sigma, E > \rightarrow < \Sigma', E' >$ and M is a model of E, a free extension M^* of M along J is isomorphic to the quotient M-mixed term algebra $T_{\Sigma'(M)/E' \cup D(M)}$.*

Proof (sketch): As shown in [22], the free extensions of a Σ, E-model M constitutes a category $Ext(M)$ consisting of Σ-morphism (extension morphism): $f : M \rightarrow M'$ as its objects and for any morphism $h : (f : M \rightarrow M') \rightarrow (f : M \rightarrow M'')$ the Σ'-morphism $\bar{h} : M' \rightarrow M''$ preserves the image of M. That is, $f' = \bar{h} \circ f$ (where M', M'' are two Σ', E'-models). Moreover, $Ext(M)$ and the category $Mod_{\Sigma'(M), E' \cup D(M)}$ are isomorphic. Since a free extension M^* of M along the morphism $J :< \Sigma, E > \rightarrow < \Sigma', E' >$ for each Σ', E'-model determines a unique free extension functor $J^{**} : Mod_{\Sigma,E} \rightarrow Mod_{\Sigma',E'}$ such that $J^{**}(M) = M^*$, such J^{**} is an initial object in $Ext(M)$. Therefore, we conclude that M^* is isomorphic to $T_{\Sigma'(M)/E' \cup D(M)}$.

The notion of free extension provides the semantic foundation for constraint equational logic programming. Let $S_A :< \mathcal{A}, V, \Phi_A, \mathcal{I} >$ be the underlying constraint system. Semantically, each program rule in a S_A-CELP program:

$$\forall x(s = t :- \wedge_{i=1}^{n} s_i = t_i, \phi)$$

should be interpreted as the logic implication:

$$\forall x : s = t :- \phi \bigwedge \{\wedge_{i=1}^{n} s_i = t_i\}$$

The intended models of a CELP program $< E, \Phi_A >$ are free extensions A^* of A along the specification morphism: $< \Sigma_A, \emptyset > \rightarrow < \Sigma', < E, \Phi_A >>$ (where Σ_A is the set of signatures of the constants and operators of A and $\Sigma_A \subseteq \Sigma'$). All these models are isomorphic to the model $T_{\Sigma'(A)}/E \cup D(A)$ where $D(A)$ is determined by the graph of operators in Σ_A. So far, we have given the semantics to constraint equational programs by freely extending the computational domain of the underlying constraint system. As to CELP program, it is an essential semantic requirement that the free extension along the specification morphism $J :< \Sigma_A, \emptyset > \rightarrow < \Sigma', < E, \Phi_A >>$ for a constraint system $< \mathcal{A}, V, \Phi_A, \mathcal{I} >$ can protect the underlying constraint system. That is, the universal morphism $\eta : A \rightarrow J^*(A^*)$ is an isomorphism. The syntactic characterization of this property was discussed in [11] in the context of parameterized specification and also in [22] for constraint logic programming. We will assume for all the CELP systems the protecting condition of free extension has been satisfied. Therefore, a CELP program $< E, \Phi_A >$ constitutes a new constraint system $< A^*, V, \Phi', I_{A^*} >$ where A^* is the free extended model $T_{\Sigma(A)}/E \cup D(A)$ of A and $\Phi' = \Phi_A \cup \Phi_E$ is interpreted by I_{A^*} as

$$[\phi \in \Phi_A]^{I_{A^*}} = [\phi]^I$$

$$[s = t \in \Phi_E]^{I_{A^*}} = \{\alpha \in V \rightarrow \bar{A^*} \mid \bar{\alpha}s =_{A^*} \bar{\alpha}t\}$$

where $\bar{\alpha}$ is the α-extended homomorphism interpreting terms as elements of A^*. The computation in constraint equational logic programming justifies the satisfaction relation between the intended model A^* and the existential closure of the query φ by computing solutions of the query constraint. In next section, we present constrained equational deduction as an operational framework for CELP.

4 Constrained Equational Deduction

4.1 Constrained Unification

When the equational theory is empty, the A-constraint equational system defines the constraint system $S_{\Sigma,A} :< A^*, V, \Phi_A \cup \Phi_E, \mathcal{I}_{A^*} >$ where $A^* = T_{\Sigma(A)}/D(A)$. Therefore, the interpretation $\mathcal{I}_{A^*}$ for equational constraints Φ_E is defined by interpreting equational constraints in terms of their $T_{\Sigma(A)}/D(A)$-solution set. By this definition, the notion of identical terms usually determined by their symbolic identity is extended to concern the " value equivalence" of the concrete elements in A. As implied by the semantic definition, this can be done by using the normal E-unification procedure by treating the elements in A as the additional new constants and the diagram of the underlying system as an equational program. However, such a diagram is not generally finite and, even it is finite, the E-unification procedure is by no means efficient. If a special-purpose constraint solver C for such A-constraints is available, one would certainly prefer to adopt it for solving constraints between concret objects. Importantly, since the free extension along the morphism $< \Sigma_A, \emptyset > \rightarrow < \Sigma_A \cup \Sigma, \emptyset >$ in this case is obviously A-protecting, we can easily combine the underlying constraint solver with normal unification (see example 2.2) for the abstract parts of the mixed terms, as shown in the following inference rules, to get a complete constrained unification algorithm CU to solve a constraint $\bigwedge_{k=1}^m \phi_k \bigwedge \bigwedge_{i=1}^n s_i = t_i$.

VT: Variable Abstraction

$$\frac{G_0 : \{s = t\} \bigwedge S}{G_1 : \exists x.\{s[\beta \longleftarrow x] = t, x = s/\beta\} \bigwedge S}$$

where s/β is a maximal pure subterm of s

SP: Spliting

$$\frac{G_0 : \{e = t\} \bigwedge S}{G_1 : \exists x.\{x = e, x = t\} \bigwedge S}$$

where e is an expression of A and t is a pure free term.

U: Unification

$$\frac{G_0 : \phi_E \bigwedge \phi_A}{G_1 : \{x_1 = t_1, \ldots, x_n = t_n\} \bigwedge \phi_A}$$

where $x_1 = t_1, \ldots, x_n = t_n$ is the solved form of ϕ_E computed by using the constraint solver for S_Σ (see example 2.2).

C: Constraint Solving

$$\frac{G_0 : \phi_E \bigwedge \phi_A}{G_1 : \phi_E \bigwedge \phi'_A}$$

where $\phi_A \longrightarrow_c \phi'_A$ such that ϕ'_A is a solved form of ϕ_A.

Definition 4.1 *A constrained solved form of a constraint:* $\bigwedge_{k=1}^m \phi_k \bigwedge \bigwedge_{i=1}^n s_i = t_i$ *is a conjunction of constraints* ϕ'_E *and* ϕ'_A*, denoted as a pair:* $< \phi'_E, \phi'_A >$*, where* ϕ'_A *is in* $A-$*solved form and* ϕ'_E *is in solved form of unification problem. That is, a set of idempotent substitution equations.*

Theorem 4.1.1 *Given a constraint system* $S_{\Sigma,A}$*, if the constraint solver of the underlying constraint system* S_A *is complete and sound,* CU *is complete and sound for transforming a constrained unification constraint* $\phi_A \wedge \phi_E$ *to its constrained solved forms.*

Proof (sketch): The proof of soundness is straightforward. To prove the completeness, we first show that VT and SP rules preserve solutions. For a constraint $\Phi_A \wedge \Phi_E$ where $\Phi_A = \bigwedge_{k=1}^m \phi_k$ and $\Phi_E = \bigwedge_{i=1}^n s_i = t_i$, by applying VT and SP rules, we can get a new equivalent constraint $\overline{\Phi_A} \wedge \overline{\Phi_{\Sigma,E}}$ where $\overline{\Phi_A}$ contains only A-objects and $\overline{\Phi_{\Sigma,E}}$ contains only

Σ-equations. For any $\alpha \in [\Phi_A \wedge \Phi_E]^{I_{A^*}}$, then $\alpha \in [\overline{\Phi_A}]^{I_A} \cap]\overline{\Phi_{\Sigma,E}}]^{I_\Sigma}$. Moreover, there is a set of idempotent substitution equations ϕ'_E which is the solved form of $\overline{\Phi_{\Sigma,E}}$. Since the constraint solver for S_A is complete, a solved form ϕ'_A of $\overline{\Phi_A}$ can be computed such that $\alpha \in [\phi'_A]^{I_A}$. Therefore, $< \phi'_E, \phi'_A >$ is the computed constrained solved form.

Constrained unification provides a general scheme for extending traditional unification to cooperate with constraint solvers, such as Boolean unification in Boolean algebra and Gaussian elimination for solving arithmetic constraints over real numbers. It is noted that, by constrained unification, only constrained solved forms are computed for solving constrained unification constraints. In general, constrained solved forms are not always good enough for verifying satisfiablity. Therefore, underlying constraint part ϕ_A and the equation part ϕ_E of a constrained solved form $< \phi_E, \phi_A >$ should communicate and be solved further to compute a uniform solved form. This problem is not solved yet because we don't know what kind of the uniform solved forms we should achieve in general. However, in some special case, the uniform solved forms for constrained unification constraints are existed and can be computed. For example, when the underlying constraint system itself is given by a special equational theory $E(A)$ together with an $E(A)$-unification procedure as the solver, such as solving equations in $T_{\{+,a,b,...\}}(V)$ by an AC-unification procedure or in $T_{\{+,*,0,1,a,b...\}}(V)$ by an Boolean unification algorithm, constrained unification can extend the underlying unfication in such a way that it can unify terms with additional free function symbols. M.Stickel first proposed a solution to the problem of extending an AC-unification algorithm into a semantic unification model for an arbitrary term algebra having some associative and commutative symbols. Boudet and Jouannaud [1] developed a general method to combine the unification in an arbitrary theory with that in the free theory. In these work, since the solved forms of underlying constraints are explicitly representable as idempotent substitution equations, two parts of a constrained solved form could be combined together and then be solved further more to generate the final solved form in idempotent substitution equations. The new unsolved problem comes from the communication of ϕ'_E and $\phi'_{E(A)}$ in the solved form $< \phi'_E, \phi'_{E(A)} >$ which happens when :

1) $\phi'_E \bigwedge \phi'_{E(A)}$ implies an impure equation $e = t$, i.e. there exists an equation $x = t \in \phi'_E$ and an equation $x = e \in \phi'_A$. Since, in this case, the free extension is A-protecting for all models of $E(A)$ (That is, the extension is conservative), a solution to the equation must be an instance of a substitution θ, mapping the $E(A)$-expression into a variable. This θ can be computed by the underlying constraint solver (the E(A)-unification procedure) by treating x as a constant. Therefore, we have the following rule:

CA: Constant Abstraction

$$\frac{G_0 : \{x = t \in \phi'_E\} \bigwedge \{x = e \in \phi'_{E(A)}\} \bigwedge S}{G_1 : \{x = a, x = t\} \in \phi'_E \bigwedge e = a \in \phi'_{E(A)} \bigwedge S}$$

where a is a free constant.

2) There is a cycle between ϕ'_E and $\phi'_{E(A)}$. That is, the occur-check problem. Breaking such a cycle, we need to compute the special solutions $\phi'_{E(A)}$ which can eliminates some variables caused the cycle. It is always necessary to request the underlying constraint system have this variable elimination ability.

VE: Variable Elimination

$$\frac{G_0 : \{x_i = t(\ldots, x_j, \ldots) \in \phi'_E\} \wedge \{x_j = e(\ldots, x_i, \ldots) \in \phi'_{E(A)}\} \wedge S}{G_1 : \sigma\phi'_E \wedge \{x_j = \sigma e \downarrow \wedge S_\sigma\} \wedge S}$$

where $S_\sigma = \{x'_i = e'_i\}$ corresponds to an $E(A)$-substitution σ, $\sigma e \downarrow$ stands for the evaluated form of the expression σe such that $x_i \notin \sigma e \downarrow$.

This approach seems also work generally when the underlying constraint system admits idempotent substitution equations as its solved forms, such as the real number constraint system in [18]. We are going to study the issue in more details. A detailed discussion on combining unifications can be found in [1] and [29].

4.2 Equational Logic Program as A Constraint System

When the underlying constraint system is empty, CELP becomes traditional equational logic programming [4]. Therefore, CELP covers equational logic programming paradigm. On the other hand, this also means that an equational logic program defines a corresponding constraint system. Given an equational logic program E, following the semantics of CELP, E defines a constraint system over free terms $S_{\Sigma,E} :< T_\Sigma/E, V, \Phi_E, \mathcal{I}_{\Sigma,E} >$ where T_Σ/E is the quotient term algebra generated by E and $\mathcal{I}_{\Sigma,E}$ is the interpretation for the equational constraint Φ_E defined as: $[s = t]^{\mathcal{I}_{\Sigma,E}} = \{\sigma \in Val_{T_\Sigma/E} \mid \bar{\sigma}s = \bar{\sigma}t\}$ where $\bar{\sigma}$ is the σ-extended homomorphism: $T_\Sigma(V) \rightarrow T_\Sigma/E$. A solved form of equational constraints is a set of idempotent equations. Given a system of equations G, its complete set of solved forms Sol_G corresponds to the complete set of E-unifiers. A general E- unification procedure constitutes the constraint solver $C_{\Sigma,E}$ of the system to compute its complete solved forms. In [30], Snyder gave a complete set of rules for computing a complete set of solved forms of a system of equations in an arbitrary equational theory E defined by a set of conditional equations. Therefore, it constitutes a complete constraint solver $C_{\Sigma,E}$ for the system $S_{\Sigma,E}$. The solver is constructed by extending the solver in S_Σ (see example 2.2) by a new rule, called lazy paramodulation, for rule application.

Lazy Paramodulation: Let E be a conditional equational program. A lazy paramodulation step is defined by following deduction rule:

$$\frac{\{t_1 = t_2\} \wedge S}{t_1/\beta = u \wedge t_2 = t_1[\beta \longleftarrow v] \wedge \{\wedge_{i=1}^{n} s_i = t_i\} \wedge S}$$

where t_1/β is a non-variable subterm of t_1 and $u = v :- \wedge_{i=1}^{n} s_i = t_i \in E$.

4.3 Constrained Deduction Model for CELP

Now, we consider the most general case of CELP. Given a constraint system $S_A :< A, V, \Phi_A, I >$ together with its constraint solver C, the computation of a S_A-CELP program $< E, \Phi_A >$ is to justify the satisfaction relation between the intended model and the existential closure of the query constraint $\phi_A \wedge \phi_E$. Since the intended model A^* is isomorphic to $T_{\Sigma(A)}/E \cup D(A)$, the program $< E, \Phi_A >$ corresponds to a pure equational logic program $E \cup D(A)$ over $T_{\Sigma(A)}(X)$ for $V \subseteq X$ formed by augmenting the program with $D(A)$ equations and treating all Σ_A symbols as normal abstract symbols and constraint $\phi \in \Phi_A$ as conjunction of normal atomic formula for each rule:$\forall \vec{x}.(\wedge_{i=1}^{n} s_i = t_i, \phi)$. Obviously, the standard model of this pure equational logic program is the same as that of the original CELP program $< E, \Phi_A >$. Solving a query constraint by $< E, \Phi_A >$ amounts to solving it by the constraint solver in $S_{\Sigma(A),E \cup D(A)}$ (see section 4.2). The lazy paramodulation step will admit constraints to provide the following rule of **constrained paramodulation:**

$$\frac{\{t_1 = t_2\} \wedge S}{t_1/\beta = u \wedge t_2 = t_1[\beta \longleftarrow v] \wedge \phi \wedge \{\wedge_{i=1}^{n} s_i = t_i\} \wedge S}$$

where t_1/β is a non-variable subterm of t_1 and $\forall x.(u = v :- \wedge_{i=1}^{n} s_i = t_i, \phi)$ is an constrained equation in the program.

For the same reason as we discussed in section 4.1, the A-protecting condition of free extensions allows us to use the constraint solver C in S_A instead of axiomatizating the algebra by its diagram. Moreover, since the constraint solver C may have some special restrictions on the forms of its acceptable input constraint and also its output solved forms may not be expressible in terms of idempotent substitution equations, the constrained deduction model for CELP should separate constraint solving in the deduction procedure, and delay call to the solver C until the required input form of a constraint is reached. Therefore, we can compute the constrained solved forms (as defined in definition 4.1) of a query constraint. This is the central idea of our constrained equational deduction model. Due to the A-protecting assumption for free extensions, a CELP program will not introduce any new relation between concret objects to violate the underlying constraint system. The underlying constraint solver then can be integrated into the deduction model for pure equational logic programming (as presented in section 4.2) to derive a constrained equational deduction model of CELP.

Theorem 4.1.2 *Let $< E, \Phi_A >$ be a CELP program over the constraint system $S_A :< A, V, \Phi_A, I >$ and C_A the complete constraint solver of S_A. The constrained unification calculus together with the constrained paramodulation rule constitutes a sound and complete constraint solver to compute the constrained solved forms for a query constraint $G : \wedge_{k=1}^{m} \phi_k \bigwedge \wedge_{i=1}^{n} s_i = t_i$.*

Proof (sketch): The proof of soundness is straightforward. The proof on completeness is based on the A-protecting assumption for free extensions. By Snyder's result, we have a complete constraint solver $C_{\Sigma,E}$ for the constraint system $S_{\Sigma,E}$ corresponding to an equational program E (see section 4.2). For CELP program $< E, \Phi_A >$, this solver can be used to compute the solved forms of the query constraint G by regarding $< E, \Phi_A >$ as the constraint system $S_{\Sigma(A),E\cup D(A)}$. Therefore, by the completeness of the constraint solver $C_{\Sigma(A),E\cup D(A)}$, for any solution $\alpha \in [G]^{I_{A^*}}$, we have $G \longrightarrow_{C_{\Sigma(A),E\cup D(A)}} G'$ such that G' is in solved form in $S_{\Sigma(A),E\cup D(A)}$ and $\alpha \in [G']^{I_{A^*}}$. Considering this derivation $\longrightarrow_{C_{\Sigma(A),E\cup D(A)}}$, if there is an equation $t = e \in G_i$ at the ith step of the deduction and t is a Σ-term, e is an A-expression, by the A- protecting assumption on free extensions, then all solutions of the equation are the $E \cup D(A)$-instances of a substitution θ such that θt and θe are $E \cup D(A)$-equivalent to a variable. Thus, solving the equation is effectively seperated in the derivation and the subproblem $e = x$ for a fresh variable x will be computed using only $D(A)$-equations. So, by separating the Σ-parts and the A-parts of each intermediate result of the derivation and replacing the applying $D(A)$ equations with calls to the constraint solver C_A, we can have a correspondent constrained deduction derivation computing G to its constrained solved form $G'' :< \Phi'_E, \Phi'_A >$ such that $[G]^{I_{A^*}} \supseteq [G'']^{I_{A^*}} \supseteq [G']^{I_{A^*}}$. Therefore, $\alpha \in [G'']^{I_{A^*}}$. Moreover, this derivation can be enumerated by using the inference steps of constrained unification and constrained paramodulation using only the constrained equations in $< E, \Phi_A >$.

5 Conclusion

In this paper, we have developed the notion of constraint equational programming. This work is a further development of the well known research on the constraint logic programming [14] where the same notion of constraint is introduced to Horn clause logic, as in our case, to Horn equational logic. As shown in [9], CLP and CELP are instances of a general free extension procedure by using the notion of *institution* as the framework of various logics. Moreover, they can be combined directly to form a very powerful paradigm of Horn clause logic with equality and constraints on which some advanced languages, such as Eqlog [10], are based. CELP constitutes a foundation for extensions of equational logic programming systems by taking the extending component as the underlying constraint system. For example, in this framework it is very natural to develop a typed equational system by using a type constraint system such as Smolka's polymorphically order-sorted type system [27] as the underlying constraint system. This motivated our future work on constrained deduction in the typed logic framework.

We also proposed constrained equational deduction system as the operational model for CELP systems. By constrained equational deduction, a wide range of extended equational deduction systems are characterized by imposing various underlying constraint systems. It is also very straightforward to extend our model as a constrained deduction model in the context of Horn clause logic with equality just by adopting the constrained SLD-resolution of CLP [27]. Actually, this rule is implied by constrained paramodulation. This work is motivated by Stickel's work on theory resolution [28], which provides a method to perform resolution proving over special theories. This idea is developed further towards a general constrained resolution model [5] [26] [2] where the notion of constraints becomes very adequate for a uniform treatment of different but related deduction systems in resolution calculi. It has been revealed that there are a lot of inherent relations between symbolic computation and the constraint satisfaction model. For example, in [18], Lassez et.al gave a comprehensive unification theory based on intensive use of the techniques from solving linear arithmetic constraints. Lassez's recent work on the general framework of *constraint querying system* [19]provides many important insights into the integration of symbolic deduction and constraints satisfaction systems by exploiting the common features of the two computational paradigms.

Some early work on the deduction model over equational rewriting systems [15], contextual rewriting [32] also incorporate the hierarchy of deduction. However, these researches are concerned with performing rewriting over particular systems which, in our case, can be systematically treated as constraint systems. It is a very stimulating area to study a general framework of *constrained rewriting* by integrating the notion of constraints into the logic of rewriting [23]. Kirchner's work on constrained equational reasoning [17] developed a way to study the completion problem within a wide range of systems by taking parts of the equational theory as constraint systems. This work is closely related with our work where we consider a more general notion of constraints. We are going to investigate programming transformation ,which concerns the completion problem, in CELP framework. We believe that many results in [17] could be extended directly to CELP. Notice that we assume the constraint systems concerned have their first order axiomatization. This restriction can be removed by adopting a more general notion of free extension [9]. Thus, we may extend the important field of equational deduction to include higher-order objects by taking the higher order unification procedure, such as higher-order transformation system for typed $\lambda - terms$ [8], as the underlying constraint system. Therefore, we hope that the research on CELP will contribute to a rich and rigorous *design space* for understanding and achieve a happy marriage of functional and logic programming.

6 Acknowledgements

We thank Prof. J-L Lassez, Dr. Meseguer for their encouragement on this research. We also thank Mr. Lock for his helpful discussions. We would like to thank all the people in the Phoenix project at Imperial college who contribute to a stimulating environment. We also thank the European Community for their funding.

References

[1] A. Boudet, J-P. Jouannaud, and M. Schmidt-Schauss. *Unification in boolean rings and abelian groups.* Proceedings of LICS'88, 1988.

[2] H-J. Burckert *A Resolution Principle for Clauses with Constraints* Proceedings of CADE 90, July.1990.

[3] R. Burstall and J. Goguen *The Semantics of Clear, A Specification Language* LNCS 86, 1980

[4] N. Dershowitz and D.A. Plaisted,*Equational Programming* in Machine Intelligence , 1986.

[5] J.Darlington, Y.Guo and H.Lock *Developing Phoenix Design Space: A Uniform Framework of Declarative Languages* Phoenix Project Report, Imperial College May, 1990.

[6] N. Dershowitz and J.P. Jouannaud. *Rewrite systems* in Van Leuven, editor, *Handbook of Theoretical Computer Science* North Holland, 1990

[7] J.H.Gallier and W.Snyder *A General Complete E-Unification Procedure* Proceedinds of RTA'87 LNCS 256 1987.

[8] W.Snyder and J.H.Gallier *Higher Order Unification Revisited: Complete Set of Transformations* Journal of Symbolic Computation 8, 1989.

[9] J. Goguen and R.Burstall *INSTITUTIONS: Abstract Model Theory for Specification and Programming* TR CSLI-85-30 SRI, 1985.

[10] J. Goguen and J. Meseguer *Equality, Types, Modules, and (why not?) Generics for Logic Programming* Journal of Logic Programminglp,Vol 2, 1984.

[11] J. Goguen and J. Meseguer *Universal Realization, Persistent interconnection and Implementation of Abstract Modules* Proceedinds of 9th ICALP, LNCS 140 , 1982

[12] G.Huet, D. C. Oppen *Equations and Rewrite Rules* In: Formal Languages: Perspectives and Open Problems , Academic Press, 1980

[13] S. Hölldobler *Foundational of Equational Logic Programming* LNCS 353 1989.

[14] J. Jaffer , J. Lassez and M. Maher *Constraint Logic Programming* Proc. of 14th ACM Symp. POPL, 1987

[15] J.P.Jouannaud, C. Kirchner and H. Kirchner *Incremental Construction of Unification Algorithms in Equational Theories* ICALP'83 LNCS 154 1983.

[16] S.Kaplan *Conditional Rewrite Rules* Theoretical Computer Science Vol.33, pp175-193, 1984.

[17] C. Kirchner and H. Kirchner *Constrained Equational Reasoning* Proceedings of The ACM International Symposium on Symbolic and Algebraic Computation. July 1989.

[18] J. L. Lassez, M. Maher, and K. Marriot. *Unification revisited.* in J. Minker, editor, *Foundations of Deductive Databases and Logic Programming* . Morgan-Kaufman, 1988

[19] J.L.Lassez.*Querying Constraints* IBM technical report, Jan 1990

[20] J.W. Lawvere *Functorial Semantics of Algebraic Theories* Proceedings, National Academy of Science USA, 1963.

[21] J. Meseguer *General Logics* SRI-I-CSL-89-5 Mar,1989.

[22] J. Goguen and J. Meseguer *Models and Equality for Logic Programming* Proceedings TAPSOFT 87, LNCS 250, 1987.

[23] J. Meseguer *Rewriting as a Unified Model of Concurrency* SRI–CSL-90-02R Feb,1990.

[24] A. Martelli and U. Montanari *An Efficient Unification Algorithm* ACM Transactions on Programming Languages and Systems 4. 1982.

[25] G.D.Plotkin *Building-In Equational Theories* In *Machine Intelligence* 7 73-90, 1972.

[26] J.W. Roach, R.Sundararajan and L.T.Waston *Replacing Unification by Constraint Satisfaction to Improve Logic Program Expressiveness* Journal of Automated Reasoning 6, 51-75, 1990.

[27] G.Smolka *Logic Programming over Polymorphically Order-Sorted Types* Ph.D Thesis Universitat Kaiserslautern 1989.

[28] M.E.Stickel *Automated Deduction by Theory Resolution* Journal of Automated Reasoning 1, 333-356 1985.

[29] M. Schmidt-Schauss *Unification in a Combination of Arbitrary Disjoint Equational Theories* Proceedings of CADE 1988.

[30] W.Snyder and C.Lynch *An Inference System for Horn Clause Logic with Equality : A Foundation for Conditional E-Unification and for Equality in Logiv Programming* Proceedings of CTRS90, June 1990

[31] J.A.Robinson and L.Wos *Maximal Models and Refutation Completeness: Semidecision Procedures in Automatic Theorem Proving* In *Word Problem* P609 –639, 1973.

[32] H. Zhang and J.L. Remy *Contextual Rewriting* Proceedings of RTA 85' LNCS 202 1985.

Higher-Order Unification, Polymorphism, and Subsorts
(Extended Abstract)

Tobias Nipkow*
University of Cambridge
Computer Laboratory
Pembroke Street
Cambridge CB2 3QG
England
Tobias.Nipkow@cl.cam.ac.uk

Abstract

This paper analyzes the problems that arise in extending Huet's higher-order unification algorithm from the simply typed λ-calculus to one with type variables. A simple, incomplete, but in practice very useful extension to Huet's algorithm is discussed. This extension takes an abstract view of types. As a particular instance we explore a type system with ML-style polymorphism enriched with a notion of *sorts*. Sorts are partially ordered and classify types, thus giving rise to an order-sorted algebra of types. *Type classes* in the functional language Haskell can be understood as sorts in this sense. Sufficient conditions on the sort structure to ensure the existence of principal types are discussed. Finally we suggest a new type system for the λ-calculus which may pave the way to a complete unification algorithm for polymorphic terms.

1 Introduction

Huet's algorithm for higher-order unification in the simply typed λ-calculus [11] forms the core of a number of theorem provers based on higher-order logic [1, 17, 14]. In this paper we discuss extensions to this unification algorithm which permit more expressive type systems: ML-style polymorphism enriched with a notion of sorts constraining type variables. It turns out that the step from the simply typed λ-calculus considered by Huet to one with type variables is nontrivial. The unification algorithm we analyze is in fact incomplete. A complete extension remains an open problem which is briefly considered at the end. On a more positive note, the introduction of sorts is completely orthogonal to the higher-order unification algorithm and causes no problems whatsoever.

It has to be stressed that this is not the first extension of Huet's algorithm to a polymorphic type system. Nadathur [14] reports on an implementation of the programming language λProlog which features polymorphism. His extensions seem to be very similar to the ones analyzed in this paper. In particular he faces the same problem with completeness that we have. However, the author is not aware of a *formal* treatment of these issues, in particular not in connection with an order-sorted type system.

All the features described in this paper have recently been implemented in the generic theorem prover Isabelle. Isabelle is generic in the sense that it can be parameterized with the intended object-logic. Its meta-logic is a fragment of higher-order logic and its inference

*Research supported by ESPRIT BRA 3245, *Logical Frameworks*.

mechanism is based on higher-order unification. Detailed descriptions of the monomorphic version of Isabelle can be found in [18, 19]. Before we go into technical details, the motivation for the extensions to the type system in presented.

Although ML-style polymorphism hardly needs to be motivated in a programming language context, some simple examples from the realm of generic theorem provers seem appropriate. The basic reason for the introduction of polymorphism is the desire to shift type-checking in many-sorted logics from the object to the meta level. A typical example is the definition of an ordinary many-sorted first-order logic. We need to model a type of formulae $form$, and an unbounded number of different types of terms like natural numbers, reals, lists etc. In a polymorphic type system, the operation of equality between terms has type $\alpha \to \alpha \to form$ where α is a type variable. The built-in type checker ensures that equality always compares terms of the same type.

In a monomorphic type system, we would have to declare a new equality $=_s$ of type $s \to s \to form$ for each new type of terms s, and give the same inference rules over and over again. This is clearly impractical. Alternatively, we can formalize types as part of the object-logic. In this case terms have to be decorated with their types, the type checking rules are part of the inference system and type checking is part of a proof. This is the approach taken in all logics defined in the monomorphic version of Isabelle. If a logic's type system is sufficiently expressive to become undecidable, for example Intuitionistic Type Theory [12], this is in fact the only possible approach. For simpler systems, however, ML-polymorphism is clearly preferable.

Having adopted polymorphism, we need to tame its power. Initially one might be tempted to declare equality and universal quantification as follows:

$$\begin{aligned} =:&\quad \alpha \to \alpha \to form \\ \forall:&\quad (\alpha \to form) \to form \end{aligned}$$

The intention is that α ranges only over different types of terms, but certainly not over formulae or arbitrary function types (if the logic is supposed to be first-order). However, there is nothing in this declaration to enforce those constraints. As a consequence, some rather surprising inferences are possible. In first-order logic $\forall$-elimination is

$$\frac{\forall x.\ P(x)}{P(t)}$$

where $x : \alpha$, $t : \alpha$, $P : \alpha \to form$ and α are variables. Using the substitution $\{\alpha \mapsto form, t \mapsto Q, P \mapsto \lambda x.x\}$ we obtain the derived rule

$$\frac{\forall x.\ x}{Q}.$$

In a first order-logic, $\forall x.\ x$ is ill-formed. This formula could only arise because of the instantiation of α by $form$. In a higher-order logic, $\forall x.\ x$ is in fact a valid formula, usually identified with falsity, and the rule expresses *ex falso quodlibet*.

In order to avoid such pitfalls, Isabelle's types are classified by so called *sorts*. The sort of a type variable defines the subset of types the variable ranges over, thus prohibiting undesirable instantiations. Details are given in Section 4. Note that a similar mechanism for restricted polymorphism is embodied in ML's *equality types* and, in a more general form, in Haskell's *type classes* [24, 16].

After fixing the basic notation in Section 2, Section 3 presents the extension to Huet's algorithm, taking an abstract view of types. Section 4 focuses on the type system, introducing polymorphism, sorts and subsorts. Finally, we speculate on a way to obtain a complete unification algorithm for the polymorphic case.

2 Preliminaries

The reader should be familiar with the basic notations and facts of equational logic as found for example in Dershowitz and Jouannaud [4]. We are broadly consistent with their notation, except that we treat substitutions as ordinary functions. Function composition is denoted by $\circ$: $(f \circ g)(x) = f(g(x))$. The rest of this section presents the notation for the typed λ-calculus with α, β and η equality. For more details see [23, 10].

For the time being we take an abstract view of types with just enough detail to describe the unification process. Section 4 gives the full description of the intended type system. This separation is possible because the unification of terms and types proceeds largely independently.

The set $\mathcal{T}$ of all *types* is a subset of the free algebra over some set of type constructors, including the function-space constructor "$\rightarrow$", generated by a set of type variables $\mathcal{V}_\mathcal{T}$. The idea is that the types themselves form a sorted algebra and $\mathcal{T}$ is the set of all "well-sorted", i.e. well-formed types. Types whose outermost constructor is "$\rightarrow$" are called *function types*, all other types are called *base types*. The letters σ, τ and γ represent types. Function types associate to the right: $\sigma \rightarrow \tau \rightarrow \gamma$ means $\sigma \rightarrow (\tau \rightarrow \gamma)$. Instead of $\sigma_1 \rightarrow \cdots \rightarrow \sigma_n \rightarrow \tau$ we write $\overline{\sigma_m} \rightarrow \tau$. The latter form is used only if τ is a base type. The type variables occurring in a type σ are denoted by $\mathcal{V}_\mathcal{T}(\sigma)$.

$\mathcal{T}$ comes with a set of *type substitutions* $\mathcal{S}_\mathcal{T}$ which is some subset of the set of all possible mappings from $\mathcal{V}_\mathcal{T}$ to $\mathcal{T}$ and contains $\{\}$, the empty map. Again, the sorted nature of types has to be taken into account. Elements of $\mathcal{S}_\mathcal{T}$ are denoted by Θ and Δ. The quasi-order $\leq$ on terms and substitutions is defined in the usual way. Correspondingly, there is a unification function $\mathcal{U} : \mathcal{T} \times \mathcal{T} \rightarrow 2^{\mathcal{S}_\mathcal{T}}$ which returns a complete set of unifiers: for any two $\sigma, \tau \in \mathcal{T}$ and any $\Delta \in \mathcal{S}_\mathcal{T}$ with $\Delta(\sigma) = \Delta(\tau)$ there is a $\Theta \in \mathcal{U}(\sigma, \tau)$ such that $\Theta \leq \Delta\ [\mathcal{V}_\mathcal{T}(\sigma, \tau)]$. For simplicity we assume that all variables in the range of a substitution returned by $\mathcal{U}$ are "new".

Terms are generated from a set of *free variables* V, a set of *bound variables* X, and a set of *constants* C by λ-abstraction and application. Free variables are denoted by F, G, and H, bound variables by x, y and z, and constants by a, b, and c. Terms are denoted by s, t, and u. The inductive definition of *typed terms* in a context Γ, which is just a mapping from X to $\mathcal{T}$, is as follows:

$$\frac{\Gamma(x) = \tau}{\Gamma \vdash x : \tau} \qquad \Gamma \vdash F^\tau : \tau \qquad \Gamma \vdash c^\tau : \tau$$

$$\frac{\Gamma \vdash s : \sigma \rightarrow \tau \quad \Gamma \vdash t : \sigma}{\Gamma \vdash (s\ t) : \tau} \qquad \frac{\Gamma \circ \{x \mapsto \sigma\} \vdash s : \tau}{\Gamma \vdash (\lambda x^\sigma . s) : \sigma \rightarrow \tau}$$

Note that free variables and constants on their own are not legal terms. They have to be tagged with some type, as in the second and third formation rule above. In particular we assume that all occurrences of a free variable in a term are decorated with the same type. There is no corresponding requirement for constants because they can be polymorphic. A term t is *well-typed* if $\{\} \vdash t : \tau$ holds for some $\tau \in \mathcal{T}$. Instead of $\{\} \vdash t : \tau$ we simply write $t : \tau$. In the sequel we consider only well-typed terms, which means in particular that there are no "loose" bound variables: x, for example, is illegal. Instead of $\lambda x_1 . \ldots \lambda x_n . s$ we write $\lambda x_1, \ldots, x_n . s$ or just $\lambda \overline{x_n} . s$. Similarly we write $t(u_1, \ldots, u_n)$ or just $t(\overline{u_n})$ instead of $(\ldots(t\ u_1)\ldots)u_n$. The free variables in t are denoted by $\mathcal{V}(t)$, the type variables by $\mathcal{V}_\mathcal{T}(t)$. Type decorations are omitted if they are not important.

We assume α, β and η conversion on terms. Relying on the strong normalization property of the typed λ-calculus we assume that terms are in β-normal form. We also ignore α conversion by working with α-equivalence classes of terms, using the generic bound variable names x and y. The η-expanded form [23] of the β-normal form of a term t is denoted by $t{\downarrow}$.

A term $\lambda \overline{x_m} . s(\overline{u_k})$ in β-normal form is called *rigid* if $s \in X \cup C$, and *flexible* if $s \in V$.

Substitutions on terms are defined as mappings from free variables to λ-terms in the usual way. They are denoted by θ and δ. Applying a type substitution to a term means applying it to all type decorations in the term.

A *unifier* of two terms s and t is a pair of substitutions $\langle\Theta,\theta\rangle$ on types and terms respectively such that $\theta(\Theta(s))$ and $\theta(\Theta(t))$ are equivalent modulo α, β and η-conversion.

3 Higher-Order Unification

The starting point for most work on higher-order unification is Huet's algorithm [11] which enumerates a complete and minimal set of unifiers with respect to the *simply typed* λ-calculus [10]. As higher-order unification is undecidable in general [9], this is the best one can hope for. Snyder and Gallier [23] have recently reformulated Huet's algorithm in terms of inference rules, which simplifies the presentation. Their version will be our reference point.

Isabelle was originally based on the simply typed λ-calculus and used Huet's algorithm. A recent extension permits *polymorphic* constants in the ML sense. As a consequence, terms may now contain type variables as well as term variables, both of which may need to be instantiated during the unification process. This is a significant departure from the original problem.

Example 3.1 Let τ be some base type and α a type variable, let $a:\tau$ be a constant and $G:\alpha$ and $F:\alpha\to\tau$ be two variables. The following matching problem is given:

$$F(G)=a \tag{1}$$

Let us first look at (some of) the infinitely many solutions to this problem. We know that any instantiation of α must be of the form

$$\tau_1\to\cdots\to\tau_n\to\gamma$$

for suitable types $\tau_1,\ldots,\tau_n$ and γ. In particular we may assume that γ is not a function type. Without any assumptions about α we get the solution $F=\lambda x.a$. In terms of Huet's algorithm this is the solution obtained after a single imitation step. In fact, for any instantiation of α where $\gamma\neq\tau$, this is the only solution.

If we also assume that $\gamma=\tau$, we get the following infinite set of solutions, indexed by n:

$$F=\lambda x.x(A_1(x),\ldots,A_n(x)),\quad G=\lambda x_1,\ldots,x_n.a$$

The A_i are new free function variables.

Further instantiations of α yield yet more solutions. For example $\alpha=\tau\to\tau$ alone has an infinite set of independent solutions:

$$F=\lambda x.x^k(a),\quad G=\lambda x.x$$

is a solution for any natural number k, where x^k is the k-fold composition of x.

This shows quite drastically that type variables introduce a new degree of freedom into unification problems. Different type instantiations give rise to completely independent sets of solutions with greatly varying cardinality (finite vs. infinite). This is not in itself surprising, but it raises the question how this new freedom can be incorporated into the unification algorithm. We shall try to "lift" Huet's algorithm to this new setting by replacing equality tests on types by unification. The following example shows why such a simple-minded extension is problematic. You need to be familiar with Huet's algorithm to follow it.

Example 3.2 We return to the unification problem in Example 3.1. As mentioned above, a single imitation step finds the solution $F = \lambda x.a$ which requires no type instantiations. Projection, however, is different. F can be a projection only if the *result type* of its argument G is τ. This means that projection is applicable for any instance of α of the form $\tau_1 \to \cdots \to \tau_n \to \tau$. Unfortunately, the set of these types does not have a finite representation in terms of function types and type variables. Hence we can either try all possible type instantiations (which is complete but completely impractical) or sacrifice completeness for efficiency. In the latter case we pick the simplest type instantiation that permits projection. In this example it means unifying α and τ. This results in the instantiation of α by τ and the single additional solution

$$F = \lambda x.x(A(x)), \quad G = \lambda x.a.$$

This is obviously incomplete. In particular, if (1) is just one in a set of equations to be solved simultaneously, some of the other equations may well require α to be instantiated by a different type. In that case we would find no solution unless we also backtrack over possible type instantiations.

Despite its incompleteness, the solution outlined in the above example has been adopted in both λProlog and Isabelle. In the sequel we present a formal treatment of this extension to Huet's algorithm.

3.1 The Algorithm

This section formalizes the idea of "lifting" Huet's algorithm. Our treatment is very close to that of Snyder and Gallier [23]. In fact, parts of our algorithm are identical to the one they present. The price we pay is the incompleteness discussed above.

The unification algorithm is presented as a collection of conditional rewrite rules on pairs $\langle\Theta, D\rangle$, where Θ is a type substitution and D a multiset of unification problems $s =^? t$. D is called a *system* in the sequel. We assume that all occurrences of a free variable in D are decorated with the same type, a property that is preserved by the inference rules below. Note that Θ merely records and accumulates the type substitutions, whereas D represents both the current multiset of unification problems and the term substitutions obtained so far.

$$\frac{s:\sigma \quad t:\tau \quad \sigma \neq \tau \quad \Delta \in \mathcal{U}(\sigma,\tau)}{\langle\Theta, \{s =^? t\} \cup D\rangle \Longrightarrow \langle\Delta \circ \Theta, \Delta(\{s =^? t\} \cup D)\rangle} \tag{T}$$

$$\langle\Theta, \{s \stackrel{?}{=} s\} \cup D\rangle \Longrightarrow \langle\Theta, D\rangle \tag{D}$$

$$\frac{t \in X \cup C}{\langle\Theta, \{\lambda\overline{x_k}.t(\overline{s_n}) =^? \lambda\overline{x_l}.F(\overline{u_m})\} \cup D\rangle \Longrightarrow \langle\Theta, \{\lambda\overline{x_l}.F(\overline{u_m}) =^? \lambda\overline{x_k}.t(\overline{s_n})\} \cup D\rangle} \tag{O}$$

$$\frac{F \notin \mathcal{V}(t) \quad F \in \mathcal{V}(D)}{\langle\Theta, \{\lambda\overline{x_k}.F(\overline{x_k}) =^? t\} \cup D\rangle \Longrightarrow \langle\Theta, \{\lambda\overline{x_k}.F(\overline{x_k}) =^? t\} \cup \{F \mapsto t\}(D)\rangle} \tag{V}$$

$$\frac{\sigma \neq \tau \quad \Delta \in \mathcal{U}(\sigma,\tau)}{\begin{array}{l}\langle\Theta, \{\lambda\overline{x_k}.c^\sigma(\overline{s_m}) =^? \lambda\overline{x_l}.c^\tau(\overline{u_n})\} \cup D\rangle \Longrightarrow \\ \langle\Delta \circ \Theta, \Delta(\{\lambda\overline{x_k}.c^\sigma(\overline{s_m}) =^? \lambda\overline{x_l}.c^\tau(\overline{u_n})\} \cup D)\rangle\end{array}} \tag{T$_S$}$$

$$\frac{t \in X \cup C}{\begin{array}{c}\langle \Theta, \{\lambda\overline{x_k}.t(\overline{s_n}) =^? \lambda\overline{x_k}.t(\overline{u_n})\} \cup D\rangle \Longrightarrow \\ \langle \Theta, \{\lambda\overline{x_k}.s_i =^? \lambda\overline{x_k}.u_i \mid i \in \{1,\ldots,n\}\} \cup D\rangle\end{array}} \quad \text{(S)}$$

$$\frac{\sigma = \overline{\sigma_{m+i}} \to \sigma' \quad \tau = \overline{\tau_{n+j}} \to \sigma'}{\begin{array}{c}\langle \Theta, \{\lambda\overline{x_k}.F^\sigma(\overline{s_m}) =^? \lambda\overline{x_l}.c^\tau(\overline{u_n})\} \cup D\rangle \Longrightarrow \\ \langle \Theta, \{F =^? \lambda\overline{x_{m+i}}.c^\tau(\overline{H_{n+j}(\overline{x_{m+i}})}), \lambda\overline{x_k}.F^\sigma(\overline{s_m}) =^? \lambda\overline{x_l}.c^\tau(\overline{u_n})\} \cup D\rangle\end{array}} \quad \text{(I)}$$

$$\frac{t \in X \cup C \quad \sigma = \overline{\sigma_{m+i}} \to \sigma' \quad \sigma_j = \overline{\tau_p} \to \tau \quad \sigma' \neq \tau \quad \Delta \in \mathcal{U}(\sigma', \tau)}{\begin{array}{c}\langle \Theta, \{\lambda\overline{x_k}.F^\sigma(\overline{s_m}) =^? \lambda\overline{x_l}.t(\overline{u_n})\} \cup D\rangle \Longrightarrow \\ \langle \Delta \circ \Theta, \Delta(\{\lambda\overline{x_k}.F^\sigma(\overline{s_m}) =^? \lambda\overline{x_l}.t(\overline{u_n})\} \cup D)\rangle\end{array}} \quad (\mathrm{T}_P)$$

$$\frac{t \in X \cup C \quad \sigma = \overline{\sigma_{m+i}} \to \tau \quad \sigma_j = \overline{\tau_p} \to \tau}{\begin{array}{c}\langle \Theta, \{\lambda\overline{x_k}.F^\sigma(\overline{s_m}) =^? \lambda\overline{x_l}.t(\overline{u_n})\} \cup D\rangle \Longrightarrow \\ \langle \Theta, \{F^\sigma =^? \lambda\overline{x_{m+i}}.x_j(\overline{H_p(\overline{x_{m+i}})}), \lambda\overline{x_k}.F^\sigma(\overline{s_m}) =^? \lambda\overline{x_l}.t(\overline{u_n})\} \cup D\rangle\end{array}} \quad \text{(P)}$$

The rules (T), (T_S), and (T_P) unify types and are new. Rule (O) orients rigid-flexible pairs. The remaining rules unify terms and are almost identical to the ones given by Snyder and Gallier. (T) unifies the types of two terms. (T_S) unifies the types of two rigid heads. (T_P) unifies the result type of a rigid term with the result type of one of its arguments. (T_S) and (T_P) may need to be applied before (S) and (P) respectively in order to make types meet. We assume that (I) and (P) are immediately followed by (V), eliminating F. The free variables H_r in (I) and (P) are assumed to be new.

Huet's insight, which turned higher-order unification from a mere curiosity into computational reality, was that flexible-flexible pairs need not be solved. This simplifies the algorithm but requires some new terminology. The following definitions are broadly consistent with the literature.

A system D is in *presolved* form if for every unification problem $s =^? t$ in D

- either $s = \lambda\overline{x_m}.F(\overline{x_m})$ and F occurs neither in t nor in the rest of D, or both s and t are flexible, and
- s and t have the same type.

If D is in presolved form, define

$$\vec{D} = \{F \mapsto t \mid (\lambda\overline{x_k}.F(\overline{x_k}) \stackrel{?}{=} t) \in D\}.$$

Like Gallier and Snyder [23] we define $\cong$ to be the least congruence on well-typed terms of the same type containing all flexible-flexible pairs. The pair $\langle\Theta, \theta\rangle$ is a *preunifier* of D if $\theta(\Theta(s))\!\downarrow \cong \theta(\Theta(t))\!\downarrow$ for all pairs $s =^? t$ in D.

The next simple lemma shows that the terminology is consistent in that presolved forms give rise to preunifiers which in turn extend to unifiers. The precise statement requires some more notation. With every base type σ we associate some arbitrary but fixed free variable F_σ, and with every type $\tau = \overline{\sigma_m} \to \sigma$ a term $\hat{u}_\tau = \lambda\overline{x_m^{\sigma_m}}.F_\sigma$ where the x_i are all distinct.

Lemma 3.1 *If D is in presolved form, $\langle\{\}, \vec{D}\rangle$ is a preunifier of D. If $\langle\Theta, \theta\rangle$ is a preunifier of D, then $\langle\Theta, \theta \cup \xi\rangle$ is a unifier of D, where*

$$\xi = \{F \mapsto \hat{u}_\sigma \mid F^\sigma \in \mathcal{V}(D) - dom(\theta)\}$$

The following soundness and completeness theorems can be proved along the same lines as their counterparts in [11, 23]:

Theorem 3.1 *If* $\langle\Theta, D\rangle \Longrightarrow^* \langle\Theta', D'\rangle$ *and* D' *is in presolved form, then* $\langle\Theta', \vec{D'}\rangle$ *is a preunifier of* D.

Theorem 3.2 *Let* $\langle\Delta, \delta\rangle$ *be a unifier of a system* D *such that no type in the range of* Δ *is a function type. Then there exists a presolved form* $\langle\Theta, D'\rangle$ *such that* $\langle\{\}, D\rangle \Longrightarrow^* \langle\Theta, D'\rangle$, $\Theta \leq \Delta\ [\mathcal{V}_T(D)]$, $\vec{D'} \leq \delta\ [\mathcal{V}(D)]$, *and* δ *is a unifier of all flexible-flexible pairs in* D'.

The completeness theorem is rather more conservative than the actual algorithm. In many practical cases it will find all solutions, although some of them require type variables to be instantiated by function types.

As with all unification algorithms expressed as rewrite rules on collections of unification problems, there are two kinds of nondeterminism during execution: the choice between different transformations that apply ("don't know") and between different unification problems they can be applied to ("don't care"). Ideally, all strategies for selecting unification problems should be equivalent with respect to completeness. For example Huet [11] and Elliot [5] show this quite explicitly for their algorithms. This result also holds with respect to the limited completeness theorem above. However, our algorithm is in general not just incomplete but also sensitive to the selection of unification problems. The point is that (T_P) commits to a particular type instantiation out of an infinite set, thus reducing the solution space. Hence the application of (T_P) should be delayed as long as possible in order to minimize incompleteness. More precisely, one can give an operational characterization of completeness: the above set of transformation rules enumerates a complete set of unifiers for a particular unification problem if (T_P) need not be applied in such a way that either γ or τ are type variables. This means that no solutions are lost if the application of (T_P) to a particular unification problem can always be delayed until the types are sufficiently instantiated.

3.2 Optimizations

In this section we briefly consider some simple optimizations of the above rather high-level algorithm. First we look at issues connected with conversion of λ-terms. In the description above, we have abstracted from all three conversion rules by considering α-equivalence classes of η-expanded β-normal forms. Here are some remarks on how to minimize the work of obtaining these normal forms:

α If De Bruijn notation [3] is used, α-conversion can be ignored completely.

β The question here is mainly how much of the normal form to compute when. As we can see from the rules, full β-normal form is not required — head-normal form will do. If all terms are in β-normal form initially, only the application of (V) entails further normalization.

η In contrast to unification for the simply typed λ-calculus, our transformation rules do not leave terms in η-expanded form because type-variables may become instantiated. Fortunately, only (S) requires the η-expansion of the head, thus minimizing the need for η-expansion. In Isabelle we have gone so far as banning the instantiation of type variables by function types during the unification process, thus removing the need for η-expansion completely (provided the input is in η-expanded form). However, preliminary experiments suggest that in practice this is a rather drastic restriction, requiring frequent explicit type instantiations by the user.

Further important optimizations concern the order in which rules are applied. In contrast to (T_P), (T) and (T_S) should be performed as soon as possible, to detect nonunifiability by type-clashes. It is in fact sufficient to apply (T) to the input because the remaining rules maintain the invariant that the two terms of a unification problem have the same type. The rules for unification of terms, simplification (S), imitation (I), and projection (P), should be tried in the order embodied in Huet's algorithm.

4 Polymorphism and Subsorts

This section fills in the details left open by the rather general discussion of types in Section 2. As indicated in that section and motivated in the introduction, type variables need to be constrained in order to capture certain restrictions, for example the first-order nature of some proof system. In Isabelle, type variables are constrained by introducing a third level into the system: so far we had terms, and types which qualify terms. Now we add *sorts* which qualify types. This is very reminiscent of *generalized type systems* [2], a relationship that needs further investigation. In contrast to generalized type systems we have a partial order on the sorts, indicating that a subclass of types identified by one sort is contained in the subclass identified by the another sort. This embodies the central ideas behind the notion of *type classes* in the functional language Haskell [24], a relationship that is made precise in [16]. Note that we are dealing with coercion on the level of sorts, not of types. Hence the system is quite different from OBJ [6], where we have a partial order between types[1].

The step from ML-style polymorphism to one with partially ordered sorts leads from a single-sorted to an *order-sorted* algebra of types. Therefore we need some basic vocabulary of order-sorted types. After that we look at order-sorted unification, which fulfills two functions in Isabelle: it is required for type checking [16] and for higher-order unification of polymorphic terms as described in Section 3.

The following definitions are consistent with [21, 22, 25]. Note that the naming conventions established in Section 2 do not all carry over. This is important to keep in mind since the order-sorted *terms* discussed below are the *types* in the λ-calculus discussed in Section 2.

An *order-sorted signature* consists of a set of sorts S, a partial order $\leq$ on S, and a set of function declarations $f : (\overline{s_n})s$. Given a fixed signature and an S-sorted set $V = \bigcup_{s \in S} V_s$ of disjoint sets of variables, a set of typed terms is defined inductively:

$$\frac{x \in V_s \quad s \leq s'}{x : s'} \qquad\qquad \frac{f : (\overline{s_n})s \quad s \leq s' \quad \forall i.\ t_i : s_i}{f(\overline{t_n}) : s'}$$

The definitions of substitutions, unifiers, complete sets of unifiers etc. are straightforward generalizations of the unsorted case and can be found for example in [21]. In the sequel the ordering $\leq$ is extended from S to S^* in the canonical componentwise way.

A signature is called *regular* [22] if every term has a least sort. Regularity is decidable for finite signatures:

Theorem 4.1 (Smolka et al. [22, 25]) *A signature* $(S, \leq, \Sigma)$ *is regular iff for every* $f \in \Sigma$ *and* $w \in S^*$ *the set* $\{s \mid \exists w' \geq w.\ f : (w')s\}$ *either is empty or contains a least element.*

Regularity is important because it implies that order-sorted unification is finitary. Order-sorted unification in non-regular signatures may be infinitary [22].

As in the section on higher-order unification, the algorithm is presented as a rewrite system on multisets of equations. The following rules are similar to those in Jouannaud and Kirchner's

[1] Which happen to be called *sorts* in OBJ, an unfortunate confusion.

survey [8]. With respect to some fixed signature define

$$\begin{aligned} s_1 \wedge s_2 &= max\{s \mid s \leq s_1 \wedge s \leq s_2\} \\ D(f,s) &= max\{w \mid \exists s'.\ f:(w)s' \wedge s' \leq s\} \end{aligned}$$

If $E = \{x_1 =^? t_1, \ldots, x_n =^? t_n\}$ such that all x_i are distinct and no x_i occurs in one of the t_j, E is in *solved form* and $\vec{E} = \{x_1 \mapsto t_1, \ldots, x_n \mapsto t_n\}$ is a most general unifier of E.

$$\{t \stackrel{?}{=} t\} \cup E \Longrightarrow E$$

$$\{f(\overline{r_n}) \stackrel{?}{=} f(\overline{t_n})\} \cup E \Longrightarrow \{r_1 \stackrel{?}{=} t_1, \ldots, r_n \stackrel{?}{=} t_n\} \cup E$$

$$\frac{t \notin V}{\{t =^? x\} \cup E \Longrightarrow \{x =^? t\} \cup E}$$

$$\frac{x \in V_s \cap \mathcal{V}(E) \quad x \notin \mathcal{V}(t) \quad t:s}{\{x =^? t\} \cup E \Longrightarrow \{x =^? t\} \cup \{x \mapsto t\}(E)}$$

$$\frac{x \in V_s \quad y \in V_{s'} \quad s < s'}{\{x =^? y\} \cup E \Longrightarrow \{y =^? x\} \cup E}$$

$$\frac{x \in V_s \quad y \in V_{s'} \quad s \not\leq s' \quad s' \not\leq s \quad z \in V_{s''} \quad s'' \in s \wedge s' \quad z \notin \{x,y\} \cup \mathcal{V}(E)}{\{x =^? y\} \cup E \Longrightarrow \{x =^? z, y =^? z\} \cup E}$$

$$\frac{x \in V_s \quad n > 0 \quad \overline{s_n} \in D(f,s) \quad \forall i.\ x_i \in V_{s_i} \wedge x_i \notin \{x\} \cup \mathcal{V}(t_1, \ldots, t_n, E)}{\{x =^? f(\overline{t_n})\} \cup E \Longrightarrow \{x =^? f(\overline{x_n}), x_1 =^? t_1, \ldots, x_n =^? t_n\} \cup E}$$

This algorithm is less deterministic but essentially equivalent to the ones given by Schmidt-Schauß [20] and Waldmann [25]. Adapting their results one can easily show:

Theorem 4.2 *For any regular signature the set of all $\vec{S}$ such that $E \Longrightarrow^* S$ and S is in solved form is a complete set of unifiers for E.*

It follows that in finite regular signatures, where $s \wedge s'$ and $D(f,s)$ must be finite, finite complete sets of unifiers always exist.

The remainder of this section focuses on the existence of single most general unifiers. Since our terms represent types, this happens to coincide with the existence of *principal types*. A lack of principal types is undesirable for two reasons: expressions may have any finite number of incomparable types, making them rather hard to comprehend from a user's point of view; finite sets of type unifiers increase the search space for higher-order unification.

Waldmann [25] characterizes *unitary signatures*, i.e. those leading to unitary unification problems[2]. He calls a partial order $(S, \leq)$ *downward complete* iff the set $s \wedge s'$ contains at most one element.

Theorem 4.3 (Waldmann [25]) *A regular signature is unitary iff it is downward complete and for all s, w, and $f:(w^0)s^0$ such that $w \leq w^0$ and $s \leq s^0$ the set*

$$W(f,w,s) = \{w' \mid \exists \overline{w}, \overline{s}.\ f:(\overline{w})\overline{s} \wedge \overline{s} \leq s \wedge w' \leq w \wedge w' \leq \overline{w}\}$$

either is empty or contains a greatest element.

[2] The characterization results by Meseguer et al. [13] do not immediately apply to our situation because they admit unsorted variables in their unification problems.

Unfortunately, the definition of the set $W(f, w, s)$ is not very intuitive, difficult to check, and hence unsuitable as a guideline for users.

The programming language Haskell [7] solves this problem by imposing a number of severe context conditions which are sufficient to guarantee principal types. Although they are not necessary, they make a lot of sense from a programming language point of view. Their essence is captured in the next lemma, which follows easily from Lemma 4.2 below.

Lemma 4.1 *A signature is unitary if it is finite, regular, downward complete,*

- injective*:* $f : (w)s$ *and* $f(w')s$ *imply* $w = w'$*, and*
- subsort reflecting*:* $f : (w')s'$ *and* $s' \leq s$ *imply the existence of* w *such that* $w' \leq w$ *and* $f : (w)s$*.*

Isabelle uses a slightly weaker criterion called *coregularity*. A signature is *coregular* if for all f and s the set $D(f, s)$ contains at most one element. The following lemma goes back to Schmidt-Schauß and is also given by Smolka et al. [22].

Lemma 4.2 *Signatures which are finite, regular, downward complete, and coregular are unitary.*

This lemma follows directly from the correctness of the unification algorithm above because downward completeness and coregularity eliminate all essential nondeterminism from that algorithm.

It is easy to see that injectivity and subsort reflection together imply coregularity but that coregularity implies neither. Nevertheless, the two criteria are equivalent in the following sense: given a coregular signature Σ, there exists an injective and subsort reflecting signature Σ' on the same sort structure but with declarations

$$\{f : (w)s \mid D_\Sigma(f, s) = \{w\}\}$$

such that $t : s$ holds w.r.t. Σ iff it holds w.r.t. Σ'. This means that the sorting judgements derivable w.r.t. Σ and Σ' are identical. Hence Σ and Σ' are interchangeable for all intents and purposes.

5 Speculation

The algorithm for higher-order unification in the presence of type variables presented in Section 3 is incomplete. The source of this incompleteness is our inability, given a type σ, to find a finite representation for the set of all types $\overline{\sigma_n} \to \sigma$. We propose a radical extension to the type system which overcomes this limitation. We introduce a "pseudo"-product type $*$ and a unit-type 1 which are related to $\to$ as follows:

$$\begin{array}{rclrcl}
\alpha * \beta \to \gamma & = & \alpha \to \beta \to \gamma & \alpha \to \beta * \gamma & = & (\alpha \to \beta) * (\alpha \to \gamma) \\
\alpha * 1 & = & \alpha & 1 * \alpha & = & \alpha \\
1 \to \alpha & = & \alpha & \alpha \to 1 & = & 1 \\
(\alpha * \beta) * \gamma & = & \alpha * (\beta * \gamma) & & &
\end{array}$$

Read from left to right, this is a terminating and confluent rewriting system. Normal forms are products of simple types. Notice that this system, together with commutativity of $*$, axiomatizes isomorphism in Cartesian Closed Categories. The unification problem for the latter and related systems, but not for the one above, has already been considered by Narendran et al. [15].

The point of this new type system is that the set of all types $\overline{\sigma_n} \to \sigma$ can be represented by $\alpha \to \sigma$, where α is a type variable. To obtain $\overline{\sigma_n} \to \sigma$, simply instantiate α with $\sigma_1 * \cdots * \sigma_n$.

The new type constructors $*$ and 1 are accompanied by two new term constructors $\langle\_,\_\rangle$ and $\langle\rangle$ and a number of equalities between λ-terms containing pairs and units.

Whether this approach leads to a complete higher-order unification algorithm for polymorphic λ-terms over these new types is not clear yet, in particular since the unification problem for the types themselves is now highly nontrivial (it contains associative unification as a special case).

Acknowledgements

The author wishes to thank Larry Paulson and David Wolfram for many discussions on higher-order unification and polymorphism, Uwe Waldmann for his patient explanations of order-sorted unification, and Annette Schumann for proof reading.

References

[1] P.B. Andrews, D.A. Miller, E.L. Cohen, F. Pfenning. Automating Higher-Order Logic. In *Automated Theorem Proving: After 25 Years*, AMS Contemporary Mathematics Series 29 (1984), 169–192.

[2] H. Barendregt. Introduction to Generalised Type Systems. To appear in *J. Functional Programming.*

[3] N. G. de Bruijn. Lambda Calculus Notation with Nameless Dummies, a Tool for Automatic Formula Manipulation, with Application to the Church-Rosser Theorem. *Indagationes Mathematicae* **34** (1972), 381–392.

[4] N. Dershowitz, J.-P. Jouannaud. Rewrite Systems. In J. van Leeuwen (editor), *Handbook of Theoretical Computer Science, Vol B: Formal Methods and Semantics*, North-Holland, to appear.

[5] C. Elliot. Higher-Order Unification with Dependent Function Types. *Proc. Rewriting Techniques and Applications*, LNCS 355 (1989), 121–136.

[6] K. Futatsugi, J.A. Goguen, J.-P. Jouannaud, J. Meseguer. Principles of OBJ2. *Proc. 12th ACM Symp. Principles of Programming Languages* (1985), 52–66.

[7] P. Hudak, P. Wadler (Eds.). *Report on the Programming Language Haskell.* Version 1.0, April 1990.

[8] J.-P. Jouannaud, C. Kirchner. *Solving Equations in Abstract Algebras: A Rule-Based Survey of Unification.* Technical report, March 1990.

[9] W. Goldfarb. The Undecidability of the Second-Order Unification Problem. *Theoretical Computer Science* **13** (1981), 225-230.

[10] J.R. Hindley, J.P. Seldin. *Introduction to Combinators and λ-Calculus*, Cambridge University Press (1986).

[11] G.P. Huet. A Unification Algorithm for Typed λ-Calculus. *Theoretical Computer Science* 1 (1975), 27–57.

[12] P. Martin-Löf. *Intuitionistic Type Theory.* Bibliopolis (1984).

[13] J. Meseguer, J.A. Goguen, G. Smolka. Order-Sorted Unification. *J. Symbolic Computation* **8** (1989), 383–413.

Second-Order Unification in the Presence of Linear Shallow Algebraic Equations[1] (Extended Abstract)

Zhenyu Qian

FB3-Dept. of Computer Science, University of Bremen,
D-2800 Bremen 33, Fed. Rep. Germany
E-mail: qian@informatik.uni-bremen.de

Abstract:

This paper presents an algorithm to compute unifiers of simply typed λ-terms w.r.t. the union of α, β and η conversion and a set of first-order equational theory E, where a λ-unification algorithm and an algorithm to check the word problem w.r.t. E are assumed to be given. If the above given algorithms are terminating and complete, then our algorithm is terminating and complete, provided that λ-terms are second-order and E is consistent, linear and shallow. An equational theory is called shallow if its axioms are all of the form $f(x_1, ..., x_m) = g(y_1, ..., y_n)$ or $f(x_1, ..., x_n) = y_1$, where f,g are function symbols, $x_1, ..., x_n$, $y_1, ..., y_n$ are variables and $m,n \geq 0$. Equations defining projections or commutativity of functions are examples of such equational theories.

1. Introduction

Combinations of first order equational theories and higher order simply typed λ-calculus have been suggested by Breazu-Tannen and Meyer [87] as a logic framework for computer programming. Although they have proved that the combinations are conservative extensions of first order algebraic theories, the combinations of computation models, which result in incorporating well-developed techniques in both areas, may be complex: Nadarhur and Miller [88] have considered the extension of logic programming with various typed λ-calculi, Breazu-Tannen [88] has got some results on the Church-Rosser property in combinations of algebras and simply typed λ-calculus.

Snyder [90] has initiated the study of computing unifiers of simply typed λ-terms w.r.t. the union of α, β and η conversion and a set of first-order equational theory E, hereafter λE-unification, by a general unification algorithm in the spirit of "universal unification". His work has provided a general theoretical foundation, but difficult to be used in implementations due to nontermination and nondeterminism. Nipkow and Qian [90] have considered a modular approach. Their algorithms are extensions of Huet's algorithm (cf. [Huet 75] and [Snyder&Gallier 89]) parameterized by the E-unification algorithm. Compared to [Snyder 90], nondeterminism has been greatly reduced.

In fact, much has been known about unifications in particular equational theories. For example, the unification algorithms are known for theories of commutativity, associativity and commutativity, Boolean rings and Abelian groups. The combinations of unification or matching algorithms for disjoint theories have been considered by Stickel [75], Kirchner [86], Herold [86], Tiden [86], Yelick [87], Schmidt-Schauß [89], Nipkow [89], etc. (Cf. Jouannaud&Kirchner [90] for a survey.)

On the other side, we have known some unification algorithms in simply typed λ-calculus. Besides Huet's algorithm, Darlington [71] gave a unification algorithm in a subset of second order logic, and

[1] The research has been partially supported by the Commission of the European Communities under the ESPRIT Programme in the PROSPECTRA Project, ref #390.

[14] G. Nadathur. *A Higher-Order Logic as the Basis for Logic Programming.* PhD Thesis, University of Pennsylvania (1987).

[15] P. Narendran, F. Pfenning, R. Statman. *On the Unification Problem for Cartesian Closed Categories.* Ergo Report 89–082, School of Computer Science, Carnegie Mellon University, September 1989.

[16] T. Nipkow, G. Snelting. *Type Classes and Overloading Resolution via Order-Sorted Unification.* Tech. Rep. 200, University of Cambridge, Computer Laboratory, August 1990.

[17] L.C. Paulson. Natural Deduction as Higher-Order Resolution. *J. Logic Programming* 3 (1986), 237–258.

[18] L.C. Paulson. Isabelle: The Next 700 Theorem Provers. In P. Odifreddi (editor), *Logic and Computer Science*, Academic Press (1990), 361–385.

[19] L.C. Paulson, T. Nipkow: *Isabelle Tutorial and User's Manual*, Tech. Rep. 189, University of Cambridge, Computer Laboratory, January 1990.

[20] M. Schmidt-Schauß. A Many-Sorted Calculus with Polymorphic Functions Based on Resolution and Paramodulation. *Proc. 9th Int. Joint Conf. Artificial Intelligence* (1985), 1162–1168.

[21] M. Schmidt-Schauß. *Computational Aspects of an Order-Sorted Logic with Term Declarations*, LNCS 395 (1989).

[22] G. Smolka, W. Nutt, J.A. Goguen, J. Meseguer. Order-Sorted Equational Computation. In H. Aït-Kaci and M. Nivat (eds.), *Resolution of Equations in Algebraic Structures Vol. 2*, Academic Press (1989), 297–367.

[23] W. Snyder, J. Gallier. Higher-Order Unification Revisited: Complete Sets of Transformations. *J. Symbolic Computation* 8 (1989), 101–140.

[24] P. Wadler, S. Blott. How to Make *ad-hoc* Polymorphism Less *ad hoc*. *Proc. 16th ACM Symp. Principles of Programming Languages* (1989), 60–76.

[25] U. Waldmann: *Unification in Order-Sorted Signatures.* Forschungsbericht 298, Universität Dortmund, Fachbereich Informatik (1989).

Huet and Lang [83] presented a match algorithm in second order simply typed λ-calculus yielding the unique complete set of minimal matches.

This paper is aiming for a modular approach where a λ-unification algorithm and an algorithm to check the word problem w.r.t. E come as as black boxes. Our algorithm is proved to be terminating (complete, resp.), if both given algorithms are terminating (complete, resp.), provided that λ-terms are second-order and E is consistent, linear and shallow.

An equational theory is called *shallow* if its axioms are all of the form $f(x_1, ..., x_m) = g(y_1, ..., y_n)$ or $f(x_1, ..., x_n) = y_1$, where f,g are function symbols, $x_1, ..., x_n, y_1, ..., y_n$ are variables and $m,n \geq 0$. Equations defining projections or commutativity of functions are examples of such equational theories. Note that shallow equational theories need neither be collapse-free nor regular.

An equational theory is called *consistent* if no equations of the form $X=Y$ with distinct variables $X, Y \in V$ are derivable.

An equation is called *linear* if no variables occur more than once on each side of this equation. An equational theory E is called *linear* if its axioms are all linear equations.

Second order terms are terms containing only variables with their types being $\alpha_1 \to ... \to \alpha_n \to \beta$, where $\alpha_1, ..., \alpha_n, \beta$ are base types.

Note that we only need an algorithm to check the word problem in equational theories. Equational unification and term rewritings, for example, are just two effective methods to do such checking.

Although we are restricted in a small class of equational theories, our results can be used to solve some practical unification problems effectively. For example, consider the operation of addition + satisfying the law of commutativity: $+(x, y) = +(y, x)$. The unifier $\theta = (y\backslash 1, F\backslash +)$ can be found for the terms $\lambda x. +(x, y)$ and $\lambda x. F(1, x)$, since $\theta(\lambda x. +(x, y)) = \lambda x. +(x, 1) = \lambda x. +(1, x) = \theta(\lambda x. +(x, y))$.

2. Preliminaries

We assume the familiarity with types and terms (i.e. λ-terms) in simply typed λ-calculus. We use C to denote the set of (function) constants, V the set of (function) variables and E the set of equations.

$\tau(t)$ denotes the type of a term t. $t[s]$ means that term s occurs as a particular subterm in term t. $t[s \leftarrow u]$ denotes the result obtained by replacing the particular subterm s in t by u.

Let O be a syntactic object. The set of free variables in O is denoted by FV(O), the set of constants in O by C(O).

An (*algebraic*) *α-term* consists of only symbols in C(E) and first order variables in V.

α, β and γ stand for arbitrary types, s, t, r, u and v for terms, X, Y, Z, F and G for free variables, x, y and z for bound variables, c, d, f, g and h for constants, σ,θ,ρ for substitutions, dom(σ) for the domain of σ, and introV(σ) for the set of introduced variables $\bigcup_{X \in dom(\sigma)} FV(\sigma(X))$, introC(σ) the set of introduced free constants $\bigcup_{X \in dom(\sigma)}(\sigma(X)) - C(E)$. Note that in σ(t), σ only substitutes the free variables in t and introduces no bound variables.

A unification problem is expressed as a *multi-equation* M, which is a set of terms of the same type. $\{t\} \cup M$ may be denoted by $t=M$. A *system* Γ is a finite set of multi-equations.

For more details, the reader is referred to Snyder [90], Snyder&Gallier [89], Nipkow&Qian [90].

3. Some informal description of our algorithm and the necessary restrictions

An equational unification algorithm in combinations of disjoint equational subtheories may consist of (i) breaking down every original unification problem into simpler subproblems of these subtheories, (ii) solving each subproblem by the given equational unification algorithm in each subtheory and (iii) conbine the solution of the original problem from solutions to subproblems. The situation of λE-unification seems to be much more complicated than this, since simply typed λ-calculus is not disjoint with equational theory E.

Consider, for example, the equation set E = {f(X)=g(X)}. Let Γ = {λx.F(g(X))=λx.f(F(x))} with F∈V be a system. σ = {F<-f} is obviously a λE-unifier. In order to find this unifier, λ-unification should be applied. However, λ-unification fails to effectively find the unifier {F<-f} or {F<-g} until we have E-reduced the system Γ into {λx.F(f(x))=λx.f(F(x))} or {λx.F(g(X))=λx.g(F(x))}.

A temptation to overcome this difficulty is first to rename the second occurrence of F, say into G∈V. The new system is {λx.F(g(x))=λx.f(G(x)), F=G}. λ-unification of λx.F(g(x))=λx.f(G(x)) would produce the solution {F<-f, G<-g}, which satisfies F=G according to E.

Unfortunately the above process does not work for unifying the system Γ' = {λx.F(g(X))=λx.f(f(x))}, where E-reducing Γ' into {λx.F(g(X))=λx.f(g(x))} is necessary before λ-unification.

Note that first E-reducing a system is not a proper solution in general because this step is too non-deterministic: one normally does not know, where and by which the equation E-reduction should be performed.

Our solution is first to replace Σ-symbols by new variables and then to do λ-unification. In this way, the above system Γ = {λx.F(g(X))=λx.f(F(x))} will be first transformed into {λx.F(G(X))=λx.H(F(x)), G=g, H=f}. Then λ-unification of λx.F(G(X))=λx.H(F(x)) may first find the substitution {G<-F, H<-F}. This substitution will transform the system further into {λx.F(F(X)), F=g, F=f}, which is expected to yield the λE-unifier {F<-f} or {F<-g}.

Similarly, the other given system Γ' = {λx.F(g(X))=λx.f(f(x))} above may be transformed into {λx.F(G(X))=λx.H(x), G=g, λx.H(x)=λx.f(f(x))}. λ-unification of λx.F(G(X))=λx.H(x) may yield {λx.H(x)<-λx.F(G(x))} and transform the system into {λx.F(G(X)), G=g, λx.F(G(x))=λx.f(f(x))}. λ-unification of λx.F(G(x))=λx.f(f(x)) may return the whole λE-unifier {F<-f,G<-f}, which satisfies G=g according to E.

To summarize, the first two steps of our algorithm should be to replace Σ-symbols by new variables and then λ-unify the resulting multi-equations.

3.1. Shallow equational theories

In order to be complete, the first step λ-unification should produce a set of substitutions, from which every possible λE-unifier of the original problem can be obtained. In other words, it should be enough for the subsequent part of our algorithm to refine the substitutions yielded by the first step λ-unification and find all λE-unifiers of the original system. Unfortunately, it is not the case in general.

For example, let E = {f(g(X))=g(f(X))} and Γ = {λx.F(g(x))=λx.g(G(x))} with F,G∈V be a given system. It seems difficult to find unifier σ = {F<-f, G<-f} by λ-unification algorithm, because it cannot use the information f(g(X))=g(f(X)). Note that in this case it does not help first to replace the Σ-symbols in Γ by new higher order variables

To avoid the above difficulty, we restrict ourselves to shallow equational theories.

3.2. Linear equational theories

There is another reason that may affect the completeness of our algorithm. For example, let $E = \{f(X,X)=g(X)\}$. Let $\Gamma = \{F(a,a))=g(a)\}$ with $F \in V$ be a given system. The λ-unification of $\{F(a,a))=g(a)\}$ yields $\sigma_1 = \{F<-\lambda xy.g(a)\}$, $\sigma_2 = \{F<-\lambda xy.g(x)\}$ and $\sigma_3 = \{F<-\lambda xy.g(y)\}$. Note that $\sigma_2 =_E \{F<-\lambda xy.f(x,x)\}$ and $\sigma_3 =_E \{F<-\lambda xy.f(y,y)\}$. But still an obvious unifier $\sigma := \{F<-f\}$ (note: $f =_{\beta\eta} \lambda xy.f(x,y)$) is missing. (One may have noticed that if we E-reduce Γ into $\Gamma' := \{F(a,a))=f(a,a)\}$, then we can find a complete set of unifiers.) Therefore, we require that the equational theories are linear. A relax of the restriction may be possible, but is beyond the scope of this paper.

3.3. Second-order λ-terms

In order to guarantee the termination of our procedure, we restrict ourselves to second order terms, since they satisfy the following property:

For any second order term $\lambda x_1...x_k.u\, t_1 ...t_n$, if some t_i, $1 \le i \le n$, is a λ-abstraction, i.e. of the form $\lambda y_1 ...y_m.s$ with $m>0$, then $u \in C$. For, second order variables in V can only have arguments of base types, therefore not λ-abstractions. Only function constants may be of higher than second order.

4. Our unification algorithm in combinations

We assume that a λ-unification algorithm in simply typed λ-calculus and an algorithm to check the word problem in algebraic theories be given.

Our algorithm is complete and/or terminating if the given algorithms used are complete and/or terminating. This means that the combination process of our algorithm causes no additional problems, except those already existing in given algorithms. Furthermore, our unification algorithm can be adjusted into a matching algorithm.

Let Γ_{orig} be the original system to be λE-unified. Let V_{orig} denote $FV(\Gamma_{orig})$.

Step 1.

(i) Unfolding terms

$$\{u[t[s_1,...,s_n]]=M\} \cup \Gamma \Rightarrow \{u[t<-Y\ s_1 ...\ s_n]=M\} \cup \{\lambda x_1...x_n.Y\, x_1 ...\, x_n = \lambda x_1...x_n.t[s_1<-x_1,...,s_n<-x_n]\} \cup \Gamma$$

if $t[s_1<-x_1,...,s_n<-x_n]$ is an α-term and $u[t[s_1,...,s_n]]=M$ is not a multi-equation generated by (i), where $Y,x_1,...,x_n \in V$ are distinct and new, $\tau(x_i) = \tau(s_i)$, $\tau(Y) = \tau(s_1)\rightarrow...\rightarrow\tau(s_n)\rightarrow\tau(t)$;

(ii) Renaming significant variables:

$$\{t[X]=M\} \cup \Gamma \Rightarrow \{t=M\}[X\backslash Y] \cup \{X=Y\} \cup \Gamma[X\backslash Y]$$

if $X \in V_{orig}$ and $t \neq X$, where $Y \in V$ is new, $\tau(X) = \tau(Y)$. ▯

As mentioned, we are going to use the λ-unification algorithm first because of higher-order variables, and the unification may fail if we do not consider possible E-reductions. In order to isolate the effects of axioms E for the moment, rule (i) abstracts all α-subterms by free variables, i.e. we just assume that the α-terms, which may be E-reduced by equations, be arbitrary.

The goal of (ii) is to rename significant variables, i.e. after applying this step, no original variables in V_{orig} occur in a non-variable term. It is a preparation step so that significant variables will not be eliminated by applying substitutions in the subsequent transformations. In the contrary, these substitutions will be registered for the significant variables.

We use the example $E = \{f(X)=g(X)\}$ and $\Gamma = \{f(F(a))=F(g(a))\}$ with $F \in V$ to illustrate transformation steps. Rule (i) unfolds the system Γ into $\{G(F(X))=F(Y),\ \lambda x_1.G(x_1)=\lambda x_1.f(x_1),\ X=a,\ Y=g(a)\}$. After renaming, we have the system $\{G(F'(X))=F'(Y),\ \lambda x_1.G(x_1)=\lambda x_1.f(x_1),\ X=a,\ Y=g(a),\ F=F'\}$.

Let Γ be the current system after step 1 and $M \in \Gamma$ an arbitrary multi-equation. It is obvious that either M contains no C(E)-symbols, or $M = \{\lambda x_1 ... x_n.Y\, x_1 ... x_n = \lambda x_1 ... x_n.t\}$ with Y a variable and t a non-variable α-term. Two subsystems of Γ, denoted by $\alpha\Gamma$ and $\lambda\Gamma$, can be constructed correspondingly: $M\text{-}V_{orig} \in \lambda\Gamma$ iff $M \in \Gamma$ contains no C(E)-symbols, $M \in \alpha\Gamma$ iff $M \in \Gamma$ contains C(E)-symbols. We assume that no subsequent transformations will change the subsystem of a multi-equation.

Step 2. λ-unification of $\lambda\Gamma$:

Let U be a set of complete λ-unifiers of $\lambda\Gamma$ obtained by the given λ-unification algorithm.
If $U=\emptyset$, then return failure, otherwise nondeterministically select $\sigma \in U$ and have $\Gamma \Rightarrow \sigma\Gamma$. []

Continuing the above example, we have $\lambda\Gamma = \{G(F'(X))=F'(Y),\ F'\}$ and $\alpha\Gamma = \{\lambda x_1.G(x_1)=\lambda x_1.f(x_1),\ X=a,\ Y=g(a)\}$ at the moment. One of the unifiers of $G(F'(X))=F'(Y)$ is $\{G \leftarrow F',\ Y \leftarrow F'(X)\}$. Then the resulting system of this transformation step is $\{F'(F'(X)),\ \lambda x_1.F'(x_1)=\lambda x_1.f(x_1),\ X=a,\ F'(X)=g(a),\ F=F'\}$ with $\lambda\Gamma = \{F'(F'(X)),\ F'\}$ and $\alpha\Gamma = \{\lambda x_1.F'(x_1)=\lambda x_1.f(x_1),\ X=a,\ F'(X)=g(a)\}$.

After successful application of step 2, the current $\lambda\Gamma$ contains only one-element multi-equations. The main problem now is to unify current $\alpha\Gamma$. We may assume w.l.g. that no α-symbols may occur in λ-unifiers in U. Then multi-equation of current $\alpha\Gamma$ must be of the form $\{\lambda x_1 ... x_n.s=\lambda x_1 ... x_n.t\}$, where s contains no C(E)-symbols, and t is a non-variable α-term with $\lambda x_1 ... x_n.t$ containing no free variables. Note that s may contain free constants. But all these free constants must disappear in the process of unifying $\lambda x_1 ... x_n.s=\lambda x_1 ... x_n.t$. Otherwise, the equation is not unifiable. Since s is of second order, s contains no bound variables of second order.

We need the following preparation for λ-unifying the equations in $\alpha\Gamma$.

Step 3: Renaming occurrences of variables:

$\{t[X]=M\} \cup \Gamma \Rightarrow \{t[X \leftarrow Y]=M\} \cup \{X=Y\} \cup \Gamma$ where Y is a new variable with $\tau(X) = \tau(Y)$. []

Step 3 renames occurrences of variables so that no variables will have more than one occurrence in $\alpha\Gamma$. It serves to separate the λ-unification and the equivalence of α-term induced by axioms E. Step 3 does some unnecessary renaming. We give the transformation in favor of simplicity. In fact, it is enough to require that an occurrence of a variable is renamed when this variable occurs twice.

We denote by $\gamma\Gamma$ the subsystem consisting of all new multi-equations X=Y generated by this step. Again, a multi-equation remains in $\gamma\Gamma$ after each of subsequent transformations.

Step 4. λ-unification of $\alpha\Gamma$:

Let U be a complete set of λ-unifiers of $\alpha\Gamma$ obtained by the given λ-unification algorithm such that IntroV(σ)=∅ and IntroC(σ)=∅ for each σ∈U.
If U=∅, then return failure, otherwise nondeterministically select σ∈U and have $\Gamma \Rightarrow \sigma\Gamma$. ▯

We can prove that we can always construct a complete set U of λ-unifiers of $\alpha\Gamma$ with IntroV(σ)=∅ and IntroC(σ)=∅ for each σ∈U from an arbitrary complete set U' of λ-unifiers of $\alpha\Gamma$.

Now for each substitution σ∈U and any X∈dom(σ), σ(X) is of the form $\lambda y_1 \ldots y_m.s$ with s consisting of only symbols in $\Sigma \cup \{y_1,\ldots,y_m\}$ and $y_1,\ldots,y_m$ being of first-order, i.e. either $s=y_j$ for some $1 \le j \le m$, or s is a Σ-term. Therefore, after the above transformation, any multi-equation in $\gamma\Gamma$ consists of terms that are either variables themselves, or a projection $\lambda y_1 \ldots y_m.y_i$, or of the form $\lambda y_1 \ldots y_m.s$ with s a α-term. To unify each equation $\lambda x_1 \ldots x_n.s = \lambda x_1 \ldots x_n.t$ in $\gamma\Gamma$, we need only to check whether $\lambda x_1 \ldots x_n.s =_E \lambda x_1 \ldots x_n.t$, or apply $\lambda x_1 \ldots x_n.s \leftarrow \lambda x_1 \ldots x_n.t$ on the current system if, say, $\lambda x_1 \ldots x_n.s$ is (the long βη-normal form of) a variable.

Step 5. (i) $\{\lambda x_1 \ldots x_n.s = \lambda x_1 \ldots x_n.t\} \cup \Gamma \Rightarrow \Gamma$ if s and t are non-variable α-terms and $s =_E t$;
(ii) $\{X = t\} \cup \Gamma \Rightarrow \{X \leftarrow t\}\Gamma$ where $X = t \in \gamma\Gamma$ and X∈V. ▯

Note that we need only an algorithm to check the word problem in the equational theory E, where no further instantiation of variables is necessary. The reason is that all necessary instantiations of the first order variables have implicitly been done non-deterministically in step 1.

We have reached the final resulting system Γ. The interpretation of the final system is very simple: If there is a multi-equation M∈Γ containing, except a possible original variable, still two distinct terms, then Γ is a failure; otherwise $\{X \leftarrow t \mid \{X=t\} \in \Gamma, X \in V_{orig}\}$ is a λE-unifier of Γ_{orig}.

References:

Breazu-Tannen, V. [88]: "Combing algebra and higher-order types." In: Proc. LICS, Edinburgh. (1988)
Breazu-Tannen, V. and Meyer, A. R. [87]: "Computable values can be classical." In: Proc. 14thPOPL, ACM. (1987)
Darlington, J. [71]: "A partial mechanization of second order logic." In: Machine Intellegence 6 (1971)
Elliot, C. [89]: "Higher-Order Unification with Dependent Function Types, RTA'89, Chapel Hill. (1989)
Herold, A. [86]: "Combination of Unification Algorithms." In: Proc. 8th CADE, ed. J. Siekmann, Springer-Verlag, LNCS 230, pp. 450-469. (1986)
Huet, G. [75]: "A Unification Algorithm for Typed λ-Calculus." In: Theoretical Computer Science 1, (1975) 27-57.
Huet, G., Lang, B. [83]: "Proving and applying program transformations expressed with second order patterns." Acta Informatica 11. (1983)
Jouannaud, J.-P. and Kirchner, C. [90]: "Solving equations in abstract algebras: a rule-based survey of unification."
Kirchner, C. [86]: "Computing Unification Algorithms." In: Proc. Conf. LICS, Cambridge, Mass. (1986) 206-216.
Nadathur, G. and Miller, D. [88]: "An Overview of λProlog." Proc. of 5th Int. Conf. Logic Programming, eds, R.A.Kowalski and K.A. Bowen, MIT Press (1988)
Nipkow,T. [89]: "Combining matching algorithms: The regular case." In: Proc. 3rd Int. Conf. RTA, LNCS 355 (1989) 343-358
Nipkow,T. and Qian Zh. [90]: "Modular Higher-Order E-Unification." Submitted.
Schmidt-Schauß,M. [89]: "Unification in a Combination of Arbitrary Disjoint Equational Theories." In: J. Symbolic Computation 8 (1989) 51-99
Snyder,W. [90]: "Higher-order E-unification." Proc. 10th Int. Conf. on Automated Deduction, LNCS (1990)
Snyder,W and Gallier,J. [89]: "Higher-Order Unification Revisited: Complete Sets of Transformations." In: J. Symbolic Computation 8 (1989) 101-140.
Stickel,M.E. [75]: "A Complete Unification algorithm for Associative-Commutative Functions." In: Proc. 4th IJCAI (1975).
Tiden,E. [86]: "Unification in combinations of collapse-free theories with disjoint sets of functions." In Proc. 8th CADE. (1986)
Yelick,K.A. [87]: "Unification in Combinations of Collapse-Free Regular Theories." In: J. Symbolic Computation 3 (1987) 153-182.

AN INFERENCE SYSTEM FOR HORN CLAUSE LOGIC WITH EQUALITY: A Foundation for Conditional E-Unification and for Logic Programming in the Presence of Equality

Wayne Snyder and Christopher Lynch
Boston University Department of Computer Science
111 Cummington St.
Boston, MA 02215
{snyder,lynch}@cs.bu.edu (617)-353-8925

Abstract: In this extended abstract of a full paper we present an inference system for Horn clause logic with equality which is complete not only refutationally but also with respect to the answer substitutions returned, without using functional reflexivity axioms or paramodulating at variable positions. It is goal directed in the sense that—as in SLD resolution—inference rules are only applied to the goal statement. We present the inference system (which subsumes those used for SLD-resolution and for general E-unification), prove its soundness and completeness, and then suggest how this approach forms the appropriate foundation for the study of inference systems (like logic programming interpreters) which return answer substitutions.

1 Introduction

There has been much interest in recent years in the integration of logic programming and functional/equational programming (see [2] for references) and in the possibility of extending the paradigm of logic programming to include the equality predicate. All previous work on this topic known to the authors either: only considers practical issues of language design; embeds the problem in an existing framework, such as a completion procedure or paramodulation theorem prover; only considers the issue of refutational completeness; assumes the existence of an E-unification procedure; only considers a subclass of Horn clause theories with equality predicate; or attempts to solve the problem by either allowing paramodulation into variable positions or adding the (prolific) axioms of equality to SLD-resolution (for a good bibliography, see [2,6,5]). As far as the authors know, there exists no proof of completeness (wrt answer substitutions) for a deduction mechanism for this logic which is goal directed in the sense that it applies inference rules only to the goal statement (as in SLD-resolution), and also avoids the use of functional reflexivity axioms, explicit axioms for equality, and paramodulating at variable positions, all of which have been rejected in the case of refutational theorem provers as generating too many irrelevant deduction steps to be practically useful.

In [4] we present such a deduction mechanism and present rigorous completeness results. In particular, we extend the work of Jean Gallier and the first author on General E-Unification [3] to the case of arbitrary sets of Horn clauses with equality. Here we shall simply summarize the results obtained. The basic approach is to extend the Herbrand-Martelli-Montanari transformations on systems of terms to account not only for the presence of arbitrary equational axioms (as in General E-Unification), but for the presence of arbitrary Horn clauses with the equality symbol among the predicates. The basic problem our inference system solves can be stated most conveniently in a logic programming framework: *Given a set P of Horn clauses with equality, a conjunction S of atoms (equational or otherwise), and a substitution θ such that $P \models \theta(S)$, is there a means of transforming the goal S, e.g.*

$$S = S_0 \implies S_1 \implies \ldots \implies S_n$$

into a simple form S_n from which we can extract a substitution σ_{S_n} such that $\sigma_{S_n} \leq_P \theta[Var(S)]$? In other words, can we compute an answer substitution σ_{S_n} which is more general than θ (modulo the congruence on terms implied by P), over the variables of S? Roughly, what we are asking for is a concrete solution to the query $P \models \exists x_1 \ldots \exists x_k . S$? in terms of (in the affirmative case) a definite answer σ which subsumes a concrete list of bindings $t_1, \ldots, t_k$. In addition, we would like this transformation process to be relatively efficient. Our inference system solves this problem and has the following features:

(i) No explicit equality axioms or functional reflexivity axioms are needed; equality is "built in" to the inference rules, as in the case of E-unification (of which our method is a natural extension);

(ii) No paramodulation into variable positions is necessary;[1] this is an important restriction which eliminates a huge number of irrelevant inference steps, as is well known from resolution theory;

(iii) The rules are "goal directed" in the sense that no inference rules are ever applied among the program clauses (as would happen in embedding this problem in a completion procedure or a paramodulation theorem prover); as in SLD-resolution we apply inferences steps in a linear fashion to the goal clause only;

The importance of (i) and (ii) is familiar from resolution theory, and the importance of (iii) is well-known as one of the central characteristics of SLD-resolution; we take it as axiomatic that inherent in the notion of "programming" is some computation mechanism which is (relatively) efficient (hence *i* and *ii*), goal-directed (hence *iii*), and produces output more informative than simply **yes** or **no** (hence the importance of completeness wrt answer substitutions). Unfortunately, in the context of the standard paramodulation inference rule, these desiderata are antagonistic: paramodulation is *not* complete with the restrictions given in (i)—(iii) above For example, if $E = \{f(a,b) \doteq g(a,b), a \doteq b\}$, then the equation $f(x,x) \doteq g(x,x)$ has a solution (i.e. is E-unifiable by) $\theta = [a/x]$, but no deduction can be found using goal-directed paramodulation at

1 In fact, we use a *lazy* form of paramodulation, so we should say "No lazy paramodulation into variables is necessary."

non-variable positions. For this reason, we give an inference system which performs lazy unification, which is complete with these restrictions; the authors know of no other inference system for this very general form of programming logic which contains all these desirable features and which is provably complete not only refutationally but also wrt the answer substitutions returned.

As in the use of transformations for unification problems, these inference rules form an abstract and mathematically elegant formalism for analysing goal-directed methods for this problem, by separating the logical issues from those of data structure and control. Unification is not built-in, as in SLD-resolution, but forms a part of the deduction process itself; in the context of equations, the increased complexity of unification necessitates a more sophisticated integration of unification and the deduction process; our formalism—which consists of simple "atomic" operations extending the Herbrand-Martelli-Montanari transformations—allows us to analyse the complex interaction between unification and deduction in Horn clause logic with equality in a clean and elegant way. Hence we hope this formalism will provide a suitable foundation for programming in the context of Horn clause logic with equality.

2 Inference Rules for Equational Horn Clause Logic

We assume the reader is familiar with the basic notions of first-order logic (with equality) in clausal form.

Definition 2.1 A *goal system* (or just *system*) is the existential closure of a conjunction of atomic formulae, represented by a multiset of atomic formulae (possibly empty and possibly including equations). (Thus, a goal system $\{A_1, \dots, A_n\}$ is interpreted as the sentence $\exists x_1 \dots \exists x_k . A_1 \wedge \dots \wedge A_n$, where $\{x_1, \dots, x_k\} = Var(A_1, \dots, A_n)$.)

Our inference system attempts to answer a query of the form $P \models \exists \overline{x}. S$ by transforming the goal S into a simple form from which an answer substitution can be trivially extracted.

Definition 2.2 Let S be a system (possibly empty) and P a program. We have the following inference rules, expressed as transformations on goal systems.

Trivial:

$$\{t \doteq t\} \cup S \implies S \tag{1}$$

Orientation:

$$\{t \doteq x\} \cup S \implies \{x \doteq t\} \cup S, \tag{2}$$

if t is not a variable.

Decomposition:

$$\{f(s_1, \ldots, s_k) \doteq f(t_1, \ldots, t_k)\} \cup S \implies \{s_1 \doteq t_1, \ldots, s_k \doteq t_k\} \cup S, \tag{3}$$

for any $k > 0$ and any $f \in \Sigma_k$.

Variable Elimination:

$$\{x \doteq t\} \cup S \implies \{x \doteq t\} \cup \sigma(S), \tag{4}$$

if $x \notin Var(t)$, $x \in Var(S)$, and $\sigma = [t/x]$.

Lazy Resolution:

$$\{R(s_1, \ldots, s_k)\} \cup S \implies \{s_1 \doteq t_1, \ldots, s_k \doteq t_k, B_1, \ldots, B_n\} \cup S, \tag{5}$$

if $R(t_1, \ldots, t_k) \leftarrow B_1, \ldots, B_n$ is a variant of a clause from P (for $n \geq 0$).

Lazy Paramodulation:

$$\{A\} \cup S \implies \{A/\alpha \doteq l, A[\alpha \leftarrow r], B_1, \ldots, B_n\} \cup S, \tag{6}$$

where $A[\alpha \leftarrow r]$ is the result of replacing the subterm of A at address α (namely A/α) by r, and A/α is a term which is is *not* a variable, and either $l \doteq r \leftarrow B_1, \ldots, B_n$ or $r \doteq l \leftarrow B_1, \ldots, B_n$ is a variant of a clause from P (for $n \geq 0$). Furthermore, if l is not a variable, we require that $Root(A/\alpha) = Root(l)$ and that (as an integral part of this rule) Decomposition is immediately applied to the equation $A/\alpha \doteq l$.

A goal system S is in *solved form* if it has the form $S = \{x_1 \doteq t_1, \ldots, x_n \doteq t_n\}$ where all the x_i are distinct and no x_i occurs in any t_j, $1 \leq i, j \leq n$. From such a system we extract a substitution $\sigma_S = [t_1/x_1, \ldots, t_n/x_n]$.

A *proof* of S from P is a sequence of transformations

$$S = S_0 \implies S_1 \implies \ldots \implies S_n$$

such that S_n is in solved form; the *computed answer substitution* of the proof is $\sigma_{S_n}|_{Var(S)}$.

Example 2.3 If $P = \{Q(f(y, g(y))), g(z) \doteq z \leftarrow b \doteq z\}$, then $P \models \theta(Q(f(x, x))$ for $\theta = [b/x]$. A proof sequence would be:

$$\begin{aligned} Q(f(x,x)) &\implies_5 f(x,x) \doteq f(y, g(y)) \\ &\implies_3 x \doteq y, x \doteq g(y) \\ &\implies_4 x \doteq y, y \doteq g(y) \\ &\implies_6 x \doteq y, y \doteq z, y \doteq z, b \doteq z \\ &\implies_2 x \doteq y, y \doteq z, y \doteq z, z \doteq b \end{aligned}$$

$$\Longrightarrow_4 \; x \doteq y, y \doteq b, y \doteq b, z \doteq b$$
$$\Longrightarrow_4 \; x \doteq b, y \doteq b, b \doteq b, z \doteq b$$
$$\Longrightarrow_1 \; x \doteq b, y \doteq b, z \doteq b$$

where the computed answer substitution is $\sigma = \theta = [b/x]$.

The soundness result is easily proved by showing that for any program P and any i, $1 \leq i \leq 6$, if $S \Longrightarrow_i S'$ and $P \models \theta(S')$ for some θ, then $P \models \theta(S)$.

Theorem 2.4 (Soundness) If

$$S = S_0 \Longrightarrow S_1 \Longrightarrow \ldots \Longrightarrow S_n$$

is a proof of S from P with computed answer substitution $\sigma = \sigma_{S_n}|_{Var(S_0)}$, then $P \models \sigma(S)$.

3 Completeness of the Inference System

In this section we sketch three completeness proofs from the full paper. In the first we extend the notion of narrowing to the case of certain kinds of equational Horn clause programs;[2] in the second we extend this to arbitrary equational Horn clause programs (conditional E-unification); and finally we show how to adapt this proof to the general case of Horn clauses with equality.

In the full paper we first discuss some basic properties of rewriting in the presence of conditional equations: the consequences of the fact that not all substitution instances of the heads of program clauses are available are developed, including a new version of the Critical Pair lemma. We then extend the notion of ground canonicity for a set of rewrite rules to the notion of *safe equational Horn clause programs*, which basically asserts that in the inductive construction of a model M_P for the program, every consequence of the program (at every level of the construction) can be justified with a rewrite proof in which no rewrite step takes place at a part of a term introduced by a substitution. Such a condition basically asserts that we can always lift a *safe* proof (where no rewrite occurs at or below a variable occurrence) using the inference rules, analogous to the way in which narrowing lifts a rewrite proof via unification. The first completeness result may be stated as follows.

2 By an *equational Horn clause* we mean a Horn clause consisting entirely of equations.

Theorem 3.1 If P is a safe set of equational Horn clauses, S a multiset of equations, and θ a substitution such that $P \models \theta(S)$, then there exists a sequence of transformations (which does not use rule (5))

$$S = S_0 \implies S_1 \implies S_2 \implies \dots \implies S_n$$

with S_n in solved form such that $\sigma_{S_n} \leq_P \theta[Var(S)]$.

The proof basically shows how to extract the requisite sequence of transformations from a "safe proof" for the fact that $P \models \theta(S)$. It is essentially a modification of the completeness theorem for basic narrowing, accounting for the presence of conditional equations.

In order to prove the completeness result for conditional equations in the general case we consider the saturation P^ω of the program P by paramodulants, and show that such an (in general infinite) set generates in the model M_{P^ω} safe proofs for every equational consequence of the program. Basically, the instances of equations from the heads of clauses form a canonical rewriting system with some special properties that allow safe proofs to be found. Thus we give a second completeness result for the general case of equational Horn clauses.

Theorem 3.2 If P is an arbitrary set of equational Horn clauses, S a multiset of equations, and θ a substitution such that $P \models \theta(S)$, then there exists a sequence of transformations (which does not use rule (5))

$$S = S_0 \implies S_1 \implies S_2 \implies \dots \implies S_n$$

with S_n in solved form such that $\sigma_{S_n} \leq_P \theta[Var(S)]$.

The proof shows that P^ω is equivalent to P, and so $P^\omega \models \theta(S)$. By the first completeness result, we have a sequence of transformations using clauses from P^ω. Since the proof is finite, in fact every clause used is from P^k for some k; we show how to transform the sequence into another sequence (with the same computed answer substitution) by demonstrating that the inference rules can calculate paramodulants "on the fly." This proof extends the general completeness proof from [3] (Theorem 6.8) to the Horn clause case.

Finally, we prove the completeness of the set of inference rules in the general case of Horn clauses with equality by translating arbitrary Horn clauses into equational Horn clauses.

Theorem 3.3 If P is an arbitrary set of Horn clauses with equality, S a multiset of equations, and θ a substitution such that $P \models \theta(S)$, then there exists a sequence of transformations

$$S = S_0 \Longrightarrow S_1 \Longrightarrow S_2 \Longrightarrow \ldots \Longrightarrow S_n$$

with S_n in solved form such that $\sigma_{S_n} \leq_P \theta[Var(S)]$.

To prove this, we first translate everything into the context of equational Horn clauses by replacing each non-equational atom A by the equation $A \doteq \mathbf{T}$, where $\mathbf{T}$ is a new constant denoting *true*. Then we appeal to the previous completeness result, observing that wlg we can assume the sequence of transformations (not involving rule (5)) which results has certain restrictions on the way equations $A \doteq \mathbf{T}$ were used in lazy paramodulation steps. But then, because of this restriction, this sequence can be immediately translated into a sequence using (1)–(6) fulfilling the conditions of the general theorem.

Finally, in the full paper we discuss various restrictions of the inference rules which preserve completeness. For example, the strategy from [1] for general E-unification, which basically requires that the equation $A/\alpha \doteq l$ formed in Lazy Paramodulation be decomposed down to variable–term equations, can easily be extended to the Horn clause case. We also discuss various putative restrictions and discuss in which cases they are complete; for example, restricting paramodulation *from* a variable (i.e., using an equation of the form $x \doteq t$) is only possible for safe programs, and is not a complete strategy in the general case.

4 Conclusion

Although the intent of the paper is to give rigorous completeness results extending naturally those found in [3], we also hope that this inference system might form a foundation for the theoretical study of deduction mechanisms (such as logic programming interpreters) which are not simply refutational in nature, but also return answer substitutions. Such inference mechanisms are a really new development in the history of proof theory, and include not only logic programming systems per se, but also E-unification procedures (note that E-unification problems can be phrased in the form of existential queries for equational programs). There has been relatively little attention paid to the unique nature of completeness wrt answer substitutions, particularly in the presence of equality. The difficulty of adding equality to logic programming systems mostly centers around the extending the notion of unification, and it seems clear that some integration of the process of unification and of deduction is necessary if such systems will ever be useful. We hope that our approach, which decomposes the necessary actions into six atomic, relatively simple rules, will allow us to cleanly analyze a wide variety of approaches to this problem, from lazy unification strategies to the use of special purpose E-unification modules.

5 Bibliography

[1] Dougherty, D., and P. Johann, "An Improved General E-Unification Method," CADE90, Kaiserlautern, Germany.

[2] Gallier, J.H., Raatz, S., "Extending SLD-Resolution to Equational Horn Clauses Using E-unification", to appear in *Journal of Logic Programming* (1987).

[3] "Complete Sets of Transformations for General E-Unification," Jean Gallier and Wayne Snyder, TCS 67 (1989) 203–260.

[4] "An Inference System for Horn Clause Logic with Equality: A foundation for conditional E-unification and Horn clause logic with equality," Wayne Snyder and Christopher Lynch, Boston University Technical Report No. BU-TR90-014.

[5] "Conditional Equational Theories and Complete Sets of Transformations," Steffen Hölldobler, Proceedings of International Conference of Fifth Generation Computer Systems (1988).

[6] Padawitz, P., *Computing in Horn Theories*, EATCS Monographs on Theoretical Computer Science, Vol. 16, Springer-Verlag, Berlin (1989).

Lecture Notes in Computer Science

For information about Vols. 1–429
please contact your bookseller or Springer-Verlag

Vol. 430: J.W. de Bakker, W.-P. de Roever, G. Rozenberg (Eds.), Stepwise Refinement of Distributed Systems. Models, Formalisms, Correctness. Proceedings, 1989. X, 808 pages. 1990.

Vol. 431: A. Arnold (Ed.), CAAP '90. Proceedings, 1990. VI, 285 pages. 1990.

Vol. 432: N. Jones (Ed.), ESOP '90. Proceedings, 1990. IX, 436 pages. 1990.

Vol. 433: W. Schröder-Preikschat, W. Zimmer (Eds.), Progress in Distributed Operating Systems and Distributed Systems Management. Proceedings, 1989. V, 206 pages. 1990.

Vol. 434: J.-J. Quisquater, J. Vandewalle (Eds.), Advances in Cryptology – EUROCRYPT '89. Proceedings, 1989. X, 710 pages. 1990.

Vol. 435: G. Brassard (Ed.), Advances in Cryptology – CRYPTO '89. Proceedings, 1989. XIII, 634 pages. 1990.

Vol. 436: B. Steinholtz, A. Sølvberg, L. Bergman (Eds.), Advanced Information Systems Engineering. Proceedings, 1990. X, 392 pages. 1990.

Vol. 437: D. Kumar (Ed.), Current Trends in SNePS – Semantic Network Processing System. Proceedings, 1989. VII, 162 pages. 1990. (Subseries LNAI).

Vol. 438: D.H. Norrie, H.W. Six (Eds.), Computer Assisted Learning – ICCAL '90. Proceedings, 1990. VII, 467 pages. 1990.

Vol. 439: P. Gorny, M. Tauber (Eds.), Visualization in Human-Computer Interaction. Proceedings, 1988. VI, 274 pages. 1990.

Vol. 440: E. Börger, H. Kleine Büning, M.M. Richter (Eds.), CSL '89. Proceedings, 1989. VI, 437 pages. 1990.

Vol. 441: T. Ito, R.H. Halstead, Jr. (Eds.), Parallel Lisp: Languages and Systems. Proceedings, 1989. XII, 364 pages. 1990.

Vol. 442: M. Main, A. Melton, M. Mislove, D. Schmidt (Eds.), Mathematical Foundations of Programming Semantics. Proceedings, 1989. VI, 439 pages. 1990.

Vol. 443: M.S. Paterson (Ed.), Automata, Languages and Programming. Proceedings, 1990. IX, 781 pages. 1990.

Vol. 444: S. Ramani, R. Chandrasekar, K.S.R. Anjaneyulu (Eds.), Knowledge Based Computer Systems. Proceedings, 1989. X, 546 pages. 1990. (Subseries LNAI).

Vol. 445: A.J.M. van Gasteren, On the Shape of Mathematical Arguments. VIII, 181 pages. 1990.

Vol. 446: L. Plümer, Termination Proofs for Logic Programs. VIII ,142 pages. 1990. (Subseries LNAI).

Vol. 447: J.R. Gilbert, R. Karlsson (Eds.), SWAT '90. 2nd Scandinavian Workshop on Algorithm Theory. Proceedings, 1990. VI, 417 pages. 1990.

Vol. 448: B. Simons, A. Spector (Eds.), Fault Tolerant Distributed Computing. VI, 298 pages. 1990.

Vol. 449: M.E. Stickel (Ed.), 10th International Conference on Automated Deduction. Proceedings, 1990. XVI, 688 pages. 1990. (Subseries LNAI).

Vol. 450: T. Asano, T. Ibaraki, H. Imai, T. Nishizeki (Eds.), Algorithms. Proceedings, 1990. VIII, 479 pages. 1990.

Vol. 451: V. Mařík, O. Stepánková, Z. Zdráhal (Eds.), Artificial Intelligence in Higher Education. Proceedings, 1989. IX, 247 pages. 1990. (Subseries LNAI).

Vol. 452: B. Rovan (Ed.), Mathematical Foundations of Computer Science 1990. Proceedings, 1990. VIII, 544 pages. 1990.

Vol. 453: J. Seberry, J. Pieprzyk (Eds.), Advances in Cryptology – AUSCRYPT '90 Proceedings, 1990. IX. 462 pages. 1990.

Vol. 454: V. Diekert, Combinatorics on Traces. XII, 165 pages. 1990.

Vol. 455: C.A. Floudas, P.M. Pardalos, A Collection of Test Problems for Constrained Global Optimization Algorithms. XIV, 180 pages. 1990.

Vol. 456: P. Deransart, J. Maluszyn´ski (Eds.), Programming Language Implementation and Logic Programming. Proceedings, 1990. VIII, 401 pages. 1990.

Vol. 457: H. Burkhart (Ed.), CONPAR '90 – VAPP IV. Proceedings, 1990. XIV, 900 pages. 1990.

Vol. 458: J.C.M. Baeten, J.W. Klop (Eds.), CONCUR '90. Proceedings, 1990. VII, 537 pages. 1990.

Vol. 459: R. Studer (Ed.), Natural Language and Logic. Proceedings, 1989. VII, 252 pages. 1990. (Subseries LNAI).

Vol. 460: J. Uhl, H.A. Schmid, A Systematic Catalogue of Reusable Abstract Data Types. XII, 344 pages. 1990.

Vol. 461: P. Deransart, M. Jourdan (Eds.), Attribute Grammars and their Applications. Proceedings, 1990. VIII, 358 pages. 1990.

Vol. 462: G. Gottlob, W. Nejdl (Eds.), Expert Systems in Engineering. Proceedings, 1990. IX, 260 pages. 1990. (Subseries LNAI).

Vol. 463: H. Kirchner, W. Wechler (Eds.), Algebraic and Logic Programming. Proceedings, 1990. VII, 386 pages. 1990.

Vol. 464: J. Dassow, J. Kelemen (Eds.), Aspects and Prospects of Theoretical Computer Science. Proceedings, 1990. VI, 298 pages. 1990.

Vol. 465: A. Fuhrmann, M. Morreau (Eds.), The Logic of Theory Change. Proceedings, 1989. X, 334 pages. 1991. (Subseries LNAI).

Vol. 466: A. Blaser (Ed.), Database Systems of the 90s. Proceedings, 1990. VIII, 334 pages. 1990.

Vol. 467: F. Long (Ed.), Software Engineering Environments. Proceedings, 1969. VI, 313 pages. 1990.

Vol. 468: S.G. Akl, F. Fiala, W.W. Koczkodaj (Eds.), Advances in Computing and Information – ICCI '90. Proceedings, 1990. VII, 529 pages. 1990.

Vol. 469: I. Guessarian (Ed.), Semantics of Systeme of Concurrent Processes. Proceedings, 1990. V, 456 pages. 1990.

Vol. 470: S. Abiteboul, P.C. Kanellakis (Eds.), ICDT '90. Proceedings, 1990. VII, 528 pages. 1990.

Vol. 471: B.C. Ooi, Efficient Query Processing in Geographic Information Systems. VIII, 208 pages. 1990.

Vol. 472: K.V. Nori, C.E. Veni Madhavan (Eds.), Foundations of Software Technology and Theoretical Computer Science. Proceedings, 1990. X, 420 pages. 1990.

Vol. 473: I.B. Damgård (Ed.), Advances in Cryptology – EUROCRYPT '90. Proceedings, 1990. VIII, 500 pages. 1991.

Vol. 474: D. Karagiannis (Ed.), Information Syetems and Artificial Intelligence: Integration Aspects. Proceedings, 1990. X, 293 pages. 1991. (Subseries LNAI).

Vol. 475: P. Schroeder-Heister (Ed.), Extensions of Logic Programming. Proceedings, 1989. VIII, 364 pages. 1991. (Subseries LNAI).

Vol. 476: M. Filgueiras, L. Damas, N. Moreira, A.P. Tomás (Eds.), Natural Language Processing. Proceedings, 1990. VII, 253 pages. 1991. (Subseries LNAI).

Vol. 477: D. Hammer (Ed.), Compiler Compilers. Proceedings, 1990. VI, 227 pages. 1991.

Vol. 478: J. van Eijck (Ed.), Logics in AI. Proceedings, 1990. IX, 562 pages. 1991. (Subseries in LNAI).

Vol. 480: C. Choffrut, M. Jantzen (Eds.), STACS 91. Proceedings, 1991. X, 549 pages. 1991.

Vol. 481: E. Lang, K.-U. Carstensen, G. Simmons, Modelling Spatial Knowledge on a Linguistic Basis. IX, 138 pages. 1991. (Subseries LNAI).

Vol. 482: Y. Kodratoff (Ed.), Machine Learning – EWSL-91. Proceedings, 1991. XI, 537 pages. 1991. (Subseries LNAI).

Vol. 483: G. Rozenberg (Ed.), Advances In Petri Nets 1990. VI, 515 pages. 1991.

Vol. 484: R. H. Möhring (Ed.), Graph-Theoretic Concepts In Computer Science. Proceedings, 1990. IX, 360 pages. 1991.

Vol. 485: K. Furukawa, H. Tanaka, T. Fullsaki (Eds.), Logic Programming '89. Proceedings, 1989. IX, 183 pages. 1991. (Subseries LNAI).

Vol. 486: J. van Leeuwen, N. Santoro (Eds.), Distributed Algorithms. Proceedings, 1990. VI, 433 pages. 1991.

Vol. 487: A. Bode (Ed.), Distributed Memory Computing. Proceedings, 1991. XI, 506 pages. 1991.

Vol. 488: R. V. Book (Ed.), Rewriting Techniques and Applications. Proceedings, 1991. VII, 458 pages. 1991.

Vol. 489: J. W. de Bakker, W. P. de Roever, G. Rozenberg (Eds.), Foundations of Object-Oriented Languages. Proceedings, 1990. VIII, 442 pages. 1991.

Vol. 490: J. A. Bergstra, L. M. G. Feljs (Eds.), Algebraic Methods 11: Theory, Tools and Applicatlons. VI, 434 pages. 1991.

Vol. 491: A. Yonezawa, T. Ito (Eds.), Concurrency: Theory, Language, and Architecture. Proceedings, 1989. VIII, 339 pages. 1991.

Vol. 492: D. Sriram, R. Logcher, S. Fukuda (Eds.), Computer-Aided Cooperative Product Development. Proceedings, 1989 VII, 630 pages. 1991.

Vol. 493: S. Abramsky, T. S. E. Maibaum (Eds.), TAPSOFT '91. Volume 1. Proceedings, 1991. VIII, 455 pages. 1991.

Vol. 494: S. Abramsky, T. S. E. Maibaum (Eds.), TAPSOFT '91. Volume 2. Proceedings, 1991. VIII, 482 pages. 1991.

Vol. 495: 9. Thalheim, J. Demetrovics, H.-D. Gerhardt (Eds.), MFDBS '91. Proceedings, 1991. VI, 395 pages. 1991.

Vol. 496: H.-P. Schwefel, R. Männer (Eds.), Parallel Problem Solving from Nature. Proceedings, 1991. XI, 485 pages. 1991.

Vol. 497: F. Dehne, F. Fiala. W.W. Koczkodaj (Eds.), Advances in Computing and Intormation - ICCI '91 Proceedings, 1991. VIII, 745 pages. 1991.

Vol. 498: R. Andersen, J. A. Bubenko jr., A. Sølvberg (Eds.), Advanced Information Systems Engineering. Proceedings, 1991. VI, 579 pages. 1991.

Vol. 499: D. Christodoulakis (Ed.), Ada: The Choice for '92. Proceedings, 1991. VI, 411 pages. 1991.

Vol. 500: M. Held, On the Computational Geometry of Pocket Machining. XII, 179 pages. 1991.

Vol. 501: M. Bidoit, H.-J. Kreowski, P. Lescanne, F. Orejas, D. Sannella (Eds.), Algebraic System Specification and Development. VIII, 98 pages. 1991.

Vol. 502: J. Bārzdiņš, D. Bjørner (Eds.), Baltic Computer Science. X, 619 pages. 1991.

Vol. 503: P. America (Ed.), Parallel Database Systems. Proceedings, 1990. VIII, 433 pages. 1991.

Vol. 504: J. W. Schmidt, A. A. Stogny (Eds.), Next Generation Information System Technology. Proceedings, 1990. IX, 450 pages. 1991.

Vol. 505: E. H. L. Aarts, J. van Leeuwen, M. Rem (Eds.), PARLE '91. Parallel Architectures and Languages Europe, Volume I. Proceedings, 1991. XV, 423 pages. 1991.

Vol. 506: E. H. L. Aarts, J. van Leeuwen, M. Rem (Eds.), PARLE '91. Parallel Architectures and Languages Europe, Volume II. Proceedings, 1991. XV, 489 pages. 1991.

Vol. 507: N. A. Sherwani, E. de Doncker, J. A. Kapenga (Eds.), Computing in the 90's. Proceedings, 1989. XIII, 441 pages. 1991.

Vol. 508: S. Sakata (Ed.), Applied Algebra, Algebraic Algorithms and Error-Correcting Codes. Proceedings, 1990. IX, 390 pages. 1991.

Vol. 509: A. Endres, H. Weber (Eds.), Software Development Environments and CASE Technology. Proceedings, 1991. VIII, 286 pages. 1991.

Vol. 510: J. Leach Albert, B. Monien, M. Rodríguez (Eds.), Automata, Languages and Programming. Proceedings, 1991. XII, 763 pages. 1991.

Vol. 511: A. C. F. Colchester, D.J. Hawkes (Eds.), Information Processing in Medical Imaging. Proceedings, 1991. XI, 512 pages. 1991.

Vol. 512: P. America (Ed.), ECOOP '91. European Conference on Object-Oriented Programming. Proceedings, 1991. X, 396 pages. 1991.

Vol. 513: N. M. Mattos, An Approach to Knowledge Base Management. IX, 247 pages. 1991. (Subseries LNAI).

Vol. 514: G. Cohen, P. Charpin (Eds.), EUROCODE '90. Proceedings, 1990. XI, 392 pages. 1991.

Vol. 515: J. P. Martins, M. Reinfrank (Eds.), Truth Maintenance Systems. Proceedings, 1990. VII, 177 pages. 1991. (Subseries LNAI).

Vol. 516: S. Kaplan, M. Okada (Eds.), Conditional and Typed Rewriting Systems. Proceedings, 1990. IX, 461 pages. 1991.